The Filamentous Fungi

Volume 2 Biosynthesis and Metabolism

The Filamentous Fungi
Volume 3 Developmental Biology
is now in preparation

The Filamentous Fungi

Volume 2 Biosynthesis and Metabolism

Edited by:
JOHN E. SMITH, D.Sc., F.I. Biol.
DAVID R. BERRY, Ph.D.
Department of Applied Microbiology, University of Strathclyde, Glasgow

EDWARD ARNOLD

First published 1976
by Edward Arnold (Publishers) Limited,
25 Hill Street, London, W1X 8LL

ISBN 0 7131 2537 3

Printed in Great Britain by J. W. Arrowsmith Ltd., Bristol

Preface

Although the fundamental biochemical processes of cells exhibit a remarkable similarity throughout the microbial, animal and plant Kingdoms, each group of organisms exhibits biochemical characteristics specific to itself. The filamentous fungi are no exception; indeed they are perhaps unique in the diversity of their biosynthetic capabilities.

The aim of this volume is to provide the reader with an up to date review of present knowledge relevant to the biosynthesis of fungal constituents and products. The title, 'Biosynthesis and Metabolism' is preferred since we wish to avoid on the one hand the more chemical and physical aspects of biochemistry and on the other the diversity of physiological responses which have been recorded in the fungi. However, it is recognized that the processes of solute uptake and energy generation are essential to biosynthesis.

It has been our aim to make each chapter a complete entity in itself rather than a survey of the most recent advances. This approach, we believe, will also be most valuable to the research worker in Universities, Research Institutes and Industry and to the postgraduate and undergraduate students to whom this book is directed.

1975

J. E. S.
D. R. B.

Contents

List of Contributors

Dr. D. R. Berry,
Department of Applied Microbiology,
University of Strathclyde,
Glasgow.

Dr. Elisabeth A. Berry,
Department of Applied Microbiology,
University of Strathclyde,
Glasgow.

Professor H. J. Blumenthal,
Department of Microbiology,
Loyola University,
Stritch School of Medicine,
2160 South First Avenue,
Maywood, Illinois 60153,
U.S.A.

Dr. A. T. Bull,
Biological Laboratory,
The University,
Canterbury, Kent.

Dr. J. D. Bu'Lock,
Department of Chemistry,
The University of Manchester,
Manchester.

Dr. M. E. Bushell,
Glaxo Research,
Sefton Park,
Stoke Poges, Bucks.

Dr. P. J. Casselton,
Department of Botany,
University of London,
Birkbeck College, London.

Professor V. W. Cochrane,
Department of Biology,
Weslyan University, Middletown,
Connecticut, U.S.A.

Professor T. W. Goodwin,
Department of Biochemistry,
The University,
Liverpool.

Professor D. H. Jennings,
Department of Botany,
The University,
P.O. Box 147, Liverpool.

Dr. J. R. Kinghorn,
Department of Genetics,
The University of Glasgow,
Glasgow.

Dr. N. J. McCorkindale,
Joint Mycology Laboratories,
Departments of Chemistry & Botany,
The University of Glasgow,
Glasgow.

Professor J. A. Pateman,
Department of Genetics,
The University of Glasgow,
Glasgow.

Professor A. H. Rose,
School of Biological Sciences,
University of Bath,
Claverton Down,
Bath.

Dr. R. F. Rosenberger,
Department of Microbiological Chemistry,
The Hebrew University,
Hadassah Medical School,
Jerusalem, Israel.

Professor G. H. N. Towers,
Department of Botany,
University of British Columbia,
Vancouver 8, Canada.

Dr. W. R. Turner,
I.C.I. Pharmaceuticals Division,
Alderley Park,
Macclesfield,
Cheshire.

Dr. L. C. Vining,
Atlantic Regional Laboratory,
National Research Council
of Canada,
1411 Oxford Street,
Halifax, Nova Scotia,
Canada.

Dr. Peter Walker,
Department of Applied
Biochemistry & Nutrition,
School of Agriculture,
Sutton Bonington,
Loughborough.

Dr. K. Watson,
Department of Biochemistry,
James Cooke University,
Townsville, Australia.

Dr. M. Woodbine,
Department of Applied
Biochemistry & Nutrition,
School of Agriculture,
Sutton Bonington,
Loughborough.

Dr. J. L. C. Wright,
Atlantic Regional Laboratory,
National Research Council
of Canada,
1411 Oxford Street,
Halifax, Nova Scotia,
Canada.

CHAPTER 1

Environmental Control of Fungal Growth

A. T. BULL and M. E. BUSHELL

1.1 Introduction

Experiments designed to explore the effects of various environmental parameters on microbial activities have long been popular with microbiologists, and in this endeavour the filamentous fungi have not been neglected. An extensive research literature concerned with both chemical and physical factors is at the disposal of the mycologist and the conclusions of the pre-continuous-flow culture era have been extensively reviewed on several occasions (Cochrane, 1958; Ainsworth & Sussman, 1965). In discussions of the environment there is a tendency to treat each factor separately but this approach is doomed to failure because it implies acceptance of independently acting factors. The true situation is quite the reverse and an appreciation of the interaction of such factors is essential for the full understanding of fungal, and indeed microbial, growth. Another concept of major significance and relevance to this discussion is that of a limiting factor which controls the rate or extent of growth. During the course of growth the nature of the limiting factor may change so that a second factor takes on the rate controlling role.

In a very cogent essay Tempest (1970) has put the case for the adoption of open or continuous-flow cultures in microbiological research. Further progress in many areas of microbiology, concludes Tempest, can be greatly facilitated by, or only follow from, the application of continuous-flow culture techniques. In our view such an approach is essential in the context of evaluating environmental effects on the growth of fungi and, of course, other protists. The advantages (and inadequacies) of continuous-flow cultures need not be reiterated here in detail, suffice it only to point out that they provide a means of controlling the growth rate while chemical and physical parameters can be varied independently and under completely definable conditions. In other words, the phenotypic variability of organisms in response to changes in the environment can be determined unequivocally.

The essential elements of chemostat theory have been presented in Volume 1 of this treatise (Righelato, 1975). The chemostat is a continuous-flow culture in which the growth rate is controlled by the rate of supply of a limiting nutrient to the organisms and is regulated at values below the maximum specific growth rate. Continuous-flow cultures also have inherent versatility of operation and, by the incorporation of multistage, branched-stream and feedback components into a system, analyses of sequential changes in growth parameters can be assessed. An interesting illustration of this facility is contained in the report on the production of a fungal α-galactosidase by Imanaka, Kaieda & Taguchi (1973). In contrast, the non-steady state characteristics of closed or batch cultures makes them very unsuitable for the analysis of environmental effects.

The influence of numerous environmental factors on fungal growth and development have been investigated and include temperature, pH, radiations of various sorts, hydrostatic pressure, ionic strength and gaseous and nutritional environments. This chapter does not attempt to embrace all these factors but will emphasize analyses made under carefully defined conditions, particularly those involving chemostasis. Furthermore, we have singled out oxygen and carbon dioxide for detailed discussion because of their obvious importance in fungal fermentation research, because recent work with fungi has not been reviewed, and finally because we believe that studies in this area of fungal physiology should be encouraged. Unfortunately, the adoption of continuous-flow techniques by those working with filamentous fungi has been noticeably slow. Therefore we will refer, when appropriate, to related studies of yeasts especially where information from such studies suggests profitable lines of investigation and the desirability for urgent application to filamentous fungi. In part we see this contribution as having an essential didactic function. Finally, because of the general inexperience of mycologists in continuous-flow techniques, we have prefaced the main discussions by a brief consideration of fermentation systems for filamentous fungi, hoping thereby to encourage their wider usage in studies of these micro-organisms.

1.2 Fermenter Systems for Filamentous Fungi: Design and Operation

Introduction

The establishment of fungal cultures, especially continuous-flow types, presents a number of technical problems which have warranted the construction of specialized equipment for laboratory and large scale fermentations. Gerhardt & Bartlett (1959) stated that accumulated growth on surfaces inside the fermenters and within pipes and valves was the most serious limitation to continuous culture of mycelial organisms. This tendency for 'wall-growth' is thought to initiate from an electrostatic attraction between the organism and the surfaces involved (Munson & Bridges, 1964). Surface growth of this type decreases the culture volume and creates a physiologically heterogeneous culture. Also growth around medium inlet pipes may cause premature depletion of nutrients or result in blockages. Care also must be taken in the design of sensing probes because a coating of fungal growth over the sensitive area will result in impaired performance.

Stirred tank reactors (STR)

This type of fermenter can be used for both batch and continuous-flow cultivation of moulds and is convenient for laboratory and large-scale

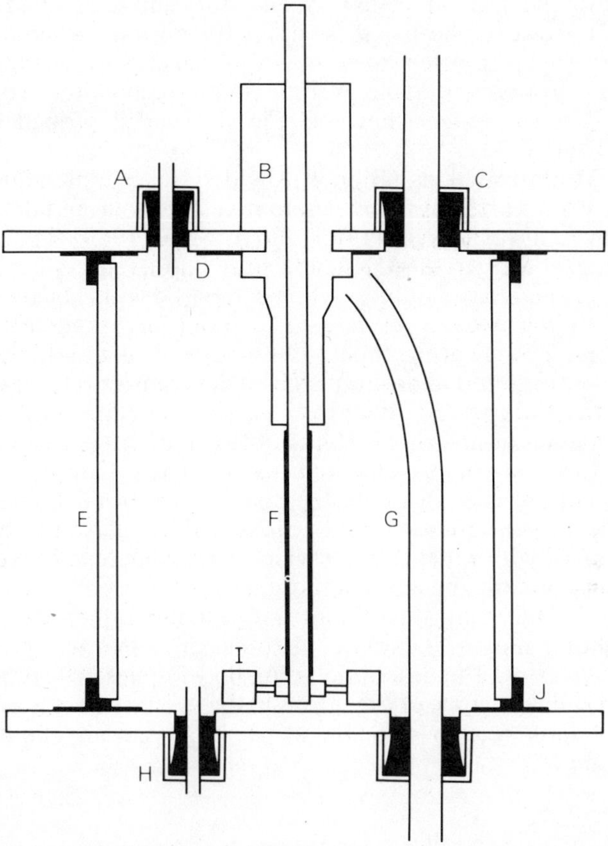

Fig. 1.1 A stirred tank reactor vessel for the continuous-flow culture of moulds. A: Inlet for air or medium; B: stirrer gland; C: outlet for sampling effluent gas; D: stainless steel top plate against the underside of which may be fitted a Teflon sheet (1 mm thick); E: Pyrex glass pipe section; F: stirrer shaft sleeved in Teflon; G: overflow pipe (normally a common effluent line for gas and spent culture); H: culture sampling line; I: paddle type stirrer; J: Neoprene rubber gasket. (From Rowley & Bull, 1973.)

experiments. A laboratory-scale apparatus suitable for the continuous-flow culture of fungi is shown in Fig. 1.1. The lines A, C, H and G are constructed from Teflon to prevent their being blocked with mycelium and to discourage surface growth. Partial blockage in the overflow lines (G) can result in the retention of mycelium in the vessel giving rise to an inadvertent feedback situation and, inevitably, to serious departures from the simple chemostat kinetics discussed in Volume 1 by Righelato (1975). Two design systems have been proposed to prevent such blockages: (1) a pinch-valve which periodically closes the overflow line allowing air pressure to build up in the vessel which, when released, has the effect of blowing any accreted mycelium out of the line (Righelato & Pirt, 1967); (2) an intermittent increase of the impeller speed which causes the shape of the culture vortex to change so that the take-off line is washed free of surface growth (Brunner & Röhr, 1972). However, both of these procedures cause transient increases in the culture oxygen transfer rate; they also cause periodic changes in the dilution rate. In order to avoid such fluctuations in culture conditions we favour the simpler system shown in Fig. 1.1 and described in detail by Rowley & Bull (1973).

In the STR illustrated, agitation is provided by a paddle impeller (I) and aeration is via a vortex system. Alternative forms of agitator have been described by Lengyel & Nyiri (1965) and by Steel (1969). The turbo-mixer used by Lengyel & Nyiri (1965) significantly improved gas transfer rates in *Penicillium chrysogenum* cultures when compared with a blade impeller. In an attempt to improve oxygen transfer without increasing hyphal damage resulting from high shearing, Steel (1969) developed a multiple rod impeller. The power required to produce a given gas transfer rate was reduced by 43% when the multiple rod rather than a turbine impeller was used. The gas transfer dynamics of mycelial cultures of fungi are little understood as the broths produced are non-Newtonian in character. One effect of such a thixotropic culture is to promote the coalescence of air bubbles, thereby reducing the oxygen solution rate. Because of this, many industrial fermentations are made with pellet cultures which approximate to Newtonian fluids and can, consequently, be aerated more easily.

Solomons (1972) has stated that a fermenter which has 'plug flow' characteristics is most suitable for microbial processes which follow zero-order reaction rates. The description plug flow implies little or no variation in the mean residence time of organisms in the vessel, whereas a continuous-flow stirred tank reactor (CSTR) has a large spread about the mean residence time.

Tower fermenters

Ross & Wilkins (1968) have published a patent describing a 'continuous microbiological process involving filamentous micro-organisms' in a cylindrical vessel. By attempting to achieve a plug-flow situation these authors sought to improve the volumetric efficiency (*VE*)* of fermentation of filamentous micro-organisms over that observed in CSTRs. Fresh medium was introduced at the bottom of the tower and the product harvested from the top; perforated disc baffles were situated at intervals in the vessel to prevent back mixing and thence give a closer approximation to plug flow. Cultures were aerated by sparging air through sintered discs and the efficiency of aeration was dependent on the morphology of the fungus. Three morphological types were identified: pellets, 'non-pellet aggregates' with diameters of about 0.1 mm and a filamentous mycelial form. Of these, the non-pellet aggregates gave the best volumetric efficiency in the fermenter. During continuous-flow growth, concentration gradients were established throughout the height of the tower giving rise to a 'heterogeneous steady state'.

A number of designs for vertical tubular reactors have been discussed recently by Greenshields & Smith (1974). The design of choice for moulds

* $VE = D\bar{x}/V$, where D is the dilution rate, $\bar{x}$ the steady-state biomass concentration and V the vessel volume.

appears to be a cylinder with an expansion chamber at the top and a pelleted growth morphology. An advantage of this system is its simplicity of construction and operation made possible by the lack of moving parts. Running costs are low as the energy requirements are far smaller than for systems requiring agitation. A biochemical engineering study of the aeration of *Aspergillus niger* cultures growing in such a tower fermenter has been made by Morris, Greenshields & Smith (1973).

Means, Savage, Reusser & Koepsell (1962) have described a novel fermenter for the continuous growth of filamentous micro-organisms which minimized congestion of process lines and head space (Fig. 1.2). The

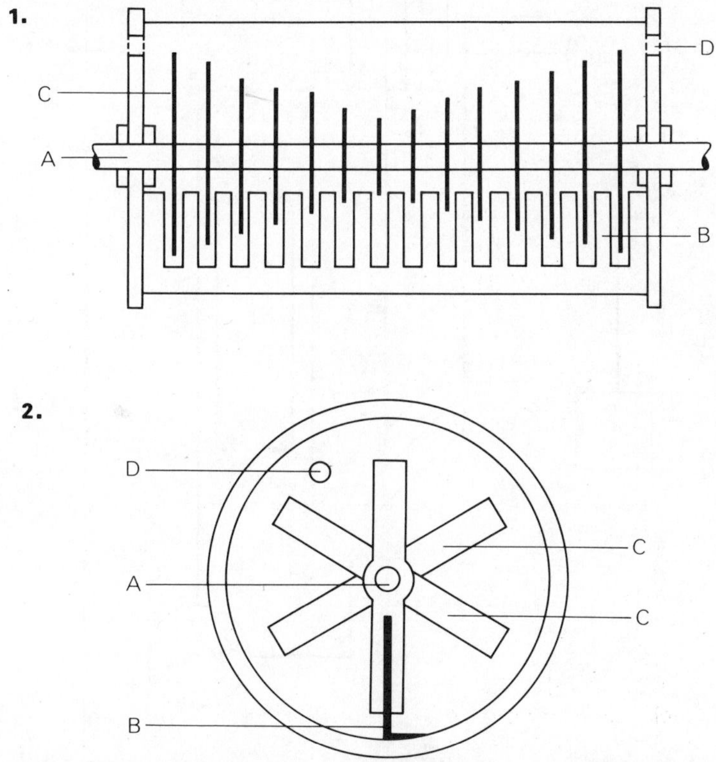

Fig. 1.2 A horizontal compartmental fermenter for the continuous-flow culture of fungi. 1, Longitudinal section; 2, cross-section. A: stirrer shaft; B: baffle plate; C: stirring blade; D: overflow port. (Redrawn from Means *et al.*, 1962.)

fermenter consisted of nine cylindrical compartments joined end on and mounted horizontally. The compartments were separated by plates with an overflow hole in the upper half. Agitation was provided by stainless steel blades mounted on a central shaft which swept passed a stainless steel comb-shaped baffle. Excellent performance of this fermenter was reported for the production of novobiocin by *Streptomyces niveus* and the bioconversion of steroids by *Septomyxa affinis*.

The cyclone column

A system for producing well mixed and aerated cultures without the use of an impeller has been described by Dawson (1963) (Fig. 1.3). Culture is pumped continuously from the base of a cylindrical reaction vessel at the rate of 6000 ml min^{-1} through a loop to the top whence it emerges as a jet,

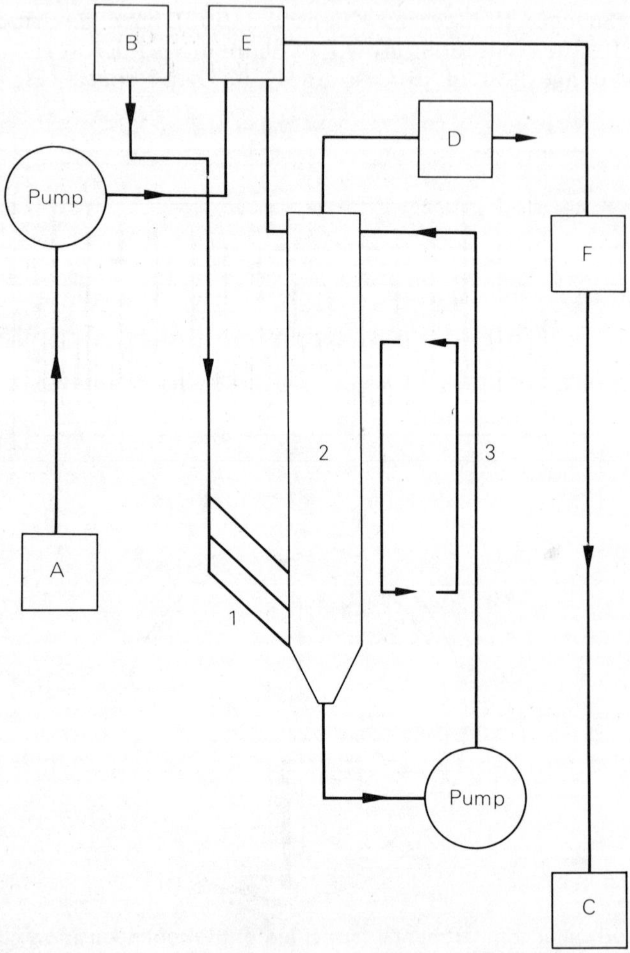

Fig. 1.3 The cyclone column fermenter. 1: Air and medium inlet ports; 2: reactor vessel; 3: culture recirculation limb. A: medium reservoir; B: air supply; C: culture receiver; D: exhaust air; E: volume control sensor; F: volume control solenoid. (Based on Dawson, 1963.)

tangentially to the walls of the reactor. The culture then descends as a curtain on the sides of the vessel until it reaches the bottom where the circulation process restarts. Good oxygen transfer rates and resistance to accreted growth and foaming were claimed, but as with the horizontal tubular reactor, little critical evaluation of the system for mould growth has been reported outside the author's laboratory.

1.3 Environmental Variables

Questions of nutrition and growth rate

It is not our intention to discuss the general nutrition of fungi; this topic is well covered in several texts and need not be reiterated. However, the need to define and control environmental variables when examining fungal nutrition does warrant brief mention. For example, the potential of carbon compounds as growth substrates may be realized or not according to the culture pH, while the requirement for organic growth factors is often a function of the medium dissolved oxygen tension (see pages 9 and 23). The requirement for water is so familiar that it is rarely studied critically and most investigations have centered on defining optimum water activity regimes for growth and sporulation. However, the requirement for hydrogen oxide *per se* is also subject to environmental control as the work of Mrtek, Crespi, Blake & Katz (1965) shows. These authors found that *Claviceps purpurea* grew in a minimal medium containing 99.6% deuterium oxide when glucose was the sole carbon source, whereas in succinate medium H_2O could be replaced only 75% by D_2O before growth was inhibited.

In this section some consideration will be given to the consequences of different nutrient limitations and concentrations on fungal growth. Use of chemostats for such purposes has clear advantages and enables unequivocal distinctions to be made between nutrient and growth rate effects. Nevertheless, batch cultures can and have been used in attempts to provide similar data. Thus, the design of media (a) in which the initial concentrations of nutrients are manipulated so that the first one to be exhausted varies and (b) which support different specific growth rates, greatly extends the range of useful batch culture operation, but in neither situation can definitive steady-state conditions be established.

One of the most thorough studies of selective exhaustion of nutrients was reported by Borrow, Jeffreys, Kessell, Lloyd, Lloyd & Nixon (1961) who used this approach to analyse the gibberellic acid fermentation. *Gibberella fujikuroi* was grown in media from which either glucose, ammonium nitrate or other nutrients would be exhausted at a desired culture biomass and the other nutrients so supplied that they would be exhausted in specific sequences thereafter. 'Balanced' growth (increasing biomass concomitant with constant rates of C, N, PO_4 Mg and K assimilation) was exponential in character and continued until the first nutrient was almost exhausted. Subsequent development of the fermentation was as follows, providing that oxygen limitation did not ensue: (1) with PO_4 and Mg exhaustion, growth continued at a linear rate until depletion of C or N ('transition phase') and endogenous metaphosphate was mobilized; (2) with N exhaustion, growth stopped but lipid and carbohydrate synthesis continued and gibberellin synthesis commenced ('storage phase'); (3) depletion of glucose in N-exhausted cultures led to a 'maintenance phase' during which endogenous lipids were metabolised and, on their eventual exhaustion, to hyphal degradation ('terminal phase').

Griffin, Timberlake & Cheney (1974) investigated the effect of growth rate on the macromolecular composition of *Achlya bisexualis* and used different nitrogen sources to vary μ over the range 0.04 h^{-1} ($t_d = 17$ h) to

$0.81\ h^{-1}$ ($t_d = 0.85$ h); the doubling time of 51 min is the fastest reported for a filamentous fungus. Griffin and his colleagues claimed that their data supported the notion of constant ribosome efficiency in protein synthesis, first proposed by Maaløe's group in Copenhagen. However, recent analyses of protein synthesis in carbon-limited chemostat cultures of *Candida utilis* (Alroy & Tannenbaum, 1973) and *Aspergillus nidulans* (M. E. Bushell & A. T. Bull, unpubl. data) question the validity of this hypothesis. The efficiency of RNA in protein synthesis increased linearly with dilution rate in both cases and reached a maximum at a D value of approximately half μ_{max}. The probable explanation of these observations is that the fraction of ribosomes involved in polysomal complexes, and thence synthesizing protein, increases with increasing growth or dilution rate; experimental support for this is afforded by studies of ribosome behaviour in bacteria (Varricchio & Monier, 1971; Harvey, 1973) and germinating spores of *Neurospora crassa* (Mirkes, 1974).

The affinities of a variety of fungi for different nutrients have been determined in chemostats. The selection of data which follows illustrate a number of physiological properties: K_s glucose, *Aspergillus nidulans*, 80–120 mg l^{-1} (Carter & Bull, 1969; M. E. Bushell & A. T. Bull, unpubl. data); K_s glucose, *Saccharomyces cerevisiae*, 108 mg l^{-1} (Leuenberger, 1971); K_s glucose, *Fusarium aquaeductuum*, 0.3 mg l^{-1} (Steensland, 1973); K_s glycerophosphate, *S.cerevisiae*, 4.6 mg l^{-1} (Francis & Hansche, 1972); K_s phosphate, *Rhodotorula rubra*, 1.0 μg l^{-1} (Button, Dunker & Morse, 1973); K_s thiamine, *Cryptococcus albidus*, 4.3×10^{-13}g (Button, 1969). The yeast and *Aspergillus* affinities for glucose represent typical values for carbon and energy substrate while the very low value reported for the sewage fungus, *F. aquaeductuum*, is consistent with its competitive colonization of polluted rivers. Affinities for vitamins are several orders greater than for macronutrients and the *Cryptococcus* data prompts the conclusion that, if thiamine is a significant factor in microbial marine ecology, it is probably below the range of current assay techniques. Francis and Hansche's paper is particularly interesting because it carefully details the selection of mutants of glycerophosphate-limited chemostats having reduced K_s values for the limiting substrate. Finally, the extremely low K_s for phosphate observed for *R. rubra* again reflects its natural growth situation in phosphate-depleted marine waters.

Carter & Bull (1969) have published results of the first systematic analysis of environmental factors (dilution rate, dissolved oxygen tension) on carbon catabolic pathways in fungi. They found functional Embden-Meyerhof-Parnas (EMP) and pentose phosphate (PP) pathways in chemostat cultures of *Aspergillus nidulans* and that the flux through the PP pathway was enhanced at dilution rates approaching μ_{max} and at low DOT values. Increased activity of the PP pathway was also characteristic of exponentially growing batch cultures and periods immediately preceding sporulation. These changes in glucose catabolism were interpreted in terms of the biosynthetic demands of the fungus in response to changing circumstances of growth. Subsequently these studies were extended to glucose-starved and glucose-maintained cultures of *A. nidulans* (Bainbridge, Bull, Pirt, Rowley & Trinci, 1971); predictably, all the glucose was catabolized via the EMP route within 10 h of terminating the glucose supply or reducing it to near the maintenance ration, a change which paralleled the complete decay of PP activity. Findings of this sort have been confirmed and extended by Smith and his coworkers with another species of *Aspergillus, A. niger* (Ng, Smith & Anderson, 1972). In a later paper Ng, Smith & McIntosh (1974) have described the growth-rate-dependent changes in enzymes of the EMP and PP pathways, the glyoxylate shunt and the tricarboxylic acid cycle under conditions of glucose and citrate limitation. Under glucose limitation the TCA cycle enzymes increased in activity as the dilution rate was raised, a result corroborating the less detailed observations in *A. nidulans* (Carter, Bull, Pirt & Rowley, 1971). For both species evidence for glucose induction of certain EMP and PP pathway enzymes was obtained. However, other environmental factors influence the operation of these catabolic routes as Hankinson's finding of nitrate-stimulated flux through the PP pathway suggests (Hankinson, 1974).

The study of fungal physiology at very low growth or dilution rates has had scant attention but the validity of a minimum specific growth rate, μ_{min}, has been substantiated by chemostat experiments on *Aspergillus nidulans* (Bainbridge *et al*; 1971) and *Penicillium chrysogenum* (Righelato, Trinci, Pirt & Peat, 1968). The μ_{min} may be defined as that growth rate below which all or part of the population ceases to grow and it becomes manifest in terms of metabolic and/or morphological differentiation. In *A. nidulans* for example, μ_{min} under glucose-limited continuous-flow culture conditions has a value of approximately 0.03 h^{-1}, *i.e.* about 15% μ_{max}; at such a dilution rate profuse conidiation occurs and ribosome efficiency approaches a minimum value (Bushell, McGetrick & Bull, 1973).

Oxygen

INTRODUCTION On the basis of present information it appears that most fungi require molecular oxygen for growth, that is to say they are strictly aerobic organisms. Obligate anaerobiosis in fungi has not been reported but facultative anaerobic growth, as distinct from anaerobic metabolism, is now well established. Fungi can grow over very wide ranges of medium oxygen tension and many of the early investigations of oxygen requirements were limited by the inability to measure very low oxygen concentrations and to establish rigorously anaerobic conditions. With hindsight it is clear to see that the quantitative relations of growth and oxygen availability would not be revealed by experiments in which other physical and chemical variables were neither monitored nor controlled. Moreover, careful quantitative measurements of growth frequently were not reported in the early literature and, as a result, conclusions on oxygen requirements have been based on semiquantitative or visual estimations. It is not surprising, therefore, to discover that much of the work attempting to define the response of fungi to aerobic, microaerophilic or anaerobic environments up to the late 1940s and 1950s is both difficult to evaluate and contradictory. Nevertheless, some important facts on fungal requirements for oxygen were revealed before this period. As early as 1900 Ternetz showed that fungi could grow under oxygen pressure from atmospheric to 10 mmHg and that growth and sporulation responded differently to changing partial pressures. Subsequently oxygen tensions of 550 mmHg and greater were found to inhibit the growth of several microfungi and yeasts without being lethal (Karsner &

Saphir, 1926). At this time also Tamiya was investigating the cytochromes of filamentous fungi (see Tabak & Cooke, 1968) and he claimed that in fungi containing cytochromes the rate of respiration was independent of the oxygen tension. The capacity of fungi to grow under low oxygen tensions is implicit in numerous early reports concerned with their distribution in soils and aquatic habitats. Among the Chytridiomycetes and Oömycetes, for example, are aquatic species ranging from those which are strictly aerobic to others which colonize stagnant, low oxygen environments; species in the latter category have a fermentative metabolism while the former have an oxidative type. Paterson (1960) analysed the parasitic growth of *Rhizosiphon anabaenae* on a cyanobacterium and found that it was dependent on the DOT of the water. Optimum growth of the chytrid occurred at sub-saturation levels but if the DOT was reduced below 120 mmHg growth was drastically reduced.

The pioneer investigations on the influence of oxygen, and other gases, on the growth of fungi have been discussed by Tabak & Cooke (1968) and the reader is directed to this comprehensive review for further information. The present discussion will concentrate on those more recent researches which attempt to define the nature of fungal responses to dissolved oxygen. Such research, still in its infancy, has had to await the development of suitable methods for oxygen measurement and environmental control.

Technical developments and considerations

MEASUREMENT OF DISSOLVED OXYGEN TENSION (DOT) The dropping mercury polarograph measures actual oxygen concentrations in a medium and was used in the earliest studies of fungal (yeast) respiration (Baumberger, 1939; Winzler, 1941). Although polarography has yielded valuable information on oxygen uptake rates and has enabled critical oxygen tensions (q.v.) to be determined it has inherent limitations: because it is a closed system the DOT of the fungal suspension changes continuously with time and experiments necessarily must be short termed. Recently an open polarographic system has been developed (Degn & Wolhrab, 1971; Degn, Lilleør & Iversen, 1973) but to date it has not been used for fungal studies.

Membrane electrodes of the applied potential or galvanic types can be used to monitor dissolved oxygen in growing populations. Galvanic probes generally have the best stability and the Mackereth version (Mackereth, 1964; Rowley, 1970) has proved very satisfactory for prolonged measurement of DOT in fungal cultures (Carter & Bull, 1971). The Mackereth electrode is not steam sterilizable, but the autoclavable galvanic probes now available tend to be rapidly overgrown by filamentous fungi and hence become ineffective. The Mackereth electrode measures the partial pressure of oxygen in a broth in equilibrium with a defined gas mixture. The calibration of membrane electrodes is not possible at DOT values below about 1 mmHg. This fact has important repercussions because the affinity of most aerobic microbes for oxygen is very great and their rates of respiration are DOT-dependent only at very low levels of oxygen. This technical limitation has considerably hampered the understanding of microbial responses to dissolved oxygen. An alternative approach is to follow changes in the redox potential (E_h) of a growing culture. From the Nernst equation it follows that the output from a redox electrode is a function of $\log_e$ DOT and, at low DOT, a small change in dissolved oxygen will appear as a large change in E_h. Unfortunately the relationship between DOT and E_h is likely to be distorted at low oxygen levels due to changes in the oxidation-reduction characteristics of culture liquors. Nevertheless, successful aeration monitoring and control of *Penicillium chrysogenum* cultures have been claimed for such a system (Lengyel & Nyiri, 1965; Nyiri & Lengyel, 1965). Details of redox potential measurements are given by Jacob (1970) while Harrison (1973) provides a valuable critique of their validity for measuring and controlling oxygen additions to cultures at DOT values below the sensitivity of polarographic probes.

Controlled growth environments

The number of published reports of fungi growing in continuous-flow culture under controlled DOT is extremely small and the feedback control system for DOT developed by Maclennan & Pirt (1966) appears to have been used in each case. With this system the DOT is controlled by varying the inflowing gas composition while maintaining the total flow rate constant. Electrically operated valves which work directly from a proportional controller are more convenient than the pneumatic type available to Maclennan and Pirt and recently Harper & Lynch (1973) have used such a valve during studies on *Mucor hiemalis*. This study, of ethylene production by *M. hiemalis*, forcibly illustrates the value of continuous-flow cultures in fungal physiology. The effect of oxygen on ethylene production in soil has been considered to be inhibitory, however, Lynch & Harper (1974) have shown that ethylene synthesis by one of the commonest soil fungi was stimulated by oxygen. A sharp linear increase in ethylene production occurred as the DOT was raised from 4 to 36 mmHg and thereafter remained constant up to air saturation. Lynch and Harper concluded that ethylene production in soil is not inhibited by oxygen *per se* but is diminished due to inadequate substrate availability under aerobic conditions.

This is not the place to enter into a discussion of oxygen supply to fungal cultures but some points need to be made, albeit briefly, because of their profound practical importance in the study of fungus–environment interactions. Fungi in submerged culture obtain oxygen in the dissolved state and the solubility of oxygen in water is low (about 200 μM at ambient temperatures and pressures); the solubility is lowered further by the presence of solutes and by increasing temperature. The specific oxygen uptake rate, q_{O_2}, of a culture is defined as the volume of O_2 utilized per unit biomass per unit time and is related to the concentration of dissolved oxygen (c_L)* and the Michaelis constant for oxygen (K_O) as follows:

$$q_{O_2} = q_{O_2(\max)} \frac{c_L}{K_O + c_L}.$$

The maximum specific oxygen uptake rate, $q_{O_2(\max)}$, is known as the specific oxygen demand of an organism under defined growth conditions and reported values for microfungi have been of the order 3 to 5 mmol g biomass^{-1} h^{-1}. Now the value of DOT (or c_L) at which the q_{O_2} falls

* The dissolved oxygen tension and the dissolved oxygen concentration are related thus, DOT $= Hc_L$, where H is the Henry's law constant.

below the maximum can be described as the critical oxygen tension, or concentration, and, clearly, under such conditions growth becomes oxygen limited. As Solomons has pointed out in a previous Chapter (Solomons, 1975) the overall transfer coefficient for oxygen is greatly affected by the apparent viscosity of a fungal culture which in turn is determined by the growth form of the fungus, *i.e.* whether yeast-like, filamentous or pelleted. Thus, very high viscosity, non-Newtonian cultures of filamentous fungi are difficult to aerate and in such cultures the apparent critical DOT increases. At the other extreme, pelleted, Newtonian cultures of fungi allow high rates of oxygen transfer to the bulk liquor but growth may be severely limited by the poor rate of oxygen diffusion into the pellet. Pirt (1966) calculated that oxygen limitations of growth would occur when the diameter of compact pellets exceeded about 150 μm, a prediction substantiated by Phillips (1966) data on *Penicillium chrysogenum* pellets. More recently Huang & Bungay (1973) have used microelectrodes to measure oxygen concentrations in and close to pellets of *Aspergillus niger.* These experiments revealed that in a dilute, air-saturated medium the oxygen gradient began to decrease about 200 μm from the pellet surface and at a depth of about 135 μm into the pellet oxygen was undetectable.

Effects of oxygen on fungal growth

AEROBIC GROWTH Cochrane (1958) concluded that, when colony development on solid media is used as a growth criterion, most fungi grow as well at oxygen partial pressures of 20 to 40 mmHg as at atmospheric pressure. However, only with the use of continuous-flow cultures have more precise descriptions of oxygen effects been possible. Thus, the steady-state specific oxygen uptake rate is given by

$$q_{O_2} = \frac{\mu}{Y_{O_2}} + m_{O_2},$$

where μ is the specific growth rate, Y_{O_2} the molar yield coefficient for oxygen and m_{O_2} the maintenance coefficient for oxygen. A plot of this equation gives a straight line for μ (or D, the dilution rate) against q_{O_2}, with the intercept on the ordinate giving the oxygen maintenance coefficient. Published data of this type for microfungi are rare but Fig. 1.4 gives plots of data obtained from glucose-limited chemostat experiments with *Aspergillus nidulans* (Carter *et al.*, 1971; M. E. Bushell and A. T. Bull, unpubl. data). The straight line obtained for *A. nidulans* 13 mel indicates that Y_{O_2} is independent of dilution rate; similar findings have been reported for *Torula utilis* (Tempest & Herbert, 1965) and *Penicillium chrysogenum* (Righelato *et al.*, 1968). However, deviation from linearity is found with a wild-type strain BWB224 and this result suggests that the yield coefficient for oxygen falls at dilution rates approaching D_{crit}. Values of 0.55, 0.60 and 0.75 mmol g biomass^{-1} h^{-1} can be calculated for the m_{O_2} of *A. nidulans* BWB224, *T. utilis* and *P. chrysogenum* respectively; the value for the hyaline mutant 13 mel of *A. nidulans* is about fourfold less (0.14 mmol g^{-1} h^{-1}).

When the effect of DOT on growth is to be analysed it is common practice to vary the pO_2 in the gas supply whilst holding the dilution rate of a chemostat culture constant. Under such conditions the steady state is

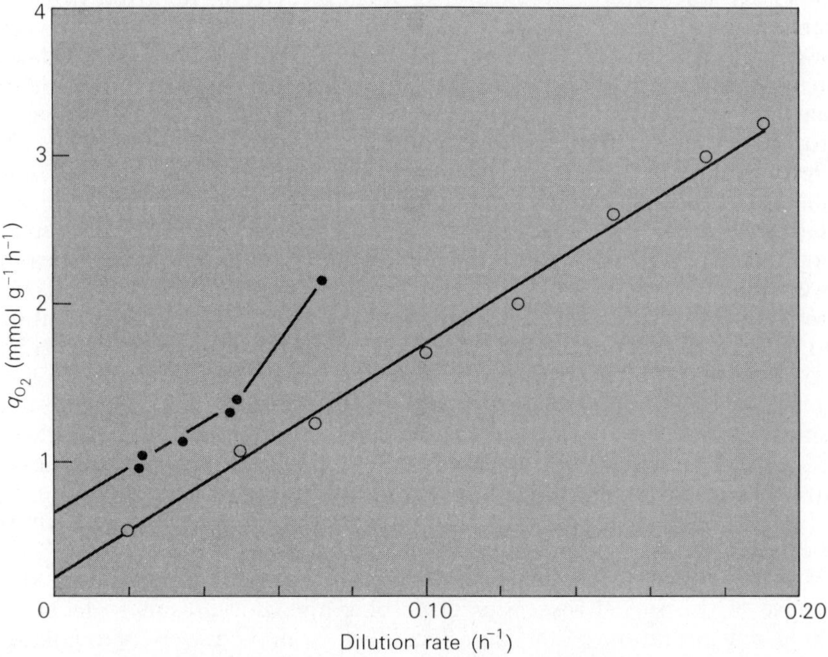

Fig. 1.4 Effect of dilution rate on the specific oxygen uptake rate of *Aspergillus nidulans*.

●, Wild type strain BWB224 ($\mu_{max} = 0.20\ h^{-1}$, $D_{crit} = 0.08\ h^{-1}$)

○, Mutant strain 13 mel ($\mu_{max} = 0.20\ h^{-}1$, $D_{crit} = 0.195\ h^{-}$).

(Data taken from Carter, Bull, Pirt & Rowley, 1971; M. E. Bushell and A. T. Bull, previously unpublished experiments 1972).

defined by the equation

$$Q' = K_L a(T_g - \text{DOT}),$$

where Q' is the oxygen uptake rate of the whole culture, K_La is the oxygen transfer coefficient and T_g the pO_2 in the inflowing gas. Harrison (1973) has pointed out that in well-regulated fermenters K_La should remain constant, thereby allowing changes in oxygen uptake rates with DOT to be sensitively detected. As far as we are aware the study of *Aspergillus nidulans* (Carter & Bull, 1971) is the only one of this type to be reported for a filamentous fungus. Carter & Bull (1971) set the dilution rate of a glucose limited chemostat at $0.05\ h^{-1}$ and used the Maclennan & Pirt (1966) method of DOT control. They found that the q_{O_2} increased by 23% as the DOT was reduced from air saturation to 10 mmHg and then fell sharply when the DOT was further reduced to 1.75 mmHg; data obtained in this way are termed '*in situ* respiration rates'. Increased respiration rates at low DOT also have been reported for *Candida utilis* (Moss *et al.*, 1969). Carter & Bull

(1971) also measured oxygen uptake rates of mycelia removed from the chemostat and supplied with an excess of all nutrients including oxygen; these data may be termed 'potential respiration rates' or the 'oxidative capacity'. From such experiments the critical oxygen tension could be measured and it was found to vary from 7 mmHg for mycelia grown at air saturation to 2.8 mmHg for mycelia grown at low DOT (2 mmHg). The different values of the *in situ* critical oxygen tension (1.75 mmHg) and those obtained by the exhaustion curve technique are quite consistent; respiration rates in the chemostat reflect glucose-limited growth and are lower than the potential respiration rates, thus the DOT would have to attain a significantly lower value before oxygen-limited respiration is imposed. These results also show that the critical DOT is itself dependent on environmental conditions and to think of this parameter in terms of a constant is invalid. The critical oxygen tensions for batch grown *A. niger* and *Penicillium chrysogenum* have been reported as 11 and 12 mmHg (Phillips & Johnson, 1961). The potential respiration rates of *A. nidulans* grown at air saturation and 2 mmHg were 1.31 (1.26) and 2.40 (1.51) mmol $O_2\ g^{-1}\ h^{-1}$ respectively; the values in parentheses are the respective *in situ* respiration rates. The higher potential and *in situ* rates under conditions of oxygen limitation accord with similar observations made of facultative anaerobic bacteria.

Harrison (1973) has concluded a recent review by suggesting that, because of the varied responses of aerobic micro-organisms to DOT, no simple interpretation of the data leading to a unified view of regulatory mechanisms is currently possible. Postulations for short-term regulation (enzyme kinetics, energy charge, reduced adenine nucleotide concentration) and longer term regulation (changes in enzyme and cytochrome concentrations) have been proposed, but the mechanism(s) of oxygen in induction or repression roles is little understood. Such understanding is almost entirely lacking with regard to filamentous fungi and research is urgently needed in this field.

ANAEROBIC GROWTH Enquiries into the anaerobic growth of fungi have been compromised by two factors: the failure to distinguish between anaerobic *growth* and anaerobic *metabolism* (fermentation), and the failure to establish rigorous anaerobic conditions. Thus, although *Blastocladia pringsheimii* apparently grew as well in nitrogen as in air it is an obligate aerobe and growth was supported by oxygen (0.7 mmHg) contaminating the N_2 (Cantino, 1956). This observation has been confirmed in subsequent more extensive studies by Held, Emerson, Fuller & Gleason (1969). Indeed, the number of authenticated reports of anaerobic growth of microfungi are sparse, although microaerophily, as we implied in the previous section, is a common phenomenon.

Results of a number of investigations have revealed that some fungi may grow anaerobically if appropriate growth factors are supplied, in other words, the organism may have impaired vitamin or sterol synthesis in the complete absence of oxygen. Early studies of *Saccharomyces cerevisiae* pointed to ergosterol and unsaturated fatty acid requirements for anaerobic growth in minimal media (Andreasen & Stier, 1954). The requirement for exogenous ergosterol has been examined further in *S. carlsbergensis* by David & Kirsop (1973); the anaerobically grown yeast contained markedly sub-aerobic concentrations of sterols but concomitantly squalene accumulation was observed. It would appear that in the absence of molecular oxygen the formation of squalene 2,3-epoxide (which subsequently cyclicizes to form the sterol nucleus, lanosterol) by the action of a mixed function oxidase, is blocked. Several species of *Mucor* can be grown in the absence of oxygen (Bartnicki-Garcia & Nickerson, 1962*a*) but in defined minimal media thiamine and nicotinic acid are essential growth factors for *M. rouxii* (Bartnicki-Garcia & Nickerson, 1961). In the latter situation the thiazole moiety completely replaces thiamine but nicotinic acid is only partly replaceable by 3-hydroxyanthranilic acid and its precursors from tryptophan; 3-hydroxy-anthranilic acid oxidase activity was absent from anaerobically grown mycelia. Bartnicki-Garcia & Nickerson (1962*b*) made two other interesting observations of anaerobically grown *M. rouxii*: only hexoses could be utilized as sources of carbon and energy; and, whereas in aerobic cultures growth rate and morphology (mycelial) were independent of spore inoculum size, under nitrogen both parameters were dependent on the inoculum size. Small inocula (less than 10^6 ml^{-1}) produced filamentous growth under nitrogen (Elmer & Nickerson, 1970).

Gunner & Alexander's data (1964) on the anaerobic growth of *Fusarium oxysporium* are much less readily interpreted. This species could grow in the complete absence of oxygen provided that the medium contained either yeast extract, MnO_2, NO_3^-, SeO_3^{2-} or Fe^{3+}. These authors concluded (1) that inorganic ions might function as electron acceptors or increase the positiveness of the redox potential, or both; (2) that organic growth factor requirements are imposed at low E_h values and that the presence of oxidants, including oxygen, alleviate the requirement. The induction of a dissimilatory nitrate reductase in *Aspergillus nidulans* in chemostats at very low DOT (B. L. A. Carter and A. T. Bull, unpublished results) may represent a similar phenomenon. In a glucose-limited, nitrate-containing medium this enzyme is repressed by high DOT and only becomes induced when the DOT is lowered to 1–2 mmHg. Moreover, at 1 mmHg nitrite accumulation in the medium is marked which suggests the involvement of nitrate in terminal electron transport.

We indicated above that *Blastocladia pringsheimii* (Chytridiomycetes: Blastocladiales) required oxygen, albeit in small concentration, for growth. Another species, *B. ramosa*, behaves as a facultative anaerobe (Held *et al.*, 1969); it lacks typical mitochondria under all growth environments and possesses cytochrome (b-type) only in trace quantities. A decreasing capacity for respiration is also seen in another group of aquatic fungi (Oömycetes : Leptomitales). Thus, species of *Saprolegnia* and *Mindeniella* have a respiratory metabolism, cristate mitochondria and aa_3, b_{564}, b_{557} and c-type cytochromes (Gleason & Unestam, 1968), whereas *Aqualinderella* totally lacks mitochondria and a cytochrome system (Held *et al.*, 1969). Neither the growth rate, molar growth yield from glucose nor the homolactic fermentative activity of *A. fermentans* were affected by oxygen. The requirements for cholesterol and a fatty acid (Held, 1970) parallel the nutritional demands of yeast under anaerobic conditions.

Finally in this section a brief comment on the development of anoxia in fungal cultures is appropriate. Several reports have been made of the recovery of microfungi from transient or long-term oxygen starvation and the implications of unwitting perturbations of this type, particularly of steady-state continuous-flow systems, will be obvious. Lengyel & Nyiri (1966) made a careful analysis of fluctuating q_{O_2} in a *Penicillium chrysogenum* fermentation and commented on the complexity of interactions between physical and biotic parameters. For example, the DOT increased when oxygen transfer from the gas to the liquid was inhibited by culture foaming, rose subsequently due to a falling q_{O_2} which, in turn, was caused by inadequate culture ventilation and the build up of CO_2 to inhibitory concentrations. Addition of antifoam removed the foam which caused the liberation of CO_2 and stimulated respiration; excess antifoam, while improving ventilation caused the q_{O_2} to fall because of impaired oxygen transfer from gas to liquid and liquid to organism. Huang & Bungay (1973) demonstrated that even the outer growing zone of pellets of *Aspergillus niger* became oxygen-depleted within almost 10 s of the air supply being interrupted.

EFFECTS OF HYPERBARIC OXYGEN Few studies have been reported on the effects of hyperbaric oxygen (gas phase in which pO_2 greater than in air) on fungi and the majority of these have referred to yeasts. Clearly it is essential to distinguish between the effects of oxygen *per se* and hydrostatic pressure; thus exposure to 100 atm. O_2 for 20 h kills *Saccharomyces cerevisiae* whereas nitrogen at the same pressure does not (Stuart, Gerschman & Stannard, 1962). Robb (1966) screened a large number of filamentous fungi against exposure to 10 atm. O_2 for varying periods and analysed the inhibitory effect on colony growth rates. She found that the lag time between return to ambient conditions and restoration of growth varied with the species and duration of exposure, generally longer exposures induced longer lags. These studies were extended by Pritchard & Hudson (1967) who suggested that the ability to survive prolonged exposure to hyperbaric oxygen was a function of induced catalase activity and the prevention of peroxide accumulation. Furthermore, they concluded that no correlation existed between survival capacity and initial catalase content, rather, the change in catalase activity in response to hyperbaric O_2 was significant. Support for this hypothesis was provided from analyses of hyperbaric O_2 toxicity during the cell cycle of *S. cerevisiae* and *Candida utilis* (Gifford & Pritchard, 1969). There were two points during the cycle at which the toxicity was nil or nearly so and these correlated with periods of catalase synthesis. Moreover, higher death rates occurred on substrates supporting rapid rates of growth in air, such as glucose, than those supporting slower rates such as ethanol.

SOME OBSERVATIONS ON FUNGAL GROWTH EFFICIENCY Growth efficiency may be considered in terms of biomass yield from a known concentration of growth-limiting substrate. For filamentous fungi the yield factor, Y (g biomass g substrate^{-1}) from sugars is usually in the range 0.40–0.50. The molar growth yield (g biomass mol substrate^{-1}) is a more useful parameter for growth efficiency studies because it enables direct comparison of substrates to be made. Typical values of $Y_{glucose}$ for aerobic, glucose-limited chemostat cultures are 77 (*Aspergillus nidulans*; Carter & Bull, 1969), 81 (*Penicillium chrysogenum*; Righelato *et al.*, 1968), 85 (*A. niger*; Ng, Smith & McIntosh, 1974) and 90 g mol^{-1} (*Fusarium aquaeductuum*; Steensland, 1973). Recently Mason & Righelato (1974) have compared the molar growth yields from glucose when *P. chrysogenum* was grown under conditions of glucose and oxygen limitation. When oxygen was limiting the $Y_{glucose}$ and Y_{O_2} were 93 and 100 g mol^{-1} compared with oxygen excess values of 73 and 40 g mol^{-1} respectively. The interest of these results lies in the fact that the energy substrate was utilized with greater efficiency at low DOT. Thornton (1965) measured the $Y_{glucose}$ for a range of fungi and found that the values fell into two distinct groups: 50–60 and 80–90 g mol^{-1}. Because of their lower molar enthalpy the molar growth yields from carboxylic acids are predictably less than those from sugars, *e.g.* $Y_{citrate}$ for *A. niger* is 27 g mol^{-1} (Ng, Smith & McIntosh, 1974). Aerobic and anaerobic molar growth yields have been determined for the facultative anaerobe *Blastocladia ramosa* (Held *et al.*, 1969); $Y_{glucose}$ (air) was 36 g mol^{-1} and $Y_{glucose}$ (H_2) was 30 g mol^{-1}, results indicative of an obligatory fermentative metabolism.

The problems of accurately determining fungal growth yields have been referred to by Solomons (1975) and the necessary experimental precautions to overcome them detailed by Stouthamer (1969). Again continuous-flow culture is the method of choice for such determinations.

Harrison & Lovelace (1971) have derived an expression such that

$$N = \frac{\mu}{Y^{ATP}} \cdot \frac{1}{q_{O_2} - m_{O_2}},$$

where Y^{ATP} is the molar yield for ATP and N the number of mols ATP produced per mol oxygen consumed. Assuming that two substrate level

Table 1.1 Yields of ATP per mole of Oxygen consumed by *Aspergillus nidulans*. Values of *N* calculated from data of Carter & Bull (1971), Carter, Bull, Pirt & Rowley (1971), M. E. Bushell and A. T. Bull (unpublished experiments) and assuming a Y^{ATP} of 10.5

	Dilution rate[a] **(h^{-1})**	***N***
BWB224	0.02	6.6
	0.05	6.5
	0.07	4.7
13 mel	0.05	6.1
	0.10	6.0
	0.15	5.8
	0.19	5.9
	Dissolved oxygen[b] **tension (mmHg)**	
BWB224	156	6.7
	30	5.8
	10	5.0
	2	5.0
	1.75	5.2

[a] DOT 156 mmHg
[b] D 0.05 h^{-}

phosphorylations and three sites of oxidative phosphorylation derive from the consumption of one molecule of glucose and six molecules of oxygen, then N equals 6.3. Similarly N equals 4.3 and 2.3 for two and one sites of oxidative phosphorylation respectively. Calculated values of N for *Aspergillus nidulans* are given in Table 1.1. Whereas a constant value of about 6 was obtained for carbon-limited strain 13 mel over a range of dilution rates, N fell to 4.7 when strain BWB224 was grown at dilution rates close to D_{crit}. The basis of this latter observation has not been examined but a partial uncoupling of oxidative phosphorylation is postulated. A less marked but progressive change in N occurred when the DOT was reduced from 156 to 1.75 mmHg.

COMPOSITION AND PRODUCT FORMATION Before concluding this discussion it is useful to consider how oxygen influences the chemical composition of and product formation by fungi. There are few systematic studies to review but some examples will illustrate the range of effects which oxygen induces. Of macromolecular constituents lipids have received most attention. A detailed turbidostat study of *Candida utilis* revealed that unsaturated fatty acids comprised about 80 to 90% of the total fatty acids, a result which was constant over a DOT range from air saturation to values too small to be monitored (Babij, Moss & Ralph, 1969). At low glucose concentrations (1 mg%) a change in DOT had little effect on fatty acid composition but at high glucose concentrations (greater than 1%) oxygen availability exerted significant qualitative and quantitative effects. Thus, a decrease in oleic acid ($C_{18:1}$) with increasing DOT was compensated by an increase in linolenic acid ($C_{18:3}$) while linoleic acid ($C_{18:2}$) remained fairly constant. The authors conclude that under highly aerated conditions the enzyme mediating multiple desaturation tends to catalyse an attack on the CH_3 end of oleic acid. Recent papers by Rogers & Stewart (1973) and Shih & Marth (1974) also include data on the lipid contents of *Saccharomyces cerevisiae* and *Aspergillus parasiticus* under various degrees of aeration. That metabolite pools vary with changes in oxygen tension is evident from the work of Brown & Johnson (1971). With decreasing DOT (75 to 1 mmHg) the concentrations of aspartate and arginine increased about 14- and 7-fold respectively while other amino acids remained constant. Such changes are likely reflections of an increasing fermentative metabolism. These data were obtained with carbon-limited cultures of *S. cerevisiae* but comparable analyses of filamentous fungi remain to be made.

Not surprisingly many studies have been made of changes in cytochrome and respiratory enzyme concentrations in response to DOT. Harrison (1972) has summarized much of this data but, although cytochrome concentrations were maximal under oxygen-limited conditions, little correlation could be made with potential respiration rates. However, the increase in potential respiration rate of *Saccharomyces cerevisiae* as a function of DOT is correlated with the synthesis of mitochondrial enzymes (Rogers & Stewart, 1973). Furthermore, the glyoxylate cycle enzymes were synthesized only at higher DOT's thereby enabling the organism to exploit the increasingly aerobic nature of its environment. In passing it should be noted that the regulation of many diverse types of enzymes may be modulated by oxygen: the secretion of α-glucosidase by *Mucor rouxii* (Flores-Carreon,

Reyes & Ruiz-Herrera, 1970) and the synthesis of ribonuclease by *Neurospora crassa* (Ueda, *et al.*, 1971) are examples of oxygen dependency.

Finally, the production of secondary metabolites by fungi is frequently influenced by oxygen availability. It is important to note that the critical dissolved oxygen tension and the minimum necessary for maximum product synthesis are distinct parameters; for the cephalosporin C fermentation the approximate values are 8 mmHg and 23 mmHg respectively (Feren & Squires, 1969). The relationship between these values may be reversed in other secondary metabolite fermentations. The rate of melanin production by *Aspergillus nidulans* is markedly DOT-dependent being maximal over the range 16 to 30 mmHg and decreasing below 16 mmHg (Rowley & Pirt, 1972). Earlier Carter & Bull (1969) had observed a strong correlation between melanogenesis and glucose catabolism via the EMP pathway and reported that the proportion of glucose metabolized via this route decreased from 78% at a DOT of 18 mmHg to 46% at a DOT of 2 mmHg. Subsequently Shih & Marth (1974) have demonstrated a similar correlation between the conditions of aeration, operation of the EMP pathway and aflatoxin production by *A. parasiticus.*

Carbon dioxide

INTRODUCTION Carbon dioxide has long been known to influence fungal morphogenesis and comprehensive reviews have been published by Tabak & Cooke (1968) and Smith & Galbraith (1971). In this Chapter discussion will be restricted to a consideration of the role of CO_2 as a regulator of fungal growth and metabolism. Since the assertion by Rockwell & Highberger (1927) that CO_2 is an essential prerequisite to the growth of bacteria, yeasts and moulds, a number of reports have described its stimulatory effects upon fungal growth, *e.g. Rhizopus* (Barinova, 1954), *Blastocladiella* (Cantino & Horenstein, 1956), *Fusarium* (Stover & Frieberg, 1958), soil fungi (Macauley & Griffin, 1969) and a number of coprophilous Pyrenomycetes (Harvey & Hodgkiss, 1972). Other authors have considered CO_2 as an inhibitor of fungal growth and Tabak & Cooke (1968) concluded that gas mixtures containing in excess of 95% CO_2 are, in general, inhibitory. In studies of the penicillin fermentation Lengyel & Nyiri (1967) found a CO_2 concentration of 1% in oxygen-saturated cultures to be optimal for growth and antibiotic production. The term 'ventilation' was coined by Nyiri & Lengyel (1968) in describing CO_2 transfer processes to distinguish them from aeration effects which are concerned with oxygen transfer and availability.

CARBON DIOXIDE FIXATION Carbon dioxide acts as a major source of carbon in a number of syntheses of fungal intermediary metabolites. Experiments with radioactively-labelled CO_2 demonstrated its incorporation into lactic, fumaric, citric and succinic acids in a number of species (Cochrane, 1958). The amino acids immediately associated with the TCA cycle have also been shown to act as acceptors of CO_2. Aspartate, glutamate and alanine all appear to be primary products of CO_2-fixing reactions in many fungi, for example, in the uredospores of *Ustilago phaseoli* (Staples & Weinstein, 1959). Similarly, arginine (Gitterman & Knight, 1952), tyrosine (Budd, 1969) and isoleucine (Hartman, Keen & Long, 1972) have been identified as radioactive products following $^{14}C—CO_2$ incorporation into *Penicillium chrysogenum*, *Neocosmospora vasinfecta* and *Verticillium albo-atrum* respectively. These results suggest that the enzymes responsible for CO_2 fixation are those which synthesize TCA cycle precursors or intermediates. The anaplerotic enzymes (see Chapter 5, this Volume) pyruvate carboxylase (EC 6.4.1.1.) and phosphoenol pyruvate carboxylase (EC 4.1.1.31.) which are capable of synthesizing oxaloacetate from CO_2 and pyruvate or phosphenol pyruvate respectively, appear to be the most likely candidates and both enzymes have been detected in Aspergillus species (Utter & Keech, 1963; Bushell & Bull, 1974*b*). Other authors have detected phosphoenol pyruvate carboxykinase (EC 4.1.1.32.) in fungi and have concluded that it operates in a reverse manner to its normal function in gluconeogenesis (*e.g.* Woronick & Johnson, 1960; Bachofen & Rast, 1967); however, the enzyme activities assayed *in vitro* have yet to be correlated with *in vivo* rates of fixation. Moreover, Hartman & Keen (1974) have argued convincingly that CO_2 is fixed in *V. albo-atrum* by anaplerotic enzymes and that phosphoenol pyruvate carboxykinase acts solely as a decarboxylase. The results of Rast & Bachofen (1967*a*, *b*) have suggested the presence of a further CO_2-fixing enzyme, β-methyl crotonyl CoA carboxylase (EC 6.5.1.5.) in *Agaricus bisporus*. The activity of this enzyme resulted in the incorporation of CO_2 into acetone.

Attempts have been made to distinguish between the utilization of gaseous CO_2 and dissolved bicarbonate as substrates for carboxylase activities. Macauley & Griffin (1969) concluded that the differential growth rates of fungi in soils of different pH could be explained by the fact that the equilibrium between gaseous CO_2 and dissolved bicarbonate is strongly pH-dependent, bicarbonate, in this case, acting as a C1 carbon source. This observation may reflect a general prerequisite for gaseous substrates to be dissolved, prior to their crossing the cell envelope. However, in the practical situation of a culture whose pH does not vary markedly, bicarbonate-mediated growth may be conveniently influenced by varying the partial pressure of gaseous CO_2 (see Ishizaki, Hirose & Shiro, 1971).

INFLUENCE OF CARBON DIOXIDE ON FUNGAL FERMENTATIONS Carbon dioxide fixation has a major role in the production of commercially important organic acids by fungi. The formation of citric acid from hexoses by *Aspergillus niger* proceeds via a split into two C3 compounds, one which is decarboxylated to C2 and the other carboxylated to C4 (Cleland & Johnson, 1954). Condensation of the C2 and C4 products then occurs to form citric acid. Itaconic acid produced by *A. terreus*, is formed in a similar manner (Bentley & Thiessen, 1957); *cis*-aconitate is formed from citrate which is then decarboxylated to form itaconate. Overman & Romano (1969) demonstrated the importance of pyruvate carboxylase in fumaric acid production by *Rhizopus nigricans*. The continued activity of the enzyme after the cessation of growth appeared to provide a carbon source for fumaric acid synthesis in the form of fixed CO_2.

In practice, ventilation parameters in organic acid fermentations are arrived at empirically. It would seem logical, however, that optimal conditions should provide a sufficient concentration of gaseous CO_2 in the culture to prevent the solubility equilibrium between dissolved bicarbonate and gaseous CO_2 becoming a limiting factor in CO_2 fixation. Nyiri & Lengyel

(1965) found an 'inhibitory critical level' of CO_2 above which penicillin production was impaired. Construction of a model system (Nyiri & Lengyel, 1968) produced data describing the effects of culture viscosity, due to mycelial concentration, buffer components and capacity, and agitation, on culture ventilation. The effect of increasing the mycelial concentration or of raising the culture viscosity with sucrose-glucose solutions appeared to inhibit the hydration of CO_2, *i.e.* to slow down the rate of bicarbonate formation. Further, it was found that increasing the impeller speed appeared to affect the mass transfer dynamics of CO_2 resulting in an increase in the rate of hydration.

A study of ventilation in continuous-flow cultures of *Aspergillus nidulans* has been made recently in our laboratory (Bushell & Bull, 1974*b* and unpublished results). Having concluded that a critical CO_2 partial pressure was required in *A. nidulans* cultures for the successful establishment of continuous-flow cultures from spore inocula (Bushell, McGetrick & Bull, 1973), the effect of manipulating ventilation parameters on biomass yield from glucose was investigated. Carbon dioxide fixation by pyruvate carboxylase and phosphoenol pyruvate carboxylase was responsible for biomass yield increases of 23% when the bicarbonate concentration in the culture broth was raised from 2.5 mM to 7.0 mM. Enzyme activity and resultant rates of CO_2 fixation were found to be strongly growth rate dependent. Pyruvate carboxylase decreased in activity as the dilution rate was increased while the phosphoenol pyruvate carboxylase activity reached a maximum at a dilution rate equivalent to approximately $1/2\mu_{max}$. Mycelium fractionation revealed that some 60% of fixed radioactive carbon appeared in the protein fraction in short-term labelling experiments. It would seem that the influence of CO_2 upon fungal growth is potentially of commercial interest and that the culture conditions which affect ventilation merit further investigation. Ishizaki and his colleagues of the Ajinomoto Co., Japan, have made an important and thorough study of this problem (see Ishizaki & Hirose, 1972) and have defined a number of culture parameters for the optimization of large-scale microbial processes.

Hydrogen ion concentration

GENERAL pH is not a unitary parameter and, as Cochrane (1958) pointed out, the response of fungi to it differs at different hydrogen ion concentrations. Thus, different points on a pH-growth plot may be interpreted in terms of effects on transport of nutrients, nutrient solubilities, enzyme reactions or surface phenomena. One is forced to conclude, therefore, that much of the data reporting pH optima for growth and fungal activities are of doubtful value because they have been obtained under ill-defined and non-steady state conditions.

As mentioned in the previous section, the culture pH influences the concentration of dissolved bicarbonate formed from gaseous CO_2. In this way, the prevailing pH and buffer capacity of the culture medium can influence fungal growth and product formation. Dissolved bicarbonate, pH and CO_2 are related by the Henderson-Hasselbalch equation:

$$\log\{HCO_3^-\} = pH - pK' + \log\{CO_2\}$$

$$K' = \frac{\{H^+\} \cdot \gamma H^+ \{HCO_3^-\} \cdot \gamma HCO_3^-}{\{CO_2\}}$$

(γH^+ and γHCO_3^- are activity coefficients for the hydrogen and bicarbonate ions respectively).

Thus, increasing the culture pH will tend to drive the equilibrium of the following equation to the left:

$$CO_2 + H_2O \rightleftharpoons H^+ + HCO_3^-.$$

Consequently, experiments to elucidate the effect of pH on growth should be designed with care as they may be difficult to assess in conditions of changing bicarbonate concentration. In this context the time-dependent variations in pH produced in fungal batch cultures should be interpreted cautiously.

The effect of varying pH in steady-state chemostat cultures of *Aspergillus nidulans* on biomass yield has been studied (M. E. Bushell & A. T. Bull, unpublished results) while keeping the dissolved bicarbonate concentration constant. The optimum pH for growth under conditions of glucose limitation and at 30°C was 6.9, the yield falling sharply at pH values on either side of the peak. In a study of melanin production by this fungus Rowley & Pirt (1972) found that pH 3.0 was the lower limit for growth and that washout from the chemostat occurred at values between 3.0 and 2.7. Culture pH also effects the growth habit of moulds. Pirt & Callow (1959) reported that increasing the pH of steady-state cultures of *Penicillium chrysogenum* above 6.0 caused a corresponding decrease in hyphal length which reached a minimum value at pH 7.0–7.4. At higher pH values extensive formation of swollen hyphae was observed. One interpretation of these results is that pH-dependent change in hyphal wall structure takes place, resulting in a decrease in the resistance of hyphae to impeller shear. Since the optimum pH for penicillin production is about 7.4, this lead Pirt & Callow (1959) to conclude that an ideal continuous-flow culture system for antibiotic production would comprise two stages, the first for growth with a pH not exceeding 7.0 and a second stage with a higher pH for penicillin production. The phenomenon of pellet formation (reviewed by Whitaker & Long, 1973) may be influenced by pH. In general, the tendency to form pellets increases as the pH value of the culture is raised.

PRACTICAL CONSIDERATION OF pH CONTROL One effect of lowering the culture pH is to decrease the propensity of fungi to grow on vessel surfaces. At pH values approaching 3.0 the accreted growth of *Aspergillus nidulans* is virtually eliminated (Rowley & Pirt, 1972). However, any advantage produced by this effect will be offset by the decreased yield, and by the unfavourable influence of low pH on the availability of metal ions when chelating agents such as citrate and EDTA are used.

Various manufactures have produced glass pH electrodes suitable for use as sensing probes in fungal cultures. The glass surface of these probes discourages the surface growth of fungi.

Mathematical modelling of the effects of pH on fermentation processes and its utility in calculating optimum conditions has been made by Andreeva & Biryukov (1973). These authors derived values for respiratory quotients and rates of penicillin production which gave good agreement with experimental data for *Penicillium chrysogenum*. They were then able to derive optimum pH values for maximum respiration rate and antibiotic production.

As the degree of sophistication and reliability of simulation techniques improves, similar studies should decrease considerably the time taken to derive optimum operating parameters for fungal fermentations.

Temperature

Closed culture systems are highly unsuitable for the elucidation of temperature effects on fungal growth. The reason for this assertion is clear: although temperature is an environmental parameter easy to control, changes in temperature produce simultaneous changes in other batch culture variables. Thus, an increase in incubation temperature within physiological ranges enhances the growth rate and Q_{10} values of 2 to 30 are common in the literature. Medium dissolved-oxygen tension is also temperature dependent and varies inversely with increasing temperature. Similarly nutritional and pH requirements for growth may be influenced by the temperature. Frequently fungi develop nutritional requirements when they are grown at temperatures approaching the upper cardinal point and the requirements of *Neurospora crassa* for riboflavin, *Saccharomyces cerevisiae* for pantothenate and *Coprinus fimetarius* for methionine at high growth temperatures clearly illustrate this phenomenon. Consequently, the large volume of data purporting to define the effects of temperature on fungal growth and activity is suspect and should be interpreted cautiously; the corollary, that the use of chemostats is necessary to evaluate temperature effects, is inescapable. One important practical point should be made here. In a continuous-flow culture the rate of medium evaporation is a polyfunctional parameter and is greatly affected by temperature. If evaporation is not accounted for or compensated, errors can be made in measurement of growth constants such as Y and m. These errors will be particularly serious at low dilution rates (below 0.05 h^{-1} according to King, Sinclair & Topiwala, 1972), dilution rates which are commonly established with fungal cultures, and at high temperatures.

Studies of temperature effects under steady-state conditions are few with respect to filamentous fungi; rather more data are available for yeasts and some will be included in this discussion. Rowley & Pirt (1972) observed that as the temperature of steady-state cultures of *Aspergillus nidulans* was raised from 23°C to 37°C the q_{O_2} increased from 1.54 to 3.24. The authors interpreted these results in terms of an increased maintenance energy requirement due to the higher turnover rates of protein and nucleic acids at the elevated temperatures. Over a similar range of temperature the specific growth rate of this fungus increased from 0.09 to 0.37 h^{-1} (Trinci, 1969) and a comparable increase in the colony growth rate, K_r, was observed. Trinci (1969) concluded that K_r was a reliable measure of the optimum temperature for fungal growth, a conclusion which has been substantiated by analyses of *Neurospora crassa* (Trinci, 1973). Temperature has a marked effect on all the growth constants: comparison of the growth of *Saccharomyces cerevisiae* at 25°C and 38°C showed that μ_{max} increased from 0.22 h^{-1} to 0.25 h^{-1}, while the substrate affinity and yield from glucose increased from 129 to 300 mg l^{-1} and decreased from 0.23 to 0.20 g g^{-1}, respectively (Jones & Hough, 1970). At the higher temperature ethanol, pyruvate and 2-oxoglutarate accumulation occurred in the medium and the level of 2-oxoglutarate dehydrogenase was reduced to nearly 10% of that at 25°C.

Thus, the primary effect of high temperature was the inhibition of 2-oxoglutarate dehydrogenase while changes in respiratory activity, metabolite concentrations and other enzymes reinforced the change to a more fermentative metabolism. A decreased yield from glucose accompanied by enhanced glycolytic activity has also been observed in filamentous fungi such as *Pythium* (Cantrell & Dowler, 1970).

The attention of several laboratories has been focussed on the biochemical bases of optimum, minimum and maximum temperatures for growth and what conclusions can be made for fungi derive almost entirely from yeast experiments. The factor determining minimum growth temperatures appears to be inactivation of solute transport systems, while temperature sensitivity of the protoplasmic membrane in the presence of metabolizable carbon substrates may be an important determinant of the maximum growth temperature (Hagler & Lewis, 1974). The interested reader is directed to the review of Farrell & Rose (1967) for a detailed assessment of this subject.

Data on temperature-induced changes in fungal composition are restricted to yeasts and some of the most interesting relate to nucleic acid and lipid. Brown & Rose (1969) made a detailed study of temperature effects on the composition of *Candida utilis* in carbon and nitrogen limited chemostats. At a fixed dilution rate dramatic increases in total RNA accompanied a reduction in temperature. When growth temperature is lowered, the attainable value of μ_{max} also is lowered so that the dilution rate becomes a larger fraction of μ_{max}, *i.e.* the yeast approaches a physiological state characteristic of unrestricted growth. The RNA content approached but did not exceed about 130 mg per g dry weight yeast, a value which may approximate to the maximum possible. An interesting corollary is that the increase in RNA concentration might determine the minimum temperature of *C. utilis* under substrate-limited conditions. Work in Rose's laboratory (Rose, 1969; Hunter & Rose, 1972) has also revealed significant changes in lipid content and composition of yeasts attending temperature reduction. The major changes involve increases in triglycerides, phospholipids and the proportion of unsaturated fatty acids.

Salt and ion effects

Tolerance of high salt environments has been looked at quite extensively but, once again few steady-state studies have been reported. When examining salt effects it is essential to know how they may be modulated by other environmental variables; the salt tolerance of microfungi, for example, is highest at the optimum growth temperature and minimal at the temperature limits. Recently Tresner & Hayes (1971) have surveyed the salt tolerance of a large number of terrestrial fungi. The effects observed were varied and included enhanced growth, increased sporulation and notable resistance of aspergilli and penicillia to 20% or more of sodium chloride. An attempt to interpret salt effects in mechanistic terms was made by Watson (1970). Working with carbon-limited chemostat cultures of *Saccharomyces cerevisiae*, Watson (1970) found that μ_{max} and $Y_{glucose}$ decreased when NaCl was increased from 0.25 M to 1.50 M. This response was due largely to a tenfold increase in the energy maintenance coefficient in the presence of the high NaCl, presumably to maintain the intracellular Na^+ concentration constant. The kinetics of glucose uptake also were altered by NaCl, much higher concentrations of sugar being required to maintain a particular uptake rate.

The effects of mono- and di-valent cations on the growth of *Dendryphiella salina* have been examined by Jennings and his colleagues. Growth was reduced by 40 and 25% in the presence of 200 mM NaCl and KCl respectively (Allaway & Jennings, 1970). Sodium ions inhibited the uptake of glucose by inhibiting its catabolism which, in turn, appeared to result from the leakage of K^+ from the mycelium. The inhibition by Na^+ was reversed almost completely by 10 mM $CaCl_2$ which reduced the loss of K^+, Na^+ and sugar alcohols from the fungus. An inhibition of glucose transport was found to be the basis of the potassium effect. Recently some interesting findings have resulted from studies of *Neurospora crassa* under magnesium-limited and magnesium-starvation conditions. Alberghina, Signorini, Trezzi & Viotti (1971) studied Mg^{2+} concentration dependency of growth rate in this fungus and reported a K_s for Mg^{2+} of 7 μmol. A significant increase in polyamine concentration occurred in mycelia of Mg-limited batch cultures as soon as intracellular levels of Mg^{2+} fell but the RNA content and growth were not affected (Viotti, Bagni, Sturani & Alberghina, 1971). These authors concluded that polyamines and magnesium may partially substitute for one another. Fluctuations in polyamine and Mg^{2+} concentration in response to dilution rate have been observed in carbon-limited chemostat cultures of *Aspergillus nidulans* (Bushell & Bull, 1974*a*). Magnesium concentration in the mycelium fell from about 100 μmol g biomass^{-1} to about 20 μmol g^{-1} as the dilution rate was raised from 0.05 to 0.18 h^{-1}. In contrast spermidine concentrations increased over this dilution rate such that the molar ratio of polyamine plus Mg^{2+} to RNA remained constant at approximately 2. Maintenance of this ratio may be critical for the stability of ribosomes or for RNA synthesis.

Very few other effects of metal and other ions on fungal growth have been analysed under steady-state conditions though a large literature reporting the results of batch experiments exists. Chelating agents now are in common use for controlling trace metal concentrations in culture media and the report of the effects of a variety of such agents on the growth of *Aspergillus niger* (Choudhary & Pirt, 1965) is noteworthy. Phenotypic changes in the respiratory chain of *Torulopsis utilis* have been induced in chemostats limited by substrates which are involved in various functional components of mitochondria (Fe, Mg, Cu, SO_4^{3-}, PO_4^{3-}). The results of such limitations on growth and in the metabolic dissection of energy conservation mechanisms have been reviewed by Light (1972). The significance of metal ions in the biosynthesis of fungal secondary metabolites has received considerable attention and the review of Weinberg (1970) should be consulted for details. Finally, reports such as that of Detroy & Ciegler (1971) indicate the role that metal ions have in fungal morphogenesis. In this case a reduction in the manganese concentration from 7.3 μmol to 0.73 μmol induces a mycelium–yeast dimorphism in *Aspergillus parasiticus*.

1.4 References

AINSWORTH, G. C. & SUSSMAN, A. S. (1965). *The fungi*, Vol. I. New York and London: Academic Press.

ALBERGHINA, F. A. M., SIGNORINI, R. C., TREZZI, F. & VIOTTI, A. (1971). Effects of magnesium on growth and morphology of *Neurospora crassa*. *Journal of Submicroscopical Cytology* **3**, 9–18.

ALLAWAY, A. E. & JENNINGS, D. H. (1970). The influence of cations on glucose transport and metabolism by, and the loss of sugar alcohols from, *Dendryphiella salina*. *New Phytologist* **69**, 581–93.

ALROY, Y. & TANNENBAUM, S. R. (1973). The influence of environmental conditions on the macromolecular composition of *Candida utilis*. *Biotechnology and Bioengineering* **15**, 239–46.

ANDREASEN, A. A. & STIER, T. J. B. (1954). Anaerobic nutrition of *Saccharomyces cerevisiae*. II. Unsaturated fatty acid requirement for growth in a defined medium. *Journal of Cellular and Comparative Physiology* **48**, 317–28.

ANDREEVA, L. N. & BIRYUKOV, V. V. (1973). Analysis of mathematical models of the effect of pH on fermentation processes and their use for calculating optimum fermentation conditions. *Biotechnology and Bioengineering Symposium* No. 4, 61–76.

BABIJ, T., MOSS, F. J. & RALPH, B. J. (1969). Effects of oxygen and glucose levels on lipid composition of the yeast *Candida utilis* grown in continuous culture. *Biotechnology and Bioengineering* **11**, 593–603.

BACHOFEN, R. & RAST, D. (1967). Carboxylierungstreaktionen in *Agaricus bisporus*. III Pyruvat und Phosphoenolpyruvat als CO_2-Acceptoren. *Archiv für Mikrobiologie* **60**, 217–34.

BAINBRIDGE, B. W., BULL, A. T., PIRT, S. J., ROWLEY, B. I. & TRINCI, A. P. J. (1971). Biochemical and morphological changes in non-growing cultures of *Aspergillus nidulans* with and without a maintenance ration of glucose. *Transactions of the British Mycological Society* **56**, 371–85.

BARINOVA, S. A. (1954). Effects of carbon dioxide on respiration in moulds. *Mikrobiologiya*, **23**, 521–6.

BARTNICKI-GARCIA, S. & NICKERSON, W. J. (1961). Thiamine and nicotinic acid: anaerobic growth factors for *Mucor rouxii*. *Journal of Bacteriology* **82**, 142–8.

BARTNICKI-GARCIA, S. & NICKERSON, W. J. (1962*a*). Induction of yeastlike development in *Mucor* by carbon dioxide. *Journal of Bacteriology* **84**, 829–40.

BARTNICKI-GARCIA, S. & NICKERSON, W. J. (1962*b*). Nutrition, growth and morphogenesis of *Mucor rouxii*. *Journal of Bacteriology* **84**, 841–58.

BAUMBERGER, J. P. (1939). The relation between 'oxidation-reduction potential' and the oxygen consumption rate of yeast cell suspensions. *Cold Spring Harbor Symposia on Quantitative Biology* **7**, 195.

BENTLEY, R. & THIESSEN, C. P. (1957). Biosynthesis of itaconic acid in *Aspergillus terreus*. I. Tracer studies with ^{14}C labelled substrates. *Journal of Biological Chemistry* **226**, 673–87.

BORROW, A., JEFFREYS, E. G., KESSELL, R. H. J., LLOYD, E. C., LLOYD, P. B. & NIXON, I. S. (1961). The metabolism of *Gibberella fujikuroi* in stirred culture. *Canadian Journal of Microbiology* **7**, 227–76.

BROWN, C. M. & JOHNSON, B. (1971). Influence of oxygen tension on the physiology of *Saccharomyces cerevisiae* in continous culture. *Antonie van Leeuwenhoek* **37**, 477–87.

BROWN, C. M. & ROSE, A. H. (1969). Fatty-acid composition of *Candida utilis* as affected by growth temperature and dissolved oxygen tension. *Journal of Bacteriology* **99**, 371–8.

BRUNNER, H. & RÖHR, M. (1972). Novel system for improved control of filamentous micro-organisms in continuous culture. *Applied Microbiology* **24**, 521–3.

BUDD, K. (1969). The assimilation of bicarbonate by *Neocosmospora vasinfecta*. *Canadian Journal of Microbiology* **15**, 389–98.

BUSHELL, M. E. & BULL, A. T. (1974*a*). Polyamine, magnesium and ribonucleic acid levels in steady state cultures of the mould *Aspergillus nidulans*. *Journal of General Microbiology* **81**, 271–2.

BUSHELL, M. E. & BULL, A. T. (1974*b*). Anaplerotic carbon dioxide fixation in steady and non-steady state fungal cultures. *Proceedings of the Society for General Microbiology* **1**, 69.

BUSHELL, M. E., MCGETRICK, A. M. T. & BULL, A. T. (1973). Chemostat culture of *Aspergillus nidulans*. Further analyses of kinetic and physiological parameters. *Proceedings of the Society for General Microbiology* **1**, 23.

BUTTON, D. K. (1969). Thiamine limited steady state growth of the yeast *Cryptococcus albidus*. *Journal of General Microbiology* **58**, 15–21.

BUTTON, D. K., DUNKER, S. E. & MORSE, M. L. (1973). Continuous culture of

Rhodotorula rubra: kinetics of phosphate-arsenate uptake, inhibition and phosphate-limited growth. *Journal of Bacteriology* **113**, 599–611.

CANTINO, E. C. (1956). The relation between cellular metabolism and morphogenesis in *Blastocladiella. Mycologia* **48**, 225–40.

CANTINO, E. C. & HORENSTEIN, E. A. (1956). The stimulatory effect of light upon growth and CO_2 fixation in *Blastocladiella. Mycologia* **48**, 777–9.

CANTRELL, H. F. & DOWLER, W. M. (1970). Relationships between temperature and glucose metabolism of *Pythium irregulare* and *P. vexans. Bacteriological Proceedings* **P153**, 147.

CARTER, B. L. A. & BULL, A. T. (1969). Studies of fungal growth and intermediary carbon metabolism under steady and non-steady state conditions. *Biotechnology and Bioengineering* **11**, 785–804.

CARTER, B. L. A. & BULL, A. T. (1971). The effect of oxygen tension in the medium on the morphology and growth kinetics of *Aspergillus nidulans. Journal of General Microbiology* **65**, 265–73.

CARTER, B. L. A., BULL, A. T., PIRT, S. J. & ROWLEY, B. I. (1971). Relationship between energy substrate utilisation and specific growth rate in *Aspergillus nidulans. Journal of Bacteriology* **108**, 309–13.

CHOUDARY, A. Q. & PIRT, S. J. (1965). Metal-complexing agents as metal buffers in media for the growth of *Aspergillus niger. Journal of General Microbiology* **41**, 99–107.

CLELAND, W. W. & JOHNSON, M. J. (1954). Tracer experiments on the mechanism of citric acid formation by *Aspergillus niger. Journal of Biological Chemistry* **208**, 679–89.

COCHRANE, V. W. (1958). *Physiology of fungi.* New York: Wiley.

DAVID, M. H. & KIRSOP, B. H. (1973). A correlation between oxygen requirements and the products of sterol synthesis in strains of *Saccharomyces cerevisiae. Journal of General Microbiology* **77**, 529–31.

DAWSON, P. S. (1963). A continuous-flow culture apparatus. The cyclone column unit. *Canadian Journal of Microbiology* **9**, 671–87.

DEGN, H., LILLEØR, M. & IVERSEN, J. J. L. (1973). The occurrence of a stepwise-decreasing respiration rate during oxidative assimilation of different substrates by resting *Klebsiella aerogenes* in a system open to oxygen. *Biochemical Journal* **136**, 1097–104.

DEGN, H. & WOHLRAB, H. (1971). Measurement of steady state values of respiration rate and oxidation levels of respiratory pigments at low oxygen tensions. A new technique. *Biochimica et Biophysica Acta* **245**, 347–55.

DETROY, R. W. & CIEGLER, A. (1971). Induction of yeast-like development in *Aspergillus parasiticus. Journal of General Microbiology* **65**, 259–64.

ELMER, G. W. & NICKERSON, W. J. (1970). Filamentous growth of *Mucor rouxii* under nitrogen. *Journal of Bacteriology* **101**, 592–602.

FARRELL, J. & ROSE, A. H. (1967). Temperature effects on micro-organisms. *Annual review of Microbiology* **21**, 101–20.

FEREN, C. J. & SQUIRES, R. W. (1969). The relationship between critical oxygen level and antibiotic synthesis of capreomycin and cephalosporin C. *Biotechnology and Bioengineering* **11**, 583–92.

FLORES-CARREON, A., REYES, E. & RUIZ-HERRERA, J. (1970). Inducible cell-wall-bound α-glucosidase in *Mucor rouxii. Biochimica et Biophysica Acta* **222**, 354–60.

FRANCIS, J. C. & HANSCHE, P. E. (1972). Directed evolution of metabolic pathways in microbial populations. *Genetics* **70**, 59–73.

GERHARDT, P. & BARTLETT, M. C. (1959). Continuous industrial fermentations. *Advances in Applied Microbiology* **1**, 215–60.

GIFFORD, G. D. & PRITCHARD, G. G. (1969). Toxicity of hyperbaric oxygen to yeasts displaying periodic enzyme synthesis *Journal of General Microbiology* **56**, 143–9.

GITTERMAN, C. O. & KNIGHT, S. G. (1952). Carbon dioxide fixation into amino acids of *Penicillium chrysogenum. Journal of Bacteriology* **64**, 223–31.

GLEASON, F. H. & UNESTAM, T. (1968). Comparative physiology of respiration in aquatic fungi. I. The Leptomitales. *Physiologia Plantarum* **21**, 556–72.

GREENSHIELDS, R. N. & SMITH, E. L. (1974). The tubular reactor in fermentation. *Process Biochemistry* **9**, 11–28.

GRIFFIN, D. H., TIMBERLAKE, W. E. & CHENEY, J. C. (1974). Regulation of macromolecular synthesis, colony development and specific growth rate of *Achlya bisexualis* during balanced growth. *Journal of General Microbiology* **80**, 381–8.

GUNNER, H. B. & ALEXANDER, M. (1964). Anaerobic growth of *Fusarium oxysporum. Journal of Bacteriology* **87**, 1309–16.

HAGLER, A. N. & LEWIS, M. J. (1974). Effect of glucose on thermal injury of yeast that may define the maximum temperature of growth. *Journal of General Microbiology* **80**, 101–9.

HANKINSON, O. (1974). Mutants of the pentose phosphate pathway in *Aspergillus nidulans*. *Journal of Bacteriology* **117**, 1121–30.

HARPER, S. H. T. & LYNCH, J. M. (1973). Modified system for the automatic control of dissolved oxygen in stirred microbial culture. *Laboratory Practice* p. 736.

HARRISON, D. E. F. (1972). Physiological effects of dissolved oxygen tension and redox potential on growing populations of micro-organisms. *Journal of Applied Chemistry & Biotechnology* **22**, 417–40.

HARRISON, D. E. F. (1973). Growth, oxygen and respiration. *CRC Critical Reviews in Microbiology* **3**, 185–228.

HARRISON, D. E. F. & LOVELACE, J. E. (1971). The effect of growth conditions on respiratory activity and growth efficiency in facultative anaerobes grown in chemostat culture. *Journal of General Microbiology* **68**, 35–43.

HARTMAN, R. E. & KEEN, N. T. (1974). The phosphoenolpyruvate carboxykinase of *Verticillium albo-atrum*. *Journal of General Microbiology* **81**, 21–6.

HARTMAN, R. E., KEEN, N. T. & LONG, M. (1972). Carbon dioxide fixation by *Verticillium albo-atrum*. *Journal of General Microbiology* **73**, 29–34.

HARVEY, R. J. (1973). Fraction of ribosomes synthesizing protein as a function of specific growth rate. *Journal of Bacteriology* **114**, 287–93.

HARVEY, R. & HODGKISS, I. J. (1972). The effect of CO_2 on the growth and sporulation of certain coprophilous Pyrenomycetes. *Transactions of the British Mycological Society* **59**, 409–18.

HELD, A. A. (1970). Nutrition and fermentative energy metabolism of the water mould *Aqualinderella fermentans*. *Mycologia* **62**, 339–58.

HELD, A. A., EMERSON, R., FULLER, M. S. & GLEASON, F. H. (1969). *Blastocladia* and *Aqualinderella*: fermentative water moulds with high carbon dioxide optima. *Science* **165**, 706–9.

HUANG, M. Y. & BUNGAY, H. R. (1973). Microprobe measurements of oxygen concentrations in mycelial pellets. *Biotechnology and Bioengineering* **15**, 1193–7.

HUNTER, K. & ROSE, A. H. (1972). Influence of growth temperature on the composition and physiology of micro-organisms. *Journal of Applied Chemistry and Biotechnology* **22**, 527–40.

IMANAKA, T., KAIEDA, T. & TAGUCHI, H. (1973). Optimization of α-galactosidase production in multistage continuous culture of mold. *Journal of Fermentation Technology* **51**, 431–9.

ISHIZAKI, A. & HIROSE, Y. (1972). Kinetics of carbon dioxide evolution in agitation systems. *Agricultural and Biological Chemistry* **37**, 1295–305.

ISHIZAKI, A., HIROSE, Y. & SHIRO, T. (1971). Studies on the ventilation in submerged fermentations. III. Behaviour of CO_2 in the submerged culture. *Agricultural and Biological Chemistry* **35**, 1860–9.

JACOB, H.-E. (1970). Redox potential. In *Methods in microbiology* Vol. 2, pp. 91–123. Edited by J. R. Norris and D. W. Ribbons. London and New York: Academic Press.

JONES, R. C. & HOUGH, J. S. (1970). The effect of temperature on the metabolism of baker's yeast growing on continuous culture. *Journal of General Microbiology* **60**, 107–16.

KARSNER, H. T. & SAPHIR, O. (1926). Influence of high partial pressure of oxygen and the growth of certain molds. *Journal of Infectious Diseases* **39**, 231–9.

KING, W. R., SINCLAIR, C. G. & TOPIWALA, H. H. (1972). Effect of evaporation losses on experimental continuous culture results. *Journal of General Microbiology* **71**, 87–92.

LENGYEL, Z. L. & NYIRI, L. (1965). An automatic aeration control system for biosynthetic processes. *Biotechnology and Bioengineering* **7**, 91–100.

LENGYEL, Z. L. & NYIRI, L. (1966). Studies on automatically aerated biosynthetic processes. II. Occurrence and eliminations of CO_2 during penicillin biosynthesis. *Biotechnology and Bioengineering* **8**, 337–52.

LENGYEL, Z. L. & NYIRI, L. (1967). Automatic aeration and the action of carbon dioxide in fermentation processes. *Industrie Chimique Belge* **32**, 798–800.

LEUENBERGER, H. G. W. (1971). Cultivation of *Saccharomyces cerevisiae* in continuous culture. I. Growth kinetics of a respiratory deficient yeast strain grown in continuous culture. *Archiv für Mikrobiologie* **79**, 176–86.

LIGHT, P. A. (1972). Influence of environment on mitochondrial function in yeast. *Journal of Applied Chemistry and Biotechnology* **22**, 509–26.

LYNCH, J. M. & HARPER, S. H. T. (1974). Formation of ethylene by a soil fungus. *Journal of General Microbiology* **80**, 187–95.

MACAULEY, B. J. & GRIFFEN, D. M. (1969). Effect of CO_2 and the bicarbonate ion on the growth of some soil fungi. *Transactions of the British Mycological Society* **53**, 223–8.

MACKERETH, F. J. H. (1964). An improved galvanic cell for determination of oxygen concentration in fluids. *Journal of Scientific Instruments* **41**, 38–41.

MACLENNAN, D. G. & PIRT, S. J. (1966). Automatic control of dissolved oxygen concentration in stirred microbial cultures. *Journal of General Microbiology* **45**, 289–302.

MASON, H. R. S. & RIGHELATO, R. C. (1974). Microbiological aspects of gas transfer problems. Discussion. *Proceedings of the Society for General Microbiology* **1**, 60.

MEANS, C. W., SAVAGE, G. M., REUSSER, F. & KOEPSELL, H. J. (1962). Design and operation of a pilot plant fermenter for the continuous propagation of filamentous micro-organisms. *Biotechnology and Bioengineering* **4**, 5–16.

MIRKES, P. E. (1974). Polysomes, ribonucleic acid and protein synthesis during germination of *Neurospora crassa* conidia. *Journal of Bacteriology* **117**, 196–202.

MORRIS, G. G., GREENSHIELDS, R. N. & SMITH, E. L. (1973). Aeration in tower fermenters containing micro-organisms. *Biotechnology and Bioengineering Symposium* **4**, 535–45.

MOSS, F. J., RICKARD, P. A. D., BEECH, G. A. & BUSH, F. F. (1969). The response by micro-organisms to steady state growth in controlled concentrations of oxygen and glucose. I. *Candida utilis*. *Biotechnology and Bioengineering* **11**, 561–80.

MRTEK, R. G., CRESPI, H. L., BLAKE, M. I. & KATZ, J. J. (1965). Effect of deuterium oxide on the saprophytic culture of *Claviceps*. I. Nutritional factors. *Journal of Pharmaceutical Science* **54**, 1450–3.

MUNSON, R. J. & BRIDGES, B. A. (1964). 'Take-over'—an unusual selection process in steady state cultures of *Escherichia coli*. *Journal of General Microbiology* **37**, 411–8.

NG, A. M. L., SMITH, J. E., MCINTOSH, A. F. (1974). Influence of dilution rate on enzyme synthesis in *Aspergillus niger* in continuous culture. *Journal of General Microbiology* **81**, 425–34.

NG, W. S., SMITH, J. E. & ANDERSON, J. G. (1972). Changes in carbon catabolic pathways during synchronous development of conidiophores of *Aspergillus niger*. *Journal of General Microbiology* **71**, 495–504.

NYIRI, L. & LENGYEL, Z. L. (1965). Studies on automatically aerated biosynthetic processes. I. The effect of agitation and CO_2 on penicillin formation in automatically aerated liquid cultures. *Biotechnology and Bioengineering* **8**, 343–54.

NYIRI, L. & LENGYEL, Z. L. (1968). Studies on ventilation of culture broths. I. Behaviour of CO_2 in model systems. *Biotechnology and Bioengineering* **10**, 133–50.

OVERMAN, S. A. & ROMANO, A. H. (1969). Pyruvate carboxylase of *Rhizopus nigricans* and its role in fumaric acid production. *Biochemical and Biophysical Research Communications* **37**, 457–63.

PATERSON, R. A. (1960). Infestation of chytridiaceous fungi on phytoplankton in relation to certain environmental factors. *Ecology* **41**, 416–24.

PHILLIPS, D. H. (1966). Oxygen transfer into mycelial pellets. *Biotechnology and Bioengineering* **8**, 456–60.

PHILLIPS, D. H. & JOHNSON, M. J. (1961). Aeration in fermentations. *Journal of Biochemical & Microbiological Technology and Engineering* **3**, 277–309.

PIRT, S. J. (1966). A theory of the mode of growth of fungi in the form of pellets in submerged culture. *Proceedings of the Royal Society, B*, **166**, 369–73.

PIRT, S. J. & CALLOW, D. S. (1959). Continuous flow culture of the filamentous mould *Penicillium chrysogenum* and the control of its morphology. *Nature, London* **184**, 307.

PRITCHARD, G. G. & HUDSON, M. A. (1967). Changes in catalase activity in higher plants and fungi treated with oxygen at high pressure. *Nature, London* **214**, 945–6.

RAST, D. & BACHOFEN, R. (1967a). Carboxylierungsreaktionen in *Agaricus bisporus*. I. Der endogene CO_2-Acceptor. *Archiv für Mikrobiologie* **57**, 392–405.

RAST, D. & BACHOFEN, R. (1967*b*). Carboxylierungsreaktionen in *Agaricus bisporus*. II. Aceton als ein CO_2-Acceptor. *Archiv für Mikrobiologie* **58**, 339–56.

RIGHELATO, R. C. (1975). Growth kinetics of mycelial fungi. In *The filamentous fungi* Vol. 1, pp. 79–103. Edited by J. E. Smith and D. R. Berry. London: Edward Arnold (Publishers) Ltd.

RIGHELATO, R. C. & PIRT, S. J. (1967). Improved control of organisms concentration in continuous cultures of filamentous micro-organisms. *Journal of Applied Bacteriology* **30**, 246–50.

RIGHELATO, R. C., TRINCI, A. P. J., PIRT, S. J. & PEAT, A. (1968). The influence of maintenance energy and growth rate on the metabolic activity, morphology and conidiation of *Penicillium chrysogenum. Journal of General Microbiology* **50**, 399–412.

ROBB, S. M. (1966). Reaction of fungi to exposure to 10 atmospheres pressure of oxygen. *Journal of General Microbiology* **45**, 17–29.

ROCKWELL, E. & HIGHBERGER, J. H. (1927). The necessity of CO_2 for the growth of bacteria, yeasts and moulds. *Journal of Infectious Diseases* **40**, 438–46.

ROGERS, P. J. & STEWART, P. R. (1973). Respiratory development in *Saccharomyces cerevisiae* grown at controlled oxygen tensions. *Journal of Bacteriology* **115**, 88–97.

ROSE, A. H. (1969). Temperature control of growth and metabolic activity. In *Fermentation advances*, pp. 157–75. Edited by D. Perlman, New York and London: Academic Press.

ROSS, N. G. & WILKINS, G. D. (1968). A continuous microbiological process involving filamentous micro-organisms. *British Patent* 1, 133, 875.

ROWLEY, B. I. (1970). Control of oxygen tension using the Mackereth electrode. In *Automation, mechanization and data handling in microbiology*, pp. 163–74. Edited by A. Baillie and R. J. Gilbert. London: Academic Press.

ROWLEY, B. I. & BULL, A. T. (1973). Chemostat for the cultivation of moulds. *Laboratory Practice* **22**, 286–9.

ROWLEY, B. I. & PIRT, S. J. (1972). Melanin production by *Aspergillus nidulans* in batch and chemostat cultures. *Journal of General Microbiology* **72**, 553–63.

SHIH, C.-N. & MARTH, E. H. (1974). Aflatoxin formation, lipid synthesis and glucose metabolism by *Aspergillus parasiticus* during incubation with and without agitation. *Biochimica et Biophysica Acta* **338**, 286–96.

SMITH, J. E. & GALBRAITH, J. C. (1971). Biochemical and physiological aspects of differentiation in the fungi. *Advances in Microbial Physiology*, **5**, 45–134.

SOLOMONS, G. L. (1972). Improvements in the design and operation of the chemostat. *Journal of Applied Chemistry and Biotechnology* **22**, 217–28.

SOLOMONS, G. L. (1975). Submerged culture production of mycelial biomass. In *The filamentous fungi* Vol. 1, pp. 249–264. Edited by J. E. Smith & D. R. Berry. London: Edward Arnold.

STAPLES, R. C. & WEINSTEIN, L. M. (1959). Dark carbon dioxide fixation by uredospores of rust fungi. *Contributions from Boyce Thompson Institute* **20**, 71–82.

STUART, B., GERSCHMAN, R. & STANNARD, J. N. (1962). Effect of high oxygen tension on potassium retentivity and colony formation of baker's yeast. *Journal of General and Comparative Physiology* **45**, 1019–30.

STEEL, R. (1969). Systems for high solid processes. In *Fermentation advances* pp. 491–514. Edited by D. Perlman. New York and London: Academic Press.

STEENSLAND, H. (1973). Continuous culture of a sewage fungus *Fusarium aquaeductuum. Archiv für Mikrobiologie* **93**, 287–94.

STOUTHAMER, A. H. (1969). Determination and significance of molar growth yields. In *Methods in microbiology* Vol. 1, pp. 629–63. Edited by J. R. Norris and D. W. Ribbons. London and New York: Academic Press.

STOVER, R. H. & FREIBERG, S. R. (1958). Effect of carbon dioxide on multiplication of *Fusarium* in soil. *Nature, London* **181**, 788–9.

TABAK, H. H. & COOKE, W. B. (1968). The effects of gaseous environments on the growth and metabolism of fungi. *The Botanical Review* **34**, 126–252.

TEMPEST, D. W. (1970). The place of continuous culture in microbiological research. *Advances in Microbial Physiology* **4**, 223–50.

TEMPEST, D. W. & HERBERT, D. (1965). Effect of dilution rate and growth limiting substrate on the metabolic activity of *Torula utilis* cultures. *Journal of General Microbiology* **41**, 143–50.

TERNETZ, C. (1900). Protoplasma bewegung und Furchkorperbildung bei *Ascophanus carneus. Jahrbuch für wissenschaftliche Botanik* **35**, 273–321.

THORNTON, D. R. (1965). The molar growth yields of a range of microorganisms. *Experientia* **21**, 384.

TRESNER, H. D. & HAYES, J. A. (1971). Sodium chloride tolerance of terrestrial fungi. *Applied Microbiology* **22**, 210–13.

TRINCI, A. P. J. (1969). A kinetic study of the growth of *Aspergillus nidulans* and other fungi. *Journal of General Microbiology* **57**, 11–24.

TRINCI, A. P. J. (1973). The hyphal growth unit of wild type and spreading colonial mutants of *Neurospora crassa. Archiv für Mikrobiologie* **91**, 127–36.

UEDA, K., TAKEBE, H., TAKAHASHI, J. & NOMOTO, M. (1971). Enlargement of the culture scale from flask to jar fermenter in the production of ribonuclease N_1 by *Neurospora crassa*: combined effect of dissolved oxygen and agitation on enzyme production. *Journal of Fermentation Technology* **49**, 981–8.

UTTER, M. F. & KEECH, D. B. (1963). Pyruvate carboxylase. I. Nature of the reaction. *Journal of Biological Chemistry* **238**, 2603–8.

VARRICCHIO, F. & MONIER, R. (1971). Ribosome patterns in *Escherichia coli* growing at various rates. *Journal of Bacteriology* **108**, 105–10.

VIOTTI, A., BAGNI, N., STURANI, E. & ALBERGHINA, F. A. M. (1971). Magnesium and polyamine levels in *Neurospora crassa* mycelia. *Biochimica et Biophysica Acta* **244**, 329–37.

WATSON, T. G. (1970). Effects of sodium chloride on steady state growth and metabolism of *Saccharomyces cerevisiae. Journal of General Microbiology* **64**, 91–9.

WEINBERG, E. D. (1970). Biosynthesis of secondary metabolites: roles of trace metals. *Advances in Microbial Physiology* **4**, 1–44.

WHITAKER, A. & LONG, P. A. (1973). Fungal pelleting. *Process Biochemistry* **8**, 27–31.

WINZLER, R. J. (1941). The respiration of baker's yeast at low oxygen tension. *Journal of Cellular and Comparative Physiology* **17**, 263–77.

WORONICK, C. L. & JOHNSON, M. J. (1960). Carbon dioxide fixation by cell-free extracts of *Aspergillus niger. Journal of Biological Chemistry* **235**, 9–15.

CHAPTER 2

Transport and Translocation in Filamentous Fungi

D. H. JENNINGS

2.1 Introduction

Fungal hyphae extend by growth at the tip. The important studies of Robertson (1959) have shown that growth at the tip depends on a suitable osmotic differential between the cytoplasm and the external environment. If this differential is suddenly changed, there is distortion of growth. We know from the studies of Robertson & Rizvi (1968) that there appears to be a constant osmotic pressure along hyphae which extend at their tip, because it is here that there is a lower wall pressure. But, of course, for growth to occur there must be absorption of solute from the external medium either by the tip itself or, if the hyphae are growing over an inert substrate, material must be translocated from those parts of the mycelium which are absorbing nutrients. In either case, if the medium contains the elements required for growth in a form which can only be made available after enzyme action, the fungus must secrete the necessary enzymes.

When the mycelium has been in contact with the substrate for a period of time, the material which the fungus is utilizing may become exhausted and further growth will require that another substrate be metabolized. As with bacteria, the metabolic processes within the hyphae become adjusted to the new substrate through various regulatory processes.

There is an implicit assumption in what has been considered above, namely that there are within the growth medium relatively high concentrations of suitable substrates for growth. But this need not be so and experience would suggest that we need to anticipate uptake systems with a high efficiency for scavenging the environment to maximize whatever nutrients are available.

In this description of the growth of a filamentous fungus, both transport of substances into the hyphae and translocation within the hyphae will be shown to be intimately involved in the process of growth. If we consider first transport of substances between the hyphae and the medium, it clearly must be a two-way process insofar as soluble compounds must enter and enzymes must be released into the medium. But as will be indicated, other solutes, particularly hydrogen and sodium ions and probably hydroxyl ions, are also released. This takes place at the boundary between the cytoplasm and the medium, the plasmalemma. There is now evidence to indicate that the plasmalemma can be thought of as not just a set of transport systems strung together in a lipid matrix but as an organized biochemical unit, in much the same way that the bacterial membrane now tends to be considered.

2.2 Ways of Studying Transport in Filamentous Fungi

There are a number of approaches to the study of transport of solutes into hyphae; none of course provide all the necessary information when used by itself.

Mutants

The ease with which mutants can be obtained has allowed the characterization of a large number of transport systems in unicellular micro-organisms. The same is true for filamentous fungi. The approach used is to find mutants which will not grow or grow only slowly in the presence of a certain solute, then show that if it is a compound which is normally metabolized, all the necessary enzymes for metabolism are present inside the mycelium. From such information, one can only conclude that these enzymes are isolated from the external medium by a barrier, the plasmalemma, and therefore the appropriate transport systems must be absent from the mutant. Alternatively, a non-metabolizable analogue of the solute in question can be used and shown that it can no longer be absorbed by the mutant. In both cases, genetic analysis will show whether or not one or more systems are present. A good example of this discriminatory power of genetic analysis comes from *Neurospora crassa* where genetic studies have confirmed the conclusions of physiological studies that there is a single transport system linking potassium uptake with loss of hydrogen and sodium ions from the cell (Slayman & Tatum, 1965*b*).

Flux analysis

The movement of a solute in either direction across the plasmalemma provides information about how the net movement of the solute into or out of the mycelium is brought about. Thus accumulation of the solute may be due to an essentially one-way flux from medium to cytoplasm or due to metabolism stimulating influx or reducing efflux. Analysis of the fluxes of potassium, hydrogen and sodium ions into and out of the mycelium of *Neurospora crassa* (Table 2.1) has provided the physiological information, referred to above, from which it has been concluded that potassium influx is very closely coupled to efflux of hydrogen and sodium ions. Although net fluxes were measured in this instance, supporting information showed that these fluxes could be considered as being essentially unidirectional.

The best procedure for obtaining information about fluxes is to pretreat mycelium with radioactive solute, wash the mycelium and place it in a new medium containing the solute which is now non-radioactive. Measurements of the rate of appearance of radioactivity in the medium can be used to calculate rates of efflux. Measurements of the change in concentration of the solute in the medium using parallel samples that do not contain the radioactive isotope gives the net flux. Subtraction of the efflux from the net flux gives the influx.

Table 2.1 The ion fluxes which have been observed to occur when mycelium of *Neurospora crassa* is put in a medium whose composition is also given. (Data from Slayman & Slayman, 1968)

Ion	Flux (pmol $cm^{-2}s^{-1}$)	Direction	Type
K^+	0.20–0.46	Outwards	Passive
K^+	15.0	Inwards	Active
Na^+	9.7	Outwards	Active
H^+	5.6	Outwards	Active
$^{36}Cl^-$	1.32	Inwards	Active?

Medium Composition	($mM\ l^{-1}$)
H^+	1.6×10^{-3}
K^+	30
Na^+	25
Cl^-	30
3,3-dimethyl glutamate	20
glucose	0.55

Flux measurement using green plant cells provide information not only about the mechanism of transport of a solute but also about compartmentation of the solute inside the cell. With filamentous fungi, it is customary to assume that the mycelium behaves as a single compartment. This assumption has not been tested rigorously. From the work of Park & Robinson (1967) it is known that fungal hyphae vacuolate as they mature and there is now plenty of evidence from electron microscopy that the cytoplasm of hyphae contains many membrane-bounded spaces, so that compartmentation of solutes is highly likely. However, in instances where the appropriate measurements have been made, it would seem that the fluxes across the membranes of these compartments are many times faster than across the plasmalemma or the solutes in question are unable to penetrate the compartments.

However, there are two instances where it is clear that fungal mycelium behaves as a two or more compartment system. Slayman & Slayman (1970) showed that above pH 8.0, potassium uptake in *Neurospora crassa* can be resolved into two distinct exponential components. There is a fast component (time constant—1.2 min) which is matched quantitatively by a rapid loss of sodium. This exchange can be attributed to ion exchange within the cell wall, since it is comparatively insensitive to low temperatures and metabolic inhibitors. The slow component is uptake into the cytoplasm.

Previous studies (Slayman & Tatum, 1965*a*) had shown that steady-state potassium exchange at pH 5.8 across the plasmalemma shows only first-order kinetics. From this it can be concluded that the cytoplasm behaves as only one compartment with respect to potassium. On the other hand, Jennings & Aynsley (1971), from an examination of steady-state fluxes of potassium at 2°C, showed that the mycelium of *Dendryphiella salina* behaves as a multi-compartment system. The fluxes are almost certainly between the cytoplasm and the medium. The probable explanation for this difference between *Neurospora crassa* and *D. salina* may be due to the bursting of hyphal tips in the former fungus during washing with distilled water (Robertson, 1959) prior to the measurement of potassium tracer efflux. If this is the case, the consequence of this phenomenon need to be examined further, particularly with respect to the measurements of fluxes of other solutes between the medium and the mycelium of *Neurospora crassa*.

In most instances flux measurements for filamentous fungi are expressed as a function of mycelial dry weight and not in terms of hyphal surface area, which is more customary for other eukaryotic cells. There are of course problems in measuring surface areas but they are not insuperable. Slayman & Slayman (1968) determined the surface area of *Neurospora crassa* from data for the inulin-impermeable space in a mycelial suspension of known volume.

Competition studies

These can be considered as being measurement of fluxes of a solute in the presence of other solutes which also combine with the transport system. Of course, the underlying idea to this approach to transport in fungi comes from a consideration of solute transport in terms of permeases. This has meant the kinetics of transport are formulated with respect to those for enzymes, particularly where two substrates compete. In actual fact, the kinetic basis of such transport competition studies is often rudimentary. Nevertheless, by studying the interaction between solutes for transport into mycelia, it has been possible to demonstrate the presence of a number of carrier systems for a particular compound.

However, not all studies are based on completely secure foundations. This is because attention is focussed on one compound, whose uptake is measured radioisotopically. Only the effects of various other structurally related solutes on the uptake of the labelled compound are studied. There is often no investigation of the uptake of competing solutes either alone or in the presence of the compound which is the focus of the study.

Electrical measurements

Only two sets of electrical measurements have been made in fungi—on hyphae of *Neurospora crassa* by C. L. Slayman and his colleagues (Slayman, 1965*a*, *b*; Slayman, Lu & Shane, 1970; Slayman & Slayman, 1962) and on the sporangiophores of *Phycomyces blakesleeanus* by Cowan, Lewis & Thain (1972) and Cowan, Thain & Lewis (1972*a*, *b*). But if the mechanism of transport of an ion into hyphae is to be fully understood, electrical measurements are essential. The biophysics of ion transport is described fully elsewhere (Lüttge & Pitman, 1975) and it is not appropriate to discuss the subject in detail here. The point to stress is that, while electrical measurements on fungi are difficult, they are technically feasible. It is salutory to note that the first unequivocal demonstration of an electrogenic pump in plant cells was made by Slayman (1965*a*, *b*). Further, although the majority of measurements on the hyphae of *Neurospora crassa* were made using those which were mature and about 20 μm wide, it is possible to insert electrodes into hyphae not far behind their tips (Slayman & Slayman, 1962).

2.3 Features of Transport of Solutes into Hyphae

Diffusion

There is not much information about the passive permeability of the outer membrane of filamentous fungi. However, two observations indicate that the permeability of the plasmalemma to potassium ions is not very different from green plant cells (Table 2.2). It should be noted that it is likely that 20%

Table 2.2 Information about passive flux of potassium from fungal hyphae with comparable data for large coenocytes of the alga *Nitella translucens*

	Flux (pmol cm^{-2} s^{-1})
Neurospora crassa (Slayman & Slayman, 1968)	
Loss into buffer (pH 5.8) from mycelium poisoned with 1.0 mM sodium azide	0.20
Loss into distilled water from untreated mycelium	0.33
Unidirectional flux extrapolated to the minimum extracellular concentrations (0.5 mM) at which cells stay in the steady-state	0.46
Dendryphiella salina (Jennings & Aynsley, 1971)	
Low temperature (2 °C) efflux	0.14
Nitella translucens (MacRobbie, 1962)	
Passive potassium efflux	0.85

of the potassium in *Dendryphiella salina* may exchange at a rate faster than that given in the table. Nothing more definite can be said at the moment because there is insufficient information about the distribution of potassium within the hyphae. Although calcium does not seem to be required as a macro-nutrient by fungi, it appears to be needed for membrane integrity. Calcium can prevent the increased loss of solutes from hyphae of *Dendryphiella salina* which is brought about by monovalent cations (Jones & Jennings, 1965; Allaway & Jennings, 1970*b*; Jennings & Austin, 1973). Other bivalent ions can perform a similar role but calcium seems the most effective. Comparable observations have been made for higher plant cells (Jennings, 1969).

As might be expected, weak acids and bases readily penetrate hyphae. MacMillan (1956) showed that this was the case for ammonia, the distribution of the solute between the medium and the mycelium of *Scopulariopsis brevicaulis* could be predicted reasonably from a knowledge of the internal and external pH and the assumption that it is the uncharged molecule which crosses the plasmalemma (Fig. 2.1). Taber (1971) examining the uptake of succinate and related acids by *Claviceps purpurea* found that it was biphasic. The first phase of 1–3 min was almost unaffected by pH (4.1 and 6.8), temperature and metabolic inhibitors. This suggests that the process is one of diffusion which supported by the linear relation between uptake over one minute and the lack of effect of equimolar concentrations of other acids,

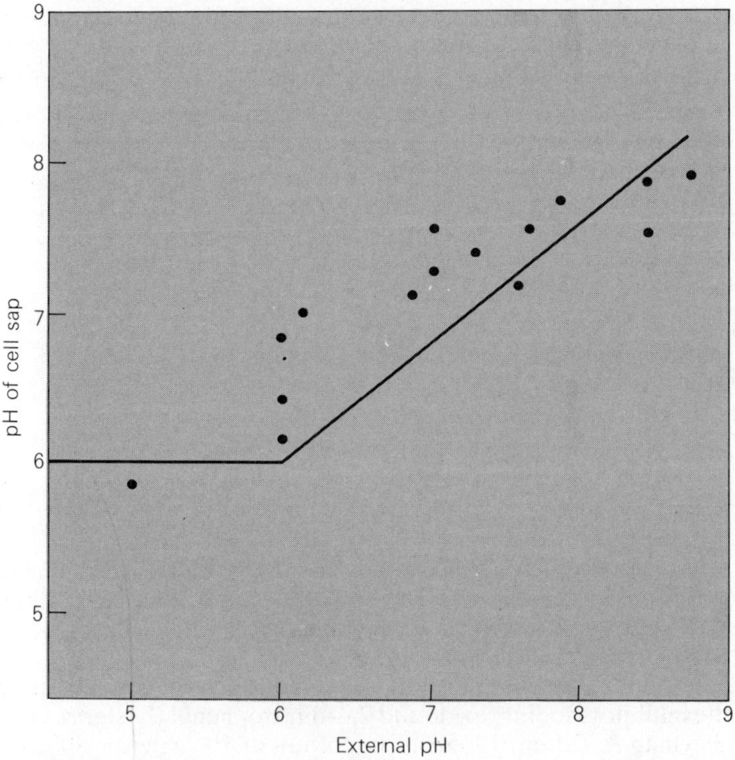

Fig. 2.1 Effect of shaking mycelium of *Scopulariopsis brevicaulis* in buffered solutions containing 0.4 mg NH_3–N cm^{-3} on the pH of the expressed sap. The line shows the calculated internal pH. From MacMillan, 1956, *Journal of Experimental Botany* **7**, 113–26; © 1956 Oxford University Press, and by permission of The Clarendon Press, Oxford.

which militates against binding of succinate to positive charges on the cell wall. The second phase of uptake continues for a longer period and is associated with metabolism.

Ion transport

It is a pity that there is no coherent body of information about the ion content of filamentous fungi. Table 2.3 gives some information for *Neurospora crassa* but it relates only to mycelium growing in a medium with a relatively low potassium concentration. In this particular instance, cations are largely balanced by phosphate. In *Fusarium oxysporium* organic acids have been shown to play an important role in balancing potassium (Shere & Jacobson, 1970*a*).

It is not clear to what extent ions contribute to the internal osmotic pressure of hyphae. It is probable that uncharged compounds can play an important role as indicated by data (Table 2.4) for *Dendryphiella salina*, which shows that the synthesis of mannitol and arabitol can make a considerable contribution to the osmotic pressure of mycelium adjusting to the presence of varying concentrations of sea water in the external medium. Much more information of this sort is required.

Table 2.3 The concentration of some of the major inorganic components of the mycelium of *Neurospora crassa* after growth in a medium whose composition is also given. (Data from Slayman & Slayman 1968)

	Medium ($mM\ l^{-1}$)	**Mycelium ($mM\ kg^{-1}$ cell water)**
K^+	0.2	56
Na^+	45.2	107
$H_2PO_4^-$	36.8	12
Poly P		69
Organic P (low molecular weight)		90
Nucleic acid P		129
Citrate	8.4	
NH_4NO_3	25.0	
$MgSO_4$	0.81	
$CaCl_2$	0.68	

POTASSIUM TRANSPORT This system has been extensively studied in *Neurospora crassa* (Lester & Hechter, 1958, 1959; Slayman, 1965*a*, *b*; Slayman, C. L., 1970; Slayman & Slayman, 1968, 1970; Slayman, C. W., 1970; Slayman & Tatum, 1964, 1965*a*, *b*).

Evidence that potassium transport is metabolically driven comes from the use of the inhibitors sodium azide and 2,4-dinitrophenol (Lester & Hechter, 1958; Slayman & Tatum, 1965*a*). The former at 10^{-4} M gives 80% inhibition of respiration and almost complete inhibition of potassium uptake. Loss of potassium of any magnitude from the mycelium only occurs at higher concentrations of the inhibitor. The system has a high degree of specificity for potassium, though it also transports rubidium. The affinity for rubidium is 40% of that for potassium at 5×10^{-3} M concentration of each ion (Slayman & Tatum, 1964). Little sodium is transported into the hyphae.

The system brings about the exchange of potassium for hydrogen or sodium ions (Slayman & Slayman, 1968). The system is probably also responsible for the rapid exchange of potassium between the mycelium and the external medium (Slayman & Tatum, 1965*a*). Such exchange has a V_{max} of 2×10^{-2} M per kg of cell water per min, which is the same as exchange of potassium for sodium and hydrogen ions. Further, a single gene change can produce mycelium in which K–K exchange and net potassium uptake are equally affected (Slayman & Tatum, 1965*b*; Slayman, C. W., 1970). The Michaelis constants for external potassium for K–K exchange and K–Na–H exchange are quite different: 1×10^{-3} M in the former instance and 11.8×10^{-3} M in the latter.

Slayman & Slayman (1970) have examined the kinetics of net potassium flux into the cytoplasm as a function of the potassium concentration and pH of the medium. At low pH (4.0–6.0) net flux is a simple exponential function of time which obeys Michaelis kinetics as a function of potassium concentration. At high pH, potassium uptake is more complex, obeying sigmoid kinetics. The data have been fitted satisfactorily by two different two-site models. In one, the transport system is thought to contain both a carrier site

Table 2.4 The contribution made by mannitol and arabitol to the excess osmocity (Δosm) within the hyphae of *Dendryphiella salina* in response to varying dilutions of sea water for 6 h. The medium contained glucose at 50 g ml^{-1}. Mycelium grown as described by Allaway & Jennings (1970*a*) and the ratio of intracellular water/dry weight taken as 2.54 (Slayman & Tatum, 1964). Osmocity of mannitol and sea water obtained from data of Weast & Selby (1967). That of arabitol taken as being the same as mannitol (Data from Jennings 1973)

Sea water in medium	Osmocity mol l^{-1}	Arabitol: mg 100 mg dry $weight^{-1}$	Arabitol: mol l^{-1} intracellular water	Arabitol: Osmocity mol l^{-1}	Mannitol: mg 100 mg dry $weight^{-1}$	Mannitol: mol l^{-1} intracellular water	Mannitol: Osmocity mol l^{-1}	Total osmocity (mol l^{-1})	Δ osm (mol l^{-1})
100	0.575	10.1	0.395	0.245	9.0	0.195	0.111	0.356	0.245
50	0.265	8.9	0.23	0.133	9.0	0.195	0.111	0.244	0.133
25	0.13	5.9	0.15	0.084	6.8	0.15	0.084	0.168	0.057
12.5	0.07	5.8	0.15	0.084	7.6	0.16	0.093	0.177	0.066
6.25	0.03	3.8	0.09	0.048	5.7	0.125	0.063	0.111	0
0	0	3.8	0.09	0.048	5.8	0.125	0.063	0.111	—

responsible for potassium uptake and a modifier site: for a hydrogen ion at low pH and a potassium ion at high pH. The other model postulates a transport system consisting of multiple subunits, each with an active site for potassium, hydrogen ions being allosteric activators.

The potassium-sodium-hydrogen system is electrogenic. Metabolic inhibitors (cyanide, carbon monoxide, dinitrophenol, azide, anoxia and low temperature) all cause rapid changes in membrane potential (Slayman, 1965*b*). Thus, within 1 min of adding 1×10^{-3} M sodium azide to the external medium, already containing 10×10^{-3} M KCl + 1×10^{-3} M $CaCl_2$ + 2% sucrose, the potential (inside negative) shifts from −226 mV to −19 mV. No significant change in membrane resistance can be detected. Slayman, Lu & Shane (1970) have also shown that the decay in potential brought about by a metabolic inhibitor (in this case, 1 mM cyanide) is paralleled by a drop in the mycelial concentration of ATP such that the voltage/time curve is superimposable upon the ATP/time curve, with rate constants for both corresponding to a half-time of 3.7 s.

The ion causing electrogenicity is hydrogen. This ion remains the only candidate after all other possibilities have been removed in experiments where it was shown that the potential could not be changed by changing the anion composition of the medium, nor by changing the sodium content of the mycelium.

The data that we now have for the potassium-sodium-hydrogen system is that it is an ATPase which differs however from the animal potassium-sodium ATPase, in that hydrogen ions play a crucial role.

ANION TRANSPORT Slayman & Slayman (1968, 1970) measured the tracer influx of chloride, sulphate and phosphate into *Neurospora crassa* (Table 2.5). The influxes are independent of whether or not potassium or sodium is the counter ion and appear to be much the same at pH 5.8 and 8.0. They are at least an eighth of the net potassium influx under the same conditions. A similar low rate of chloride uptake was found in *Neocosmospora vasinfecta* (Budd, 1969). Miller & Budd (1971) have pointed out that this chloride uptake, though small, is probably active, if the internal potential is like that of *Neurospora crassa*, namely negative. The presence of glucose in the

Table 2.5 Anion influxes into low potassium mycelium of *Neurospora crassa* (Slayman & Slayman, 1968, 1970)

Tracer anion	Concentration (potassium salt) (mM)	Initial anion influx (pmol cm^{-2} s^{-1})
pH 5.8		
^{36}Cl	30	1.32
$^{35}SO_4$	15	0.13
$^{32}PO_4$	27.4	0.39
pH 8.0		
^{36}Cl	30	1.39
$^{35}SO_4$	15	0.061
$^{32}PO_4$	15.5	0.37

external medium inhibits net chloride uptake. Both influx and efflux of the ion appear to be affected.

Rates of uptake of anions lower than those of cations are not necessarily the rule. Shere & Jacobson (1970*b*) have shown that when *Fusarium oxysporium* is grown on a low phosphate medium, the rate of phosphate uptake is equal to that of potassium. However, when the fungus is grown in a medium rich in phosphate, only a small fluctuating uptake is observed. Potassium uptake is much retarded. This observation may help to explain the data for *Neurospora crassa*, where there is a high phosphorus content but low flux into the mycelium from the experimental solution. It is likely that under these conditions, phosphate uptake is being repressed by some phosphate compound within the cell, probably orthophosphate. This is known to occur in yeast (Holzer, 1953).

There is more information about sulphate transport from other studies. Marzluff (1970*a*, *b*, 1972) has shown that *Neurospora crassa* contains two separate sulphate transport systems which are encoded by unlinked genes. The so-called system I which occurs in conidia and has a K_m of 10^{-4} M and system II which has a K_m of 8×10^{-6} M and which replaces system I when mycelial outgrowth occurs. Preincubation of mycelium in medium containing inorganic sulphate leads to a reduction in uptake, indicating feedback inhibition of uptake. The flux rates measured by Slayman & Slayman (1968, 1970) may be under conditions where feedback inhibition is occurring.

Transport of organic solutes

Considerations of organic solute transport in fungi are helped by a simple model which conceives transport in terms of a carrier to which the solute can combine and which can move across the membrane either free or in combination with the solute. Once across the membrane, the carrier, if the solute is bound into it, may release the solute into the cytoplasm. This basic system can be varied in a number of ways and two variations seem to be present in filamentous fungi.

ALTERATION OF THE AFFINITY OF THE CARRIER FOR THE SOLUTE ON ONE OR OTHER SIDE OF THE MEMBRANE If the affinity of the carrier is the same on both sides of the membrane, solute will move across the membrane until the concentration is the same on both sides of the membrane. Under these conditions movement of solute into the hyphae can be considered as being brought about by facilitated diffusion. If, however, the affinity of the carrier when it reaches the cytoplasmic side of the membrane is reduced, then at equilibrium there will be a greater concentration of solute inside the hyphae than in the external medium. Accumulation will have occurred. For this change to take place some metabolic process must be involved either directly or indirectly. Because of this we can speak of the process as being active. If metabolic participation is only indirect, the process is spoken of as being secondarily active. This possibility will be further considered later.

This relationship between facilitated diffusion and active transport has been shown for monosaccharide transport in *Saccharomyces cerevisiae* (van Steveninck & Rothstein, 1965) and the same situation could hold for glucose transport in *Neurospora crassa*. There are two systems for glucose in this fungus (Scarborough, 1970*a*, *b*; Schneider & Wiley, 1971*a*, *b*, *c*). They have K_m's of $1–7 \times 10^{-5}$ M (high affinity system) and $8–25 \times 10^{-3}$ M (low affinity system). The high affinity system is repressed by growth in glucose. When this occurs, only the low affinity system is operative and this brings about facilitated diffusion of 3-*O*-methyl glucose, and presumably glucose, into the hyphae. Active transport of sorbose is also repressed. It appears that it is some metabolic product of glucose, not glucose itself, which represses the high affinity system, since 3-*O*-methyl glucose has no effect. It is not unreasonable to postulate that the same carrier is involved in both instances, the process which is repressed being the metabolic step which changes the affinity of the carrier on the outside of the membrane. Genetic analysis could decide whether or not this is the case.

EFFECT OF IONS ON ORGANIC SOLUTE TRANSPORT Transport can be brought about by a difference of affinity of the carrier for solute on the two sides of a membrane. This can happen as a consequence of some metabolic reaction causing a lower affinity on the inside of the membrane of higher affinity on the outside. There is now evidence that ions may be involved in maintaining this difference of affinity. Certain studies (Schultz & Curran, 1970; Heinz, 1972) have shown that the active transport of sugars and amino acids in animal cells involves sodium and possibly potassium ions. Transport of the sugar or amino acid is inhibited if the intracellular sodium and/or extracellular potassium is increased. It seems that the movement of the organic solute is as a rule accompanied by the parallel movement of Na^+ and the antiparallel movement of K^+ ions.

There seems to be general agreement about the model of transport. All the proposed models involve a protein carrier which is also capable of forming a complex with the activating cation. For the moment, attention will be confined to Na^+. The carrier can of course bind a sugar molecule, but the affinity of the carrier for it is increased when sodium is also bound. If it is assumed that there is a relatively high concentration of Na^+ in the medium, the cation will combine with the carrier on the outside of the membrane, increasing the affinity of the carrier for the sugar. The sugar can then be accumulated in the cell if there is some way of removing that Na^+ which is transported across the cell membrane when it reaches the cell interior. This is because the removal of the Na^+ will decrease the affinity of the carrier for the sugar on the inside of the membrane and thus tend to prevent movement of the sugar back out of the cell. In animal cells, the concentration of Na^+ in the cytoplasm is kept low by the action of the sodium pump which drives these ions out of the cell against their electrochemical gradient.

This is not the only possible way in which Na^+ might bring about accumulation of a sugar, but the model just presented is conceptually the most useful. Readers are referred to the paper by Heinz, Geck & Wilbrandt (1972) for a full analysis of possible models. In all cases, however, there must be some mechanism for keeping a low level of Na^+ within the cytoplasm. Thermodynamically, the accumulation of sugar (or amino acid), is driven by the flux of Na^+ down its electrochemical potential gradient and the consequent dissipation of energy inherent in this gradient. The cell does work in maintaining this gradient by pumping sodium ions out of the cell. The active accumulation of sugar by such a system is said to be *secondary* since it is not coupled directly to metabolism. It is driven by the primary active process, the pumping of Na^+. The secondary status of the active system means of course that phosphorylation of the sugar *during* transport does not take place.

There is now compelling evidence from yeasts that accumulation of amino acids and certain sugars is brought about by a secondary active transport system, the important activating ions being K^+ and, unlike animal cells, H^+ (Seaston, Inkson & Eddy, 1973). The role of H^+ in sugar transport in yeast could be said to be not entirely unexpected since there is very good evidence that H^+ is pumped out of the cells in exchange for K^+ (Jennings, 1975).

Attention has been drawn to the very well documented evidence for K^+–H^+ exchange in *Neurospora crassa* (see page 38). Therefore, the occurrence of ion-driven organic solute transport can be reasonably expected in filamentous fungi. Hunter & Segal (1973) have presented evidence that a proton gradient can drive amino acid transport in *Penicillium chrysogenum.* Their evidence for this comes from the fact that a variety of weak acids at or below their pK_a are potent inhibitors of L-leucine transport. Although the inhibitors caused a decrease in cellular ATP levels, there was no constant correlation between inhibition and reduction in ATP levels.

Allaway & Jennings (1970*a*, *b*; 1971) have presented evidence that glucose transport in *Dendryphiella salina* is affected by K^+ and Na^+ and they indicated that Na^+ might stimulate transport. It should be noted that this fungus comes from marine habitats, so that involvement of Na^+ is not unexpected.

DIFFERENTIAL MOBILITY OF THE CARRIER ACROSS THE MEMBRANE Accumulation will also occur if the loaded carrier travels more slowly back across the membrane. In one sense, what is happening here is that the process of solute movement is tending to take place in one direction only, though as will be shown later it cannot be considered as simply as this. The situation certainly applies to amino acid transport in *Neurospora crassa.*

Neurospora crassa contains five amino acid transport systems (Table 2.6). Regulation of transport is brought about by the concentration of amino acids within the mycelium and this inhibition is called transinhibition. Where it has been examined in detail in systems I, II and V it is found to be system specific. The ability of an amino acid to transinhibit is highly correlated with its affinity for the system. Amino acids with high affinity are effective transinhibitors; those with lower affinity are less effective.

Pall (1971) has indicated two possible ways in which transinhibition might occur. In the first, transinhibition is suggested to occur through the protein involved in transport having an allosteric binding site for the amino acids concerned. Binding of the amino acids could inhibit the activity of the protein and therefore transport. By this mechanism, the transport system would have two binding sites, one binding the amino acid prior to transport into the hyphae and another binding transinhibiting amino acids inside the hyphae.

On the other hand, and this seems more likely, transport and transinhibition may be determined by a single site. In this case, transinhibition will be caused by the binding of the appropriate amino acid to the active site of the

Table 2.6 Major amino acid transport systems in *Neurospora crassa* (Pall, 1970) The amino acids in this table are only a partial listing of the amino acids having affinity for the different transport systems. In most cases, amino acids with similar properties to those listed will also have affinity. Affinity constants (K_m or K_i) are all expressed in μM

	System I **L-Neutral amino acids**	**System II** **D- or L-basic, neutral and acidic amino acids**	**System III** **L-Basic amino acids**	**System IV** **D- or L-acidic amino acids**
Amino acids transported and affinity constants (μm)	L-Tryptophan (60) L-Leucine (110) L-Phenylalanine (50)	L-Arginine (0.2) L-Phenylalanine (2) D-Phenylalanine (25) Glycine (7) L-Aspartic acid (1200)	L-Arginine (2.4) L-Lysine (4.8)	L-Cysteic acid (7) L-Aspartic acid (13) L-Glutamic acid (16)
Other amino acids showing affinity	L-Valine L-Alanine Glycine L-Histidine L-Serine	L-Lysine L-Leucine α-Aminoisobutyric acid β-Alanine L-Histidine	L-Ornithine L-Canavanine L-Histidine (low affinity)	D-Aspartic acid D-Glutamic acid

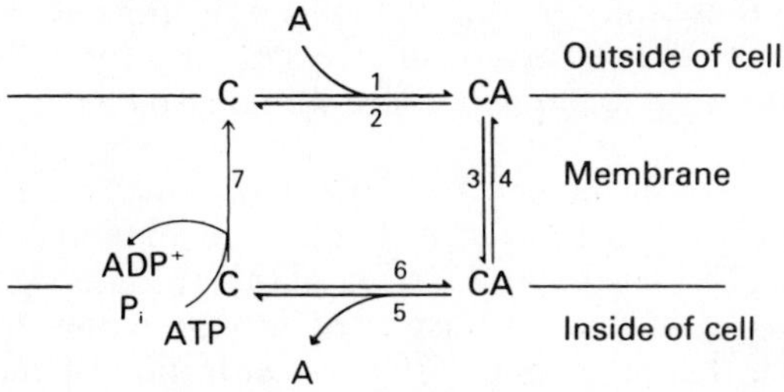

Fig. 2.2 A model for system—specific transinhibition (Pall, 1971). A = amino acid, C = carrier protein.

carrier when that site is oriented towards the hyphal interior. Pall (1971) has proposed a model on this basis in which the binding of the amino acid to the carrier prevents the carrier returning to the outside of the membrane. A diagram of this model is shown in Fig. 2.2. The features of the various reactions are as follows:

1 and 2. Allows reversible association of the amino acid with the carrier. 3. Brings about orientation of the carrier from the outside to the inside of the membrane. 4. The reverse of 3 but very much slower. 5 and 6. Allows reversible dissociation of the amino acid from the carrier. 7. Brings about recycling of the carrier by a process involving ATP or some other energy-rich compound.

The slowness of reaction 4 compared with 3 is equivalent to the carrier-amino acid complex being in a lower energy state when oriented to the inside of the hypha than when oriented towards the outside. When an amino acid which brings about transinhibition accumulates inside the hypha, the equilibrium between the free carrier and carrier-amino acid complex is shifted towards the latter. Thus the concentration of free carrier is reduced. This leads to an inhibition of the energy-dependent recycling process. Recycling cannot occur via reaction 4 because of its slowness.

An important aspect of this model concerns reaction 4. The slowness of this reaction leads to transinhibition and also a low rate of exchange diffusion. The results of Wiley & Matchett (1968) show little or no exchange diffusion of tryptophan in *Neurospora crassa.* On the other hand, if reaction 4 is fast, there will be no transinhibition but considerable exchange diffusion.

Of course, if amino acid transport is driven by ion gradients, an energy-dependent recycling process will not be involved. However, there is little problem about reconciling the above model with transport driven in this way. The only necessary assumption (and this is implicit in the above model) is that the unloaded carrier travels back to the outside of the membrane at a much higher rate than when it is loaded.

Control of transport

A priori considerations would lead us to expect that transport in filamentous fungi ought to be subject to regulatory processes. Under natural conditions and indeed to an extent on solid and in batch culture, the growing mycelium will be exposed both in space and time to a range of nutrients which may also vary in concentration. If transport can be related to internal metabolism, it is more likely that these nutrients will be efficiently utilized. Also metabolism will not become unbalanced as a result of excess absorption of any particular compound. If this were to happen, there would also be detrimental osmotic consequences.

In one particular instance, the interaction between transport and metabolism has been treated as an example of regulation of hyphal osmotic pressure. Jennings & Austin (1973) showed that the non-metabolized sugar 3-*O*-methyl glucose is actively transported into the hyphae of *Dendryphiella salina* and when this takes place, the concentration of the other soluble carbohydrates (predominantly mannitol and arabitol) adjusts so that the total soluble carbohydrate, which includes the 3-*O*-methyl glucose absorbed, remains unchanged. The drop in mannitol and arabitol concentration is accompanied by an increase in polysaccharide hydrolysable with 1.0 N sulphuric acid. Jennings & Austin (1973) suggest that the simplest hypothesis to explain their observations is that transport of each molecule of sugar into the mycelium is accompanied by the concomitant conversion of either a mannitol or an arabitol molecule into polysaccharide or some other insoluble compound.

In bacteria the ability of the phosphotransferase transport system to bring about both facilitated diffusion and active transport has been interpreted as a mechanism by which a cell can utilize substrates sequentially (Roseman, 1969). When bringing about facilitated diffusion, the system has broad specificity and when one substrate in the medium is utilized another can penetrate the cell membrane and induce the proteins necessary to convert the transport system to the condition where it brings about active accumulation of that solute. It would be of interest to find out whether the glucose transport system of *Neurospora crassa* behaves in this matter with respect to some sugar, known to require a period of induction before it can be metabolized.

Romano & Kornberg (1968, 1969) have shown that the utilization of a whole range of carbohydrates, glucose, fructose, maltose, lactose, galactose, mannose, sorbitol and glycerol (but not sucrose) by *Aspergillus nidulans* is inhibited by acetate. Using mutants, they were able to demonstrate that the effect of acetate requires the presence in the cells of the enzyme that catalyses the formation of acetyl CoA from acetate. The lack of inhibition of sucrose utilization indicated that acetate was having its effect on transport. This was confirmed by the finding that acetate inhibits the uptake of 2-deoxyglucose which is phosphorylated by the cell but the phosphorylated sugar is not metabolized. Jennings & Austin (1973) have shown that acetate inhibits the uptake of 3-*O*-methyl glucose by *Dendryphiella salina*. The inhibitory agent produced from acetate in *Aspergillus nidulans* appears to be acetyl CoA and it must act on at least three sugar transport systems, since Mark & Romano (1971) have shown from competitive studies that this fungus contains separate carriers for D-glucose, D-galactose and D-fructose. These authors also point out that, either each separate carrier has a common biochemical feature that renders it sensitive to the acetate effect, or that sugar transport involves an additional element that is distinct from the stereospecific carrier. But whatever the nature of the agent it can be seen as being a mechanism whereby the two fungi use acetate and sugar sequentially if the two compounds are present together in the medium.

Table 2.7 Regulation of transport in filamentous fungi

Solute transported	Fungus	K_m for transport system (M)	Feedback inhibition	Repression by	Reference
Ammonia Methylamine Ethylamine	*Penicillium chrysogenum*	2.5×10^{-7} 10^{-5} 10^{-4}	Principally asparagine glutamine		Hackett *et al.* (1970)
Sulphate	*Neurospora crassa*	10^{-6}	Some metabolite early in pathway of sulphate metabolism	Methionine	Marzluff (1970*a*, *b*, 1972, 1973)
	P. chrysogenum	$3–7\times10^{-5}$	Methionine cystine		Segel & Johnson (1961) Yamamoto & Segel (1966)
Choline-*O*-sulphate	*P. chrysogenum* *P. expansum* *P. notatum* *N. crassa* *Aspergillus nidulans*	$10^{-4}–3\times10^{-4}$	Choline-*O*-sulphate methionine	Growth in media containing sulphur compounds	Bellenger *et al.* (1968)
Glucose	*N. crassa*	$1–7\times10^{-6}$		Glucose	Schneider & Wiley (1971*a*, *b*, *c*)
Tryptophan	*N. crassa*	5×10^{-5}		Tryptophan Leucine Phenylalanine	Wiley & Matchett (1966, 1968)
Methionine	*N. crassa*		Methionine	Growth in media containing sulphur compounds	Pall (1971)

It is not surprising to find that a considerable number of transport systems are subject to feedback inhibition (Table 2.7). Though in nearly all cases there has been no kinetic examination of the situation, it is highly likely that they are further examples of transinhibition.

On the basis of experience with bacteria, repression of synthesis of transport systems in fungi ought to be expected. However, the evidence for this occurring is not always easy to demonstrate unequivocally. Care needs to be exercised in separating repression of synthesis from feedback inhibition. Cycloheximide has been used on a number of occasions to examine whether removal of those conditions causing repression allows the transport protein to be re-synthesized. Failure of the rate of uptake to increase in the presence of cycloheximide is taken to mean that the relevant transport protein is not being synthesized. But this sort of conclusion should be viewed with caution, since it is known that cycloheximide can have other effects, particularly directly on transport itself (Reilly, Fuhrmann & Rothstein, 1970). Particular care is needed when using cycloheximide to investigate the transport of amino acids, since inhibition of protein synthesis can cause a build-up of free amino acids within the hyphae with consequent transinhibition.

An example of the complicated effects of cycloheximide in transport comes from a study of L-leucine transport in *Penicillium chrysogenum* (Hunter & Segel, 1973; Hunter, Norberg & Segel, 1973). When nitrogen-starved mycelium is incubated with relatively high concentrations of leucine there is uptake followed by loss of the amino acid. Cycloheximide inhibits this loss. The inhibitor also reduces the initial rate of uptake but this inhibition only occurs when phosphate is present in the medium. The inhibition can be relieved by calcium and cycloheximide promotes the uptake of this cation. The effect of calcium may be on proton gradients across the cell membrane.

One of the more convincing cases of repression of a transport protein comes from studies on the tryptophan-uptake system (System I) of *Neurospora crassa* by Wiley & Matchett (1968) and Wiley (1970). In particular, Wiley has isolated a tryptophan-binding protein from germinated conidia by cold osmotic shock, the procedure being similar to that used for bacterial cells (Heppel, 1969). The germinated conidia so treated lose 90% of their ability to transport tryptophan without loss of viability. There is good evidence that the tryptophan-binding protein is closely involved in tryptophan transport and it is significant that a decreased capacity for binding tryptophan is observed in shock fluids from cells repressed for tryptophan uptake. This seems good evidence that the number of carrier proteins is reduced as a result of repression.

A particularly interesting situation with respect to transport regulation appears to be emerging from studies on the effect of ammonia on a number of enzyme and uptake systems in *Aspergillus nidulans*. The systems repressed by ammonium include nitrate reductase, xanthine dehydrogenase, acetamidase, formamidase and adenine dinucleotide phosphate-glutamate dehydrogenase (NADP-GDH), L-glutamate, urea and thiourea uptake and extracellular protease (for references see Pateman, Kinghorn, Dunn & Forbes, 1973). There appear to be two types of ammonium-regulated systems. One, the ammonia uptake system is determined by the intracellular ammonia concentration while the levels of the enzymes and other uptake systems listed above are determined by the extracellular ammonium concentration.

Pateman and his colleagues believe that NADP-GDH is located in the cell membrane where it complexes with extracellular ammonium. This so-called first regulatory complex determines the level of L-glutamate uptake, thiourea uptake, nitrate reductase, xanthine dehydrogenase and protease by repression or inhibition or both. NADP-GDH can also combine with intracellular ammonium at another site to form a second type of regulatory complex which determines the ammonium repression or inhibition or both of ammonium uptake. The situation is summarized in Fig. 2.3. It would seem

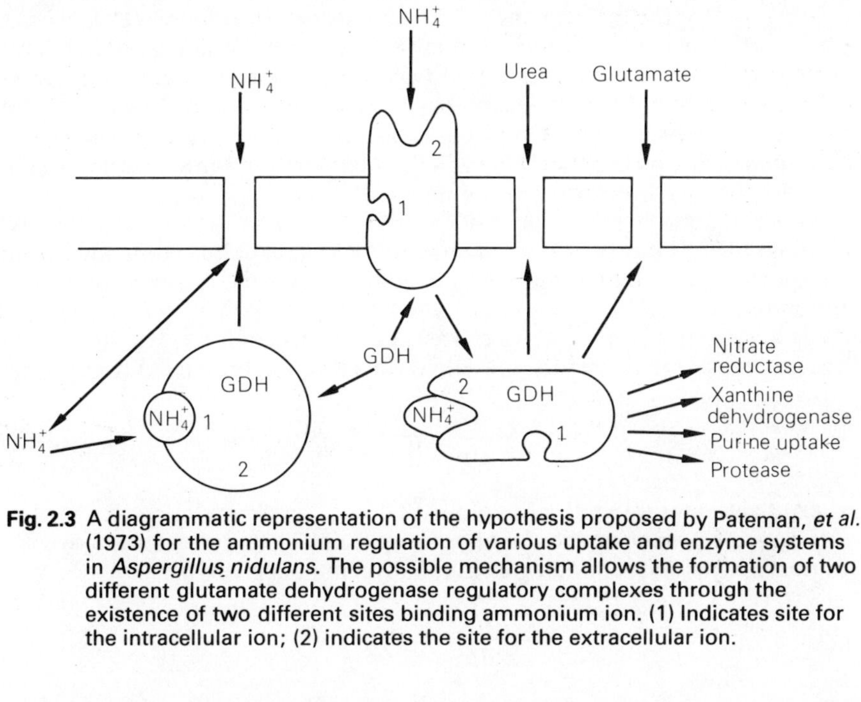

Fig. 2.3 A diagrammatic representation of the hypothesis proposed by Pateman, *et al.* (1973) for the ammonium regulation of various uptake and enzyme systems in *Aspergillus nidulans*. The possible mechanism allows the formation of two different glutamate dehydrogenase regulatory complexes through the existence of two different sites binding ammonium ion. (1) Indicates site for the intracellular ion; (2) indicates the site for the extracellular ion.

that according to Pateman and his colleagues NADP-GDH must be released from the cell membrane before second type regulation can occur. This seems somewhat unlikely, especially since it does not appear that control via glutamate and not the second type regulatory complex has been ruled out. Nevertheless, the information is of considerable interest because it indicates how a number of transport and metabolic activities might be regulated in an integrated manner.

Fluxes and growth

Very little thought has yet been devoted to the relationship between transport and growth. Van Uden (1969) has reviewed the situation but little consideration was given to the changing surface area of the growing cell. I shall indicate ways of taking this into account but before doing this we need to be clear about what we want to know. Initially, the key question is 'What is the relationship between the rates of uptake or net flux which are measured in resting mycelium and those which are present in growing mycelium?' Other questions can of course be asked but this question seems to be initially the most important.

It is difficult to answer because flux measurement must of necessity be made with resting mycelium. Growing mycelium will confound any measurements because, during the time that a difference in concentration of solute or a difference in radioactivity can be measured, there can be a difference in surface area of the mycelium. However, one comparison of value can be made very easily and that concerns maintenance feed rates for glucose and net fluxes for the same sugar. The former figure is obtained for growing mycelium in a chemostat. We have a value of 0.029 g glucose g dry weight^{-1} h^{-1} for the maintenance ratio for *Aspergillus nidulans* (Bainbridge, Bull, Pirt, Rowley & Trinci, 1971). Schneider & Wiley (1971*a*) have estimated the V_{max} for the high-affinity glucose transport system in *Neurospora crassa* to be 4.6×10^{-8} M mg dry weight^{-1} min^{-1}, which is 0.049 g of glucose g dry weight^{-1} h^{-1}. Thus if the same system were present in *Aspergillus nidulans*, it would be working at half maximal rate. More information of this sort is needed for other nutrients.

Another approach is to consider growth in a more theoretical way. Jennings (1969) has provided equations for a plant cell relating growth to transport. These equations can be adapted for the growth of fungal mycelium.

For a filamentous species, we can assume that the radius (r) of the hyphae remains constant. Therefore the rate of increase in volume (v) is given by:

$$\frac{dv}{dt} = \pi r^2 \frac{dl}{dt} \qquad (1)$$

where l is the total length of hyphae.

If we are considering the site of uptake of solute, the rate of change of the number (N) of solute molecules in the mycelium is given by

$$\frac{dN}{dt} = J \cdot 2\pi rl \qquad (2)$$

where J is the flux of solute across the membrane. The rate of change of the internal concentration (C) of solute is then given by

$$\frac{dC}{dt} = \frac{V\, dN/dt - N\, dV/dt}{V^2} \qquad (3)$$

since $C = N/V$.

Equation (*3*) can be expanded using equations (*1*) and (*2*)

$$\frac{dC}{dt} = J\frac{2\pi rl}{V} - \frac{N}{V^2} \cdot \pi r^2 \frac{dl}{dt} \qquad (4)$$

$$= \frac{2J}{r} - \frac{C}{l} \cdot \frac{dl}{dt} \qquad (5)$$

This can be rearranged

$$J = \frac{r \cdot C}{2 \cdot l} \cdot \frac{dl}{dt} + \frac{r}{2} \cdot \frac{dC}{dt} \tag{6}$$

Another way of considering the rate of increase in length is to assume that growth is controlled by an enzyme reaction involving the solute within the cell. Thus we get

$$\frac{dl}{dt} = \frac{V_{max} \cdot C}{C + K_m} \tag{7}$$

where V_{max} is the maximum rate of growth (increase in length) and K_m is the Michaelis constant for the enzyme reaction.
Substituting equation (*7*) in (*6*) we get

$$J = \frac{r}{2}\left[\frac{dC}{dt} + \frac{C}{l} \cdot \frac{V_{max} \cdot C}{C + K_m}\right] \tag{8}$$

As it stands, this equation is not immediately very useful. However, data has been obtained using *Dendryphiella salina* which indicate both that the equation can be simplified and allow a calculation of J, the net flux of solute. The data concern potassium and have been obtained with batch cultures. The biological basis of the situation is represented in Fig. 2.4 (see p. 57). In essentials, glucose metabolism and therefore growth is dependent upon the mycelial potassium concentration (Allaway & Jennings, 1970*a*, *b*), so the model just described applies here.

First, it seems likely that for any one potassium concentration in the medium, the concentration of the ion in the mycelium during growth changes only slowly. Thus we can put $dC/dt = 0$. Equation (*8*) reduces to

$$J = \frac{r}{2} \cdot \frac{C}{l} \cdot \frac{V_{max} \cdot C}{C + K_m} \tag{9}$$

Second, under our particular experimental conditions where potassium is not limiting growth, the maximum rate of growth is 8.5×10^{-10} mg dry weight s^{-1} which works out at an increase in length of 11.1 cm s^{-1}, assuming a hyphal diameter of 5 μm, a fresh/dry weight ratio of 2.55 and a specific gravity for the mycelium of unity.

In one particular instance, in a medium containing 2.5×10^{-3} M potassium chloride we obtained 41.3 mg of mycelium containing 0.76 mg of potassium (Jennings & Aynsley, 1971). One can take a number of values for the K_m of the enzyme reaction catalysed by potassium and the flux values obtained by substituting the appropriate figures in equation (*9*), using the assumptions listed in the previous paragraph, are given in Table 2.8.

Not too much should be made of these data in themselves except to point out that the figures indicate that reasonable rates of growth can be maintained by rather low flux rates. They are much lower for instance than the initial net fluxes determined for low potassium mycelium of *Neurospora crassa* (Slayman & Slayman, 1968 and Table 2.1). This would indicate that once a certain concentration of potassium is achieved in the mycelium transport comes under internal regulation.

Table 2.8 The values for net flux of potassium obtained using equation 9 (in text) and chosen values for K_m.

Chosen K_m (M)	Net flux (pmol cm^{-2} s^{-1})
0.5	0.13
10^{-1}	0.32
10^{-2}	0.47

The data used for the calculations are somewhat crude. A much more refined experimental analysis is required, possibly involving continuous culture. Further, it would be sensible to modify the equations to take account of the specific growth rate. This is a relatively easy task and would lead to equations of general applicability. Nevertheless, I think there is sufficient information above to indicate the sort of thinking that needs to be carried out with respect to relating transport to growth.

2.4 Translocation

Translocation is the process of movement of materials along fungal hyphae. It is tempting to assume that since hyphae are linear structures there must be extensive translocation. But caution must be exercised about this assumption. In the first place, there is evidence that septal blocking can occur quite early on in hyphal development (Trinci & Collinge, 1973) and indeed in certain fungi, *e.g. Geotrichum candidum* the septal structure may be such that only a small proportion of the total area of the septum allows direct contact from one cell to the next (Hashimoto, Morgan & Conti, 1973).

Table 2.9 Movement of materials in fungal hyphae

Process	Type of evidence for the occurrence of the process
Nuclear migration	Genetic analysis Visual
Movement of mitochondria	Genetic analysis Use of 'tagged' organelles
Movement of viruses	Pathological symptoms
Movement of vesicles to tips of hyphae	Combined light and electron microscope studies
Cyclosis	Visual
Bulk flow of protoplasm	Visual Theoretical considerations
Movement of solutes	Tracer studies Chemical analysis Natural history, *e.g.* observation of growth from food source over an inert surface

Second, such an assumption is not entirely in keeping with fungal growth under natural conditions. Many filamentous fungi do not grow away from nutrients but towards them. It is not difficult to conceive of a situation where the tip and the first millimetre or so behind it absorb all the essential nutrients and the hyphae then subsequently go into physiological decline with eventual autolysis. One of the problems that occurs in mycology is that much of the work is done with petri dishes. Here the conditions may be very unlike those in nature not only in terms of the structure of the substratum but because nutrients can so readily diffuse in agar towards the mycelium. In soil, for instance, where organic matter has to be broken down before it can be utilized, a situation akin to that which has been mentioned above is quite likely. Of course the situation assumes that hyphal tips possess all the necessary transport systems.

On the other hand, there is definitive information about movement of materials in hyphae. The various processes are listed in Table 2.9.

Nuclear migration

It is true to say that more is known about this process than any other process which brings about movement of material in hyphae. A nucleus, because of its genetic material, can produce detectable morphological and biochemical changes which are ascribable only to the presence of that nucleus or other nuclei derived from it at a certain locus in the mycelium. Thus these morphological and biochemical changes at different parts of a mycelium can be used as markers for the movement of a nucleus through that mycelium. Table 2.10 summarizes the studies that have been made. It can be seen that the rate of nuclear migration is in all cases greater than the rate of hyphal extension indicating that movement of nuclei cannot be explained on the basis of hyphal growth.

Nuclear migration might be brought about by either a specific force acting solely on the nucleus or a more generalized force bringing about both nuclear and cytoplasmic movement. As we shall see it is difficult to discount the latter, especially in the sense that whatever force brings about nuclear migration may also bring about movement of other cytoplasmic components.

It is the evidence of electron microscopy which has provided positive suggestions regarding a possible mechanism of nuclear migration. The relevant information comes from studies on the short-range nuclear movements which occur in Basidiomycetes. Two types of movement have been distinguished (Girbardt, 1968; Snider, 1968; Niederpruem, 1969). One occurs after nuclear division when the two daughter nuclei move apart and the other maintains the central position of the nuclei in the cells. Girbardt (1968) using a combined light and electron microscopical approach showed that microtubules are involved in the short-range movements in *Polystictus versicolor*. He also showed that oscillation of nuclei in a longitudinal direction occurs independently of cytoplasmic streaming. Subsequently, Raudaskoski (1972, 1973) demonstrated by electron microscopy the presence of microtubules scattered in the hyphae around interphase nuclei in mycelium in which intracellular migration is occurring. Further, during hyphal fusion, microtubules close to an interphase nucleus were found to be connected with electron-dense structures thought to be centres of assembly of microtubules.

Table 2.10 Relative rates of nuclear migration (NM) and hyphal tip growth (TG). (From Snider, 1965)

Fungus	Temperature (°C)	Rates(mm h^{-1}) NM	TG	Relative Rate (NM/TG)	References
Basidiomycetes					
Coprinus lagopus	22	1.51	0.15	10.0	Buller, 1931
	28	1.0	0.25	4.0	Swiezynski & Day, 1960
Coprinus macrorhizus	30	3.2	0.15	21.3	Kimura, 1954
Coprinus radiatus	23	0.58	0.09	6.4	Prévost, 1962
Cyathus stercoreus	R.T.	0.37	0.16	2.2	Fulton, 1950
Schizophyllum commune	22	0.5	0.13	3.8	Snider & Raper, 1958
	32	3.0	0.22	13.6	Snider & Raper, 1958
Ascomycetes					
Gelasinospora tetrasperma	Not given	4.0	2.00	2.0	Dowding & Buller, 1940
	Not given	10.5	*c.* 3.00	3.5	Dowding & Bakerspigel, 1954
	30–33	(40.0)	(*c.* 3.00)	(13.3)	Dowding, 1958
Ascobolus stercorarius	22	(20.0)	(*c.* 1.50)	(13.3)	Snider, unpublished results

These observations by themselves justify the assumption that microtubules are involved but it is also important to realize that the observed rates of movement are in the range where one can reasonably expect contractile systems to be involved.

Movement of mitochondria

It is important to know whether or not when nuclei migrate to one hypha from another there is also movement of cytoplasm. The evidence is conflicting. Thus Snider (1968) using *Schizophyllum commune* found that the receptor of light-induction of fruiting did not travel with the nucleus, though this assumes that the receptor molecule is located within the cytoplasm. However, another test carried out by Casselton & Condit (1972) who used mitochondrial genes as markers would suggest that there is no movement of cytoplasm. On the other hand Jinks (1959) has produced genetic evidence that in *Aspergillus* that there is transfer of cytoplasmic organelles and more emphatically Watrud & Ellingboe (1973*a*, *b*) have produced visual evidence that this is the case.

The latter two workers grew *Schizophyllum commune* on a medium containing 1.0 g l^{-1} cobalt chloride hexahydrate. The cobalt was slowly absorbed from the medium and incorporated in the mitochondria so that they were both denser than normal and could be visualized by light microscopy. Movement of mitochondria from one mycelium to another was demonstrated both by isolation of the mitochondria by density gradient centrifugation and by visual observation. The results that they obtained suggest that, as with nuclei, there is a dual regulation of mitochondrial transfer by both A and B factors. This would suggest that the phenomenon of mitochondrial transfer parallels that of nuclear migration. However, transfer occurs with common AB matings where nuclear migration is limited. Furthermore, observations of anastomoses of fully compatible matings showed that mitochondrial transfer was evident prior to the completion of development of clamp connections indicating that mitochondrial transfer could occur prior to nuclear migration. Movement of mitochondria did not appear to be accompanied by mass cytoplasmic movement.

It is unfortunate that no rates of movement of mitochondria were reported in these studies. This information is badly needed.

Movement of viruses

There is now plenty of evidence that viruses infect fungi (Hollings & Stone, 1971) and can move through the hyphae. No figures for the rate of movement are available. However, Last, Hollings & Stone (1967) investigating the movement through the mycelium of the mushroom *Agaricus bisporus* of three out of five viruses that infect the fungus, showed that the rate of movement as detected by disease symptoms was related to the size of the virus. The topic of translocation of viruses deserves considerable study.

Movement of vesicles to the tips of hyphae

There is now considerable evidence from electron microscopy supported by light microscopy to suggest strongly that vesicles move to the hyphal tip from the Golgi apparatus which are located some distance back along the hypha. It is believed that these vesicles contain wall materials which are inserted in the wall where extension growth is occurring (Bartnicki-Garcia, 1973).

Two hypotheses have been proposed for the way these vesicles move to the tip. Both hypotheses rest on the observations of Slayman & Slayman (1962) that the membrane potential of *Neurospora crassa* decreases as one moves from the mature parts of a hypha to its tip. Bartnicki-Garcia (1973) proposed that the vesicles move to the tip by electrophoresis driven by the potential gradient along the hypha. Jennings (1973) believes that the drop in potential as one moves to the tip is a consequence of the fact that there is also a drop in the number of potassium pumps (ATPases) at the outer membrane. He produced in support of this hypothesis evidence of changes with time of potassium/sodium ratios in growing mycelium of *Neurospora crassa* and of *Dendryphiella salina* (Table 2.11). The net effect of this presumed distribution could be the presence of a standing gradient osmotic flow (Diamond & Bossert, 1967) in the hypha, so that there will be a bulk flow of liquid to the tip. This bulk flow of liquid would have a directional effect on the movement of the vesicles. Jennings, Thornton, Galpin & Coggins (1974) have reported light-microscopical histochemical evidence in favour of this hypothesis.

Cyclosis and protoplasmic streaming

When considering movement of protoplasm it is essential to distinguish between cyclosis or movement of cytoplasm driven by shearing forces which are metabolically actuated as in the coenocytic alga *Nitella* (Allen, 1974)

Table 2.11 Potassium/sodium ratios in mycelium of *Neurospora crassa* (calculated from data in Fig. 1 of Slayman & Tatum, 1964) and *Dendryphiella salina* (Jennings & Aynsley, 1971) at different times during growth in media containing 36.8 mM potassium and 8.4 mM sodium and 0.11 mM potassium and 7.6 mM sodium respectively

Neurospora crassa[a]		***Dendryphiella salina***[b]	
Time (h)	**K/Na**	**Time (h)**	**K/Na**
0	8.5	10	0.0049
2	6.5	24	0.11
4	4.5	30	0.42
6	15.0	48	0.25
8	12.0	54	0.085[c]
10	12.5		
12	14.0		
14	16.5		

[a] End of lag period at 3½ h.
[b] End of lag period at 7½ h.
[c] All the potassium in the medium depleted by this time.

and bulk flow of cytoplasm driven either by pressure or by evaporative water loss.

Cyclosis has been shown to occur in sporangiophores of *Phycomyces blakesleeanus* (Bergman *et al.*, 1969). It has a rate of about 2.5×10^{-6} m s^{-1} upwards and about 3.0×10^{-6} m s^{-1} downwards. In an atmosphere of nitrogen, streaming in the upper part of the sporangiophore stops in 1–2 min but streaming in the lower part may continue for 30–40 min. Movement of cytoplasm in the outer mantle of a hypha (Buller, 1933) is also likely to be cyclosis.

Bulk flow of protoplasm in hyphae is well authenticated (Arthur, 1897; Schröter, 1905; Buller, 1933). The evidence that it is brought about by differences in water potential comes from studies which have shown that streaming occurs when there is evaporation from one part of the mycelium or local applications of high concentrations of salt or sugar. It is in keeping with this that septal bulging is observed to occur in the direction of streaming (Buller, 1933) while it has also been shown that density of cytoplasm increases toward a septum in the same direction (Isaac, 1964). If, as we suppose, streaming occurs as a result of a suitable water potential gradient, evaporation should cause bulk movement of solutes along hyphae. This will be considered again in the following section but it is appropriate to point out here that there is good evidence that evaporation increases the rate of translocation of material into basidiomycete sporophores (Plunkett, 1958; Littlefield, Wilcoxson & Sudia, 1965).

Jennings *et al.* (1974) have called bulk flow of protoplasm driven by a water potential gradient, osmotic flow. They point out that osmotic flow may occur over short distances and the movement of vesicles to the tips of hyphae can be instanced as a possible example of short-distance movement.

Movement of solutes

There is considerable evidence from studies using radioactive tracers that solutes can move along fungal hyphae (Wilcoxson & Sudia, 1968). Unfortunately, there is almost no information about rates of movement which is vital if conclusions are to be reached about the mechanism of movement. Much of the difficulty comes from the procedure which has been used.

Nearly all studies have been based on the use of petri dishes. The fungus is allowed to grow across a diffusion barrier, in nearly all instances the mycelium has been allowed to colonise the plate. Tracer is presented to mycelium on one side of the diffusion barrier and appearance on the other side determined. There are two immediate criticisms of this procedure. First, by allowing the mycelium to spread over the plate one may be dealing with older hyphae which are highly vacuolate and indeed, if nutrients are at a low

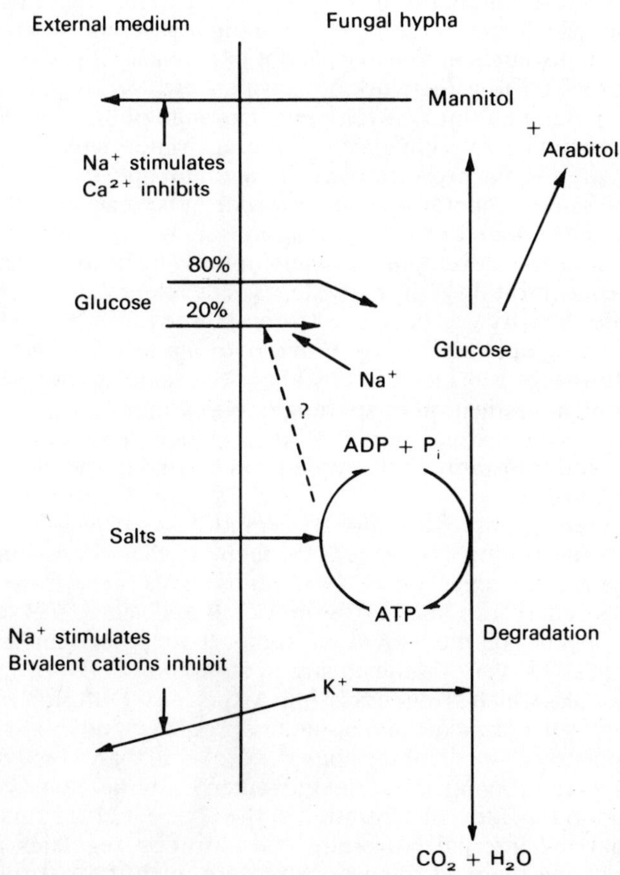

Fig. 2.4 A diagram summarizing the known effects of cations on the uptake and metabolism of glucose by *Dendryphiella salina* and upon the permeability characteristics of the hyphae (from Allaway & Jennings, 1970*b*).

level in the substratum, conceivably without contents because of autolysis. This has been shown to occur in shake culture over a similar period of time (Trinci & Righelato, 1970; Holligan & Jennings, 1972). Second, the geometry of the situation makes analysis difficult. Ideally the movement of radioactivity should be considered in terms of movement in two dimensions out from a point source so that a comparison can be made with the rate of transport if it were to occur by diffusion. Mathematical solutions are available for situations similar to that under consideration which can extend to the case where the mycelium is still growing (Crank, 1956). Of course, there are further problems if one is dealing with a substance which is metabolized, particularly carbohydrates which breakdown to give radioactive carbon dioxide. The gas will diffuse rapidly across the plate and if fixed by the mycelium which has not yet received translocate, give rise to spurious results. On the other hand, the presence of an absorbant for carbon dioxide can set up anomalous water potential gradients. There is a great deal to be said for studies involving solutes which are not metabolized.

In view of these comments, it is not surprising to find that the two sets of information which are reasonably unambiguous come from studies of translocation of solutes in fungal cells not in contact with agar.

The first to refer to is the work of Cowan, Lewis & Thain (1972) on the uptake of potassium into developing sporangiophores of *Phycomyces blakesleeanus* under the influence of different water activities ($w_a = P/P_0$, where P is the vapour pressure over the medium and P_0 is that over pure water at the same temperature and pressure). It is clear from these studies that water movement caused by evaporation brought about potassium movement into the developing sporangiophore from the mycelium. The potassium concentration at the early stages of development was increased by 50% as water activity was decreased from 0.98 to 0.60. Parallel studies on isolated sporangiophores, using a microporometer, showed that water uptake at this stage was increased by about the same amount.

Microscopic investigation of sporangiophores showed two distinct types of streaming—strand streaming of constant velocity which is unaffected by external w_a and transpirational streaming occurring in the inner regions of the sporangiophore.

As the sporangiophore develops it becomes less permeable to water loss due to cuticularization. However, throughout the development of the sporangiophore, the apical portion maintains virtually the same concentration of potassium (0.125 M l^{-1}). Cowan, Lewis & Thain (1972) suggest that, since the velocity of movement of radioactive potassium to the apex (2.1×10^{-6} m s^{-1}) is very similar to strand streaming (2.8×10^{-6} m s^{-1}), it is the latter process which brings potassium to the apex. Diffusion is unlikely to be important, since the expected boundary profile was not observed. In any case, the appearance of ^{42}K at the apex is more rapid than would be expected if diffusion were bringing about the movement of potassium.

A constant potassium concentration at the apex could of course be due to the fact that the internal potassium concentration regulates the rate of growth occurring there. If this were so, there might be a variable rate of transport; the constant concentration of potassium being maintained by the growth rate. Under these circumstances, transpiration might be causing some of the movement. The point is difficult to test, since although water activity had no significant effect on the rate of volume growth of sporangiophores, the standard deviations were large.

The other study, which is of value, concerns estimated rates of translocation in *Serpula lacrymans*. Jennings *et al.* (1974) using the data of Butler (1957), who made measurements of the rate of growth of the fungus over acid-washed brick, calculated that there must be a rate of translocation of 1.27×10^4 pmol of hexose $cm^{-2}\ s^{-1}$ and if a 10% solution moves it does so with a velocity of 3×10^{-6} cm s^{-1}. The thinking behind the calculations was as follows. The growth of the fungus can be considered as the extension of a bundle of hyphae, each of which is growing from a food base. If one can calculate the rate of production of dry matter, one can get a measure of the rate of movement of dry matter through a unit cross-sectional area of hyphae in unit time. It was assumed that hexose is the carbon compound which forms the bulk of the translocate.

Pressure-driven bulk flow seems to be the most likely mechanism to account for these calculated values. There seems to be no problem of maintaining a suitable pressure gradient and further calculations showed that diffusion cannot bring about the calculated values of rates of translocation without having impossible concentration gradients. The bulk flow could of course be driven by evaporation but this seems unlikely, since growth of *Serpula lacrymans* is best under conditions of high relative humidity (Brown, Fahim & Hutchinson, 1968) though it must be admitted that the water relations of the growth of this fungus have yet to be categorized in proper quantitative terms.

Two pieces of information provide more positive evidence that translocation is pressure driven (C. R. Coggins and D. H. Jennings unpublished results). First, studies on the rate of drop formation at the tips of hyphae growing from agar over an inert surface—the situation at microscopic level which is the same as 'weeping' at the macroscopic level—have shown that drops either stop growing or decrease in size when the hyphae are cut or presented with inhibitor at a place some distance from the tips. Table 2.12 shows some data concerning the application of sodium azide. Second, chemical analysis has shown that the mycelium growing on the food base

Table 2.12 Effect of 10^{-4} M sodium azide on the size of droplets produced at hyphal tips of *Serpula lacrymans*

Hours after addition of water or azide to centre of colony	Increase in droplet size ($\mu m^3 \times 10^{-3}$) Water control	10^{-4} M azide
1	2.6	4.7
2	12.5	−1.5
4	22.8	−11.8
6	35.5	−29.8
Initial drop size	$10.7 \times 10^5\ \mu m^3$	

(wood) can contain very high concentrations of soluble carbohydrate, particularly arabitol. There can be as much as 15 mg of arabitol per 100 mg dry weight of mycelium.

It can therefore be envisaged that the fungus, as it grows on the wood, very effectively breaks down the cellulose absorbing the soluble carbohydrate produced, converting it to arabitol. The internal concentration which develops generates the pressure necessary to drive the bulk flow of liquid along the hyphae. The large amounts of rigid wall material in the strands built up from bundles of hyphae (Falk, 1912) can be envisaged as being there to maintain pressure gradients in the hyphae in which translocation is occurring.

It should be emphasized that the translocational flux which has been calculated is several orders of magnitude greater than known fluxes of glucose across the outer membranes of cells. Therefore, there must be a considerable area of loading for the system and so translocation at the rate indicated will only occur when the mycelium has spread over a sizeable area of the food-base.

2.5 References

ALLAWAY, A. E. & JENNINGS, D. H. (1970*a*). The influence of cations on glucose uptake by the fungus *Dendryphiella salina. New Phytologist* **69**, 567–79.

ALLAWAY, A. E. & JENNINGS, D. H. (1970*b*). The influence of cations on glucose transport and metabolism by, and the loss of sugar alcohols from, the fungus *Dendryphiella salina. New Phytologist* **69**, 581–93.

ALLAWAY, A. E. & JENNINGS, D. H. (1971). The effect of cations on glucose utilisation by, and on the growth of, the fungus *Dendryphiella salina. New Phytologist* **70**, 511–8.

ALLEN, R. D. (1974). Some new insights concerning cytoplasmic transport. *Symposium of the Society for Experimental Biology* **28**, 15–26.

ARTHUR, J. C. (1897). The movement of protoplasm in coenocytic hyphae. *Annals of Botany* **11**, 491–507.

BAINBRIDGE, B. W., BULL, A. T., PIRT, S. J., ROWLEY, B. I. & TRINCI, A. P. J. (1971). Biochemical and structural changes in non-growing maintained autolysing cultures of *Aspergillus nidulans. Transactions of the British Mycological Society* **56**, 371–85.

BARTNICKI-GARCIA, S. (1973). Fundamental aspects of hyphal morphogenesis. *Symposium of the Society for General Microbiology* **23**, 245–67.

BELLENGER, N., NISSEN, P. W., WOOD, T. C. & SEGEL, I. H. (1968). Specificity and control of choline-*O*-sulfate transport in filamentous fungi. *Journal of Bacteriology* **96**, 1574–85.

BERGMAN, K., BURKE, P. V., CERDA-OLMEDO, E., DAVID, C. N., DELBRÜCK, M., FOSTER, K. W., GOODELL, E. W., HEISENBERG, M., MEISSNER, G., ZALOKAR, M., DENNISON, D. S. & SHROPSHIRE, Jr, W. (1969). *Phycomyces. Bacteriological Reviews* **33**, 99–157.

BROWN, J., FAHIM, M. N. & HUTCHINSON, S. A. (1968). Some effects of atmospheric humidity on the growth of *Serpula lacrymans. Transactions of the British Mycological Society* **51**, 507–10.

BUDD, K. (1969). Potassium transport in non-growing mycelium of *Neucosmospora vasinfecta. Journal of General Microbiology* **59**, 229–38.

BULLER, A. H. R. (1931). *Researches on fungi*, IV. London: Longmans Green.

BULLER A. H. R. (1933). *Researches on fungi*, V. London: Longmans Green.

BUTLER, G. M. (1957). The behaviour of mycelial strands in *Merulius lacrymans* (Wulf.) Fr. 1. Strand development from a food-base through a non-nutrient medium. *Annals of Botany* NS **21**, 527–37.

CASSELTON, L. A. & CONDIT, A. (1972). A mitochondrial mutant of *Coprinus lagopus. Journal of General Microbiology* **72**, 521–7.

COWAN, M. C., LEWIS, B. G. & THAIN, J. F. (1972). Uptake of potassium by the developing sporangiophore of *Phycomyces blakesleeanus. Transactions of the British Mycological Society* **58**, 113–26.

COWAN, M. C., THAIN, J. F. & LEWIS, B. G. (1972*a*). Mechanism of translocation of potassium in sporangiophores of *Phycomyces blakesleeanus* in an aqueous environment. *Transactions of the British Mycological Society* **58**, 91–102.

COWAN, M. C., THAIN, J. F. & LEWIS, B. G. (1972*b*). The electrical mobility of potassium in the sporangiophores of *Phycomyces blakesleeanus. Transactions of the British Mycological Society* **58**, 161–4.

CRANK, J. (1970). *The mathematics of diffusion.* Oxford: Clarendon Press.

DIAMOND, J. M. & BOSSERT, W. H. (1967). Standing gradient osmotic flow. A mechanism for coupling of water and solute transport in epithelia. *Journal of General Physiology* **50**, 2061–83.

DOWDING, E. S. (1958). Nuclear streaming in *Gelasinospora. Canadian Journal of Microbiology* **4**, 295–301.

DOWDING, E. S. & BAKERSPIGEL, A. (1954). The migrating nucleus. *Canadian Journal of Microbiology* **1**, 68–78.

DOWDING, E. S. & BULLER, A. H. R. (1940). Nuclear migration in *Gelasinospora. Mycologia* **32**, 471–88.

FALK, R. (1912). Die *Merulius* faüle des bauholzes. *Hausschwammforschung* **4**, 1–405.

FULTON, W. (1950). Unilateral nuclear migration and the interactions of haploid mycelium in the fungus *Cyanthus stercoreus. Proceedings of the National Academy of Sciences, U.S.A.* **36**, 306–12.

GIRBARDT, M. (1968). Ultrastructure and dynamics of the moving nucleus. *Symposium of the Society for Experimental Biology* **22**, 249–60.

HACKETT, S. L., SKYE, G. E., BURTON, C. & SEGEL, I. H. (1970). Characterisation of an ammonium transport system in filamentous fungi with methyl ammonium-^{14}C as the substrate. *Journal of Biological Chemistry* **245**, 4241–50.

HASHIMOTO, T., MORGAN, J. & CONTI, S. F. (1973). Morphogenesis and ultrastructure of *Geotrichum candidum* septa. *Journal of Bacteriology* **116**, 447–55.

HEINZ, E. (1972). *Na-linked transport of organic solutes.* Berlin, Heidelberg and New York: Springer-Verlag.

HEINZ, E., GECK, P. & WILBRANDT, W. (1972). Coupling in secondary active transport. Activation of transport by cotransport and/or counter-transport with the fluxes of other solutes. *Biochimica et Biophysica Acta* **225**, 442–61.

HEPPEL, L. A. (1969). The effect of osmotic shock on release of bacterial proteins and on active transport. *Journal of General Physiology* **54**, 953–1133.

HOLLIGAN, P. M. & JENNINGS, D. H. (1972). Carbohydrate metabolism in the fungus *Dendryphiella salina.* 1. Changes in the levels of soluble carbohydrates during growth. *New Phytologist* **71**, 569–82.

HOLLINGS, M. & STONE, O. M. (1971). Viruses that infect fungi. *Annual Review of Phytopathology* **9**, 93–118.

HOLZER, H. (1953). Zur penetration von orthophosphat in lebende hefezellin. *Biochemische Zeitschrift* **324**, 144–55.

HUNTER, D. R. & SEGEL, I. H. (1973). Effect of weak acids on amino acid transport by *Penicillium chrysogenum*: evidence for a proton or charge gradient as the driving force. *Journal of Bacteriology* **113**, 1184–92.

HUNTER, D. R., NORBERG, C. L. & SEGEL, I. H. (1973). Effect of cycloheximide on L-leucine transport by *Penicillium chrysogenum*: involvement of calcium. *Journal of Bacteriology* **114**, 956–60.

ISAAC, P. K. (1964). Cytoplasmic streaming in filamentous fungi. *Canadian Journal of Botany* **42**, 787–92.

JENNINGS, D. H. (1969). The physiology of the uptake of ions by the growing plant cell. In *Ecological aspects of the mineral nutrition of plants.* Edited by I. H. Rorison, 261–79. Oxford: Blackwell Scientific Publications.

JENNINGS, D. H. (1973). Cations and filamentous fungi: invasion of the sea and hyphal functioning. In *Ion transport in plants.* Edited by W. P. Anderson 323–35. London: Academic Press.

JENNINGS, D. H. (1975). Transport in fungi. In *Encyclopedia of Plant Physiology.* Vol. 2. Edited by U. Lüttge and M. G. Pitman. Berlin: Springer-Verlag (in Press).

JENNINGS, D. H. & AUSTIN, S. (1973). The stimulatory effect of the non-metabolised sugar 3-*O*-methyl glucose on the conversion of mannitol and arabitol to polysaccharide and other insoluble compounds in the fungus *Dendryphiella salina. Journal of General Microbiology* **75**, 287–94.

JENNINGS, D. H. & AYNSLEY, J. S. (1971). Compartmentation and low temperature fluxes of potassium in mycelium of *Dendryphiella salina. New Phytologist* **70**, 713–23.

JENNINGS, D. H., THORNTON, J. D., GALPIN, M. F. J. & COGGINGS, C. R. (1974). Translocation in fungi. *Symposium of the Society for Experimental Biology* **28**, 137–54.

JINKS, J. L. (1959). Lethal suppressive cytoplasm in aged clones of *Aspergillus glaucus. Journal of General Microbiology* **21**, 397–408.

JONES, E. B. G. & JENNINGS, D. H. (1965). The effect of cations on the growth of fungi. *New Phytologist* **64**, 86–100.

KIMURA, K. (1954). Diploidisation in the Hymenomycetes. II. Nuclear behaviour in the Buller phenomenon. *Biological Journal of Okayama University* **4**, 1–59.

LAST, F. T., HOLLINGS, M. & STONE, O. M. (1967). Some effects of cultural treatments on virus diseases of cultivated mushroom, *Agaricus bisporus. Annals of Applied Biology* **59**, 451–62.

LESTER, G. & HECHTER, O. (1958). Dissociation of rubidium uptake by *Neurospora crassa* into entry and binding phases. *Proceedings of the National Academy of Sciences, U.S.A.* **44**, 1141–9.

LESTER, G. & HECHTER, O. (1959). The relationship of sodium, potassium and deoxycorticosterone in *Neurospora crassa. Proceedings of the National Academy of Sciences, U.S.A.* **45**, 1792–801.

LITTLEFIELD, L. J., WILCOXSON, R. D. & SUDIA, T. W. (1965). Translocation in sporophores of *Lentinus tigrinus. American Journal of Botany* **52**, 599–605.

LÜTTGE, U. & PITMAN, M. G. (1975). *Encyclopedia of Plant Physiology*. Vol. 2. Berlin, Springer-Verlag (in press).

MacMILLAN, A. (1956). The entry of ammonia into fungal cells. *Journal of Experimental Botany* **7**, 113–26.

MacROBBIE, E. A. C. (1962). Ionic relations of *Nitella translucens. Journal of General Physiology* **45**, 861–78.

MARK, C. G., & ROMANO, A. H. (1971). Properties of the hexose transport systems of *Aspergillus nidulans. Biochimica et Biophysica Acta* **249**, 216–26.

MARZLUF, G. A. (1970*a*). Genetical and biochemical studies of distinct sulphate permease species in different developmental stages of *Neurospora crassa. Archives of Biochemistry and Biophysics* **138**, 254–63.

MARZLUF, G. A. (1970*b*). Genetic and metabolic controls for sulphate metabolism in *Neurospora crassa*: isolation and study of chromate-resistant and sulfate transport-negative mutants. *Journal of Bacteriology* **102**, 716–21.

MARZLUF, G. A. (1972). Control of the synthesis, activity and turnover of enzymes of sulfur metabolism in *Neurospora crassa. Archives of Biochemistry and Biophysics* **150**, 714–24.

MARZLUF, G. A. (1973). Regulation of sulfate transport in *Neurospora* by trans-inhibition and by inositol depletion. *Archives of Biochemistry and Biophysics* **156**, 244–54.

MILLER, A. G. & BUDD, K. (1971). Chloride uptake by mycelium of *Neocosmospora vasinfecta* and its inhibition by glucose. *Journal of General Physiology* **66**, 243–5.

NIEDERPRUEM, D. J. (1969). Direct studies of nuclear movements in *Schizophyllum commune. Archiv für Mikrobiologie* **64**, 387–95.

PALL, M. L. (1970). Amino acid transport in *Neurospora crassa.* III. Acidic amino acid transport. *Biochimica et Biophysica Acta* **211**, 513–20.

PALL, M. L. (1971). Amino acid transport in *Neurospora crassa.* IV. Properties and regulation of a methionine transport system. *Biochimica et Biophysica Acta* **233**, 201–14.

PARK, D. & ROBINSON, P. M. (1967). A fungal hormone controlling internal water distribution normally associated with cell ageing in fungi. *Symposium of the Society for Experimental Biology* **21**, 323–36.

PATEMAN, J. A., KINGHORN, J. R., DUNN, E. & FORBES, E. (1973). Ammonium regulation in *Aspergillus nidulans. Journal of Bacteriology* **114**, 943–50.

PLUNKETT, B. E. (1958). Translocation and pileus formation in *Polyporus brumalis. Annals of Botany NS* 237–50.

PRÉVOST, G. (1962). Etude genetique d'un Basidiomycete: *Coprinus radiatus* Fr. ex Bolt. *These*, Universite de Paris.

RAUDASKOSKI, M. (1972). Occurrence of microtubules in the hyphae of *Schizophyllum commune* during intracellular nuclear migration. *Archiv für Mikrobiologie* **86**, 91–100.

RAUDASKOSKI, M. (1973). Light and electron microscope study of unilateral mating between a secondary mutant and wild-type strain of *Schizophyllum commune. Protoplasma* **76**, 35–48.

REILLY, C., FUHRMANN, G-F. & ROTHSTEIN, A. (1970). The inhibition of K^+ and phosphate uptake in yeast by cycloheximide. *Biochimica et Biophysica Acta* **203**, 583–5.

ROBERTSON, N. F. (1959).. Experimental control of hyphal branching and branch forms in hyphomycetous fungi. *Journal of the Linnean Society, London* **56**, 207–11.

ROBERTSON, N. F. & RIZVI, S. R. H. (1968). Some observations on the water relations of the hyphae of *Neurospora crassa. Annals of Botany NS* **32**, 279–91.

ROMANO, A. H. & KORNBERG, H. L. (1968). Regulation of sugar utilisation by *Aspergillus nidulans*. *Biochimica et Biophysica Acta* **158**, 491–3.

ROMANO, A. H. & KORNBERG, H. L. (1969). Regulation of sugar uptake by *Aspergillus nidulans*. *Proceedings of the Royal Society, Series B*, **173**, 475–90.

ROSEMAN, S. (1969). The transport of carbohydrates by a bacterial phosphotransferase system. *Journal of General Physiology* **54**, 138S–86S.

SCARBOROUGH, G. A. (1970*a*). Sugar transport in *Neurospora crassa*. *Journal of Biological Chemistry* **245**, 1694–8.

SCARBOROUGH, G. A. (1970*b*). Sugar transport in *Neurospora crassa*. II. A second glucose transport system. *Journal of Biological Chemistry* **245**, 3985–7.

SCHNEIDER, R. P. & WILEY, W. R. (1971*a*). Kinetic characteristics of the two glucose transport systems in *Neurospora crassa*. *Journal of Bacteriology* **106**, 479–86.

SCHNEIDER, R. P. & WILEY, W. R. (1971*b*). Regulation of sugar transport in *Neurospora crassa*. *Journal of Bacteriology* **106**, 487–92.

SCHNEIDER, R. P. & WILEY, W. R. (1971*c*). Transcription and degradation of messenger ribonucleic acid for a glucose transport system in *Neurospora*. *Journal of Biological Chemistry* **246**, 4784–9.

SCHRÖTER, A. (1905). Über protoplasmaströ munz bei Mucorineen. *Flora* **45**, 1–30.

SCHULTZ, S. G. & CURRAN, P. F. (1970). Coupled transport of sodium and organic solutes. *Physiological Reviews* **50**, 637–718.

SEASTON, A., INKSON, C. & EDDY, A. A. (1973). The absorption of protons with specific amino acids and carbohydrates by yeast. *Biochemical Journal* **134**, 1031–43.

SEGEL, I. H. & JOHNSON, M. J. (1961). Accumulation of intracellular inorganic sulfate by *Penicillium chrysogenum*. *Journal of Bacteriology* **81**, 91–106.

SHERE, S. M. & JACOBSON, L. (1970*a*). Mineral uptake in *Fusarium oxysporium* f. sp. *vasinfectum*. *Physiologia Plantarum* **23**, 51–62.

SHERE, S. M. & JACOBSON, L. (1970*b*). The influence of phosphate uptake on cation uptake in *Fusarium oxysporium* f. sp. *vasinfectum*. *Physiologia Plantarum* **23**, 294–303.

SLAYMAN, C. L. (1965*a*). Electrical properties of *Neurospora crassa*. Effects of external cations on the intracellular potential. *Journal of General Physiology* **49**, 69–92.

SLAYMAN, C. L. (1965*b*). Electrical properties of *Neurospora crassa*. Respiration and the intracellular potential. *Journal of General Physiology* **49**, 93–116.

SLAYMAN, C. L. (1970). Movement of ions and electrogenesis in micro-organisms. *American Zoologist* **10**, 377–92.

SLAYMAN, C. L. & SLAYMAN, C. W. (1962). Measurement of membrane potentials in *Neurospora*. *Science* **136**, 876–7.

SLAYMAN, C. L. & SLAYMAN, C. W. (1968). Net uptake of potassium in *Neurospora*. Exchange for sodium and hydrogen ions. *Journal of General Physiology* **52**, 424–43.

SLAYMAN, C. L., LU, C.Y-H. & SHANE, L. (1970). Correlated changes in membrane potential and ATP concentrations in *Neurospora*. *Nature, London* **226**, 274–6.

SLAYMAN, C. W. (1970). Net potassium transport in *Neurospora*. Properties of a transport mutant. *Biochimica et Biophysica Acta* **211**, 502–12.

SLAYMAN, C. W. (1970). Net potassium transport in *Neurospora*. Properties of a transport mutant. *Biochimica et Biophysica Acta* **211**, 502–12.

SLAYMAN, C. W. & SLAYMAN, C. L. (1970). Potassium transport in *Neurospora*. Evidence for a multisite carrier at high pH. *Journal of General Physiology* **55**, 758–86.

SLAYMAN, C. W. & TATUM, E. L. (1964). Potassium transport in *Neurospora*. I. Intracellular sodium and potassium concentrations and cation requirements for growth. *Biochimica et Biophysica Acta* **88**, 578–92.

SLAYMAN, C. W. & TATUM, E. L. (1965*a*). Potassium transport in *Neurospora*. II. Measurement of steady-state potassium fluxes. *Biochimica et Biophysica Acta* **102**, 149–60.

SLAYMAN, C. W. & TATUM, E. L. (1965*b*). Potassium transport in *Neurospora*. III. Isolation of a transport mutant. *Biochimica et Biophysica Acta* **109**, 184–93.

SNIDER, P. J. (1965). Incompatibility and nuclear migration. In *Incompatibility in fungi*. Edited by K. Esser and J. R. Raper, pp. 52–70. Berlin: Springer-Verlag.

SNIDER, P. J. (1968). Nuclear movements in *Schizophyllum*. *Symposium of the Society for Experimental Biology* **22**, 261–83.

SNIDER, P. J. & RAPER, J. R. (1958). Nuclear migration in the Basidiomycete *Schizophyllum commune*. *American Journal of Botany* **45**, 538–46.

STEVENINCK, J. van & ROTHSTEIN, A. (1965). Sugar transport and metal binding in yeast. *Journal of General Physiology* **49**, 235–46.

SWIEZYNSKI, K. M. & DAY, P. R. (1960). Heterokaryon formation in *Coprinus lagopus*. *Genetical Research* **1**, 129–39.

TABER, W. A. (1971). Uptake of ^{14}C-labelled succinate, L(+)-dihydroxysuccinate, L-monohydroxysuccinate, citrate, alpha-ketoglutarate and D-glucose by washed mycelium of *Claviceps purpurea*. *Mycologia* **63**, 290–307.

TRINCI, A. P. J. & COLLINGE, A. J. (1973). Structure and plugging of septa of wild type and spreading colonial mutants of *Neurospora crassa*. *Archiv für Mikrobiologie* **91**, 355–64.

TRINCI, A. P. J. & RIGHELATO, R. C. (1970). Changes in constituents and ultrastructure during autolysis of glucose-starved *Penicillium chrysogenum*. *Journal of General Microbiology* **60**, 239–50.

UDEN, N. van (1969). Kinetics of nutrient-limited growth. *Annual Review of Microbiology* **23**, 473–86.

WATRUD, L. S. & ELLINGBOE, A. H. (1973*a*). Cobalt as a mitochondrial density marker in a study of cytoplasmic exchange during mating of *Schizophyllum commune*. *Journal of Cell Biology* **59**, 127–33.

WATRUD, L. S. & ELLINGBOE, A. H. (1973*b*). Use of cobalt as a mitochondrial vital stain to study cytoplasmic exchange in matings of the Basidiomycete *Schizophyllum commune*. *Journal of Bacteriology* **115**, 1151–8.

WEAST, R. C. & SELBY, S. M. (1967). *Handbook of chemistry and physics*. Cleveland: The Chemical Rubber Co.

WILCOXSON, R. D. & SUDIA, T. W. (1968). Translocation in fungi. *Botanical Reviews* **34**, 32–50.

WILEY, W. R. (1970). Tryptophan transport in *Neurospora crassa*: a tryptophan binding protein released by cold osmotic shock. *Journal of Bacteriology* **103**, 656–62.

WILEY, W. R. & MATCHETT, W. H. (1966). Tryptophan transport in *Neurospora crassa*. I. Specificity and kinetics. *Journal of Bacteriology* **92**, 1698–705.

WILEY, W. R. & MATCHETT, W. H. (1968). Tryptophan transport in *Neurospora crassa*. II. Metabolic control. *Journal of Bacteriology* **95**, 959–66.

YAMAMOTO, L. A. & SEGEL, I. H. (1966). The inorganic sulfate transport system of *Penicillium chrysogenum*. *Archives of Biochemistry and Biophysics* **114**, 523–38.

CHAPTER 3

Glycolysis

V. W. COCHRANE

3.1 Introduction

Although in its original usage the term glycolysis denoted the conversion of glycogen to lactic acid in muscle, the term is now used more broadly, if less precisely, to mean the entire family of metabolic sequences by which sugars are converted to small molecules which can enter the pathways of terminal oxidation or biosynthesis. (Perhaps the term 'glucolysis' would be preferable for this broader usage.) The principal known metabolic sequences are the Embden–Meyerhof (EM), the pentose phosphate (PP) and the Entner–Doudoroff (ED) pathways. In addition, the metabolism of gluconate requires brief comment. Gluconeogenesis, the synthesis of glucose and polymers of glucose from smaller molecules, utilizes many of the reactions of the EM pathway, and the process is intimately related to regulation of crucial enzymes of the EM system. Finally, in cells or tissues with more than one pathway operative, the question of how they are balanced in normal growth and in development is of great current interest.

In this broader usage of the term, glycolysis can occur under aerobic or anaerobic conditions. Energetically, of course, the aerobic process, normally involving the tricarboxylic acid cycle and oxidative phosphorylation, is far more efficient than any anaerobic metabolic sequence.

3.2 The Embden–Meyerhof Pathway

Figure 3.1 summarizes our knowledge of the principal reactions of the EM pathway. From the glucose level to that of pyruvate, the reactions can be summarized:

$$\underset{\text{Glucose}}{C_6H_{12}O_6} \longrightarrow \underset{\text{Pyruvate}}{2CH_3COCOOH} + 4[H]$$

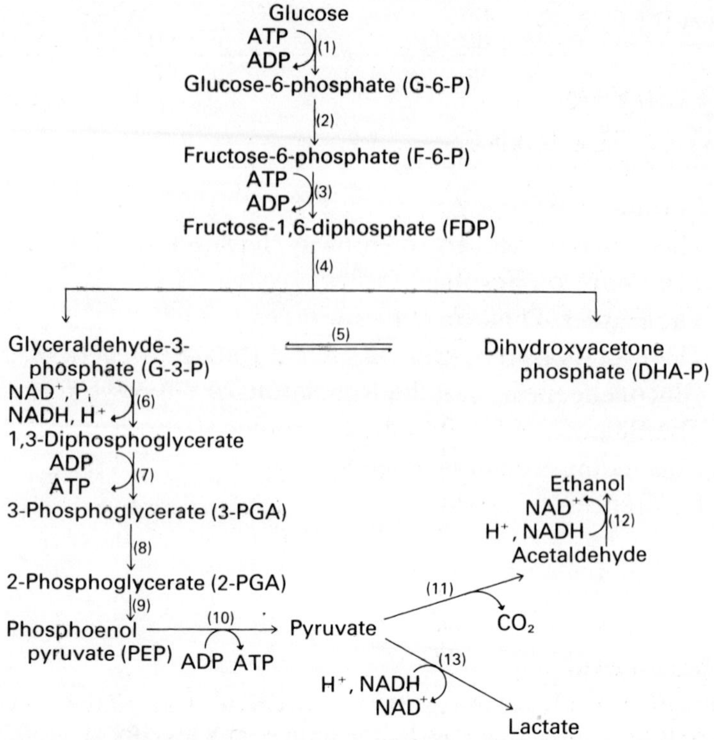

Fig. 3.1 Reactions of the Embden–Meyerhof (EM) pathway. Although most reactions are more or less reversible, arrows have, except for the triose phosphate isomerase reaction (5), been drawn only in the forward direction. Enzymes catalysing the numbered reactions are: (1) hexokinase, (2) phosphoglucoisomerase, (3) phosphofructokinase (PFK), (4) aldolase, (5) triose phosphate isomerase, (6) glyceraldehyde-3-phosphate dehydrogenase, (7) phosphoglycerate kinase, (8) phosphoglyceromutase, (9) enolase, (10) pyruvate kinase (PK), (11) pyruvate decarboxylase, (12) ethanol dehydrogenase (ADH), (13) lactate dehydrogenase (LDH). Other abbreviations: ATP = adenosine triphosphate, ADP = adenosine diphosphate, P_i = inorganic phosphate, NAD^+ = nicotinamide adenine dinucleotide, oxidized form, NADH = nicotinamide adenine dinucleotide, reduced form.

In lactic acid fermentation (muscle, some bacteria and fungi), the reducing equivalents of the above equation are used to reduce pyruvate to lactate, while in yeast, the other branch of Fig. 3.1, pyruvate is decarboxylated to acetaldehyde and CO_2, and the reducing equivalents of the first equation used to reduce acetaldehyde to ethanol. In either case, as shown in Fig. 3.1, the *net* formation of ATP from ADP is 2 moles ATP per mole glucose; by comparison, if pyruvate is completely oxidized to CO_2 aerobically, with oxidative phosphorylation, the net synthesis of ATP is 36 moles per mole of glucose oxidized.

Although many of the pioneer studies on the EM system employed yeast as the experimental organism and although the pathway was well established in yeast as early as 1933, the study of glycolysis in the filamentous fungi progressed rather more slowly; until 1956, for example, it was maintained by some that the EM pattern, involving phosphorylated intermediates, was absent in the fungi. Part of the reason for this slow progress is, of course, the relatively small number of biologists, then and now, interested in the filamentous fungi. In addition, the fungi present some technical problems, particularly the inhomogeneity of cellular preparations (Smith & Galbraith, 1971), and the difficulty of preparing active cellular and cell-free systems (Cochrane, 1958; Blumenthal, 1965). Lack of interest and technical difficulties are reflected in the fact that relatively few fungi have been intensively investigated and perhaps in the fact that no genuinely new and distinctive glycolytic pathways have been unequivocally established in the group.

However, by the time of the review by Blumenthal (1965) it had become clear that many fungi carry out a typical glycolysis using the Embden–Meyerhof (EM) sequence of reactions. Evidence came from product balances, from the study of single enzymatic reactions, and from isotope distributions in metabolic products. This evidence was most convincing for the lactate-producing lower Phycomycetes, and for *Fusarium lini*, *Aspergillus* spp., *Rhizopus* sp., and *Penicillium* spp. (Cochrane, 1958; Blumenthal, 1965).

Since 1965, enough more information has appeared so that we may safely conclude that the EM system is functional in both anaerobic and aerobic metabolism of fungi; this does not of course exclude the possibility of particular enzyme deficiencies in still to be investigated species.

Anaerobic growth is not common in the fungi (Tabak & Cooke, 1968), but it is definitely established in some of the lower Phycomycetes (Gleason, 1968; Emerson & Held, 1969; Held, Emerson, Fuller & Gleason, 1969) and in *Mucor* spp. (Flores-Carréon, Reyes & Ruíz-Herrera, 1969; Elmer & Nickerson, 1970; Clark-Walker, 1972).

Anaerobic metabolism among fungi which require oxygen for growth is fairly common but by no means universal. From still limited evidence it appears that there are three patterns of anaerobic glycolysis from the standpoint of the final product.

In the first of these, known particularly in the Chytridiomycetes and Oömycetes, lactic acid is the sole product of anaerobic fermentation of glucose (Golueke, 1957; Gleason & Price, 1969; Held, 1970; LéJohn, 1971); the group therefore resembles the homofermentative lactic acid bacteria. The lactic acid produced by these fungi is the L(−)isomer (Gleason, Nolan, Wilson & Emerson, 1966). Aeration results in the replacement of lactate by CO_2 as the principal product in *Sapromyces elongatus* (Golueke, 1957) and *Blastocladiella emersonii* (Khuow & McCurdy, 1969), but lactate accumulation in *Aqualinderella fermentans* is unaffected by aeration (Held, 1970).

The second fermentation pattern is that which yields ethanol and CO_2 as the principal products, as in yeast. *Fusarium lini* converts glucose stoichiometrically to these products, and the labelling pattern from (^{14}C) glucose is consistent with the operation of the EM pathway (Cochrane, 1956; Heath, Nasser & Koffler, 1956). A mutant of *Neurospora crassa* able to grow anaerobically also produces the theoretical ethanol yield from glucose (Howell, Zuiches & Munkres, 1971). However, a strain of the wild type of this fungus has been reported to contain detectable but low levels of lactate dehydrogenase (Bianchi, Purohit & Turian, 1971).

The third pattern is that of *Rhizopus* sp. (the culture studied has been called *R. oryzae* and *Rhizopus* MX—its identity is uncertain). Superficially, this fermentation resembles that of the heterolactic bacteria, in that ethanol, CO_2 and lactic acid are all formed in substantial yield. The resemblance is only superficial: neither the relative amounts of the three end-products nor the product labelling from (^{14}C) glucose are as in the bacteria (Gibbs & Gastel, 1953; Margulies & Vishniac, 1961). In *Rhizopus oryzae* the lactic acid formed is the D(+)isomer (Obayashi, Yorifuji, Yamagata, Ijichi & Kanie, 1966).

Macroconidia of *Fusarium solani* f. sp. *phaseoli* do not metabolize glucose anaerobically, although extracts contain detectable amounts of all of the enzymes of the EM pathway (Cochrane & Cochrane, 1966). A similar phenomenon in bacteria has been suggested to result from the absence of systems able to re-oxidize NADH (Hill & Mills, 1954; Cochrane, 1955).

A second line of evidence which may be offered in support of the generalization that the EM pathway is common in fungi is provided by many studies of the activity of enzymes of the pathway; to those reviewed by Cochrane (1958) and Blumenthal (1965) may be added *Achlya ambisexualis* (Warren & Mullins, 1969), *Venturia inaequalis* (Van Skoyoc, Kuć & Williams, 1970), *Polyporus brumalis* (Casselton, 1966), *Colletotrichum gloeosporioides* and *Glomerella musarum* (Greene, 1969), *Rhizoctonia solani* and *Sclerotium bataticola* (Van Etten, Molitoris & Gottlieb, 1966), and *Verticillium albo-atrum* (Malca, Erwin, Sims, Long & Jones, 1968).

The reader should be aware that these determinations of enzymatic activities have been made with crude extracts, and that such extracts may conceal artifacts which affect results either qualitatively or quantitatively. The pyruvate kinase of *Neurospora crassa* has been purified (Kapoor & Tronsgaard, 1972), as have the lactate dehydrogenases of an *Allomyces* hybrid (Purohit & Turian, 1972) and *Pythium debaryanum* (LéJohn, 1971).

A third line of evidence for the EM system in fungi comes from studies on the distribution of ^{14}C from specifically labelled (^{14}C) glucose in metabolic products. This evidence is reviewed later in the section on the balance of pathways (p. 83).

From these data, it can be generalized, first, that we know of no anaerobic metabolic system in the filamentous fungi which does not rely on the reactions of the EM pathway. That is, there is as yet no evidence for the operation of anaerobic non-EM systems such as those known in bacteria (Doelle, 1969). Secondly, the EM system is at least very common in the filamentous fungi and is involved in aerobic as well as anaerobic catabolism of sugars.

3.3 The Pentose Phosphate Pathway

This system of reactions, the details of which were worked out during the 1950s, has also been called the phosphogluconate oxidation pathway and the hexose monophosphate shunt. The term pentose phosphate (PP) pathway recognizes the central position of several pentose phosphates in the system; however, as discussed below, there are other ways to synthesize pentoses, and the principal role of the pathway is probably provision of reduced nicotinamide adenine dinucleotide phosphate (NADPH) for synthetic reactions.

The PP pathway is of a quite different type than is the EM sequence. In the latter, one can write down a starting point, glucose or glycogen, and a final product, lactate or ethanol and CO_2 (more generally, pyruvate can be considered the final product of the EM system). Intermediates may be drawn off—triose phosphate in glycerol synthesis or 3-PGA in serine formation, for example—but the end-product is never in doubt.

This is not the situation in the PP pathway. A variety of compounds form and may be drawn upon by the cell for a variety of reactions, including degradation *via* the EM pathway. There is, in short, no one end-product; instead, a pool of triose, pentose, hexose, and heptose phosphates is formed by the system.

For this reason, the PP pathway is described here as a series of four sets of reactions, rather than as a single unitary sequence. The division is somewhat arbitrary, but has the advantage of avoiding a misleading oversimplification of what is in fact a quite complex metabolic situation. These four 'phases' are, in brief:

1. The oxidative phase, in which glucose is phosphorylated and converted first to 6-phosphogluconate (6-PG) and then to ribulose-5-phosphate (Ru-5-P) and CO_2, with concomitant reduction of $NADP^+$ to NADPH.
2. The non-oxidative conversion of Ru-5-P to xylulose-5-phosphate (Xu-5-P) and ribose-5-phosphate (R-5-P).
3. The non-oxidative reactions carried out by transketolase.
4. The non-oxidative reactions catalyzed by transaldolase.

The oxidative phase is depicted in Fig. 3.2. Equilibria strongly favour the formation of Ru-5-P, largely because of the hydrolysis of the lactone in reaction (3).

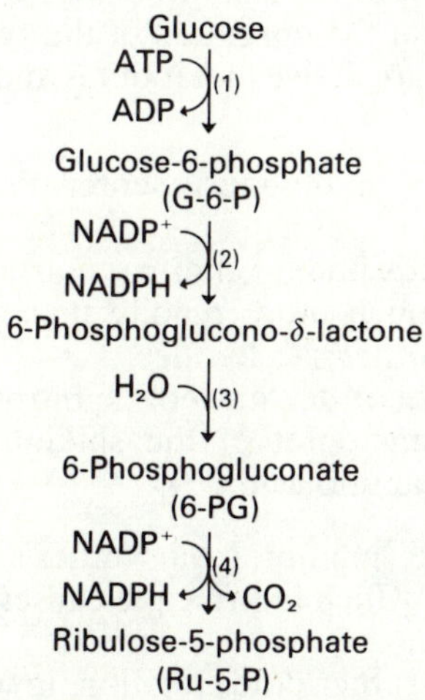

Fig. 3.2 The conversion of glucose to ribulose-5-phosphate in the pentose phosphate pathway. The enzymes catalysing the numbered reactions are: (1) hexokinase, (2) glucose-6-phosphate dehydrogenase, (3) gluconolactonase, (4) 6-phosphogluconate dehydrogenase.

In the second phase, two isomerization reactions convert Ru-5-P to other pentose phosphates, *viz*:

(1) Ru-5-P ⇄ R-5-P, catalyzed by phosphoribose isomerase.
(2) Ru-5-P ⇄ Xu-5-P, catalyzed by phosphoketopentose isomerase.

Equilibria in these reactions are not far from unity.

In the third phase, the enzyme transketolase catalyzes a number of transfer reactions, in all of which a 2-carbon keto fragment ('active glycolaldehyde') is transferred from a donor keto sugar to an acceptor aldehyde. Physiologically, the important reaction is:

R-5-P + Xu-5-P ⇄ sedoheptulose-7-phosphate (S-7-P)
+ glyceraldehyde-3-phosphate (G-3-P).

It is important, in order to appreciate the complexity of this system, to bear in mind that transketolase can act on other sugars of appropriate structure, *e.g.*

F-6-P + G-3-P ⇄ Xu-5-P + erythrose-4-phosphate (E-4-P)

F-6-P + R-5-P ⇄ S-7-P + E-4-P

Again equilibria are near unity, and the reactions are freely reversible.

The last phase, dependent on the activity of transaldolase, is normally the reaction:

S-7-P + G-3-P ⇄ F-6-P + E-4-P

Transaldolase is specific for a class of reactions, transferring a dihydroxyacetone residue from a ketose phosphate to an aldose phosphate.

Since all of the reactions of the last three phases are freely reversible, it is clear, as noted earlier, that the operation of the system generates a pool of sugar phosphates. These are active metabolites and may be used in a variety of ways, *e.g.*

1. F-6-P and G-3-P can obviously enter the EM pathway and be metabolized to pyruvate.
2. The same compounds under other circumstances could be converted back to G-6-P (F-6-P much more readily than G-3-P) and used in the synthesis of glycogen or other saccharides.
3. R-5-P may be drawn off for nucleotide biosynthesis.
4. E-4-P is the starting point of the shikimic acid pathway for the biosynthesis of aromatic amino acids.

Still other reactions, *e.g.* mannitol biosynthesis in fungi, can easily be envisaged, as can a cycle which in principle causes a complete oxidation of glucose (see below).

We have to visualize a rather fluid situation, in which the demands of the cell at any one time determine the flow of pentose phosphate carbon. This is important in connection with efforts, considered later, to estimate the balance of pathways by tracing out the fate of isotopically labelled glucose carbon atoms.

This fluidity, or open-endedness, may not hold in all cases. Certain of the obligately aerobic yeasts of the genus *Rhodotorula* appear to lack phosphofructokinase and therefore to be unable to utilize the EM pathway from glucose. It has been suggested that in these forms the PP pathway is the route by which all glucose carbon flows (Brady & Chambliss, 1967; Höfer, Betz & Becker, 1970).

It should be recognized that, in principle, glucose could be completely oxidized to CO_2 by the PP reaction sequence. It is only necessary to assume that all pentose carbon winds up, by the pathways just described, in F-6-P and G-3-P, and that these are totally resynthesized to G-6-P. Then, if the G-6-P so formed re-enters the PP pathway, and the cycle is repeated over and over, one may write, for each six turns, the equation (Hollman & Touster, 1964):

$$6\text{G-6-P} + 12\text{NADP}^+ \longrightarrow 5\text{G-6-P} + 6\text{CO}_2 + 12\text{NADPH} + 12\text{H}^+ + \text{P}_i$$

Or, cancelling common terms:

$$\text{G-6-P} + 12\text{NADP}^+ \longrightarrow 6\text{CO}_2 + 12\text{NADPH} + 12\text{H}^+ + \text{P}_i$$

An *in vitro* system of this type can be reconstituted (Couri & Racker, 1959), but it seems unlikely that any cell routinely uses this completely oxidative cycle, primarily because there is as yet no evidence that NADPH oxidation can be coupled to phosphorylation; hence, it would seem that such a cell would rapidly exhaust its ATP stores. Further, it is not clear how the many biosyntheses based on intermediates of the EM pathway and the tricarboxylic acid cycle could be provided under this metabolic regimen.

An anaerobic utilization of part of the hexose monophosphate pathway is a well-known metabolic system in heterofermentative lactic acid bacteria such as *Leuconostoc mesenteroides*. However, in the bacterial system the two dehydrogenases are linked to NAD^+, not $NADP^+$, and the transketolase and transaldolase reactions do not form part of the system. Instead, pentose phosphate is cleaved directly, by phosphoketolase, to G-3-P and acetyl phosphate; the ultimate fermentation products from glucose are thus CO_2, lactate, and acetate in roughly equimolar amounts. This heterolactic system is mentioned here because there is limited evidence of something similar to phosphoketolase activity in fungi metabolizing pentose (p. 74), and because there is one report (Hochster, 1957) of an NAD-dependent G-6-P dehydrogenase in *Aspergillus flavus-oryzae*.

Occurrence of the PP pathway in fungi

Methods for the demonstration of this pathway do not depend on product balances, since there is no one end-product. Instead, two criteria have been used:

1. Catalysis of pathway reactions by cell-free extracts of fungi.
2. The distribution of ^{14}C from specifically labelled (^{14}C) glucose in CO_2 or in other metabolites.

The second of the criteria is discussed later (p. 81), in connection with the problem of the balance of pathways during growth and development. It is suggested there that such data may be highly suggestive but that full confidence in their interpretation is easier if corroborative evidence from other kinds of experiments is available.

Evidence from extracts is straightforward as far as the two dehydrogenases are concerned—their assay poses no serious problem. For the rest of the PP pathway, a number of approaches have been used. In the simplest, the entire system is assayed by measuring the formation of hexose or triose phosphates from R-5-P (Casselton, 1966; Greene, 1969). Although this method has the advantage of simplicity, it is preferable to measure key enzymes specifically (Van Skoyoc, Kuć & Williams, 1970) or to assay for specific intermediates (Sih, Hamilton & Knight, 1957).

By the time of the review by Blumenthal (1965) it was clear that the PP pathway is widespread in the higher fungi; evidence from the Phycomycetes was less clear. Since then, additional evidence has been reported from lower Phycomycetes (Belsky & Goldstein, 1964; Warren & Mullins, 1969), Ascomycetes (Van Skoyoc, Kuć & Williams, 1970), Basidiomycetes (Meloche, 1961; Casselton, 1966, 1967; Mitchell & Shaw, 1968; Williams & Shaw, 1968; Holligan & Jennings, 1972), and Fungi Imperfecti (Malca *et al.*, 1968; Greene, 1969). Although the evidence varies somewhat in conclusiveness, the pathway appears to be quite general. The two dehydrogenases of the pathway in *Neurospora crassa* have been highly purified, and their kinetics have been reported (Scott & Tatum, 1970; Lechner, Fuscaldo & Bazinet, 1971; Scott & Abramsky, 1973*a*). Two negative findings, on spores of *Tilletia caries* (Newburgh & Cheldelin, 1958) and mycelium of *Caldariomyces fumago* (Ramachandran & Gottlieb, 1963), propose that there is an Entner–Doudoroff (ED) pathway rather than the PP system; these are discussed in connection with the ED pathway below.

Significance of the PP pathway

At first glance, it might appear that the major role of the PP pathway might be in the provision of starting materials for biosynthesis, *e.g.* pentose for nucleic acids. However, it should be stressed that it is not at all necessary that the entire PP pathway operate in order for the sugar phosphates to be synthesized; they can also be formed non-oxidatively from intermediates, specifically F-6-P and G-3-P, of the EM pathway. It is only necessary that transketolase and transaldolase be present; Horecker (1962) proposed that these enzymes are very common or even ubiquitous. Evidence from the filamentous fungi which directly bears on this question is not at hand; in the yeast-like *Candida albicans* labelling studies indicate that ribose is formed oxidatively (David & Renaut, 1955; Horecker, 1962), but the situation in organisms generally is that either an oxidative or a non-oxidative route of pentose synthesis can be used (Hollman & Touster, 1964).

Lehninger (1970) sums up the current view on the significance of the PP pathway thus: 'Its primary purpose in most cells is to generate reducing power in the extra-mitochondrial cytoplasm in the form of NADPH.' In organisms generally, it appears that biosynthetic reductions require NADPH rather than NADH; in the fungi, for example, there are both NAD- and NADP-linked glutamate dehydrogenases, the latter predominating under conditions in which glutamate must be synthesized (Kato, Koike, Yamada, Yamada & Tanaka, 1962; Sanwal & Lata, 1962).

The most nearly decisive evidence for an essential role of the PP pathway in providing NADPH for biosyntheses comes from studies on mutants of *Neurospora crassa* which show a correlation between altered morphology (colonial growth habit) and altered G-6-P and/or 6-PG dehydrogenases. The *col*-2 mutant, for example, forms a G-6-P dehydrogenase which is physically different from that of the wild type (Brody & Tatum, 1966; Lechner, Fuscaldo & Bazinet, 1971); correspondingly, the level of NADPH is abnormally low (Brody, 1970). And, cultivation in media which derepress other NADPH-forming dehydrogenases effects a partial reversion of morphology toward normal, non-colonial, morphology (Fuscaldo, Lechner & Bazinet, 1971). Other morphological mutants also have an altered G-6-P dehydrogenase (Scott & Tatum, 1970), or both G-6-P and 6-PG dehydrogenases (Lechner, Fuscaldo & Bazinet, 1971; Scott & Abramsky, 1973*a*, *b*).

These results implicate the PP pathway as normally providing the NADPH necessary for cell-wall biosynthesis. Although it is not yet possible to prove that the cause of abnormal growth is a shortage of NADPH (Scott & Abramsky, 1973*b*), it is reasonable to suggest that fatty-acid synthesis, necessary for construction of some essential membrane, is disrupted by too low a level of NADPH. It is significant that colonial mutants have an abnormally low content of unsaturated fatty acids, for the formation of which NADPH is required (Brody & Nyc, 1970).

Finally, if the PP pathway has as its main function the provision of NADPH, it might be expected that the rate of the operation of the reactions is controlled by the cellular level of the reduced co-enzyme. This is the case in animal tissues (Eggleston & Krebs, 1974), and the same suggestion has been made for micro-organisms in general (Eagon, 1963) and for yeast in particular (Osmond & ap Rees, 1969).

The catabolism of pentoses

The only extensive studies of aerobic pentose metabolism in fungi are those of S. G. Knight and co-workers, paralleled by the work of Horecker and others on *Candida* spp. (Horecker, 1962). In *Penicillium chrysogenum* xylose is converted to Xu-5-P by a sequence of three reactions which are similar to those reported in animal cells but differ significantly from bacterial systems (Hollman & Touster, 1964). Chiang & Knight (1959, 1960*a*) provide evidence for the reactions:

$$\text{D-xylose} + \text{NADPH} + \text{H}^+ \rightarrow \text{xylitol} + \text{NADP}^+$$

$$\text{xylitol} + \text{NAD}^+ \rightarrow \text{D-xylulose} + \text{NADH} + \text{H}^+$$

$$\text{D-xylulose} + \text{ATP} \rightarrow \text{D-xylulose-5-phosphate} + \text{ADP}$$

$$\text{Sum: } \text{D-xylose} + \text{NADPH} + \text{NAD}^+ + \text{ATP} \rightarrow \text{D-xylulose-5-phosphate} + \text{NADP}^+ + \text{NADH}$$

The three enzymes involved—two polyol dehydrogenases and a kinase—may differ somewhat in substrate specificity according to species, but are very similar in *P. chrysogenum* and *Candida* spp. (Chiang & Knight, 1959; Horecker, 1962).

Of fourteen fungi able to grow on xylose—including *Penicillium* spp., *Aspergillus* spp., *Rhizopus nigricans*, and *Neurospora crassa*—all have the two dehydrogenases found in *P. chrysogenum* (Chiang & Knight, 1960*b*).

In both *Penicillium chrysogenum* and *Candida* spp. L-arabinose also is converted ultimately to D-xylulose-5-phosphate, although the complete reaction sequence is not certain (Chiang & Knight, 1961; Horecker, 1962).

The anaerobic breakdown of pentose by *Lactobacillus* spp. is distinctive by reason of the participation of the enzyme phosphoketolase, which effects a cleavage of pentose to acetyl phosphate and G-3-P. The former is converted to free acetate, and G-3-P is metabolized *via* EM reactions to lactate, giving the balance, in the case of xylose:

$$\text{xylose} \rightarrow \text{acetate} + \text{lactate}$$

Label from xylose-1-^{14}C is incorporated only into the methyl group of acetate.

For *Fusarium lini*, Gibbs *et al.* (1954) reported preliminary evidence for a comparable reaction in the anaerobic metabolism of xylose. The pentose is metabolized as follows:

$$\text{xylose} \rightarrow \text{acetate} + \text{ethanol} + CO_2$$

The evidence suggesting the involvement of phosphoketolase is that essentially all of the label from xylose-1-^{14}C appears in the methyl carbon of acetate. The authors postulate a C_2–C_3 split of pentose phosphate by phosphoketolase, as in *Lactobacillus* spp., followed by the conversion of G-3-P *via* EM reactions known in *F. lini* (Cochrane, 1956) to ethanol and CO_2. The labelling evidence alone seems insufficient to establish the hypothesis, but the reaction is worth further investigation.

The formation of xylonic acid from xylose has been reported to be carried out by several fungi: *Trametes versicolor* and *Fomes igniarius* (Lyr, 1962), *Polyporus obtusus* (Ruelius, Kerwin & Janssen, 1968), and *Penicillium corylophilum* (Ikeda & Yamada, 1963). Lyr (1962) characterized the enzyme responsible as a xylose oxidase. The relationship of this enzyme to other oxidases is in dispute (Lyr, 1962; Ruelius, Kerwin & Janssen, 1968); the well-known glucose oxidase of *Aspergillus niger* is specific for glucose (Bentley, 1963).

3.4 The Entner–Doudoroff Pathway

The Entner–Doudoroff (ED) pathway is used by some bacteria in the anaerobic breakdown of glucose, by others only in gluconate metabolism (Zablotny & Fraenkel, 1967). In the ED system, 6-PG is formed from glucose as in the PP pathway, but is then converted to 2-keto-3-deoxy-6-phosphogluconate (KDPG). This is then split, by KDPG aldolase, to pyruvate and G-3-P. Since the G-3-P is then converted to pyruvate by EM reactions, the product balance is the same as in the EM pathway, *viz.*

$$\text{Glucose} \rightarrow 2\,\text{pyruvate} + 4[\text{H}]$$

However, the mode of formation of the pyruvate means that its carboxyl carbon arises from both carbons 1 and 4 of glucose. Pyruvate formed in the

EM reaction has its carboxyl carbon derived from carbons 3 and 4 of glucose (Blumenthal, 1965). Thus, in the sole operation of the ED pathway, isotope from glucose-1-^{14}C appears in the carboxyl of pyruvate; no other known system yields this result.

The first report of the ED pathway in fungi is that of Newburgh & Cheldelin (1958), working with chlamydospores (teliospores) of *Tilletia caries*. The method used was to measure over time the appearance of ^{14}C from spot-labelled glucose in the CO_2 produced aerobically during glucose oxidation. Results were consistent with, but were not claimed to prove, the operation of the ED pathway. For reasons given below (p. 81), it seems best to regard this as suggestive rather than conclusive; it is to be hoped that enzymatic studies will be made on this material.

The other report of the ED pathway is that of Ramachandran & Gottlieb (1963) on *Caldariomyces fumago*. The evidence offered is in part negative, and as such not convincing. The positive evidence is that carboxyl-labelled pyruvate was formed from glucose-1-^{14}C. Like the work on *Tilletia caries*, this evidence provides sufficient reason for further study of the enzymatic capacities of the organism.

Aspergillus niger forms gluconic acid from glucose and presumably metabolizes the gluconate further (see below), but two studies (McDowell & DeHertogh, 1958; Lakshminarayana, Modi & Shah, 1969) failed to produce evidence for the ED pathway.

3.5 The Metabolism of Gluconate and Related Reactions

The high yields of gluconate from glucose in some fungal cultures has long been known (Cochrane, 1958). The enzyme responsible in *Aspergillus niger*—glucose oxidase, or glucose aerodehydrogenase—has been studied extensively (Bentley, 1963); it is the basis for a useful method of glucose analysis. The overall reaction is:

$$\text{D-glucose} + O_2 + H_2O \rightarrow \text{D-gluconic acid} + H_2O_2$$

Notatin, from *Penicillium notatum*, is also a glucose oxidase but it differs physically from the *A. niger* enzyme (Bodmann & Walter, 1965).

Two proposals have been made for the further metabolism of gluconate by *Aspergillus niger*. Lakshminarayana, Modi & Shah (1969) propose that it is phosphorylated to 6-PG and further metabolized by the PP system; gluconate-adapted cells were found to have gluconokinase and 6-PG dehydrogenase.

In contrast, Elzainy, Hassan & Allam (1973) have provided some evidence for a non-phosphorylative pathway of gluconate catabolism, *viz.*

$$\text{Gluconate} \rightarrow \text{2-keto-3-deoxygluconate (KDG)}$$

$$\text{KDG} \rightarrow \text{pyruvate} + \text{glyceraldehyde}$$

The enzymes involved are gluconate dehydratase and 2-keto-3-deoxygluconate aldolase. The system is of course analogous to the ED pathway discussed above, and might be responsible for the results on fungi discussed in the previous section.

In some bacteria gluconate metabolism involves, among others, the compounds 2-keto- and 5-ketogluconate. These have been identified in fungi—the 2-keto acid in *Penicillium brevi-compactum* (Simonart & Godin, 1951), the 5-keto derivative in *Aspergillus niger* (Martin & Steel, 1955). 2-Ketogluconate can be reduced to gluconate by *Aspergillus nidulans* (De Ley & Defloor, 1959). Direct phosphorylation of 2-ketogluconate, known in bacteria, does not occur in any of 15 fungi studied by De Ley & Vandamme (1955).

Two other oxidases deserve mention, although they cannot now be identified as part of any metabolic sequence. *Polyporus obtusus* forms a 'carbohydrate oxidase' oxidizing glucose at carbon-2:

$$\text{Glucose} + O_2 \rightarrow \text{glucosone} + H_2O_2$$

The enzyme also acts on D-xylose, D-gluconolactone, and L-sorbose (Janssen & Ruelius, 1968; Ruelius, Kerwin & Janssen, 1968), and is presumably identical with the sorbose oxidase of *Trametes sanguinaria*, (Yamada, Iizuka, Aida & Uemura, 1966).

The second oxidase is the galactose oxidase of *Polyporus circinatus*, which oxidizes galactose and galactosides at carbon-6, for example:

$$\text{D-galactose} + O_2 \rightarrow \text{D-galacto-hexodialdose} + H_2O_2$$

The description by Avigad, Amaral, Asensio & Horecker (1962) corrects in several respects an earlier report on the same enzyme (Cooper, Smith, Bacila & Medina, 1959).

3.6 Gluconeogenesis and the Regulation of Glycolysis

Gluconeogenesis

Gluconeogenesis, the conversion of pyruvate to glucose (or a polysaccharide), is well known in animal cells; the process as currently understood is compared in Fig. 3.3 with the EM system of glycolysis, with which it shares many reversible enzymatic reactions. The salient features of the interaction shown in Fig. 3.3 can be summarized:

1. Three EM reactions—those catalyzed by hexokinase, phosphofructokinase, and pyruvate kinase—are irreversible, as is the phosphorolysis of glycogen. The other steps of the pathway are reversible and at these points the same enzymes function in both gluconeogenesis and glycolysis.
2. The special reactions of gluconeogenesis are essentially ways in which to get around the irreversible steps of glycolysis, using different enzyme systems.
3. Energy, as ATP and GTP, and reducing power, as NADH, are used in the conversion of pyruvate to glucose; if glycogen is formed, energy from UTP is also required.

The scheme of Fig. 3.3 is somewhat simplified. Thus, the conversion of pyruvate to PEP requires, in animal cells, both extra- and intramitochondrial enzyme systems and a mitochondrion-cytoplasm shuttle involving the formation and oxidation of malate (Lehninger, 1970).

With respect to the fungi, there are two obvious questions, *viz.*

1. Does gluconeogenesis occur in at least approximately the same way as it does in animal cells?

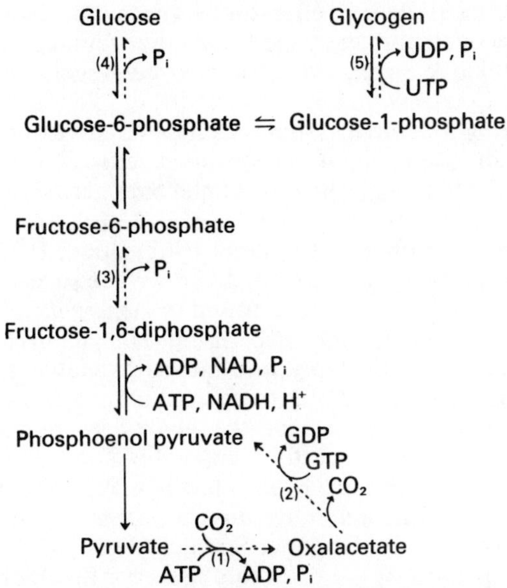

Fig. 3.3 Glycolysis and gluconeogenesis in animal cells. Reactions which are only reversed in gluconeogenesis are shown by paired solid arrows, irreversible reactions of glycolysis by single solid arrows, and the special by-pass reactions of gluconeogenesis by dotted arrows. Omitted steps (see Fig. 3.1) are reversible. Inputs and outputs are shown only for gluconeogenesis. The principal enzymes unique to gluconeogenesis are numbered: (1) pyruvate carboxylase (PC), (2) phosphoenolpyruvate carboxykinase (PEPCK), (3) fructose-1,6-diphosphatase (FDPase), (4) glucose-6-phosphatase, (5) combined action of UDP-glucose pyrophosphorylase and glycogen synthetase. Abbreviations: GTP, GDP = guanosine triphosphate and diphosphate, respectively; UTP, UDP = uridine triphosphate and diphosphate, respectively; and P_i = inorganic phosphate.

2. How are essentially opposing reactions, *e.g.* those of phosphofructokinase and fructose-1,6-diphosphatase, regulated?

The evidence for gluconeogenesis in the fungi is in the first instance that many are able to grow with small molecules such as acetate, lactate, or amino acids as the sole source of carbon (Cochrane, 1958; Perlman, 1965). Since these, like all fungi, synthesize cell wall polysaccharides, a process at least analogous to gluconeogenesis must be operative. Secondly, oxidative assimilation during the metabolism of small molecules also suggests gluconeogenesis; spores of *Fusarium solani* have been shown (Cochrane, Cochrane, Vogel & Coles, 1963) to metabolize 2-carbon compounds thus:

$$CH_3COOH + O_2 \rightarrow (CH_2O) + CO_2 + H_2O$$

$$2CH_3CH_2OH + 3O_2 \rightarrow 3(CH_2O) + CO_2 + 3H_2O$$

Isotopic studies of this system show that carbon from acetate is incorporated into polysaccharide (Cochrane *et al.*, 1963*b*).

Finally, glycogen in *Neurospora crassa* and *Blastocladiella emersonii* is synthesized, as in animals, from uridine diphosphate glucose (Camargo,

Meuser & Sonneborn, 1969; Téllez-Iñón, Terenzi & Torres, 1969). The breakdown of glycogen, too, seems to follow the animal pattern (Shepherd, Rosenthal, Lundblad & Segel, 1969; Shepherd & Segel, 1969; Téllez-Iñón & Torres, 1970).

Fig. 3.3, referring to animal systems, designates pyruvate carboxylase as the key enzyme in generating 4-carbon intermediates which can be converted to PEP in the first major process of gluconeogenesis. The enzyme has been reported in a number of fungi (Bloom & Johnson, 1962; Stan & Schormüller, 1968; Overman & Romano, 1969; Budd, 1971; Hartmann & Keen, 1974*a*); its regulation is discussed below. The succeeding enzyme in the animal system, PEPCK, has been found in *Aspergillus niger* (Woronick & Johnson, 1960) and *Verticillium albo-atrum* (Hartmann & Keen, 1974*b*).

Pyruvate kinase is not the only one means available to fungi for the synthesis of 4-carbon intermediates from smaller molecules. In *Neurospora crassa* there is good evidence that the glyoxylate bypass (Chapter 5) synthesizes 4-carbon acids which then enter the gluconeogenesis pathway *via* PEPCK (Flavell & Fincham, 1968; Flavell & Woodward, 1971). However, fumaric acid in *Rhizopus nigricans*, is probably not formed by the glyoxylate system (Romano, Bright & Scott, 1967).

Finally, the malic enzyme is a third candidate for the operative system in synthesizing 4-carbon acids from pyruvate; the enzyme in *Neurospora crassa* is competitively inhibited in both aspartate and oxalacetate, consistent with a biosynthetic function (Zink, 1967; Zink & Shaw, 1968).

The next critical step in gluconeogenesis is the action of FDPase (Fig. 3.3). This step has not been studied in the filamentous fungi, but the enzyme is known in *Candida utilis* (Pontremoli & Horecker, 1971).

Like FDPase, glucose-6-phosphatase has been little studied in the fungi, and it is not certain that the few reports available refer to this enzyme specifically or whether non-specific phosphatases have been confused with it (Nordlie, 1971).

Thus, there is evidence that gluconeogenesis occurs in the fungi, that there are available mechanisms for the synthesis of PEP from 2- and 3-carbon intermediates, and that glycogen synthesis strongly resembles that in mammalian cells. Whether one or different systems are responsible for PEP formation, and the course of the reaction from PEP to the glucose level, are essentially unknown for the fungi, and what follows on regulation is based on the assumption that the entire system will turn out to be much like that already known in higher animals.

The regulation of glycolysis and gluconeogenesis

In this section, we are concerned primarily with regulatory activities affecting the function of pre-existing enzymes. Many of the initial enzymes of glycolysis, especially those functioning in the metabolism of uncommon substrates like xylose and gluconate, are inducible (De Ley & Defloor, 1959; Kornfeld & Knight, 1962; Lakshminarayan, Modi & Shah, 1969; Elzainy, Hassan & Allan, 1973), and of course feedback repression is well known in the fungi (Zalokar, 1965). These mechanisms, however, operate at the transcription level and control the synthesis of enzymes rather than their activity.

The principal mechanism of control of the reactions of glycolysis and gluconeogenesis rests on the allosteric nature of certain key enzymes. Particular small molecules, not necessarily either substrate or product, are believed to bind to non-catalytic sites on the enzyme protein and, by inducing conformational changes, increase (activate) or decrease (inhibit) enzyme activity (Newsholme & Start, 1973). The 'energy charge', essentially the ATP : AMP ratio, is especially important (Atkinson, 1969), but other small molecules such as citrate and acetyl-CoA may also modulate enzyme activity.

From a teleological point of view, it is intuitively obvious that a cell undergoing active growth or differentiation would benefit from some mechanism which would maintain or accelerate glycolysis and other energy-yielding processes. Conversely, a cell with fewer demands and an abundant supply of carbon could, if glycolysis is checked and gluconeogenesis stimulated, store reserves for future use. This appears to be the principal selective advantage of the type of metabolic control under discussion.

Our understanding of regulatory phenomena in filamentous fungi is at present not extensive, and only in part consistent with the much wider knowledge of regulation in bacterial and animal cells.

Phosphofructokinase (PFK) and pyruvate kinase (PK) are two glycolytic enzymes known in other systems to be under metabolic control; conditions favouring gluconeogenesis reduce their activity. The PFK of *Candida albicans* is inhibited by ATP and activated by AMP as expected (Chattaway, Bishop, Holmes & Odds, 1973), but the same enzyme in *Neurospora crassa*, surprisingly, is inhibited by ATP, ADP, and AMP, as well as by citrate and P_i (Tsao & Madley, 1972; Tsao, Smith & Borondy, 1969). *Aspergillus niger*, under conditions favouring glycolysis, accumulates glycolytic intermediates in amounts consistent with a control function for PFK (Smith & Ng, 1972).

Pyruvate kinase (PK) is regulated in *Coprinus lagopus*, both by the carbon source and by specific modulators (Stewart & Moore, 1971); the data are consistent with a minimal activity of PK during gluconeogenesis.

In the gluconeogenic pathway, as discussed earlier, it is not certain which enzymes are of most importance in fungi; in particular, the mechanism for the generation of PEP from pyruvate is in doubt. Pyruvate carboxylase (PC) is controlled, in the directions expected, by aspartate and acetyl-coenzyme A in *Verticillium albo-atrum* (Hartmann & Keen, 1974*a*), by acetyl-coenzyme A in *Rhizopus nigricans* (Overman & Romano, 1969), and by energy charge and aspartate in *Penicillium camemberti* (Stan & Schormüller, 1968). Data presented by Hartmann & Keen (1974*b*) indicate that PEPCK of *Verticillium albo-atrum* is regulated so that maximal activity prevails under conditions favouring gluconeogenesis.

Other enzymes which may be involved in the synthesis of 4-carbon intermediates from pyruvate or acetate and may also be under at least some degree of metabolic control include the malic enzyme of *Fusarium oxysporum* (Zink, 1974) and *Neurospora crassa* (Zink & Shaw, 1968), and malate synthase and isocitric lyase of *N. crassa* (Flavell & Woodward, 1971).

In *Pythium debaryanum* and other Oömycetes, although not in other Phycomycetes, guanosine triphosphate (GTP) is an allosteric inhibitor of lactic dehydrogenase; the effect is to divert pyruvate into gluconeogenesis when the level of cellular carbohydrate is low, and into lactate when the carbohydrate level is high (LéJohn, 1971; Wang & LéJohn, 1974).

In a rather different kind of regulation, both glycogen synthetase and glycogen phosphorylase of *Neurospora crassa* exist in two interconvertible forms, as in animal cells (Flawia & Torres, 1972*a*, *b*). Cyclic AMP is an effector in converting inactive phosphorylase *b* to active phosphorylase *a* (Téllez-Iñón & Torres, 1970, 1973). However, Camargo, Meuser & Sonneborn (1969) were unable to find in *Blastocladiella emersonii* the two forms of glycogen synthetase reported from *N. crassa* (Téllez-Iñón, Terenzi & Torres, 1969).

At this time it is not possible to make any extensive generalizations as to regulation in the fungi, except that mechanisms at least comparable to those in other organisms surely exist. The most important kind of research needed is probably to trace out the exact path of gluconeogenesis in at least a few fungi; too much of our concept of metabolic regulation is fundamentally based on analogy.

3.7 The Balance of Co-existing Glycolytic Pathways

The near universal occurrence of both the EM and the PP pathways in fungi raises two questions, *viz.*

1. For a given organism, what share of carbohydrate metabolism is attributable to each pathway?
2. Are developmental events marked by or even caused by shifts in the balance of pathways?

These have proved to be complex and difficult problems, and it must be admitted that the already large literature is inconclusive in many respects. Detailed coverage of this literature is not possible; this review can at best serve to introduce the principal problems.

The problem of method

Four types of method are available for the investigation of the relative rates of competing pathways:

1. Estimation of the specific activities of known enzymes of the pathways known to be operative.
2. Indirect estimation of pathway activities by tracing the fate of specifically isotope-labelled substrate atoms.
3. Chemical estimation of the concentrations of glycolytic intermediates.
4. Determination of the metabolic consequences of specific genetic blocks in the operation of one or more pathways.

The first of these is methodologically fairly straightforward, but subject to the reservation, raised especially by Wright (1968), that most enzymes appear to be present in cells at concentrations far exceeding the level at which metabolism would be limited. At the very least, this underlines the desirability of focussing attention on those enzymes thought to be limiting to a pathway. Doubling or halving of the concentration of an enzyme present in tenfold excess is not likely to be very important.

Isotopic methods have been very appealing because of their apparent simplicity, and the literature on these is very large. Much of this work has been done by measuring the relative outputs of $^{14}CO_2$ from glucose-1-^{14}C and glucose-6-^{14}C. In the EM pathway, carbons 1 and 6 of glucose both wind up as the methyl group of pyruvate; thus, a 'pure' EM system always yields a C_1/C_6 ratio of unity. Conversely, in the PP pathway carbon-1 of glucose is split off in the oxidative decarboxylation of 6-PG, and the ratio is therefore well above unity until the point of complete oxidation of glucose.

Unfortunately for the simple theory, pentose phosphate formed in the PP pathway is converted to F-6-P and G-3-P. The F-6-P formed is, so far as we know, in equilibrium with G-6-P. Thus, if recycling of pentoses to G-6-P occurs, there is a dilution of the G-6-P pool by unlabelled compound arising from the PP pathway. If carbon-2 of substrate glucose is labelled, the recycled G-6-P will be labelled in positions 1 and 3 after one recycling, and in positions 1, 2 and 3 equally after a large number of cycles (Wood, Katz & Landau, 1963).

Other possible reactions which may limit the value of isotope methods include diversion of triose carbon to fermentation products (Katz & Wood, 1960), reversal of the transketolase and transaldolase reactions (Katz & Rognstad, 1967), diversion of glucose to glycogen synthesis (Katz & Wood, 1960), and recycling through all or part of the tricarboxylic acid cycle.

These difficulties do not necessarily mean that the 'C1/C6' method is useless, but do suggest that each problem should be studied rather carefully before a method is adopted, and that assumptions should be made clear. As suggested earlier, it seems highly desirable, even necessary, that isotope methods be supplemented with other kinds of evidence. In general, very well-known systems can often be studied by methods which take account of known complicating reactions. This is the case in some animal systems, for which quite reliable isotope methods have been devised (Rognstad & Katz, 1966; Katz & Rognstad, 1967; Heath, 1968; Heath & Threlfall, 1968; Nakayama & Weser, 1972).

Wang (1972) has described in detail the 'radiorespirometric method', in which $^{14}CO_2$ yield from glucose labelled in all six positions is assayed over time. Wang states the assumptions very clearly; the most important are:

1. Only two pathways can be present, not more.
2. There must be negligible resynthesis of hexose from triose formed.
3. Randomization of hexose label by transketolase and transaldolase must be negligible.

Another family of isotope methods is based on the analysis of label in products other than CO_2. The rationale and the limitations are described by Blumenthal (1965); the most useful method depends very heavily on the specific organism under study, so that generalization is difficult. Such methods often help to improve confidence in conclusions based on $^{14}CO_2$ data (Cochrane *et al.*, 1963*a*).

The use of genetically blocked mutants for the study of glycolysis has been quite successful in bacteria (Fraenkel & Vinopal, 1973). Apart from the aforementioned studies on colonial and anaerobic mutants in *Neurospora crassa*, the approach has not had the attention in work on fungi which it undoubtedly deserves.

The balance of pathways in non-differentiating cells

Blumenthal (1965) reviewed much of the work on the balance of pathways in cells which are either resting or growing without differentiation, with the general conclusion that most fungi use the EM pathway primarily, the PP sequence to some less important degree. In some—*Penicillium charlesii*, *Ustilago maydis*, and spores of *Tilletia controversa*—the relationships appear to be reversed, with the PP pathway the more important. Both the flax rust fungus, *Melampsora lini* (Mitchell & Shaw, 1968; Williams & Shaw, 1968) and *Phymatotrichum omnivorum* (Gunesakaran, 1972) similarly appear to have both pathways, with the EM the more important quantitatively.

The provision of nitrate nitrogen to cells metabolizing labelled glucose usually increases the apparent activity of the PP pathway. This may be credited to a simple demand for more NADPH for the biosyntheses involved in growth—ammonium nitrogen has the same effect in yeast (Holzer, 1961)—or to the fact that nitrate reduction has the effect of regenerating $NADP^+$ from NADPH (Holligan & Jennings, 1972). In *Aspergillus nidulans*, however, the effect of nitrate appears to be a more subtle affair than either of these (Hankinson & Cove, 1974).

Dimorphic fungi—those able to grow either in a yeast-like or a mycelial form—show distinct changes in metabolism and in the apparent contribution of the EM and PP pathways to total glucose catabolism (Chattaway *et al.*, 1973; Friedenthal, Roselino & Passeron, 1973; Friedenthal, Epstein & Passeron, 1974), but no general theory can be constructed, and it is probable that no explanation framed solely in terms of glycolytic mechanisms will suffice.

The balance of pathways in spore germination

Spore germination is the clearest and most easily manipulated developmental stage in mycelial fungi, and metabolic changes associated with it have received considerable attention. It is well established that overall respiratory activity increases dramatically in most fungi upon germination, the rust fungi being the principal exception (Sussman, 1966). For glycolysis—as well as other systems—two questions have been foremost:

1. Do particular enzymes appear *de novo* during germination, having been absent from the ungerminated spore?
2. Is there a change in the balance of pathways during the conversion of ungerminated spores to young mycelium?

Only two claims of *de novo* synthesis of glycolytic enzymes have been made, those of Bhatnagar & Krishnan (1960) for *Aspergillus niger* conidia and of Caltrider & Gottlieb (1963) for teliospores of *Ustilago maydis.* Internal evidence suggests that there were difficult technical problems in both studies, and a suspension of judgment seems advisable. Other recent reports on a wide variety of fungi indicate that particular EM and PP enzymes may increase dramatically during germination, but that at least low levels of the enzyme are present in the ungerminated spore (Caltrider & Gottlieb, 1963; Cochrane & Cochrane, 1966; Niederpruem, Updike & Henry, 1965; Ohmori & Gottlieb, 1965; Rudolph & Furch, 1970).

With regard to the second question, on changes in the balance of pathways, there are two reports (Cochrane, 1966; Aitken & Niederpruem, 1973) of changes in the contribution of carbon-1 of glucose to CO_2 relative to that of other glucose carbon atoms, but in neither is a claim made for a definitive change in the balance of pathways. Studies on *Neurospora sitophila* ascospores (Budd, Sussman & Eilers, 1966; Eilers, Ikuma & Sussman, 1966) and on *Phycomyces blakesleeanus* sporangiospores (Delvaux, 1973) provide no evidence for a change in pathways during spore germination.

This second generalization is, however, challenged by the results of Newburgh & Cheldelin (1958): radiorespirometric data indicated that mycelium of *Tilletia caries* uses the EM system predominantly, with some flow of carbon through the PP pathway, but that ungerminated teliospores use the ED pathway, possibly as the sole glycolytic mechanism. The data are very striking, and it is hoped that studies of this organism can be renewed.

With the exceptions noted, it therefore appears that as a general rule both of the questions posed about spore germination can be answered in the negative at this time. The data base for generalization is, however, rather slender.

The balance of pathways in sporulation

Sporulation is another well-marked developmental phase in fungi. Unlike spore germination, however, it is very difficult to get homogeneous cell preparations for metabolic studies (Smith & Anderson, 1973), and this feature has been a serious limitation on research in the field.

Historically, sporulation has been viewed, following Klebs, as a consequence of some check to vegetative growth (Cochrane, 1958); recent work has complicated this hypothesis but has not seriously undermined it (Morton, 1961; Righelato, Trinci, Pirt & Peat, 1968; Galbraith & Smith, 1969). Usually growth is checked by exhaustion of a nutrient, and herein lies one of the problems involved in determining the balance of glycolytic pathways as a factor in sporulation. For, it seems obvious that a drastic change in nutrient availability may of itself cause changes in glycolytic and other pathways, changes which are coincident with but causally unrelated to sporulation. This problem of experimental design, which applies also to the induction of sporulation by adding either nutrients or metabolic poisons, severely limits the decisiveness of the experimental observations, no matter how striking they are.

Other experimental approaches to the relation of metabolism to sporulation are available, in particular the induction of sporulation in susceptible species by light (McDowell & DeHertogh, 1968; Smith & Galbraith, 1971; Wilson & Huisingh, 1972), and the use of asporogenous mutants (Turian, 1966; Martinelli & Clutterbuck, 1971; Smith & Anderson, 1973). These approaches have been relatively less well investigated than have methods based on the manipulation of nutritional factors, but are arguably more promising.

Two major investigations of metabolic pathways as they relate to sporulation have been conducted in recent years: studies of *Neurospora crassa* by G. Turian and associates (Turian & Bianchi, 1972), and of *Aspergillus niger* by J. E. Smith and others (Smith & Anderson, 1973).

In the work on *Neurospora crassa*, reliance has been placed primarily on manipulation of nutritional conditions to induce or repress sporulation, followed by assay of catabolic enzymes or their products. The principal conclusion is that mycelial (vegetative) growth is associated with a 'fermentative' metabolism, marked by vigorous ethanol formation and high activity of ethanol dehydrogenase. Conidiation, on the other hand, appears to be associated with an 'oxidative' metabolism. The simplest hypothesis to explain these data is of course outside the realm of glycolysis, relating to the disposition—to ethanol or to products of the tricarboxylic acid cycle—of pyruvate formed in glycolysis. Turian & Bianchi (1972), however, suggest that conidiating cultures have a higher rate of gluconeogenesis than do vegetative, and that the PP pathway may be more active during or just before sporulation than it is in vegetative mycelium.

An aconidial mutant of *Neurospora crassa* forms acetaldehyde and ethanol in media which normally encourage conidiation (Oulevey-Matikian & Turian, 1968); this is consistent with the main hypothesis. A mutant which is able to grow anaerobically, using the EM pathway, nevertheless requires oxygen for sporulation (Howell, Zuiches & Munkres, 1971); this too suggests a fundamental shift in respiratory mechanisms coincident with and required for sporulation.

Studies on the sporulation of *Aspergillus niger* by J. E. Smith and associates (Smith & Galbraith, 1971; Smith & Anderson, 1973) have employed a wider range of techniques to differentiate between the metabolism of sporulating and that of non-sporulating cultures. Insofar as glycolysis is a factor, most of their results are consistent with a higher activity of the PP pathway just prior to sporulation (Ng, Smith & Anderson, 1972); however, when glutamate is added to induce sporulation, the EM pathway is the more active (Smith, Valenzuela-Perez & Ng, 1971).

An association of the PP pathway with sporulation in *Neurospora crassa* and *Aspergillus niger* is consistent with the results of radiorespirometric studies on *Endothia parasitica* (McDowell & DeHertogh, 1968) and *Aspergillus nidulans* (Carter & Bull, 1969).

There appears to be as yet no strong reason to believe that changes in glycolytic pathways are *determinative* in sporulation; they may be only incidental to other changes. Sporulation is a complex process, and involves, besides shifts in the balance of glycolytic pathways, definite changes in protein and amino-acid metabolism, nucleic acids and other organic phosphorus compounds, lipid metabolism, and activity of tricarboxylic acid cycle enzymes (Smith & Galbraith, 1971; Ng, Smith & McIntosh, 1973*a*, *b*). The association of conidiogenesis with aerial growth may involve still more subtle physiological changes (Morton, 1961).

3.8 Concluding Remarks

A review of glycolysis, as of any other metabolic system, in the fungi brings home the fragmentary and incomplete nature of our knowledge. The literature in a sense has more good beginnings than good endings; promising but still tentative conclusions abound. Some aspects of glycolysis appear to be quite well understood; others such as regulation, pentose catabolism, gluconeogenesis, and the metabolism of sugar acids, obviously need much more study.

Although many fungi have been studied in one aspect of glycolysis or another, there are two unmet needs. One is simply the need for studies on groups which have been relatively neglected, *e.g.* the Peronosporales, Mucorales, and Ustilaginales. Some of these have, of course, been neglected because they pose special difficulties, but the effort should be made nevertheless. The second need is in a sense antithetical to the first: detailed and in-depth studies of one single organism, or a few organisms, with the purpose of developing the kind of picture that animal biochemists have of the liver cell and bacteriologists of *Escherichia coli.*

Neurospora crassa is suggested, for three reasons, as a suitable organism for such a study. First, we already know a good deal about its metabolism. Second, its genetics are well understood, and methods for the induction and mapping of mutations have been worked out. Finally, it grows rapidly in a synthetic medium, and enzyme extraction from mycelium is relatively easy. The drawback to *N. crassa* is its formation of NADase, complicating some standard enzyme assays; this does not appear to be an insuperable objection.

3.9 References

AITKEN, W. B. & NIEDERPRUEM, D. J. (1973). Isotopic studies of carbohydrate metabolism during basidiospore germination in *Schizophyllum commune.* II. Changes in specifically labelled glucose and sugar alcohol utilization. *Archiv für Mikrobiologie* **88**, 331–44.

ATKINSON, D. E. (1969). Regulation of enzyme function. *Annual Review of Microbiology* **23**, 47–68.

AVIGAD, G., AMARAL, D., ASENSIO, C. & HORECKER, B. L. (1962). The D-galactose oxidase of *Polyporus circinatus. Journal of Biological Chemistry* **237**, 2736–43.

BELSKY, M. & GOLDSTEIN, S. (1964). Glucose metabolism of *Thraustochytrium roseum,* a non-filamentous marine Phycomycete. *Archiv für Mikrobiologie* **49**, 375–82.

BENTLEY, R. (1963). Glucose oxidase. In *The enzymes,* 2nd edition. Edited by P. D. Boyer, M. Lardy and K. Myrbäck, Vol. 7, 567–86. New York & London: Academic Press.

BHATNAGAR, G. M. & KRISHNAN, P. S. (1960). Enzymatic studies in spores of *Aspergillus niger.* III. Enzymes of Embden–Meyerhof–Parnas pathway in germinating spores. *Archiv für Mikrobiologie* **37**, 211–4.

BIANCHI, D. E., PUROHIT, K. & TURIAN, G. (1971). Lactic dehydrogenase activity and cellular localization of several dehydrogenases in *Neurospora* and *Allomyces. Archiv für Mikrobiologie* **75**, 163–70.

BLOOM, SANDRA J. & JOHNSON, M. J. (1962). The pyruvate carboxylase of *Aspergillus niger. Journal of Biological Chemistry* **237**, 2718–20.

BLUMENTHAL, H. J. (1965). Glycolysis. In *The fungi.* Edited by G. C. Ainsworth and A. S. Sussman, Vol. I, 229–68. New York & London: Academic Press.

BODMANN, O. & WALTER, M. (1965). Die glucose-oxydasen aus *Penicillium notatum* (notatin) und *Aspergillus niger* (nigeran). I. Isolierung und einige molekulare Eigenschaften. *Biochimica et Biophysica Acta* **110**, 496–506.

BRADY, R. J. & CHAMBLISS, G. H. (1967). The lack of phosphofructokinase in several species of *Rhodotorula. Biochemical and Biophysical Research Communication* **29**, 898–903.

BRODY, S. (1970). Correlation between reduced nicotinamide adenine dinucleotide phosphate levels and morphological changes in *Neurospora crassa. Journal of Bacteriology* **101**, 802–7.

BRODY, S. & NYC, J. F. (1970). Altered fatty acid distribution in mutants of *Neurospora crassa. Journal of Bacteriology* **104**, 780–6.

BRODY, S. & TATUM, E. L. (1966). The primary biochemical effect of a morphological mutation in *Neurospora crassa. Proceedings of the National Academy of Sciences U.S.A.* **56**, 1290–7.

BUDD, K. (1971). Bicarbonate fixation by cell-free extracts and by mycelium of *Neocosmopara vasinfecta. Journal of General Microbiology* **67**, 99–106.

BUDD, K., SUSSMAN, A. S. & EILERS, F. I. (1966). Glucose-^{14}C metabolism of dormant and activated ascospores of *Neuro-*

spora. Journal of Bacteriology **91**, 551–61.

CALTRIDER, P. G. & GOTTLIEB, D. (1963). Respiratory activity and enzymes for glucose catabolism in fungus spores. *Phytopathology* **53**, 1021–30.

CAMARGO, E. P., MEUSER, R. & SONNEBORN, D. (1969). Kinetic analyses of the regulation of glycogen synthetase activity in zoospores and growing cells of the water mold, *Blastocladiella emersonii. Journal of Biological Chemistry* **244**, 5910–19.

CARTER, B. L. A. & BULL, A. T. (1969). Studies of fungal growth and intermediary carbon metabolism under steady and non-steady state conditions. *Biotechnology and Bioengineering* **11**, 785–804.

CASSELTON, P. J. (1966). Enzymes of the Embden–Meyerhof and pentose phosphate pathways in *Polyporus brumalis* extracts. *Journal of Experimental Botany* **17**, 579–89.

CASSELTON, P. J. (1967). Oxygen uptake by cell-free preparations of Basidiomycetes. *Transactions of the British Mycological Society* **50**, 377–84.

CHATTAWAY, F. W., BISHOP, R., HOLMES, M. R. & ODDS, F. C. (1973). Enzyme activities associated with carbohydrate synthesis and breakdown in the yeast and mycelial forms of *Candida albicans. Journal of General Microbiology* **75**, 97–109.

CHIANG, C. & KNIGHT, S. G. (1959). D-xylose metabolism by cell-free extracts of *Penicillium chrysogenum. Biochimica et Biophysica Acta* **35**, 454–63.

CHIANG, C. & KNIGHT, S. G. (1960*a*). A new pathway of pentose metabolism. *Biochemical and Biophysical Research Communications* **3**, 554–9.

CHIANG, C. & KNIGHT, S. G. (1960*b*). Metabolism of D-xylose by moulds. *Nature, London* **188**, 79–81.

CHIANG, C. & KNIGHT, S. G. (1961). L-Arabinose metabolism by cell-free extracts of *Penicillium chrysogenum. Biochimica et Biophysica Acta* **46**, 271–8.

CLARK-WALKER, G. D. (1972). Development of respiration and mitochondria in *Mucor genevensis* after anaerobic growth: absence of glucose repression. *Journal of Bacteriology* **109**, 399–408.

COCHRANE, V. W. (1955). The metabolism of species of Streptomyces. VIII. Reactions of the Embden–Meyerhof–Parnas sequence in *Streptomyces coelicolor. Journal of Bacteriology* **69**, 256–63.

COCHRANE, V. W. (1956). The anaerobic dissimilation of glucose by *Fusarium lini. Mycologia* **48**, 1–12.

COCHRANE, V. W. (1958). *Physiology of fungi.* 524p. New York: John Wiley & Sons.

COCHRANE, V. W. (1966). Respiration and spore germination. In *The fungus spore*, ed. M. F. Madelin, pp. 201–215. London: Butterworths.

COCHRANE, V. W. & COCHRANE, J. C. (1966). Spore germination and carbon metabolism in *Fusarium solani.* V. Changes in anaerobic metabolism and related enzyme activities during development. *Plant Physiology* **41**, 810–14.

COCHRANE, V. W., COCHRANE, J. C., VOGEL, J. M. & COLES, R. S., Jr (1963*b*). Spore germination and carbon metabolism in *Fusarium solani.* IV. Metabolism of ethanol and acetate. *Journal of Bacteriology* **86**, 312–19.

COCHRANE, V. W., BERRY, S. J., SIMON, F. G., COCHRANE, J. C., COLLINS, C. B., LEVY, J. A. & HOLMES, P. K. (1963*a*). Spore germination and carbon metabolism in *Fusarium solani.* III. Carbohydrate respiration in relation to germination. *Plant Physiology* **38**, 533–41.

COOPER, J. A. D., SMITH, W., BACILA, M. & MEDINA, H. (1959). Galactose oxidase from *Polyporus circinatus* Fr. *Journal of Biological Chemistry* **234**, 445–8.

COURI, D. & RACKER, E. (1959). The oxidative pentose phosphate cycle. V. Complete oxidation of glucose-6-phosphate in a reconstructed system of the oxidative pentose phosphate cycle. *Archives of Biochemistry and Biophysics* **83**, 195–205.

DAVID, S. & RENAUT, J. (1955). Répartition de la radioactivité sur le D-ribose biosynthetique. *Biochimica et Biophysica Acta* **16**, 548–9.

DE LEY, J. & DEFLOOR, J. (1959). 2-Ketogluconoreductase in micro-organisms. *Biochimica et Biophysica Acta* **33**, 47–54.

DE LEY, J. & VANDAMME, J. (1955). The metabolism of sodium 2-keto-D-gluconate by micro-organisms. *Journal of General Microbiology* **12**, 162–71.

DELVAUX, E. (1973). Some aspects of germination induction in *Phycomyces blakesleeanus* by an ammonium acetate pretreatment. *Archiv für Mikrobiologie* **88**, 273–84.

DOELLE, H. W. (1969). *Bacterial metabolism.* 486 p. New York & London: Academic Press.

EAGON, R. G. (1963). Rate-limiting effects of pyridine nucleotides on carbohydrate catabolic pathways of micro-organisms. *Biochemical and Biophysical Research Communications* **12**, 274–9.

EGGLESTON, L. V. & KREBS, H. A. (1974). Regulation of the pentose phosphate cycle. *Biochemical Journal* **138**, 425–35.

EILERS, F. I., IKUMA, H. & SUSSMAN, A. S. (1970). Changes in metabolic intermediates during activation of *Neurospora* ascospores. *Canadian Journal of Microbiology* **16**, 1351–6.

ELMER, G. W. & NICKERSON, W. J. (1970). Filamentous growth of *Mucor rouxii* under nitrogen. *Journal of Bacteriology* **101**, 592–4.

ELZAINY, T. A., HASSAN, M. M. & ALLAM, A. M. (1973). New pathway for non-phosphorylated degradation of gluconate by *Aspergillus niger. Journal of Bacteriology* **114**, 457–9.

EMERSON, R. & HELD, A. A. (1969). *Aqualinderella fermentans* gen. et sp. n., a Phycomycete adapted to stagnant waters. II. Isolation, cultural characteristics, and gas relations. *American Journal of Botany* **56**, 1103–20.

FLAVELL, R. B. & FINCHAM, J. R. S. (1968). Acetate non-utilizing mutants of *Neurospora crassa*. II. Biochemical deficiencies and the role of certain enzymes. *Journal of Bacteriology* **95**, 1063–8.

FLAVELL, R. B. & WOODWARD, D. O. (1971). Metabolic role, regulation of synthesis, cellular localization, and genetic control of the glyoxylate cycle in *Neurospora crassa. Journal of Bacteriology* **105**, 200–10.

FLAWIÁ, M. M. & TORRES, H. N. (1972*a*). Adenylate cyclase activity in *Neurospora crassa*. I. General properties. *Journal of Biological Chemistry* **247**, 6873–9.

FLAWIÁ, M. M. & TORRES, H. N. (1972*b*). Adenylate cyclase activity in *Neurospora crassa*. II. Kinetics. *Journal of Biological Chemistry* **247**, 6880–3.

FLORES-CARRÉON, A., REYES, E. & RUIZ-HERRERA, J. (1969). Influence of oxygen on maltose metabolism by *Mucor rouxii. Journal of General Microbiology* **59**, 13–19.

FRAENKEL, D. G. & VINOPAL, R. T. (1973). Carbohydrate metabolism in bacteria. *Annual Review of Microbiology* **27**, 69–100.

FRIEDENTHAL, M., EPSTEIN, A. & PASSERON, S. (1974). Effect of potassium cyanide, glucose and anaerobiosis on morphogenesis of *Mucor rouxii. Journal of General Microbiology* **82**, 15–24.

FRIEDENTHAL, M., ROSELINO, E. & PASSERON, S. (1973). Multiple molecular forms of pyruvate kinase from *Mucor rouxii.* Immunological relationships among the three isozymes and nutritional factors affecting the enzymatic pattern. *European Journal of Biochemistry* **35**, 148–58.

FUSCALDO, K. E., LECHNER, J. F. & BAZINET, G. (1971). Genetic and biochemical studies of the hexose monophosphate shunt in *Neurospora crassa*. I. The influence of genetic defects in the pathway on colonial morphology. *Canadian Journal of Microbiology* **17**, 783–8.

GALBRAITH, J. C. & SMITH, J. E. (1969). Sporulation of *Aspergillus niger* in submerged liquid culture. *Journal of General Microbiology* **59**, 31–45.

GIBBS, M., COCHRANE, V. W., PAEGE, L. M. & WOLIN, H. (1954). Fermentation of D-xylose-1-^{14}C by *Fusarium lini* Bolley. *Archives of Biochemistry and Biophysics* **50**, 237–42.

GIBBS, M. & GASTEL, R. (1953). Glucose dissimilation by *Rhizopus. Archives of Biochemistry and Biophysics* **43**, 33–8.

GLEASON, F. H. (1968). Nutritional comparisons in the Leptomitales. *American Journal of Botany* **55**, 1003–10.

GLEASON, F. H. & PRICE, J. S. (1969). Lactic acid fermentation in lower fungi. *Mycologia* **61**, 945–56.

GLEASON, F. H., NOLAN, R. A., WILSON, A. C. & EMERSON, R. (1966). D(−)-Lactate dehydrogenase in lower fungi. *Science* **152**, 1272–3.

GOLUEKE, C. G. (1957). Comparative studies of the physiology of *Sapromyces* and related genera. *Journal of Bacteriology* **74**, 337–43.

GREENE, G. L. (1969). Enzymes of glucose catabolism pathways in *Colletotrichum* and *Gloeosporium. Mycologia* **61**, 902–14.

GUNASEKARAN, M. (1972). Physiological studies on *Phymatotrichum omnivorum*. I. Pathways of glucose catabolism. *Archiv für Mikrobiologie* **83**, 328–31.

HANKINSON, O. & COVE, D. J. (1974). Regulation of the pentose phosphate pathway in the fungus *Aspergillus nidulans.* The effect of growth with nitrate. *Journal of Biological Chemistry* **249**, 2344–53.

HARTMAN, R. E. & KEEN, N. T. (1974*a*). The pyruvate carboxylase of *Verticillium albo-atrum. Journal of General Microbiology* **81**, 15–19.

HARTMAN, R. E. & KEEN, N. T. (1974*b*). The phosphoenolpyruvate carboxykinase of *Verticillium albo-atrum. Journal of General Microbiology* **81**, 21–6.

HEATH, D. F. (1968). The redistribution of carbon label by the reactions involved in glycolysis, gluconeogenesis and the tricarboxylic acid cycle in rat liver. *Biochemical Journal* **110**, 313–35.

HEATH, D. F. & THRELFALL, C. J. (1968). The interaction of glycolysis, gluconeogenesis and the tricarboxylic acid cycle in rat liver *in vivo*. *Biochemical Journal* **110**, 337–62.

HEATH, E. C., NASSER, O. & KOFFLER, H. (1956). Biochemistry of filamentous fungi. III. Alternative routes for the breakdown of glucose by *Fusarium lini*. *Archives of Biochemistry and Biophysics* **64**, 80–7.

HELD, A. A. (1970). Nutrition and fermentative energy metabolism of the water mold *Aqualinderella fermentans*. *Mycologia* **62**, 339–58.

HELD, A. A., EMERSON, R., FULLER, M. S. & GLEASON, F. H. (1969). *Blastocladia* and *Aqualinderella*: fermentative water molds with high carbon dioxide optima. *Science* **165**, 706–9.

HILL, R. L. & MILLS, R. C. (1954). The anaerobic glucose metabolism of *Bacterium tularense*. *Archives of Biochemistry and Biophysics* **53**, 174–83.

HOCHSTER, R. M. (1957). Pyridine nucleotide specificities and rates of formation of glucose-6-phosphate and 6-phosphogluconate dehydrogenases in *Aspergillus flavus-oryzae*. *Archives of Biochemistry and Biophysics* **66**, 499–501.

HÖFER, M., BETZ, A. & BECKER, J.-U. (1970). Metabolism of the obligately aerobic yeast *Rhodotorula gracilis*. I. Changes in metabolite concentrations following D-glucose and D-xylose addition to the cell suspension. *Archiv für Mikrobiology* **71**, 99–110.

HOLLIGAN, P. M. & JENNINGS, D. H. (1972). Carbohydrate metabolism in the fungus *Dendryphiella salina*. III. The effect of the nitrogen source on the metabolism of [1-^{14}C]- and [6-^{14}C]-glucose. *New Phytologist* **71**, 1119–33.

HOLLMAN, S. & TOUSTER, O. (1964). *Non-glycolytic pathways of metabolism of glucose*. 276 p. New York and London: Academic Press.

HOLZER, H. (1961). Regulation of carbohydrate metabolism by enzyme competition. *Cold Spring Harbor Symposia on Quantitative Biology* **26**, 277–88.

HORECKER, B. L. (1962). *Pentose metabolism in bacteria*. 100 p. New York: John Wiley and Sons.

HOWELL, N., ZUICHES, C. A. & MUNKRES, K. D. (1971). Mitochondrial biogenesis in *Neurospora crassa*. I. An ultrastructural and biochemical investigation of the effects of anaerobiosis and chloramphenicol inhibition. *Journal of Cell Biology* **50**, 721–36.

IKEDA, S. & YAMADA, K. (1963). Xylonic acid fermentation. *Nippon Nogei Kagaku Kaishi* **37**, 514–17, 518–23.

JANSSEN, F. W. & RUELIUS, H. W. (1968). Carbohydrate oxidase, a novel enzyme from *Polyporus obtusus*. II. Specificity and characterization of reaction products. *Biochimica et Biophysica Acta* **167**, 501–10.

KAPOOR, M. & TRONSGAARD, T. M. (1972). Pyruvate kinase of *Neurospora crassa*: purification and some properties. *Canadian Journal of Microbiology* **18**, 805–15.

KATO, K., KOIKE, S., YAMADA, K., YAMADA, H. & TANAKA, S. (1962). Di- and triphosphopyridine nucleotide-linked glutamic dehydrogenases of *Piricularia oryzae* and their behavior in glutamate media. *Archives of Biochemistry and Biophysics* **98**, 346–7.

KATZ, J. & ROGNSTAD, R. (1967). The labeling of pentose phosphate from glucose-^{14}C and estimation of the rates of transaldolase, transketolase, the contribution of the pentose cycle, and ribose-phosphate synthesis. *Biochemistry* **6**, 2227–47.

KATZ, J. & WOOD, H. G. (1960). The use of glucose-^{14}C for the evaluation of the pathways of glucose metabolism. *Journal of Biological Chemistry* **235**, 2165–77.

KHUOW, B. T. & MCCURDY, H. D. (1969). Tricarboxylic acid cycle enzymes and morphogenesis in *Blastocladiella emersonii*. *Journal of Bacteriology* **99**, 197–205.

KORNFELD, J. M. & KNIGHT, S. G. (1962). Induction of xylose reductase in germinating spores of *Penicillium chrysogenum*. *Mycologia* **54**, 407–14.

LAKSHMINARAYANA, K., MODI, V. V. & SHAH, V. K. (1969). Studies on gluconate metabolism in *Aspergillus niger*. II. Comparative studies on the enzyme make-up of the adapted and parent strains of *Aspergillus niger*. *Archiv für Mikrobiologie* **66**, 396–405.

LECHNER, J. F., FUSCALDO, K. E. & BAZINET, G. (1971). Genetic and biochemical studies of the hexose monophosphate pathway in *Neurospora crassa*. II. Characterization of biochemical defects of the morphological mutants *colonial* 2 and *colonial* 3. *Canadian Journal of Microbiology* **17**, 789–94.

LEHNINGER, A. L. (1970). *Biochemistry*. 833 p. New York: Worth Publishers, Inc.

LéJOHN, H. B. (1971). D(−)Lactate dehydrogenase in fungi. *Journal of Biological Chemistry* **246**, 2116–26.

LYR, H. (1962). Nachweis einer Xylose-Oxidase (Xylose-O_2-Transhydrogenase) bei höheren Pilzen. *Enzymologia* **24**, 69–80.

MALCA, I., ERWIN, D. C., SIMS, J. J., LONG, M. & JONES, B. (1968). Respiratory and enzymatic studies with *Verticillium alboatrum* from conidia. *Phytopathology* **58**, 348–53.

MARGULIES, M. & VISHNIAC, W. (1961). Dissimilation of glucose by the MX strain of *Rhizopus*. *Journal of Bacteriology* **81**, 1–9.

MARTIN, S. M. & STEEL, R. (1955). Effect of phosphate on production of organic acids by *Aspergillus niger*. *Canadian Journal of Microbiology* **1**, 470–2.

MARTINELLI, S. D. & CLUTTERBUCK, A. J. (1971). A quantitative survey of conidiation mutants in *Aspergillus nidulans*. *Journal of General Microbiology* **69**, 261–8.

McDOWELL, L. L. & DeHERTOGH, A. A. (1968). Metabolism of sporulation in filamentous fungi. I. Glucose and acetate oxidation in sporulating and nonsporulating cultures of *Endothia parasitica*. *Canadian Journal of Botany* **46**, 449–51.

MELOCHE, H. P. (1961). The metabolism of ribose-5-phosphate by cell-free extracts of *Lactarius tomentosus*. *Biochimica et Biophysica Acta* **51**, 586–9.

MITCHELL, D. & SHAW, M. (1968). Metabolism of glucose-^{14}C, pyruvate-^{14}C and mannitol-^{14}C by *Melampsora lini*. II. Conversion to soluble products. *Canadian Journal of Botany* **46**, 453–60.

MORTON, A. G. (1961). The induction of sporulation in mould fungi. *Proceedings of the Royal Society, Series B* **153**, 548–69.

NAKAYAMA, H. & WESER, E. (1972). Adaptation of small bowel after intestinal reaction: increase in the pentose phosphate pathway. *Biochimica et Biophysica acta* **279**, 416-23.

NEWBURGH, R. W. & CHELDELIN, V. H. (1958). Glucose oxidation in mycelia and spores of the wheat smut fungus *Tilletia caries*. *Journal of Bacteriology* **76**, 308–11.

NEWSHOLME, E. A. & START, C. (1973). *Regulation in metabolism*. 349 p. New York: John Wiley & Sons.

NG, A. M. L., SMITH, J. E. & MCINTOSH, A. F. (1973*a*). Conidiation of *Aspergillus niger* in continuous culture. *Archiv für Mikrobiologie* **88**, 119–26.

NG, A. M. L., SMITH, J. E. & McINTOSH, A. F. (1973*b*). Changes in activity of tricarboxylic acid cycle and glyoxylate cycle enzymes during synchronous development of *Aspergillus niger*. *Transactions of the British Mycological Society* **61**, 13–20.

NG, W. S., SMITH, J. E. & ANDERSON, J. G. (1972). Changes in carbon catabolic pathways during synchronous development of *Aspergillus niger*. *Journal of General Microbiology* **71**, 495–504.

NIEDERPRUEM, D. J., UPDIKE, J. & HENRY, L. (1965). Hexose monophosphate metabolism during basidiospore germination in *Schizophyllum commune*. *Journal of Bacteriology* **89**, 908–9.

NORDLIE, R. C. (1971). Glucose-6-phosphatase, hydrolytic and synthetic activities. In *The enzymes*, 3rd edition. Edited by P. D. Boyer, Vol. 4, 543–608. New York and London: Academic Press.

OBAYASHI, A., YORIFUJI, H., YAMAGATA, T., IJICHI, T. & KANIE, M. (1966). Respiration in organic acid-forming molds. I. Purification of cytochrome c, coenzyme Q_9 and L-lactic dehydrogenase from lactate-forming *Rhizopus oryzae*. *Agricultural and Biological Chemistry* (*Tokyo*) **30**, 717–24.

OHMORI, K. & GOTTLIEB, D. (1965). Development of respiratory enzyme activities during spore germination. *Phytopathology* **55**, 1328–36.

OSMOND, C. B. & ap REES, T. (1969). Control of the pentose-phosphate pathway in yeast. *Biochimica et Biophysica Acta* **184**, 35–42.

OULEVEY-MATIKIAN, N. & TURIAN, G. (1968). Contrôle métabolique et aspects ultrastructuraux de la conidiation (macro-microconidies) de *Neurospora crassa*. *Archiv für Mikrobiologie* **60**, 35–58.

OVERMAN, S. A. & ROMANO, A. H. (1969). Pyruvate carboxylase of *Rhizopus nigricans* and its role in fumaric acid production. *Biochemical and Biophysical Research Communications* **37**, 457–63.

PERLMAN, D. (1965). The chemical environment for fungal growth. 2. Carbon sources. In *The Fungi*, ed. G. C. Ainsworth and A. S. Sussman, Vol. I, 479–89. New York and London: Academic Press.

PONTREMOLI, S. & HORECKER, B. L. (1971). Fructose-1,6-diphosphatases. In *The enzymes*, 3rd edition. Edited by P. D.

Boyer, Vol. 4, 611–46. New York and London: Academic Press.

PUROHIT, K. & TURIAN, G. (1972). D(−) Lactate dehydrogenase from *Allomyces.* Partial purification and allosteric properties. *Archiv für Mikrobiologie* **84**, 287–300.

RAMACHANDRAN, S. & GOTTLIEB, D. (1963). Pathways of glucose catabolism in *Caldariomyces fumago. Biochimica et Biophysica Acta* **69**, 74–84.

RIGHELATO, R. C., TRINCI, A. P. J., PIRT, S. J. & PEAT, A. (1968). The influence of maintenance energy and growth rate on the metabolic activity, morphology and conidiation of *Penicillium chrysogenum. Journal of General Microbiology* **50**, 399–412.

ROGNSTAD, R. & KATZ, J. (1966). The balance of pyridine nucleotides and ATP in adipose tissue. *Proceedings of the National Academy of Sciences, U.S.A.* **55**, 1148–56.

ROMANO, A. H., BRIGHT, M. M. & SCOTT, W. E. (1967). Mechanism of fumaric acid accumulation in *Rhizopus nigricans. Journal of Bacteriology* **93**, 600–4.

RUDOLPH, H. & FURCH, B. (1970). Untersuchungen zur Aktivierung von Sporenhomogenaten durch Wärmeaktivierung. *Archiv für Mikrobiologie* **72**, 175–81.

RUELIUS, H. W., KERWIN, R. M. & JANSSEN, F. W. (1968). Carbohydrate oxidase, a novel enzyme from *Polyporus obtusus.* I. Isolation and purification. *Biochimica et Biophysica Acta* **167**, 493–500.

SANWAL, B. D. & LATA, M. (1962). Concurrent regulation of glutamic dehydrogenases of *Neurospora. Archives of Biochemistry and Biophysics* **97**, 582–8.

SCOTT, W. A. & ABRAMSKY, T. (1973*a*). Neurospora 6-phosphogluconate dehydrogenase. I. Purification and properties of the wild type enzyme. *Journal of Biological Chemistry* **248**, 3535–41.

SCOTT, W. A. & ABRAMSKY, T. (1973*b*). Neurospora 6-phosphogluconate dehydrogenase. II. Properties of two purified mutant enzymes. *Journal of Biological Chemistry* **248**, 3542–5.

SCOTT, W. A. & TATUM, E. L. (1970). Glucose-6-phosphate dehydrogenase and Neurospora morphology. *Proceedings of the National Academy of Sciences U.S.A.* **66**, 515–22.

SHEPHERD, D., ROSENTHAL, S., LUNDBALD, G. T. & SEGEL, I. H. (1969). *Neurospora crassa* glycogen phosphorylase: characterization and kinetics via a new radiochemical assay for phosphorolysis. *Archives of Biochemistry and Biophysics* **135**, 334–40.

SHEPHERD, D. & SEGEL, I. H. (1969). Glycogen phosphorylase of *Neurospora crassa. Archives of Biochemistry and Biophysics* **131**, 609–20.

SIH, C. J., HAMILTON, P. B. & KNIGHT, S. G. (1957). Demonstration of the pentose cycle reactions in *Penicillium chrysogenum. Journal of Bacteriology* **73**, 447–51.

SIMONART, P. & GODIN, P. (1951). Formation of pentoses, 2-keto-gluconic acid and glucuronic acid by *Penicillium brevicompactum. Bulletin de la Société Chimique Belge* **60**, 446–8.

SMITH, J. E. & ANDERSON, J. G. (1973). Differentiation in the Aspergilli. In *Microbial growth, Symposia of the Society for General Microbiology* **23**, 295–337.

SMITH, J. E. & GALBRAITH, J. C. (1971). Biochemical and physiological aspects of differentiation in fungi. *Advances in Microbial Physiology* **5**, 45–134.

SMITH, J. E. & NG, W. S. (1972). Fluorometric determination of glycolytic intermediates and adenylates during sequential changes in replacement culture of *Aspergillus niger. Canadian Journal of Microbiology* **18**, 1657–64.

SMITH, J. E., VALENZUELA-PEREZ, J. & NG, W. S. (1971). Changes in activities of the Embden–Meyerhof–Parnas and pentose phosphate pathways during the growth cycle of *Aspergillus niger. Transactions of the British Mycological Society* **57**, 93–101.

STAN, H.-J. & SCHÖRMULLER, J. (1968). Regulation of pyruvate carboxylase of *Penicillium camemberti* by the adenylate system. *Biochemical and Biophysical Research Communications* **32**, 289–94.

STEWART, G. R. & MOORE, D. (1971). Factors affecting the level and activity of pyruvate kinase from *Coprinus lagopus sensu* Buller. *Journal of General Microbiology* **66**, 361–70.

SUSSMAN, A. S. (1966). *Dormancy and spore germination.* In *The fungi.* Edited by G. C. Ainsworth and A. S. Sussman, Vol. II, 733–64. New York and London: Academic Press.

TABAK, H. H. & COOKE, W. B. (1968). The effects of gaseous environment on the growth and metabolism of fungi. *Botanical Review* **34**, 124–52.

TÉLLEZ-IÑÓN, M. T. & TORRES, H. N. (1970). Interconvertible forms of glycogen phosphorylase in *Neurospora crassa. Proceedings of the National Academy of Sciences, U.S.A.* **66**, 459–63.

TÉLLEZ-IÑÓN, M. T. & TORRES, H. N. (1973). Regulation of glycogen phosphorylase *a* phosphatase in *Neurospora crassa. Biochimica et Biophysica Acta* **297**, 399–412.

TELLEZ-INÑÓN, M. T., TERENZI, H. & TORRES, H. N. (1969). Interconvertible forms of glycogen synthetase in *Neurospora crassa. Biochimica et Biophysica Acta* **191**, 765–8.

TSAO, M. U. & MADLEY, T. I. (1972). Kinetic properties of phosphofructokinase of *Neurospora crassa. Biochimica et Biophysica Acta* **258**, 99–105.

TSAO, M. U., SMITH, M. W. & BORONDY, P. E. (1969). Metabolic response of *Neurospora crassa* to environmental change. *Microbios* **1**, 37–43.

TURIAN, G. (1966). Morphogenesis in Ascomycetes. In *The fungi.* Edited by G. C. Ainsworth and A. S. Sussman, Vol. II, 339–85. New York and London: Academic Press.

TURIAN, G. & BIANCHI, D. E. (1972). Conidiation in *Neurospora. Botanical Review* **38**, 119–54.

VAN ETTEN, J. L., MOLITORIS, H. P. & GOTTLIEB, D. (1966). Changes in fungi with age. II. Respiration and respiratory enzymes of *Rhizoctonia solani* and *Sclerotium bataticola. Journal of Bacteriology* **91**, 169–76.

VAN SCOYOC, S. W., KUĆ, J. & WILLIAMS, E. B. (1970). Respiratory enzymes in *Venturia inaequalis. Canadian Journal of Microbiology* **16**, 181–5.

WANG, C. H. (1972). Radiorespirometric methods. In *Methods in microbiology.* Edited by J. R. Norris and D. W. Ribbons, Vol. 6B, 185–230. New York and London: Academic Press.

WANG, H. S. & LÉJOHN, H. B. (1974). Analogy and homology of the dehydrogenases of Oömycetes. II. Regulation by GTP of D(−)lactic dehydrogenases and isozyme patterns. *Canadian Journal of Microbiology* **20**, 575–80.

WARREN, C. O. & MULLINS, J. T. (1969). Respiratory mechanisms in *Achlya ambisexualis. American Journal of Botany* **56**, 1135–42.

WILLIAMS, P. G. & SHAW, M. (1968). Metabolism of glucose-^{14}C, pyruvate-^{14}C, and mannitol-^{14}C by *Melampsora lini.* I. Uptake. *Canadian Journal of Botany* **46**, 435–40.

WILSON, D. M. & HUISINGH, D. (1972). Respiration, peroxidase, and diphenol oxidase isozymes in reproductive and non-reproductive growth of *Glomerella cingulata. Canadian Journal of Microbiology* **18**, 1525–9.

WOOD, H. G., KATZ, J. & LANDAU, B. R. (1963). Estimation of pathways of carbohydrate metabolism. *Biochemische Zeitschrift* **338**, 809–47.

WORONICK, C. L. & JOHNSON, M. J. (1960). Carbon dioxide fixation by cell-free extracts of *Aspergillus niger. Journal of Biological Chemistry* **235**, 9–15.

WRIGHT, B. E. (1968). An analysis of metabolism underlying differentiation in *Dictyostelium discoideum. Journal of Cellular Physiology* **72**, Suppl. 1, 145–60.

YAMADA, K., IIZUKA, K., AIDA, K. & UEMURA, T. (1966). L-Sorbose oxidase from *Trametes sanguinaria. Agricultural and Biological Chemistry* (*Tokyo*) **30**, 97–8.

ZABLOTNY, R. & FRAENKEL, D. G. (1967). Glucose and gluconate metabolism in a mutant of *Escherichia coli* lacking gluconate-6-phosphate dehydrase. *Journal of Bacteriology* **93**, 1579–81.

ZALOKAR, M. (1965). Integration of cellular metabolism. In *The fungi.* Edited by G. C. Ainsworth and A. S. Sussman, Vol. 1, 377–426. New York and London: Academic Press.

ZINK, M. W. (1967). Regulation of the 'malic' enzyme in *Neurospora crassa. Canadian Journal of Microbiology* **13**, 1211–21.

ZINK, M. W. (1974). Mechanism of regulation of the malic enzyme from *Fusarium. Canadian Journal of Microbiology* **20**, 443–54.

ZINK, M. W. & SHAW, D. A. (1968). Regulation of 'malic' isozymes and malic dehydrogenases in *Neurospora crassa. Canadian Journal of Microbiology* **14**, 907–12.

CHAPTER 4

The Biochemistry and Biogenesis of Mitochondria

K. WATSON

4.1 Introduction

Electron microscopy has revealed that fungal cells show an abundance of mitochondrial structures which, in general, resemble those observed in plant and animal cells (Hawker, 1965). However, acceptance of the generalization that fungal mitochondria have biochemical properties which are identical to those of other eukaryotic mitochondria requires exercise of caution. Studies of the biochemical properties of fungal mitochondria have almost exclusively been confined to the yeasts, especially *Saccharomyces* and *Candida*, and the filamentous fungi *Aspergillus* spp. and *Neurospora crassa.*

Yeast mitochondria have been extensively studied in recent years, particularly with regard to their biogenesis and several reviews have been published (Ashwell & Work, 1970; Linnane & Haslam, 1970; Linnane, Haslam, Lukins & Nagley, 1972; Schatz & Mason, 1974). In this chapter no attempt will be made to cover yeast mitochondria and their biogenesis except on a comparative basis with work on other fungal mitochondria.

The intention of this section is to summarize the more recent literature on the isolation and biochemical properties of fungal mitochondria, primarily those of the filamentous fungi. The use of cytoplasmic mutants, the role of mitochondria in cellular differentiation and the biogenesis of fungal mitochondria are also discussed.

4.2 Isolation of Fungal Mitochondria

The difficulties encountered in the isolation of functional mitochondria, that is mitochondria capable of coupling oxidation to phosphorylation, has been a major drawback in past work on fungal mitochondria. The drastic techniques employed for the disruption of fungal hyphae have generally led to ill-defined particulate fractions which, although having respiratory activity, were devoid of phosphorylation ability.

This situation may be contrasted with studies on yeast. Mitochondria have been isolated in good yields and in a relatively intact state by the use of snail gut enzyme to digest away the cell wall. The resulting protoplasts or spheroplasts may be disrupted by a combination of osmotic and mechanical means to yield relatively intact mitochondria as judged by both biochemical criteria and electron microscopy (Duell, Inoue & Utter, 1964; Ohnishi, Kawaguchi & Hagihara, 1966; Kovac, Bednarova & Greksak, 1968).

One of the earliest reports of oxidative phosphorylation in fungal mitochondria (Iwasa, 1960*a*) described the preparation of a mitochondrial fraction from *Aspergillus oryzae* by homogenization of hyphae in a high-speed blender. The mitochondria had a maximum P/O ratio of 1.6 with succinate as substrate. However, the preparation required the addition of a nucleotide fraction, prepared from mycelial mats, for maximum phosphorylation. Furthermore, other substrates of the tricarboxylic acid cycle were either not oxidised or only weakly oxidized (Iwasa, 1960*b*).

Weiss (1965) compared oxidative phosphorylation in *Neurospora crassa* mitochondria isolated by disruption of protoplasts, following snail enzyme digestion of cell walls, with mitochondria isolated by grinding hyphae and conidia with sand in a mortar. Results indicated that, with succinate as substrate, the mechanical procedure gave inconsistent preparations with P/O ratios of less than one while the enzymatic procedure yielded mitochondria which gave P/O ratios consistently greater than one.

In recent years, the isolation of functional mitochondria from fungi has been improved by (a) the introduction of mechanical and enzymatic methods which lead to good cell breakage without having too disruptive an effect on the integrity of the mitochondria and (b) the realization that the isolation of functional mitochondria requires careful control of conditions such as osmolarity, pH and the presence of bovine serum albumin and EDTA.

Relatively intact mitochondria with high respiratory control ratios have been isolated from *Aspergillus niger* (Watson & Smith, 1967*a*) and *A. oryzae* (Watson, Paton & Smith, 1969) by the use of an all-glass homogenizer. The inclusion of bovine serum albumin, EDTA and the use of an incubation buffer of pH 6.5 were found to be necessary for the observation of high respiratory control ratios. This simple technique has been successfully applied to the isolation of mitochondria from other *Aspergillus* species. Mitochondria exhibiting respiratory control have been isolated from a citric acid producing strain of *A. niger* (Ahmed, Smith & Anderson, 1972) and an itaconic acid producing strain of *A. terreus* (Nowakowska-Waszczuk, 1973). Ahmed, Smith & Anderson (1972) emphasized that thorough washing of the mycelium with buffer was necessary to remove the citric acid accumulated in the medium during growth.

The main disadvantage of these methods is that only small amounts of mycelium (1–2 g wet weight) can be disrupted at any one time and only between 10–15 g wet weight of mycelium can be handled. However, the yields of mitochondrial protein are comparable with other mechanical methods described for the preparation of fungal mitochondria (Table 4.1). Grinding of mycelia from *Aspergillus niger*, with sand in a mortar, followed by differential centrifugation has also been found to give mitochondria with good respiratory control ratios. However, the large increase in respiration

Table 4.1 Methods for the isolation of fungal mitochondria

Organism	Method	Yield (mg protein/g wet weight mycelium)	References
Aspergillus niger	Glass-homogenizer	1–2	Watson & Smith (1967*a*)
	Mortar and sand	1–2	Watson (1968)
	High speed Vir-Tis homogenizer		Higgins & Friend (1968)
Aspergillus oryzae	Glass-homogenizer	1–2	Watson, Paton & Smith (1969)
	High-speed blender	0.8–1	Kawakita (1970)
Neurospora crassa	Mortar and sand		Luck (1963)
	Snail-gut enzyme		Weiss (1965)
	Micro-mill		Hall & Greenawalt (1967)
	Snail-gut enzyme		Greenawalt, Hall & Wallis (1967)
	Grind-mill	2–3	Weiss *et al.* (1970)
	Snail-gut enzyme	3–4	Lambowitz, Smith & Slayman (1972*b*)

on the addition of exogenous cytochrome c is indicative of structural damage (Watson, 1968).

Mitochondria capable of carrying out oxidative phosphorylation with most of the common intermediates of the tricarboxylic acid cycle have been isolated from *Neurospora crassa* by the use of a micro-mill (Hall & Greenawalt, 1967). An enzymatic procedure similar to that used for the preparation of yeast mitochondria has also been described (Greenawalt, Hall & Wallis, 1967). In the latter method enzymatic digestion of the cell wall is followed by controlled gentle homogenization of the resulting protoplasts. Both procedures yield mitochondria with some respiratory control.

A further improvement in these methods was the introduction of a grind-mill for the disruption of *Neurospora crassa* hyphae (Weiss, von Jagow, Klingenberg & Bucher, 1970). Large amounts of hyphae can be disrupted by this method and, moreover, high yields of intact mitochondria were reported using this technique. A modified snail enzyme procedure has recently been reported by Lambowitz, Smith & Slayman (1972*a*). This method gives very good yields of mitochondrial protein and the mitochondria show respiratory control with a wide range of substrates.

Kawakita (1970) has described a mechanical method in which disruption of large amounts of *Aspergillus oryzae* mycelia (60–80 g wet weight) between two stainless steel rollers yielded mitochondria showing some degree of respiratory control.

The isolation of comparatively intact mitochondria from *Neurospora crassa* and *Aspergillus* species appears now to be well established (Table 4.1). Methods, both mechanical and enzymatic, have been described for small- or large-scale preparations. The addition of bovine serum albumin and EDTA to the isolation and assay media appears to be necessary for the observation of respiratory control. Where necessary, mitochondria may be purified by sucrose density gradient centrifugation (Luck, 1963). The purification step is particularly critical in studies which require minimum contamination by cytoplasmic components, for example in investigations of mitochondrial ribosomes and protein synthesis. This procedure is now widely adopted by many laboratories. In general, a continuous or discontinuous gradient consisting of 20–70% w/v sucrose, EDTA and appropriate buffer may be used with equally good results.

Most authors have stated that best results were obtained using cells in the logarithmic or early stationary growth phases. This was particularly critical for the methods utilizing enzymatic digestion of cell walls (Greenawalt, Hall & Wallis, 1967; Lambowitz, Smith & Slayman 1972*a*). It must be recognized that the properties of fungal mitochondria may be markedly influenced by growth conditions and age of mycelium (see p. 110).

4.3 Biochemical Properties

Oxidative phosphorylation

The biochemical properties of fungal mitochondria have only recently been elucidated, although the presence of an active tricarboxylic acid cycle and respiratory chain in fungi have been long established (for reviews see, Niederpruem, 1965; Lindenmayer, 1965).

It is now generally accepted that mitochondria from fungi, at least in the case of *Aspergillus* species and *Neurospora crassa*, are biochemically similar to those of yeasts, plants and animals.

Mitochondria capable of oxidizing all the common intermediates of the tricarboxylic acid cycle have been isolated from *Aspergillus niger* (Watson & Smith, 1967*a*, *b*; Ahmed, Smith & Anderson, 1972) and *A. oryzae* (Watson, Paton & Smith 1969; Kawakita, 1970) (Fig. 4.1). These studies indicated the presence of three phosphorylation sites in the case of NAD-linked substrates such as pyruvate-malate and citrate, two phosphorylation sites with succinate and one phosphorylation site with ascorbate-TMPD as substrate. The P/O and respiratory control ratios were somewhat lower than those observed in mammalian mitochondria but were similar to those observed in *Saccharomyces carlsbergensis* (Ohnishi, Kawaguchi & Hagihara, 1966*a*). Phosphorylation efficiencies with NAD-linked substrates were, however, significantly higher in *Aspergillus* mitochondria as compared with the yeast. Typical polarographic traces of oxygen uptake and phosphorylation efficiencies in *A. niger* mitochondria utilizing different substrates are illustrated in Fig. 4.1.

The response of the electron transport chain of *Aspergillus niger* to cyanide, antimycin A, 2,4-dinitrophenol and oligomycin was essentially the same as that of mitochondria from other sources. Kawakita (1970) reported that mitochondria from *A. oryzae* were slightly more resistant to antimycin A and cyanide than mammalian mitochondria. In this respect the mitochondria resembled those of higher plants (Hackett, 1961; Wiskich & Bonner, 1963).

Oxidative phosphorylation in wild-type *Neurospora crassa* was first demonstrated by Hall & Greenawalt (1964, 1967). These initial studies showed that *N. crassa* mitochondria were capable of oxidation coupled to phosphorylation with a number of tricarboxylic acid cycle intermediates.

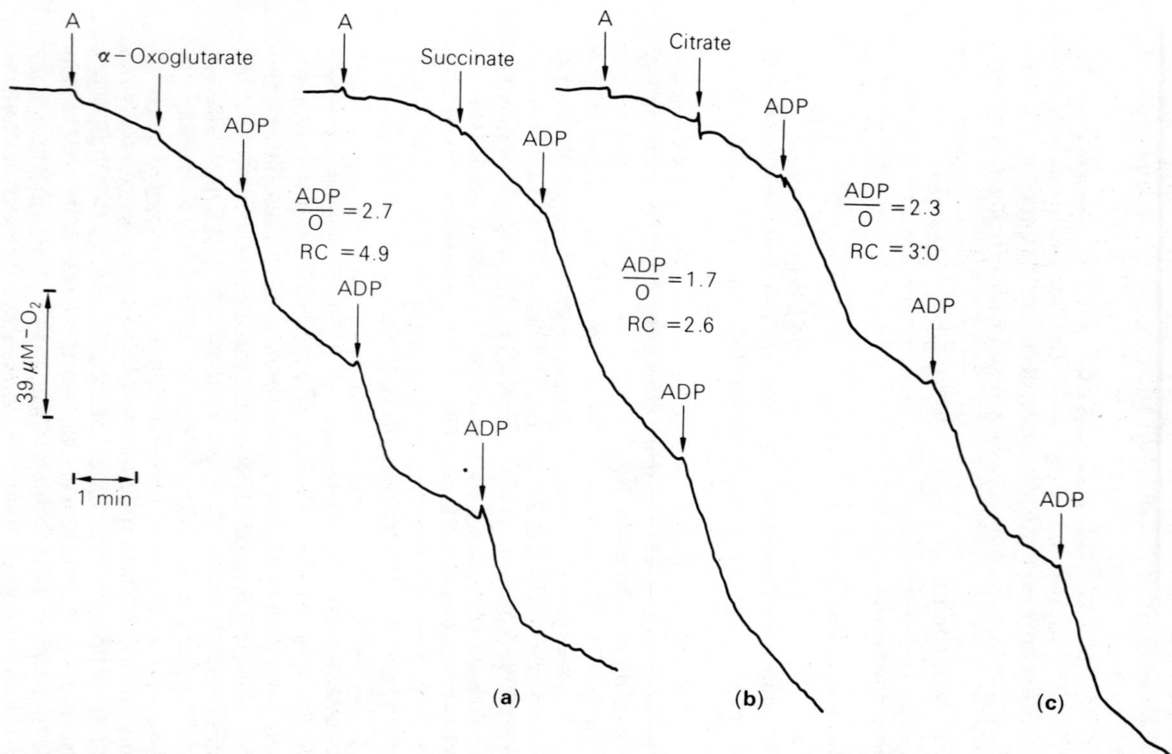

Fig. 4.1 Oxygen uptake and phosphorylation efficiences in *Aspergillus niger* mitochondria with different substrates. Oxygen uptake was measured polarographically in a reaction medium containing 0.5 M mannitol, 0.5 mM EDTA, 5 mM KCl, 5 mM potassium phosphate and 0.2% w/v bovine serum albumin. The final pH was 6.5. Additions were made as indicated in the graphs. The substrate concentration in each case was 5 mM. At A, 1.2 mg mitochondrial protein was added followed by substrate and ADP, 161 nmoles. ADP/O and RC (respiratory control) ratios were calculated by the method of Chance & Williams (1955). (Watson & Smith, unpublished observations.)

Respiratory control ratios were low, but phosphorylation efficiencies were comparable to those of mammalian mitochondria.

More recent studies by Weiss *et al.* (1970) and von Jagow, Weiss & Klingenberg (1973), using mitochondria with respiratory control ratios of 2–3 with succinate and NAD-linked substrates, have confirmed these observations. The ADP/O ratios calculated graphically from polarographic traces, agreed fairly well with those determined manometrically by Hall & Greenwalt (1967). Lambowitz, Smith & Slayman (1972*a*), using a procedure employing snail gut enzyme, have also reported the isolation of *Neurospora crassa* mitochondria with good respiratory control ratios. Phosphorylation efficiencies with α-ketoglutarate (P/O ratio of 2.1), citrate (1.6) and pyruvate-malate (1.5) were significantly lower than those reported by Hall & Greenawalt (1967) for the same substrates, 3.3, 3.0 and 1.9 respectively. It is difficult to reconcile low or high P/O ratios determined on isolated mitochondria with *in vivo* phosphorylation efficiencies. This is particularly the case with mitochondria isolated from fungal mycelium. It is not unlikely that the mitochondria may have suffered some structural damage as the result of the breakage of the tough hyphal walls.

A more attractive explanation for the low P/O ratios obtained by Lambowitz, Smith & Slayman, (1972*a*) comes from work on yeast mitochondria and the concept of site I phosphorylation. The reported absence of site I phosphorylation in *Saccharomyces* yeast (Ohnishi, Kawaguchi & Hagihara, 1966; Ohnishi, Sottocasa & Ernster, 1966; Kovac, Bednarova & Greksak, 1968) has now been clarified by recent work which has shown that three sites of phosphorylation are only observed in cells harvested at late stationary growth phase (Mackler & Haynes, 1973). The mitochondrial energy-transfer chain is apparently not fully developed in exponentially growing cells of *Saccharomyces* yeast, resulting in a defect in the phosphorylation site localized between NADH and cytochrome b. A similar situation may be operative in the case of *Neurospora crassa* cells. It is interesting to note that Lambowitz, Smith & Slayman (1972*a*) used cells which were harvested in the late exponential growth phase. The P/O ratios of less than two with pyruvate-malate (1.5) and citrate (1.6) as substrates, suggests the absence of one phosphorylation site (see also p. 100). Furthermore, the isolated mitochondria were reported to be relatively insensitive to respiratory inhibition by rotenone (10 μM) and amytal (2 μM), levels which are maximally inhibitory to mammalian mitochondria. On the other hand, Weiss *et al.* (1970) have reported that *N. crassa* mitochondria are sensitive to inhibition by rotenone (10 μM) with pyruvate-malate as substrates. The P/O ratios in these experiments were in the range 1.7–2.3, suggesting a more fully developed mitochondrial energy-transfer system. In a more recent study, it was reported that the levels of cytochromes aa_3 and cytochrome b increased with the age of the cells and correlated well with increases in phosphorylation efficiency (von Jagow, Weiss & Klingenberg, 1973).

A possible correlation between the loss of rotenone-sensitive respiration and loss of phosphorylation efficiency has been suggested for mitochondria from the yeast *Torulopsis utilis* (Katz, Kilpatrick & Chance, 1971). On the other hand, a direct relationship between loss and acquisition of rotenone sensitivity and site I phosphorylation has been questioned by some workers (*e.g.* Clegg & Garland, 1971). Nevertheless, it is clear that conditions of cell growth and age of cells are extremely important parameters and must be taken into consideration when discussing biochemical properties of fungal mitochondria.

Respiratory components

The respiratory components of the electron transport chain of fungi have only recently been quantitatively determined. The very early observation by Boulter & Derbyshire (1957) of the presence of a cytochrome system in fungi resembling that of a reduced yeast suspension has been confirmed.

Quantitative data are limited to studies on *Aspergillus* species and *Neurospora crassa*. Table 4.2 summarizes the available data on the respiratory components of mitochondria isolated from *A. niger*, *A. oryzae* and *N. crassa*. For comparison, the respiratory components of the yeast *Saccharomyces carlsbergensis* are included. Mitochondria from *N. crassa* are characterized by a high level of all cytochromes, particularly cytochromes aa_3 and b when compared with *Aspergillus* or *S. carlsbergensis*. Conversely, the ubiquinone level is lower than in *A. niger* and considerably lower than that of *S. carlsbergensis* (see also p. 100). It would be appropriate at this stage to re-emphasize that the level of respiratory components in fungal mitochondria is markedly influenced by growth conditions and age of cells. Direct comparison of such data, therefore, requires careful interpretation.

Table 4.2 Respiratory components of mitochondria from filamentous fungi and yeast

(Concentrations expressed as nmol/mg^{-1} protein)

Component	***Aspergillus niger***[a]	***Aspergillus oryzae***[b]	***Neurospora crassa***[c]	***Saccharomyces carlsbergensis***[d]
Cytochromes aa_3	0.15	0.20	0.38	0.15
Cytochrome b	0.19	0.13	0.38	0.28
Cytochromes $c + c_1$	0.39	0.20	1.6	0.65
Ubiquinone	2.9		1.5[e]	5.4
NAD	7.0		12.0	

References: [a]Watson & Smith (1968); [b]Wakiyama & Ogura (1972); [c]Lambowitz, Smith & Slayman (1972*b*); [d]Ohnishi *et al* (1967); [e]Drabikowska & Kruszewska (1972).

Detailed studies of the b-type cytochromes of *Aspergillus oryzae* (Wakiyama & Ogura, 1972) and *Neurospora crassa* (Lambowitz, Smith & Slayman, 1972*b*; von Jagow & Klingenberg, 1972; von Jagow, Weiss & Klingenberg, 1973) have been reported. In *A. oryzae*, three b-type cytochromes, in approximately equal proportions, were observed absorbing at 556, 559.5 and 564 nm respectively. Two b-type cytochromes, present in equimolar amounts, have been characterized in *N. crassa* mitochondria with absorption maxima at 556 and 562 nm. The analogy between these b-type cytochromes and the cytochromes designated as b_K and b_T found in animal mitochondria (Sato, Wilson & Chance, 1971; Wikstrom, 1972) has been discussed by these authors.

NADH oxidation

A characteristic property of fungal mitochondria is their ability to oxidize exogenous NADH at a rapid rate. Moreover, this oxidation is coupled to the synthesis of two molecules of ATP. Fungal mitochondria are thus distinguished from mammalian mitochondria which normally do not oxidize exogenous NADH (Lehninger, 1951). Under certain conditions, mitochondria from pigeon heart (Blanchaer, Lundquist & Griffith, 1966) and rat liver (Cereijo-Santalo, 1966) will oxidize NADH, but without concomitant ATP formation.

This ability to carry out oxidative phosphorylation with exogenous NADH as substrate is not confined to fungal mitochondria, but appears to be a common feature of plant mitochondria (Wiskich & Bonner, 1963; Douce, Mannella & Bonner, 1973). Oxidative phosphorylation with exogenous NADH as substrate has been reported for mitochondria isolated from *Aspergillus niger* (Watson & Smith, 1967*a*, *b*; Ahmed, Smith & Anderson, 1972), *A. oryzae* (Watson, Paton & Smith, 1969; Kawakita, 1970), *A. terreus* (Nowakowska–Waszczuk, 1973) and *Neurospora crassa* (Hall & Greenawalt, 1967; Weiss *et al.*, 1970; Lambowitz, Smith & Slayman, 1972*a*).

Oxidative phosphorylation of exogenous NADH by fungal mitochondria is anomalous in that the pathway is relatively insensitive to inhibition by rotenone and amytal, as compared with the oxidation of NAD-linked substrates. This phenomenon has been observed in mitochondria from *Aspergillus niger* (Watson & Smith, 1967*a*, *b*), *A. oryzae* (Kawakita, 1970) and *Neurospora crassa* (Weiss *et al.*, 1970; Lambowitz, Smith & Slayman, 1972*a*). The effect of rotenone on oxidative phosphorylation in *A. niger* mitochondria with NADH and α-oxoglutarate as subtrates is shown in Fig. 4.2.

Higgins & Friend (1968) have reported that oxidation of exogenous NADH by *Aspergillus niger* mitochondria was sensitive to inhibition by rotenone (7 μM) and high levels of amytal (10 mM). Although the mitochondrial preparations showed some degree of respiratory control the oxidation of exogenous NADH was not coupled to phosphorylation. It is difficult, therefore, to draw any conclusions from these results and those in which oxidation of NADH is coupled to ATP synthesis.

Studies by von Jagow & Klingenberg (1970) on *Saccharomyces* mitochondria provided evidence for the existence of two NADH dehydrogenases with different localizations on the inner mitochondrial membrane. One pathway, for the oxidation of exogenous NADH, was localized on the outside of the inner mitochondrial membrane. The other pathway, associated with the oxidation of NADH generated endogenously was localized on the inside of the inner mitochondrial membrane. Both pathways in *Saccharomyces* yeasts, in which site I phosphorylation is absent (see also p. 101), are apparently rotenone insensitive. In the case of *Torulopsis* yeasts, in which site I phosphorylation is present, the latter pathway is rotenone sensitive (Ohnishi, Sottocasa & Ernster, 1966). Pathways for the oxidation of exogenous and endogenous NADH in *Neurospora crassa* are similar to *Torulopsis* and different from *Saccharomyces* (Weiss *et al.*, 1970). A second pathway for the oxidation of exogenous NADH has recently been shown to occur in plant mitochondria (Douce, Mannella & Bonner, 1973) and is

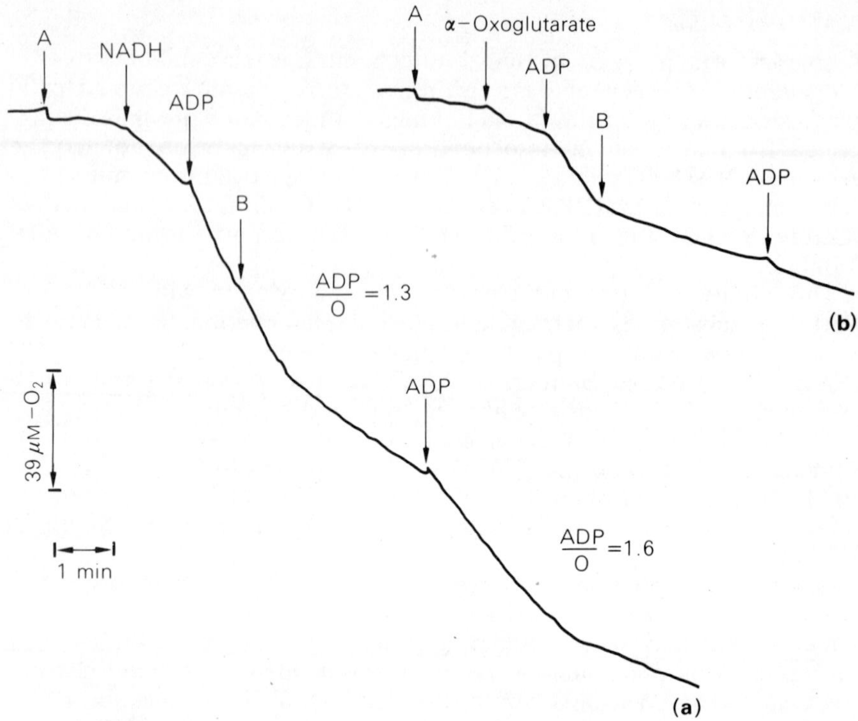

Fig. 4.2 The effect of rotenone on oxidative phosphorylation in *Aspergillus niger* mitochondria with NADH and α-oxoglutarate as substrates. Reaction conditions were similar to that given in Fig. 4.1. The following additions were made: (a) A, 1.0 mg mitochondrial protein; NADH, 2 mM; rotenone 16 μM; ADP, 176 nmoles. (b) Additions were the same as in (a) except the substrate was α-oxoglutarate, 5 mM (Watson & Smith, unpublished results.)

associated with the mitochondrial outer membrane. It would be of interest to establish the existence, or otherwise, of this second pathway in fungal mitochondria.

4.4 Mutants

Poky mutants of Neurospora crassa

The utilization of cytoplasmic mutants in the elucidation of the mechanism of oxidative phosphorylation and mitochondrial biogenesis in yeast has been comprehensively reviewed (Roodyn & Wilkie, 1968; Linnane & Haslam, 1970; Coen, Deutsch, Netter, Petrochillo & Slonimski, 1969; Perlman & Mahler, 1970; Linnane *et al.*, 1972). It is apparent from recent work that the exploitation of fungal mutants of cytoplasmic origin, other than yeast, may provide an important tool for investigations of mitochondrial electron transport, oxidative phosphorylation and biogensis.

The most extensively studied of these mutants are the cytoplasmic mutants of *Neurospora crassa* which have abnormal cytochrome spectra (Mitchell & Mitchell, 1952; Haskins, Tissieres, Mitchell & Mitchell, 1954;

Tissieres & Mitchell, 1954). One of these mutants, *poky* or *mi-1*, has very low levels of cytochromes aa_3 and b but contains an alternative oxidative pathway. This pathway is insensitive to the usual inhibitors of respiration, azide, cyanide and antimycin A (Tissieres, Mitchell & Haskins, 1953; Lambowitz & Slayman, 1971) which enables the mutant to respire at rates faster than the wild type. Recent quantitative analyses of the electron transport components of wild-type and *poky* mutants of *N. crassa* have been made (Rifkin & Luck, 1971; Lambowitz, Slayman, Slayman & Bonner, 1972; von Jagow & Klingenberg, 1972; Drabikowska & Kruszewska, 1972; von Jagow, Weiss & Klingenberg, 1973; Drabikowska, Kosmakos & Brodie, 1974). These studies have confirmed the original reports on the low levels of cytochromes aa_3 and b and the high levels of cytochrome c in *poky* mitochondria (Haskins *et al.*, 1954; Tissieres, Mitchell & Haskins, 1953). In addition to cytochrome c abnormally high levels of ubiquinone were observed in the *mi-1* (von Jagow & Klingenberg, 1972; von Jagow, Weiss & Klingenberg, 1973; Drabikowska & Kruszewska, 1972), *mi-3* (von Jagow, Weiss, & Klingenberg, 1973) and *mi-4* (Drabikowska & Kruszewska, 1972) mutants. It has been suggested that ubiquinone plays a role in both oxidase pathways, namely the cyanide-sensitive cytochrome-mediated pathway and the cyanide-insensitive pathway (von Jagow, Weiss & Klingenberg, 1973). These authors have proposed that ubiquinone is localized at the branch point of the two oxidative pathways.

Oxidative phosphorylation has only recently been demonstrated in the *poky* mutant (Eakin & Mitchell, 1970; Lambowitz, Smith & Slayman, 1972*a*; von Jagow, Weiss & Klingenberg, 1973; Drabikowska, Kosmakos & Brodie, 1974). While there is general agreement that the cyanide-insensitive pathway is not mediated by the cytochrome system and does not include a phosphorylation site, there is some uncertainty as to the number of phosphorylation sites associated with the cytochrome system.

Lambowitz, Smith & Slayman (1972*a*) interpreted their results as indicative of the absence of site I phosphorylation in *poky* mitochondria. This conclusion was based on the measured P/O ratios and the small reduction (Less than 10%) of the endogenous NAD by succinate oxidation in *poky* mitochondria. It must be pointed out, however, that even in the wild type, P/O ratios of much less than 2 were observed with NAD-linked substrates and the reduction (approximately 27%) of endogenous NAD by succinate was also relatively low. Additionally, both the *poky* and wild-type mitochondria were comparatively insensitive to rotenone (10 μM) and amytal (2 mM), which are known inhibitors of the NADH dehydrogenase region of the mitochondrial respiratory chain. It is suggested, therefore, that the mitochondrial energy-transfer systems in both the wild-type and *poky* cells were not fully developed in these studies in which the cells were harvested at late exponential growth phase. It has been noted that *poky* cells become more efficient both in terms of an increase in cytochromes (Haskins *et al.*, 1954; Eakin & Mitchell, 1970; Rifkin & Luck, 1971) and phosphorylation ability (von Jagow, Weiss & Klingenberg, 1973) with increasing age of the culture. A somewhat similar phenomenon is observed in yeast cells in which the mitochondrial energy transfer system is not fully developed (absence of site I phosphorylation) in the exponential growth phase. Fully functional mitochondria, with all three phosphorylation sites are not observed until cells have reached the stationary growth phase (Macklet & Haynes, 1973).

In another recent study of oxidative phosphorylation in the *poky* mutant, Drabikowska *et al.* (1974) have concluded that only two sites of phosphorylation, site I and site III, are present in mitochondria isolated from young hyphae. Evidence for site I was based on the observation of the energy-linked reduction of NAD (approximately 57%) by succinate plus ATP, and the evidence for site III on the phosphorylation associated with the oxidation of ascorbate/TMPD. However, very low P/O ratios were obtained, 0.4 with succinate and ascorbate/TMPD and 0.8 with citrate as substrates. Rotenone (20 μM) was reported to have no effect on oxidation or phosphorylation with citrate as substrate. By contrast, von Jagow, Weiss & Klingenberg (1973) have concluded from their studies that there are three sites of phosphorylation in wild-type and *poky* mitochondria. They attributed the low ADP/O ratios of *poky* mitochondria to the high rates of respiration associated with the cyanide-insensitive non-phosphorylating pathway.

Despite these extensive studies, it is evident that further work on the electron transport chain of the *poky* mutant is called for. The potential of studies involving mutants with lesions at different points along the mitochondrial electron transport chain is apparent. It is surprising, therefore, to find that, apart from the *poky* mutants, very few studies on mitochondria from nuclear and cytoplasmic mutants of fungi have been made. A list of some of these mutants which have been studied is presented in Table 4.3. Undoubtedly, the reasons for this situation are the inherent difficulties in selecting mutants in which the respiratory chain is altered or defective (Bertrand & Pittenger, 1972). Several researchers have now published methods which enable the selection of respiratory mutants of *Neurospora crassa* (Gillie, 1970; Edwards, Kwiecinski & Horstmann, 1973). Such mutants would be expected to be extremely useful for elucidating the pathway and mechanism of mitochondrial electron transport and energy conservation.

Other mutants

The isolation of respiratory-defective cytoplasmic mutants of fungi, apart from yeast and *Neurospora*, has only recently been reported. The isolation of a mitochondrial mutant of the Basidiomycete, *Coprinus lagopus* has been communicated (Casselton & Candit, 1972). The mutation was shown to be inherited cytoplasmically and to result in a defective cytochrome spectrum. The cytochrome spectrum showed the presence of cytochromes b and c and the absence of cytochromes aa_3 and was thus reminiscent of that of the *poky*, *mi-3*, mutant of *N. crassa* (Mitchell, Mitchell & Tissieres, 1953).

Some species of *Mucor* exhibit dimorphism, the yeast-like form being favoured when the cells are grown anaerobically or aerobically in the presence of 2-phenethanol (Terenzi & Storck, 1969) and other antibiotics (Fisher, Pring, Richmond & Michael, 1973). Storck & Morrill (1971) have studied stable mutants of *M. bacilliformis* which are unable to grow filamentously and which grow exclusively in a yeast-like form even under aerobic conditions. These mutants were shown to be respiratory deficient and to lack cytochromes aa_3 and cytochrome oxidase activity. The presence of cytochrome c, however, was detected in the isolated mitochondria. The

Table 4.3 Respiratory defective mutants of fungi

Organism	Mutagenic origin	Properties	References
Neurospora crassa			
mi-1 (poky)	Spontaneous	Slow growth	Mitchell & Mitchell (1952)
		Low cytochromes aa_3 and b	Mitchel, Mitchell & Tissieres (1953); Haskins *et al.* (1953).
		Cyanide and antimycin A-insensitive respiration	Tissieres, Mitchell & Haskins (1953); Eakin & Mitchell (1970); Lambowitz & Slayman (1971)
		High ubiquinone content	Drabikowska & Kruszewska (1972); von Jagow Weiss & Klingenberg (1973)
		Low oxidative phosphorylation	Eakin & Mitchell (1970); Drabikowska *et al.* (1974); Lambowitz, Smith & Slayman (1972*a*)
mi-3	Spontaneous	Slow growing	Mitchell & Mitchell (1952)
		Low cytochromes aa_3	Mitchell, Mitchell & Tissieres (1953); Haskins *et al.* (1953)
		Cyanide and antimycin A-insensitive respiration	Tissieres, Mitchell & Haskins (1953); von Jagow, Weiss & Klingenberg (1973)
		High ubiquinone content	von Jagow, Weiss & Klingenberg (1973)
mi-4	Spontaneous	Slow growth	Pittenger (1956)
		Abnormal cytochromes	
		High ubiquinone content	Drabikowska & Kruszewska (1972)
{ *cn-1*, *cn-2*, *resp-1* }	Inositol-less death and tetrazolium salts	Abnormal cytochromes and respiration	Edwards, Kwiecinski & Horstmann (1973)
{ *abn-1*, *abn-2* }	Spontaneous	Slow growth	Diacumakos, Garnjobst & Tatum (1965)
		Abnormal cytochromes	
Aspergillus nidulans			
cs67	Nitrosoguanidine	Cold-sensitive	Waldron & Roberts (1973)
		Abnormal cytochromes	
—	Oligomycin	Oligomycin-resistant	Rowlands & Turner (1973)
		Abnormal cytochromes	
Coprinus lagopus			
asc-10	Nitrosoguanidine	Unable to grow on acetate	Casselton & Candit (1971)
		Abnormal cytochromes	
Mucor bacilliformis	Spontaneous	Yeast-like growth	Storck & Morrill (1971)
		Abnormal cytochromes	

Mucor mutant was thus somewhat analagous to the well-documented cytoplasmic *petite* mutant of yeast. It is noteworthy, in this respect that 2-phenethanol has been shown to induce *petite* formation in *Saccharomyces cerevisiae* (Wilkie & Maroudas, 1969). Further studies on these respiratory-deficient mutants of *Mucor* would be rewarding, just as the cytoplasmic *petite* mutants of yeast have proved to be invaluable in studies on the biochemistry, genetics and biogenesis of yeast mitochondria (Roodyn & Wilkie, 1968; Schatz, 1968; Linnane & Haslam, 1970; Perlman & Mahler, 1970; Linnane *et al.*, 1972).

Several very interesting mutants of *Aspergillus nidulans* have recently been characterized. They include an oligomycin-resistant mutant which showed non-Mendelian behaviour and it was concluded that the mutation was of cytoplasmic origin (Rowlands & Turner, 1973). Oligomycin-resistant mutants of yeast have been isolated (Avner & Griffiths, 1970; Stuart, 1970; Wakabayashi & Gunge, 1970). These mutants have been further subdivided into two classes on the basis of their cross-resistance to a variety of inhibitors and uncouplers of mitochondrial energy-transfer reactions (Avner & Griffiths, 1973*a*, *b*). Class I mutants are specifically resistant to oligomycin and rutamycin while class II mutants show cross resistance to a range of inhibitors and uncouplers. The oligomycin resistance has been shown to be expressed *in vitro* at the inner mitochondrial membrane and to result in an increased resistance of the mitochondrial ATPase to oligomycin (Griffiths, Avner, Lancashire & Turner, 1972). However, no gross differences in inner-membrane morphology are observed between normal and oligomycin-resistant mutants (Watson & Linnane, 1972). The study of fungal mutants resistant to antibiotics such as oligomycin which inhibit mitochondrial ATP synthesis, provides a biochemical and genetical means of approaching the problem of mitochondrial oxidative phosphorylation.

A cold sensitive mutant of *Aspergillus nidulans* unable to grow at 20°C or below, has been characterized as a mutation of cytoplasmic origin (Waldron & Roberts, 1973). The biochemical properties of this mutant have, as yet, not been communicated although preliminary data suggests a defective cytochrome spectrum (Waldron & Roberts, 1973).

4.5 Biogenesis of Fungal Mitochondria

Over ten years ago Nass & Nass (1963) demonstrated the presence of DNAase-susceptible filaments in liver mitochondria. In the succeeding year the isolation of DNA species with buoyant densities distinct from nuclear DNA was described in mitochondria isolated from *Neurospora crassa* (Luck & Reich, 1964) and *Saccharomyces cerevisiae* (Schatz, Haslbrunner & Tuppy, 1964). Subsequently, mitochondria were shown to have a protein-synthesizing system which was distinguished in many respects from the cytoplasmic system (for review see Ashwell & Work, 1970). It was evident, however, that the information content of the mitochondrial DNA was insufficient to code for the entire mitochondrial protein-synthesizing apparatus as well as all the mitochondrial proteins (Borst & Kroon, 1969).

A voluminous literature now exists concerning the biogenesis of mitochondria and it would be beyond the scope of this chapter to cover fields adequately treated in various reviews (Ashwell & Work, 1970; Linnane & Haslam, 1970; Rabinowitz & Swift, 1970; Borst, 1972; Linnane *et al.*, 1972;

Schatz & Mason 1974). Only the very recent literature, covering specific aspects of mitochondrial biogenesis relevant to fungal mitochondria, are discussed in this section.

The following generalizations appear now to be well established and generally accepted by the majority of workers in this field: (a) the mitochondrion possesses a distinct protein-synthesizing system, complete with DNA, RNA and ribosomes; (b) the biogenesis of the mitochondrion requires the combined operation of the mitochondrial and cytoplasmic synthesizing systems; and (c) the proteins synthesized by the mitochondrion are localized exclusively in the inner mitochondrial membrane and constitute approximately 10–15% of the total mitochondrial protein.

Mitochondrial DNA and RNA

Apart from *Neurospora crassa* and *Saccharomyces cerevisiae*, unique species of mitochondrial DNA have been identified in fourteen species of fungi belonging to the Zygomycetes, Ascomycetes and Basidiomycetes (Villa & Storck, 1968). The base composition, as moles per cent of guanine plus cytosine, of the mitochondrial DNAs varied from 28% for *Schizophyllum commune* to 44% for *Gelasinospora tetrasperma*. No gross differences were noted in the percentage guanine plus cytosine between mitochondrial DNA and nuclear DNA in the Zygomycetes, whereas average differences of 15% and 19% were found for Ascomycetes and Basidiomycetes respectively. These results prompted the authors to suggest the possibility that the analysis of the base composition of mitochondrial and nuclear DNA may have a phylogenetic significance.

Physicochemical properties of mitochondrial ribosomal RNA have been reported for the filamentous fungi, *Trichoderma viride, Neurospora crassa* and *Aspergillus nidulans* (Edelman, Verma, Herzog, Calun & Littauer, 1971). The mitochondrial ribosomal RNAs were clearly distinguishable from those of cytoplasmic and bacterial origin. The authors further concluded that there was a considerable heterogeneity in the physical properties among the fungal ribosomal RNA species themselves.

The respiratory-deficient cytoplasmic mutant of *Saccharomyces cerevisiae* has been shown to lack mitochondrial RNA species (Wintersberger, 1967) and an active mitochondrial protein-synthesizing system (Kuzela & Grecna, 1969; Schatz & Saltzgaber, 1969). In the extreme case of *neutral petites*, a complete loss of mitochondrial DNA is also observed (Goldring, Grossman, Kruprick, Cryer, & Marmur, 1970; Nagley & Linnane, 1970; Michaelis, Douglass, Tsai, & Criddle, 1971). For a more detailed discussion of the cytoplasmic *petite* mutant see review by Linnane *et al.* (1972).

The *mi-1* or *poky* mutant of *Neurospora crassa* shows properties similar to those of the *petite* mutant. It has a defective mitochondrial protein-synthesizing system (Sebald, Machleidt & Otto, 1968; Neupert, Massinger & Pfaller, 1971; Rifkin & Luck, 1971). These latter studies have been extended by recent work of Kuriyama & Luck (1974). Evidence was presented that in wild-type *Neurospora*, mitochondrial ribosomal RNAs and their precursor were methylated, while in *poky* mitochondria, by contrast, the mitochondrial RNAs were undermethylated. The ribosomal RNA defect in *poky* was suggested to be due either to (a) a primary defect in methylation leading to low rates of synthesis of small subunits, or (b) a primary defect in the ribosomal RNA precursor or in ribosome assembly resulting in undermethylation of the RNAs.

The mitochondrial ribosome of *Neurospora crassa* was, until recently, firmly established as a species sedimenting at 73 S (Kuntzel, 1969; Borst & Grivell, 1971). A reappraisal of this conclusion is warranted in the light of more recent data which suggests that the native mitochondrial ribosome of *N. crassa* has a sedimentation coefficient of 80 S (Datema, Agsterribe & Kroon, 1974). Evidence was presented that both 80 S and 73 S ribosomes were of mitochondrial origin. The 80 S ribosome could be converted to the 73 S form by ageing or by incubation with the post-ribosomal supernatant fraction of lysed mitochondria.

The presence of 80 S ribosomes in yeast mitochondria has been previously reported by various workers. Schmidt (1970) described the presence of both 73 S and 80 S ribosomes in yeast mitochondria. The particles sedimenting at 80 S were regarded as a cytoplasmic contaminant which could be removed by washing the mitochondria with EDTA. Morimoto, Scragg, Nekhorochoff, Villa & Halvorson (1971) also presented evidence that mitochondrial ribosomes of *Saccharomyces cerevisiae* sediment at 80 S. These observations were reiterated in a subsequent paper (Morimoto & Halvorson, 1971). The presence of 80 S ribosomes in isolated yeast mitochondria has also been reported by Kellems & Butow (1972). The cytoplasmic nature of these ribosomes was demonstrated by their insensitivity to inhibition by ethidium bromide and chloramphenicol and because this class of 80 S ribosome could be isolated from a cytoplasmic *petite* mutant lacking mitochondrial DNA. These ribosomes could, however, be distinguished from cytoplasmic ribosomes by their having greater stability to dissociation by high concentrations of KCl. The authors raised the intriguing possibility that the 80 S ribosome, although of cytoplasmic origin, may be an integral part of the mitochondrial protein-synthesizing system.

Proteins synthesized by the mitochondrion

The separation and partial purification of the inner and outer mitochondrial membranes was first achieved in mammalian mitochondria (Parsons, Thompson, Wilson & Chance, 1967; Sottocasa, Ernster, Kuylenstierna & Bergstrand, 1967; Schnaitman & Greenawalt, 1968). Following these procedures for membrane separation, it was subsequently demonstrated that isolated mitochondria incorporate radioactive amino acids only into the inner membrane proteins (Beattie, Basford & Koritz, 1967; Neupert, Brdiczka & Bucher, 1967). It was thus concluded that the proteins of the mitochondrial outer membrane are products of the cytoplasmic protein-synthesizing system.

The separation of the inner and outer membranes of mitochondria from *Neurospora crassa* (Cassady & Wagner, 1971; Neupert & Ludwig, 1971) and *Saccharomyces cerevisiae* (Bandlow, 1972) has also been accomplished in recent years. These studies have confirmed the observation that, in the presence of cycloheximide, radioactive amino acids are incorporated exclusively into the inner membrane proteins. Under these experimental conditions, the outer membrane proteins were not labelled. It is interesting to note that the outer membrane of fungal mitochondria is distinguished from that of mammalian mitochondria by the absence of monoamine oxidase activity. Kynurenine hydroxylase, on the other hand, was found to be characteristically associated with the outer mitochondrial membrane of *N. crassa* (Cassady & Wagner, 1971) and *S. cerevisiae* (Bandlow, 1972).

Evidence has accumulated from recent studies with *Saccharomyces cerevisiae* and *Neurospora crassa* that at least some of the subunits of cytochrome oxidase (Schatz *et al.*, 1972; Tzagoloff, Rubin & Sierra, 1973; Sebald, Machleight & Otto, 1973; Rowe, Landsman & Woodward, 1974), cytochrome b (Weiss, 1972) and mitochondrial ATPase (Tzagoloff, Rubin & Sierra, 1973) are products of mitochondrial protein synthesis.

Cycloheximide-resistant protein synthesis of mitochondrial membrane components has recently been reported in *Aspergillus nidulans* cells labelled *in vivo* with ^{14}C-leucine (Turner, 1973). Polyacrylamide gel electrophoresis of labelled mitochondrial membrane proteins revealed a pattern similar to that observed in *Neurospora crassa* and *Saccharomyces cerevisiae.* It would be of some interest to establish if these bands correspond to subunits of mitochondrial ATPase and/or cytochrome oxidase or to some other, as yet unidentified, mitochondrial component.

It would be premature to draw final conclusions from these studies, based almost exclusively on a few micro-organisms, as to the exact nature and function of the proteins synthesized by the mitochondrion. There is no reason to believe that proteins synthesized by mitochondria from different organisms would be produced in equivalent amounts or, indeed would be the same proteins. For example, the information content of mitochondrial DNA from *Neurospora crassa* and *Saccharomyces cerevisiae* is potentially much greater than that of DNA of animal mitochondria (Borst, 1972). The physiological state of the organism may also be a determining factor in influencing the nature and amount of protein synthesized by the mitochondrion.

Anaerobic growth

The ability of filamentous fungi to grow under strict anaerobic conditions is limited to a very few species, and these are mainly restricted to *Mucor* and *Fusarium* (Curtis, 1969). Reports on the growth of fungi under anaerobic conditions are thus of some interest. Two reports on anaerobic growth of fungi, relevant to mitochondrial biogenesis, have appeared in the recent literature.

Facultative anaerobic mutants of *Neurospora crassa* designated An^+, have been described (Howell, Zuiches & Munkres, 1971). Anaerobic growth of these mutants leads to a reduction of cytochrome oxidase activity and a decrease in the number of recognizable mitochondria with fewer cristae. Morphological changes in the cellular membrane components were also noted. The absence of glucose repression of respiratory development distinguishes *N. crassa* from the yeast system, in which glucose strongly represses respiratory activity (Slonimski, 1953; Yotsuyanagi, 1962; Polakis, Bartley & Meek, 1964; Jayaraman, Cotman & Mahler, 1966).

Many *Mucor* species exhibit dimorphism and can grow either filamentously or yeast-like. Cell populations of the latter form are particularly observed in the absence of oxygen (Bartnicki-Garcia & Nickerson, 1962; Elmer & Nickerson, 1970; see also p. 102). In anaerobically grown yeast-like cells of *M. genevesis*, mitochondrial profiles with cristae are still visible in electron micrographs, although there is a complete loss of the respiratory cytochromes (Clark-Walker, 1972). As was the case in anaerobically grown *Neurospora crassa*, glucose repression of respiratory development was not observed. Nevertheless, it is clear that anaerobically grown *M. genevesis* and *N. crassa* are not unlike the yeast system in which cristate mitochondria are still visible under the extreme conditions of anaerobiosis, glucose repression and lipid depletion (Damsky, Nelson & Claude, 1969; Plattner & Schatz, 1969; Watson, Haslam & Linnane, 1970; Watson, Haslam, Veitch & Linnane, 1971).

Formation of mitochondria

Evidence for the formation of mitochondria from pre-existing organelles was first demonstrated by Luck (1963) utilizing a choline requiring auxotroph of *Neurospora crassa.* Cells labelled with ^{3}H-choline were allowed to undergo several divisions in unlabelled media. Quantitative autoradiography showed that the mitochondria were randomly labelled, suggesting that newly formed mitochondria arose from the growth and division of pre-existing ones. Similar conclusions were reached based on choline labelling studies, with mitochondrial density changes observed in isopycnic sucrose density gradient centrifugations (Luck, 1965).

The suggestion of *de novo* formation of mitochondria appears now to be untenable. Early electron microscopy observations of the absence of mitochondrial profiles in anaerobically grown yeast cells (Polakis, Bartley & Meek, 1964; Wallace, Huang & Linnane, 1968) have now been shown to be the result of inadequate fixation procedures (Damsky, Nelson & Claude, 1969; Plattner & Schatz, 1969; Watson, Haslam & Linnane, 1970; Watson *et al.*, 1971).

The formation of mitochondria from non-mitochondrial membranes, on the other hand, has remained an intriguing alternative to current concepts of mitochondrial biogenesis. Robertson (1960, 1964) was one of the earliest advocates of the existence of interrelationships among cellular membranes, including the origin of mitochondria from the endoplasmic reticulum.

The inner and outer membranes of the mitochondrion differ, both structurally and enzymatically (Parsons *et al.*, 1967; Sottocasa *et al.*, 1967; Schnaitman & Greenawalt, 1968; Ernster & Kuylenstierna, 1970). On the other hand, an impressive body of evidence has now accumulated relating the outer membrane of mitochondria and the endoplasmic reticulum (Parsons *et al.*, 1967; Sottocasa *et al.*, 1967; Schnaitman, 1969; Ernster & Kuylenstierna, 1970). Electron micrographs suggesting a close association of endoplasmic reticulum with the outer mitochondrial membrane have been observed by many workers in diverse organisms (André, 1962; Robertson, 1960; Sandborn, Koen, McNabb & Moore, 1964; Ruby, Dyer & Salko, 1969; Kessel, 1971; Morré, Merritt & Lembi, 1971).

Malhotra (1970) has presented electron micrographs of *Neurospora crassa* which show a close association between the mitochondrial outer membrane and invaginations of the plasma membrane. Some of these invaginations take the form of complex whorls which have been termed mesosomes in view of their resemblance to bacterial mesosomes (Malhotra, 1966). The function of these associations is not known, but an analogy between the mesosomal membranes of *Neurospora* and the endoplasmic reticulum of higher animal and plant cells has been proposed (Malhotra, 1970; Gupta & Malhotra, 1971).

Recent electron microscopy studies of fungal hyphae by Bracker & Grove (1971) have shown several different types of intimate associations between endoplasmic reticulum-like membranes and the outer membrane of fungal mitochondria. Smooth endoplasmic reticulum membranes were found to predominate in these close associations. The functional implications of these associations were discussed is some detail by Bracker & Grove (1971). Electron micrographs showing association and continuity of endoplasmic reticulum with the outer membrane of fungal mitochondria are illustrated in Fig. 4.3 and Fig. 4.4 respectively. These authors further observed that in *Aspergillus niger* hyphae, ribosome-like granules were seen aligned along the periphery of the inner mitochondrial membrane, analogous to that of ribosomes on rough endoplasmic reticulum membranes. The alignment of mitochondrial ribosomes along the periphery of the inner mitochondrial membrane has also recently been observed in *Saccharomyces cerevisiae* (Watson, 1972). The significance of such an arrangement of the mitochondrial ribosomes in unknown.

It is now well established that approximately 10–15% of the proteins of the inner membrane are synthesized by the mitochondrion, but all the proteins of the outer membrane are apparently synthesized in the cytoplasm (Linnane & Haslam, 1970; Borst, 1972). The majority of the inner membrane proteins (80–90%) are thus either synthesized in the cytoplasm and then transported into the mitochondrion, or require the interaction of both the mitochondrial and the cytoplasmic protein-synthesizing systems.

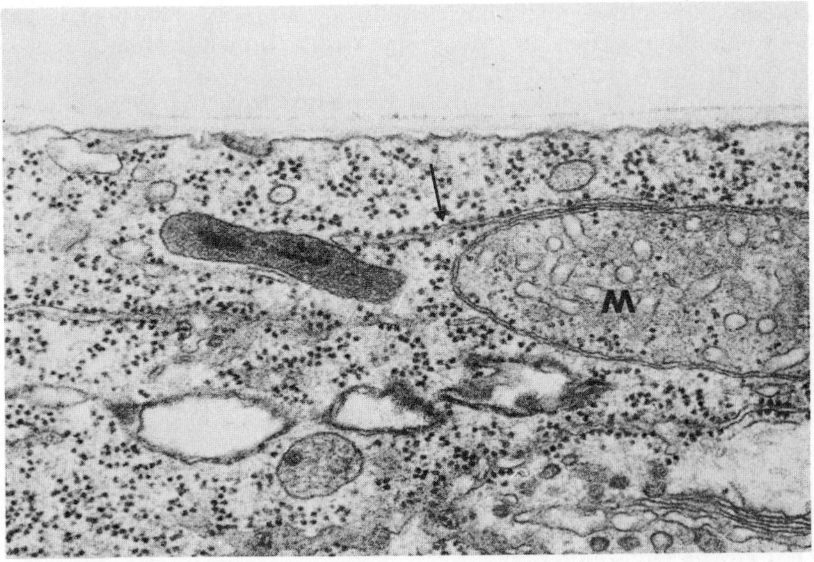

Fig. 4.3 Electron micrograph showing a close association of endoplasmic reticulum (arrow) with the outer membrane of mitochondria (M) in *Pythium aphanidermatum*. ×5 200. (Micrograph courtesy of Dr C. E. Bracker, from Bracker & Grove, 1971, Protoplasm, **73**, 15–34).

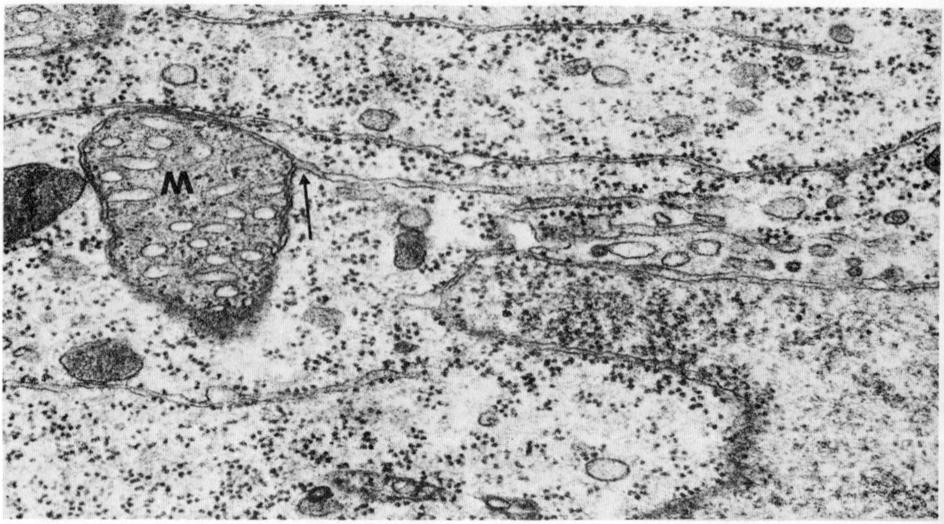

Fig. 4.4 Continuity of smooth endoplasmic reticulum (arrow) with the outer membrane of mitochondria (M) in *Pythium aphanidermatum*. ×5 000. *(Micrograph courtesy of Dr C. E. Bracker, from Bracker & Grove, 1971, Protoplasma,* **73**, 15–34).

The endoplasmic reticulum is a logical candidate as the site for synthesis of proteins of the outer membrane, as well as the majority of the proteins of the inner membrane. Thus far, cytochrome c is the only mitochondrial protein for which there is evidence of its being a microsomal product which is transported into the mitochondrion (Kadenbach, 1966, 1969; Gonzalez-Cadavid & Campbell, 1967). The formation of certain phospholipids by the endoplasmic reticulum and subsequent transfer to the mitochondrion have also been proposed (McMurray & Dawson, 1969; Wirtz & Zilversmit, 1969; Jungalwala & Dawson, 1970). However, it must be taken into consideration that endoplasmic reticulum is not abundant in fungal cells and that the majority of ribosomes are apparently free in the cytoplasm (Hawker, 1965; Bracker, 1967). Despite these observations, it is tempting to speculate that the close association of endoplasmic reticulum with the outer membrane is not merely coincidental, but has functional significance. These associations, together with the peripheral arrangement of the mitochondrial ribosomes, may be interpreted as facilitating the exchange and transfer of materials between these two cell components. The elucidation of the exact nature of these interactions remains the prime objective in understanding the problem of the biogenesis of mitochondria.

4.6 Cellular Differentiation

Germination

It is a common observation that fungal spores show a marked increase in respiratory activity at the onset of germination. Since the initial studies by

Goddard (1935) and Goddard & Smith (1938) with ascospores of *Neurospora tetrasperma*, numerous reports have appeared in the literature pertaining to an increased respiratory rate accompanying fungal spore germination (Allen, 1965; Sussman & Halvorson, 1966; Madelin, 1966; Sussman, 1969).

Spore activation may result from various treatments, including heat and chemical action (Sussman & Halvorson, 1966). Evidence for the role of critical metabolites, such as ATP and acetyl-CoA, in the activation of spores of *Penicillium roqueforti* has been presented (Lawrence & Bailey, 1970). These authors concluded that induction of the tricarboxylic acid cycle, required for the synthesis of these key intermediates, was a prerequisite for spore germination.

Despite extensive studies, the precise role of the mitochondrion in the germination process is not yet clearly understood. The ultrastructural changes associated with spore germination, in particular the role of the mitochondrion, have been comprehensively reviewed by Smith, Gull, Anderson & Deans (1975). Observations are consistent with the generally accepted concept that spore germination is an energy-requiring process and that an active respiratory system is obligatory. Increase in the ratio of mitochondrial DNA to total cellular DNA (Mannino & Greenawalt, 1972) is in keeping with the observation of increased numbers of mitochondria at the onset of germination (Bracker, 1967; Bartnicki-Garcia, Nelson & Cota-Robles, 1968).

Alternatively, it has been demonstrated that mitochondrial DNA synthesis is apparently not required for germination of spores of *Botrydiplodia theobromae* (Dunkle, van Etten & Brambl, 1972). Ethidium bromide, a specific inhibitor of mitochondrial DNA synthesis (Goldring *et al.*, 1970), was found to have no effect on nuclear DNA synthesis or on the extent of spore germination. It was concluded that, although mitochondrial DNA synthesis was not required for germination, it was probably essential for continued vegetative growth.

Sporulation

Recent studies on the biochemical changes accompanying sporogenesis in filamentous fungi have been conducted on *Neurospora crassa* (Turian & Bianchi, 1972) and *Aspergillus niger* (Galbraith & Smith, 1969*a*, *b*; Valenzuela-Perez & Smith, 1971; Smith, Valenzuela-Perez & Ng, 1971; Ng, Smith, & McIntosh, 1974). These studies have indicated the importance of the glycolytic, glyoxylate and tricarboxylic acid cycles to fungal sporulation.

Although the initiation of sporulation involves the utilization of large amounts of energy, the fungus spore itself has very little metabolic activity (Sussman & Halvorson, 1966; Madelin, 1966). It is thus an apparent contradiction that numerous mitochondria are to be found in nearly all fungal spores, for example in *Aspergillus* sp. (Tanaka, 1966; Tsukahara, Yamada & Itagati, 1966; Florence, Denison & Allen, 1972), *Neurospora* (Sussman, 1966) and other species (see Madelin, 1966; Bracker, 1967). The single giant mitochondrion in the spore of the water fungus *Blastocladiella emersonii* makes this fungal spore, so far, unique (Cantino, Lovett, Leak & Lythgoe, 1963).

Respiratory enzymes which have been found to be less active in fungal spores (as compared with vegetative cells) include succinate dehydrogenase (Zalokar, 1959), isocitrate dehydrogenase (Gottleib & Caltrider, 1963), malate dehydrogenase (Gottlieb & Caltrider, 1963), NADH and cytochrome c reductase (White & Ledingham, 1961) and cytochrome oxidase (White & Ledingham, 1961). Despite the low activity of cytochrome oxidase in uredospores in *Puccinia graminis*, White & Ledingham (1961) have concluded that the enzyme was not limiting. It is notable that cytochrome oxidase activity has been found to be higher in conidia as opposed to mycelia in *Neurospora crassa* (Holton, 1960) and *Glomerella singulata* (Sussman & Marker, 1953). In a study of conidiation in *N. crassa*, Weiss (1965) and Weiss & Turian (1966) have observed very little difference in the specific activities of succinate dehydrogenase, cytochrome oxidase and succinate cytochrome c reductase in mycelial and conidial cultures. Moreover, the concomitant production of ATP via oxidative phosphorylation remained high throughout conidial development. It seems surprising, in view of the fact that fungal spores have a very low respiratory activity, that the observations of relatively high cytochrome oxidase activity have never been adequately substantiated.

The isolation of a natural sulphydryl reagent from sporulating gill tissue of the mushroom, *Agaricus bisporus*, which inhibits mitochondrial respiration is suggestive (Weaver, Rajagopalan, Handler, Jeffs, Byrne & Rosenthal, 1970). The compound, γ-L-glutaminyl-4-hydroxybenzene was shown to inhibit succinate oxidase and succinate dehydrogenase activities in mitochondria isolated from *A. bisporus* and rat liver (Vogel & Weaver, 1972). NADH oxidase, by contrast, was not inhibited. The depression of mitochondrial energy production as a possible control of induction of dormancy in the spores is discussed by the authors.

Although significant progress has been made in recent years, the nature of the control mechanism(s) which stimulate and depress mitochondrial function at the onset of spore germination and dormancy, remains an open question. Nothing is known about the function of mitochondria in the fungus spore although recently Smith *et al.* (1975) have considered that the mitochondria in fungal spores may be involved in the establishment of polarity during germination.

Virtually all biochemical studies involving mitochondria from filamentous fungi have utilized batch cultivation techniques. The distinct advantages of continuous culture have not yet been exploited. However, the recent studies by Ng, Smith & McIntosh (1974) indicate how continuous cultivation techniques can be used to study enzyme regulation in a manner not possible with conventional batch methods. Using either glucose or citrate limiting conditions Ng, Smith & McIntosh (1974) were able to show that certain enzymes of the tricarboxylic acid cycle varied in activity according to dilution rate or growth rate. In particular, the specific activity of aconitase, NAD(P) isocitrate dehydrogenase and malate dehydrogenase increased with increasing dilution rate. Under citrate limitation citrate synthase remained relatively constant with increasing dilution rate whereas under glucose limitation citrate synthase activity decreased with increasing dilution rate. Clearly, these studies support the concepts set out in Chapter 1 and show how important the type of cultural system can be in understanding how the environment impinges on fungal metabolism.

4.7 References

AHMED, S. A., SMITH, J. E. & ANDERSON, J. G. (1972). Mitochondrial activity during citric acid production by *Aspergillus niger*. *Transactions of the British Mycological Society* **59**, 51–61.

ALLEN, P. J. (1965). Metabolic aspects of spore germination in fungi. *Annual Review of Phytopathology* **3**, 313-42.

ANDRÉ, J. (1962). Contribution à la connaissance du chondriome, Étude de ses modification ultrastructurales pendant la spermatogenese. *Journal of Ultrastructural Research Supplement* **3**, 1–185.

ASHWELL, M. & WORK, T. S. (1970). The biogenesis of mitochondria. *Annual Review of Biochemistry* **39**, 251–90.

AVNER, P. & GRIFFITHS, D. E. (1970). Oligomycin-resistant mutants in yeast. *Federation of European Biochemical Societies Letters* **10**, 202–7.

AVNER, P. & GRIFFITHS, D. E. (1973*a*). Studies on energy-linked reactions. Isolation and characterization of oligomycin-resistant mutants of *Saccharomyces cerevisiae*. *European Journal of Biochemistry* **32**, 301–11.

AVNER, P. & GRIFFITHS, D. E. (1973*b*). Studies on energy-linked reactions. Genetic analysis of oligomycin-resistant mutants of *Saccharomyces cerevisiae*. *European Journal of Biochemistry* **32**, 312–21.

BANDLOW, W. (1972). Membrane separation and biogenesis of the outer membrane of yeast mitochondria. *Biochimica et Biophysica Acta* **282**, 105-22.

BARTNICKI-GARCIA, S. & NICKERSON, W. J. (1962). Nutrition, growth and morphogenesis of *Mucor rouxii*. *Journal of Bacteriology* **84**, 841–58.

BARTNICKI-GARCIA, S., NELSON, N. & COTA-ROBLES, E. (1968). Electron microscopy of spore germination and cell wall function in *Mucor rouxii*. *Archiv für Mikrobiologie* **63**, 242–55.

BEATTIE, D. S., BASFORD, R. E. & KORITZ, S. B. (1967). The inner membrane as the site of the *in vitro* incorporation of L-^{14}C-leucine into mitochondrial protein. *Biochemistry* **6**, 3099–106.

BERTRAND, H. & PITTENGER, T. H. (1972). Isolation and classification of extranuclear mutants of *Neurospora crassa*. *Genetics* **71**, 521–33.

BLANCHAER, M. C., LUNDQUIST, C.-G. & GRIFFITH, T. J. (1966). Factors influencing the utilization of reduced nicotinamide adenine dinucleotide by pigeon heart mitochondria. *Canadian Journal of Biochemistry* **44**, 105–17.

BORST, P. (1972). Mitochondrial nucleic acids. *Annual Review of Biochemistry* **41**, 333–76.

BORST, P. & GRIVELL, L. A. (1971). Mitochondrial ribosomes. *Federation of European Biochemical Societies Letters* **13**, 73–106.

BORST, P. & KROON, A. M. (1969). Mitochondrial DNA: physico-chemical properties, replication and genetic function. *International Review of Cytology* **26**, 108–90.

BOULTER, D. & DERBYSHIRE, E. (1957). Cytochromes of fungi. *Journal of Experimental Botany* **8**, 313–8.

BRACKER, C. E. (1967). Ultrastructure of fungi. *Annual Review of Phytopathology* **5**, 343–74.

BRACKER, C. E. & GROVE, S. N. (1971). Continuity between endomembranes and outer mitochondrial membranes in fungi. *Protoplasma* **73**, 15–34.

CANTINO, E., LOVETT, J. S., LEAK, L. V. & LYTHGOE, J. (1963). The single mitochondrion, fine structure and germination of the spore of *Blastocladiella emersonii*. *Journal of General Microbiology* **31**, 393–404.

CASSADY, W. E. & WAGNER, R. P. (1971). Separation of mitochondrial membranes of *Neurospora crassa*. *Journal of Cell Biology* **49**, 536–41.

CASSELTON, L. A. & CANDIT, A. (1972). A mitochondrial mutant of *Coprinus lagopus*. *Journal of General Microbiology* **72**, 521–7.

CEREIJO-SANTALO, R. (1966). The role of magnesium in the oxidation of external reduced nicotinamide adenine dinucleotide by rat liver mitochondria. *Canadian Journal of Biochemistry* **44**, 67–76.

CHANCE, B. & WILLIAMS, S. R. (1955). Respiratory enzymes in oxidative phosphorylation. I. Kinetics of oxygen utilization. *Journal of Biological Chemistry* **217**, 383–484.

CLARK-WALKER, G. D. (1972). Development of respiration and mitochondria in *Mucor genevesis* after anaerobic growth: absence of glucose repression. *Journal of Bacteriology* **109**, 399–428.

CLEGG, R. A. & GARLAND, P. B. (1971). Non-haem iron and the dissociation of piriericidin A sensitivity from site I energy conservation in mitochondria from *Torulopsis utilis*. *Biochemical Journal* **124**, 135–54.

COEN, D., DEUTSCH, J., NETTER, P., PETROCHILLO, E. & SLONIMSKI, P. P. (1969). Mitochondrial genetics. I.

Methodology and phenemenology. *Symposium of the Society for Experimental Biology* **24**, 449–96.

CURTIS, P. J. (1969). Anaerobic growth of fungi. *Transactions of the British Mycological Society* **53**, 299–302.

DAMSKY, C. H., NELSON, W. M. & CLAUDE, A. (1969). Mitochondria in anaerobically grown lipid-limited brewer's yeast. *Journal of Cell Biology* **43**, 174–9.

DATEMA, R., AGSTERIBBE, E. & KROON, A. M. (1974). The mitochondrial ribosomes of *Neurospora crassa.* I. On the occurrence of 80 S ribosomes. *Biochimica et Biophysica Acta* **335**, 386–95.

DIACUMAKOS, E. G., GARNJOBST, L., & TATUM, E. L. (1965). A cytoplasmic character in *Neurospora crassa.* The role of nuclei and mitochondria. *Journal of Cell Biology* **26**, 427–43.

DOUCE, R., MANNELLA, C. A. & BONNER, W. D. (1973). The external NADH dehydrogenase of intact plant mitochondria. *Biochimica et Biophysica Acta* **292**, 105–16.

DRABIKOWSKA, A. & KRUSZEWSKA, A. (1972). Ubiquinone function in *Neurospora crassa. Journal of Bacteriology* **112**, 1112–7.

DRABIKOWSKA, A., KOSMAKOS, F. C. & BRODIE, A. F. (1974). Studies of respiratory components and oxidative phosphorylation in mitochondria of *mi-1 Neurospora crassa. Journal of Bacteriology* **117**, 733–40.

DUELL, E., INOUE, S. & UTTER, M. F. (1964). Isolation and properties of intact mitochondria from spheroplasts of yeast. *Journal of Bacteriology* **88**, 1762–73.

DUNKLE, L., VAN ETTEN, J. L. & BRAMBL, R. M. (1972). Mitochondrial DNA synthesis during fungal spore germination. *Archiv für Mikrobiologie* **85**, 224–32.

EAKIN, R. T. & MITCHELL, H. R. (1970). Alterations of the respiratory system of *Neurospora crassa* by the *mi-1* mutation. *Journal of Bacteriology* **104**, 74–8.

EDELMAN, M., VERMA, I. M., HERZOG, R., GALUN, E. & LITTAUER, U. Z. (1971). Physico-chemical properties of mitochondrial ribosomal RNA from fungi. *European Journal of Biochemistry* **19**, 372–8.

EDWARDS, D. K., KWIECINSKI, F. & HORSTMANN, J. (1973). Selection of respiratory mutants of *Neurospora crassa. Journal of Bacteriology* **114**, 164–8.

ELMER, G. W. & NICKERSON, W. J. (1970). Nutritional requirements for growth and yeast-like development of *Mucor rouxii* under carbon dioxide. *Journal of Bacteriology* **101**, 595–602.

ERNSTER, L. & KUYLENSTIERNA, B. (1970). Outer membrane of mitochondria. In *Membranes of mitochondria and chloroplasts,* pp. 172–212. Edited by E. Racker, New York: Van Nostrand-Reinhold.

FISHER, D. J., PRING, R. J., RICHMOND, D. V. & MICHAEL, A. (1973). The production of yeast-like forms of *Mucor* by antibiotics and fungicides. *Journal of General Microbiology* **75**, xiv.

FLORANCE, E. R., DENISON, W. C. & ALLEN, T. C. (1972). Ultrastructure of dormant and germinating conidia of *Aspergillus nidulans. Mycologia* **64**, 115–23.

GALBRAITH, J. C. & SMITH, J. E. (1969*a*). Sporulation of *Aspergillus niger* in submerged liquid culture. *Journal of General Microbiology* **59**, 31–45.

GALBRAITH, J. C. & SMITH, J. E. (1969*b*). Changes in the activity of the tricarboxylic acid cycle and glyoxylate cycle during the initiation of conidiation of *Aspergillus niger. Canadian Journal of Microbiology* **15**, 1207–12.

GILLIE, O. J. (1970). Methods for the study of nuclear and cytoplasmic variation in respiratory activity of *Neurospora crassa*, and the discovery of three new genes. *Journal of General Microbiology* **61**, 379–95.

GODDARD, D. R. (1935). The reversible heat activation inducing germination and increased respiration in the ascospores of *Neurospora tetrasperma. Journal of General Physiology* **19**, 45–69.

GODDARD, D. R. & SMITH, P. E. (1938). Respiratory block in the dormant spores of *Neurospora tetrasperma. Plant Physiology* **24**, 241–64.

GOLDRING, E. S., GROSSMAN, L. I., KRUPNICK, D., CRYER, D. R. & MARMUR, J. (1970). The petite mutation in yeast. Loss of mitochondrial deoxyribonucleic acid during induction of petites with ethidium bromide. *Journal of Molecular Biology* **52**, 323–35.

GONZALEZ-CADAVID, N. F. & CAMPBELL, P. N. (1967). The biosynthesis of cytochrome c. Sequence of incorporation *in vivo* of ^{14}C-lysine into cytochrome c and total protein of rat liver subcellular fractions. *Biochemical Journal.* **105**, 443–50.

GOTTLIEB, D. & CALTRIDER, P. G. (1963). Synthesis of enzymes during the germination of fungus spores. *Nature, London* **197**, 916–7.

GREENAWALT, J. W., HALL, D. O. & WALLIS, O. C. (1967). Preparation and properties of *Neurospora* mitochondria. In *Methods in enzymology*, Vol. 10, pp. 142–7. Edited by R. W. Estabrook and M. E. Pullman. London: Academic Press.

GRIFFITHS, D. E., AVNER, P. R., LANCASHIRE, W. E. & TURNER, J. R. (1972). In *International symposium on the biochemistry and biophysics of mitochondrial membranes* pp. 605–21. Edited by E. Carafoli, A. Lehninger, S. Papa and N. Siliprandi. London: Academic Press.

GUPTA, P. D. & MALHOTRA, S. K. (1971). Lanthanum as a tracer for infoldings of plasma membrane in *Neurospora. Cytobios* **4**, 21–7.

HACKETT, D. F. (1961). Respiratory mechanisms in higher plants. *Annual Review of Plant Physiology* **10**, 113–46.

HALL, D. O. & GREENAWALT, J. W. (1964). Oxidative phosphorylation by isolated mitochondria of *Neurospora crassa. Biochemical and Biophysical Research Communications* **17**, 565–9.

HALL, D. O. & GREENAWALT, J. W. (1967). The preparation and biochemical properties of mitochondria from *Neurospora crassa. Journal of General Microbiology* **48**, 419–30.

HASKINS, F. A., TISSIERES, A., MITCHELL, H. K. & MITCHELL, M. B. (1954). Cytochromes and the succinic acid oxidase system of *poky* strains of *Neurospora crassa. Journal of Biological Chemistry* **200**, 819–26.

HAWKER, L. E. (1965). Fine structure of fungi as revealed by electron microscopy. *Biology Reviews* **40**, 59–92.

HIGGINS, E. S. & FRIEND, W. H. (1968). Comparative mitochondrial function in two strains of *Aspergillus niger. Canadian Journal of Biochemistry* **46**, 1515–21.

HOLTON, R. W. (1960). Studies on pyruvate metabolism and cytochrome system in *Neurospora tetrasperma. Plant Physiology* **35**, 757–66.

HOWELL, N., ZUICHES, C. A. & MUNKRES, K. D. (1971). Mitochondrial biogenesis in *Neurospora crassa.* I. An ultrastructural and biochemical investigation of the effects of anaerobiosis and chloramphenicol inhibition. *Journal of Cell Biology* **50**, 721–36.

IWASA, K. (1960*a*). Phophorus metabolism of fungal cells. I. Oxidative phosphorylation by mitochondria from *Aspergillus oryzae. Journal of Biochemistry, Tokyo* **47**, 445–53.

IWASA, K. (1960*b*). Phosphorus metabolism of fungal cells. II. Some properties of the oxidative phosphorylation system of mitochondria from *Aspergillus oryzae. Journal of Biochemistry, Tokyo* **47**, 584–91.

JAYARAMAN, J., COTMAN, C. & MAHLER, H. R. (1966). Biochemical correlates of respiratory deficiency. VII. Glucose repression. *Archives of Biochemistry and Biophysics* **116**, 224–51.

JUNGALWALA, F. B. & DAWSON, R. M. (1970). Phospholipid biosynthesis and exchange in isolated liver cells. *Biochemical Journal* **117**, 481–90.

KADENBACH, B. (1966). Synthesis of mitochondrial proteins: demonstration of a transfer of proteins from microsomes into mitochondria. *Biochimica et Biophysica Acta* **134**, 430–42.

KADENBACH, B. (1969). A quantitative study of the biosynthesis of cytochrome c. *European Journal of Biochemistry* **10**, 312–7.

KATZ, R., KILPATRICK, Z. & CHANCE, B. (1971). Acquisition and loss of rotenone sensitivity in *Torulopsis utilis. European Journal of Biochemistry* **21**, 301–7.

KAWAKITA, M. (1970). Studies on the respiratory system of *Aspergillus oryzae. II.* Preparation and some properties of mitochondria from mycelium. *Journal of Biochemistry, Tokyo* **68**, 625–31.

KELLEMS, R. & BUTOW, R. A. (1972). Cytoplasmic-type 80 S ribosomes associated with yeast mitochondria. I. Evidence for ribosome binding sites on yeast mitochondria. *Journal of Biological Chemistry* **247**, 8053–60.

KESSEL, R. G. (1971). Cytodifferentiation in the *Rana pipiens* oöcyte. II. Intramitochondrial yolk. *Zeitschrift für Zellforschung* **112**, 313–32.

KOVAC, L., BEDNAROVA, H. & GREKSAK, M. (1968). Oxidative phosphorylation in yeast I. Isolation and properties of phosphorylating mitochondria from stationary phase cells. *Biochimica et Biophysica Acta* **153**, 32–42.

KUNTZEL, H. (1969). Proteins of mitochondrial and cytoplasmic ribosomes from *Neurospora crassa. Nature, London* **222**, 142–6.

KURIYAMA, Y. & LUCK, D. J. L. (1974). Methylation and processing of mitochondrial ribosomal RNAs in *poky* and wild-type *Neurospora crassa. Journal of Molecular Biology* **83**, 253–66.

KUZELA, S. & GRECNA, E. (1969). Lack of amino acid incorporation by isolated mitochondria from respiratory-deficient cytoplasmic yeast mutants. *Experientia* **25**, 776–7.

LAMBOWITZ, A. M. & SLAYMAN, C. W. (1971). Cyanide-resistant respiration in *Neurospora crassa*. *Journal of Bacteriology* **108**, 1087–96.

LAMBOWITZ, A. M., SMITH, E. W. & SLAYMAN, C. W. (1972*a*). Oxidative phosphorylation in *Neurospora crassa* mitochondria. Studies on wild-type, *poky* and chloramphenicol-induced wild-type. *Journal of Biological Chemistry* **247**, 4859–65.

LAMBOWITZ, A. M., SMITH, E. W. & SLAYMAN, C. W. (1972*b*). Electron transport in *Neurospora* mitochondria. Studies on wild-type and *poky*. *Journal of Biological Chemistry* **247**, 4850–8.

LAMBOWITZ, A. M., SLAYMAN, C. W., SLAYMAN, C. L. & BONNER, W. D. (1972). The electron transport components of wild-type and *poky* strains of *Neurospora crassa*. *Journal of Biological Chemistry* **247**, 1536–45.

LAWRENCE, R. C. & BAILEY, R. W. (1970). Evidence for the role of the citric acid cycle in the activation of spores of *Penicillium roqueforti*. *Biochimica et Biophysica Acta* **208**, 77–86.

LEHNINGER, A. L. (1951). Phosphorylation coupled to the oxidation of dihydrophosphopyridine nucleotide. *Journal of Biological Chemistry* **190**, 345–59.

LINDENMAYER, A. (1965). Carbohydrate metabolism. 3. Terminal oxidation and electron transport. In *The fungi*, Vol. 1, pp. 301–47. Edited by G. C. Ainsworth and A. S. Sussman, Academic Press: London.

LINNANE, A. W. & HASLAM, J. M. (1970). The biogenesis of yeast mitochondria. In *Current topics in cellular regulation*, Vol. 2, pp. 101–72. Edited by B. L. Horecker and E. L. Stadtman. London: Academic Press.

LINNANE, A. W., HASLAM, J. M., LUKENS, H. B. & NAGLEY, P. (1972). Biogenesis of mitochondria in micro-organisms. *Annual Review of Microbiology* **26**, 163–98.

LUCK, D. J. L. (1963). Formation of mitochondria in *Neurospora crassa*. A quantitative radioautographic study. *Journal of Cell Biology* **16**, 483–99.

LUCK, D. J. L. (1965). Formation of mitochondria in *Neurospora crassa*. A study based on mitochondrial density changes. *Journal of Cell Biology* **24**, 461–70.

LUCK, D. J. L. & REICH, E. (1964). DNA of mitochondria of *Neurospora crassa*. *Proceedings of the National Academy of Sciences, U.S.A.* **52**, 931–8.

MACKLER, B. & HAYNES, B. (1973). Studies of oxidative phosphorylation in *Saccharomyces cerevisiae* and *Saccharomyces carlsbergenesis*. *Biochimica et Biophysica Acta* **292**, 88–91.

MADELIN, M. F. Ed. (1966). *The fungus spore*. London: Butterworths.

MALHOTRA, S. K. (1966). Mesosome-like structures in mitochondria of *poky Neurospora*. *Nature, London* **219**, 1267–8.

MALHOTRA, S. K. (1970). Organization of the cellular membranes. In *Progress in biophysics and molecular biology*, Vol. 20, pp. 67–131. Edited by J. A. V. Butler and D. Noble. Oxford: Pergamon Press.

MANNINO, R. J. & GREENAWALT, J. W. (1972). Mitochondrial nucleic acid changes accompanying differentiation in *Neurospora crassa*. *Journal of Cell Biology* **55**, 164a.

MCMURRAY, C. & DAWSON, R. M. C. (1969). Phospholipid exchange reactions within the liver cell. *Biochemical Journal* **112**, 91–108.

MICHAELIS, G., DOUGLASS, S., TSAI, M.-J. & CRIDDLE, R. S. (1971). Mitochondrial DNA and suppressiveness of petite mutants in *Saccharomyces cerevisiae*. *Biochemical Genetics* **5**, 487–95.

MITCHELL, H. K. & MITCHELL, M. B. (1952). A case of maternal inheritance in *Neurospora crassa*. *Proceedings of the National Academy of Sciences, U.S.A.* **38**, 442–9.

MITCHELL, H. K., MITCHELL, M. B. & TISSIERES, A. (1953). Mendelian and non-mendelian factors affecting the cytochrome system of *Neurospora crassa*. *Proceedings of the National Academy of Sciences, U.S.A.* **39**, 606–13.

MORIMOTO, H. & HALVORSON, H. O. (1971). Characterization of mitochondrial ribosomes from yeast. *Proceedings of the National Academy of Sciences, U.S.A.* **68**, 324–8.

MORIMOTO, H., SCRAGG, A. H., NEKHOROCHEFF, J., VILLA, V. & HALVORSON, H. O. (1971). Comparison of the protein-synthesizing systems from mitochondria and cytoplasm of yeast. In *Autonomy and biogenesis of mitochondria and chloroplasts*, pp. 282–92. Edited by N. K. Boardman, A. W. Linnane and R. M. Smillie. Amsterdam: North-Holland Press.

MORRÉ, D. J., MERRITT, W. D. & LEMBI, C. A. (1971). Connections between mitochondria and endoplasmic reticulum in rat liver and onion stem. *Protoplasma* **73**, 43–9.

NAGLEY, P. & LINNANE, A. W. (1970). Mitochondrial DNA deficient mutants of yeast. *Biochemical and Biophysical Research Communications* **39**, 989–96.

NASS, M. M. K. & NASS, K. (1963). Intramitochondrial fibres with DNA characteristics, I. Fixation and electron staining reactions. *Journal of Cell Biology* **19**, 593–611.

NEUPERT, W. & LUDWIG, G. D. (1971).Sites of biosynthesis of outer and inner membrane proteins of *Neurospora crassa* mitochondria. *European Journal of Biochemistry* **19**, 523–32.

NEUPERT, W., BRDICZKA, D. & BUCHER, T. (1967). Incorporation of amino acids into the outer and inner membranes of isolated rat liver mitochondria. *Biochemical and Biophysical Research Communications* **27**, 488–93.

NEUPERT, W., MASSINGER, P. & PFALLER, A. (1971). Amino acid incorporation into mitochondrial ribosomes of *Neurospora crassa* wild-type and *mi-1* mutant. In *Autonomy and biogenesis of mitochondria and chloroplasts*. pp. 328–38. Edited by N. K. Boardman, A. W. Linnane and R. M. Smillie. Amsterdam: North-Holland Press.

NG, A. M. L., SMITH, J. E. & McINTOSH, A. F. (1973). Changes in activity of tricarboxylic acid cycle and glyoxylate cycle enzymes during synchronous development of *Aspergillus niger. Transactions of the British Mycological Society* **61**, 13–20.

NG, A. M. L., SMITH, J. E. & McINTOSH, A. F. (1974). Influence of dilution rate on enzyme synthesis in *Aspergillus niger* in continuous culture. *Journal of General Microbiology* **81**, 425–34.

NIEDERPRUEM, D. J. (1965). Carbohydrate metabolism. 2. Tricarboxylic acid cycle. In *The fungi*, Vol. 1, pp. 269–300. Edited by G. C. Ainsworth and A. S. Sussman, London: Academic Press.

NOWAKOWSKA-WASZCZUK, A. (1973). Utilization of some tricarboxylic acid cycle intermediates by mitochondria and growing mycelium of *Aspergillus terreus. Journal of General Microbiology* **79**, 19–29.

OHNISHI, T., KAWAGUCHI, K. & HAGIHARA, B. (1966*a*). Preparation and some properties of yeast mitochondria. *Journal of Biological Chemistry* **241**, 1797–806.

OHNISHI, T., SOTTOCASA, G. L. & ERNSTER, L. (1966*b*). Current approaches to the mechanism of energy-coupling in the respiratory chain. Studies on yeast mitochondria. *Bulletin Societe Chimique et Biologique, Paris* **48**, 1189–203.

OHNISHI, T., KROGER, A., HELDT, H. W., PFAFF, E. & KLINGENBERG, M. (1967). The response of the respiratory chain and adenine nucleotide system to oxidative phosphorylation in yeast mitochondria. *European Journal of Biochemistry* **1**, 301–11.

PARSONS, D. F., THOMPSON, W., WILSON, D. & CHANCE, B. (1967). Improvements in the procedure for purification of mitochondrial outer membrane. In *Mitochondrial structure and compartmentation*, pp. 29–73. Edited by E. Quagliariello, S. Papa, E. C. Slater and J. M. Tager. Bari: Adriatica Editrice.

PERLMAN, P. S. & MAHLER, H. R. (1970). Formation of yeast mitochondria. III. Biochemical properties of mitochondria isolated from a cytoplasmic petite mutant. *Journal of Bioenergetics* **1**, 113–38.

PITTENGER, T. H. (1956). Synergism of two cytoplasmically inherited mutants in *Neurospora crassa. Proceedings of the National Academy of Sciences, U.S.A.* **42**, 747–52.

PLATTNER, H. & SCHATZ, G. (1969). Promitochondria of anaerobically grown yeast. III. Morphology. *Biochemistry* **8**, 339–49.

POLAKIS, E. S., BARTLEY, W. & MEEK, G. A. (1964). Changes in the structure and enzyme activity of *Saccharomyces cerevisiae* in response to change in the environment. *Biochemical Journal* **90**, 369–74.

RABINOWITZ, M. & SWIFT, H. (1970). Mitochondrial nucleic acids and their relation to the biogenesis of mitochondria. *Physiological Reviews* **50**, 376–427.

RIFKIN, M. R. & LUCK, D. J. L. (1971). Defective production of mitochondrial ribosomes in the *poky* mutant of *Neurospora crassa. Proceedings of the National Academy of Sciences, U.S.A.* **68**, 287–90.

ROBERTSON, J. D. (1960). Interrelationships between mitochondria and the endoplasmic reticulum bearing on a theory of the origin of mitochondria. *Journal of Physiology* **153**, 58–9P.

ROBERTSON, J. D. (1964). Unit membranes. A review with recent new studies of experimental alterations and a new subunit structure in the synaptic membranes. In *Cellular membranes in development* pp. 1–81. Edited by M. Locke, London: Academic Press.

ROODYN, D. B. & WILKIE, D. (1968). *The biogenesis of mitochondria.* London: Methuen.

ROWE, M. J., LANSMAN, R. A. & WOODWARD, D. O. (1974). A comparison of mitochondrially synthesized proteins from whole mitochondria and cytochrome oxidase in *Neurospora. European Journal of Biochemistry* **41**, 25–30.

ROWLANDS, R. T. & TURNER, G. (1973). Nuclear and extranuclear inheritance of oligomycin-resistance in *Aspergillus nidulans. Journal of General Microbiology* **75**, xix.

RUBY, J. R., DYER, R. F. & SKALKO, R. G. (1969). Continuities between mitochondria and endoplasmic reticulum in the mammalian ovary. *Zeitschrift für Zellforschung* **97**, 30–7.

SANDBORN, E., KOEN, P. F., MCNABB, J. D., & MOORE, G. (1964). Cytoplasmic microtubules in mammalian cells. *Journal of Ultrastructural Research* **11**, 123–38.

SATO, N., WILSON, D. F. & CHANCE, B. (1971). The spectral properties of the b cytochromes in intact mitochondria. *Biochimica et Biophysica Acta* **253**, 88–97.

SCHATZ, G. (1968). Impaired binding of mitochondrial ATPase in the cytoplasmic petite mutant of *Saccharomyces cerevisiae. Journal of Biological Chemistry* **243**, 2192–9.

SCHATZ, G. & MASON, T. L. (1974). The biosynthesis of mitochondrial proteins. *Annual Review of Biochemistry*, **43**, 51–88.

SCHATZ, G. & SALTZGABER, J. (1969). Protein synthesis in yeast promitochondria. *Biochemical and Biophysical Research Communications* **37**, 996–1001.

SCHATZ, G., HASLBRUNNER, E. & TUPPY, H. (1964). Deoxyribonucleic acid associated with yeast mitochondria. *Biochimica et Biophysica Acta* **15**, 127–32.

SCHATZ, G., GROOT, G. S. P., MASON, T., ROUSLIN, W., WHARTON, D. C. & SALTZGABER, J. (1972). Biogenesis of mitochondrial inner membranes in baker's yeast. *Federation Proceedings* **31**, 21–9.

SCHMIDT, H. (1970). Characterization of a 72 S mitochondrial ribosome from *Saccharomyces cerevisiae. European Journal of Biochemistry* **17**, 278–83.

SCHNAITMAN, C. A. (1969). Comparison of rat-liver mitochondrial and microsomal membrane proteins. *Proceedings of the National Academy of Sciences, U.S.A.* **63**, 412–9.

SCHNAITMAN, C. A. & GREENAWALT, J. W. (1968). Enzymatic properties of the inner and outer membranes of rat-liver mitochondria. *Journal of Cell Biology* **38**, 158–75.

SEBALD, W., MACHLEIDT, W. & OTTO, J. (1973). Products of mitochondrial protein synthesis in *Neurospora crassa. European Journal of Biochemistry* **38**, 312–24.

SEBALD, W., BUCHER, T., OLBRICH, B. & KAUDWITZ, F. (1968). Electrophoretic pattern of amino acid incorporation *in vitro* into the insoluble mitochondrial protein of *Neurospora crassa* wild-type and *mi-1* mutant. *Federation of European Biochemical Societies Letters* **1**, 235–9.

SMITH, J. E., VALENZUELA-PEREZ, J. & NG, W. S. (1971). Changes in the activities of enzymes of the Embden–Meyerhof–Parnas and pentose phosphate pathways during the growth cycle of *Aspergillus niger. Transactions of the British Mycological Society* **57**, 93–101.

SMITH, J. E., GULL, K., ANDERSON, J. S. & DEANS, S. (1975). Organelle changes during spore germination. In *The Fungal Spore.* Edited by B. Hess and D. Weber. Wiley & Sons, New York (in press).

SLONIMSKI, P. P. (1953). *Formation des enzymes respiratoires chez la levure.* Paris: Masson.

SOTTOCASA, G. L., ERNSTER, L., KUYLENSTIERNA, B. & BERGSTRAND, A. (1967). Occurrence of an NADH-cytochrome c reductase system in the outer membrane of rat-liver mitochondria. In *Mitochondrial structure and compartmentation*, pp. 74–119. Edited by E. Quagliariello, S. Papa, E. C. Slater & J, M. Tager, Bari: Adriatica Editrice.

STORCK, R. & MORRILL, R. C. (1971). Respiratory-deficient yeast-like mutant of *Mucor. Biochemical Genetics* **5**, 467–79.

STUART, K. (1970). Cytoplasmic inheritance of oligomycin and rutamycin resistance in yeast. *Biochemical and Biophysical Research Communications* **39**, 1045–51.

SUSSMAN, A. S. (1966). Types of dormancy as represented by conidia and ascospores of *Neurospora.* In *The fungus spore*, pp. 235–56. Edited by M. F. Madelin. London: Butterworths.

SUSSMAN, A. S. (1969). The dormancy and germination of fungus spores. *Symposium Society of Experimental Biology* **23**, 99–121.

SUSSMAN, A. S. & HALVORSON, H. O. (1966). *Spores, their dormancy and germination.* New York: Harper & Row.

SUSSMAN, A. S. & MARKER, C. L. (1953). The development of tyrosinase and cytochrome oxidase activities in mutants of *Glommerella cingulata, Archives of Biochemistry and Biophysics* **45**, 31–40.

TANAKA, K. (1966). Change in the ultrastructure of *Aspergillus oryzae* conidia during germination. *Journal of General and Applied Microbiology* **12**, 239–46.

TERENZI, H. F. & STORCK, R. (1969). Stimulation of fermentation and yeast-like morphology in *Mucor rouxii* by phenethyl alcohol. *Journal of Bacteriology* **97**, 1248–61.

TISSIERES. A. & MITCHELL, H. K. (1954). Cytochromes and respiratory activity in some slow growing strains of *Neurospora*. *Journal of Biological Chemistry* **208**, 241–9.

TISSIERES, A., MITCHELL, H. K. & HASKINS, F. A. (1953). Studies on the respiratory system of the *poky* strain *Neurospora*. *Journal of Biological Chemistry* **205**, 423–33.

TSUKAHARA, T., YAMAHA, M. & ITAGATI, T. (1966). Micromorphology of conidiospores of *Aspergillus niger* by electron microscopy. *Japanese Journal of Microbiology* **10**, 93–106.

TURIAN, G. & BIANCHI, D. E. (1972). Conidiation in *Neurospora*. *Botanical Reviews* **38**, 119–54.

TURNER, G. (1973). Cycloheximide-resistant synthesis of mitochondrial membrane components in *Aspergillus nidulans*. *European Journal of Biochemistry* **40**, 201–6.

TZAGOLOFF, A. RUBIN, M. S. & SIERRA, M. F. (1973). Biosynthesis of mitochondrial enzymes. *Biochimica et Biophysica Acta* **301**, 71–104.

VALENZUELA-PEREZ, J. & SMITH, J. E. (1971). Role of glycolysis in sporulation of *Aspergillus niger* in submerged culture. *Transactions of the British Mycological Society* **57**, 111–19.

VILLA, V. D. & STORCK, R. (1968). Nucleotide composition of nuclear and mitochondrial deoxyribonucleic acid in fungi. *Journal of Bacteriology* **96**, 184–90.

VOGEL, F. S. & WEAVER, R. F. (1972). Concerning the induction of dormancy in the spores of *Agaricus bisporus*. *Experimental Cell Research* **75**, 95–104.

VON JAGOW, G. & KLINGENBERG, M. (1970). Pathways of hydrogen in mitochondria of *Saccharomyces carlsbergensis*. *European Journal of Biochemistry* **12**, 583–92.

VON JAGOW, G. & KLINGENBERG, M. (1972). Close correlation between antimycin titered cytochrome b_T content in mitochondria of chloramphenicol treated *Neurospora crassa*. *Federation of European Biochemical Societies Letters* **24**, 278–82.

VON JAGOW, G., WEISS, H. & KLINGENBERG, M. (1973). Comparison of the respiratory chain of *Neurospora crassa* wild-type and the *mi*-mutants *mi-1* and *mi-3*. *European Journal of Biochemistry* **33**, 140–57.

WAKABAYASHI, K. & GUNGE, N. (1970). Extrachromosomal inheritance of oligomycin resistance in yeast. *Federation of European Societies Letters* **6**, 302–4.

WAKIYAMA, S. & OGURA, Y. (1972). Studies on the respiratory system of *Aspergillus oryzae*. IV. Low temperature difference spectra of mitochondria from mycelia grown in the presence and absence of chloramphenicol. *Journal of Biochemistry, Tokyo*, 295–300.

WALDRON, C. & ROBERTS, C. F. (1973). Cytoplasmic inheritance of a cold-sensitive mutant in *Aspergillus nidulans*. *Journal of General Microbiology* **78**, 379–81.

WALLACE, P., HUANG, M. & LINNANE, A. W. (1968). The biogenesis of mitochondria. II. The influence of medium composition on the cytology of anaerobically grown *Saccharomyces cerevisiae*. *Journal of Cell Biology* **37**, 207–20.

WATSON, K. (1968). PhD Thesis. University of Strathclyde, Glasgow.

WATSON, K. (1972). The organization of ribosomal granules within mitochondrial structures of aerobic and anaerobic cells of *Saccharomyces cerevisiae*. *Journal of Cell Biology* **55**, 721–6.

WATSON, K. & LINNANE, A. W. (1972). Headpiece-stalk particles lining membranes of mitochondria isolated from normal and oligomycin-resistant mutants of *Saccharomyces cerevisiae*. *Journal of Bioenergetics* **3**, 235–43.

WATSON, K. & SMITH, J. E. (1967*a*). Oxidative phosphorylation and respiratory control in mitochondria from *Aspergillus niger*. *Biochemical Journal* **104**, 332–9.

WATSON, K. & SMITH, J. E. (1967*b*). Rotenone and amytal insensitive coupled oxidation of NADH by mitochondria from *Aspergillus niger*. *Journal of Biochemistry, Tokyo* **61**, 527–30.

WATSON, K. & SMITH, J. E. (1968). Respiratory components of *Aspergillus niger* mitochondria. *Journal of Bacteriology* **96**, 1546–50.

WATSON, K., HASLAM, J. M. & LINNANE, A. W. (1970). Biogenesis of mitochondria. XIII. The isolation of mitochondrial structures from anaerobically grown *Saccharomyces cerevisiae*. *Journal of Cell Biology* **46**, 88–96.

WATSON, K., PATON, W. P. & SMITH, J. E. (1969). Oxidative phosphorylation and respiratory control in mitochondria from *Aspergillus oryzae*. *Canadian Journal of Microbiology* **15**, 975–81.

WATSON, K., HASLAM, J. M., VEITCH, B. & LINNANE, A. W. (1971). Mitochondrial precursors in anaerobically grown yeast. In *Autonomy and biogenesis of mitochondria and chloroplasts* pp. 162–74. Edited by N. K. Boardman, A. W. Linnane and R. M. Smillie, Amsterdam: North-Holland Press.

WEAVER, R. F., RAJAGOPALAN, K. V., HANDLER, P., JEFFS, P., BYRNE, W. L. & ROSENTHAL, D. (1970). Isolation of γ-L-glutaminyl-4-hydroxybenzene and γ-L-glutaminyl-3,4-benzoquinone: a natural sulfhydryl reagent from sporulating gill tissues of the mushroom, *Agaricus bisporus*. *Proceedings of the National Academy of Sciences, U.S.A.* **67**, 1050-6.

WEISS, B. (1965). An electron microscope and biochemical study of *Neurospora crassa* during development. *Journal of General Microbiology* **39**, 85–94.

WEISS, B. & TURIAN, G. (1966). A study of conidiation in *Neurospora crassa*. *Journal of General Microbiology* **44**, 407–18.

WEISS, H. (1972). Cytochrome b in *Neurospora crassa* mitochondria. A membrane protein containing subunits of cytoplasmic and mitochondrial origin. *European Journal of Biochemistry* **30**, 469–78.

WEISS, H., VON JAGOW, G., KLINGENBERG, M. & BUCHER, T. (1970). Characterization of *Neurospora crassa* mitochondria prepared with a grind-mill. *European Journal of Biochemistry* **14**, 75–82.

WHITE, G. A. & LEDINGHAM, G. A. (1961). Studies on the cytochrome oxidase and oxidation pathway in uredospores of wheat stem rust. *Canadian Journal of Botany* **39**, 1131–48.

WIKSTROM, M. K. F. (1972). The cytochrome b complex and energy conservation: a possible relation between proton transfer and oxidative phosphorylation. In *Biochemistry and biophysics of mitochondrial membranes*, pp. 147–64. Edited by G. F. Azzone, E. Carafoli, A. Lehninger, E. Quagliariello and N. Siliprandi, London: Academic Press.

WILKIE, D. & MAROUDAS, N. G. (1969). Induction of cytoplasmic respiratory-deficiency in yeast by phenethyl alcohol. *Genetics Research* **13**, 107–16.

WINTERSBERGER, E. (1967). A distinct class of ribosomal RNA components in yeast mitochondria as revealed by gradient centrifugation and by DNA-RNA hybridization. *Zeitschrift für Physiologies Chemie* **348**, 1701–4.

WIRTZ, K. W. A. & ZILVERSMIT, D. B. (1969). Participation of soluble liver proteins in the exchange of membrane phospholipids. *Biochimica et Biophysica Acta* **193**, 105–16.

WISKICH, J. T. & BONNER, W. D. (1963). Preparation and properties of sweet potato mitochondria. *Plant Physiology* **38**, 594–604.

YOTSUYANAGI, Y. (1962). Études sur le chondriome de la levure. II. Chondriosomes des mutants a deficience respiratoire. *Journal of Ultrastructural Research* **7**, 141–58.

ZALOKAR, M. (1959). Enzyme activity and cell differentiation in *Neurospora*. *American Journal of Botany* **46**, 555–69.

CHAPTER 5

Anaplerotic Pathways

P. J. CASSELTON

5.1 Introduction

The concept of anaplerotic pathways was developed by Kornberg (1966) to draw attention to enzymatic reactions which compensate for the removal of intermediates from the central metabolic routes during biosynthesis, thus ensuring that the provision of energy is not interrupted. Such reactions occur in fungi mainly because several tricarboxylic acid (TCA) cycle acids are precursors of biosynthetic pathways (Krebs, Gurin & Eggleston, 1952) even though the cycle as a whole is the major route of terminal respiration (Kornberg, 1959; Niederpruem, 1965). Carbon dioxide fixation yielding oxalacetate can act as an anaplerotic pathway to the TCA cycle when metabolism of the available substrate readily yields pyruvate (Fig. 5.1). However, carbon dioxide fixation need not always occur (Hartman, Keen & Long, 1972). If present, the glyoxylate cycle (Kornberg & Krebs, 1957) will provide for the net synthesis of C4 dicarboxylic acids from acetyl CoA (Fig. 5.1). The simultaneous operation of the glyoxylate and TCA cycles can be detected radiochemically. Although succinate is a precursor of malate in the TCA cycle, malate initially labels more rapidly when both cycles are active in a micro-organism to which radioactive acetate is supplied (Kornberg & Elsden, 1961). For example, Collins & Kornberg (1960) obtained this result with mycelium of *Aspergillus niger* which had been grown in a medium containing acetate as carbon source. The present review considers some aspects of anaplerotic metabolism that have been studied in filamentous fungi.

5.2 Enzymes of Carbon Dioxide Fixation

Whilst a number of enzymes could explain carbon dioxide fixation by heterotrophic organisms, there is considerable agreement that pyruvate and phosphoenolpyruvate carboxylases are involved in the anaplerotic synthesis of TCA cycle acids and that phosphoenolpyruvate carboxykinase is normally a decarboxylase synthesizing phosphoenolpyruvate (Scrutton, 1971; Utter & Kolenbrander, 1972; Scrutton & Young, 1972). Accordingly, mutants of *Neurospora* and *Aspergillus* which lack pyruvate carboxylase require a supplement of a C4 acid, such as succinate, before growth can occur with sucrose as carbon source (Flavell & Woodward, 1971*a*; Skinner & Armitt, 1972; Beever, 1973). Mutants of *Neurospora* which are phosphoenolpyruvate carboxykinase deficient are unable to utilize acetate, whilst normal in their growth with sucrose (Flavell & Fincham, 1968*a*, *b*). Both pyruvate carboxylase and phosphoenolpyruvate carboxykinase have been detected in extracts from several filamentous fungi (Table 5.1). Phosphoenolpyruvate carboxylase has not been reported.

Although pyruvate carboxylases obtained from filamentous fungi have been studied, none has received the attention that has been given to the enzyme from *Saccharomyces*. All appear to contain biotin, at least as evidenced by avidin inhibition, and to require a monovalent ion, usually K^+, as well as a divalent ion, usually Mn^{2+} or Mg^{2+}. The yeast enzyme also

Table 5.1 Summary of some reported occurrences of anaplerotic and related enzymes in the mycelia or spores of filamentous fungi

	Pyruvate Carboxylase (E.C. 6.4.1.1.)[a] Pyruvate + ATP + HCO_3^- $\rightleftharpoons$ oxalacetate + ADP + P_i	**Phosphoenolpyruvate carboxykinase (cf. E.C. 4.1.1.32)**[b] Phosphoenolpyruvate + CO_2 + ADP $\rightleftharpoons$ oxalacetate + ATP	**Malate Enzyme (E.C. 1.1.1.40)**[c] L-malate + NAD(NADP) $\rightleftharpoons$ pyruvate + CO_2 + $NADH_2$
Neurospora crassa	Beever (1973)	Flavell & Fincham (1968*b*) Beever & Fincham (1973)	Zink (1972)
Aspergillus niger	Woronick & Johnson (1960) Bloom & Johnson (1962) Feir & Suzuki (1969)	Woronick & Johnson (1960)	
Aspergillus nidulans	Skinner & Armitt (1972)		
Penicillium spp.	Schormüller & Stan (1970)		Caltrider & Gottlieb (1963)
Verticillium albo-atrum	Hartman & Keen (1973)	Hartman & Keen (1973)	
Puccinia graminis			Caltrider & Gottlieb (1963)
Uromyces phaseoli			Caltrider & Gottlieb (1963) Rick & Mirocha (1968)
Agaricus bisporus	Bachofen & Rast (1968)	Bachofen & Rast (1968)	Bachofen & Rast (1968)
Coprinus lagopus sensu Buller (= *C. cinereus*)		Casselton (unpublished results)	
Shizophyllum commune	Tachibana, Siode & Hanai (1967)		
Rhizopus nigricans	Overman & Romano (1969)		

[a] Pyruvate carboxylase also reported from *Endothia parasitica* (McDowell, personal communication *Neocosmospora vasinfecta* (Budd, 1971) and *Helminthosporium cynodontis* (Clarke & Hartman, 1973).

[b] Phosphoenolpyruvate carboxylase also reported from *Helminthosporium cynodontis* (Clarke & Hartman, 1972).

[c] Malate enzyme also reported from *Ustilago maydis* (Caltrider & Gottlieb, 1963; Rick & Mirocha 1968).

contains bound Zn^{2+} (Scrutton, Young & Utter, 1970). The kinetics of the enzyme from *Penicillium camembertii* suggested that activity would depend on the energy charge of the adenylate pool (Stan, 1972). MgATP and MgADP competed for a binding site on the enzyme, as predicted by Atkinson (1969). Acetyl-CoA allosterically activated this enzyme (Schormüller & Stan, 1970) as well as that from *Saccharomyces* (Cazzulo & Stoppani, 1968; Cooper & Benedict, 1968). Nevertheless, the enzymes from *Aspergillus niger* (Feir & Suzuki, 1969) and *Neurospora crassa* (Beever, 1973) were unaffected by acetyl-CoA while that from *Rhizopus nigricans* (Overman & Romano, 1969) resembled the mammalian enzyme in having an absolute requirement for it. Acetyl-CoA did not activate the enzyme from *Verticillium albo-atrum* but did reverse the inhibitory effect of L-aspartate (Hartman & Keen, 1974*a*). L-Aspartate inhibition is characteristic of pyruvate carboxylases being non-competitive with substrate for the *Aspergillus* enzyme (Scrutton & Young, 1972). Although further study is required, it is not unreasonable to postulate a regulatory advantage for

	Isocitrate Lyase (E.C. 4.1.3.1)[d] Isocitrate ⇌ glyoxylate + succinate	**Malate Synthase (E.C. 4.1.3.2)**[e] acetyl-CoA + glyoxylate + H_2O → L-malate + CoA	**Acetyl-CoA Synthase (E.C. 6.2.1.1)** Acetate + ATP + CoA → acetyl-CoA + AMP + PP_i
Neurospora crassa	Turian (1961) Flavell & Fincham (1968*b*)	Turian (1961) Flavell & Fincham (1968*b*)	Flavell & Fincham (1968*b*)
Aspergillus niger	Olson (1954) Collins & Kornberg (1960) Galbraith & Smith (1969)	Collins & Kornberg (1960) Galbraith & Smith (1969)	
Aspergillus nidulans	Armitt, Roberts & Kornberg (1970)	Armitt, Roberts & Kornberg (1971)	
Penicillium spp.	Olson (1954) Gottlieb & Ramachandran (1960) Ohmori & Gottlieb (1965)	Gottlieb & Ramachandran (1960)	
Verticillium albo-atrum	Tokunaga *et al.* (1969)	Tokunaga *et al.* (1969)	
Puccinia graminis	Caltrider, Ramachandran & Gottlieb (1963)	Caltrider, Ramachandran & Gottlieb (1963)	
Uromyces phaseoli	Caltrider, Ramachandran & Gottlieb (1963)	Caltrider, Ramachandran & Gottlieb (1963)	
Agaricus bisporus			
Coprinus lagopus sensu Buller (= *C. cinereus*)	O'Sullivan & Casselton (1973)	O'Sullivan & Casselton (1973)	Todd & Casselton (1973)
Shizophyllum commune	Cotter *et al.* (1970)	Cotter *et al.* (1970)	
Rhizopus nigricans	Olson (1954) Wegener & Romano (1964)	Wegener, Schell & Romano (1967)	

[d] Isocitrate lyase also reported from *Sclerotium rolfsii* (Maxell & Bateman, 1968), *Sclerotinia sclerotiorium* (Corsini & Le Tourneau, 1973), *Fusarium solani* (Cochrane, 1966), *Microsporum gypseum* (Leighton, Stock & Kelln, 1970). *Trichoderma viride* (Ohmori & Gottlieb, 1965), *Melampsora lini* (Frear & Johnson, 1961), *Allomyces* (Turian, 1961), *Blastocladiella emersonii* (McCurdy & Cantino, 1960) and *Pythium* (Sietsma, 1971).

[e] Malate synthase also reported from *Fusarium solani* (Cochrane, 1966), *Microsporum gypseum* (Leighton, Stock & Kelln, 1970) and *Melampsora lini* (Frear & Johnson, 1961).

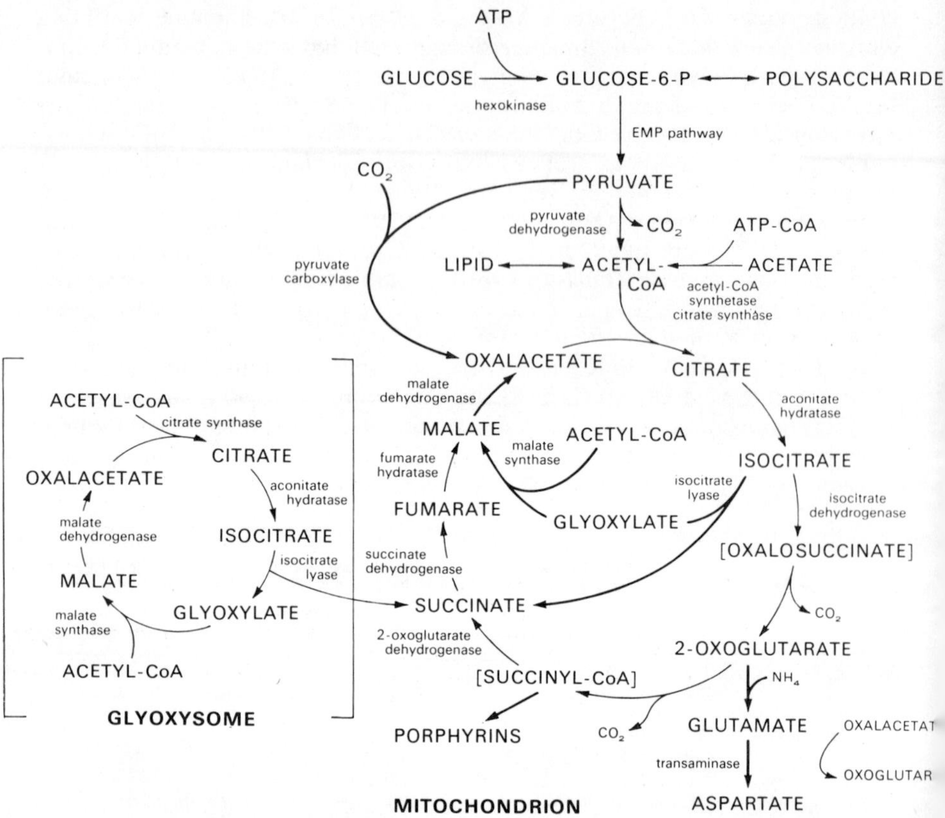

Fig. 5.1 Diagram to illustrate how carbon dioxide fixation or the glyoxylate cycle can compensate for the removal of TCA cycle intermediates during amino acid and porphyrin biosynthesis. In organisms containing glyoxysomes, succinate is passed to the mitochondrion.

enzyme activation when acetyl-CoA is accumulating and for inhibition by a substance metabolically close to oxalacetate (Fig. 5.1). Indeed Buston, Moss & Tyrrell (1966) observed that carbon dioxide fixation by *Chaetomium globosum* mycelium was inhibited by L-aspartate. Other observations include inhibition of the pyruvate carboxylase from *Verticillium* by both $NADP^+$ and diphosphoglycerate (Hartman & Keen, 1974*a*) and of the yeast enzyme by NADH (Cazzulo & Stoppani, 1969). Variation in the regulatory properties of pyruvate carboxylases, in contrast to the basic similarity of their catalytic properties, is possibly an example of enzyme evolution (Scrutton & Young, 1972). The specific activity of pyruvate carboxylase in extracts of *Neocosmospora vasinfecta*, but not of *Verticillium* (Hartman & Keen, 1974*a*), was two to three times greater when the fungus was cultured with glucose rather than acetate (Budd, 1971).

What little is known about the regulation of the phosphoenolpyruvate carboxykinases of filamentous fungi indicates a different role from that of the pyruvate carboxylases. The former enzyme increased in specific activity during growth of *Verticillium* with acetate or malate as carbon source. *In vitro*, it was inhibited by $NADP^+$, whereas acetyl-CoA and L-aspartate had no effect (Hartman & Keen, 1974*b*). Phosphoenolpyruvate carboxykinase activity increased after transfer of *Neurospora* mycelium from sucrose medium into acetate medium (Flavell & Fincham, 1968*b*). Beever (personal communication) believes this was due to removal of repression by a glycolytic intermediate. Although a certain amount is known about the malate enzymes of both *N. crassa* (Zink, 1972) and *Fusarium oxysporum* (Zink & Katz, 1973), it is unlikely that these enzymes are involved in anaplerotic carbon dioxide fixation. Malate utilization and NADPH generation are more likely functions (Scrutton, 1971). Nevertheless, Rick & Mirocha (1968) believed that carbon dioxide fixation by uredospores was due to a malate enzyme. This related to the observation that dark carbon dioxide fixation by rust infected bean plants was greater than that of uninfected controls (Mirocha & Rick, 1967). Other data suggested oxalacetate was the first product of fixation in uredospores (Gottlieb, 1966).

5.3 Enzymes of the Glyoxylate Cycle

Isocitrate lyase was known to occur in certain filamentous fungi even before its involvement in the glyoxylate cycle was suggested by Kornberg and Krebs in 1957 (Olson, 1954). Subsequently the detection of both isocitrate lyase and malate synthase in cell-free extracts of fungi representing all the major taxonomic groups has been reported (Table 5.1). In the case of *Aspergillus niger*, it was confirmed that the observed activities were adequate to account for the rate of growth of the fungus in acetate medium (Collins & Kornberg, 1960). Acetyl-CoA synthetase, also required when acetate is metabolized, has received less attention. The enzyme assay procedures have recently been reviewed by Reeves, Rabin, Wegener & Ajl (1971).

Regulation

Mutants unable to utilize acetate as carbon source have been obtained in *Neurospora crassa* (Flavell & Fincham, 1968*a*, *b*), *Aspergillus nidulans* (Armitt, Roberts & Kornberg, 1970; 1971; Skinner & Armitt, 1972) and *Coprinus cinereus* (Todd & Casselton, 1973). Some of these mutations have been associated with deficiencies of particular enzymes. However, reversion of a single gene mutation to show partial restoration of growth on acetate coupled with the occurrence of an enzyme showing abnormal properties, as compared to those of wild type, would best indicate a structural gene coding for the enzyme (Ashworth & Kornberg, 1964). Such observations have been made for both the isocitrate lyase (Leckie & Fincham, 1971) and the phosphoenolpyruvate carboxykinase (Beever & Fincham, 1973) of *Neurospora*. Some of the *Aspergillus* mutants, lacking either isocitrate lyase or malate synthase, also reverted to being capable of growth with acetate and were then found to possess both enzyme activities. In contrast to the situation in bacteria, close genetic linkage between the structural genes coding for isocitrate lyase and malate synthase was not observed (Armitt, Roberts & Kornberg, 1970, 1971).

Considerable differences in the specific activity of the glyoxylate cycle enzymes can be obtained by transferring a vegetative mycelium from a carbohydrate medium into one containing acetate. For example, twenty to forty-fold increases of activity were observed in two hours using *Neurospora* (Flavell & Woodward, 1970*b*) and comparable changes can be obtained in *Coprinus*, allbeit over longer time intervals (O'Sullivan & Casselton, 1973). However, isocitrate lyase occurs in ungerminated spores of *Verticillium albo-atrum* with only traces of malate synthase (Tokunaga *et al.*, 1969) and so-called 'constitutive' glyoxylate cycles have been described from several spores (Gottlieb, 1966). Germination in acetate medium usually causes some alteration of enzyme activity in such cases. For example, three-fold increases of isocitrate lyase activity were recorded in *Fusarium oxysporum* (Cochrane, 1966) and *Microsporum gypseum* (Leighton, Stock & Kelln, 1970), utilization of acetate being strongly pH dependent (Jarvis & Johnson, 1947).

Cycloheximide inhibited all of the enzymatic changes normally associated with transfer of *Neurospora* to acetate medium (Flavell & Woodward, 1970*b*). Nevertheless, it is difficult to distinguish between induction and derepression of the glyoxylate cycle (Metzenberg, 1972). Kobr, Turian & Zimmerman (1965) reported that isocitrate lyase activity was increased in *Neurospora* grown for four days in sucrose medium containing acetate. This might indicate induction by acetate. Conversely, the synthesis of glyoxylate cycle enzymes in mycelium transferred from sucrose to acetate medium was prevented by the addition of sucrose (Flavell & Woodward, 1971*a*). Observations on mutants of *Neurospora* (Flavell & Fincham, 1968*b*; Flavell & Woodward, 1971*a*) and *Coprinus* (Casselton & Casselton, unpublished results) also suggest that acetate is not an inducer. In which case, a conventional interpretation would be that metabolism of acetate leads to removal of the repressor, or repressors, of the genes controlling the synthesis of the enzymes. The occurrence of quite high levels of glyoxylate cycle enzymes in some spores favours this derepression hypothesis. Armitt, Roberts & Kornberg (1971) suggested that malate synthase was co-ordinately regulated with isocitrate lyase in *Aspergillus nidulans*. It had previously been established that the increase in activity of the latter preceded the resumption of growth by mycelium transferred from sucrose medium into acetate medium (Armitt, Roberts & Kornberg, 1970). However, even though it does not appear to be essential for the conidiation of *Neurospora* (Flavell & Fincham 1968*a*, *b*), elevated levels of isocitrate lyase activity during asexual reproduction of both *Neurospora* and *Aspergillus* have been reported, apparently often in the virtual absence of malate synthase activity (Smith & Galbraith, 1971). Further, in an extensive study of regulation in *Neurospora* by Flavell & Woodward (1970*a*, *b*; 1971*a*, *b*), co-ordinate control was only observed in some transfer experiments. Co-ordinate control of neither the TCA cycle enzymes nor the glyoxylate cycle enzymes was invariably observed in all situations. It was concluded that the former, which were repressed during growth with sucrose as compared to acetate, were regulated largely by the energy charge and only to a lesser extent by specific metabolic regulators (Flavell & Woodward, 1970*a*).

Glutamate suppressed the synthesis of isocitrate lyase in acetate medium. Isocitrate lyase was also partially repressed or inactivated in *Neurospora* grown in acetate medium as compared to mycelium transferred to acetate from sucrose. Both glyoxylate cycle enzymes were elevated, to a level intermediate between those normally observed in sucrose and acetate media respectively, in wild-type mycelium grown with glutamate or casamino acids as the sole source of carbon and in mutants lacking pyruvate carboxylase activity grown in sucrose medium with a limiting succinate supplement. Similar elevated levels of the enzymes were found in amino acid requiring auxotrophs transferred to sucrose medium after an initial period of growth in sucrose medium containing appropriate supplementation. The TCA cycle was repressed under the latter conditions. It was concluded that synthesis of the glyoxylate cycle enzymes was regulated by a glycolytic intermediate or its derivative (Flavell & Woodward, 1971*a*). Other interpretations are possible, particularly in the absence of reliable data as to the intracellular concentrations of the metabolites being postulated as regulators. For example, a C4 acid may repress isocitrate lyase in *Neurospora* (Beever, personal communication). Since the activities of isocitrate lyase and malate synthase can vary independently, these enzymes are presumably regulated by different regulators, even if these regulators are often in a constant ratio to one another (Flavell & Woodward, 1970*b*). Similar problems have arisen in relation to the regulation of the glutamate dehydrogenases of these fungi (Casselton, 1969).

The repression of isocitrate lyase in carbohydrate containing media has been established for two Basidiomycetes (Casselton, Fawole & Casselton, 1969; Cotter, La Clave, Wegener & Niederpruem, 1970), it being assumed that carbohydrate is utilized preferentially to acetate. However, addition of fluoroacetate did slightly derepress acetyl-CoA synthetase in *Coprinus* grown in the presence of low glucose concentrations (Casselton & Casselton, unpublished results). Romano & Kornberg (1968, 1969) found that acetate actually inhibited glucose uptake by the Ascomycete *Aspergillus nidulans*, though not of a mutant lacking acetyl-CoA synthetase. This effect was reflected in the activities of isocitrate lyase and malate synthase observed in mycelia grown in media containing both acetate and glucose (Armitt, Roberts & Kornberg, 1970; 1971).

A comprehensive study has been made of one Phycomycete. Both isocitrate lyase (Wegener & Romano, 1964) and malate synthase (Wegener, Schell & Romano, 1967) were present in *Rhizopus nigricans* grown in medium containing casein hydrolysate, but enzyme synthesis lagged behind growth showing neither enzyme was essential. Both enzymes were repressed by addition of glucose unless Zn^{2+} was also added. Transfer to acetate medium led to increased levels of both enzymes, whether the mycelium was previously glucose repressed or not. The apparent inducing effect of acetate was increased by Zn^{2+}. The results are probably best interpreted as enzyme control by repression/depression with any factor that stimulates growth promoting relief, as for example with an increase in available nitrogen. There is evidence that Zn^{2+} stimulates RNA synthesis (Wegener & Romano, 1963). Zn^{2+} was without effect on the synthesis of malate synthase in the presence of glycollate, which exceeded that caused by acetate alone or in the presence of zinc. The malate synthase produced in glycollate medium was less sensitive to thermal inactivation than that produced in acetate medium. Thus, glycollate may be a true inducer of a different enzyme (Wegener, Reeves, Rabin & Ajl, 1968). Two malate synthases can be present in

Escherichia coli (Vanderwinkel, Liard, Ramos & Wiame, 1963) but not apparently *Aspergillus nidulans* (Armitt, Roberts & Kornberg, 1971). The metabolism of glycollate by bacteria is well understood (Kornberg, 1966) but does not appear to have been further investigated in *Rhizopus. Zygorrhynchus moelleri* can grow on glycollate (Casselton, unpublished results). The activity of isocitrate lyase and of aconitate hydratase decreased when *Pythium* was grown in the presence of cholesterol, whereas most enzymes of the TCA cycle were increased (Sietsma, 1971).

The control of the activity of isocitrate lyase has been studied. Phosphoenolpyruvate, considered to be the regulator of isocitrate lyase in *Escherichia coli* (Ashworth & Kornberg, 1963), is a non-competitive inhibitor of the enzyme from *Neurospora crassa* (Sjogren & Romano, 1967) and *Rhizopus* (Romano, Bright & Scott, 1967). Isocitrate lyase has also been purified from *Neurospora* by Turian & Kobr (1965), Leckie & Fincham (1971) and Johanson, Hill & McFadden (1972). Johanson, Hill and McFadden (personal communication) also found that activity of the enzyme *in vitro* was inhibited in excess of 90% by succinate, glyoxylate and phosphoenolpyruvate. Significant inhibition also occurred with fumarate, malate and fructose 1,6-diphosphate. There was apparently no effect of energy charge. The idea that *N. crassa* produces two isoenzymes of isocitrate lyase (Sjorgen & Romano, 1967; Flavell & Woodward, 1971*a*) may have arisen from an experimental artefact (Rougemont & Kobr, 1973).

It is clear that operation of the glyoxylate cycle depends on enzymes in addition to isocitrate lyase and malate synthase. The identity of these is related to the question of the intracellular localization of the cycle.

Intracellular localization

The glyoxylate cycle operates in bacteria as a bypass mechanism for two of the oxidative reactions of the TCA cycle (Fig. 5.1; Kornberg & Madsen, 1957). However, the glyoxylate cycle of some angiosperm seedlings, which show a substantial conversion of lipid into carbohydrate during germination, was located in a microbody, the glyoxysome (Breidenbach, Kahn & Beevers, 1968), this separation, perhaps, being the basis of a regulatory system (Beevers, 1969). These glyoxysomes resemble peroxisomes in containing catalase. However, glyoxysomes can convert fatty acids into succinate. The transfer of this succinate to the mitochondria is sufficient to support all the biosynthetic activities requiring TCA cycle acids, including the initiation of gluconeogenesis by the decarboxylation of oxalacetate (Tolbert, 1971). A comparable biochemical situation may be uredospore germination. Although it has been questioned whether a substantial conversion of lipid into carbohydrate occurs (Daly, Knoche & Wiese, 1967), these spores undoubtedly utilize lipid reserves, contain the enzymes of the glyoxylate cycle and readily utilize exogenous acetate (Shu, Tanner & Ledingham, 1954; Frear & Johnson, 1961; Caltrider, Ramachandran & Gottlieb, 1963; Williams & Ledingham, 1964; Staples & Wynn, 1965). The addition of non-labelled glyoxylate reduced the radioactivity of malate synthesized from labelled acetate, as expected if malate synthase was active (Staples, 1962). Microbodies resembling the glyoxysomes of angiosperms have been described in sections of both haustoria and sporelings of rusts examined by electron microscopy (Coffey, Palevitz & Allen, 1972; Mendgen 1973*a*). These disappeared from *Uromyces phaseoli* during formation of the first haustorium, in parallel with the number of lipid bodies (Mendgen, 1973*a*). Catalase could not be demonstrated histochemically. Rust microbodies have not been isolated for *in vitro* examination. Microbodies have also been observed in sections of *Sclerotinia sclerotiorum* (Maxwell, Williams & Maxwell, 1970), *Fusarium oxysporum* (Wergin, 1972) and *Coprinus cinereus* (McLaughlin, 1973).

The current position regarding the extraction of a particulate glyoxylate cycle from fungi is unsatisfactory. This is hardly surprising as even the preparation of fully active mitochondria from these organisms presents problems (Khouw & McCurdy, 1969; Weiss, von Jagow, Klingenberg & Bücher, 1970). The sedimentation of glyoxylate cycle enzymes from cell-free extracts of *Neurospora* (Combépine, 1969; Kobr, Vanderhaeghe & Combépine, 1969) and *Coprinus cinereus* (Casselton, Fawole & Casselton, 1969) has been reported. Cotter *et al.* (1970) found only soluble activities in *Schizophyllum commune.* Similar discrepancies have occurred with yeasts (Avers, 1971). The proportion of bound isocitrate lyase in *N. crassa* extracts varied with the carbon source used to grow the mycelium. A twenty-fold increase in soluble activity resulting from growth with acetate compared with only a seven to ten-fold increase in particulate activity (Kobr, Vanderhaeghe & Combépine, 1969). This could reflect cytoplasmic synthesis of the enzyme followed by association with a glyoxysome-like particle. The distributions of isocitrate lyase and malate synthase in a sucrose density gradient were identical after centrifugation; around a median density of 1.22 g cm^{-3}. By contrast, succinate dehydrogenase and NAD isocitrate dehydrogenase were both located around 1.16–1.19 g cm^{-3}. Subsequently, Kobr & Vanderhaeghe (1973) established that NAD malate dehydrogenase activity was only associated with the mitochondrial enzymes and that the equilibrium density of both the mitochondria and the glyoxysome-like particles was affected by the nature of the carbon source used to culture the fungus. The apparent density of the latter was also related to the tonicity of the suspending medium. The distribution of catalase was not reported. This enzyme occurs in cell-free extracts of *Uromyces phaseoli* (Mendgen, 1973*b*) and *Coprinus cinereus* (Hardy & Casselton, unpublished results) even though it cannot be detected histochemically in the microbodies of either. O'Sullivan and Casselton (1973) obtained evidence for a glyoxysome-like particle in *C. cinereus* by sucrose-density gradient centrifugation of cell-free extracts.

Even if it is finally demonstrated that the functional glyoxylate cycle of filamentous fungi is located in the microbodies observed with the electron microscope, the fact that the glyoxysome like particles of *Neurospora crassa* lack malate dehydrogenase activity suggests that the inter-relations of the microbodies, the mitochondria and the cytosol would necessarily be different to those in Angiosperms. Both succinate and malate would have to pass to the mitochondria (Fig. 5.2). Benveniste & Munkres (1970) reported that a cytoplasmic isoenzyme of malate dehydrogenase was subject to glucose repression in *N. crassa* and increased in activity after exposure of mycelium to acetate. This enzyme could provide substrate for the phosphoenolpyruvate carboxykinase, opening the way to carbohydrate synthesis (Fig. 5.2).

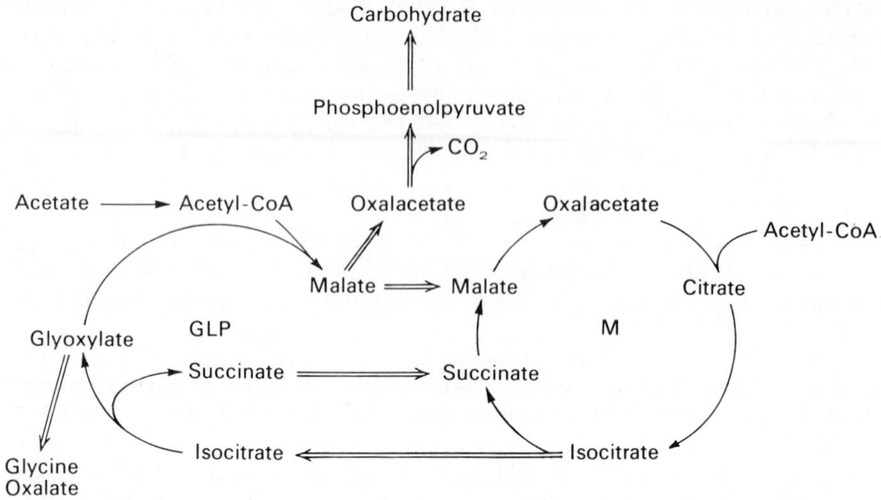

Fig. 5.2 Diagram to illustrate the possible metabolism of acetate in a fungus containing mitochondria (M) and glyoxysome like particles (GLP) which lack any duplication of TCA cycle enzymes. Both malate and succinate must be transferred to the mitochondria to provide a net synthesis of TCA cycle intermediates (cytosol reactions and intraparticular transfers shown by double lines; the scheme does not account for the derepression of NADP isocitrate dehydrogenase in the presence of acetate.)

5.4 Anaplerotic Metabolism and Acid Accumulation

The growth of many fungi is sometimes accompanied by the accumulation of an organic acid in the medium (Cochrane, 1958; Tabak & Cooke, 1968). Therefore, since anaplerotic pathways function to maintain the concentration of metabolic intermediates during biosynthesis, the study of anaplerotic pathways overlaps with that of organic acid accumulation mechanisms (Corsini & Le Tourneau, 1973). For example, Kornberg (1959) suggested that the glyoxylate cycle could account for the data concerning accumulation of fumaric acid by certain strains of *Rhizopus nigricans*. This was investigated by Romano, Bright & Scott (1967). Acid production was known to occur under aerobic conditions in media of high glucose concentration, probably mainly after the available nitrogen had been utilized. Since isocitrate lyase was found to be strongly repressed while glucose remained in the medium, the glyoxylate cycle was excluded as a mechanism for acid synthesis. The available evidence was in agreement with the occurrence of a C3 plus C1 condensation (cf. Margulies & Vischniac, 1961). Overman & Romano (1969) implicated pyruvate carboxylase. Acid production by *R. nigricans* appears to be a continuation of the normal anaplerotic pathway under conditions where, due to limiting nitrogen supply, cell synthesis is restricted. The excess C4 acid synthesized accumulated primarily as fumarate.

The accumulation of citric acid was first studied by Wehmer at the end of the nineteenth century. Despite the commercial utilization of *Aspergillus niger*, some of the quantitative aspects of acid production still remain obscure (Meyrath, 1967). The recent isolation of fully functional mitochondria during acid production indicated that the metabolic block causing citrate to accumulate is probably located outside the TCA cycle (Ahmed, Smith & Anderson, 1972). Wold & Suzuki (1973) proposed that abnormal *c*AMP metabolism was responsible. *c*AMP concentrations greater than 10^{-6} M increased the rate of citrate accumulation in experiments. Nevertheless, since the activities of the glyoxylate cycle enzymes were extremely low (Ahmed, Smith & Anderson, 1972) and there is evidence for carbon dioxide fixation (Martin, Wilson & Burris, 1950; Cleland & Johnson, 1954), the anaplerotic reaction during citrate production is probably catalysed by pyruvate carboxylase.

Other examples of acid accumulation have been studied. The production of malate by *Schizophyllum commune* involves carbon dioxide fixation (Tachibana, Siode & Hanai, 1967). In the case of oxalate accumulation by *Sclerotium rolfsii*, Maxwell & Bateman (1968) proposed that isocitrate lyase coupled to an NAD^+-dependent glyoxylate oxidoreductase which oxidized glyoxylate to oxalate. Although this pathway also produces succinate, no net synthesis of TCA cycle acid is involved and it would not be regarded as anaplerotic (Fig. 5.2; Kornberg, 1966). These reactions cannot occur in *Sclerotinia sclerotiorum* because isocitrate lyase was virtually absent under conditions favouring oxalate accumulation (Corsini & Le Tourneau, 1973). However, since certain strains of *Aspergillus niger* can apparently form oxalate by splitting oxalacetate (Hayaishi, Shimazono, Katagiri & Saito, 1956), the same acid may be synthesized by different mechanisms. Accumulations of organic acids in the fruit-bodies of *Agaricus spp.* have also been reported (Cochrane, 1958; Le Roux, 1962; Piquemal, Baldy, De Serres & Latché, 1970). Although the overproduction of organic acids may sometimes result from abnormalities in metabolic control brought about by certain environmental situations like high carbohydrate concentrations in the medium, re-utilization of an excreted acid has been observed and resulted in a form of diauxic growth (Zink & Katz, 1973).

5.5 Conclusions

There is no doubt that the growth of fungal mycelia often requires the fixation of carbon dioxide (Cochrane, 1958). This provides for the anaplerotic synthesis of C4 intermediates and direct carboxylations such as occur during purine synthesis (Hartman, Keen & Long, 1972). Carbon dioxide fixation by germinating spores is also very common, although not universal (Gottlieb, 1966). Anaplerotic metabolism has been implicated in some cases. For example, Tsay, Nishi & Yanagita (1965) found that the fixation of radioactive carbon dioxide by conidiospores of *Aspergillus niger* labelled the dicarboxylic acids before the nucleic acids and proteins. It is also clear that the glyoxylate cycle operates in fungi under suitable conditions, notably laboratory culture in acetate medium. As to its operation in at least some of the spores which do not require a supply of external nutrients for germination, it probably occurs in uredospores (Caltrider, Ramachandran & Gottliebs, 1963; Reisener & Jäger, 1967). In addition, germination of *Neurospora crassa* ascospores was much reduced if they carried mutations preventing growth of the mycelium with acetate as carbon source (Flavell &

Fincham, 1968*a*) even though germination apparently involves a switch from lipid to carbohydrate utilization (Lingappa & Sussman, 1959). Further study would obviously be justified.

A full understanding of anaplerotic metabolism in filamentous fungi would include its occurrence during all stages of differentiation. The realization that more than one enzyme complement may be capable of providing the biosynthetic potential for asexual reproduction (Turian & Bianchi, 1972; Smith & Anderson, 1973) suggests that much work remains to be done.

5.6 References

AHMED, S. A., SMITH, J. E. & ANDERSON, J. G. (1972). Mitochondrial activity during citric acid production by *Aspergillus niger*. *Transactions of the British Mycological Society* **59**, 51–61.

ARMITT, S., ROBERTS, C. F. & KORNBERG, H. L. (1970). The role of isocitrate lyase in *Aspergillus nidulans*. *FEBS Letters* **7**, 231–4.

ARMITT, S., ROBERTS, C. F. & KORNBERG, H. L. (1971). Mutants of *Aspergillus nidulans* lacking malate synthase. *FEBS Letters* **12**, 276–8.

ASHWORTH, J. M. & KORNBERG, H. L. (1963). Fine control of the glyoxylate cycle by allosteric inhibition of isocitrate lyase. *Biochimica et Biophysica Acta* **73**, 519–22.

ASHWORTH, J. M. & KORNBERG, H. L. (1964). The role of isocitrate lyase in *Escherichia coli*. *Biochimica et Biophysica Acta* **89**, 383–4.

ATKINSON, D. E. (1969). Regulation of enzyme function. *Annual Review of Microbiology* **23**, 47–68.

AVERS, C. J. (1971). Peroxisomes of yeast and other fungi. *Subcellular Biochemistry* **1**, 25–37.

BACHOFEN, R. & RAST, D. (1968). Carboxylierungsreaktionen in *Agaricus bisporus*. III. Pyruvat und Phosphoenolpyruvat als CO_2-Acceptoren. *Archiv für Mikrobiologie* **60**, 217–34.

BEEVER, R. E. (1973). Pyruvate carboxylase and *N. crassa suc* mutants. *Neurospora Newsletter* **20**, 15–16.

BEEVER, R. E. & FINCHAM, J. R. S. (1973). Acetate-nonutilizing mutants of *Neurospora crassa: acu-6*, the structural gene for PEP carboxykinase and inter-allelic complementation at the *acu-6* locus. *Molecular and General Genetics* **126**, 217–26.

BEEVERS, H. (1969). Glyoxysomes of castor bean endosperm and their relation to gluconeogenesis. *Annals of the New York Academy of Sciences* **168**, 313–24.

BENVENISTE, K. & MUNKRES, K. D. (1970). Cytoplasmic and mitochondrial malate dehydrogenases of *Neurospora*. Regulatory and enzymic properties. *Biochimica et Biophysica Acta* **220**, 161–77.

BLOOM, S. J. & JOHNSON, M. J. (1962). The pyruvate carboxylase of *Aspergillus niger*. *Journal of Biological Chemistry* **237**. 2718–20.

BREIDENBACH, R. W., KAHN, A. & BEEVERS, H. (1968). Characterisation of glyoxysomes from castor bean endosperm. *Plant Physiology* **43**, 705–13.

BUDD, K. (1971). Bicarbonate fixation by cell-free extracts and by mycelium of *Neocosmospora vasinfecta*. *Journal of General Microbiology* **67**, 99–106.

BUSTON, H. W., MOSS, M. O. & TYRRELL, D. (1966). The influence of carbon dioxide on growth and sporulation of *Chaetomium globosum*. *Transactions of the British Mycological Society* **49**, 387–96.

CALTRIDER, P. G. & GOTTLIEB, D. (1963). Respiratory activity and enzymes for glucose catabolism in fungus spores. *Phytopathology* **53**, 1021–30.

CALTRIDER, P. G., RAMACHANDRAN, S. & GOTTLIEB, D. (1963). Metabolism during germination and function of glyoxylate enzymes in uredospores of rust fungi. *Phytopathology* **53**, 86–92.

CASSELTON, P. J. (1969). Concurrent regulation of two enzymes in fungi. *Science Progress, Oxford* **57**, 207–27.

CASSELTON, P. J., FAWOLE, M. O. & CASSELTON, L. A. (1969). Isocitrate lyase in *Coprinus lagopus* (sensu Buller). *Canadian Journal of Microbiology* **15**, 637–40.

CAZZULO, J. J. & STOPPANI, A. O. M. (1968). The regulation of yeast pyruvate carboxylase by acetyl-coenzyme A and L-aspartate. *Archives of Biochemistry and Biophysics* **127**, 563–7.

CAZZULO, J. J. & STOPPANI, A. O. M. (1969). Effects of adenosine phosphates and nicotinamide nucleotides on pyruvate carboxylase from baker's yeast. *Biochemical Journal* **112**, 755–62.

CLARKE, G. A. & HARTMAN, R. E. (1972). Carbon dioxide fixation by *Helminthosporium cynodontis. Abstracts of the Annual Meeting of the American Society for Microbiology* p. 156.

CLARKE, G. A. & HARTMAN, R. E. (1973). Pyruvate carboxylase of *Helminthosporium cynodontis. Abstracts of the Annual Meeting of the Americal Society for Microbiology* p. 140.

CLELAND, W. W. & JOHNSON, M. J. (1954). Tracer experiments on the mechanism of citric acid formation by *Aspergillus niger. Journal of Biological Chemistry* **208**, 679–89.

COCHRANE, V. W. (1958). *Physiology of fungi.* New York: John Wiley.

COCHRANE, V. W. (1966). Respiration and spore germination. In *The fungus spore,* pp. 201–213. Edited by M. F. Madelin, London: Butterworths.

COFFEY, M. D., PALEVITZ, B. A. & ALLEN, P. J. (1972). The fine structure of two rust fungi, *Puccinia helianthi* and *Melampsora lini. Canadian Journal of Botany* **50**, 231–40.

COLLINS, J. F. & KORNBERG, H. L. (1960). The metabolism of C2 compounds in micro-organisms. 4. Synthesis of cell materials from acetate by *Aspergillus niger. Biochemical Journal* **77**, 430–8.

COMBÉPINE, G. (1969). Localisation intracellulaire comparée entre la transaminase alanine-glyoxylate et la malate synthétase chez *Neurospora crassa. Comptes Rendus Hebdomadaires des Séances de l'Académie des Sciences, Paris D* **268**, 2694–6.

COOPER, T. G. & BENEDICT, C. R. (1968). Regulation of pyruvate carboxylase by coenzyme A and acyl coenzyme A thioesters. *Biochemistry* **7**, 3032–6.

CORSINI, D. L. & LE TOURNEAU, D. (1973). Organic acid metabolism in *Sclerotinia sclerotiorum. Archiv für Mikrobiologie* **90**, 59–64.

COTTER, D. A., LA CLAVE, A. J., WEGENER, W. S. & NIEDERPRUEM, D. J. (1970). CO_2 control of fruiting in *Schizophyllum commune*: noninvolvement of sustained isocitrate lyase derepression. *Canadian Journal of Microbiology* **16**, 605–8.

DALY, J. M., KNOCHE, H. W. & WIESE, M. V. (1967). Carbohydrate and lipid metabolism during germination of uredospores of *Puccinia graminis tritici, Plant Physiology* **42**, 1633–42.

FEIR, H. A. & SUZUKI, I. (1969). Pyruvate carboxylase of *Aspergillus niger:* kinetic study of a biotin-containing carboxylase. *Canadian Journal of Biochemistry* **47**, 697–710.

FLAVELL, R. B. & FINCHAM, J. R. S. (1968*a*). Acetate-nonutilising mutants of *Neurospora crassa.* I. Mutant isolation, complementation studies, and linkage relationships. *Journal of Bacteriology* **95**, 1056–62.

FLAVELL, R. B. & FINCHAM, J. R. S. (1968*b*). Acetate-nonutilising mutants of *Neurospora crassa.* II. Biochemical deficiencies and the roles of certain enzymes. *Journal of Bacteriology* **95**, 1063–8.

FLAVELL, R. B. & WOODWARD, D. O. (1970*a*). The regulation of synthesis of Krebs cycle enzymes in *Neurospora* by catabolite and end-product repression. *European Journal of Biochemistry* **13**, 548–53.

FLAVELL, R. B. & WOODWARD, D. O. (1970*b*). The concurrent regulation of metabolically related enzymes. The Krebs cycle and glyoxylate shunt enzymes in *Neurospora. European Journal of Biochemistry* **17**, 284–91.

FLAVELL, R. B. & WOODWARD, D. O. (1971*a*). Metabolic role, regulation of synthesis, cellular localisation, and genetic control of the glyoxylate cycle enzymes in *Neurospora crassa. Journal of Bacteriology* **105**, 200–10.

FLAVELL, R. B. & WOODWARD, D. O. (1971*b*). Selective inhibition of enzyme synthesis under conditions of respiratory inhibition. *Journal of Bacteriology* **107**, 853–63.

FREAR, D. S. & JOHNSON, M. A. (1961). Enzymes of the glyoxylate cycle in germinating uredospores of *Melampsora lini* (pers.) lév. *Biochimica et Biophysica Acta* **47**, 419–21.

GALBRAITH, J. C. &. SMITH, J. E. (1969). Changes in activity of certain enzymes of the tricarboxylic acid cycle and the glyoxylate cycle during the initiation of conidiation of *Aspergillus niger, Canadian Journal of Microbiology* **15**, 1207–12.

GOTTLIEB, D. (1966). Biosynthetic processes in germinating spores. In *The fungus spore,* pp. 217–233, edited by M. F. Madelin, London: Butterworths.

GOTTLIEB, D. & RAMACHANDRAN, S. (1960). The nature of production of the glyoxylate pathway enzymes in germinating spores of *Penicillium oxalicum*. *Mycologia* **52**, 599–607.

HARTMAN, R. E. & KEEN, N. T. (1973). Enzymes catalysing anaplerotic carbon dioxide fixation in *Verticillium albo-atrum*. *Phytopathology* **63**, 947–53.

HARTMAN, R. E. & KEEN, N. T. (1974*a*). The pyruvate carboxylase of *Verticillium albo-atrum*. *Journal of General Microbiology* **81**, 15–9.

HARTMAN, R. E. & KEEN, N. T. (1974*b*). The phosphoenolpyruvate carboxykinase of *Verticillium albo-atrum*. *Journal of General Microbiology* **81**, 21–6.

HARTMAN, R. E., KEEN, N. T. & LONG, M. (1972). Carbon dioxide fixation by *Verticillium albo-atrum*. *Journal of General Microbiology* **73**, 29–34.

HAYAISHI, O., SHIMAZONO, H., KATAGIRI, M. & SAITO, Y. (1956). Enzymatic formation of oxalate and acetate from oxaloacetate. *Journal of the American Chemical Society* **78**, 5126–7.

JARVIS, F. G. & JOHNSON, M. J. (1947). The role of the constituents of synthetic media for penicillin production. *Journal of the American Chemical Society* **69**, 3010–17.

JOHANSON, R. A., HILL, J. M. & McFADDEN, B. A. (1972). Isocitrate lyase-1 from *Neurospora crassa*. *Federation Proceedings* **31**, 475.

KHOUW, B. T. & McCURDY, H. D. (1969). Tricarboxylic acid cycle enzymes and morphogenesis in *Blastocladiella emersonii*. *Journal of Bacteriology* **99**, 197–205.

KOBR, M. J. & VANDERHAEGHE, F. (1973). Changes in density of organelles from *Neurospora*. *Experientia* **29**, 1221–3.

KOBR, M. J., TURIAN, G., ZIMMERMAN, E. J. (1965). Changes in enzymes regulating isocitrate breakdown in *Neurospora crassa*. *Archiv für Mikrobiologie* **52**, 169–77.

KOBR, M. J., VANDERHAEGHE, F. & COMBÉPINE, G. (1969). Particulate enzymes of the glyoxylate cycle in *Neurospora crassa*. *Biochemical and Biophysical Research Communications* **37**, 640–5.

KORNBERG, H. L. (1959). Aspects of terminal respiration in micro-organisms. *Annual Review of Microbiology* **13**, 49–78.

KORNBERG, H. L. (1966). Anaplerotic sequences and their role in metabolism. In *Essays in biochemistry*, Vol. 2. Edited by P. N. Campbell and G. D. Greville. London: Academic Press.

KORNBERG, H. L. & ELSDEN, S. R. (1961). The metabolism of 2-carbon compounds by microorganisms. *Advances in Enzymology* **23**, 401–70.

KORNBERG, H. L. & KREBS, H. A. (1957). Synthesis of cell constituents from C2-units by a modified tricarboxylic acid cycle. *Nature, London* **179**, 988-91.

KORNBERG, H. L. & MADSEN, N. B. (1957). Synthesis of C4-dicarboxylic acids from acetate by a 'glyoxylate bypass' of the tricarboxylic acid cycle. *Biochimica et Biophysica Acta* **24**, 651–3.

KREBS, H. A., GURIN, S. & EGGLESTON, L. V. (1952). The pathway of oxidation of acetate in baker's yeast. *Biochemical Journal* **51**, 614–28.

LE ROUX, P. (1962). Metabolism des acides organiques dans les carpophores d' *Agaricus campestris*. *Annales de Physiologie Végétale, Paris* **4**, 149–60.

LECKIE, B. J. & FINCHAM, J. R. S. (1971). A structural gene for *Neurospora crassa* isocitrate lyase. *Journal of General Microbiology* **65**, 35–43.

LEIGHTON, T. J., STOCK, J. J. & KELLN, R. A. (1970). Macroconidial germination in *Microsporum gypseum*. *Journal of Bacteriology* **103**, 439–46.

LINGAPPA, B. T. & SUSSMAN, A. S. (1959). Endogenous substrates of dormant, activated and germinating ascospores of *Neurospora tetrasperma*. *Plant Physiology* **34**, 466–72.

McCURDY, H. D. & CANTINO, E. C. (1960). Isocitritase, glycine-alanine transaminase, and development in *Blastocladiella emersonii*. *Plant Physiology* **35**, 463–76.

McLAUGHLIN, D. J. (1973). Ultrastructure of sterigma growth and basidiospore formation in *Coprinus* and *Boletus*. *Canadian Journal of Botany* **51**, 145–50.

MARGULIEṢ, M. & VISHNIAC, W. (1961). Dissimilation of glucose by the MX strain of *Rhizopus*. *Journal of Bacteriology* **81**, 1–9.

MARTIN, S. M., WILSON, P. W. & BURRIS, R. H. (1950). Citric acid formation from $^{14}CO_2$ by *Aspergillus niger*. *Archives of Biochemistry*, **26**, 103–11.

MAXWELL, D. P. & BATEMAN, D. F. (1968). Oxalic acid biosynthesis by *Sclerotium rolfsii*. *Phytopathology* **58**, 1635–42.

MAXWELL, D. P., WILLIAMS, P. H. & MAXWELL, M. D. (1970). Microbodies and lipid bodies in the hyphal tips of *Sclerotinia sclerotiorum*. *Canadian Journal of Botany* **48**, 1689–91.

MENDGEN, K. (1973*a*). Feinbau der Infektionsstrukturen von *Uromyces phaseoli. Phytopathologische Zeitschrift* **78**, 109–20.

MENDGEN, K. (1973*b*). Microbodies (glyoxysomes) in infection structures of *Uromyces phaseoli. Protoplasma* **78**, 477–82.

METZENBERG, R. L. (1972). Genetic regulatory systems in *Neurospora. Annual Review of Genetics* **6**, 111–32.

MEYRATH, J. (1967). Citric acid production. *Process Biochemistry* **2**, 25–7.

MIROCHA, C. J. & RICK, P. D. (1967). Carbon dioxide fixation in the dark as a nutritional factor in parasitism. In *The dynamic role of molecular constituents in plant parasite interaction*, pp. 121–141 Edited by C. J. Mirocha and I. Uritani, St Paul: Bruce Publishing Co.

NIEDERPRUEM, D. J. (1965). Tricarboxylic acid cycle. In *The fungi*, Vol. I, pp. 269–300. Edited by G. C. Ainsworth and A. S. Sussman. New York: Academic Press.

OHMORI, K. & GOTTLIEB, D. (1965). Development of respiratory enzyme activities during spore germination. *Phytopathology* **55**, 1328–36.

OLSON, J. A. (1954). The D-isocitric lyase system: the formation of glyoxylic and succinic acids from D-isocitric acid. *Nature, London* **174**, 695–6.

O'SULLIVAN, J. & CASSELTON, P. J. (1973). The subcellular localisation of glyoxylate cycle enzymes in *Coprinus lagopus* (sensu Buller). *Journal of General Microbiology* **75**, 333–7.

OVERMAN, S. A. & ROMANO, A. H. (1969). Pyruvate carboxylase of *Rhizopus nigricans* and its role in fumaric acid production. *Biochemical and Biophysical Research Communications* **37**, 457–63.

PIQUEMAL, M., BALDY, P., DE SERRES, M. & LATCHÉ, J. (1970). Etude, en fonction du temps de croissance, de quelques constitants du mycélium d'*Agaricus bisporus* Lge, cultivé sur milieu semi-synthetique. *Comptes Rendus Hebdomadaires des Seánces de L'Académie des Sciences, Paris D* **271**, 2316–9.

REEVES, H. C., RABIN, R., WEGENER, W. S. & AJL, S. J. (1971). Assays of enzymes of the tricarboxylic acid and glyoxylate cycles. In *Methods in microbiology*, Vol. 6A, pp. 425–462. Edited by J. R. Norris and D. W. Ribbons. London: Academic Press.

REISENER, H.-J. & JÄGER, K. (1967). Untersuchungen über die Quantitative Bedeutung einiger Stoffwechselwege in den Uredosporen von *Puccinia graminis* var. tritici. *Planta, Berlin* **72**, 265–83.

RICK, P. D. & MIROCHA, C. J. (1968). Fixation of carbon dioxide in the dark by the malic enzyme of bean and oat stem rust uredospores. *Plant Physiology* **43**, 201–7.

ROMANO, A. H. & KORNBERG, H. L. (1968). Regulation of sugar utilisation by *Aspergillus nidulans. Biochimica et Biophysica Acta* **158**, 491–3.

ROMANO, A. H. & KORNBERG, H. L. (1969). Regulation of sugar uptake by *Aspergillus nidulans. Proceedings of the Royal Society, Series B* **173**, 475–90.

ROMANO, A. H., BRIGHT, M. M. & SCOTT, W. E. (1967). Mechanism of fumaric acid accumulation in *Rhizopus nigricans. Journal of Bacteriology* **93**, 600–4.

ROUGEMONT, A. & KOBR, M. J. (1973). Isocitrate lyase-2 from *N. crassa. Neurospora Newsletter* **20**, 28–9.

SCHORMÜLLER, J. & STAN, H.-J. (1970). Stoffwechseluntersuchungen an lebensmitteltechnologische wichtigen Mikroorganismen. Pyruvatcarboxylase aus *Penicillium camemberti* var *candidum*. 3. Allosterische Eigenschaften des Enzyms. *Zeitschrift für Lebensmitteluntersuchung und Forschung* **142**, 321–30.

SCRUTTON, M. C. (1971). Assay of enzymes of CO_2 metabolism. In *Methods in microbiology*, Vol. 6A, pp. 479–541. Edited by J. R. Norris and D. W. Ribbons. London: Academic Press.

SCRUTTON, M. C. & YOUNG, M. R. (1972). Pyruvate carboxylase. In *The enzymes*, Vol. 6, pp. 1–35. Edited by P. D. Boyer. New York: Academic Press.

SCRUTTON, M. C., YOUNG, M. R. & UTTER, M. F. (1970). Pyruvate carboxylase from baker's yeast. The presence of bound zinc. *Journal of Biological Chemistry* **245**, 6220–7.

SHU, P., TANNER, K. G. & LEDINGHAM, G. A. (1954). Studies on the respiration of resting and germinating uredospores of wheat stem rust. *Canadian Journal of Botany* **32**, 16–23.

SIETSMA, J. H. (1971). Some effects of cholesterol on the metabolism of *Pythium sp.* PRL 2142. *Biochimica et Biophysica Acta* **244**, 178–85.

SJOGREN, R. E. & ROMANO, A. H. (1967). Evidence for multiple forms of isocitrate lyase in *Neurospora crassa. Journal of Bacteriology* **93**, 1638–43.

SKINNER, V. M. & ARMITT, S. (1972). Mutants of *Aspergillus nidulans* lacking pyruvate carboxylase. *FEBS Letters* **20**, 16-8.

SMITH, J. E. & ANDERSON, J. G. (1973). Differentiation in the Aspergilli. *Symposium of the Society for General Microbiology* **23**, 295–337.

SMITH, J. E. & GALBRAITH, J. C. (1971). Biochemical and physiological aspects of differentiation in the fungi. *Advances in Microbiol Physiology* **5**, 45–134.

STAN, H.-J. (1972). Pyruvate carboxylase from *Penicillium camemberti*. 5. Regulation of the enzyme activity by the ratio of ATP and ADP. *International Journal of Biochemistry* **3**, 573–82.

STAPLES, R. C. (1962). Initial products of acetate utilization by bean rust uredospores. *Contributions of the Boyce Thompson Institute* **21**, 487–97.

STAPLES, R. C. & WYNN, W. K. (1965). The physiology of uredospores of the rust fungi. *The Botanical Review* **31**, 537–64.

TABAK, H. H. & COOKE, W. B. (1968). The effects of gaseous environments on the growth and metabolism of fungi. *The Botanical Review* **34**, 126–252.

TACHIBANA, S., SIODE, J. & HANAI, T. (1967). Studies on CO_2-fixing fermentation. XIV. On the relationship between L-malate formation and carbohydrate metabolism of cell-free extracts of *Schizophyllum commune*. *Journal of Fermentation Technology* **45**, 1130–8.

TODD, N. & CASSELTON, L. A. (1973). Non-complementation and recessiveness as properties of missense suppressor genes in the fungus *Coprinus*. *Journal of General Microbiology* **77**, 197–207.

TOKUNAGA, J., MALCA, I., SIMS, J. J., ERWIN, D. C. & KEEN, N. T. (1969). Respiratory enzymes in the spores of *Verticillium albo-atrum*. *Phytopathology* **59**, 1829–32.

TOLBERT, N. E. (1971). Microbodies, peroxisomes and glyoxysomes. *Annual Review of Plant Physiology* **22**, 45–74.

TSAY, Y., NISHI, A. & YANAGITA, T. (1965). Carbon dioxide fixation in *Aspergillus oryzae* conidia at the initial phase of germination. *The Journal of Biochemistry* **58**, 487–93.

TURIAN, G. (1961). Cycle glyoxylique, transaminase alanine-glyoxylate et différenciation sexuelle chez *Allomyces* et *Neurospora*. *Pathologia et Microbiologia* **24**, 819–39.

TURIAN, G. & BIANCHI, D. E. (1972). Conidiation in *Neurospora*. *The Botanical Review* **38**, 119–54.

TURIAN, G. & KOBR, M. (1965). Isocitrate lyase from *Neurospora crassa*. *Biochimica et Biophysica Acta* **99**, 178–80.

UTTER, M. F. & KOLENBRANDER, H. M. (1972). Formation of oxalacetate by CO_2 fixation on phosphoenolpyruvate. In *The enzymes*, Vol. 6, pp. 117–168 Edited by P. D. Boyer New York: Academic Press.

VANDERWINKEL, E., LIARD, P., RAMOS, F. & WIAME, J. M. (1963). Genetic control of the regulation of isocitritase and malate synthase in *Escherichia coli* K. 12. *Biochemical and Biophysical Research Communications* **12**, 157–62.

WEGENER, W. S. & ROMANO, A. H. (1963). Zinc stimulation of RNA and protein synthesis in *Rhizopus nigricans*. *Science* **142**, 1669–70.

WEGENER, W. S. & ROMANO, A. H. (1964). Control of isocitratase formation in *Rhizopus nigricans*. *Journal of Bacteriology* **87**, 156–61.

WEGENER, W. S., SCHELL, J. E. & ROMANO, A. H. (1967). Control of malate synthase formation in *Rhizopus nigricans*. *Journal of Bacteriology* **94**, 1951–6.

WEGENER, W. S., REEVES, H. C., RABIN, R. & AJL, S. J. (1968). Alternate pathways of metabolism of short-chain fatty acids. *Bacteriological Reviews* **32**, 1–26.

WEISS, H., VON JAGOW, G., KLINGENBERG, M. & BÜCHER, T. (1970). Characterisation of *Neurospora crassa* mitochondria prepared with a grind-mill. *European Journal of Biochemistry* **14**, 75–82.

WERGIN, W. P. (1972). Ultrastructural comparison of microbodies in pathogenic and saprophytic hyphae of *Fusarium oxysporum* f. sp. *lycopersici*. *Phytopathology* **62**, 1045–51.

WILLIAMS, P. G. & LEDINGHAM, G. A. (1964). Fine structure of wheat stem rust uredospores. *Canadian Journal of Botany* **42**, 1503–8.

WOLD, W. S. M. & SUZUKI, I. (1973). Cyclic AMP and citric acid accumulation by *Aspergillus niger*. *Biochemical and Biophysical Research Communications* **50**, 237–44.

WORONICK, C. L. & JOHNSON, M. J. (1960). Carbon dioxide fixation by cell-free extracts of *Aspergillus niger*. *Journal of Biological Chemistry* **235**, 9–15.

ZINK, M. W. (1972). Regulation of the two 'malic' enzymes in *Neurospora crassa*. *Canadian Journal of Microbiology* **18**, 611–7.

ZINK, M. W. & KATZ, J. S. (1973). Malic enzyme of *Fusarium oxysporum*. *Canadian Journal of Microbiology* **19**, 1187–96.

CHAPTER 6

The Biosynthesis of Fatty Acids

P. WALKER and M. WOODBINE

6.1 Introduction

Micro-organisms classified as Protista may be further divided into the Prokaryota including bacteria and blue-green algae, and Eukaryota including other algae, with fungi and protozoa. The terms 'lower protists' and 'higher protists' may be substituted for prokaryotic and eukaryotic micro-organisms (Pelczar & Reid, 1972).

Apart from the differences in the form of the chromosome and nucleus, the membranes and lipids of filamentous fungi and other eukaryotes differ considerably from prokaryotes and on these distinctions the division of the two classes is made (Stanier, Douderoff & Adelberg, 1971; Stanier, 1970). These differences are reflected in the fatty acid composition of the lipids of the organism, bacteria having a large number and variety of fatty acids, many of which are uncommon, or unknown, in fungi and other eukaryotes (O'Leary, 1967). Perhaps the greatest difference of all is the fact that bacteria and other prokaryotic cells do not synthesize polyunsaturated (polyenoic) fatty acids (although some blue-green algae may perform this synthesis), whereas these fatty acids are ubiquitous in eukaryotic cells (Shaw, 1966).

Although eukaryotic micro-organisms may show differences in lipid composition within the group, these are only variations of a common pattern, characteristic of eukaryotic cells. Therefore, results of studies on fatty acid biosynthesis, in particular eukaryotic micro-organisms are likely to be applicable to other eukaryotic organisms. However, because of the large differences between the two classes, extreme caution must be taken when applying the results of studies with prokaryotic micro-organisms to eukaryotes.

Most studies on microbial fatty acid biosynthesis have used yeasts and bacteria and, to a lesser extent, algae and protozoa. Comparatively little work has been done with filamentous fungi. This account of fatty acid biosynthesis discusses studies using the higher protists and other eukaryotic cells, makes comparisons with results from bacteria but emphasizes implications for the filamentous fungi.

Historical aspects

The real concept of the change of carbohydrate to fat resulted from the studies of Lawes & Gilbert (1860). Several theories were suggested for the mechanism of fatty acid biosynthesis, among which was Emil Fischer's 'hexose condensation theory', whereby sugar molecules combined their carbon chains, *e.g.* three C_6 units to give C_{18}, the sugar chains being condensed and the hydroxyl groups reduced (Fischer, 1890).

In contrast, Nencki (1878) suggested that two molecules of acetaldehyde after condensation to alcohol might be rearranged to form butyric acid in yeast. This view was supported by Hoppe-Seyler (1879*a, b*), Magnus-Levy (1902) and Raper (1907), while Smedley (1911) showed that crotonaldehyde could be similarly condensed to yield straight-chain unsaturated aldehydes. Raper (1907) expressed the views held at the time by stating, 'the formation of fatty acids in animals from carbohydrates . . . suggest that these fatty acids are produced by the condensation of some highly reactive substance containing two carbon atoms and formed in the decomposition of sugar'.

A modification of the concepts of Raper was proposed by Smedley & Lubrzynska (1913), whereby acetaldehyde condenses with pyruvic acid to form a higher keto-acid from which an aldehyde could be formed by decarboxylation or a fatty acid by oxidation.

Haehn (1921) and Haehn & Kintoff (1923; 1925*a, b*) stimulated the synthesis of fatty acids in the yeast *Endomycopsis vernalis* (*Trichosporon pullulans*) by addition of acetaldehyde to the medium. The addition of compounds which might yield acetaldehyde such as lactic and pyruvic acid, also caused fat production. The same observations were made with the mould *Aspergillus niger* (Terroine & Bonnett, 1927) and much later with *Neurospora crassa* (Ottke, Tatum, Zabin & Bloch, 1951).

It was also observed that acetate, or some form of it, was the precursor of fatty acids in *Fusarium solani* (Weiss, Fiore & Nord, 1947), *Clostridium butyricum* (Wood, Brown & Werkman, 1945) and *Saccharomyces cerevisiae* (White & Werkman, 1947). Such work as this, together with the isotopic work by Sonderhoff & Thomas (1937), which demonstrated that labelled acetic acid gave greater amounts of labelled fatty acids than a carbohydrate substrate, showed that C_2 compounds are the 'building blocks' in fatty acid biosynthesis. During this period fatty acid synthesis was regarded as the reverse of fatty acid oxidation and the processes of oxidation and biosynthesis were complementary, with acetate or a fatty acid and phosphate as functional building blocks.

The extensive work of Lipmann and Lynen indicated that a high-energy level, or 'active', acetate was involved in biosynthesis and the discovery of coenzyme A (CoA) by Lipmann (1945) was followed by the elucidation of 'active' acetate as the acetyl mercaptoester of coenzyme A, or acetyl-CoA, formed as a thioester of CoA and acetic acid (Lynen & Reichert, 1951;

Lynen, Reichert & Rueff, 1951). The work of Wakil (1958) and Wakil & Ganguly (1959) then showed that malonic acid was involved in the biosynthesis of long-chain fatty acids with acetyl-CoA supplying the first two atoms in the long-chain sequence and the rest via malonyl-CoA.

For a more detailed discussion of the development of general ideas of the biosynthesis of fatty acids, from the earliest view to the concept of the fatty acid cycle, see Woodbine (1959, 1964), Deuel (1951, 1955, 1957) and Bloch (1960).

6.2 The Biosynthesis of Straight-Chain Saturated Fatty Acids

When the mechanism of the β-oxidation of fatty acids was elucidated, it was assumed that the mechanism of fatty acid synthesis occurred by simple reversal of this pathway (Lynen, 1961), until Gibson, Titchener & Wakil (1958) found that fatty acid synthesis in extracts of avian liver required ATP and bicarbonate, which could not be explained by the catabolic pathway. Wakil (1961) subsequently showed a requirement for acetyl-CoA, ATP, HCO_3^-, MN^{2+} and NADPH in avian liver systems, the bicarbonate being required for the carboxylation of acetyl-CoA to malonyl-CoA which acts as the actual donor for chain elongation to palmitic acid (Wakil, 1958).

Ganguly (1960) isolated two distinct enzyme fractions from plant mitochondria, the first fraction catalysing the carboxylation of acetyl-CoA into malonyl CoA, the second the conversion of acetyl-CoA and malonyl-CoA into palmitate in the presence of NADPH as hydrogen donor:

$$CH_3COSCoA + HCO_3^- + ATP \rightleftharpoons HO_2CCH_2COSCoA + ADP + P_i \qquad (1)$$

$$\begin{aligned} &CH_3COSCoA + 7HO_2CCH_2COSCoA + 14NADPH + 14H^+ \rightarrow \\ &\quad CH_3[CH_2]_{14} \cdot CO_2H + 7CO_2 + 14NADP^+ + 8CoASH + 6H_2O \end{aligned} \qquad (2)$$

The first reaction is catalysed by a biotin enzyme, acetyl-CoA carboxylase, the second by a group of enzymes known as the fatty acid synthetase system (Wakil & Gibson, 1960).

Acetyl-CoA carboxylase

The first and rate limiting step in the *de novo* synthesis of long-chain fatty acids in prokaryotes and eukaryotes is the ATP- and Mn^{2+}-dependent carboxylation of acetyl-CoA to malonyl-CoA, catalysed by acetyl-CoA carboxylase (Vagelos, 1964) (Equation *1*).

The participation of biotin in the reaction was shown by the use of avidin, a raw egg-white protein which inhibits the reaction by complexing with biotin (Wakil & Gibson, 1960; Lynen, 1961) which is bound to the enzyme through the ε-amino group of lysine (Wright *et al.*, 1952) and so prevents the carboxylation of acetyl-CoA and fatty acid synthesis.

Acetyl-CoA carboxylase is an accessory to, but not part of, the fatty acid synthetase system of fungi and animals (Wakil & Gibson, 1960; Lynen, 1961) and it has been shown that fatty acid synthesis is stimulated by isocitrate or citrate in enzyme preparations from yeast and animals by exercising allosteric control on acetyl-CoA carboxylase and not the fatty acid synthetase (Rasmussen & Klein, 1968*a*, *b*, *c*; Goodridge, 1973).

Although the incompletely activated enzyme appears to control the rate of fatty acid biosynthesis, the citrate-activated enzyme generated malonyl-CoA at a rate equal to the capacity of the fatty acid synthetase to convert malonyl-CoA to fatty acid (Chang, Seidman, Tebor & Lane, 1967).

Fatty acid synthetase

The second step in fatty acid synthesis is the conversion of malonyl-CoA to higher fatty acids, requiring acetyl-CoA and NADPH and catalysed by an enzyme complex known as fatty acid synthetase.

Although the fatty acid synthetase system for the *de novo* synthesis of fatty acids is present in all micro-organisms, plants and animals, synthetase systems do differ in their type of organization. Those from bacteria (Lennarz, Light & Bloch, 1962; Wakil, Pugh & Sauer, 1964) and plants (Nagai & Bloch, 1967) may be readily dissociated into their component enzymes, whereas in animal livers and yeast, the complex has been isolated in a homogeneous form and the component protein fractionated as a single multi-enzyme unit (Lynen, 1961). This multienzyme complex has been shown to be present in the moulds *Penicillium patulum* (Lynen, 1972) and *Pythium debaryanum* (Law & Burton, 1973). Attempts to fractionate the complex by the use of enzymes and detergents led to loss of activity in *Pythium debaryanum* (Law & Burton, 1973) and yeast enzyme complex (Lynen, 1961).

The moiecular weight of the enzyme complex of pigeon liver was calculated as 4.5×10^5 (Yang, Butterworth, Bock & Porter, 1967; Butterworth, Yang, Bock & Porter, 1967), that of yeast, 2.3×10^6 (Lynen, 1967*a*) and that of *Pythium debaryanum*, 4.0×10^6 (Law & Burton, 1973). In *Saccharomyces cerevisiae* the synthetase is found in a particle distinct from the ribosomes, sediments at high centrifugation speed and is non-membranous (Klein, Volkmann & Chao, 1967; Klein, Volkmann & Leaffer, 1967; Klein, Volkmann & Weibel, 1967) and this is true of the synthetase systems of birds and mammals (*e.g.* Philips *et al.*, 1970; Dils & Popják, 1962).

It is of interest to note that the alga, *Euglena gracilis*, contains the particulate and soluble fatty acid synthetase systems (Nagai & Bloch, 1967), but the significance of the activity of either system depends on the stage of development of the chloroplasts where the soluble synthetase is located (Ernst-Fonberg, 1973).

Yeast fatty acid synthetase consists of one of each of the seven synthetase enzymes responsible for the reactions of fatty acid synthesis and two sulphydryl groups, identified as having a carrier function and designated as the 'central' and 'peripheral' groups to denote their relative positions within the complex (Lynen, 1967*a*). The 'peripheral' group has been identified as belonging to a cysteine residue of the condensing enzyme component of the complex and the 'central' thiol group as 4′-phosphopantetheine, the acyl carrier protein component (ACP) (see later) of the yeast synthetase which is bound to the enzyme through a phosphodiester linkage with the hydroxyl group of a serine residue, similar to that found in *Esherichia coli* ACP (Lynen, 1967*b*).

The yeast enzyme complex appears to occur as a 'trimer', each containing one 'peripheral' thiol group and one 'central' thiol group (Lynen, 1967*a*), *i.e.* there are three sets of seven enzymes.

Acyl carrier protein

Since no free intermediates accumulate during synthesis of fatty acids and as the complex is stimulated by thiols and inhibited by sulphydryl binding agents, protein-bound acyl derivatives are likely intermediates (Lynen, 1961). In *Escherichia coli* it has been found that the acyl intermediates are bound as thioesters to a low molecular weight (about 9000) protein, termed acyl carrier protein (ACP) (Wakil, Pugh & Sauer, 1964). This protein contains 4′-phosphopantetheine as a prosthetic group, identical with that of CoA, on to which the intermediates of fatty acid synthesis are attached by the sulphydryl moeity of the group, as in CoA derivatives (Pugh & Wakil, 1965; Majerus, Alberts & Vagelos, 1964). In *E. coli*, the 4′-phosphopantetheine of ACP is transferred from CoA (Alberts & Vagelos, 1966; Prescott, Elovson & Vagelos, 1969).

The structure of ACP has been described as 'proteinated' CoA and because of the binding of the prosthetic group to the carrier protein rather than ADP, as in CoA, many binding sites for the enzymes of the complex are created (Willecke, Ritter & Lynen, 1969).

Studies of synthesised *Escherichia coli* ACP have shown that omission of more than three residues of the single polypeptide chain that comprises ACP causes a significant reduction in activity of the protein. Arginine at position 6 of the chain and the NH_2 terminus have a critical biological function (Hancock, Marshall & Vagelos, 1973). Proteins having properties similar to *E. coli* ACP and containing 4′-phosphopantetheine have been shown to be present in the fatty acid synthetase systems of other bacteria (see *e.g.* Matsumura & Stumpf, 1968), yeast (Lynen, Oesterhelt, Schweizer & Willecke, 1968) and pigeon-liver (Phillips *et al.*, 1970). Moreover, *E. coli* ACP has been shown to function as an acyl carrier in other synthetase systems where the presence of ACP-like proteins had not been shown, such as in higher-plant tissues (Nagai & Bloch, 1968) and *Bacillus subtilis* (Butterworth & Bloch, 1970), suggesting that ACP functions as an acyl carrier generally in prokaryotic and eukaryotic cells.

In yeast, it has been shown that 4′-phosphopantetheine is attached to a structural protein component within the enzyme complex analogous to the ACP of bacteria and plants (Willecke, Ritter & Lynen, 1969). According to Lynen (Lynen *et al.*, 1968), the ACP is the core of the multi-enzyme complex around which are grouped the six main enzymes (Fig. 6.1). The sixteen-bonded unit of the ACP (Fig. 6.2) is a flexible arm of about 20 Å length when fully extended, but may shorten by folding. The pantetheine unit rotates so that the intermediates bond covalently to the thiol group pass to and from the active sites of the enzymes surrounding the ACP. These enzymes have little freedom of motion in the rigid complex. In contrast, bacterial and plant ACPs are not firmly attached to the individual enzymes involved and so the fatty acyl derivatives may shift and serve as substrates in subsequent reactions. Support for Lynen's theory was demonstrated by electron micrographs of purified synthetase, which showed single, ovoid particles and a structure of three interlocking subunits, the structural unit containing seven proteins and suggesting the enzymes were arranged around the carrier protein containing the 'central' sulphydryl group (Lynen, 1967*a*).

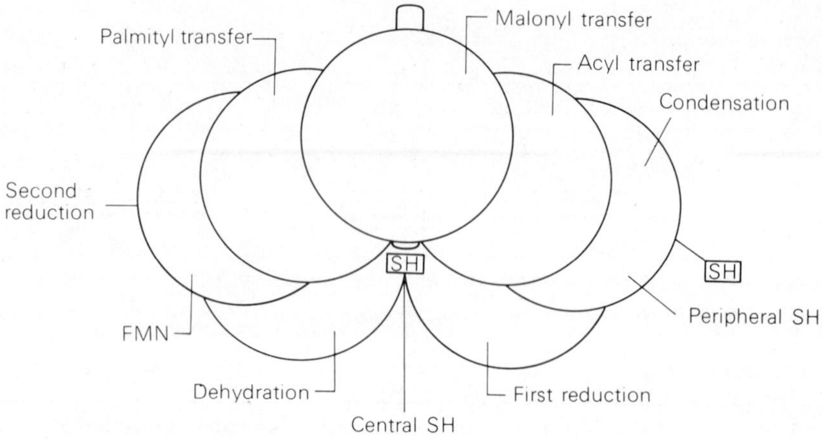

Fig. 6.1 Yeast fatty acid synthetase (Lynen, 1967*b*).

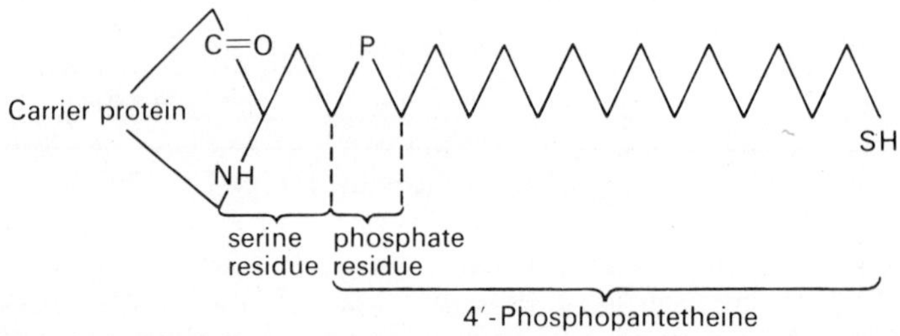

Fig. 6.2 The flexible bonding system of the ACP prosthetic group.

There are other groups on the synthetase complex onto which acetyl and malonyl groups are bound (Lynen, 1967*a*) and it has been shown that acyl transfer from acyl-CoA derivatives to the central sulphydryl group in the complex initially occurs at non-thiol groups (Lynen *et al.*, 1968; Nixon, Phillips, Abramovitz & Porter, 1970) identified as the hydroxyl groups of serine residues. Malonate is attached to a non-thiol group before transfer to the 'central' thiol group (Schweizer *et al.*, 1970) and Lynen (1972) has shown that the acetyl group is bound to a non-thiol group at some stage besides attachment to the 'central' site and then transfer to the 'peripheral' site.

The reactions involved in fatty acid biosynthesis

Malonyl-CoA supplies all the carbon atoms of the long-chain fatty acids with the exception of the two methyl terminal carbons (Wakil & Ganguly, 1959) which are supplied by a short-chain acyl primer, normally acetyl-CoA (Lynen, 1961), although other short-chain acyl derivatives may be used as primer (Bressler & Wakil, 1961), resulting in the formation of even-chain fatty acids, usually palmitate. Substitution of propionyl or iso-propionyl CoA for acetyl-CoA results in the formation of an odd or a branched-chain fatty acid (Bressler & Wakil, 1961; Lynen, 1961).

A summary of the reactions catalysed by yeast fatty acid synthetase has been provided by Lynen (Lynen *et al.*, 1968). The first reaction of fatty acid synthesis involves the transfer of the acetate residue of acetyl-CoA to the 'central' (designated HS_C) and then to the 'peripheral' (designated HS_P) sulphydryl group of the enzyme complex, liberating CoA.

The bracketed enzymes refer to the individual enzymes catalysing the corresponding reaction in bacteria, where the substrates are linked to ACP.

$$
CH_3COSCoA + \begin{array}{l} HS_C \\ \quad\diagdown \\ \qquad \text{Enzyme} \\ \quad\diagup \\ HS_P \end{array} \xrightleftharpoons{[\text{Acetyl transacylase}]} \begin{array}{l} HS_C \\ \quad\diagdown \\ \qquad \text{Enzyme} + CoASH \\ \quad\diagup \\ CH_3COS_P \end{array} \qquad (3)
$$

Next, a malonyl group is similarly transferred from malonyl-CoA to the 'central' thiol group, again liberating CoA.

$$
HO_2CCH_2COSCoA + \begin{array}{l} HS_C \\ \quad\diagdown \\ \qquad \text{Enzyme} \\ \quad\diagup \\ CH_3COS_P \end{array} \xrightleftharpoons{[\text{Malonyl transacylase}]} \begin{array}{l} HO_2CCH_2COS_C \\ \quad\diagdown \\ \qquad \text{Enzyme} + CoASH \\ \quad\diagup \\ CH_3COS_P \end{array} \qquad (4)
$$

A condensation reaction occurs between the enzyme-bound acetyl and malonyl groups to give enzyme-bound acetoacetate, covalently bound to the 'central' thiol group, with the release of carbon dioxide.

$$
\begin{array}{l} HO_2CCH_2COS_C \\ \quad\diagdown \\ \qquad \text{Enzyme} \\ \quad\diagup \\ CH_3COS_P \end{array} \xrightleftharpoons{[\beta\text{-Ketoacyl-ACP synthetase}]} \begin{array}{l} CH_3COCH_2COS_C \\ \quad\diagdown \\ \qquad \text{Enzyme} + CO_2 \\ \quad\diagup \\ HS_P \end{array} \qquad (5)
$$

Acetoacetyl-enzyme

The acetoacetyl group then undergoes reduction to D(−)-β-hydroxyacyl-enzyme,

$$\begin{matrix} CH_3COCH_2COS_C \\ \diagdown \\ \text{Enzyme} + NADPH + H^+ \\ \diagup \\ HS_P \end{matrix} \xrightleftharpoons[\text{[}\beta\text{-Ketoacyl-ACP reductase]}]{} \begin{matrix} CH_3CHCH_2COS_C \\ | \quad\quad \diagdown \\ OH \quad \text{Enzyme} + NADP \\ \diagup \\ HS_P \end{matrix} \quad (6)$$

D(−)-β-Hydroxybutyryl-enzyme

dehydration to the α,β-unsaturated acyl-enzyme,

$$\begin{matrix} CH_3CHCH_2COS_C \\ | \quad\quad \diagdown \\ OH \quad \text{Enzyme} \\ \diagup \\ HS_P \end{matrix} \xrightleftharpoons[]{\text{[}\beta\text{-Hydroxyacyl-ACP dehydrase]}} \begin{matrix} CH_3CH=CHCOS_C \\ \diagdown \\ \text{Enzyme} + H_2O \\ \diagup \\ HS_P \end{matrix} \quad (7)$$

Crotonyl-enzyme

and then a second reduction using NADPH and enzyme-bound flavin mononucleotide (FMN) as hydrogen carrier, resulting in the formation of butyrate.

$$\begin{matrix} CH_3CH=CHCOS_C \\ \diagdown \\ \text{Enzyme} + NADPH + H^+ \\ \diagup \\ HS_P \end{matrix} \xrightarrow[\text{[Enoyl-ACP reductase]}]{FMN} \begin{matrix} CH_3CH_2CH_2COS_C \\ \diagdown \\ \text{Enzyme} + NADP \\ \diagup \\ HS_P \end{matrix} \quad (8)$$

Butyryl-enzyme

The 'central' thiol group is then vacated for other chain-lengthening sequences by transfer of the butyryl derivative to the 'peripheral' group and a new malonyl residue is introduced on to the 'central' group. Condensation between the butyryl and malonyl groups takes place and the reaction cycle

Table 6.1 The major differences between fungal and bacterial mechanisms of fatty acid synthesis

Fungi	Bacteria
Fatty acid synthetase is a multienzyme, particulate complex	Fatty acid synthetase consists of individual, soluble enzymes
Intermediates in fatty acid synthesis are linked to the 'central' SH group analogous to bacterial ACP. The 'peripheral' SH group is used initially for acetyl-CoA binding and then the termination of the cycle	Intermediates of fatty acid synthesis are linked to ACP
Intermediates are firmly attached to the enzymes involved, via the SH groups	Intermediates are not firmly attached to individual enzymes and may migrate to serve as substrates in subsequent reactions
The second reduction in biosynthesis (Eqn 8) requires NADPH and FMN as coenzymes	The second reduction does not require FMN
Unsaturated fatty acids are formed by desaturation and not directly by the synthetase	Synthetase may form a C18 monounsaturated fatty acid entirely independent of O_2 in some bacteria
Polyunsaturated fatty acids are formed by desaturation	Polyunsaturated fatty acids are not formed

commences again. When a long-chain acyl group of sixteen or eighteen carbon atoms has been synthesised, it is transferred from the 'central' thiol group to CoA by action of a transferase enzyme, liberating a long-chain acyl-CoA and the free enzyme complex. The complex may then react with further molecules of acetyl-CoA and malonyl-CoA to initiate the sequence again. Gas-radiochromatography has shown that the fatty acid products of the multi-enzyme complex of *Pythium debaryanum* were a mixture of free and esterified palmitic and stearic acids (Law & Burton, 1973).

The mechanism of long-chain fatty acid formation in bacteria is similar to that found in the fungi and the major differences between the two systems are in Table 6.1. Table 6.2 indicates the major differences between fatty acid synthesis and β-oxidation, although the basic mechanism is essentially the same for both.

Elongation of the fatty acid chain

All fatty acid synthesising systems produce predominantly sixteen or eighteen carbon atom saturated fatty acids. But longer-chain fatty acids exist in filamentous fungi, bacteria and other organisms (*e.g.* fish) and so alternative mechanisms for their production must exist.

Two enzyme systems have been isolated which increase the chain length of preformed long-chain saturated or unsaturated fatty acids by the addition of C_2 units. One of these, found in the mitochondria of animals (Harlan & Wakil, 1963) and plants (Barron, Squires & Stumpf, 1961), uses acetyl-CoA as the source of C_2 units. The second system found in the microsomal

Table 6.2 The major differences between fatty acid biosynthesis and β-oxidation

Fatty acid biosynthesis	β-oxidation
Malonate is the source of additional C_2 units	Acetate is the source of additional C_2 units
Acyl carrier is ACP and substrates are attached to 4′-phosphopantetheine	Acyl carrier is CoA
Redox coenzymes are NADP, NADPH (and FMN in yeast)	Redox coenzymes are NAD and NADH
β-Hydroxyacyl derivative is the D(−) stereoisomer	β-Hydroxyacyl derivative is the L(+) stereoisomer
Site of synthesis is in the soluble fraction of the cell	Site of oxidation is in the mitochondria

fraction of mammalian cells, *e.g.* liver microsomes (Guchhait, Putz & Porter, 1966), uses C_2 units from malonyl-CoA for the elongation of acyl-CoA derivatives containing ten or more carbon atoms (Stoffel & Ach, 1964; Nugteren, 1965).

Enzymes for chain elongation must occur in fungi and other eukaryotic micro-organisms, although they have been little studied, since there is labelling of longer-chain fatty acids after the incorporation of (^{14}C)-palmitate into the growth medium of fungi, algae and protozoa (Erwin, Hulanicka & Bloch, 1964).

Elongation by the mitochondria

Because only acetyl-CoA is used as the C_2 unit, it is difficult to understand how the initial condensation reaction occurs without the presence of malonyl-CoA. The enzyme taking part in this condensation may be dependent on pyridoxal phosphate for activity (Wakil, 1961) and may be connected with activation of acetyl-CoA by formation of a Schiff's base so that the reaction is equivalent to that after carboxylation (Goodwin & Mercer, 1972). Another explanation could be that there is initial formation of a β-ketoacyl-CoA during condensation, followed by reduction, dehydration and further reduction to form a fully saturated acyl-CoA containing two more carbon atoms than the original primer (Seubert, Lamberts, Kramer & Ohly, 1968).

Elongation of long-chain fatty acids requires an acyl primer, usually palmityl-CoA and acetyl-CoA, NADPH and NADH. Synthesis is insensitive to avidin and therefore independent of malonyl-CoA (Harlan & Wakil, 1963).

In the perfused rat liver, the chain elongation mechanism uses only palmitate synthesized *de novo* and little or no pre-formed palmitate (Wadke *et al.*, 1973) and cis-vaccenate is formed by elongation of palmitoleyl-CoA, introduction of the double bond occurring at the C_{16} and not the C_{18} stage (Holloway & Wakil, 1964). This pathway also occurs in the mitochondria of beef heart, which elongates acids from 6–20 carbon atoms (Dahlen & Porter, 1968).

For mitochondrial elongation, palmitate and stearate pass through the inner membrane of the mitochondria as their CoA and carnitine derivatives (Greville & Tubbs, 1968). Although L-palmitylcarnitine is a more effective substrate primer than palmityl-CoA for chain elongation in intact guinea pig mitochondria, in fact palmityl-CoA is the immediate precursor for fatty acid chain elongation (Warshaw & Kimura, 1973*a*).

Elongation by the microsomes

The microsomal fraction of liver contains an enzyme system which catalyses the elongation of saturated and unsaturated acyl-CoA thioesters containing ten or more C units by the addition of C_2 units from malonyl-CoA in the presence of NADH (Lorch, Abraham & Chaikoff, 1963; Guchhait, Putz & Porter, 1966) but are unable to synthesize fatty acids *de novo* (Warshaw & Kimura, 1973*b*). Unsaturated fatty acids are elongated more rapidly than the corresponding saturated acids, the higher the degree of unsaturation, the higher the rate of elongation, suggesting that this system may be responsible for the synthesis of long-chain polyunsaturated acids, especially since the products are usually C_{18} and C_{20} fatty acids with a large portion unsaturated (Nugteren, 1965).

6.3 The Biosynthesis of Monounsaturated Fatty Acids

Fatty acids containing a single double bond are commonly found in eukaryotes and prokaryotes. Palmitoleic and oleic acids are the most common monoenes in fungi and other eukaryotes, whereas in bacteria many positional isomers occur which vary from species to species (O'Leary, 1967).

Double bond incorporation into fatty acids proceeds by at least two mechanisms. One pathway is anaerobic, whereby a double bond is introduced into a medium-chain-length fatty acid, followed by chain elongation. This pathway is unused by fungi and other higher protists (O'Leary, 1967). The other is direct desaturation of a pre-formed fatty acid using molecular oxygen by enzymes known as oxygenases. The introduction of other double bonds into a monoenoic fatty acid occurs only by the aerobic pathway.

Anaerobic pathway

Since a large number of micro-organisms are able to synthesise monoenoic fatty acids on fat-free media under anaerobic conditions and, as mentioned earlier, *Escherichia coli* fatty acid synthetase, in contrast to fungal and mammalian synthetases, produces unsaturated as well as saturated fatty acids, studies were made with obligate anaerobes (*cf Clostridium*, Goldfine & Bloch, 1961) and the facultative anaerobe *E. coli* (Bloch, 1962) to elucidate the mechanism of this pathway.

The acyl chains at the ten carbon atom stage (D(−)-β-hydroxydecanoyl ACP) are diverted from the sequence leading to saturated fatty acids to one leading to monounsaturated fatty acids (Brock, Kass & Bloch, 1967). β-Hydroxydecanoyl-ACP dehydrase is the enzyme that causes the branching by catalysing the dehydration of the hydroxydecanoyl-ACP intermediate to *cis*-3-decanoyl-ACP, besides the usual *trans*-2-decanoyl ACP. Since enoyl-ACP reductase of the synthetase system cannot reduce the *cis*-3-decanoyl-ACP, the double bond remains intact and is elongated to an unsaturated fatty acid.

Aerobic pathway

Bloomfield & Bloch (1960) demonstrated that anaerobically maintained cells of *Saccharomyces cerevisiae* only synthesised saturated fatty acids, whereas unsaturated fatty acids were formed when the cells were grown aerobically, indicating that there existed an independent pathway requiring molecular oxygen for the biosynthesis of monounsaturated fatty acids. They isolated a particulate enzyme system from the microsomal fraction of the cell which was able to desaturate palmityl-CoA directly to palmitoleyl-CoA and stearyl-CoA to oleyl-CoA, requiring molecular oxygen and NADH or NADPH, a system also found in *Candida utilis* (Yuan & Bloch, 1961). Similar enzymes have been isolated from the mould *Penicillium chrysogenum* (Bennett & Quackenbush, 1969; Bloch *et al.*, 1961), bacteria (Fulco & Bloch, 1964) and animals (Jones, Holloway, Peluffo & Wakil, 1969). These systems catalyse the insertion of a double bond into the 9,10-position of long-chain acyl-CoA compounds for which they are specific and so are unable to use acyl-ACP or the free saturated fatty acid (Bloomfield & Bloch, 1960).

The stearyl-CoA desaturase system purified from hen liver microsomes requires the presence of micellar dispersions of phospholipid triacylglycerol and free fatty acids for activity (Jones *et al.*, 1969). When fractionated, it yields several subparticles requiring NADH, cytochrome b_5 reductase and cytochrome b_5 for oleic acid production, which is similar to the microsomal NADPH-cytochrome P_{450} system concerned with hydroxylation, but which is unable to desaturate (Holloway & Wakil, 1970). A cyanide-sensitive factor has also been implicated in this system (Oshino, Imai & Sato, 1966).

Similar desaturase systems exist in algae and plants but there are found to be differences when comparisons are made with those in fungi and animals. In spinach and photoauxotropic *Euglena*, the system is fully soluble and associated with the chloroplasts. *Euglena gracilis* will only desaturate stearyl-CoA when grown heterotrophically and stearyl-CoA or stearyl-ACP when grown photosynthetically, the system requiring NADPH, ferridoxin-NADP reductase, ferridoxin, molecular oxygen and the desaturase (Nagai & Bloch, 1968).

In a series of experiments (Fulco, 1969, 1970, 1972*a*, *b*), it was shown that a number of species of *Bacillus* desaturated palmitic acid to *cis*-5-hexadecanoic acid in a reaction that is both oxygen and iron-dependent and temperature-regulated. The *in vivo* substrate specificities and positional specificity of double bond insertion for six bacilli desaturases were also demonstrated (Quint & Fulco, 1973).

6.4 The Biosynthesis of Polyunsaturated Fatty Acids

Long-chain fatty acids containing two or more double bonds are found both in abundance and variety in fungi and other eukaryotic cells (Shaw, 1966). Fungi usually only synthesize linoleic, α-linolenic, and in some cases, γ-linolenic acids (Shaw, 1965). By contrast, polyunsaturated fatty acids do not generally occur in bacteria (O'Leary, 1967), although *Bacillus licheniformis* produces small amounts of polyunsaturates under certain environmental conditions (Fulco, 1970) and blue-green algae are able to synthesize linoleic acid and sometimes α-linolenic acid (Levin, Lennarz & Bloch, 1964). The double bonds of the polyunsaturates are always separated by a methylene group and the acids known as 'methylene-interrupted' polyunsaturated fatty acids. It is easier to recognize the polyenoic acids by numbering the double bonds from the methyl end rather than the carboxyl end of the chain and the designation ω categorizes the terminal portion (Schlenk & Sand, 1967). The three most important families of polyunsaturated fatty acids are $\omega 9$, $\omega 6$ and $\omega 3$, and the first members of each family are oleic, linoleic and α-linolenic respectively (Gurr & James, 1971).

Mechanism of synthesis

Studies using *Penicillium chrysogenum* (Bennett & Quackenbush, 1969), *Torulopsis utilis* (Yuan & Bloch, 1961) and other eukaryotic microorganisms, such as members of the algae (Erwin, Hulanicka & Bloch, 1964), have shown that polyenoic fatty acids are formed by a series of alternate desaturations and elongations of the monoenoic fatty acid, oleic acid.

Enzyme systems converting monoenoic fatty acids to polyenoic fatty acids, isolated from *Torulopsis utilis* (Yuan & Bloch, 1961), *Chlorella vulgaris* (Harris & James, 1965) and rat liver (Nugteren, 1962), show that, like the oxidative desaturations leading to monoenoic fatty acids, these further desaturations involve the CoA thioester of oleic acid as substrate, oxygen, NADH or NADPH as cofactors and are firmly particle bound. In *Chlorella vulgaris*, the system is located in the chloroplasts (Harris & James, 1965), whereas in *Torulopsis utilis* the system is found in the microsomal pellet and supernatant (Talamo, Chang & Bloch, 1973). Additional cofactors required for the mechanism of polyenoic fatty acid synthesis, but not monoenoic fatty acid synthesis, are located in the supernatant fraction (Meyer & Bloch, 1963; Talamo, Chang & Bloch, 1973). Sterols have been suggested as the precursors of these cofactors in the protozoan *Tetrahymena setifera* (Erwin, Beach & Holz, 1966).

It is thought that an alternative pathway may exist for producing polyunsaturated fatty acids involving lipid-linked fatty acid as substrate. Experiments with *Chlorella* and *Euglena* which can synthesize animal (*e.g.* arachidonic) and plant (*e.g.* α-linolenic) polyenoic fatty acids, have shown that the animal-type fatty acids accumulate in the phospholipids and the plant-type fatty acids in the galactolipids. Labelling experiments have shown that desaturation of the fatty acids of the lipids takes place after *de novo* synthesis of the lipid molecules and not desaturation of the fatty acids and then incorporation into the lipids (Nichols, James & Breuer, 1967; Safford & Nichols, 1970). Renkonen & Bloch (1969) have shown that *Euglena* uses oleyl-ACP for synthesis of galactolipids but oleyl-CoA for synthesis of phospholipids.

Observations such as these seem to implicate certain complex lipids in the mechanism of desaturation. It has been shown that there is evidence in support of a desaturating pathway in *Neurospora crassa* involving oleyl lipid (Baker & Lynen, 1971) and that in *Chlorella vulgaris* (Gurr, Robinson & James, 1969) and *Torulopsis utilis* (Talamo, Chang & Bloch, 1973) labelled oleyl-phosphatidylcholine catalysed the formation of labelled linoleyl-phosphatidylcholine. This evidence suggests that there is an interplay of two systems, whereby oleyl-CoA may be desaturated to linoleyl-CoA and used for producing other polyunsaturated fatty acids by desaturation, or the oleyl-CoA may synthesise lipid-linked oleyl-CoA, further desaturation occurring on the lipid (Renkonen & Bloch, 1969).

The conversion of one polyenoic fatty acid into another of the same family by alternate desaturation and elongations may proceed by two alternative pathways. In the $\omega6$ family, synthesis is from linoleic acid and all additional double bonds are introduced towards the carboxyl end of the chain. The organism concerned can usually form linoleic acid from oleic acid, but in vertebrates, since the second and subsequent double bonds are introduced between the first double bond and the carboxyl group, the inability of animals to desaturate oleic acid toward the methyl end of the chain means that linoleic acid becomes a dietary requirement or an 'essential fatty acid' (Mead, 1968; Gurr & James, 1971). When linoleic acid is present, arachidonic acid may be produced by desaturation of linoleic to γ-linolenic acid followed by chain elongation, the system found in animals, such as rats (Nugteren, 1962), or linoleic acid may be chain elongated to 11,14-eicosadienoic acid followed by desaturation towards the carboxyl end, such as in the alga, *Euglena gracilis* (Hulanicka, Erwin & Bloch, 1964).

Interconversion between the $\omega9$, $\omega6$ and $\omega3$ families of polyenoic fatty acids has been shown in the filamentous fungi (Erwin & Bloch, 1963) and other higher protists such as the yeasts (Bloch *et al.*, 1961), whereby oleic acid (18 : 1 $\omega9$) is desaturated to linoleic acid (18 : 2 $\omega6$) and then to α-linolenic acid (18 : 3 $\omega3$).

Some *Phycomycetes* characteristically contain γ-lineolic acid (18 : 3 $\omega6$) (Shaw, 1965) and earlier evidence suggested that it is formed by the desaturation of oleate in *Phycomyces blakesleeanus* (Bernhard, Abisch & Wagner, 1957).

6.5 The Biosynthesis of Branched-Chain, Cyclopropane and Hydroxy Fatty Acids

There are no reports of the identification of cyclopropane and branched-chain fatty acids in the filamentous fungi and very few reports of hydroxy fatty acids (*e.g.* Morris, 1968), although they are found abundantly in bacteria (O'Leary, 1967). For this reason, the biosynthesis of these acids will not be discussed.

6.6 The Influence of Environmental Parameters on Fungal Fatty Acid Biosynthesis

The fatty acid composition of fungi is very easily modified depending on the conditions imposed during culture. Such conditions include the medium composition, aeration, temperature, pH and period of incubation.

Medium composition

Increasing the concentration of the carbon source in the medium has been reported to cause an increase in the level of saturated fatty acids, such as in the mould *Gibberella fujikuroi* (Borrow *et al.*, 1961), the yeast *Saccharomyces cerevisiae* (Brown & Johnson, 1970) and in *Escherichia coli* (Marr & Ingraham, 1962). However, completely opposite effects have been observed with other micro-organisms, such as the moulds *Aspergillus nidulans* (Singh

& Datt, 1957), *Blakeslea trispora* (Dedyukhina & Bekhtereva, 1969) and the yeast *Candida utilis* (Babij, Moss & Ralph, 1969).

Many micro-organisms are able to oxidize *n*-alkanes as carbon source to the corresponding fatty acid. In this way, the composition of the fatty acids may be altered depending on the chain length of the *n*-alkane, such as with *Nocardia* species (Davis, 1964) and *Candida* species (Ratledge, 1970). Supplementation of the carbon source by the addition of small amounts of fatty acids may also influence the fatty acids synthesised (Kamio, Kanegasaki & Takahashi, 1970).

An increase in the concentration of nitrogen in the medium has been shown to increase the saturated fatty acids of some filamentous fungi (Gregory & Woodbine, 1953; Singh & Datt, 1958). However, as with the effect of carbon concentration, this is not true for all micro-organisms, as is the case with *Escherichia coli* (Marr & Ingraham, 1962) and *Rhodotorula gracilis* (Enebo & Iwamoto, 1966), which produce higher levels of unsaturated fatty acids with increasing concentration of nitrogen. Kates & Hagen (1964) have demonstrated that in *Serratia* the source of nitrogen also has an effect on the fatty acid composition.

Aeration

Increased oxygen tension has been reported to elevate the unsaturated fatty acid content in many fungi, such as in *Mucor* species (Sumner, Morgan & Evans, 1969) and *Candida utilis* (Babij, Moss & Ralph, 1969).

As mentioned previously, the biosynthesis of unsaturated acids is regulated by desaturase enzymes requiring molecular oxygen as a cofactor and, therefore, the oxygen availability within certain limits becomes rate-limiting for unsaturated fatty acid production.

Temperature

There are conflicting reports concerning the effect of the temperature of incubation on the fatty acid content of fungi. The majority suggest that the lower the temperature the greater the increase in unsaturated acids (see *e.g.* Gregory & Woodbine, 1953; Kates & Baxter, 1962). In some fungi, however, an increase in unsaturation occurs at higher growth temperatures (Gad & Hassan, 1964; Shaw, 1966), while in others there is little relationship between temperature and fatty acid composition (Bowman & Mumma, 1967). It has been suggested that the degree of unsaturation may be a physiological response to the changing environmental temperature facilitating the survival of the organism (Caldwell & Vernberg, 1970).

Kates & Baxter (1962) proposed that the rates of both synthesis *and* degradation of unsaturated acids are temperature dependent and at lower temperatures their synthesis is less retarded than their degradation. Evidence for this postulate has been produced by Meyer & Bloch (1963) who demonstrated that the ability of enzyme systems from *Torulopsis utilis* to produce unsaturated fatty acids was dependent on the temperature at which the yeast was grown and Chang & Fulco (1973) showed that the lower the temperature of incubation of *Bacillus licheniformis*, the greater the rate and extent of desaturation of palmitate. The rate of temperature-mediated inactivation of six desaturases of *Bacillus* has been determined by Quint & Fulco (1973).

The temperature mechanism is complicated by the fact that a decreasing incubation temperature increases oxygen solubility and, as mentioned earlier, oxygen tension is a significant factor in the relative amounts of saturated and unsaturated acids produced.

Desaturase enzymes producing polyunsaturated fatty acids do not operate above the mesophilic temperature range, which may be one reason why prokaryotes but never eukaryotes are found in truly thermophilic environments (Brock, 1967).

Hydrogen ion concentration

Wide changes in the percentages of free fatty acids can be made by altering the pH of the medium (Dyatlovitskaya *et al.*, 1969). An increase in unsaturated fatty acids with increasing pH has been reported in many fungi (Prill, Wenck & Peterson, 1935; Singh & Walker, 1956; Kessell, 1968).

Period of incubation

In general, increasing incubation time, as with increasing temperature, increases the degree of saturation in fungi (Van Etten & Gottlieb, 1965; Shaw, 1966; Sumner, Morgan & Evans, 1969; Soumalainen & Keranen, 1968). However, this mechanism is not true for all fungi, since *Cephalosporium* shows a general increase in unsaturated fatty acids with culture age (Gellerman & Schlenk, 1965).

The fact that there is a tendency for oxygen to become limiting in older cultures must also be taken into account (O'Leary, 1967).

For further information concerning the effect of cultural conditions on fatty acid biosynthesis of filamentous fungi, see Walker (1975), Shaw (1974) and Taylor (1974).

The authors are grateful to Steven Shaw, Mrs J. Wortley and to our library and secretarial colleagues in preparing this contribution.

6.7 References

ALBERTS, A. W. & VAGELOS, P. R. (1966). Acyl carrier protein. VIII. Studies of acyl carrier protein and coenzyme A in *Escherichia coli* pantothenate or β-alanine auxotrophs. *The Journal of Biological Chemistry* **241**, 5201–4.

BABIJ, T., MOSS, F. J. & RALPH, B. J. (1969). Effects of oxygen and glucose levels on lipid composition of yeast *Candida utilis* grown in continuous culture. *Biotechnology and Bioengineering* **11**, 593–603.

BAKER, N. & LYNEN, F. (1971). Factors involved in fatty acyl CoA desaturation by fungal microsomes. The relative roles of acyl CoA and phospholipids as substrates. *European Journal of Biochemistry* **19**, 200–10.

BARRON, E. J., SQUIRES, C. L. & STUMPF, P. K. (1961). Fat metabolism in higher plants. XV. Enzymic synthesis of fatty acids by an extract of avocado mesocarp. *The Journal of Biological Chemistry* **236**, 2610–4.

BENNETT, A. S. & QUACKENBUSH, F. W. (1969). Synthesis of unsaturated fatty acids by *Penicillium chrysogenum*. *Archives of Biochemistry and Biophysics* **130**, 567–72.

BERNHARD, K., ABISCH, L. & WAGNER, H. (1957). Versuche zur Aufklärung der Fettsäurebildungen bei Mikroorganismen. *Helvetica Chimica Acta* **40**, 1292–8.

BLOCH, K. (1960). *Lipid metabolism*. New York: Wiley.

BLOCH, K. (1962). Oxygen and biosynthetic patterns. *Federation Proceedings* **21**, 1058–63.

BLOCH, K., BARONOWSKY, P., GOLDFINE, H., LENNARZ, W. J., LIGHT, R., NORRIS, A. T. & SCHEUERBRANDT, G. (1961). Lipid metabolism. Biosynthesis and metabolism of unsaturated fatty acids. *Federation Proceedings* **20**, 921–7.

BLOOMFIELD, D. K. & BLOCH, K. (1960). The formation of Δ^9-unsaturated fatty acids. *The Journal of Biological Chemistry* **235**, 337–45.

BORROW, A., JEFFREYS, E. G., KESSELL, R. H. J., LLOYD, E. C., LLOYD, P. B. & NIXON, I. S. (1961). The metabolism of *Gibberella fujikuroi* in stirred culture. *Canadian Journal of Microbiology* **7**, 227–76.

BOWMAN, R. D. & MUMMA, R. O. (1967). The lipids of *Pythium ultimum*. *Biochimica et Biophysica Acta* **144**, 501–10.

BRESSLER, R. & WAKIL, S. J. (1961). Studies on the mechanism of fatty acid synthesis. IX. The conversion of malonyl-coenzyme A to long chain fatty acids. *The Journal of Biological Chemistry* **236**, 1643–51.

BROCK, D. J. H., KASS, L. R. & BLOCH, K. (1967). β-Hydroxydecanoyl thioester dehydrase. II. Mode of action. *The Journal of Biological Chemistry* **242**, 4432–40.

BROCK, T. D. (1967). Life at high temperatures. *Science* **158**, 1012–19.

BROWN, C. M. & JOHNSON, B. (1970). Influence of the concentration of glucose and galactose on the physiology of *Saccharomyces cerevisiae* in continuous culture. *The Journal of General Microbiology* **64**, 279–87.

BUTTERWORTH, P. H. W. & BLOCH, K. (1970). Comparative aspects of fatty acid synthesis in *Bacillus subtilis* and *Escherichia coli*. *European Journal of Biochemistry* **12**, 496–501.

BUTTERWORTH, P. H. W., YANG, P. C., BOCK, R. M. & PORTER, J. W. (1967). The partial dissociation and the reassociation of the pigeon liver fatty acid synthetase complex. *The Journal of Biological Chemistry* **242**, 3508–16.

CALDWELL, R. S. & VERNBERG, F. J. (1970). The influence of acclimation temperature on the lipid composition of fish gill mitochondria. *Comparative Biochemistry and Physiology* **34**, 174–91.

CHANG, H. C., SEIDMAN, I., TEEBOR, G. & LANE, M. D. (1967). Liver acetyl-CoA carboxylase and fatty acid synthetase: relative activities in the normal state and in hereditary obesity. *Biochemical and Biophysical Research Communications* **28**, 682–6.

CHANG, N. & FULCO, A. F. (1973). The effects of temperature and fatty acid structure on lipid metabolism in *Bacillus licheniformis* 9259. *Biochimica et Biophysica Acta* **296**, 287–99.

DAHLEN, J. V. & PORTER, J. W. (1968). Studies on the synthesis of fatty acids by a beef heart mitochondrial enzyme system. *Archives of Biochemistry and Biophysics* **127**, 207–23.

DAVIS, J. B. (1964). Microbial incorporation of fatty acids derived from *n*-alkanes into glycerides and waxes. *Applied Microbiology* **12**, 210–4.

DEDYUKHINA, E. G. & BEKHTEREVA, M. N. (1969). Formation of lipids by *Blakeslea trispora* as a function of the composition of the nutrient medium. *Mikrobiologiya* **38**(5), 653–8.

DEUEL, H. J. (1951, 1955, 1957). *The lipids*, 3 vols. New York: Interscience.

DILS, R. & POPJÁK, G. (1962). Biosynthesis of fatty acids in cell-free preparations. 5. Synthesis of fatty acids from acetate in extracts of lactating rat mammary gland. *The Biochemical Journal* **83**, 41–51.

DYATLOVITSKAYA, E. V., GRESHNYKH, U. P., ZHDANNIKOVA, E., KOZLOVA, L. I. & BERGELSON, L. D. (1969). pH effect on the lipid composition of *Candida* yeast grown on *n*-alkanes. *Prikladnaya Biokhimiya i Mikrobiologiya* **5**, 511–4.

ENEBO, L. & IWAMOTO, H. (1966). Effects of cultivation temperature on fatty acid composition in *Rhodotorula gracilis*. *Acta Chemica Scandinavica* **20**, 439–43.

ERNST-FONBERG, M. L. (1973). Fatty acid synthetase in *Euglena gracilis* var. *bacillarius*. Characterisation of an acyl carrier protein dependent system. *Biochemistry* **12**, 2449–55.

ERWIN, J. & BLOCH, K. (1963). Lipid metabolism of a ciliated protozoan. *The Journal of Biological Chemistry* **238**, 1618–24.

ERWIN, J., BEACH, D. & HOLZ, G. G. (1966). Effect of dietary cholesterol on unsaturated fatty acid biosynthesis in a ciliated protozoan. *Biochimica et Biophysica Acta* **125**, 614–6.

ERWIN, J., HULANICKA, D. & BLOCH, K. (1964). Comparative aspects of unsaturated fatty acid synthesis. *Comparative Biochemistry and Physiology* **12**, 191–207.

FISCHER, E. (1890). Synthesen in der Zuckergruppe. *Berichte der Deutschen Chemischen Gesellschaft* **23**, 2114–41.

FULCO, A. J. (1969). The biosynthesis of unsaturated fatty acids by bacilli. I. Temperature induction of the desaturation reaction. *The Journal of Biological Chemistry* **244**, 889–95.

FULCO, A. J. (1970). The biosynthesis of unsaturated fatty acids by bacilli. II. Temperature-dependent biosynthesis of polyunsaturated fatty acids. *The Journal of Biological Chemistry* **245**, 2985–90.

FULCO, A. J. (1972*a*). The biosynthesis of unsaturated fatty acids by bacilli. III. Uptake and utilisation of exogenous palmitate. *The Journal of Biological Chemistry* **247**, 3503–10.

FULCO, A. J. (1972*b*). The biosynthesis of unsaturated fatty acids by bacilli. IV. Temperature-mediated control mechanisms. *The Journal of Biological Chemistry* **247**, 3511–9.

FULCO, A. J. & BLOCH, K. (1964). Cofactor requirements for the formation of Δ^9-unsaturated fatty acids in *Mycobacterium phlei. The Journal of Biological Chemistry* **239**, 993–7.

GAD, A. M. & HASSAN, M. M. (1964). Chemistry of mould fat. VI. Influence of temperature and incubation on carbohydrate fat conversion in the mould *Aspergillus fischeri. Journal of Chemistry of the United Arab Republic* **7**, 31–41.

GANGULY, J. (1960). Studies on the mechanism of fatty acid synthesis. III. Biosynthesis of fatty acids from malonyl CoA. *Biochimica et Biophysica Acta* **40**, 110–18.

GELLERMAN, J. L. & SCHLENK, A. (1965). Preparation of fatty acids labelled with ^{14}C from *Ochromonas danica. The Journal of Protozoology* **12**, 178–89.

GIBSON, D. M., TITCHENER, E. B. & WAKIL, S. J. (1958). Studies on the mechanism of fatty acid synthesis. V. Bicarbonate requirement for the synthesis of long-chain fatty acids. *Biochimica et Biophysica Acta* **30**, 376–83.

GOLDFINE, H. & BLOCH, K. (1961). On the origin of unsaturated fatty acids in *Clostridia. The Journal of Biological Chemistry* **236**, 2596–601.

GOODRIDGE, A. G. (1973). Regulation of fatty acid synthesis in isolated hepatocytes. Evidence for a physiological role for long-chain fatty acyl coenzyme A and citrate. *The Journal of Biological Chemistry* **248**, 4318–26.

GOODWIN, T. W. & MERCER, E. I. (1972). *Introduction to plant biochemistry.* London: Pergamon Press.

GREGORY, M. & WOODBINE, M. (1953). Microbiological synthesis of fat. *Journal of Experimental Botany* **4**, 314–18.

GREVILLE, G. D. & TUBBS, P. K. (1968). The catabolism of long chain fatty acids in mammalian tissues. In *Essays in biochemistry*, pp. 155–212, Vol. 4, edited by P. N. Campbell and G. D. Greville. London & New York: Academic Press Inc.

GUCHHAIT, R. B., PUTZ, G. R. & PORTER, J. W. (1966). Synthesis of long-chain fatty acids by microsomes of pigeon liver. *Archives of Biochemistry and Biophysics* **117**, 541–9.

GURR, M. I. & JAMES, A. T. (1971). *Lipid biochemistry: an introduction.* London: Chapman & Hall.

GURR, M. I., ROBINSON, M. P. & JAMES, A. T. (1969). The mechanism of formation of polyunsaturated fatty acids by photosynthetic tissue. The tight coupling of oleate desaturation with phospholipid synthesis in *Chlorella vulgaris. European Journal of Biochemistry* **9**, 70–8.

HAEHN, H. (1921). Uber die Möglichkeit der Fettsynthese durch Pilz-bezw. Hefeenzyme. *Zeitschrift für technische Biologie* **9**, 217–24.

HAEHN, H. & KINTOFF, W. (1923). Uber den chemischen Mechanismus bei der Fettbildung in der lebenden Zelle. *Berichte der Deutschen Chemischen Gesellschaft* **56**, 439–45.

HAEHN, H. & KINTOFF, W. (1925*a*). Beitrag über den chemischen Mechanismus der Fettbildung aus Zucker. *Chemie der Zelle und Gewebe* **12**, 115–56.

HAEHN, H. & KINTOFF, W. (1925*b*). Uber die biochemische Fettbildung aus Zucker. *Wochenschrift für Brauerei* **42**, 218–29.

HANCOCK, W. S., MARSHALL, G. R. & VAGELOS, P. R. (1973). Acyl carrier protein. XX. Chemical synthesis and characterisation of analogues of acyl carrier protein. *The Journal of Biological Chemistry* **248**, 2424–34.

HARLAN, W. R. & WAKIL, S. J. (1963). Synthesis of fatty acids in animal tissues. I. Incorporation of ^{14}C-acetyl coenzyme A into a variety of long-chain fatty acids by subcellular particles. *The Journal of Biological Chemistry* **238**, 3216–23.

HARRIS, R. V. & JAMES, A. T. (1965). Linoleic and α-linolenic acid biosynthesis in plant leaves and a green alga. *Biochimica et Biophysica Acta* **106**, 456–64.

HOLLOWAY, P. W. & WAKIL, S. J. (1964). Synthesis of fatty acids in animal tissues. II. The occurrence and biosynthesis of cis-vaccenic acid. *The Journal of Biological Chemistry* **239**, 2489–95.

HOLLOWAY, P. W. & WAKIL, S. J. (1970). Requirement for reduced diphosphopyridine nucleotide-cytochrome b_5 reductase in stearyl-coenzyme A desaturation. *The Journal of Biological Chemistry* **245**, 1862–5.

HOPPE-SEYLER, F. (1879*a*). Uber Gährungsprocesse. Synthese bei Gährungen. *Zeitschrift für physiologische Chemie* **3**, 351–61.

HOPPE-SEYLER, F. (1879*b*). Uber Lecithin in der Hefe. *Zeitschrift für physiologische Chemie* **3**, 374–80.

HULANICKA, D., ERWIN, J. & BLOCH, K. (1964). Lipid metabolism of *Euglena gracilis*. *The Journal of Biological Chemistry* **239**, 2778–87.

JONES, P. D., HOLLOWAY, P. W., PELUFFO, R. O. & WAKIL, S. J. (1969). A requirement for lipids by the microsomal stearyl coenzyme A desaturase. *The Journal of Biological Chemistry* **244**, 744–54.

KAMIO, Y., KANEGASAKI, S. & TAKAHASHI, H. (1970). Fatty acid and aldehyde compositions in phospholipids of *Selenomonas ruminantium* with reference to growth conditions. *Journal of General and Applied Microbiology* **16**, 29–37.

KATES, M. & BAXTER, R. M. (1962). Lipid composition of mesophilic and psychrophilic yeasts (*Candida* species) as influenced by environmental temperature. *Canadian Journal of Biochemistry and Physiology* **40**, 1213–27.

KATES, M. & HAGEN, P. O. (1964). Influence of temperature on fatty acid composition of psychrophilic and mesophilic *Serratia* species. *Canadian Journal of Biochemistry* **42**, 481–8.

KESSELL, R. H. J. (1968). Fatty acids of *Rhodotorula gracilis*: fat production in submerged culture and the particular effect of pH value. *Journal of Applied Bacteriology* **31**, 220–31.

KLEIN, H. P., VOLKMANN, C. M. & CHAO, F. C. (1967). Fatty acid synthetase of *Saccharomyces cerevisiae. Journal of Bacteriology* **93**, 1966–71.

KLEIN, H. P., VOLKMANN, C. M. & LEAFFER, M. A. (1967). Subcellular sites involved in lipid synthesis in *Saccharomyces cerevisiae. Journal of Bacteriology* **94**, 61–5.

KLEIN, H. P., VOLKMANN, C. M. & WEIBEL, J. (1967). Membranes of *Saccharomyces cerevisiae. Journal of Bacteriology* **94**, 475–81.

LAW, S. W. J. & BURTON, D. N. (1973). Fatty acid synthetase from *Pythium debaryanum. Canadian Journal of Biochemistry* **51**, 241–8.

LAWES, G. B. & GILBERT, J. H. (1860). On the composition of oxen, sheep and pigs and of their increase whilst fattening. *Journal of the Royal Agricultural Society of England* **21**, 433–88.

LENNARZ, W. J., LIGHT, R. J. & BLOCH, K. (1962). A fatty acid synthetase from *E. coli. Proceedings of the National Academy of Sciences, U.S.A.* **48**, 840–6.

LEVIN, E., LENNARZ, W. J. & BLOCH, K. (1964). Occurrence and localisation of α-linolenic acid containing galactolipids in the photosynthetic apparatus of *Anabaena variablis. Biochimica et Biophysica Acta* **84**, 471–4.

LIPMANN, F. C. (1945). Acetylation of sulfanilamide by liver homogenates and extracts. *The Journal of Biological Chemistry* **160**, 173–90.

LORCH, E., ABRAHAM, S. & CHAIKOFF, I. L. (1963). Fatty acid synthesis by complex systems. The possibility of regulation by microsomes. *Biochimica et Biophysica Acta* **70**, 726–41.

LYNEN, F. (1961). Biosynthesis of saturated fatty acids. *Federation Proceedings* **20**, 941–51.

LYNEN, F. (1967*a*). The role of biotin-dependent carboxylations in biosynthetic reactions. *The Biochemical Journal* **102**, 381–400.

LYNEN, F. (1967*b*). Multienzyme complex of fatty acid synthetase. In *Organisational biosynthesis* pp. 243–266. Edited by H. J. Vogel, J. O. Lampen and V. Bryson, New York & London: Academic Press.

LYNEN, F. (1972). Enzyme systems for fatty acid synthesis. *The Biochemical Journal* **128**, 1P.

LYNEN, F. & REICHERT, E. (1951). Zur chemischen Strukter der 'activierten Essigsäure'. *Angewandte Chemie* **63**, 47–8.

LYNEN, F., REICHERT, E. & RUEFF, L. (1951). Zum biologischen Abbau der Essigsäure. VI. 'Aktivierte Essigsäure', ihre Isolierung aus Hefe und ihre chemische Natur. *Annalen der Chemie* **574**, 1–32.

LYNEN, F., OESTERHELT, D., SCHWEIZER, E. & WILLECKE, K. (1968). The biosynthesis of fatty acids. In *Cellular compartmentalisation and control of fatty acid metabolism.* pp. 1–24. Edited by F. C. Gran, London & New York: Academic Press.

MAGNUS-LEVY, A. (1902). Uber den Aufbau der höher Fettsäuren aus Zucker. *Archives für Anatomie und Physiologie* 365–9.

MAJERUS, P. W., ALBERTS, A. W. & VAGELOS, P. R. (1964). The acyl carrier protein of fatty acid synthesis: purification, physical properties, and substrate binding site. *Proceedings of the National Academy of Sciences, U.S.A.* **51**, 1231–8.

MARR, A. G. & INGRAHAM, J. L. (1962). The effect of temperature on the composition of fatty acids in *Escherichia coli*. *Journal of Bacteriology* **84**, 1260–7.

MATSUMURA, S. & STUMPF, P. K. (1968). Fat metabolism in higher plants. Partial primary structure of spinach acyl carrier protein. *Archives of Biochemistry and Biophysics* **125**, 932–41.

MEAD, J. F. (1968). The metabolism of the polyunsaturated fatty acids. *Progress in the Chemistry of Fats and other Lipids* **9**, 161–92.

MEYER, F. & BLOCH, K. (1963). Effect of temperature on the enzymatic synthesis of unsaturated fatty acids in *Torulopsis utilis*. *Biochimica et Biophysica Acta* **77**, 671–2.

MORRIS, L. J. (1968). Fatty acid composition of *Claviceps* species. Occurrence of (+)-threo-9,10,dihydroxystearic acid. *Lipids* **3**, 260–1.

NAGAI, J. & BLOCH, K. (1967). Elongation of acyl carrier protein derivatives by bacterial and plant extracts. *The Journal of Biological Chemistry* **242**, 357–62.

NAGAI, J. & BLOCH, K. (1968). Enzymatic desaturation of stearyl acyl carrier protein. *The Journal of Biological Chemistry* **243**, 4626–33.

NENCKI, M. (1878). Uber den chemischen Mechanisms der Fäulniss. *Journal für praktische Chemie* **17**, 105–24.

NICHOLS, B. W., JAMES, A. T. & BREUER, J. (1967). Interrelationships between fatty acid biosynthesis and acyl-lipid synthesis in *Chlorella vulgaris*. *The Biochemical Journal* **104**, 486–96.

NIXON, J. E., PHILLIPS, G. T., ABRAMOVITZ, A. S. & PORTER, J. W. (1970). Intermediates of fatty acid synthesis: sites of binding to the pigeon liver fatty acid synthetase. *Archives of Biochemistry and Biophysics* **138**, 372–9.

NUGTEREN, D. H. (1962). Conversion *in vitro* of linoleic acid into γ-linolenic acid by rat-liver enzymes. *Biochimica et Biophysica Acta* **60**, 656–7.

NUGTEREN, D. H. (1965). The enzymic chain elongation of fatty acids by rat-liver microsomes. *Biochimica et Biophysica Acta* **106**, 280–90.

O'LEARY, W. M. (1967). *The chemistry and metabolism of microbial lipids*. Cleveland and New York: The World Publishing Company.

OSHINO, N., IMAI, Y. & SATO, R. (1966). Electron-transfer mechanisms associated with fatty acid desaturation catalysed by liver microsomes. *Biochimica et Biophysica Acta* **128**, 13–28.

OTTKE, R. C., TATUM, E. L., ZABIN, I. & BLOCH, K. (1951). Isotopic acetate and isovalerate in the synthesis of ergosterol by *Neurospora*. *The Journal of Biological Chemistry* **189**, 429–33.

PELCZAR, M. J. & REID, R. D. (1972). *Microbiology*, 3rd edition. New York: McGraw-Hill.

PHILLIPS, G. T., NIXON, J. E., DORSEY, J. A., BUTTERWORTH, P. H. W., CHESTERTON, C. J. & PORTER, J. W. (1970). The mechanism of synthesis of fatty acids by the pigeon liver enzyme system. *Archives of Biochemistry and Biophysics* **138**, 380–91.

PRESCOTT, D. J., ELOVSON, J. & VAGELOS, P. R. (1969). Acyl carrier protein. XI. The specificity of acyl carrier protein synthetase. *The Journal of Biological Chemistry* **244**, 4517–21.

PRILL, E. A., WENCK, P. R. & PETERSON, W. H. (1935). III. The chemistry of mould tissue. VI. Factors influencing the amount and nature of fat produced by *Aspergillus fischeri*. *The Biochemical Journal* **29**, 21–33.

PUGH, E. L. & WAKIL, S. J. (1965). Studies on the mechanism of fatty acid synthesis. XIV. The prosthetic group of acyl carrier protein and the mode of its attachment to the protein. *The Journal of Biological Chemistry* **240**, 4727–33.

QUINT, J. F. & FULCO, A. J. (1973). The biosynthesis of unsaturated fatty acids by bacilli. V. *In vivo* substrate specificities of fatty acid desaturases. *The Journal of Biological Chemistry* **248**, 6885–95.

RAPER, H. S. (1907). The condensation of acetaldehyde and its relation to the biochemical synthesis of fatty acids. *Journal of the Chemical Society* **91**, 1831–8.

RASMUSSEN, R. K. & KLEIN, H. P. (1968*a*). Activation of fatty acid synthesis in cell-free extracts of *Saccharomyces cerevisiae*. *Journal of Bacteriology* **95**, 157–61.

RASMUSSEN, R. K. & KLEIN, H. P. (1968b). Effects of metals on acetyl-coenzyme A carboxylase activity of *Saccharomyces cerevisiae*. *Journal of Bacteriology* **95**, 727–8.

RASMUSSEN, R. K. & KLEIN, H. P. (1968*c*). Mechanism of α-glycerophosphate regulation of acetyl-coenzyme A carboxylase of *Saccharomyces cerevisiae*. *Journal of Bacteriology* **95**, 1090–3.

RATLEDGE, C. (1970). Microbial conversions on *n*-alkanes to fatty acids: a new attempt to obtain economical microbial fats and fatty acids. *Chemistry and Industry* 843–54.

RENKONEN, O. & BLOCH, K. (1969). Biosynthesis of monogalactosyl diglycerides in photoauxotrophic *Euglena gracilis. The Journal of Biological Chemistry* **244**, 4899–903.

SAFFORD, R. & NICHOLS, B. W. (1970). Positional distribution of fatty acids in monogalactosyl diglyceride fractions from leaves and algae. Structural and metabolic studies. *Biochimica et Biophysica Acta* **210**, 57–64.

SCHLENK, H. & SAND, D. M. (1967). A new group of essential fatty acids and their comparison with other polyenoic fatty acids. *Biochimica et Biophysica Acta* **144**, 305–20.

SCHWEIZER, E., PICCININI, F., DUBA, C., GÜNTHER, S., RITTER, E. & LYNEN, F. (1970). Die Malonyl-Bindungsstellen des Fettsäure synthetase-Komplexes aus Hefe. *European Journal of Biochemistry* **15**, 483–99.

SEUBERT, W., LAMBERTS, I., KRAMER, R. & OHLY, B. (1968). On the mechanism of malonyl-CoA-independent fatty acid synthesis. I. The mechanism of elongation of long-chain fatty acids by acetyl-CoA. *Biochimica et Biophysica Acta* **164**, 498–517.

SHAW, R. (1965). The occurrence of γ-linolenic acid in fungi. *Biochimica et Biophysica Acta* **98**, 230–7.

SHAW, R. (1966). The polyunsaturated fatty acids of micro-organisms. *Advances in Lipid Research* **4**, 107–74.

SHAW, S. (1974). B.Sc. Honours dissertation, University of Nottingham.

SINGH, J. & DATT, I. (1957). Synthesis of fat by *Aspergillus nidulans*: Part I. Changes in the composition of *A. nidulans* fat with sugar concentration in the medium. *Journal of Scientific and Industrial Research* **16C**, 164–5.

SINGH, J. & DATT, I. (1958). Synthesis of fat by *Aspergillus nidulans*: Part II. Changes in the composition of *A. nidulans* fat with nitrogen concentration in the medium. *Journal of Scientific and Industrial Research* **17C**, 7–9.

SINGH, J. & WALKER, T. K. (1956). Changes in the composition of the fat of *Aspergillus nidulans* with age of culture. *The Biochemical Journal* **62**, 286–9.

SMEDLEY, I. (1911). The condensation of crotonaldehyde. *Journal of the Chemical Society* **99**, 1627–33.

SMEDLEY, I. & LUBRZYNSKA, E. (1913). The biochemical synthesis of fatty acids. *The Biochemical Journal* **7**, 364–74.

SONDERHOFF, R. & THOMAS, H. (1937). Die enzymatische Dehydrienung der Frideutero-essigsäure. *Annalen der Chemie* **530**, 195–213.

SOUMALAINEN, H. & KERANEN, A. J. A. (1968). The fatty acid composition of baker's and brewer's yeast. *Chemistry and Physics of Lipids* **2**, 296–315.

STANIER, R. Y. (1970). Some aspects of the biology of cells and their possible evolutionary significance. *Symposium of the Society for General Microbiology* **20**, 1–38.

STANIER, R. Y., DOUDEROFF, M. & ADELBERG, E. A. (1971). *General microbiology*, 3rd edition. London: Macmillan.

STOFFEL, W. & ACH, K. L. (1964). Der Stoffwechsel der ungesättigten Fettsäuren. II. Eigenschaften der kettenverlängernden Enzyms zur Frage der Biohydrogenierung der ungesättigten Fettsäuren. *Hoppe Seylerzeitschrift für Physiologische Chemie* **337**, 123–32.

SUMNER, J. L., MORGAN, E. D. & EVANS, H. C. (1969). The effect of growth temperature on the fatty acid composition of fungi in the order *Mucorales. Canadian Journal of Microbiology* **15**, 515–20.

TALAMO, B., CHANG, N. & BLOCH, K. (1973). Desaturation of oleyl phospholipid to linoleyl phospholipid in *Torulopsis utilis. The Journal of Biological Chemistry* **248**, 2738–42.

TAYLOR, P. W. (1974). M.Sc. Thesis, University of Nottingham.

TERROINE, E. F. & BONNET, R. (1927). Formation of fats at the expense of sugars in micro-organisms. *Bulletin de la société de chimie biologique* **9**, 588–96.

VAGELOS, P. R. (1964). Lipid metabolism. *Annual Review of Biochemistry* **33**, 139–72.

VAN ETTEN, J. L. & GOTTLIEB, D. (1965). Biochemical changes during the growth of fungi. II. Ergosterol and fatty acids in *Penicillium atrovenetum. Journal of Bacteriology* **89**, 409–14.

WADKE, M., BRUNENGRABER, H., LOWENSTEIN, J. M., DOLHUN, J. J. & ARSENAULT, G. P. (1973). Fatty acid synthesis by the liver perfused with deuterated and tritiated water. *Biochemistry* **12**, 2619–24.

WAKIL, S. J. (1958). A malonic acid derivative as an intermediate in fatty acid synthesis. *Journal of the American Chemical Society* **80**, 6465.

WAKIL, S. J. (1961). Mechanism of fatty acid synthesis. *Journal of Lipid Research* **2**, 1–24.

WAKIL, S. J. & GANGULY, J. (1959). On the mechanism of fatty acid synthesis. *Jour-*

nal of the American Chemical Society **81**, 2597–8.

WAKIL, S. J. & GIBSON, D. M. (1960). Studies on the mechanism of fatty acid synthesis. VIII. The participation of protein-bound biotin in the biosynthesis of fatty acids. *Biochimica et Biophysica Acta* **41**, 122–9.

WAKIL, S. J., PUGH, E. L. & SAUER, F. (1964). The mechanism of fatty acid synthesis. *Proceedings of the National Academy of Sciences, U.S.A.* **52**, 106–14.

WALKER, P. J. (1975). Ph.D. Thesis, University of Nottingham.

WARSHAW, J. B. & KIMURA, R. A. (1973*a*). Palmityl-CoA and palmitylcarnitine interactions in mitochondrial fatty acid elongation. *Archives of Biochemistry and Biophysics* **157**, 44–9.

WARSHAW, J. B. & KIMURA, R. E. (1973*b*). Changes in hepatic microsomal fatty acid synthesis during development of the rat. *Biologia Neonatorum* **22**, 133–40.

WEISS, S., FIORE, J. V. & NORD, F. F. (1947). On the structure and possible functions of a pigment of *Fusarium solani* D_2 purple. *Archives of Biochemistry* **15**, 326–8.

WHITE, A. G. C. & WERKMAN, C. H. (1947). Assimilation of acetate by yeast. *Archives of Biochemistry* **13**, 27–32.

WILLECKE, K., RITTER, E. & LYNEN, F. (1969). Isolation of an acyl carrier protein component from the multienzyme complex of yeast fatty acid synthetase. *European Journal of Biochemistry* **8**, 503–9.

WOOD, H. G., BROWN, R. W. & WERKMAN, C. H. (1945). The mechanism of butyl alcohol fermentation studied with the heavy carbon acetic and butyric acids. *Archives of Biochemistry* **6**, 243–60.

WOODBINE, M. (1959). Microbial fat: micro-organisms as potential fat producers. *Progress in Industrial Microbiology* **1**, 179–245.

WOODBINE, M. (1964). Biogenesis of fatty acids. In *Fatty acids*, Vol. 3, pp. 1967–1982. Edited by K. S. Markley, New York: Interscience.

WRIGHT, L. D., CRESSON, E. L., SKEGGS, H. R., WOOD, T. R., PECK, R. L., WOLF, D. E. & FOLKERS, K. (1952). Isolation of crystalline biocytin from yeast extract. *Journal of the American Chemical Society* **74**, 1996–9.

YANG, P. C., BUTTERWORTH, P. H. W., BOCK, R. M. & PORTER, J. W. (1967). Further studies on the properties of the pigeon liver fatty acid synthetase. *The Journal of Biological Chemistry* **242**, 3501–7.

YUAN, C. & BLOCH, K. (1961). Conversion of oleic acid to linoleic acid. *The Journal of Biological Chemistry* **236**, 1277–9.

CHAPTER 7

Nitrogen Metabolism

J. A. PATEMAN and J. R. KINGHORN

7.1 Introduction

This article is concerned with various aspects of the metabolism of simple nitrogen compounds and amino acids by filamentous fungi. Inevitably, such a vast field cannot be comprehensively covered and certain areas such as the early physiological studies are at most briefly mentioned.

A great deal of more recent knowledge of fungal metabolism derives from combined biochemical and genetical studies concentrated largely on a few species of yeasts and the two Ascomycetes *Neurospora crassa* and *Aspergillus nidulans.* As a result there is considerable emphasis in this article on the *Neurospora* and *Aspergillus* work, and the yeast studies are dealt with elsewhere. It is obvious that considerable metabolic differences must exist throughout such a large and diverse group of organisms as the fungi. Nevertheless, the studies in depth on the few well investigated species probably provide a general view of how the fungi manage some of their metabolic affairs.

7.2 Utilization of Simple Nitrogen Compounds

Nitrate

Most fungi can use nitrate as a sole or significant source of nitrogen although there are reported exceptions in all the major groups. It is usually difficult to be sure of the basis of a failure to use nitrate from the reports in the literature. It may be that the culture medium is deficient in some other component or the culture conditions may result in the accumulation of toxic amounts of nitrite from the nitrate. This latter possibility is due to the ability of some fungi to use nitrate as a hydrogen acceptor under anaerobic conditions and the resultant accumulation of nitrite (Walker & Nicholas, 1961).

The only known metabolic route for the utilization of nitrate nitrogen is by reduction first to nitrite and subsequently to ammonia. The production of the necessary enzymes, nitrate reductase, nitrite reductase and hydroxylamine reductase, is dependent on the presence of nitrate in those species investigated (see section 7.6). Nitrate will be present in most fungal environments and since it is readily utilized it is probably an important source of nitrogen for the majority of fungi.

Nitrite

Nitrite is formed from nitrate and further reduced to ammonia, and thus in one sense is utilized by all fungi which can use nitrate. It is probable that in this process the intracellular concentration of nitrite is low due to nitrite reductase activity which can be induced by nitrate (see later). This is clearly shown in *Aspergillus nidulans* where normal strains grown on nitrate do not accumulate intracellular nitrite or excrete it into the medium. A mutant, *niiA*, which possesses nitrate reductase, but lacks nitrite reductase, accumulates nitrite and excretes it in large quantities on medium containing nitrate (Pateman & Cove, 1967). Although nitrite is toxic to many fungi and bacteria, it can be used as a nitrogen source by some fungi, *e.g. Aspergillus nidulans, A. niger, Coprinus* sp., *Fusarium niveum, Neurospora crassa, N. sitophila, Penicillium* sp., *Phymatotrichum omnivorum, Scopulariopsis brevicaulis* and *Ustilago maydis.* The relative toxicity of nitrite to some fungi and most bacteria may be partly determined by pH since it is less toxic if the medium is kept at pH 6–8. There are a number of ways in which the highly reactive nitrite ion might be toxic. It can deaminate amino acids and other important molecules, or it may interfere with sulphur metabolism due to its similarity to the sulphite ion. In *Aspergillus nidulans* mutant strains which are unable to synthesize nitrite reductase are very sensitive to nitrite. This is largely due to the inability to use sulphate or sulphite as a source of sulphate in the presence of nitrite, since nitrite inactivates sulphite reductase. The nitrite toxicity is partially relieved by the presence of relatively reduced sulphur compounds such as thiosulphate (Pateman, unpublished work). In some fungi nitrite toxicity may result in the accumulation of pyruvate (Nord & Mull, 1945).

Ammonia

As with nitrate, it is likely that the great majority of fungi can use ammonia as a sole nitrogen source. This has been shown in *Alternaria* sp., *Aspergillus* sp., *Botrytis* sp., *Cladosporium* sp., *Coprinus* sp., *Diplodia* sp., *Mucor* sp., *Neurospora* sp., *Penicillium* sp., *Ustilago* sp. *etc.* (Morton & MacMillan, 1954; Pateman, Rever & Cove, 1967; Lewis & Fincham, 1970*a*). The form in which the ammonia is supplied is important. Ammonium chloride, ammonium nitrate and ammonium sulphate are utilized poorly or not at all by some fungi. It was shown by Morton & MacMillan (1954) that this is due to pH effects in the medium. When these salts are used the pH of the medium rapidly drops and consequently growth stops. If the medium is kept at about pH 5 by the addition of organic acid salts such as malate or tartrate there is good growth and complete assimilation of the ammonia. This is most readily achieved, at least in *A. nidulans*, by using ammonium tartrate as the nitrogen source. In the case of ammonium nitrate the sharp drop in the pH of the medium is due to the preferential use of the ammonia which occurs in many fungi (Morton & MacMillan, 1954). This ammonia repression of the nitrate utilization pathway is a good example of the widespread phenomenon of ammonia repression which is discussed in section 7.13.

Urea

Although urea is neither an inorganic molecule nor an amino acid, some aspects of its metabolism are closely concerned with ammonia. Consequently, it is briefly dealt with in various sections. Urea can be utilized as a source of nitrogen by such fungi as *Aspergillus* sp., *Coprinus* sp., *Neurospora* sp., *Penicillium* sp., and *Ustilago* sp. It is probable that most fungi can use it as a source of nitrogen, but not of carbon. Urea is converted to ammonia and carbon dioxide by the enzyme urease. It has been shown in *Aspergillus nidulans* (Darlington & Scazzocchio, 1967) and *Neurospora crassa* (Kolmark, 1969*a, b*; Haysman & Branch Hoire, 1971) that mutant strains which lack urease cannot utilize urea as a nitrogen source. In *A. nidulans* urease is the final enzyme in the pathway responsible for the oxidative degradation of purines to ammonia. This is an important salvage and detoxifying pathway, since purines are toxic at quite low concentrations. It is likely that in most fungi an important function of urease is to utilize urea resulting from purine breakdown and arginine breakdown (see section 7.9).

7.3 Transport of Simple Nitrogen Compounds

The utilization of a compound usually implies that it is taken into the cells by some transport mechanism. Despite their importance, comparatively little work has been done on the transport of nitrate, nitrite and ammonia in fungi or in bacteria. The main reason for this is the lack of suitable radioactive isotopes of nitrogen, which severely limits the resolution of uptake studies. However, some evidence has been obtained about these systems using direct determinations of the ^{15}N-labelled substrates or the ammonia analogue methylammonia.

Nitrate

Nitrate transport or assimilation has been studied in a number of fungi (Morton & MacMillan, 1954; Goldsmith, Livoni, Norberg & Segel, 1973; Schloemer & Garrett, 1974*a*). The situation is very unclear with as yet, to our knowledge, no absolutely unequivocal demonstration that a nitrate transport system other than diffusion exists in any fungus. The basic difficulty is to distinguish between some form of nitrate transport system and simple diffusion down a concentration gradient into the cell created by the action of nitrate reductase. There are two main ways in which this difficulty might be resolved: the use of mutants lacking nitrate reductase activity and/or nitrate uptake activity; the demonstration that nitrate can be transported against a concentration gradient.

Schloemer & Garrett (1974*a*) claim to have demonstrated a nitrate uptake system in *Neurospora crassa* which is induced by nitrate and nitrite but not repressed by ammonia. This system is apparently independent of nitrate reduction, since it is present in the *nit-1*, *nit-2* and *nit-3* mutants which lack nitrate reductase activity. There are several reasons why this claim must be treated with reserve at present. The nitrate uptake assay is usually based on measurement of nitrate disappearance from the medium and not accumulation in the cells. The nitrate uptake rate in the nitrate reductase minus mutants apparently does not reach significant levels until 3–6 h after induction, a remarkably slow response if there really is an uptake system independent of nitrate reduction. These authors also claim that the uptake system can concentrate nitrate 50-fold compared with the external concentration. This finding, if substantiated, would certainly indicate the existence of an active transport system for nitrate in *N. crassa*, but no experimental details are given.

A similar type of investigation of nitrate uptake has been carried out in *Penicillium chrysogenum* by Goldsmith *et al.* (1973). Nitrate uptake by the cells was assayed by the measurement of nitrate disappearance from the medium. The authors consider that probably there is a nitrate-uptake system independent of nitrate reduction. However the same difficulties apply here as in the *Neurospora crassa* system, and the experimental evidence does not unequivocally demonstrate the existence of an independent nitrate-uptake system.

In *Aspergillus nidulans* uptake studies utilizing ^{15}N-labelled substrates have been done with nitrate and nitrite (Pateman, unpubl. work) and ammonia (Pateman, Dunn, Kinghorn & Forbes, 1974). Essentially the assay technique consists of incubating cells with ^{15}N-enriched substrate, *e.g.* $K^{15}NO_3$ and determining the percentage enrichment of ^{15}N in samples of cells taken at various times. This was done with the normal strain and the mutant *nia-D*, which lacks nitrate reductase activity. The results in Fig. 7.1 are for cells grown on 5 mM urea and then incubated in 2 mM potassium nitrate (30% ^{15}N enrichment) with and without 5 mM ammonium chloride. Further experimental details were essentially the same as those given for uptake of ^{15}N ammonia in Pateman *et al.* (1974). The main findings were:

1. There is no significant uptake of nitrate by the normal strain during the first 40 min, thereafter there is an increase until a maximum rate of about 1.5% enrichment/g wet weight cells/h is reached (in this system 1% enrichment = 12 μmol nitrate taken up/g wet weight cells/h). This nitrate uptake is correlated with nitrate reductase activity which is first detectable about 30 min after addition of nitrate, and reaches a maximum after about 1 h.

2. There was no uptake of nitrate by the normal strain when 5 mM ammonia was present with the 2 mM nitrate. This also is correlated with the virtual absence of nitrate reductase in these conditions.

3. The mutant which lacked nitrate reductase activity did not take up nitrate to a significant degree in the presence or absence of ammonia. Thus in *A. nidulans* there is no indication of a nitrate-transport system independent of nitrate reduction, although it is possible that some mechanism of

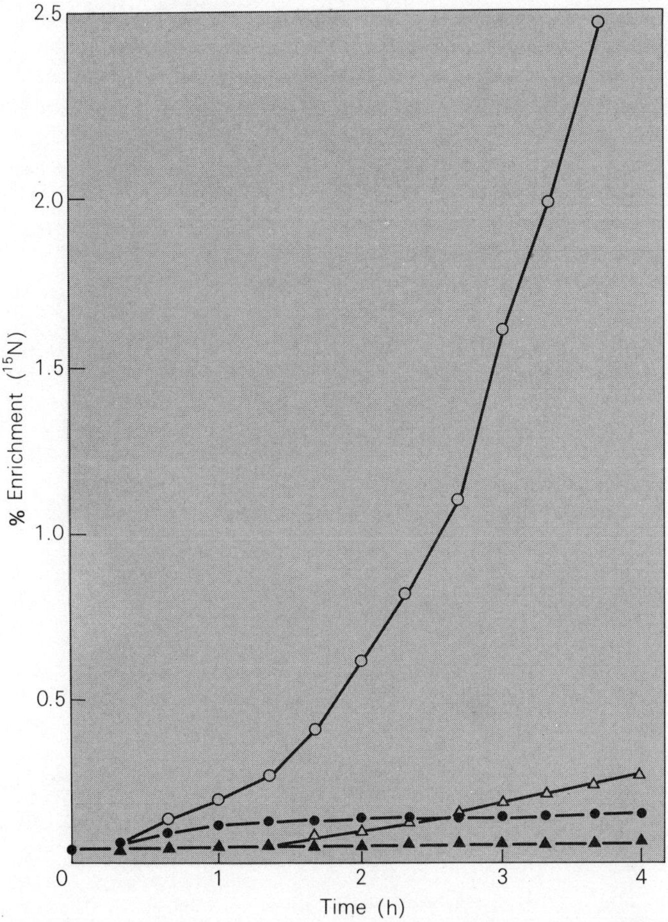

Fig. 7.1 Uptake of ^{15}N-enriched potassium nitrate by *Aspergillus nidulans*. Prototrophic, *nia-D*$^+$, and mutant, *nia-D*, cells were grown on 5 mM urea and transferred to 2 mM potassium nitrate (30% ^{15}N enrichment) with and without 5 mM ammonium chloride. Samples of cells were taken at 20 min intervals and assayed for ^{15}N enrichment. ○—○, *nia-D*$^+$ on nitrate. ●—●, *nia-D*$^+$ on nitrate plus ammonia. △—△, *nia-D* on nitrate. ▲—▲, *nia-D* on nitrate plus ammonia.

facilitated diffusion operates together with the nitrate reduction pathway. It seems unlikely that two similar fungi such as *Neurospora crassa* and *A. nidulans* should differ markedly with respect to nitrate transport; the apparent discrepancies may be resolved by further work.

Nitrite

A nitrite uptake system in *Neurospora crassa* has been reported by Schloemer & Garrett (1974*b*). As with the nitrate system previously mentioned, the uptake assay is based on the disappearance of the substrate, nitrite, from the medium. It was not possible to demonstrate the concentration of nitrite against a gradient or the independence of nitrite uptake from

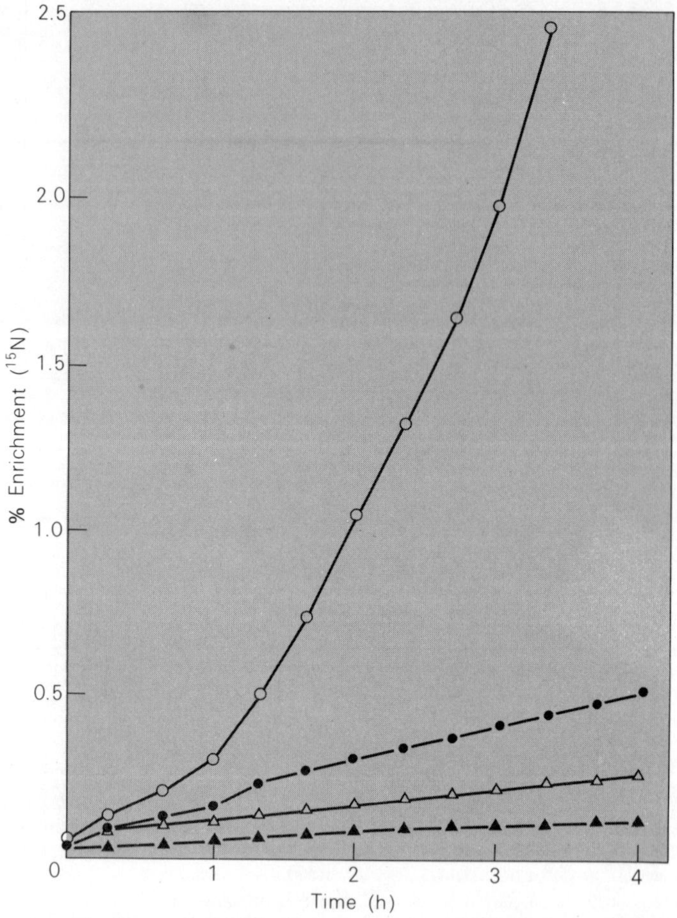

Fig. 7.2 Uptake of ^{15}N-enriched potassium nitrite by *Aspergillus nidulans*. Prototrophic, *nii-A*$^+$, and mutant, *niiA*$^-$, cells were grown on 5 mM urea transferred to 2 mM potassium nitrite (30% ^{15}N enrichment) with and without 5 mM ammonium chloride. Samples of cells were taken at 20 min intervals and assayed for ^{15}N enrichment. ○—○, *nii-A*$^+$ on nitrite. ●—●, *nii-A*$^+$ on nitrite plus ammonia. △—△, *nii-A*$^-$ on nitrite. ▲—▲, *nii-A*$^-$ on nitrite plus ammonia.

nitrite reduction, since no mutants lacking nitrite reductase activity were available. There are therefore the same doubts about the interpretation of these results as with nitrate uptake in *N. crassa*.

Some results from uptake studies using ^{15}N-labelled nitrite in *Aspergillus nidulans* are given in Fig. 7.2. The normal strain and the mutant *niiA*, which lacks nitrite reductase activity, were grown on 5 mM urea and then incubated in 2 mM potassium nitrite (30% ^{15}N enrichment) with and without 5 mM ammonium chloride. There was no indication of nitrite uptake independent of nitrite reduction by nitrite reductase.

Ammonia

There have been a number of reports of ammonia transport in fungi (MacMillan, 1956; Hackette, Skye, Burton & Segel, 1970; Pateman *et al.*, 1974). Ammonia transport in *Penicillium chrysogenum* has been studied by Hackette *et al.* (1970) using the analogue methylammonium labelled with ^{14}C. There are good reasons for supposing that methylammonium uptake is mediated by the ammonia-transport system so the labelled analogue can be used to characterize the system. The level of methylammonia transport is low in nitrogen sufficient cells; nitrogen starvation by depression and/or deinhibition results in increases of up to 800-fold in the rate of transport. It is also claimed that feedback inhibition by intracellular glutamine and asparagine occurs (Hackette *et al.*, 1970).

In *Aspergillus nidulans* an active transport system for ammonia and methylammonia has been characterized using ^{15}N-labelled ammonia and ^{14}C-labelled methylammonia (Pateman *et al.*, 1974). The system has a K_m of less than 5×10^{-5} M for ammonia and a K_m of 2×10^{-5} M for methylammonia and can concentrate methylammonia at least 120-fold over the extracellular concentration. Several classes of mutants are known which exhibit abnormal levels of ammonia transport. Mutations in the *meaA* and *mod-meaA* genes result in low transport rates and altered K_ms for ammonia and methylammonia, and consequently these genes are thought to determine membrane proteins. Mutation in the *amrA* gene results in low transport levels under certain growth conditions only, and this gene is thought to play a regulatory role in ammonia transport. The regulation of the ammonia transport system in *A. nidulans* is complicated and not fully understood. The available evidence suggests that the intracellular concentration of ammonia regulates transport both by inhibition and by repression of the synthesis of some protein component of the uptake system (Pateman, Kinghorn, Dunn & Forbes, 1973; Pateman *et al.*, 1974*a*).

Urea

A transport system, which is specific for urea and thiourea, has been reported for *Aspergillus nidulans* (Dunn & Pateman, 1972). Mutants resistant to the toxic urea analogue, thiourea, have been isolated and these are defective in urea and thiourea transport, but normal for ammonia and amino acid transport. The urea transport system is regulated by ammonia control and this is discussed later.

7.4 Enzymes Concerned with the Reduction of Nitrate to Ammonia

Nitrate is reduced via nitrite and hydroxylamine to ammonia in a series of steps which are essentially electron transfer reactions. There is still some doubt about the possible existence of an organic reductive route from nitrite to ammonia, but it is generally accepted that the inorganic reductive pathway is the important one in fungi. The enzymes involved contain a number of cofactors and metals, and utilize NADPH as a hydrogen donor (Table 7.1).

It is probable that one or more intermediate compounds are formed between nitrite and hydroxylamine at the +1 oxidation state for the nitrogen atom. The nature of the postulated intermediate(s) is uncertain: nitroxyl

Table 7.1 Summary of nitrate reduction

Possible pathway	Hydrogen donor	Cofactors	Oxidation/ reduction state of N	Enzyme	Mutants lacking enzyme activity: *Aspergillus nidulans*	Mutants lacking enzyme activity: *Neurospora crassa*
NO_3			+5			
↓ +2e	NADPH	FAD, Fe, Mo		Nitrate reductase	*niaD, cnxA, cnxB, cnxC, cnxE, cnxF, cnxG, cnxH, nirA*	*nit-1*, *nit-2*, *nit-3*
NO_2			+3			
↓ +2e	NADPH	FAD, Cu, Fe		Nitrite reductase	*nii ribo A*, *nirA*	
(NOH)? ($\rightarrow N_2O_2 \rightarrow N_2O \rightarrow$ (NOH)?)			+1			
↓ +2e				Hyponitrite reductase?		
NH_2OH			−1			
↓	NADPH	FAD, Mg, Mn		Hydroxylamine reductase	*nii ribo A*, *nirA*	
NH_3			−3			

nitroxyl	NOH
hyponitrite	N_2O_2
nitrous oxide	N_2O

(NOH), hyponitrite and nitrous oxide have all been considered as possibilities (Nicholas, 1963). There are several further possibilities that at present cannot be ruled out. It may be that, *in vivo*, one or more intermediates are not present in the free form, but are bound to some organic compound or are never released in a free form from an enzyme aggregate. These ideas are attractive since the postulated intermediates between nitrite and hydroxylamine are unstable or toxic or both. Another possibility is that nitrite is not reduced completely to ammonia, but some intermediate after nitrite is used for amino acid biosynthesis. It has, for example, been suggested that hydroxylamine could be used to form oximes which could then be reduced to amino acids (Saris & Virtanen, 1957*a*, *b*).

Nitrate reductase

There are two types of nitrate reductase, assimilatory and respiratory. These will be considered separately.

ASSIMILATORY ENZYME Nitrate reductase (NAD[P]H-nitrate oxidoreductase, E.C.1.6.6.3) from *Neurospora crassa*, was partially purified and characterized by Nason & Evans (1953) and shown to contain FAD. Subsequently, Nicholas and coworkers (Nicholas & Nason, 1954; Nicholas, 1963) demonstrated that molybdenum is a functional constituent of nitrate reductase. Kinsry & McElroy (1958) found that nitrate reductase activity was closely correlated with cytochrome c reductase activity (NADPH-cytochrome c oxidoreductase, E.C.1.6.2.3) during nitrate induction of the cells and during purification of nitrate reductase activity in *N. crassa*. They suggested that a single enzyme protein was involved in both catalytic activities. In *Aspergillus nidulans* the properties of various mutants (Pateman, Rever & Cove, 1967) and the enzyme activities of nitrate reductase during purification (MacDonald, Cove & Coddington, 1974) provide strong evidence that flavin and molybdenum are involved in nitrate reduction and that nitrate reductase can show a range of NADPH-diaphorase activities including NADPH-cytochrome c reductase. The reduction of nitrate may be considered as an electron transfer chain confined to a single complex enzyme molecule. The probable situation is represented diagrammatically in Fig. 7.3. A pair of electrons are donated by NADPH to FAD, then possibly to iron in a cytochrome, then to two molybdenum atoms which are reduced from Mo^{6+} to Mo^{5+} and finally to nitrate which is reduced to nitrite. It is apparent that nitrate reductase can donate electrons from some intermediate point on the chain, probably the flavin, to suitable electron acceptors, and this is the observed NADPH-diaphorase activity. The molecule can also accept electrons from reduced violegen dyes at an intermediate point, probably molybdenum, and thus show benzyl violegen-nitrate oxidoreductase activity both in *A. nidulans* (Patemen, Rever & Cove, 1967) and in *N. crassa* (Sorger, 1965).

The obvious complexity of the nitrate reductase molecule and the reactions it mediates are accompanied by a complex genetical determination of the protein. Combined biochemical and genetic studies of nitrate reductase have been done particularly in *Neurospora crassa*, *Aspergillus nidulans* and *Ustilago maydis*. The work in *A. nidulans* is described in most detail, since it is more extensive at the present time and the situation in the other fungi is probably similar. Cove & Pateman (1963) described mutants

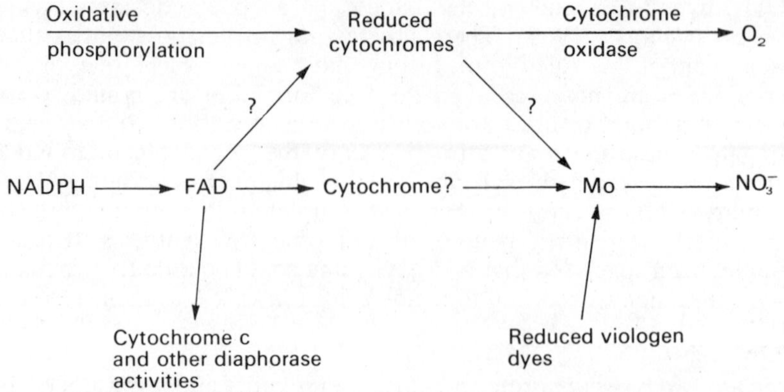

Fig. 7.3 Electron transfer steps associated with nitrate reduction in fungi. Assimilatory nitrate reduction involves electron transfer on the molecule direct from NADPH to nitrate. The possible involvement of iron in a cytochrome intermediate between FAD and molybdenum is indicated by a query.

of *A. nidulans* which were unable to use nitrate as a sole nitrogen source. Mutants which are unable to grow on nitrate because they lack normal nitrate reductase activity map in seven separate genes in *A. nidulans* (Table 7.2) (Pateman *et al.*, 1964). Mutation in *nia-D*, the structural gene for the nitrate reductase protein, leads to the reduction or absence of nitrate reductase activity alone. Mutation in at least five genes, *cnxA,B,C, cnxE, cnxF, cnxG* and *cnxH*, results in the inability to grow on hypoxanthine or nitrate, and in the reduction or loss of both nitrate reductase and xanthine dehydrogenase activities. Some of the *nia-D* mutants and all the *cnx* mutants retain the NADPH-cytochrome c reductase activity associated with nitrate reductase. The status of *cnxA, cnxB* and *cnxC* is uncertain. They may represent separate genes or complementation groups within a complex structural gene. These *cnx* genes are thought to be responsible for the production and incorporation into the nitrate reductase and xanthine dehydrogenase molecules of a protein molybdenum cofactor which is essential for both enzyme activities (Pateman, Rever & Cove, 1967). The *cnxE* mutants are repaired by high concentrations of molybdenum in the medium and can then grow on nitrate or hypoxanthine as sources of nitrogen and possess nitrate reductase or xanthine dehydrogenase activity. It has been suggested that the *cnxE* gene is the structural gene for an enzyme responsible for the incorporation of molybdenum into the nitrate reductase molecule (Arst, MacDonald & Cove, 1970). On the basis of sucrose density-gradient centrifugation it has been suggested that the *nia-D* gene specifies a protein of molecular weight 90–100 000 and the *cnx* genes a low molecular weight protein molybdenum cofactor of possibly 10–20 000 molecular weight. The *nia-D* gene product, both as a monomer and dimer, possesses NADPH-cytochrome c reductase activity but not nitrate reductase activity. The complete nitrate reductase molecule is a dimer of two *nia-D* gene products and one or more molecules of the cofactor specified by the *cnx* genes. Only this aggregate of about 200 000 MW possesses nitrate reductase activity in addition to cytochrome c reductase activity. In terms of

Table 7.2 Enzyme activities of the prototroph and mutant strains grown on various nitrogen sources (From Pateman, Rever & Cove (1967), *Biochemical Journal* **104**, 106)

Cell-free extracts were prepared from mycelium grown as follows: (1) 5 mM-urea as initial and only nitrogen source, mycelium harvested at 20 h, (2) 5 mM-urea as initial nitrogen source, 10 mM-$NaNO_3$ added at 15 h., mycelium harvested at 20 h; (3) 5 mM-urea as initial nitrogen source 10 mM-NH_4NO_3 added at 15 h, mycelium harvested at 20 h. The activities of nitrate reductase (NO_3R), nitrite reductase (NO_2R), hydroxylamine reductase (NH_2OHR), cytochrome *c* reductase (CR), benzyl viologen reductase (BVR), iodophenylnitrophenylphenyltetrazolium reductase (INTR) and potassium ferricyanide reductase (FCN R) were determined and are given in nmoles of substrate transformed min^{-1} mg^{-1} of protein.

Enzyme activities

(1) Mycelium grown on urea

	NO_3 R	NO_2 R	NH_2OH R	BV R	CR	INT R	FCN R
Prototroph	4	2	3	66	268	151	383
Prototroph	4	1	15	30	169	162	466
cnxA5	0	199	319	627	508	710	2631
cnxB11	0	225	345	655	380	906	3200
cnxC2	0	100	120	280	480	798	3580
cnxE2	<1	10	22	80	209	—	—
cnxE3	0	153	161	1110	828	—	—
cnxE4	<1	268	352	1800	1890	—	—
cnxE13	0	238	281	1260	1210	—	—
cnxE14	0	4	26	210	248	265	1313
cnxF1	0	108	157	668	565	—	—
cnxF2	0	224	253	924	985	—	—
cnxF8	0	108	205	813	716	833	3800
cnxG2	2	<1	15	22	153	—	—
cnxG4	0	174	252	816	1110	543	2605
cnxH3	<1	192	284	1195	1290	706	3050
niaD17	0	137	246	1030	470	1204	4450
niiA1	1	3	9	114	89	114	424
niiB2	2	3	24	36	82	146	387
niiriboA1	3	0	13	74	209	—	—
niiriboA1[a]	2	1	7	103	87	—	—

(2) Mycelium grown on urea + $NaNO_3$

	NO_3 R	NO_2 R	NH_2OH R	BV R	CR	INT R	FCN R
Prototroph	75	136	146	524	528	777	2644
Prototroph	91	151	224	550	907	972	3000
cnxA5	0	261	330	667	327	758	2820
cnxB11	0	239	258	659	352	860	3200
cnxC2	0	152	143	381	741	952	3571
cnxE2	4	269	340	1400	1180	—	—
cnxE3	0	131	209	1260	971	—	—
cnxE4	<1	272	340	1630	1970	—	—
cnxE13	0	257	307	1540	1360	—	—
cnxE14	0	276	381	1605	1960	1375	4700
cnxF1	0	156	227	722	737	—	—
cnxF2	0	195	219	1063	897	—	—
cnxF8	0	148	223	840	475	714	3043
cnxG2	2	212	254	643	975	—	—
cnxG4	0	184	247	894	1060	732	3250
cnxH3	<1	143	167	794	780	844	3150
niaD17	0	102	139	530	537	1200	4190
niiA1	76	2	21	309	327	457	857
niiB2	2	2	21	44	90	159	462
niiriboA1	72	33	92	320	805	—	—
niiriboA1[a]	113	217	314	577	506	—	—

(3) Mycelium grown on urea + NH_4NO_3

	NO_3 R	NO_2 R	NH_2OH R	BV R	CR	INT R	FCN R
Prototroph	20	35	45	189	484	375	875
Prototroph	9	10	26	171	254	—	—
cnxA5	0	9	33	80	119	155	610
cnxB11	0	17	46	124	136	245	920
cnxC2	—	—	—	—	—	—	—
cnxE2	1	22	43	174	263	—	—
cnxE3	0	15	12	139	303	—	—
cnxE4	0	18	67	—	631	—	—
cnxE13	0	7	46	199	226	—	—
cnxE14	0	52	92	447	370	—	—
cnxF1	0	66	14	151	364	—	—
cnxF2	0	25	36	224	435	—	—
cnxF8	0	4	11	74	148	304	756
cnxG2	2	37	69	271	502	—	—
cnxG4	0	25	49	284	624	—	—
cnxH3	0	25	47	239	365	—	—
niaD17	0	19	44	193	444	—	—
niiA1	6	<1	14	54	108	166	538
niiB2	0	4	17	36	90	154	344
niiriboA1	5	3	15	89	342	—	—
niiriboA1[a]	13	38	70	184	172	—	—

[a] All media supplemented with 2.5 μg of riboflavine ml^{-1}

the diagram in Fig. 7.3 mutation in *nia-D* can result in failure of one or more of the electron transfer steps through abnormalities of the *nia-D* protein. Mutation in the *cnx* genes only affects the electron transfer mediated by the molybdenum cofactor with the loss of nitrate reduction, but production of normal *nia-D* protein and its associated NADPH-diaphorase activities. There is another gene in *A. nidulans* mutation in which can result in failure to utilize nitrate and loss of nitrate reductase activity. Mutations in the *nirA* gene result in the non-inducibility of both nitrate and nitrite reductases, while xanthine dehydrogenase activity is not affected (Pateman & Cove, 1967). Since *nirA* is a regulatory gene concerned with nitrate induction its properties are discussed later.

The situation is less clear with regard to mutants of *Neurospora crassa* unable to use nitrate and lacking normal nitrate reductase activity. Mutations in four different genes, *nit-1*, *nit-2*, *nit-3* and *nit-4*, unable to grow on nitrate were described by Blakeley & Srb (1962), and Sorger & Giles (1965) report a fifth gene, *nit-5*. It is not clear from the paper of Sorger & Giles (1965) which of the *nit* genes in *N. crassa* correspond to the *nia-D* structural gene and the *cnx* cofactor genes in *Aspergillus nidulans*. They claim that *nit-2* grows poorly on hypoxanthine, which would make it analogous to the *cnx* genes. However, MacDonald, Cove & Coddington (1974) quote unpublished work by A. Coddington which suggests that *nit-1* is equivalent to the *cnx* genes and *nit*-3 corresponds to the structural gene *nia-D* of *A. nidulans*. Clearly the situation in *N. crassa* is somewhat similar to *A. nidulans* but needs further work.

Mutants of *Ustilago maydis* unable to utilize nitrate and lacking normal nitrate reductase activity fall into six complementation groups (Lewis & Fincham, 1970*a*, *b*). One of these, *narA*, probably represents the structural gene for the nitrate reductase protein analogous to the *nia-D* gene in *Aspergillus nidulans*. A temperature-sensitive *narA* mutant produces a temperature sensitive nitrate reductase. Another group of mutants, *narC*, unable to grow on nitrate or nitrite, is apparently similar to the *nirA* mutants of *A. nidulans*. At present nothing further is known about the other four groups of mutants.

DISSIMILATORY ENZYME A dissimilatory nitrate reductase activity is present in *Neurospora crassa* grown with a limited oxygen supply (Walker & Nicholas, 1961). The enzyme requires iron as well as molybdenum for activity, reduced methyl or benzyl viologen is the most effective hydrogen donor, and under these conditions there is no flavin requirement. The purified dissimilatory enzyme does not possess NADPH-cytochrome c reductase activity (Nicholas & Wilson, 1964). It is difficult to tell from the literature if there are two distinct forms of nitrate reductase in fungi, as is usually assumed. It is possible, at least in *N. crassa* and *Aspergillus nidulans*, that under conditions of low oxygen tension and consequent low cytochrome oxidase activity there is a build-up of reduced cytochromes, particularly reduced cytochrome c. In the presence of nitrate the reduced cytochromes would donate electrons probably to molybdenum in the nitrate-induced nitrate reductase enzyme, with the consequent reduction of nitrate. Thus nitrate would substitute for oxygen as the terminal electron acceptor in respiration. This possibility is made more likely by the fact that the assimilatory enzyme in both *N. crassa* and *A. nidulans* can accept electrons from reduced viologen dyes and then reduce nitrate (Sorger, 1965; Pateman, Rever & Cove, 1967). The situation with respect to electron transfer in both enzymes is summarized in Fig. 7.3. The uncertainty about the existence of iron in the assimilatory nitrate reductase molecule and the reduction of cytochromes in dissimilatory nitrate reduction is indicated by queries. It may be that under conditions of low oxygen and available nitrate there is a special dissimilatory nitrate reductase formed which indirectly, via external cytochromes, or directly, can transfer electrons from reduced pyridine nucleotides (NADH, NADPH) to nitrate. However, the matter requires further investigation and could probably be best settled by studying apparent dissimilatory nitrate reduction in various classes of mutants of *N. crassa* and *A. nidulans* lacking normal assimilatory nitrate reductase activity.

Nitrite reductase

A nitrite reductase activity was demonstrated in *Neurospora crassa*, by Nason, Abraham & Auerbach (1954). Later it was purified and shown to contain FAD, copper and iron (Nicholas, Medina & Jones, 1960). These authors proposed that the copper is required for the terminal electron transfer to nitrite, univalent copper reduces nitrite non-enzymatically and the divalent copper is enzymatically reduced by the nitrite reductase.

$$\text{NADPH} \rightarrow \text{FAD} \rightarrow \text{cytochrome} \rightarrow \text{Cu} \rightarrow NO_2$$

The precise nature of the reduction product(s) fron nitrite is unknown (see Table 7.1).

In *Aspergillus nidulans* nitrate reductase activity (NADPH-nitrite oxidoreductase 1.6.6.4) is induced in cells grown in the presence of nitrate or nitrite (Table 7.2) (Pateman *et al.*, 1964; Pateman, Rever & Cove, 1967). The properties of various mutant strains demonstrate that this nitrite reductase, like the nitrate reductase in *A. nidulans*, possesses NADPH-cytochrome c reductase activity and other NADPH-diaphorase activities. Mutation in three genes, *niiA*, *nii ribo A* and *nirA* can result in low levels or the absence of nitrite reductase activity. Mutants of the *niiA* gene possess normal nitrate inducible nitrate reductase but lack nitrite reductase and hydroxylamine reductase activities.

It is probable that the product of the *niiA* gene can mediate the reduction of nitrite to ammonia. The evidence for this is circumstantial but strong. In an original series of mutation experiments designed to produce mutants unable to utilize nitrate, a total of 98 were unable to use nitrate or nitrite and complementation tests showed that all these mutants were either *niiA* or *nirA*, together with one *nii riboA* mutant (Cove & Pateman, 1963). Since then a number of workers have made further large numbers of mutants unable to utilize nitrite (Pateman, unpubl. work; Cove, pers. comm.). Complementation tests proved that all these were either $niiA^-$ or $nirA^-$. Thus, in a total of at least 100 to 200 non-nitrite utilizing mutants, only one probable candidate has been discovered for a structural gene determining a catalytic protein for the reduction of nitrite to ammonia, and that is *niiA*. If another structural gene existed, which determined some other protein essential for nitrite reduction, then mutation in that gene should result in a frequently occurring class of non-nitrite utilizing mutants. The fact that no further class of non-nitrite utilizing mutant has been discovered in extensive selection experiments strongly indicates that a single protein, determined by the *niiA* gene, catalyses the reduction of nitrite to ammonia in *A. nidulans*. It has also been postulated that ammonia is the product of the reduction of nitrite by nitrite reductase without the formation of free intermediates in various higher plants (Joy & Hageman, 1966; Ingle, Joy & Hageman, 1966).

It is probable that the *nii riboA* gene is involved in the synthesis of a flavin cofactor or its incorporation into the nitrite reductase molecule. Mutation in the *nii riboA* gene results in poor growth on nitrate or nitrite medium without a riboflavin supplement and low nitrite reductase and hydroxylamine reductase activities (Pateman, Rever & Cove, 1967). Growth is normal on nitrate or nitrite with a riboflavin supplement and the enzyme activities are fully restored. None of the *niiA* or *nirA* mutants are supplemented on nitrate or nitrite by riboflavin. Thus there is every indication that the *nii riboA* gene is involved in the flavin cofactor status of nitrite reductase in an analogous fashion to the role of the *cnx* genes in the molybdenum cofactor status of nitrate reductase. It seems that the *nii riboA* gene is not involved in any way with the flavin cofactor of nitrate reductase since nitrate reductase activity is not dependent on a riboflavin growth supplement in the *nii riboA* mutant.

Hyponitrite reductase

Hyponitrite has been suggested as a possible intermediate in the reduction of nitrite to ammonia. Hyponitrite reductase, an enzyme which uses NADH as an electron donor to reduce hyponitrite to ammonia, has been demonstrated in *Neurospora crassa* (Medina & Nicholas, 1957*a*). This enzyme is apparently distinct from the nitrite reductase in *N. crassa* (Medina & Nicholas, 1957*b*).

Hydroxylamine reductase

Hydroxylamine reductase (NADPH-oxidoreductase E.C.1.6.6.4) has been demonstrated in *Neurospora crassa* (Nicholas, 1959*a*, *b*), and in *Aspergillus nidulans* (Pateman, Rever & Cove. 1967). As with nitrate and nitrite reductases, the level of hydroxylamine reductase activity is greatly increased by growth on nitrate or nitrite. In *N. crassa* the enzyme is a flavoprotein inhibited by metal chelating agents and uses NADH as the electron donor to reduce hydroxylamine to ammonia. Manganese- or magnesium-deficient cells have very low hydroxylamine reductase activity, but neither metal accumulates in purified enzyme, nor does addition of these metals to enzyme preparations increase activity. It is suggested that the two metals may be necessary for the formation of the enzyme molecule *in vivo*.

In *A. nidulans* the electron donor is NADPH not NADH as in *N. crassa*. The enzyme activity *in vitro* is strongly dependent on added FAD, like nitrite reductase activity and to a lesser extent nitrate reductase activity. As described previously, hydroxylamine reductase activity is associated with nitrite reductase activity. The *niiA* gene may be regarded as a structural gene for a protein which catalyses the reduction of both nitrite and hydroxylamine using NADPH as the electron donor. The fact that *nii riboA* mutants lack normal hydroxylamine reductase activity unless grown with a riboflavin supplement implies that flavin is also a necessary component for hydroxylamine reductase whether or not it is the same protein as nitrite reductase. It is not possible to be sure if *A. nidulans* can use hydroxylamine directly as a nitrogen source. Hydroxylamine is extremely toxic to *A. nidulans* and rather unstable. The concentration (10^{-4} M) at which it is not toxic is not sufficient to act as a significant nitrogen source either alone or together with a limiting concentration of another nitrogen source (Pateman, unpubl. work).

Although it is not strictly relevant in this section, some mutants in *Neurospora crassa* concerned with nitrite utilization deserve mention. Silver & McElroy (1954) described five classes of mutants which were unable to grow normally on nitrite. Two of these mutants excreted nitrite and hydroxylamine when grown in the presence of nitrate. A third mutant required pyridoxine for normal growth; on a low pyridoxine supplement plus nitrate the mutant grew poorly, excreted nitrite and had low nitrite reductase activity. In an attempt to explain the characteristics of these mutants Silver & McElroy (1954) suggested that pyridoxine plays the following role in nitrite reduction. The reduction of nitrite results in free hydroxylamine, which reacts with pyridoxal phosphate to form pyridoxal oxime phosphate. The oxime is reduced to pyridoxamine, which then acts as a donor in transamination with keto acids. The end result is the regeneration of pyridoxal phosphate and the production of amino acids whose amino nitrogen derives from hydroxylamine and not ammonia. However, subsequent attempts to reduce pyridoxal oxime phosphate with *N. crassa* cell preparations were not successful and other classes of pyridoxine requiring mutants do not require more pyridoxine when grown on nitrate (Nicholas, 1959*a*, *b*). Thus there is no direct evidence at present for a reductive pathway other than the well-established direct route from nitrate to ammonia.

7.5 Oxidation of Inorganic Nitrogen Compounds

A number of fungi can oxidize ammonia to nitrite and/or nitrite to nitrate, although this is in no way comparable to the oxidations carried out by *Nitrosomonas* and *Nitrobacter.* Eylar & Schmidt (1959) surveyed about a thousand strains of fungi for the ability to oxidize ammonia. Some thirteen strains of *Aspergillus flavus* were able to oxidize ammonia to a significant degree. However, this level of activity was low compared with bacterial oxidation, and very much less than the rate of reduction of nitrate to ammonia achieved by most fungi. When *A flavus* was grown on peptone, amino acids or ammonia as nitrogen sources, then β-nitropropionic acid, bound hydroxylamine, nitrite and nitrate were produced. Also the fungus contained a peroxidase capable of producing nitrite from β-nitropropionic acid. A number of other *Aspergillus* species: *A. aureus, A. catatae, A. wentii,* and various *Penicillium* species can oxidize nitrite to nitrate when grown on a peptone medium. In *Aspergillus wentii* the oxidation of ammonia, hydroxylamine and nitrite to nitrate has been demonstrated in cell-free extracts. The oxidation rates were increased by the addition of NADP and cytochrome c, and the oxidation of hydroxylamine to nitrite and of nitrite to nitrate is apparently catalysed by cytochrome c reductases (Alleem, Lees & Lyric, 1964). It seems from the work on oxidation *in vitro* that the inorganic pathway via hydroxylamine and nitrite is likely to be the important route in the small minority of fungi capable of oxidizing ammonia.

7.6 Regulation of Nitrate Reduction in *Aspergillus nidulans*

This section will consist largely of an account of the genetical studies of nitrate reduction in *Aspergillus nidulans*, since these have been more extensive and consequently the situation is better understood than in any other fungus.

Nitrate induction and ammonia repression

In *A. nidulans* the enzyme activities necessary for nitrate reduction are nitrate reductase and nitrite reductase. These two enzyme activities and the associated hydroxylamine reductase activity are induced by nitrate and nitrite and repressed by ammonia (Cove & Pateman, 1963; Pateman *et al.*, 1964; Pateman, Rever & Cove, 1967). It is probable that both nitrate induction and ammonia repression act by determining the rate of enzyme synthesis, but it is not known if this is at the transcription or translation level or both. There is also some evidence that nitrate reductase *in vivo* may be degraded in the absence of nitrate and/or the presence of ammonia, with consequent loss of activity (Hynes, 1973*a*). The genes known to play some role in nitrate reduction are: *nia-D*, the structural gene for nitrate reductase; *niiA*, the probable structural gene for nitrite and hydroxylamine reductase; the five *cnx* (cofactor, nitrate reductase, xanthine dehydrogenase) genes *cnxABC, cnxE, cnxF, cnxG, cnxH* concerned with a cofactor common to nitrate reductase and xanthine dehydrogenase; *nii riboA* concerned in some way with the flavin cofactors of nitrate and nitrite reductases; *nirA* (previously *niiB*), probably a regulator gene concerned with the nitrate induction of nitrate and nitrite reductases. Recessive mutations in the *nia-D* and *cnx* genes result in the failure to grow on nitrate; recessive mutations in the *niiA*, *nii riboA* and *nirA* genes result in failure to grow on nitrate or nitrite. A number of genes which affect the regulation of the pathway by ammonia are described later.

The regulatory role of nitrate reductase

The majority of mutations in the *nia-D* and *cnx* genes not only result in the loss of nitrate reductase activity, they also result in the constitutive synthesis of nitrite reductase in the absence of nitrate (Table 7.2). Furthermore, some *nia-D* gene mutants and all *cnx* mutants retain the cytochrome c reductase activity possessed by the nitrate reductase molecule (Table 7.2). This cytochrome c reductase protein, is also synthesized constitutively in the absence of nitrate. Thus, when nitrate reductase activity is lost as a result of mutation in any one of six different genes, the synthesis of nitrate reductase itself—as represented by its cytochrome c reductase moiety—and of nitrite reductase, becomes constitutive. There are two main hypotheses to account for this remarkable situation:

1. The apparent constitutive synthesis of the two nitrate induced enzyme proteins might be due to the accumulation of a low concentration of nitrate in cells that lack nitrate reductase activity. This nitrate might come from the oxidation of ammonia or the concentration of trace amounts from the medium. There are two main arguments against this: (a) at low concentrations nitrite is a comparable inducer to nitrate (Pateman, Rever & Cove, 1967) and *niiA* mutants which lack nitrite reductase activity would convert the nitrate to nitrite. Thus, as a result of nitrite induction, *niiA* mutants

should appear constitutive for nitrate reductase activity. However all *niiA* mutants are fully nitrate inducible for nitrate reductase (Table 7.2); (b) in the normal prototrophic strains of *A. nidulans* ammonia repression is more effective than nitrate induction in that the presence of equimolar amounts of ammonia and nitrate results in low levels of nitrate and nitrite reductase. Also the constitutive enzyme synthesis of the *nia-D* and *cnx* mutants is repressed by ammonia. Thus it is possible to observe the effect of various combined nitrate and ammonia concentrations in the medium on nitrite reductase levels in the prototroph and the mutants. If the *nia-D* and *cnx* mutants were constitutive because of intracellular nitrate accumulation then the mutants should respond to nitrate in the presence of ammonia by as much or more enzyme synthesis as the prototroph. However the mutants have significantly lower levels of enzyme in these conditions (Cove & Pateman, 1969). This is strong evidence that the *nia-D* and *cnx* mutants are not constitutive because of intracellular concentration of nitrate.

2. The alternative and favoured hypothesis is that nitrate reductase has two functions, one catalytic and the other regulatory. It is based on the observations that mutations in the *nia-D* and *cnx* genes (but not in the *niiA* gene) can simultaneously result in: (a) loss of nitrate reductase activity; (b) constitutive synthesis of all the enzyme proteins of the nitrate reduction pathway in the absence of ammonia; (c) altered response to nitrate induction in the presence of ammonia.

The regulatory gene nirA

A common class of recessive mutations in the *nirA* gene result in the loss of all the enzyme activities necessary for the reduction of nitrate. A second type of mutation can occur in the *nirA* gene and result in constitutive synthesis of the nitrate and nitrite reductases in the absence of nitrate. These mutations were originally made in a *niiA* mutant using a selection method which depended on the intracellular accumulation of nitrite by *niiA* mutants when supplied with nitrate (Pateman & Cove, 1967). The constitutive mutant, $nirA^c$, possesses the following characteristics: (a) the $nirA^c$ mutation is in the same gene as the $nirA^-$ mutations; (b) the $nirA^c$ mutation affects the level of all the enzymes necessary for nitrate reduction; (c) the $nirA^c$ is semi-dominant (*i.e.* intermediate in phenotype) with respect to the induction of nitrate reductase and recessive for nitrite reductase activity in the heterozygous diploid genotype, $nirA^c/nirA^+$; (d) the constitutive enzyme synthesis in $nirA^c$ is repressed by ammonia in a similar fashion to the ammonia repression of the nitrate induced enzyme synthesis in the prototroph $nirA^+$ (Table 7.3). In order to explain these results it was proposed that the *nirA* gene produced a regulator product which was necessary for the synthesis of the nitrate reducing enzymes. The simplest type of model which accounts for the properties of the regulatory gene *nirA* is given by Pateman & Cove (1967). However, this model does not take into consideration the probable involvement of nitrate reductase itself or ammonia in the regulation of the nitrate reduction pathway.

A possible model for the regulation of nitrate reduction

A model proposed by Cove & Pateman (1969) which illustrates possible mechanisms for the control of nitrate and nitrite reductase is given in Fig.

Table 7.3 Enzyme activities of *Aspergillus nidulans* mycelium of various genotypes grown in the presence and absence of nitrate (From Pateman & Cove, (1967), *Nature* **215**, 1235)

	Enzyme activities of mycelium grown on urea			Enzyme activities of mycelium grown on urea + sodium nitrate		
Genotype	**Nitrate reductase**	**Nitrite reductase**	**Hydroxylamine reductase**	**Nitrate reductase**	**Nitrite reductase**	**Hydroxylamine reductase**
nir^+	4.2	2.2	2.6	75	136	146
nir^+, $niiA^-$	1	3	9	76	2	21
nir^c, $niiA^-$	61	1	5	52	1	10
nir^c	124	60	100	125	237	348
nir^c/nir^c	107	71	108	140	267	451
nir^c/nir^+	41	8	29	121	85	119
nir^+/nir^+	4	1	18	80	192	204
nir^c/nir^-	58	30	47	108	109	149

All enzyme activities are expressed as nmoles of substrate transformed min^{-1} mg^{-1} of protein

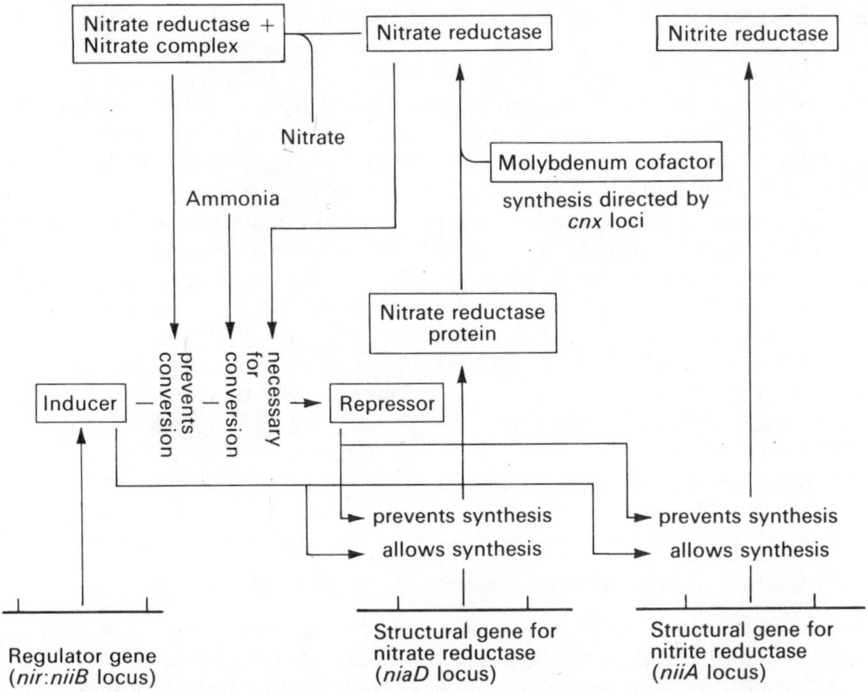

Fig. 7.4 Model illustrating the possible mechanisms of the control of nitrate and nitrite production in *Aspergillus nidulans.* The *nir* gene produces a substance which is necessary for expression of the structural genes. This substance is converted to a repressor, which prevents the expression of the structural genes, either by ammonium or by nitrate reductase when it is not complexed with nitrate. Complexed with nitrate, nitrate reductase interferes with the conversion of the substance to a repressor. (From Cove & Pateman (1969).) Journal of Bacteriology **97**, 1374.

7.4. Although it is rather complex it is the simplest type of model which takes into account the involvement of nitrate, ammonia, nitrate reductase and the *nirA* gene product in the regulation of the pathway. There is no direct biochemical evidence as yet concerning the mechanisms of the postulated regulatory functions of nitrate reductase and the *nirA* gene product. Therefore, the relationships suggested in the model between the various control components are speculative. They represent just one of a number of possible ways of explaining the experimental observations. There are several points which are worth further discussion.

The postulated requirement of the *nirA* gene product (*i.e.* positive action) for synthesis of the enzyme proteins is based on the lack of all enzyme activity in the recessive *nirA*$^-$ mutants and the semi-dominance of the *nirA*^{c} mutant. The *nirA*$^-$ mutants would produce either no regulatory protein or an abnormal form unable to mediate induction of enzyme synthesis. The *nirA*^{c} mutant would produce a regulatory protein whose positive function in enzyme induction was not prevented by the nitrate reductase molecule and the haploid *nirA*^{c} would thus be constitutive. An interaction between the *nirA*^{c} regulatory protein and ammonia must still occur since the *nirA*^{c}

mutant is repressed by ammonia. The semi-dominance of the *nirA*^{c} mutant in the heterozygous diploid *nirA*^{c}/*nirA*$^+$ may be explained by a repressor function (*i.e.* negative action) of the *nirA* regulatory protein after interaction with nitrate reductase or ammonia, as illustrated in Fig. 7.4. Alternatively, it may be that the *nirA* regulatory protein has only a positive inducing function and the semi-dominance of *nirA*^{c} is a gene dosage effect, as suggested by Cove & Pateman (1969). This would require a variant, but not fundamentally different, form of the model in which both ammonia and nitrate reductase could interact with the *nirA* regulatory protein to prevent induction. The postulated positive action of the *nirA* gene product is analogous to that of the *C* gene regulatory protein in the L-arabinose system in *Escherichia coli* (Englesberg, Irr, Power & Lee, 1965) and in direct contrast to the repressor function (negative action) of the *i* gene regulatory protein in the lactose system of *E. coli.* The nature of the control system, if one exists, for the synthesis of the molybdenum cofactor determined by the *cnx* genes is an intriguing but unresolved question. It is certainly possible that some form of genetic regulation exists since the requirement for the cofactor would increase at least a 100-fold after the induction of nitrate reductase and xanthine dehydrogenase by nitrate and purines respectively.

Regulation by ammonia of nitrate reductase activity

There is good evidence in such fungi as *Neurospora crassa, Aspergillus nidulans* and *Ustilago maydis* that ammonia represses the synthesis of nitrate reductase. There is also evidence in *N. crassa* and *U. maydis* that ammonia can inhibit or inactivate in some way nitrate reductase activity *in vivo* (Lewis & Fincham, 1970*a*; Subramanian & Sorger, 1972*b*). The essential findings are the same in both fungi and are exemplified in Fig. 7.5 taken from Subramanian & Sorger (1972*b*). The induction of nitrate reductase by nitrate is inhibited by cycloheximide and actinomycin D, which indicates that *de novo* protein synthesis is necessary for the production of enzyme activity. If cells with a high level of nitrate reductase activity are transferred to medium lacking nitrogen or containing ammonia there is a rapid decrease in nitrate reductase activity. A simple explanation of this could be that the nitrate reductase protein is unstable and that constant resynthesis is required to maintain the level of enzyme activity (Cove, 1966). However, nitrate reductase activity is more stable in cell-free enzyme preparations than in the *in vivo* conditions described. Also, if nitrate induced cells are transfered to nitrogen free or ammonia medium—which also contains cycloheximide—the decrease in nitrate reductase activity is much less rapid or is abolished. Thus protein synthesis seems to be necessary both for the production of nitrate reductase activity and for the rapid loss of activity *in vivo* in the absence of nitrogen or the presence of ammonia. The most plausible explanation is that a specific protein is responsible for the inactivation of nitrate reductase, while the enzyme on its own is quite stable. This controlling protein might be a specific protease breaking up the nitrate reductase molecule, or it might complex with and inhibit the enzyme. The controlling protein would either be induced by ammonia or would be subject to repression by some metabolite whose intracellular concentration dropped rapidly when extracellular ammonia was available. There are precedents for this sort of situation in the case of rat liver tyrosinase (Kenny, 1967), and in

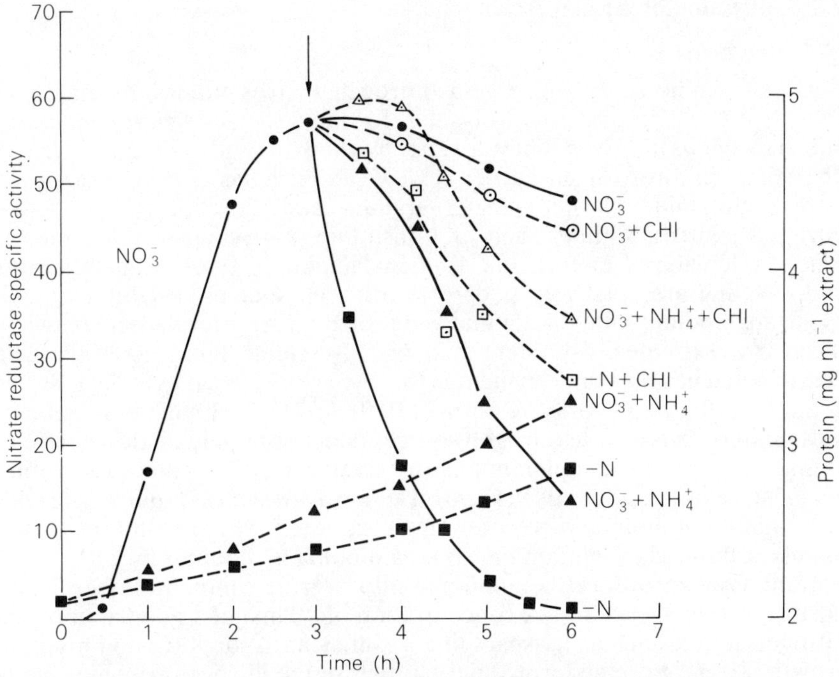

Fig. 7.5 Effect of various culture conditions on the stability of nitrate reductase in wild type mycelia in the presence and absence of cycloheximide. Ammonia-grown wild type mycelia were fully induced by incubation in medium containing 20 mM nitrate. At the time indicated by arrow, the cultures were divided into three groups, each of which received a different treatment. Each of these treatments was in turn carried out in the presence (open symbols), absence (full symbols) of cycloheximide (CHI). One set of mycelia were transferred to medium containing no nitrogen source (□—□); to another set of cultures in nitrate medium, ammonium tartrate was added at a final concentration of 20 mM (△—△); a third set of cultures was kept in nitrate medium as before (○—○). At various time intervals after the treatment, the specific activity of nitrate-reductase of the mycelia remaining in the absence (full symbols connected by solid lines) and presence (open symbols connected by dotted lines) of cycloheximide, and the protein content of the mycelia (full symbols connected by dotted lines) were determined. (From Subramanian & Sorger (1972*b*), Journal of Bacteriology **110**, 540).

yeast where arginase complexed with arginine inhibits the activity of ornithine transcarbamylase (Messenguy & Wiame, 1969). The controlling protein would need to be synthesized very rapidly in the appropriate conditions in order to cause rapid loss of nitrate reductase activity. Thus the synthesis of the controlling protein itself would be very sensitive to protein synthesis inhibitors like cycloheximide, and the presence of the antibiotic would preserve the nitrate reductase activity after transfer to nitrogen-free or ammonia-containing medium. The situation is probably analogous to feedback inhibition in bacteria and would provide a fast-acting regulation of nitrate reductase activity to complement the slower acting repression of enzyme synthesis.

7.7 Utilization of Amino Acids

L-Amino acids

Certain L-amino acids can serve as nitrogen sources either individually or collectively, *e.g.* in casein hydrolysates, for fungi such as *Neurospora crassa* and *Aspergillus nidulans.* They are effective both in solid or liquid and in still or shaken culture (for an earlier review see Nicholas, 1965). The amino acids L-glutamine, L-asparagine, L-arginine and L-proline are good sole nitrogen sources. Others such as L-histidine, L-glycine, L-isoleucine, L-lysine, L-leucine, L-methionine, L-phenylalanine, L-tyrosine, L-threonine and L-valine are relatively poor sole nitrogen sources. L-Glutamate, L-aspartate (sodium salts), L-alanine and L-orthine are intermediate between these two extremes. L-Cysteic acid and L-cysteine are very toxic. The possible mechanisms underlying this toxicity are discussed by Adiga, Sastry & Sarma (1962) and Adiga & Sarma (1970, 1971). Although utilization of most amino acids as sole nitrogen sources depends largely on the release of ammonia, this is not completely so. For example, it is likely that some amino acids can be utilized directly via central amino acids such as L-glutamate with only minimum ammonia release. However, even where amino acids are mainly utilized via L-glutamate, some ammonia is required by certain important steps incorporating ammonia into organic compounds (see Kinghorn & Pateman, 1975*c*). A concentration of 10 mM of a good amino acid nitrogen source such as L-asparagine is sufficient to support good mycelial growth (10–12 g pressed wet weight/l) before the nitrogen becomes limiting. Proteins, such as found in milk, can be hydrolysed by extracellular proteases to a mixture of amino acids and therefore are good sole nitrogen sources. Moreover, although certain amino acids are poor nitrogen sources, they can be utilized efficiently as supplements or for repairing amino acid auxotrophy.

A. nidulans and *N. crassa* can use amino acids as the sole nitrogen and carbon source, albeit very poorly in most cases. In general, amino acids which can be degraded to intermediates of carbon metabolism by one or two reactions (*e.g.* L-glutamate to α-oxoglutarate) are better sources than those which require more elaborate degradation. Although single amino acids are poor carbon and nitrogen sources, mixtures of amino acids such as protein digests are quite good. An important factor when an amino acid is utilized as a sole source of carbon is the consequent production of a large amount of ammonia. The amount of reduced carbon required for respiration is far higher than the nitrogen requirement for growth. This results in excess ammonia production and can lead to growth inhibition as a result of ammonia toxicity or high pH levels.

D-Amino Acids

All the following statements are based on unpublished work with *Aspergillus nidulans* and *Neurospora crassa* by J. R. Kinghorn. All the D-amino acids are extremely poor sole nitrogen sources for *A. nidulans* and *N. crassa.* These include D-asparagine, D-glutamine and D-arginine, the L-isomers of which are good nitrogen sources. Moreover, in *A. nidulans* the D-isomers of asparagine, lysine, arginine and ornithine do not repair asparagine, lysine, arginine and ornithine auxotrophy respectively. D-Serine and D-phenylalanine are toxic for mycelium while D-glutamine and D-asparagine appear only to inhibit conidiation. It is likely that fungi can use D-amino acids to some extent since they possess D-amino acid oxidases (Ohnishi, MacLeod & Horowitz, 1962). All D-amino acids are extremely poor as combined nitrogen and carbon sources.

7.8 Transport of Amino Acids

Amino acids constitute one of the largest and most diverse groups of substances which need to be transported through the fungal cell membrane. Such transport may be partly facilitated diffusion but in most investigated cases is an active transport process. Kinetic and genetic studies have demonstrated the existence of a number of active systems in fungi varying in their specificities and affinities for a wide range of amino acids. The activity of these transport systems may be dependent on many factors: the age of the culture and therefore stage of development, *e.g.* conidia, young or old mycelial cells; the external environment, *e.g.* the composition and concentration of exogenous inorganic nitrogen sources and/or amino acids; and the pH, which affects the ionization and electric charge of many substrates.

Kinetic studies, biochemical analysis of transport components and genetic techniques have all been used in the study of transport processes in micro-organisms; for a description of transport systems in micro-organisms, see recent reviews by Kaback (1970), Lin (1970), Heppel (1971), Slayman (1973) and Chapter 2. So far, kinetic studies and genetic analysis have provided most of the information in fungi. Transport mutants have been isolated by a variety of methods, the most important of these being the selection of mutants resistant to toxic amino acid analogues such as *p*-fluorophenylalanine. Many toxic analogues are transported into the cell by one of the normal amino acid transport systems. Consequently, a mutation which reduces uptake efficiency may confer resistance by keeping the toxic analogue out of the cell. Fungal transport studies have centred around *Neurospora crassa, Aspergillus nidulans* and *Penicillium chrysogenum*, and information from transport systems and mutations which affect amino acid transport is summarized in Tables 7.4, 7.5 and 7.6. Comparisons of the data on transport systems from various sources must be made with caution since the techniques of cultivation, the composition of media, the age and stage of development of cells, the transport assay methods, the units in which transport rates are expressed, all vary considerably from one laboratory to another. In spite of these difficulties some general patterns are discernible.

In *Neurospora crassa* the studies of Pall and others (see Table 7.4 for references and also Chapter 2) have shown the existence of four major transport systems: system I (for neutral amino acids) and system III (for basic amino acids). The level of activity of these two systems is high in young rapidly growing cells on nutrient sufficient media. System II (for neutral, basic and acidic amino acids) and system IV (for acidic amino acids) are most active in older cells which are deficient in carbon or nitrogen. The interpretation of essentially kinetic data by Pall is to some extent supported by the characteristics of mutations which result in a loss of these systems. These are *mtr* (for system I), *su-mtr, Pm G* (for system II) and *bat* and others (for system III). No mutants have been isolated which are defective in acidic

Table 7.4 Amino acid transport in *Neurospora crassa*

Transport system	Specificity	Conditions	Transport mutants
Neutral aliphatic and aromatic L-amino acids [Transport System I—Pall, 1969]	Takes up L-tryptophan, L-leucine, L-valine, L-phenylalanine, L-alanine, L-glycine, L-histidine and L-serine (Pall, 1969; DeBusk & DeBusk, 1965; Stadler, 1966; Lester, 1966; Wiley & Matchett, 1966; Brink, 1972). Also, neutrally charged L-aspartate and L-glutamate at low pH in conidia (Wolfinbarger & DeBusk, 1972; Wolfinbarger & Kay, 1973*a*; Wolfinbarger, Jervis & DeBusk, 1971). A neutral amino acid protein (Wiley, 1970) and KCl-extractable protein (Stuart & DeBusk, 1971) have been claimed to be involved in neutral amino acid transport.	Found in young growing mycelial cells (Pall, 1969)	*Mtr* impaired transport of L-tryptophan, L-phenylalanine, L-tyrosine, L-methionine, L-valine, L-leucine and L-histidine (Lester, 1966; Stadler, 1966). *PmN, neu*[a] and *neu*[t] are probably allelic with *mtr* and defective in certain neutral amino acids tested (Wolfinbarger & DeBusk, 1971; Woodward, Read & Woodward, 1967; Magill, Sweeney & Woodward, 1972)
Acidic, basic and neutral L-amino acids [Transport System II—Pall, 1969]	Takes up a wide range of amino acids including L-arginine, D- and L-phenylalanine, L-glycine, L-lysine, L-leucine, L-aminobutyric, L-alanine, L–histidine and L-asparagine (Pall, 1969, 1970*b*; Sanchez, Martinez & Mora, 1972). Also neutrally charged L-aspartate and L-glutamate at low pH in conidia (Wolfinbarger & DeBusk, 1972; Wolfinbarger, Jervis & DeBusk, 1971; Wolfinbarger & Kay, 1973*a*).	Found in older cells which have been starved of carbon and/or nitrogen (Pall, 1969). Probably regulated by ammonia repression (Sanchez *et al.*, 1972).	*Su-mtr*: These are reported to have increased transport of System II (Stadler, 1967). *PmG,* defective in the general amino acid system (Rao & DeBusk, 1975).

Basic L-amino acids [Transport System III—Pall, 1970*a*]	Takes up L-arginine, L-ornithine, L-lysine and canavanine (Bauerle & Garner, 1964; Roess & DeBusk, 1968; Pall, 1970*a*; Wolfinbarger & DeBusk, 1971).	Found in young rapidly growing mycelial cells (Bauerle & Garner, 1964) and germinated conidia (Pall, 1970*a*)	*bat* defective in basic amino acid transport (Thwaites & Pendyala, 1969; Pall, 1970*a*). *PmB* defective in L-lysine and L-arginine (Roess & DeBusk, 1968; Wolfinbarger & DeBusk, 1971). *hlp-1, bm-1, bas-9, CR-10* are also defective in basic amino acid transport (Roess & DeBusk, 1968; Choke, 1969; Sanchez *et al.* 1972; Magill, Sweeney & Woodward, 1972 respectively)
Acidic D- and L-amino acids [Transport System IV—Pall, 1970*b*]	Takes up anionically charged species of L-aspartate, D-aspartate, L-glutamate, D-glutamate and L-cysteate (Pall, 1970*b*). The presence of this system is questioned by Wolfinbarger & Kay (1973*a*). Their studies indicate that acidic amino acids are taken up primarily, if not exclusively by the neutral or general system. Furthermore they described a dicarboxylic acid transport system in young mycelial cells which takes up L-aspartate (Wolfinbarger & Kay, 1973*b*).	High activity in mycelial cells which have been carbon, sulphur or nitrogen starved (Pall, 1970*b*).	None

Table 7.5 Amino acid transport in *Aspergillus nidulans*

Transport system	Specificity	Conditions	Transport mutants
Acidic L-amino acids	Takes up L-glutamate, L-aspartate and L-cysteate (Robinson, Anthony & Drabble, 1973*a, b;* Kinghorn & Pateman, 1972, 1975*b*; Pateman, Kinghorn & Dunn, 1974; Hynes, 1973*b*).	Found in germinated conidia (Robinson, Anthony & Drabble, 1973*a, b*) and young mycelial cells (Pateman, Kinghorn & Dunn, 1974)	*aauA1* defective in L-glutamate and L-aspartate transport (Kinghorn & Pateman, 1975*b*; Pateman, Kinghorn & Dunn, 1974). *aauB, aauC, aauD* (Kinghorn & Pateman, 1975*b*) and *fpaD* (Sinha, 1969) are defective in the transport of a number of amino acids.
Basic L-amino acids	Takes up L-arginine and possibly L-lysine (Cybis & Weglenski, 1969)	Found in young mycelial cells (Cybis & Weglenski, 1969)	No mutants

amino acid transport (system IV) in *N. crassa*. However, mutants have been isolated which show reduced transport of all or a number of these systems and these are difficult to interpret in terms of Pall's grouping. These include *fpr–1* (Kinsey & Stadler, 1969), *mod-5* (St Lawrence, Maling, Altwerger & Rachmeler, 1964), *nap* (Jacobson & Metzenberg, 1968), *UM300* (Davis & Zimmerman, 1965) and *un-t* (55701) (Kappy & Metzenberg, 1965, 1967; Jacobson & Metzenberg, 1968).

In *Aspergillus nidulans* there appears to be a system specific for acidic amino acids in germinated conidia and young mycelial cells (Table 7.5). The specificity of the system is confirmed by a mutation at the *aauA* locus which results in low transport rates for L-glutamate and L-aspartate alone (Pateman, Kinghorn & Dunn, 1974; Kinghorn & Pateman, 1975*b*). A transport system for basic amino acids has been described but is not extensively characterized. Mutations have been described in *A. nidulans* which result in the simultaneous loss of several different amino acid transport activities. These mutations map at the following loci: *aauB, aauC, aauD* (Kinghorn & Pateman, 1975*b*) and *fpaD* (Sinha, 1969). It is likely that *aauB, aauC* and *aauD* determine membrane proteins which are important structural components shared by several or all amino acid transport sites. The *aauA* gene may determine a specific component such as a carrier protein for the acidic amino acid transport site.

Segel and coworkers have investigated the specificity of transport systems in *Penicillium chrysogenum* (Table 7.6). These kinetic studies suggest the existence of specific systems for L-methionine (Benko, Wood & Segel, 1967), L-cysteine and L-cystine (Skye & Segel, 1970), L-lysine, L-arginine and L-proline (Hunter & Segel, 1971). In addition, group-specific systems have been found, for acidic amino acids in cells which have been carbon or nitrogen starved (Hunter & Segel, 1971) and for basic amino acids in nutrient-sufficient media (Hunter & Segal, 1971). Furthermore, Benko, Wood & Segel (1969) report a general and relatively non-specific system for L-amino acids including L-glutamate and L-aspartate in nitrogen- or carbon-starved cells.

In summary, a number of general points can be made at the present time about amino acid transport in filamentous fungi. The kinetic evidence of Segel and coworkers suggests that *Penicillium chrysogenum* possesses a number of highly specific systems for individual amino acids. Such specific transport systems are well known in bacteria and yeast (Slayman, 1973, and references therein). Transport systems whose specificity is limited to the acidic or basic or neutral groups of amino acids have been reported for all three fungi. Unlike the situation in bacteria, but similar to yeast, some filamentous fungi also have a non-specific general amino acid transport system. Such a general system has been found in *Arthrobotyris conoides* (Gupta & Pramer, 1970), *Neurospora crassa* (Pall, 1969, 1970*a, b*), *P. chrysogenum* ((Benko, Wood & Segel, 1969), *P. griseofulvum* (Whitaker & Morton, 1971) and *Botrytis fabae* (Jones & Watson, 1962; Jones, 1963). In some fungi aromatic amino acids are transported by non-specific systems which can also transport aliphatic neutral amino acids. This is in contrast to the situation in bacteria where there is usually a specific transport system for aromatic amino acids.

Table 7.6 Amino acid transport in *Penicillium chrysogenum*.

Transport system	Specificity	Conditions	Transport mutants
L-Cysteine and L-cystine	Relatively specific for L-cystine and L-cysteine (Skye & Segel, 1970)	Young mycelial cells (Skye & Segel, 1970)	None
L-Methionine	Relatively specific system for L-methionine (Benko, Wood & Segel, 1967)	Mycelial cells which have been sulphur starved (Benko, Wood & Segel, 1967)	None
L-Proline	Specific for L-proline (Hunter & Segel, 1971	Mycelial cells which have been nitrogen or carbon starved (Hunter & Segel, 1971)	None
Basic L-amino acids	Highly specific for L-arginine, L-lysine and low specificity for L-ornithine and L-histidine (Hunter & Segel, 1971)	Found in nutrient sufficient mycelia (Hunter & Segel, 1971)	None
Acidic L-amino acids	Specific for L-glutamate and L-aspartate (Hunter & Segel, 1971)	Found in mycelial cells which have been nitrogen or carbon starved (Hunter & Segel, 1971)	None
General system, acidic, basic and neutral L-amino acids	Takes up L-methionine, L-leucine and L-phenylalanine (Benko, Wood & Segel, 1969) and neutrally charged acidic L-amino acids (Hunter & Segel, 1971)	Nitrogen starved mycelial cells (Benko, Wood & Segel, 1969)	None

The study of the regulation of amino acid transport systems is fraught with difficulties. In general, transport levels can only be assayed by function and not by direct analysis of transport proteins. This makes it difficult to differentiate between the regulation of transport activity, the regulation of the synthesis of transport systems and the functional integration of transport systems into the cell membrane. The level of transport systems is often affected by the composition of the growth medium. A general transport system has been reported to occur in older cells of *Neurospora crassa* (system II) and *Penicillium chrystogenum*, which are carbon and/or nitrogen starved. The evidence of Hunter & Segel (1971) suggests that in *P. chrysogenum* this general system is subject to feedback inhibition ('transinhibition') by the intracellular concentration of the various substrates, and this is more evident under conditions in which amino acids are transported faster than they are metabolized. Although many of the transport systems in bacteria and yeast, as well as fungi, are subject to feedback inhibition the mechanism is largely unknown. A possible explanation is offered by Bradfield *et al.* (1970). In *P. chrysogenum* the specific transport systems for L-methionine and L-cystine are regulated by the availability of sulphur. In *N. crassa* the specific systems for L-cysteine, L-arginine and L-lysine appear to be constitutive (defined in this context as active in cells in nutrient sufficient media without the presence of the specific substrate), while the systems for L-proline and acidic amino acids are found only to develop to a high level in nitrogen or carbon deprived cells. More extensive studies in *Aspergillus nidulans* have shown that the acidic amino acid transport system is regulated by ammonia repression and some form of carbon control (Hynes, 1973*b*; Kinghorn & Pateman, 1974*a*; Pateman, Kinghorn & Dunn, 1974). There is some evidence for changes in amino acid transport systems during development. Tisdale & DeBusk (1970) have shown developmental differences, including specific systems for certain stages in development, in amino acid transport levels in *N. crassa*.

Several features are apparent from the genetic analysis of transport systems. In *Aspergillus nidulans* and *Neurospora crassa* a large number of genes are known to be involved in some way in the determination of amino acid transport. There are a number of possible functions for these genes. It is probable that some determine proteins which have a specific function in the transport of an individual amino acid, others will determine proteins with a transport function specific for one or more groups of amino acids. But in no case so far in fungi has a specific transport function for a membrane protein been directly demonstrated, although membrane protein changes in amino acid transport mutants of *N. crassa* are known (Wiley, 1970; Stuart & DeBusk, 1971). There are other probable functions for genes with an apparent involvement in amino acid transport. They might determine enzymes concerned with some aspect of amino acid metabolism or affecting energy systems coupled with amino acid transport, or some of them may be regulatory genes controlling the level of one or more transport systems. Most mutations affecting transport in fungi are recessive. Some dominant or partially dominant transport mutants have been described such as *aauC* and *aauD* (Kinghorn & Pateman, 1975*b*) and *fpaD* (Sinha, 1969). These could be regulatory genes. Alternatively, it may be that the heterozygous diploid containing the normal and mutant alleles specifies two forms of membrane protein and various combinations of these are built into the transport sites.

Thus the cell membrane of the heterozygous diploid could be a mosaic of several kinds of transport site, each with various degrees of efficiency, this being apparent as complete or partial dominance of the mutant.

7.9 Amino Acid Catabolism

Although the biosynthesis and regulation of amino acid biosynthesis in fungi has been well documented, relatively little is known about amino acid breakdown systems. The main pathways found in bacteria and mammalian tissues such as liver are covered in reviews by Thiamin (1963), Greenberg (1969*b*) and Rodwell (1969*b*). In bacteria certain amino acids such as L-glutamate, L-aspartate and L-alanine, are deaminated or transaminated to give directly such carbon metabolites as α-oxoglutarate, oxalacetate and pyruvate, respectively. With other amino acids such as L-threonine, L-arginine and L-histidine there is considerable breakdown before carbon metabolites are formed and this breakdown is usually coupled to oxidative reactions yielding energy. In some cases there may be little or no energy yield, *e.g.* the opening of the aromatic ring of aromatic amino acids such as L-tyrosine, which is catalyzed by oxygenases (Hayaishi, 1962). Similar patterns of breakdown may exist in the fungi, although in few cases has this been definitively shown. Examples of known catabolic steps in the fungi are given in the following discussion, together with a section on protein degradation, since protein is a direct source of amino acids in the wild. The regulation of amino acid catabolism in fungi may be partly similar to that in bacteria, where a number of steps are regulated by catabolite repression (Jacoby, 1964). Moreover, amino acid breakdown may also, or alternatively, be regulated by ammonia repression (see section 7.13).

L-GLUTAMATE The breakdown of L-glutamate is of central importance since a number of amino acids, *e.g.* L-glutamine, L-histidine, L-arginine, L-ornithine and L-proline, are themselves broken down via L-glutamate. The major route of L-glutamate catabolism is via NAD-linked glutamate dehydrogenase (NAD-GDH). This enzyme has been found in a number of fungi (LéJohn, 1971), in particular *Neurospora crassa* (Kapoor & Grover, 1970*a, b*; Strickland, 1971; Grover & Kapoor, 1973), *Aspergillus nidulans* (Kinghorn & Pateman, 1974*b*; Hynes, 1974*a*) and *Coprinus lagopus* (Fawole & Casselton, 1972; Stewart & Moore, 1974). NAD-GDH mainly serves a catabolic function and high enzyme activity is found in fungi when L-glutamate is the main or only carbon source. These studies indicate that NAD-GDH is at least partly regulated by glucose or some effector derived from glucose (Kinghorn & Pateman, 1974*b*). In *A. nidulans* and *N. crassa* NAD-GDH is a different protein determined by a different structural gene *gdhB* (Kinghorn & Pateman, 1973*a*, 1975*c*) from that of NADP-GDH.

L-ASPARTATE L-aspartate is probably utilized via glutamate-oxaloacetate aminotransferase (GOT). In *Aspergillus nidulans* there are two forms of GOT, one of which may be regulated by the carbon status of the cell (Kinghorn, unpubl. work) (see L-aspartate synthesis).

L-ALANINE Little is known definitely about alanine utilization. It is either a reversal of the normal aminating activity of glutamate-pyruvate aminotransferase (GPT) or a different enzyme with alanine α-oxoglutarate aminotransferase activity (AOT) yielding pyruvate and L-glutamate. High activity of GOT and GPT (or AOT) are found in cells grown on amino acids as the main carbon and nitrogen source (Kinghorn, unpubl. work). This is associated with high levels of NAD-GDH activity which would provide large amounts of α-oxoglutarate as an acceptor for transamination.

L-THREONINE AND L-GLYCINE The studies of the utilization of L-threonine as a carbon and nitrogen source in *Penicillium chrysogenum* and *P. janthinellum* have revealed two reactions mediating the conversion of L-threonine to L-glycine and acetyl-CoA (Willetts, 1972). The first, NAD threonine dehydrogenase, converts L-threonine to 2-amino-3-oxobutyrate, which is cleaved to acetyl-CoA and L-glycine by the second enzyme, 2-amino-3-oxobutyrate ligase. Both enzymes are inducible. The acetyl-CoA is further metabolized via the tricarboxylic acid and glyoxylate cycles.

Further enzymatic evidence of Willetts (1972) suggested that L-glycine is initially deaminated by glycine-pyruvate aminotransferase to glyoxylate, possibly a reversal of L-glycine biosynthesis (alanine-glyoxylate aminotransferase), with the glyoxylate subsequently metabolized via the glycerate pathway in a sequence essentially similar to that found in bacteria (Kornberg, 1958; Kornberg & Elsen, 1961). Moreover, Willetts (1972) presented direct evidence against L-glycine breakdown via L-glycine decarboxylase or by conversion to L-serine.

L-ASPARAGINE AND L-GLUTAMINE It is likely that the major route of L-asparagine and L-glutamine breakdown is deamination by amidases to give L-aspartate and L-glutamate respectively. Asparaginase has been demonstrated in a number of fungi such as *Penicillium camembertii* (Dox, 1909), *Aspergillus niger* (Bach, 1928; Schmalfuss & Mothes, 1930), *A terreus* (De–Angeli *et al.*, 1970). *A. nidulans* (Drainas and Pateman, unpubl. work), and in culture filtrates of a number of species of *Fusarium, Penicillium, Verticillium, Hypomyces* and *Nectaria* (Arima, Sakamoto & Araki-Tamura, 1972; Imada, Igarasi, Nakamama & Isono, 1972).

It is well known that asparaginase from certain micro-organisms, especially organisms of the Enterobacteriaceae, possess anti-leukaemia properties. Asparaginases, from fungi, have mainly not been tested for this, with the following exceptions. Scheetz, Whelan & Wriston (1971) showed that *Fusarium tricinctum* asparaginase did not have anti-lymphoma activity. De-Angeli *et al.* (1970) reported that the asparaginase from *Aspergillus terreus* suppressed Walker 256 ascites carcimoma in rats.

Less is known about glutaminase activity in fungi. It has only been looked for in culture filtrates and has been found in relatively few cases, the exceptions being two strains of the family Moniliaceae and a few Ascomycete fungi (Imada *et al.*, 1972). Little is known about the regulation of these systems, but it is at least possible that they are regulated by ammonia and/or carbon repression, as has been found in the bacterium *Lactobacillus* (Ravel, Norton, Humphreys & Shire, 1962). In addition to glutaminase there is, in *Neurospora crassa*, glutamine aminotransferase, which is responsible for L-glutamine breakdown. α-Oxoglutaramate is formed from L-glutamine and this is readily deaminated by a ω-amidase to give α-oxoglutarate. Glutamine aminotransferase acts mainly in the one direction (*i.e.* glutamine catabolism), and can be separated from asparaginase aminotransferase (see L-asparagine synthesis) and ω-amidase (Meister, 1953; Monder & Meister, 1959).

It is of peripheral interest here that other amides such as acetamide and formamide can be used as nitrogen and in some cases carbon sources (Hynes & Pateman, 1970*c*). These amides are hydrolysed by specific amidases which do not hydrolyse L-asparagine or L-glutamine. The acetamidase and formamidase systems are regulated by ammonia and/or carbon repression (Hynes, 1970; Hynes & Pateman, 1970*a*, *b*).

L-ARGININE, L-ORNITHINE AND L-PROLINE The general utilization of these amino acids proceeds by similar means (Fig. 7.6). Two enzymes, arginase and ornithine-γ-aminotransferase, are required for the breakdown of L-arginine to L-ornithine and urea, and subsequently of L-ornithine to glutamate-γ-semialdehyde in *Neurospora crassa* (Fincham, 1953; Castaneda Martusielli & Mora, 1967; Davis, Lawless & Port, 1971; Davis & Mora, 1968; Morgan, 1970*a, b*; Mora, Salceda & Sanchez, 1972), and *Aspergillus nidulans* (Cybis & Weglenski, 1972; Cybis, Piotrowska & Weglenski, 1972*a, b*; Bartnik & Weglenski, 1973). Mutants of *A. nidulans (otaA)*, which have lost ornithine-γ-aminotransferase activity, are able to grow on L-arginine but not on L-ornithine as a sole nitrogen source (Piotrowska, Sawachi & Weglenski, 1969). The oxidation of glutamate-γ-semialdehyde is discussed below with proline degradation. It has been shown in *A. nidulans* that arginase and ornithine-γ-aminotransferase are induced by L-arginine (Cybis & Weglenski, 1972) and repressed by ammonium and/or glucose (Bartnik, Guzewska & Weglenski, 1973; Bartnik & Weglenski, 1973). More recently Bartnik & Weglenski (1974) have presented evidence that arginase and ornithine-γ-aminotransferase synthesis are probably regulated by positive control.

The first step in L-proline breakdown in fungi is catalysed by proline oxidase and results in an equilibrium mixture of Δ′ pyrroline-5-carboxylate and glutamate-γ-semialdehyde. Mutants of *Aspergillus nidulans (prnB)* lacking this activity are unable to grow on proline as sole nitrogen source, but can use L-arginine or L-ornithine (Arst, unpubl. work). Glutamate-γ-semialdehyde (or perhaps γ′ pyrroline-5-carboxylate) derived from L-proline, L-arginine or L-ornithine breakdown is oxidized by the action of Δ′ pyrroline-5-carboxylate dehydrogenase to L-glutamate. Mutants, *prnA*, blocked at this step, are unable to utilize L-ornithine or L-proline as a sole nitrogen source but can use L-arginine. The reason that *prnA* and *otaA* mutants can utilize L-arginine is probably the production of ammonia from urea by urease activity. Conversely, mutants which lack urease activity can utilize L-arginine and in this case ammonia is derived from L-ornithine via L-glutamate. Mutants of *N. crassa* which lack arginase activity (*aga*) fail to grow on L-arginine since neither urea nor L-ornithine is made.

L-TRYPTOPHAN There are a number of pathways for L-tryptophan breakdown in bacteria (for a review, see Greenberg, 1969*b*). In *Neurospora crassa* L-tryptophan is broken down either by the kynurenine-anthranilate pathway via formylkynurenine and kynurenine (Jacoby, 1954), 3-hydroxykynurenine (Yanofsky & Bonner, 1950*a*) and anthranilate or 3-hydroxyanthranilate (Jacoby & Bonner, 1953; Yanofsky & Bonner,

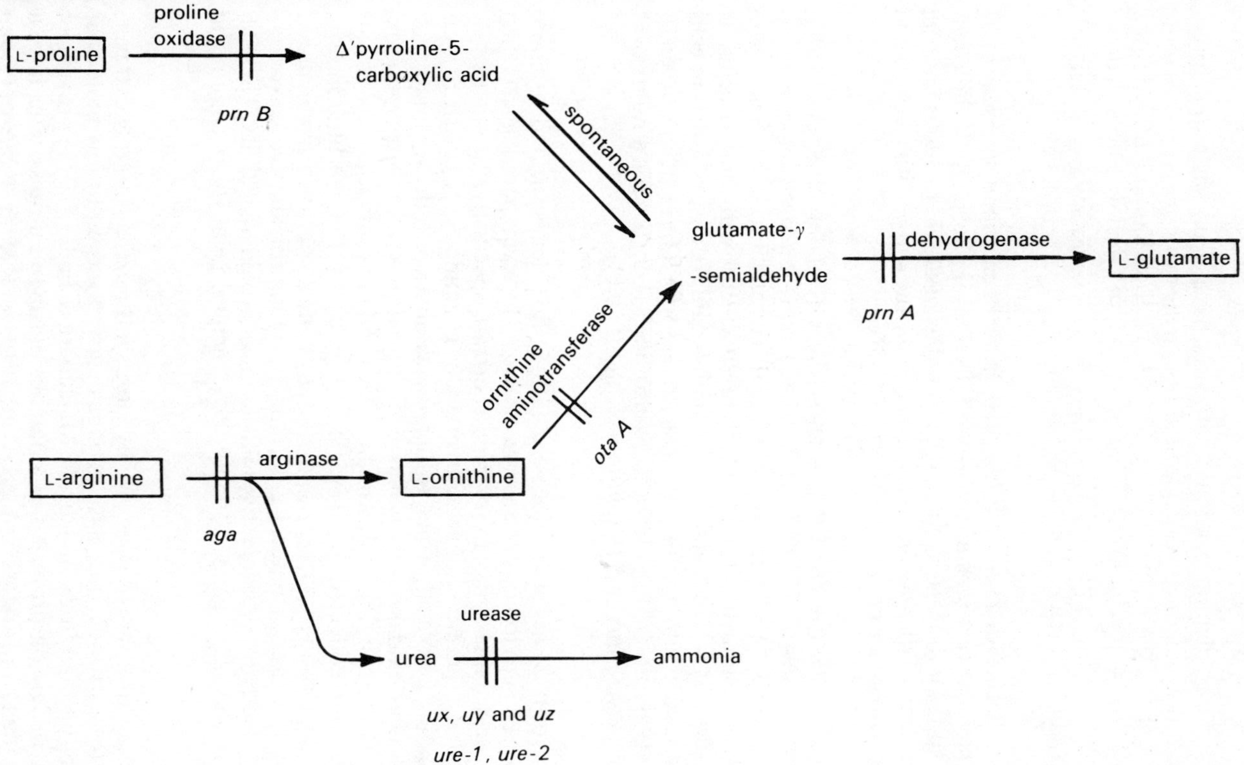

Fig. 7.6 The pathways of L-arginine, L-ornithine and L-proline degradation in *Neurospora crassa* and *Aspergillus nidulans* showing the steps blocked by various mutants. Further information on the mutants *prnA* and *prnB* from Arst & MacDonald (1974); *otaA* from Davis & Mora (1968), Piotrowska, Sawacki & Weglenski (1969); *aga* from Morgan (1970*a,b*), Davis, Lawless & Port, (1971), Castaneda, Martusielli & Mora (1967), Mora, Salceda & Sanchez (1972); *ux, uy* and *uz* from Darlington & Scazzocchio (1967); *ure-1, ure-2* from Kolmark (1969*a, b*), Haysman & Branch Hoire (1971) and Davis (1970).

1950*b*) or by the quinolinic pathway via kynurenic acid or xanthurenic acid (Jacoby & Bonner, 1956).

L-LYSINE Four distinct catabolic pathways have been suggested for L-lysine breakdown in micro-organisms although the evidence is weak in some cases a number of intermediates have still to be found and identified (for a review see Rodwell, 1969*b*). In *Neurospora crassa* L-lysine is broken down via α-keto-E-amino caproate, Δ' piperideine-2-carboxylate and Δ' pipecolate (Schweet, Holden & Lowry, 1954*a, b*).

L-SERINE Serine deaminating enzymes have been found in *Neurospora crassa* which convert serine to pyruvate. Pyridoxal phosphate is the coenzyme (Yanofsky & Reissig, 1953). The studies of Dekker (1960) indicate that there are two distinct enzymes in *N. crassa*, one active with L-serine and the other with D-serine.

L-AMINO ACID OXIDASE Bender and others (Bender & Krebs, 1950; Thayer & Horowitz, 1951; Burton, 1952) reported the presence of an L-amino acid oxidase in the mycelium and culture filtrate of *Neurospora crassa*. This system has broad substrate specificity with amino acids as possible substrates. High activity of L-amino acid oxidase is found in cells grown on media with limiting nitrogen or cells which have been starved of nitrogen. L-amino acid oxidases have also been reported for a number of species of *Aspergillus* and *Penicillium* (Knight, 1948).

D-AMINO ACID OXIDASES Horowitz (1944) first demonstrated D-amino acid oxidase activity in *Neurospora crassa*. The evidence suggests that D-amino acid oxidase has fairly wide substrate specificity. It is generally assumed that D-amino acids are converted, by D-amino acid oxidase, to their corresponding keto acid, which is then transaminated to the L-amino acid. D-Amino acid oxidase activity has been found in several Aspergilli and Penicillia (Emerson, Puziss & Knight, 1950) and in twenty wild-type strains of *Neurospora* collected from various regions of the world (Ohnishi, Macleod & Horowitz, 1962) and therefore must perform some useful metabolic function in natural environments. Ohnishi and his coworkers (1962) isolated mutants defective in D-amino acid oxidase activity. These all mapped in the same gene, possibly the structural gene for D-amino acid oxidase.

EXTRACELLULAR PROTEASES Many fungi synthesize and secrete extracellular proteases which hydrolyse proteins such as casein to low molecular weight peptides and/or amino acids. Matsubara & Feder (1971) provide an extensive review on proteases in a number of fungi. These systems have a wide specificity and are of commercial importance. Recent studies of Cohen and Drucker have shown that extracellular protease activity in *Aspergillus* and *Neurospora* is high with protein as sole nitrogen and carbon source and low when a combination of low molecular weight metabolites such as ammonium, sulphur and glucose are present (Cohen, 1972, 1973a, b; Cohen, Morris & Drucker, 1975; Drucker, 1972, 1973).

7.10 Amino Acid Biosynthesis

Most fungi can grow on media containing only inorganic nitrogen, or require a supplement of only a small number of amino acids. Under these conditions they synthesize some or all of the twenty-one or more essential L-amino acids required for protein synthesis and other purposes. Most of the biosynthetic pathways in micro-organisms have been extensively covered by various reviews (Umbarger & Davis, 1962; Greenberg, 1969*a*; Rodwell, 1969*a*). The review of Esser & Kuenen (1967) should be followed by readers wishing to be aware of the genetical contribution to the elucidation of amino acid biosynthesis in the fungi. Amino acid biosynthesis in filamentous fungi closely parallels that found in yeast and bacteria, but there are a few notable exceptions, *e.g.* L-lysine biosynthesis. Therefore, figures showing biosynthetic steps, which have actually been demonstrated in the fungi, together with a selection of literature references, are given rather than exhaustive description. This, it is hoped, will provide a lead-in to work on amino acid biosynthesis in higher fungi.

Most of the steps of biosynthesis which have been elucidated have been first observed in the bread mould *Neurospora crassa* and, perhaps to a lesser extent, *Aspergillus nidulans, A. niger, Penicillium chrysogenum* and *Ustilago maydis*. Evidence has been obtained in one or more of the following ways: (a) by the study of the characteristics of amino acid auxotrophic mutants, *e.g.* accumulation of an intermediate, or response to a product [lists of amino acid auxotrophs in *A. nidulans* and *N. crassa* are given by Clutterbuck (1974) and Barratt & Ogata (1972) respectively]; (b) detection of enzymic activity; (c) isolation and purification of the enzyme system; and (d) tracer experiments with isotopically labelled molecules.

The amino acids have been divided, rather arbitrarily, on the basis of their origin, *e.g.* tricarboxylic acid (TCA) cycle etc.

Group I *Amino acids from TCA intermediates:*
- (a) L-glutamate, L-aspartate and L-glutamine
- (b) L-lysine

Group II *Amino acids from glycolytic intermediates:*
- (a) L-alanine
- (b) L-valine, L-isoleucine and L-leucine
- (c) L-aromatic amino acids, L-tyrosine, L-phenylalanine and L-tryptophan
- (d) L-serine, L-cysteine and L-glycine

Group III *Amino acids from glutamate:*
- (a) L-proline
- (b) L-ornithine, L-citrulline and L-arginine

Group IV *Amino acids from aspartate:*
- (a) L-asparagine
- (b) L-threonine, L-homoserine, L-methionine, L-cysteine

Group V *Amino acids from imidazole glycerophosphate:*
- (a) L-histidine

Fig. 7.7 Group I Biosynthesis of amino acids from tricarboxylic acid intermediates
Group Ia: L-glutamate, L-glutamine and L-aspartate

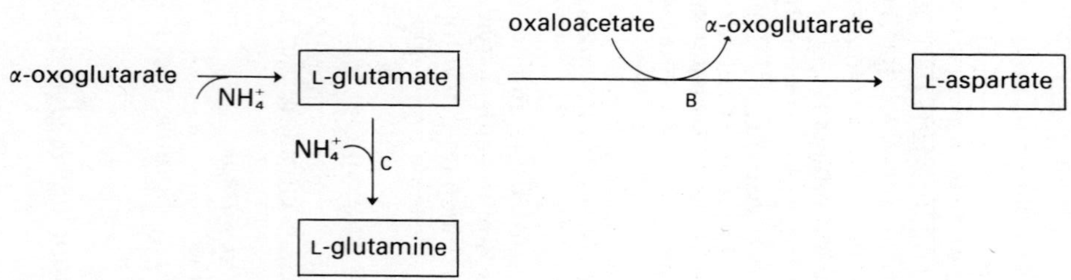

Group Ib: L-lysine

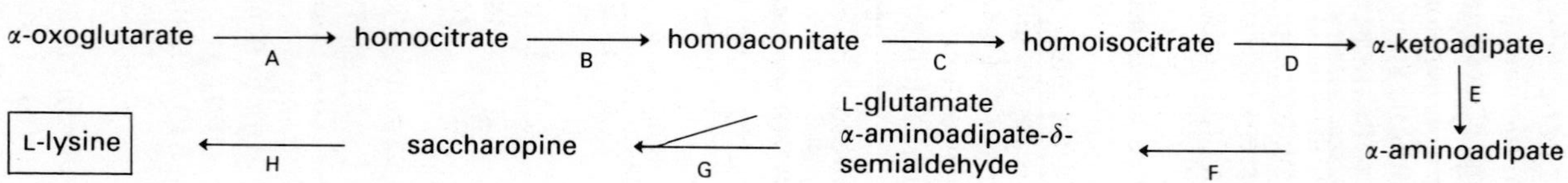

References for fig. 7.7 (Group I)

Step	Enzyme system	Organism	References
Group Ia			
A	NADP L-glutamate dehydrogenase	*Neurospora crassa*	Fincham (1962); Burke & Pateman (1962); Pateman & Fincham (1964); Arkin & Grossowicz (1970); Strickland (1971); Wootton *et al.* (1973)
		Aspergillus nidulans	Pateman (1969); Arst & MacDonald (1973); Kinghorn & Pateman (1973*a*, 1974*a*, 1975*b*); Hynes (1974*a*)
		Other fungi	Goldin & Frieden (1971); LéJohn (1971); Fawole & Casselton (1972); Stewart & Moore (1974)
B	Glutamate-oxaloacetate aminotransferase	*N. crassa*	Fincham & Boulter (1956); Munkres (1965*a, b*); Munkres, Giles & Case (1965); Kitto *et al.* (1967).
		A. nidulans	Kinghorn (unpubl. work)
C	Glutamine synthetase	*N. crassa*	Kapoor & Bray (1968); Kapoor, Bray & Ward (1969); Pateman (1969)
		A. nidulans	Pateman (1969)
Group Ib			
A	Homocitrate condensing enzyme	*N. crassa*	Andersson-Kotto *et al.* (1954); Abelson & Vogel (1955); Hogg & Broquist (1968)
B	Homoaconitase	*N. crassa*	Broquist (1971)
C	Homoaconitase	*N. crassa*	Broquist (1971)
D	Homoisocitrate dehydrogenase	*N. crassa*	Broquist (1971)
E	Glutamate α-ketoadipate aminotransferase	*N. crassa*	Mitchell & Houlahan (1948); Broquist (1971)
F	Aminoadipate reductase system	*N. crassa*	Abelson & Vogel (1955); Broquist (1971)
G	Aminoadipate semialdehyde-glutamate reductase	*N. crassa*	Sanders & Broquist (1966)
H	Saccharopine dehydrogenase	*N. crassa*	Broquist & Stiffey (1959); Trupin & Broquist (1965); Sanders & Broquist (1966)

Amino acids synthesized from intermediates of the TCA cycle

L-GLUTAMATE The primary amination step in filamentous fungi is catalysed by NADP-linked glutamate dehydrogenase with the synthesis of L-glutamate (Fig. 7.7 Ia Step A). Other main routes of primary amination, such as L-aspartate and L-glutamate synthetase found in bacteria and yeasts, have not yet been demonstrated in filamentous fungi. NADP-GDH has been identified in a number of fungi (Léjohn, 1971), including *Neurospora crassa* (Fincham, 1962) and *Aspergillus nidulans* (Pateman, 1969) and can constitute about 0.5% of the total soluble protein in cells grown on inorganic nitrogen sources.

L-ASPARTATE Oxaloacetate, another intermediate of the tricarboxylic acid cycle, acts as an amino acceptor with L-glutamate as the donor to form L-aspartate (Fig. 7.7, Ia Step B). This is carried out by glutamate-oxaloacetate aminotransferase (transaminase) and in fungi is probably tightly bound to its co-enzyme pyridoxal phosphate (Fincham & Boulter, 1956). Munkres, in a series of communications (Munkres, 1965*a*, *b*; Munkres, Giles & Case, 1965), presented evidence that glutamate-oxaloacetate aminotransferase, GOT, and malate dehydrogenase, MDH, activities reside in the same protein. This view was challenged by Kitto and coworkers (Kitto *et al.*, 1967), whose results suggest that two forms of both MDH and GOT are present in *Neurospora crassa* and that all four are distinct physical entities rather than MDH and GOT activities being associated with a single protein. Subsequent work by Kinghorn & Pateman (unpubl. work) indicates that MDH and GOT are separate enzymes in *Aspergillus nidulans*, and suggests that one form of GOT is important in the synthesis and the other in the breakdown of L-aspartate.

L-GLUTAMINE Glutamine synthetase catalyses the amination of L-glutamate with the formation of L-glutamine (Fig. 7.7, Ia Step C). Kapoor, Bray & Ward (1969) have shown that glutamine synthetase from *Neurospora crassa* is a tetrameric enzyme of molecular weight 350 000–360 000. Pateman (1969) found high levels of activity in cells of *Aspergillus nidulans* and *N. crassa* after growth on L-glutamate, but low activity on all other nitrogen sources with or without L-glutamate. From his data he postulated that glutamine synthetase is regulated by L-glutamine repression and not induced by L-glutamate. Since L-glutamine serves as a donor of ammonia in a number of reactions leading to the biosynthesis of carbamyl phosphate, L-histidine, L-tryptophan, purine and pyrimidines, Kapoor & Bray (1968) and Kapoor *et al.* (1969) investigated the inhibition effect of metabolic end-products on glutamine synthetase activity of *N. crassa*. They found that it is subject to feedback inhibition by a number of end-products at the termini of certain metabolic routes whose origin can be traced to L-glutamine, *e.g.* adenosine monophosphate, cytidine triphosphate, guanosine triphosphate, nicotinamide, cytidine triphosphate, guanosine triphosphate, nicotinamide-adenine dinucleotide, L-glycine, L-histidine and anthranilic acid. Since L-glutamine serves as an ammonia donor in the biosynthesis of a number of nitrogenous substances, L-glutamine is at an important point of divergence of a number of biosynthetic pathways (see Shapiro & Stadtman (1970) for a review of glutamine synthetase in micro-organisms, including *N. crassa*).

L-LYSINE The biosynthesis of L-lysine is rather unusual in microorganisms since there are two quite independent synthetic pathways. In bacteria and lower fungi, L-lysine is synthesized via diaminopimelic acid (DAP pathway), whereas in higher fungi its synthesis proceeds via α-aminoadipic acid (AAA pathway). The distribution patterns of these pathways in fungi are described by Vogel (1960, 1961), Rodwell (1969*a*), LéJohn (1971) and the assay procedures by Broquist (1971). So far no organism has been found in which both pathways exist together. It appears that, in general, when cellulose is the major component of the cell wall, the DAP pathway is the pathway of choice. On the other hand, when chitin is the main component of the cell wall as in fungi, the AAA pathway is used. The AAA pathway is described in Fig. 7.7 (Ib). Lysine auxotrophs have been isolated in *Aspergillus nidulans* (Aspen & Meister, 1962) and *Neurospora crassa* (see Barratt & Ogata, 1972). Preliminary studies indicate that the initial enzyme of this pathway (homocitrate condensing enzyme) is subject to end-product control by lysine in *N. crassa* (Hogg & Broquist, 1968).

Amino acids derived from glycolytic intermediates

L-ALANINE L-alanine is probably formed mainly by transamination from L-glutamate and pyruvate by the action of glutamate-pyruvate aminotransferase (GPT) in *Neurospora crassa* (Fincham & Boulter, 1956) and *Aspergillus nidulans* (Kinghorn, unpubl. work) (Fig. 7.8, IIa). Burk & Pateman (1962) found that the NADP L-glutamate dehydrogenase of *N. crassa* also has some NADP L-alanine dehydrogenase activity.

L-ISOLEUCINE AND L-VALINE An interesting feature of the biosynthesis of L-valine and L-isoleucine from pyruvate and α-ketobutyrate respectively, is that it is carried out in four steps by enzymes with dual substrate specificities (Fig. 7.8, IIb Steps B, C, D, E). For example, the same aminotransferase catalyses the synthesis of both L-isoleucine and L-valine (Step E). Extensive studies by Wagner and colleagues with *Neurospora crassa* (Wagner, Bergquist & Barbee, 1965) and Kiritani, Nasise, Bergquist & Wagner (1965), provide evidence that the synthesis of these amino acids occurs in the presence of a complex of four enzymes which is bound or oriented on a particulate fraction of the cell. Recent work (Bergquist, LaBrie & Wagner, 1969; LaBrie, Leiter, Bergquist & Wagner, 1972) has shown that mitochondria can synthesize L-valine when pyruvate and cofactors associated with L-valine synthesis are present. This system, then, may afford the opportunity to study the relationship of localization to regulation. The initial studies of Olshan & Gross (1974) show that the synthesis of the enzymes of L-isoleucine and L-valine synthesis is determined by the product of the *leu3*$^{+}$ gene and the effector, α-isopropylmalate (an intermediate in L-leucine synthesis). Mutants which are specifically blocked in the synthesis, of L-valine and L-isoleucine have been found in *N. crassa* (Wagner, Bergquist, Barbee & Kiritani, 1965).

α-Ketobutyrate is synthesized by threonine dehydratase from L-threonine (Step A). Structural gene mutations for this stage of L-isoleucine synthesis have been found in *Aspergillus nidulans* (Pees, 1966; MacDonald, Arst & Cove, 1974).

Fig. 7.8 Group II Biosynthesis of amino acids from glycolytic intermediates

Group IIa: L-alanine

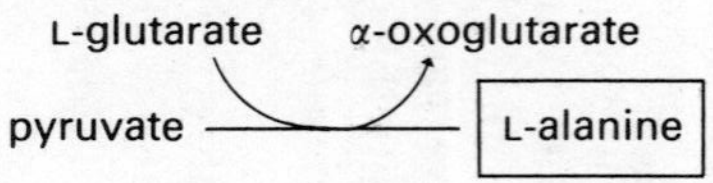

Step	Enzyme system	Organism	References
A	Glutamate-pyruvate aminotransferase	*Neurospora crassa*	Abelson & Vogel (1955); Fincham & Boulter (1956)
		Aspergillus nidulans	Kinghorn (unpubl. work)

Group IIb: L-leucine, L-isoleucine and L-valine

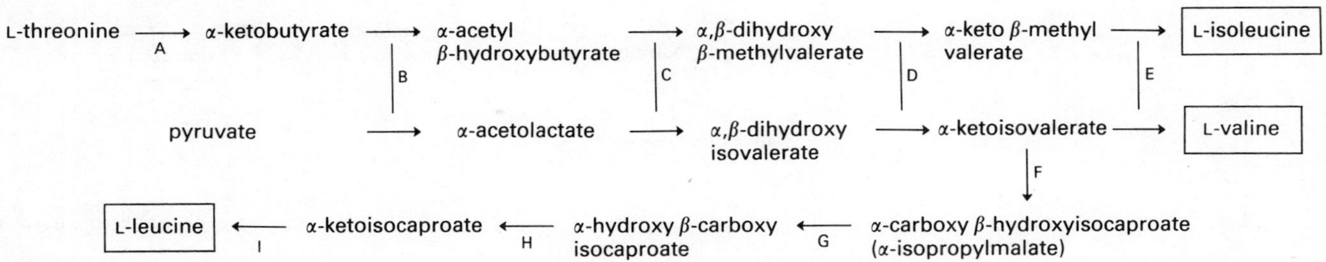

Step	Enzyme system	Organism	References
Group IIb			
A	Threonine dehydratase	*A. nidulans*	Macdonald, Arst & Cove (1974)
B	α-aceto α-hydroxyacid synthetase	*N. crassa*	Abelson & Vogel (1955); Kiritani *et al.* (1965); Wagner, Bergquist & Barbee (1965); Wagner *et al.* 1965); LaBrie *et al.* (1972)
C	Acetohydroxyacid isomeroreductase	*N. crassa*	Abelson & Vogel (1955); Kiritani *et al.* (1965); Wagner, Bergquist & Barbee (1965); Wagner *et al.* (1965); LaBrie *et al.* (1972)
D	α,β-dyhydroxyacid dehydratase	*N. crassa*	Abelson & Vogel (1955); Kiritani *et al.* (1965); Wagner, Bergquist & Barbee (1965); Wagner *et al.* (1965); LaBrie *et al.* (1972)
E	Valine aminotransferase	*N. crassa*	Abelson & Vogel (1955); Kiritani *et al.* (1965); Wagner, Bergquist & Barbee (1965); Wagner *et al.* (1965); LaBrie *et al.* (1972)
F	α-carboxy β-hydroxy isocaproate synthetase	*N. crassa*	Gross (1965, 1969); Webster, Nelson & Gross (1965); Webseter & Gross (1965)
G	α-carboxy β-hydroxy isocaproate isomerase	*N. crassa*	Jungwirth *et al.* (1963); Gross (1962, 1969)
H	α-hydroxy β-carboxy isocaproate dehydrogenase	*N. crassa*	Jungwirth *et al.* (1963); Gross (1965)
I	Leucine aminotransferase	*N. crassa*	Fincham & Boulter (1956); Gross (1969)

Fig. 7.8 continued
Group IIc: aromatic amino acids

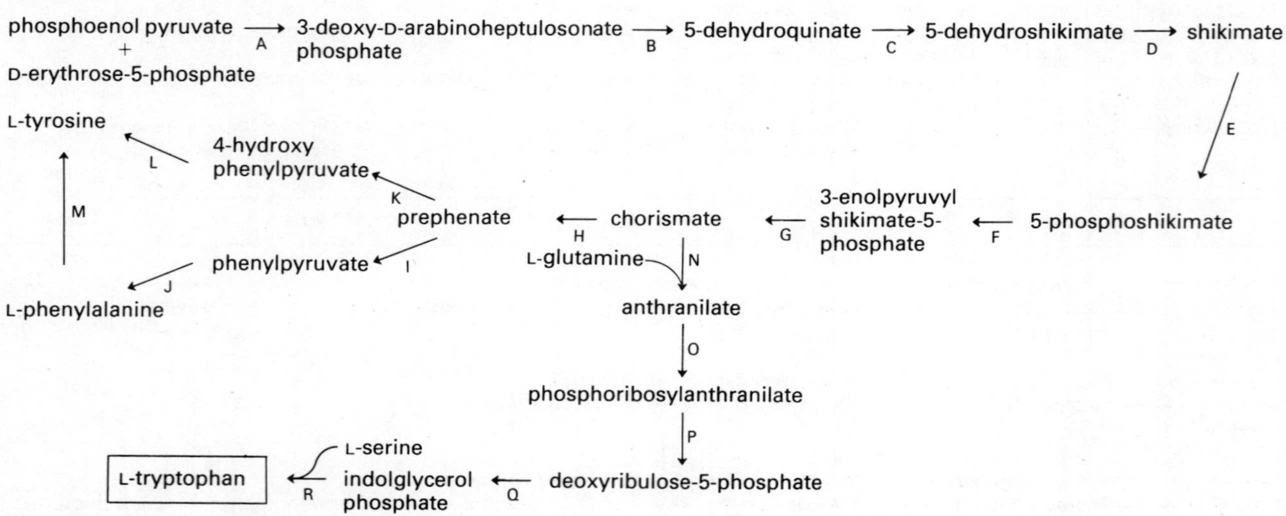

Step	Enzyme system	Organism	References
A	3 deoxy-D-arabinoheptulosonic-7-phosphate synthase	*N. crassa*	Doy (1968); Halsall & Doy (1969); Halsall, Catcheside & Doy (1971); Halsall & Catcheside (1971)
B	Dehydroquinate synthetase	*N. crassa*	Gordon, Haskins & Mitchell (1950); Giles *et al.* (1967*a, b*); Case & Giles (1968, 1971); Burgoyne, Case & Giles (1969)
C	5 dehydroquinate dehydratase	*N. crassa*	Gordon *et al.* (1950); Giles *et al.* (1967*a, b*); Case & Giles (1968, 1971); Burgoyne, Case & Giles (1969)
D	Shikimate dehydrogenase	*N. crassa*	Tatum & Perkins (1950); Tatum, Ehrensvärd & Garnjobst (1954); Case and Giles (1968, 1971); Burgoyne, Case & Giles (1969)
E	Shikimate phosphatase	*N. crassa*	DeMoss (1965*a*); Edwards & Jackman (1965); Case & Giles (1968, 1971); Burgoyne, Case & Giles (1969)
F	3-enolpyruvyl shikimate 7-phosphate synthase	*N. crassa*	DeMoss (1965*a*); Edwards & Jackman (1965); Case & Giles (1968, 1971); Burgoyne, Case & Giles (1969)
G	Chorismate synthetase	*N. crassa*	DeMoss (1965*a*); Edwards & Jackman (1965); Burgoyne, Case & Giles (1969); Gaertner & Cole (1973)
H	Chorismate mutase	*N. crassa*	Metzenberg & Mitchell (1956); Colburn & Tatum (1965); Baker (1966, 1968); Lingens, Goebel & Vesseler (1967); Woodin & Nishioka (1973)
I	Prephenic dehydratase	*N. crassa*	Colburn & Tatum (1965); Baker (1968)
J	Phenylpyruvate aminotransferase	*N. crassa*	Colburn & Tatum (1965); Baker (1968)
K	Prephenate dehydrogenase	*N. crassa*	Seecof & Wagner (1959); Colburn & Tatum (1965); Edwards & Jackman (1965); Baker (1968); Catcheside (1969); El-Eryani (1969)
L	4 hydoxyphenyl pyruvate aminotransferase	*N. crassa*	Colburn & Tatum (1965); Baker (1968); El-Eryani (1969)
M	Phenylalanine hydroxylase	*A. nidulans*	Sinha (1967)
N	Anthranilic acid synthetase	*N. crassa*	DeMoss (1962, 1965*b*); DeMoss & Wegman (1965); Arroyo-Begovich & DeMoss (1969, 1973)
O	Anthranilic PP ribose-phosphoribosyl transferase	*N. crassa*	Wegman & DeMoss (1965)
P	Phosphoribosylanthranilate isomerase	*N. crassa*	Lester (1963, 1968, 1971); DeMoss & Wegman (1965); DeMoss, Jackson & Chalmers (1967)
Q	Indole glycerophosphate synthetase	*N. crassa*	Lester (1963, 1968, 1971); Wegman & DeMoss (1965)
R	Tryptophan synthetase	*N. crassa*	DeMoss & Bonner (1959); Mohler & Suskind (1960); DeMoss (1962); Tsai, Tsai & Yu (1973); Yu, Kula & Tsai (1973)

Fig. 7.8 continued
Group IId: L-serine, L-cysteine and L-glycine
Non-phosphorylated pathway

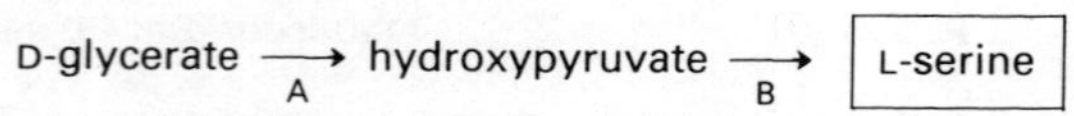

Phosphorylated pathway

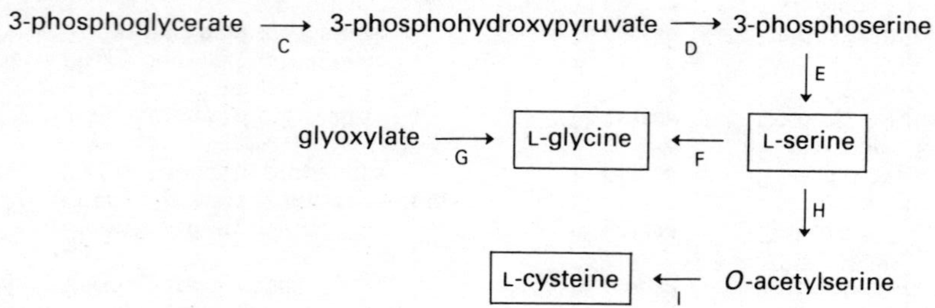

Step	Enzyme system	Organism	References
A	NAD glycerate dehydrogenase	*N. crassa*	Sojka & Garner (1967)
B	Aminotransferase	*N. crassa*	Sojka & Garner (1967)
C	Phosphoglycerate dehydrogenase	*N. crassa*	Sojka & Garner (1967)
D	Transaminase	*N. crassa*	Sojka & Garner (1967)
E	Phosphatase	*N. crassa*	Sojka & Garner (1967)
F	Serine hydroxymethyl transferase	*N. crassa*	Tatum & Perkins (1950); Sakami (1955); Combépine & Turian (1965)
G	Alanine glyoxylate aminotransferase	*N. crassa*	Turian (1961)
		A. niger	Frank & Jilge (1961); Galbraith & Smith (1969)
H	Serine transacetylase	*N. crassa*	Abelson & Vogel (1955); Wiebers & Garner (1967*a, b*)
		A. nidulans	Pieniazek *et al.* (1974)
I	*O*-acetylserine sulphydrylase	*A. nidulans*	Pieniazek *et al.* (1974)

L-LEUCINE The synthesis of L-leucine is less well characterized, but is thought to proceed via α-ketoisovalerate by a four-step reaction (Fig. 7.8, IIb Steps F. G, H, I). The first three steps are exclusive to L-leucine biosynthesis, whereas the fourth, an aminotransferase step, is not. The first step involves condensation of α-ketoisovalerate and acetyl-CoA producing α-carboxy β-hydroxyisocaproate (α-isopropyl malate) by a synthetase (Step F). This enzyme has now been purified from *Neurospora crassa* and its amino acid composition determined (Webster, Nelson & Gross, 1965). The isomerase which catalyses Step G is a polymeric protein and appears to be composed of two different polypeptide subunits which are specified by two genes (Gross, 1962). It may be that isopropyl maleic acid is an intermediate in this reaction (Jungwirth, Gross, Margolin & Umbarger, 1963). The regulation of L-leucine biosynthesis is complex and cannot be detailed here, except to say that L-leucine exerts control by both feedback inhibition and repression of the first enzyme, a synthetase (Step F). Also, the product of the synthetase activity, α-isopropyl malate, acts as an inducer of the isomerase and dehydrogenase (Gross, 1965, 1969; Kashmiri & Gross, 1970).

AROMATIC AMINO ACIDS The biosynthetic pathway for L-tyrosine, L-phenylalanine and L-tryptophan appears to be similar in all micro-organisms investigated so far. Detailed studies, however, reveal the diversity in the control mechanisms, isozyme patterns and aggregation patterns. The first part of the shikimic acid pathway culminates in the formation of chorismate. The biosynthesis of chorismate is initiated by the condensation of phosphoenol pyruvate and D-erythrose-5-phosphate to 3 deoxy-D-arabino heptulosonic-7-phosphate (DAPH) by DAPH synthetase (Fig. 7.8 IIc Step A). Doy (1968), from his studies on *Neurospora crassa*, concluded that this enzyme is allosteric and activated by its two substrates and inhibited by the end-products of aromatic biosynthesis (L-phenylalanine, L-tyrosine and L-tryptophan). Three genes determine this first step and there is evidence that these specify three distinct synthetases, each subject to a distinct form of allosteric inhibition by the aromatic end-products (Halsall & Doy, 1969; Halsall & Catcheside, 1971). The rest of the enzymes responsible for the synthesis of chorismate, a branch point intermediate, are constitutive, metabolic regulation occurring mainly by allosteric inhibition at the first step. DAPH is converted to shikimic acid, the starting material for the benzene ring, by three steps, B, C, D. The synthesis proceeds via 5-phosphoshikimate, 3-enol pyruvylshikimate-7-phosphate to chorismate Steps E, F, G). At this point the pathway branches, at the end of which three different amino acids (L-phenylalanine, L-tyrosine and L-tryptophan) are formed. It is interesting that mutants affecting steps B, C, D, E and F map in a tight cluster (*arom* gene cluster in *N. crassa* (Giles *et al.*, 1967*a, b*; Case & Giles, 1968; Rines, Case & Giles, 1969). Centrifugation analysis by Giles and others (Burgoyne, Case & Giles 1969; Case & Giles, 1971; Jacobson, Hart, Doy & Giles, 1972) revealed that all five activities resided in a protein complex of molecular weight of around 230 000. There are thought to be five polypeptides in this complex, one corresponding to each enzymatic activity. The state of aggregation of these enzymes has been investigated for *Rhizopus stolonifer, Phycomyces* *nitens, Absidia glauca, Aspergillus nidulans, Coprinus lagopus* and *Ustilago maydis.* These studies show that the five enzymes sediment together (Ahmed & Giles, 1969).

L-PHENYLALANINE AND L-TYROSINE Prephenic acid is a precursor common to both L-phenylalanine and L-tyrosine, and arises by the rearrangement of chorismate catalysed by chorismate mutase (Fig. 7.8 IIe, Step H). In *Neurospora crassa,* there is only one mutase present (Baker, 1968), in *Penicillium chrysogenum* and *P. duponti* there are two and three isozyme forms respectively. These isozymes are inhibited by L-tyrosine and L-phenylalanine and activated by L-tryptophan (Woodin & Nishioka, 1973). The conversion of prephenate to L-phenylalanine proceeds through phenylpyruvic acid catalysed by prephenate dehydratase (Step I) and an aminotransferase (Step J) respectively. L-Tyrosine is synthesized from prephenate via p-hydroxyphenylpyruvic acid by prephenate dehydrogenase (Step K) and an aminotransferase (Step L) respectively. Certain micro-organisms possess phenylalanine hydroxylase which carries out the direct hydroxylation of phenylalanine to tyrosine (Step M). This enzyme has been reported in *Aspergillus nidulans* (Sinha, 1967).

L-TRYPTOPHAN The biosynthesis of L-tryptophan branches off from L-tyrosine and L-phenylalanine biosynthesis at chorismate from which five steps (Fig. 7.8, IIc, Steps N, O, P, Q, R) are required for the formation of L-tryptophan in *Neurospora crassa* (Wegman & DeMoss, 1965). Four unlinked genes specify the enzymes which catalyse this conversion in *N. crassa* (Ahmed & Catcheside, 1960; DeMoss & Wegman, 1965) and *Aspergillus nidulans* (Roberts, 1967), unlike those in *Escherichia coli* and *Salmonella typhimurium,* which are arranged in operons. Three enzymes, anthranilate synthetase, phosphoribosylanthranilate isomerase and indole glycerolphosphate synthetase are components of a multi-enzyme complex which is specified by two genes (DeMoss & Wegman, 1965). From studies of sedimentation behaviour of the five enzymes Hutter & DeMoss (1967) recognized a number of patterns of enzyme association, and these were used to evaluate proposed phylogenetic relationships in the fungi.

On the question of regulation of L-tryptophan synthesis this may be subject to several controls (Carsiotis *et al.*, 1970; Lester, 1971) in *Neurospora crassa.* Only one has been unequivocally demonstrated; feedback inhibition of anthranilic acid synthetase (Step N) and anthranilate phosphoribosyl transferase (Step O) by L-tryptophan as demonstrated *in vitro* (DeMoss, 1965*b*) and *in vivo* (Lester, 1963, 1968).

L-SERINE, L-GLYCINE AND L-CYSTEINE Sojka & Garner (1967) have shown that two pathways of serine biosynthesis, found in bacteria and plants, are present in *Neurospora crassa.* These are (1) the non-phosphorylated pathway in which D-glycerate is converted to hydroxy pyruvate and then aminated to yield L-serine, and (2) the phosphorylated pathway in which 3-phosphoglycerate is converted to phospho-hydroxy pyruvate and then aminated to yield phosphoserine, which is converted to L-serine by a specific phosphatase (Fig. 7.8 IId). Moreover, their data suggest that the relative activities of the two pathways can be regulated by altering the carbon source.

L-Serine is converted to L-glycine by serine hydroxymethyl-transferase (Step F). More recently an alternative reaction for glycine biosynthesis has been described for *Neurospora crassa* (Turian, 1961), *Aspergillus niger* (Frank & Jilge, 1961; Galbraith & Smith, 1969), *Blastocladiella* (McCurdy & Cantino, 1960), *A. nidulans* (Kinghorn, unpubl. work). In this reaction glyoxylate is converted to L-glycine by transamination with L-alanine. The enzyme which catalyses this reaction is alanine glyoxylate aminotransferase (AGT, Step G). L-Serine is converted to L-cysteine by a two-step mechanism (H, I) (Pieniazek, Bol, Balbin & Stepien, 1974).

Biosynthesis of amino acids from L-glutamate

L-PROLINE Most of the available evidence supports the view that L-proline is synthesized from exogenously supplied L-arginine or L-ornithine via L-glutamate semialdehyde. Synthesis from the normally operative glutamate pathway proceeds via glutamate semialdehyde (Fig. 7.9, IIIa Step A) (Vogel & Bonner, 1954). This is in rapid non-enzymic equilibrium with Δ-pyrroline-5-carboxylate (Step B) (Vogel & Davis, 1952), which is reduced to L-proline by an NADP-linked reductase (Step C) (Yura & Vogel, 1955, 1959; Yura, 1959).

L-ORNITHINE, L-CITRULLINE AND L-ARGININE Meister (1965), suggested a re-investigation of the L-ornithine biosynthetic pathway, since some uncertainty existed in fungi about the presence of acetylated or non-acetylated pathways. However, mutants defective in a number of these steps indicate that the acetylated pathway (Fig. 7.9, IIIb) is important in the biosynthesis and the non-acetylated pathway, via glutamate-γ-semialdehyde, in the breakdown, as discussed before. Formation of L-arginine from L-ornithine proceeds via L-citrulline and arginosuccinate. In contrast to the situation in yeast, detailed analysis of the regulation of L-arginine synthesis in filamentous fungi is not so well advanced (Morgan, 1966; Cybis, Piotrowska & Weglenski, 1972*b*; Barthelmess *et al.*, 1974). Nevertheless, Davis and colleagues (Subramanian, Weiss & Davis, 1973; Weiss & Davis, 1973) have gone far in showing the complexity of regulation and compartmentation of L-arginine synthesis in the fungal cell (see section 7.12).

Amino acids from L-aspartate

L-ASPARAGINE In contrast to the situation in bacteria and mammals, knowledge of L-asparagine biosynthesis in fungi is rather meagre. It is assumed that fungal asparagine synthetase is similar to that found in bacteria (Fig. 7.10, IVa Step A). Asparagine auxotrophs have been found for *Neurospora crassa* (Tamenbaun, Garnjobst & Tatum, 1954) and *Aspergillus nidulans* (Arst, Kinghorn, unpubl. work), and these mutants may be deficient in asparagine synthetase activity. Another route of L-asparagine synthesis has been shown to occur in *N. crassa* (Monder & Meister, 1959). This involves the amination of α-ketosuccinamate with L-glutamine as the amino donor by asparagine aminotransferase, to give L-asparagine.

L-HOMOSERINE AND L-THREONINE The pathway of L-homoserine and L-threonine synthesis has been demonstrated in *N. crassa*. The enzyme that catalyses the conversion of *O*-phosphohomoserine to L-threonine,

Fig. 7.9 Group III Biosynthesis of amino acids from L-glutamate

Group III (a) L-proline

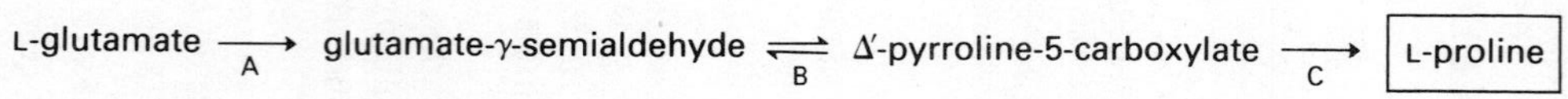

Group III (b) L-ornithine, L-citrulline and L-arginine

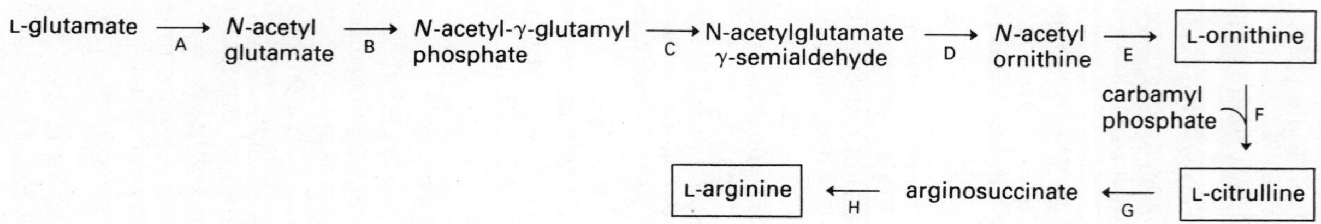

References for Fig. 7.9 (Group IIIa and b)

Step	Enzyme system	Organism	References
Group IIIa			
A	Glutamate-γ-semialdehyde system	*Neurospora crassa*	Vogel & Bonner (1954); Vogel & Kopac (1959); Fincham (1953)
		Aspergillus nidulans	Weglenski (1966); Piotrowska, Sawacki & Weglenski (1969)
B	Δ′-pyrroline-5-carboxylate formation (spontaneous)	*N. crassa*	Vogel & Davis (1952)
C	Δ′-pyrroline-5-carboxylate reductase	*N. crassa*	Abelson & Vogel (1955); Yura & Vogel (1955, 1959)
Group IIIb			
A	Glutamate-N-acetyl transferase	*N. crassa*	Vogel & Bonner (1954); Abelson & Vogel (1955); Morgan (1966)
		A. nidulans	Cybis, Piotrowska & Weglenski (1972*b*)
B	*N*-acetyl-γ-glutamokinase	*N. crassa*	Vogel & Vogel (1963)
C	*N*-acetyglutamic semialdehyde dehydrogenase	*N. crassa*	Vogel & Vogel (1963)
D	*N*-acetyl ornithine aminotransferase	*N. crassa*	Fincham (1953); Morgan (1965)
E	*N*-acetyl ornithine deacetylase	*N. crassa*	De Deken (1963)
F	Ornithine carbamyl transferase	*N. crassa*	Davis (1962); Davis & Thwaites (1963); Barthelmess *et al.* (1974)
		A. nidulans	Cybis, Piotrowska & Weglenski (1970, 1972*a*)
G	Arginosuccinate synthetase	*N. crassa*	Srb & Horowitz (1944); Newmeyer (1962); Wampler & Fairley (1967); Barthelmess *et al.* (1974)
H	Arginosuccinase	*N. crassa*	Fincham & Boylen (1957); Bainbridge, Dalton & Walpole (1966); Cohen & Bishop (1966); Barthelmess *et al.* (1974)

Fig. 7.10 Group IV Biosynthesis of amino acids from aspartate

IVa: L-aspargine

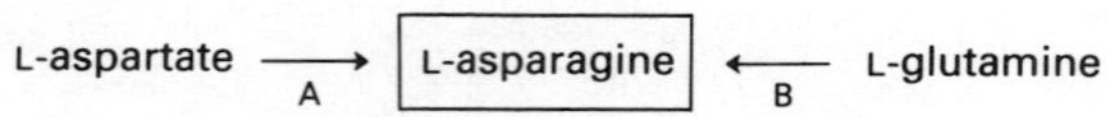

IVb; L-homoserine, L-methionine, L-threonine and L-cysteine

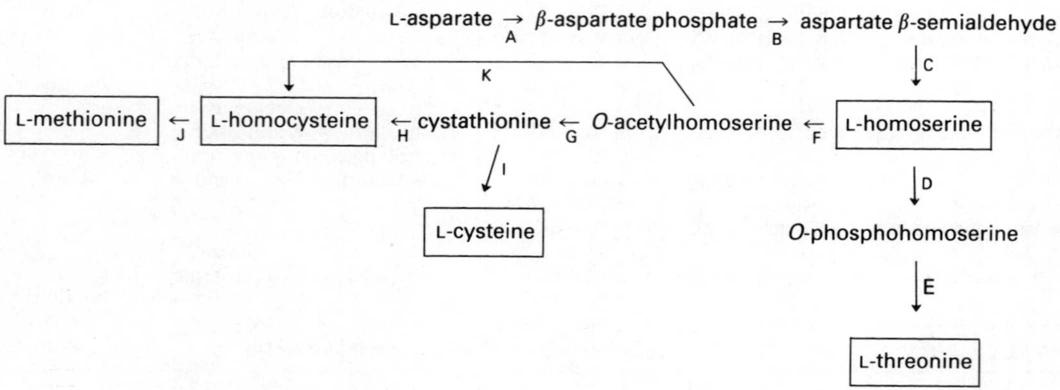

References for Fig. 7.10 (Group IV)

Step	Enzyme system	Organism	References
Group IVa			
A	Asparagine synthetase	*Aspergillus nidulans*	Drainas & Kinghorn (unpubl. work)
B	Asparagine aminotransferase	*Neurospora crassa*	Meister (1953); Richlerich-Van Baerle & Foldstein (1957); Monder & Meister (1959)
Group IVb			
A	β-aspartokinase	*N. crassa*	Abelson & Vogel (1955)
B	Aspartic semialdehyde dehydrogenase	*N. crassa*	Abelson & Vogel (1955)
C	Homoserine dehydrogenase	*N. crassa*	Jenkins & Woodward (1970)
D	Homoserine kinase	*N. crassa*	Teas, Horowitz & Fling (1948); Kaplan & Flavin (1965*b*)
E	Threonine synthetase	*N. crassa*	Flavin & Slaughter (1960*a,b*)
F	Homoserine transacetylase	*N. crassa*	Flavin & Slaughter (1965); Nagai & Flavin (1966); Kerr (1971)
G	Cystathionine-γ-synthetase	*N. crassa*	Kaplan & Flavin (1965*a, b*)
H	β-cystathionase	*N. crassa*	Flavin & Slaughter (1964, 1967*b*)
		P. baarnense	Bouvier (1973)
I	γ-cystathionase	*N. crassa*	Flavin & Segel (1964); Flavin & Slaughter (1964, 1967*b*); Wiebers & Garner (1964); Kerr & Flavin (1970); Pieniazek *et al.* (1974)
J	Methionine formation	*N. crassa*	Kerr & Flavin (1970)
		Penicillium baarnense	Bouvier (1973)
K	*O*-acetyl homoserine sulphydrylase	*N. crassa*	Flavin & Slaughter (1967*a*); Wiebers & Garner (1967*a, b*); Kerr (1971)

threonine synthetase, has been purified 500-fold (Flavin & Slaughter, 1960*a*). As discussed before, L-threonine is an important precursor of L-isoleucine.

L-METHIONINE There may be two biosynthetic routes in *Neurospora crassa* for the synthesis of L-homocysteine, an intermediate in L-methionine synthesis, from L-homoserine. In the first, L-homoserine or *O*-acetyl homoserine reacts directly with H_2S to form homocysteine (Fig. 7.10, IVb Step K). The second involves a number of steps via cystathionine. Flavin & Slaughter (1964) have shown that there are two cystathionine cleavage enzymes in *N. crassa*, one which cleaves the γ-linkage to yield L-cysteine (Step I). The γ-cleavage system has been purified 400-fold in *N. crassa* (Flavin & Segel, 1964). β-Cystathionase has not been purified and indeed has not been well demonstrated in fungi (Step H). There is therefore considerable doubt as to the importance of this in the biosynthesis of L-homocysteine and L-methionine. It may be of importance in the biosynthesis of cystathionine and L-cysteine (Pieniazek *et al.*, 1974). Moreover, cystathionine may be an important sulphur sink for fungi. Although bacteria can form cystathionine from L-cysteine, this step has not been demonstrated in the fungi (Dulavier-Klutchko & Flavin, 1965). This is supported by the fact that L-methionine auxotrophs of *Aspergillis nidulans* do not respond to L-cysteine.

Amino acids from imidazole glycerophosphate

L-HISTIDINE The general pattern of L-histidine biosynthesis emerged from the study of imidazole substances accumulated in histidine auxotrophs of *Neurospora* and *Penicillium* (Ames & Mitchell, 1952, 1955); three of these imidazole derivatives were found to be phosphate esters (Ames, Mitchell & Mitchell, 1953). (See Fig. 7.11 for the later stages of L-histidine synthesis.) In *N. crassa* (Ahmed & Catcheside, 1960) there are seven genes determining the enzymes, which catalyse the ten reactions necessary for histidine biosynthesis. These genes are distributed at random throughout the genome, unlike the position in *Salmonella typhimurium*, where they are arranged in an operon. One of the genes, *hist-3*, in *N. crassa* (Catcheside, 1960; Webber & Case, 1960; Ahmed, Case & Giles, 1964; Ahmed, 1968), and the equivalent gene in *Aspergillus nidulans* (Berlyn, 1967), probably specify a protein with three enzyme activities, the second, third and tenth steps. At first this was interpreted as an operon of three genes controlling three enzyme activities. However, later genetical and biochemical evidence suggested that, like the *arom* region (discussed before), a multifunctional protein aggregate catalyses all three activities and these activities can be differently affected by mutation.

7.11 NAD and NADP Linked L-Glutamate Dehydrogenases

A number of fungi including *Aspergillus nidulans* and *Neurospora crassa* can synthesize both L-glutamate: NADP oxidoreductase (NADP-GDH), which has a synthetic aminating function, and L-glutamate: NAD oxidoreductase (NAD-GDH), which has a degrading deaminating function (LèJohn, 1971).

The structural gene for NADP-GDH in *Neurospora crassa* is *am-1* and a considerable number of *am-1* mutants affecting the enzyme molecule have

Fig. 7.11 Group V Biosynthesis of L-histidine from imidazoleglycerol phosphate

$$\text{imidazoleglycerol phosphate} \xrightarrow[A]{} \text{imidazoleacetol phosphate (1AP)} \xrightarrow[B]{} \text{L-histidinol phosphate (HP)}$$

$$\downarrow C$$

$$\text{L-histidine} \xleftarrow[D]{} \text{L-histidinol}$$

References for Fig. 7.11 (Group V)

Step	Enzyme system	Organism	References
A	Dehydratase	*N. crassa*	Ames & Mitchell (1952, 1955); Ames (1957*b*)
B	Aminotransferase	*N. crassa*	Ames & Horecker (1956)
C	Phosphatase	*N. crassa*	Ames (1957*a*)
D	Histidinol dehydrogenase	*N. crassa*	Ames, Garry & Herzenberg (1960); Creaser *et al.* (1965, 1967); Creaser & Garner (1969); Minson & Creaser (1969)

been studied (Fincham, 1962). The enzyme molecule is a hexamer with a molecular weight of 280 000, each identical subunit of which contains 467 amino acids, and the complete sequence is now known (Wootton, Chambers, Taylor & Fincham, 1973, and J. R. S. Fincham, pers. commun.). The structural gene, *gdhA*, for NADP-GDH in *A. nidulans* is known (Kinghorn & Pateman, 1974*c*). The level of NADP-GDH activity in *A. nidulans* is markedly affected by both nitrogen and carbon sources utilized by the cells. The synthesis of NADP-GDH is repressed by L-glutamate when glucose is present as a carbon source (Pateman, 1969). If the glucose is replaced by amino acids or the cells are carbon starved, there is a rapid loss of NADP-GDH activity (Hynes, 1974*a*; Kinghorn & Pateman, 1974*a*). The protein synthesis inhibitor cycloheximide prevents this loss of NADP-GDH activity. This suggests that the loss of NADP-GDH activity is caused by an inactivating protein, possibly a protease, and that glucose in some way prevents the synthesis of the inactivating protein (Hynes, 1974*a*).

The structural gene for NAD-GDH in *Aspergillus nidulans* is probably *gdhB*. Mutations in *gdhB* result in loss of NAD-GDH activity (Kinghorn & Pateman, 1973*a*; 1975*c*). The structural gene is not known in *N. crassa*, the report by Ahmed & Sanwal (1967) is probably incorrect. The level of NAD-GDH activity in *A. nidulans* is in some way regulated by a carbon metabolite (Kinghorn & Pateman, 1974*b*). NAD-GDH activity is low in cells when glucose is present. In carbon-starved cells, or when amino acids are the carbon source, there is a 5 to 8-fold increase in NAD-GDH activity. The increase in NAD-GDH activity on transfer from glucose medium to medium without glucose is not inhibited by cycloheximide. This implies that *de novo* protein synthesis is not required for the appearance of enzyme activity (Hynes, 1974*a*). The factors affecting the level of activity of the two GDHs are obviously complex, especially when the role of NADP-GDH in ammonia regulation is taken into account. A review covering more extensive information available is given by Goldin & Frieden (1971).

There is some indication that the situation may be rather different in Basidiomycetes (Casselton, 1969; Fawole & Casselton, 1972). It has been shown that NAD-GDH alone serves both aminating and deaminating functions in *Coprinus lagopus*, since it is the major enzyme present under all growth conditions in the vegetative mycelium. Under normal growth conditions NADP-GDH activity only increases during sporophore development and is only present in the pileus. It is suggested that NADP-GDH is reserved for specific developmental functions—possibly autolysis—in the cap (Stewart & Moore, 1974).

7.12 Amino Acid Pools

Fungi, in common with other micro-organisms, contain a number of free amino acids (for a review of amino acid pool studies in a number of fungi, see Holden, 1962). It is outside the scope of this article to discuss these studies in depth. Instead, a few general points will be made about some salient features. One important aspect which is emerging is that the concentration of free intracellular L-glutamate is higher than any other amino acid in a number of fungi examined (Holden, 1962), including *Neurospora crassa* (DeBusk & DeBusk, 1967; Wootton, 1967), *Aspergillus nidulans* (Robinson, Anthony & Drabble, 1973*b*; Kinghorn, unpub. work) and *Penicillium* *chrysogenum* (Hunter & Segel, 1971). In general, L-alanine, L-aspartate and L-glutamine pool sizes are present in higher concentrations than the rest of the amino acids. Second, and not unexpectedly, the free amino acid pool of young mycelial cells can show marked changes in composition in response to environmental conditions, *e.g.* nitrogen or carbon starvation (Wootton, 1967; Hunter & Segel, 1971). Furthermore, growth on a particular amino acid as sole nitrogen source increases the free intracellular concentration of that amino acid in most cases, with the exception of L-glutamate. For example, mycelial cells grown with L-aspartate as the sole nitrogen source have an L-aspartate pool size ten times that found in urea or ammonium grown cells (Kinghorn, unpub. work). Third, evidence accumulated over the past few years suggests that amino acids are not distributed homogeneously within living cells, although little is known about their degree of compartmentation (for reviews, see Oaks & Bidwell, 1970; Mortimore, Woodside & Henry, 1972). Recently, the elegant studies of Davis and colleagues (Weiss, 1973; Weiss & Davis, 1973; Subramanian *et al.*, 1973) demonstrate that over 98% of L-arginine and L-ornithine are found in a vesicle, an organelle which is readily distinguishable from mitochondria. This vesicle fraction also contains substantial amounts of L-histidine and L-lysine. Similarly, Brooke & DeBusk (1973) have evidence of cellular compartmentation of aromatic amino acids in *N. crassa*. These studies are of obvious significance in the regulation by amino acids of certain anabolic and catabolic pathways and the channelling of nitrogenous metabolites. However, it is as yet too early to be more precise about the role of metabolite compartmentation in regulation.

7.13 Metabolic Regulation by Ammonia

The effect of ammonia upon the enzymes of the nitrate reduction pathway has been discussed in previous sections. However, nitrate reduction is only one of the areas of metabolism that are in some sense regulated by ammonia. In this section we discuss what is known of the regulatory role of ammonia in *Aspergillus nidulans*, the species in which the majority of work in this field has been done. A number of enzyme and uptake systems in *A. nidulans* are regulated by ammonia. The level of all these systems is minimal if prototrophic (normal) cells are grown, or held, in the presence of ammonia. It is not known whether this effect of ammonia is on protein synthesis, is an inhibition of activity, or is a combination of both. However, it is usually referrred to as ammonia repression and for convenience will be so called here. Repression by ammonia is widespread and very important in the whole area of inorganic and simple organic nitrogen metabolism in an analogous fashion to catabolite repression in the carbon metabolism of many prokaryotes. Systems repressed by ammonia include nitrate reduction (Pateman & Cove, 1967); purine degradation (Scazzocchio & Darlington, 1968); amide degradation (Hynes & Pateman, 1970*a, b*); amino acid transport systems (Benko, Wood & Segel 1969; Hunter & Segel, 1971; Pateman, Kinghorn & Dunn, 1974); urea transport (Dunn & Pateman, 1972; Pateman *et al.*, 1973); purine transport (Arst & Cove, 1969); ammonia transport (Hackette *et al.*, 1970; Pateman *et al.*, 1973; Pateman *et al.*, 1974); L-arginine (Bartnik, Guzewska & Weglenski, 1973; Bartnik & Weglenski, 1973, 1974); L-histidine catabolism (Hynes, 1974*b*); and extracellular protease (Cohen, 1972). Mutations in a number of genes can affect the regulation of these systems. The properties of these mutants will be described in separate sections and a hypothesis to explain their diverse characteristics put forward in a final section.

Mutations affecting ammonia regulation: meaA, mod meaA and meaB genes

It is probable that the ammonia analogue methylammonia can to some extent repress the synthesis of the ammonia regulated systems in an analogous fashion to ammonia itself, although this is difficult to prove because of methylammonia toxicity. Methylammonia can serve as a poor nitrogen source at low concentrations (less than 5 mM) but is toxic at high concentrations (100 mM). It is possible that repression of enzyme synthesis contributes to methylammonia toxicity, but repression is certainly only a minor part of its toxic action. The target(s) of methylammonia toxicity and the basis of methylammonia resistance in the various classes of mutants to be described are not known. However, it is easy to select for mutant strains resistant to the usual toxic level of methylammonia and this procedure has produced several types of mutants with abnormal ammonia regulation.

It was shown by Arst & Cove (1969) that mutations in two genes *meaA* and *meaB* can result in simultaneous resistance to methylammonia and derepression for several ammonia repressed activities (see Table 7.7). the *meaA* gene probably determines some membrane component which affects ammonia and methylammonia transport (Arst & Page, 1973; Pateman *et al.*, 1974). Mutations in a modifier gene *mod meaA*, when combined in a double mutant *meaA mod meaA* (previously called DER-3, see Table 7.3), result in altered K_ms for both ammonia and methylammonia transport (Pateman *et al.*, 1974). This strongly indicates that the two genes *meaA* and *mod meaA* in some way determine a structural component(s) of the ammonia transport sites. It may be of significance that both *meaA* and the double mutant *meaA mod meaA* also efflux ammonia considerably faster than the normal strain under conditions which result in a high intracellular concentration of ammonia (Pateman, unpubl. work). It is easy to understand why a reduction in transport efficiency should confer resistance to the toxic analogue methylammonia. It is less clear why mutation affecting a membrane component should also result in ammonia derepression.

The function of the *meaB* gene is unknown at present. In general *meaB* mutants show a rather lower level of methylammonia resistance than *meaA* mutants. It has been shown (Arst & Cove, 1973) that *meaB* mutants are also altered in their growth responses to a number of amino acid analogues, and this suggests that the *meaB* gene may possibly be concerned with some aspect of amino acid uptake.

The amrA gene

A mutant *amrA1* was obtained by selection for poor growth on ammonia as a sole nitrogen source (Pateman *et al.*, 1973). This mutation in the *amrA* gene simultaneously results in ammonia derepression for some but not all ammonia-regulated systems, but it does not confer methylammonia resistance (Table 7.7). The *amrA1* possesses a low level of ammonia transport and of methylammonia transport under certain conditions (Pateman *et al.*, 1974). It also effluxes ammonia faster than the normal strain under certain

Table 7.7 Intracellular ammonium concentration and ammonium regulation[a]

Genotype	Thiourea uptake		L-glutamate uptake		Methylammonium uptake		Intracellular ammonium (mM)	
	N-free	10 mM NH_4^+	N-free	10 mM NH_4^+	N-free	10 mM NH_4^+	N-free	10 mM NH_4^+
Wild type	0.97 (0.22)	0.09	1.07 (0.21)	0.08 (0.02)	9.1 (1.1)	5.2 (0.3)	0.13 (0.02)	0.17 (0.04)
meaA8	0.84	0.83	1.09	0.56	4.5 (1.3)	4.4	0.18	0.15
meaB6	0.79	0.12	1.2	0.08	12.9	8.6	0.23	0.55
DER3	0.71	0.85	1.01	0.89	2.0 (0.48)	1.4	0.20	0.22
amrA1	0.46 (0.032)	0.14	0.51 (0.17)	0.29	8.5	4.3	0.15	1.4 (0.55)
xprD1	0.17 (0.014)	0.15	1.08	1.1	4.8	1.7	0.55	1.3 (0.41)
gdhA1	1.20	0.95	1.14	0.31	12.6	8.3	0.10	1.5 (0.51)

[a] N-free and 10 mM NH_4^+ refer to 4 h of incubation in these conditions before the uptake or ammonium assay. Uptake activities are expressed as nmol min^{-1} mg^{-1} (dry weight) of cells. Standard deviation in brackets.

conditions (Pateman, unpublished work). The function of the *amrA* gene is unknown, perhaps the best guess is that it in some way determines a membrane component, since the *amrA1* mutant shows altered transport properties.

The xprD gene

The *xprD1* mutant was obtained by Cohen (1972) and is ammonia derepressed for extracellular protease production. It is not known if ammonia regulates the synthesis, activation, or the release of protease. However, *xprD1* results in the production of extracellular protease in the presence of ammonia, which does not occur in the normal strain. The mutant is ammonia derepressed for several other ammonia regulated systems and also has altered levels of thiourea and methylammonia transport (Table 7.7). The function of the *xprD* gene is discussed later, together with the *areA* gene, since it is possible that they are in fact the same gene (Arst & Cove, 1973).

The gdhA gene

Mutations in the *gdhA* gene result in the loss of NADP glutamate dehydrogenase activity (Arst & MacDonald, 1973; Kinghorn & Pateman, 1973*a*), and it has been shown that *gdhA* is the structural gene for the enzyme (Kinghorn & Pateman 1975*a*). The *gdhA* mutants are derepressed for a number of ammonia regulated systems (Table 7.7). They are not resistant to methylammonia but rather more sensitive than the normal strain. They are sensitive to high external ammonia concentrations (200 mM), a fact which has been utilized in selection experiments. The basis of this sensitivity to ammonia is not known—it does not appear to be simply due to the loss of glutamate dehydrogenase activity.

The *gdhA* mutants are unique in that they are the only class of mutants exhibiting abnormalities of ammonia regulation where the primary gene function—determination of the NADP glutamate dehydrogenase protein—is known. There are two main hypotheses to explain the involvement of NADP glutamate dehydrogenase in ammonia regulation. It might be that ammonia itself is not the true effector whose concentration determines the level of the various enzymes and uptake systems. Instead, the true effector might be some metabolite, *e.g.* L-glutamate or L-glutamine produced from ammonia via glutamate dehydrogenase activity. Thus lack of glutamate dehydrogenase activity would result in low levels of the true effector and consequent derepression. There are two powerful arguments against this hypothesis. The only likely candidates for the role of effector are L-glutamate and L-glutamine and both are readily taken up by cells, yet neither results in effective repression of the systems under discussion. This is difficult to understand if they themselves, or anything derived from them, were the true effector. Furthermore, it appears that it is the extracellular concentration of ammonia, not the intracellular concentration, which determines the level of many of the ammonia regulated systems (Pateman *et al.*, 1973, and see later section). This too is difficult to understand if the true effector were some product derived from intracellular ammonia.

The alternative hypothesis is that the NADP glutamate dehydrogenase protein has some kind of function in ammonia regulation in addition to its catalytic activity. We consider this the more probable state of affairs, and a possible model for ammonia regulation which incorporates a regulatory role for NADP glutamate dehydrogenase is given later.

In *Aspergillus nidulans*, as in many other fungi, there is a NAD glutamate dehydrogenase in addition to the NADP specific enzyme. The NAD glutamate dehydrogenase does not appear to be involved in ammonia regulation, and *gdhB* mutants which lack NAD glutamate dehydrogenase activity are in no way abnormal with regard to ammonia regulation (Kinghorn & Pateman, 1973*a*, 1975*c*).

The areA gene

As a consequence of ammonia regulation the toxicity of a number of poisonous analogues is reduced by ammonia. An example of this is chlorate, which is toxic to *Aspergillus nidulans*, because in some sense it is an analogue of nitrate. In general, it appears that chlorate is only toxic to cells which possess nitrate reductase activity, although there are exceptions to this. In any event ammonia probably protects cells from chlorate poisoning by the repression of the synthesis of both nitrate reductase and other systems. Arst & Cove (1973) obtained two chlorate resistant strains which were the result of mutations in a gene which they called *areA*. These two mutants, *areA*^{r}*-1* and *areA*^{r}*-2*, grow poorly on most nitrogen sources including nitrate, nitrite, amino acids, purines and amides. The growth characteristics of the *areA*^{r} strains are recessive in heterokaryous and heterozygous diploids with the normal allele *areA*$^+$. The *areA*^{r}*-1* strain possesses low levels of nitrate and nitrite reductase and of urate oxidase, the enzymes necessary for nitrate and uric acid utilization. These properties of the *areA*^{r} strains suggest that the *areA* gene might be a gene involved in ammonia regulation. The mutant alleles would result in the gene product repressing the ammonia regulated enzymes and uptake systems even in the absence of ammonia, and thus growth would only occur when ammonia itself was the nitrogen source. This hypothesis is strengthened by the finding that the mutants *xprD-1* and *amdT-102* are probably allelic with *areA*^{r}*-1*. The mutant *xprD-1* is ammonia derepressed for various systems, and *amdT-102* is derepressed for acetamidase (Hynes & Pateman, 1970*a*, *b*; Hynes, 1972). Thus mutations in the gene *areA* can apparently result in either ammonia derepression or repression in the absence of ammonia. The probable explanation is that the *areA* gene determines a regulatory protein which is necessary for allowing the synthesis of a number of enzymes of nitrogen metabolism, *i.e.* a positive action regulatory protein functioning as a component of ammonia regulation.

An important characteristic of any apparent regulatory gene such as *areA* is the phenotype of double mutants carrying an *areA*^{r} allele and a mutant allele at some other gene which also determines abnormal control characteristics. The double mutants *areA*^{r}*-1 nirA*^{c}*-1* and *areA*^{r}*-1 gdhA-10* and *areA*^{r}*-1 meaA-8* and *areA*^{r}*-1 meaB-6* have been made and tested as far as possible for various growth characteristics and some enzyme activities (Arst & Cove, 1973). There are considerable practical difficulties involved in many of the tests but in general it seems the *areA*^{r}*-1* allele is epistatic to the other mutations and the phenotype of the double mutants resembles the phenotype of the single *areA*^{r}*-1* mutant.

The tamA and tanA genes

The toxic effects of the ammonia analogue methylammonia, the urea analogue thiourea (Dunn & Pateman, 1972) and the asparagine analogue aspartic hydroxamate (Pateman, Kinghorn & Dunn, 1974), are like those of chlorate relieved by ammonia. Consequently, there is a possibility that simultaneous resistance to two or more such toxic analogues might result from mutation altering the properties of a regulatory molecule concerned with ammonia repression. A number of resistant mutants were obtained by selecting for simultaneous resistance to the three analogues chlorate, thiourea and aspartic hydroxamate (Kinghorn & Pateman, 1973*b*, 1975*d*). These triply resistant mutants were genetically complex, each strain carrying at least two and often three mutations. The analysis of these mutant strains is still not complete but enough is known to suggest that they probably are the result of mutations in regulatory genes.

Mutations in the *tanA* gene are resistant to thiourea, aspartic hydroxamate and methylammonia with nitrate as the nitrogen source, but are not triply resistant with amino acids such as alanine as the nitrogen source. The *tanA′* mutations are recessive in heterozygous diploids and show low levels of thiourea and methylammonia transport. The double mutant *tanA′-1 gdhA-1* is phenotypically similar to *tanA-1* alone, *i.e.* it is not derepressed like *gdhA-1*. The *tanA* mutants have normal NADP glutamate dehydrogenase activity and grow normally on all nitrogen sources. The *tanA* gene is located on chromosome III and may be the same gene as *areA*. The only characteristic which is difficult to understand is the ability of *tanA-1* to grow on all nitrogen sources. Mutations which result in multiple resistance to toxic analogues because of repression in the absence of ammonia should also result in poor utilization of nitrogen sources such as nitrate.

The *tamA′* mutants are resistant to thiourea, aspartic hydroxamate and methylammonia with alanine and some other amino acids as nitrogen sources, but not with nitrate. The *tamA′* mutants are recessive in heterozygous diploids and have low levels of thiourea and methylammonia transport. The double mutant *tamA′-1 gdhA-1* has a repressed phenotype like *tamA′-1*. The *tamA′* mutants grow normally on all nitrogen sources and the *tamA* gene is located on chromosome VI, unlike *areA* on chromosome III. The *tamA′* mutants have two further unique characteristics, they efflux ammonia particularly when using amino acids as nitrogen sources, and they have low levels (about 10%) of NADP glutamate dehydrogenase activity. This second instance of a connection between ammonia regulation and NADP glutamate dehydrogenase is intriguing but at present unexplained. The most plausible explanation of the diverse properties of the *tamA′* mutants is that the *tamA* gene determines another component of the ammonia regulatory system in an analogous fashion to the *areA* gene (Kinghorn & Pateman, 1975*d*).

A model for ammonia regulation

It is evident from the brief survey of the more important characteristics of the numerous genes involved in ammonia regulation that the system is extraordinarily complex. In such a situation it is both instructive and necessary to consider the simplest set of assumptions concerning the functions of the genes which could plausibly explain the experimental observations. One such scheme or 'model' for ammonia regulation is given below. It is derived from the hypothesis already presented by Pateman *et al.* (1973), but extended to take into account the existence of the probable regulatory genes *areA* and *tamA*. It is only one of a number of possible schemes, but it serves as a guide amidst a welter of data and, hopefully, has heuristic value.

There are four main features which need to be incorporated into a general scheme of ammonia regulation: (1) the differential effect of extracellular and intracellular ammonia on ammonia repression. It can be seen from Table 7.7 (Pateman *et al.*, 1973) that the thiourea and glutamate uptake systems of normal cells are fully derepressed in cells which have been nitrogen starved. In contrast, cells held in 10 mM ammonia are fully repressed, with the specific activities of the two uptake systems some 10 to 20-fold lower. The concentration of the intracellular ammonia pool is similar, about 100 to 200 μM, after the two treatments. Therefore normal cells can be either fully repressed or fully derepressed for the two uptake systems and yet have the same low level of intracellular ammonia. This indicates that, in some way, normal cells monitor and respond to the extracellular ammonia concentration with respect to the regulation by ammonia of glutamate and thiourea uptake. When the external concentration of ammonia is high (50 mM) the intracellular pool is also high and the level of methylammonia uptake is low. A high intracellular ammonia pool derived from urea by urease activity also results in a low level of methylammonia uptake (Table 7.8) (Pateman *et al.*, 1973). This indicates that it is the concentration of the intracellular ammonia pool which determines the level of the methylammonia uptake system, which is in fact the uptake system for ammonia itself. However, the *gdhA* mutants are considerably derepressed with respect to regulation by intracellular ammonia of methylammonia uptake. The *gdhA* mutants—which lack NADP glutamate dehydrogenase—are unique in their response to intracellular ammonia, unlike the normal strain or any of the other mutants; (2) the role of NADP glutamate dehydrogenase in ammonia regulation; (3) the roles of the probable regulatory genes *areA* and *tamA*; (4) the roles of the various classes of ammonia derepressed mutants other than *gdhA*.

The main features of our hypothesis illustrated in Fig. 7.12 are: (a) normal cells of *Aspergillus nidulans* monitor separately the extracellular and intracellular ammonia concentrations with respect to ammonia regulation; (b) the extracellular ammonia concentration determines the level of a number of enzyme and uptake systems, notably nitrate reductase, urea uptake and glutamate uptake; (c) the intracellular ammonia concentration determines the level of the ammonia uptake system; (d) NADP glutamate dehydrogenase located in an ammonia monitoring site in the cell membrane can complex with extracellular but not intracellular ammonia. It is this complex of enzyme and external ammonia which serves as the measure of external ammonia for the cell. The concentration of this complex indirectly determines the level of nitrate reductase, urea uptake and glutamate uptake, *etc.*; (e) NADP glutamate dehydrogenase can complex with intracellular ammonia. This second type of enzyme and ammonia complex serves as the measure of internal ammonia for the cell. The concentration of this second

Table 7.8 Intracellular ammonium concentration and methylammonium uptake

Genotype	Treatment[a]	Methylammonium uptake[b]	Intracellular ammonium (mM)
Wild type	N-free	9.0 (1.10)	0.13
	10 mM urea	5.2 (1.10)	1.05
	20 mM urea	2.3 (0.98)	1.43
	50 mM urea	0.35 (0.40)	3.50
	100 mM urea	0.02	9.30
	50 mM NH_4^+	0.05	2.30
gdhA	N-free	11.6	0.12
	10 mM urea	8.7	3.80
	20 mM urea	5.3	7.90
	50 mM urea	1.7	8.41
	50 mM NH_4^+	1.66	5.60
meaA8	50 mM urea	0.51	2.36
	50 mM NH_4^+	0.18	3.26
meaB6	50 mM urea	1.03	2.84
	50 mM NH_4^+	1.28	1.96
DER3	50 mM urea	0.18	2.50
	50 mM NH_4^+	0.25	2.28
amrA1	50 mM urea	0.03	9.60
	50 mM NH_4^+		
xprD1	50 mM urea	0.54	2.56
	50 mM NH_4^+	0.53	2.88

[a] Treatment refers to 4 h of incubation in the presence of the nitrogen source before the uptake or ammonium assay.
[b] Figures in parentheses refer to L-glutamate uptake in these treatments. Uptake activity expressed as nmol min^{-1} mg^{-1} (dry weight) of cells.

type of complex indirectly determines the level of the ammonia uptake system; (f) a number of cell membrane components are essential both for the function of a number of metabolically unrelated transport systems such as urea and glutamate, and also for the function of the external ammonia monitoring site. Mutation in the genes determining such structural membrane components could simultaneously result in the malfunction of a number of transport systems and the ammonia monitoring sites. Thus genes such as *meaA*, *meaB*, *amrA* and possibly *xprD* which show both transport and ammonia regulation abnormalities (Table 7.7) would be interpreted as structural genes for cell membrane proteins; (g) a control pathway converts information concerning the external and internal concentration of ammonia—which is in the form of the NADP glutamate dehydrogenase complexes—into action with respect to synthesis of the regulated enzyme and uptake systems. The probable regulatory genes *areA* and *tamA* determine proteins which mediate steps in this control pathway.

There are several points which merit further brief discussion. It seems clear from the intracellular ammonia pool data and the derepression characteristics of the *gdhA* mutants that cells of *Aspergillus nidulans* distinguish between external and internal ammonia with respect to regulation. A monitoring mechanism located in the cell membrane itself is the simplest type that can be envisaged for distinguishing between the external

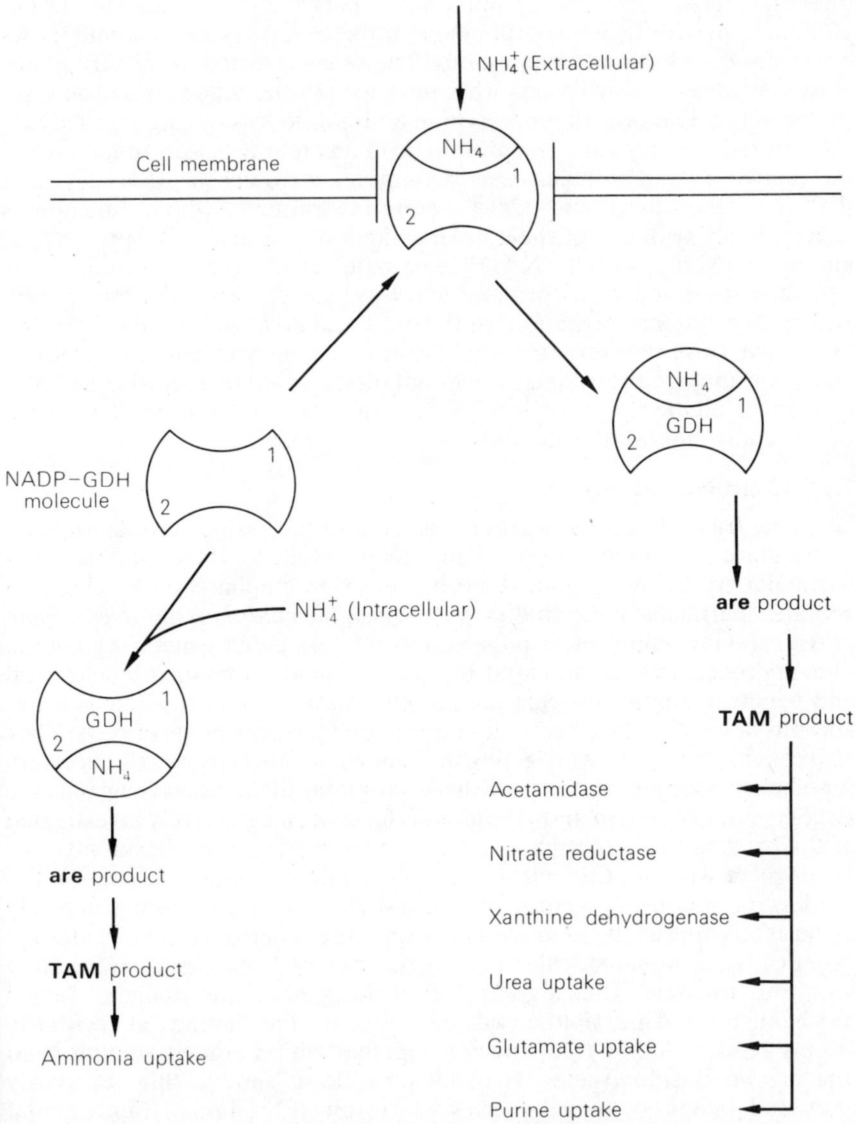

Fig. 7.12 A model for ammonia regulation. The NADP-GDH molecule can complex with extracellular ammonia and this complex is the first intermediate in a control pathway which determined the level of nitrate reductase, urea uptake, *etc.* NADP-GDH can also form a second type of complex with intracellular ammonia, and this is the first intermediate in a control pathway which determines the level of ammonia uptake. Thus the cell monitors the concentration of internal and external ammonia indirectly by the interaction of NADP-GDH with ammonia. The *are* and *tam* gene products mediate later reactions in the control pathways.

and internal concentrations of a metabolite. Alternatively, some form of internal compartmentalism of ammonia, in part at least mediated by NADP glutamate dehydrogenase, could explain the observations. It would, however, require additional assumptions. The role suggested for NADP glutamate dehydrogenase explains why mutations in the *gdhA* structural gene can result in ammonia derepression in addition to loss of catalytic activity. The altered or absent enzyme protein no longer interacts with ammonia and the regulatory system behaves as if ammonia were absent. It is suggested that the genes *areA* and *tamA* determine proteins whose function is necessary for synthesis of the uptake and enzyme systems. The presence of ammonia together with the NADP glutamate dehydrogenase protein inactivates the *areA* and *tamA* proteins and results in repression of the enzyme and uptake systems. Mutations in the *areA* and *tamA* genes which resulted in inactive or absent proteins would lead to low enzyme and uptake levels and toxic analogue resistance. Such mutations would be recessive and also epistatic to all the classes of derepressed mutants. These are the characteristics of known mutants of the *areA* and *tamA* genes.

7.14 Concluding Remarks

We have tried to present a general account of the various topics, together with sufficient documentation to allow deeper study for those interested in a particular field. Throughout there has been an emphasis on biochemical genetics, particularly the studies with *Aspergillus nidulans* and *Neurospora crassa*, and including where possible information about genetic regulation. Our approach was inevitable at the present time. Combined biochemical and genetical studies provide unique information about the biochemistry and physiology of the structural and functional components of cells. They also provide a way of investigating the subtle regulatory processes necessary for efficient metabolism and growth. Amongst the filamentous fungi only the Ascomycetes *N. crassa* and *A. nidulans* have been extensively investigated in this way so far, although a start has been made on the Basidomycetes *Ustilago maydis* and *Coprinus lagopus.* It requires a tremendous amount of spadework before an organism becomes sufficiently well-known genetically to be really useful. It seems unlikely that any other fungi will be studied genetically on the same scale as *N. crassa* or even *A. nidulans*, at least for a long time to come. Thus a great deal of our general knowledge of fungal biochemistry and physiology will be based on the findings in two fairly closely related Ascomycete species, together with a limited amount from one or two Basidomycetes. It is not possible to know if this necessarily restricted sample of fungal species will result in the loss of fundamental knowledge, although it will certainly leave many intriguing byways unexplored.

The metabolic control systems of fungi are of interest in their own right [see the reviews of Gross, 1969; Calvo & Fink, 1971; Fink, 1971; Metzenberg, 1972]. There is also the added incentive that the study of such lower eukaryotes may serve as some guide to control systems in higher eukaryotes, including the mammals. Even with the limited amount known at present about control systems in fungi, a few general features are worth mentioning.

The complexity of the control systems is something of a surprise. In the nitrate reduction pathway in *Aspergillus nidulans* an enzyme, nitrate reductase, and a regulatory gene product are both involved in a complex relationship with the ammonia regulatory system. This ammonia regulatory system itself involves another enzyme, NADP-GDH, and several regulatory and other gene products. Other systems such as L-arginine utilization and purine degradation are probably equally complex. Perhaps it was naïve to expect control systems to be simple when metabolic pathways are branched and interrelated in complicated ways. In any event the lesson from simple eukaryotes is that metabolic control systems are complex and require sophisticated genetical analysis to dissect out their component parts.

In bacteria the genes determining the enzyme proteins which catalyse the steps in a pathway are often close together in the genome, forming operons. This is not the case in fungi. The great majority of genes determining sequential or closely related metabolic steps are scattered throughout the genome. There are exceptions, for example some of the genes determining the enzymes for L-histidine and aromatic amino acid synthesis are clustered in *Neurospora crassa* and *Aspergillus nidulans.* However, it is clear that 'operon' organization, if it exists, is not as important in filamentous fungi as it is in bacteria.

A feature of most, but not all, regulatory genes in bacteria is the negative action of the regulatory gene product (repressor). The situation is quite different in filamentous fungi. There are at least seven systems—including those described here—in *Aspergillus nidulans* where there is strong evidence for positive control, and there is as yet no good evidence for negative control in either *A. nidulans* or *Neurospora crassa* (Metzenberg, 1972). The reason for the marked difference between bacteria and filamentous fungi with respect to operon organization and regulatory genes is unknown. It may be related to the unicellular as contrasted to the filamentous habit, although bacteria and fungi often occupy similar types of habitat. Alternatively, it may be the organization of eukaryote genetic material into chromosomes and nuclei with a bounding nuclear membrane as compared to the naked bacterial genome. The resolution of such problems lies in the future.

7.15 References

ABELSON, P. H. & VOGEL, H. J. (1955). Amino acid biosynthesis in *Torulopsis utilis* and *Neurospora crassa. Journal of Biological Chemistry* **213**, 355–64.

ADIGA, P. R. & SARMA, P. S. (1970). Cysteine toxicity in *Neurospora crassa*: comparison of counteraction by sulphur amino acids and iron. *Indian Journal of Biochemistry* **7**, 141–4.

ADIGA, P. R. & SARMA, P. S. (1971). Cysteine toxicity in *Neurospora crassa*: the mechanism of counteraction by amino acids. *Indian Journal of Biochemistry and Biophysics* **8**, 9–15.

ADIGA, P. R., SASTRY, K. S. & SARMA, P. S. (1962). Amino acid interlationships in cysteine toxicity in *Neurospora crassa. Journal of General Microbiology* **29**, 149–55.

AHMED, A. (1968). Organisation of the histidine-3 region of *Neurospora. Molecular and General Genetics* **103**, 185–93.

AHMED, M. & CATCHESIDE, D. G. (1960). Physiological diversity amongst tryptophan mutants in *Neurospora crassa. Heredity* **15**, 55–64.

AHMED, S. I. & GILES, N. H. (1969). Organization of enzymes in the common aromatic synthetic pathway: evidence for aggregation in fungi. *Journal of Bacteriology* **99**, 231–7.

AHMED, S. I. & SANWAL, B. D. (1967). A structural gene for NAD-GDH in *Neurospora crassa. Genetics* **55**, 359–64.

AHMED, A., CASE, M. E. & GILES, N. H. (1964). The nature of complementation among mutants in the histidine-3 region of *Neurospora crassa. Brookhaven Symposium Biology* **17**, 53–65.

ALLEEM, M. I. H., LEES, H. & LYRIC, R. (1964). Ammonium oxidation by cell-free extracts of *Aspergillus wentii. Canadian Journal of Biochemistry* **42**, 989–98.

AMES, B. N. (1957*a*). The biosynthesis of histidine: L-histidinol phosphate phosphatase. *Journal of Biological Chemistry* **226**, 583–93.

AMES, B. N. (1957*b*). The biosynthesis of histidine: D-erythrose-imidazole-glycerol phosphate dehydrase. *Journal of Biological Chemistry* **228**, 131–43.

AMES, B. N. & HORECKER, B. L. (1956). The biosynthesis of histidine: imidazole-acetol phosphate-transaminase. *Journal of Biological Chemistry* **220**, 113–28.

AMES, B. N. & MITCHELL, H. K. (1952). The paper chromatography of imidazoles. *Journal of American Chemical Society* **74**, 252–3.

AMES, B. N. & MITCHELL, H. K. (1955). The biosynthesis of histidine: imidazole-glycerol phosphate, imidazole-acetol phosphate and histidinol phosphate. *Journal of Biological Chemistry* **212**, 687–96.

AMES, B. N., GARRY, B. J. & HERZENBERG, J. (1960). The genetic control of the enzymes of histine biosynthesis in *Salmonella typhimurium. Journal of General Microbiology* **22**, 369–78.

AMES, B. N., MITCHELL, H. K. & MITCHELL, M. B. (1953). Some new naturally occurring imidazoles related to the biosynthesis of histidine. *Journal of American Chemical Society* **75**, 1015–8.

ANDERSSON-KOTTO, J., EHRENSVARD, G., HOGSTROM, G., REIO, L. & SALUSTE, E. (1954). Amino acid formation and utilisation in *Neurospora. Journal of Biological Chemistry* **210**, 455–63.

ARIMA, K., SAKAMOTO, T. & ARAKI-TAMURA, G. (1972). Production of extracellular L-asparaginases by micro-organisms. *Agricultural and Biological Chemistry* **36**, 356–61.

ARKIN, H. & GROSSOWICZ, M. (1970). Inhibition by D-glutamate of growth and glutamate dehydrogenase activity of *Neurospora crassa. Journal of General Microbiology* **61**, 255–61.

ARROYO-BEGOVICH, A. & DEMOSS, J. A. (1969). *In vitro* formation of an active multienzyme complex in the tryptophan pathway of *Neurospora crassa. Proceedings of the National Academy of Sciences, U.S.A.* **64**, 1072–8.

ARROYO-BEGOVICH, A. & DEMOSS, J. A. (1973). The isolation of the components of the anthranilate synthetase complex from *Neurospora crassa. Journal of Biological Chemistry* **248**, 1262–7.

ARST, H. N. & COVE, D. J. (1969). Methylammonium resistance in *Aspergillus nidulans. Journal of Bacteriology* **98**, 1284–93.

ARST, H. N. & COVE, D. J. (1973). Nitrogen metabolite repression in *Aspergillus nidulans. Molecular and General Genetics* **126**, 111–41.

ARST, H. N. & MACDONALD, K. (1973). A mutant of *Aspergillus nidulans* lacking NADP-linked glutamate dehydrogenase. *Molecular and General Genetics* **122**, 261–5.

ARST, H. N. & MACDONALD, D. W. (1974). Control of proline breakdown in *Aspergillus nidulans. Heredity* **33**, 137.

ARST, H. N. & PAGE, M. (1973). Mutants of *Aspergillus nidulans* altered in the transport of methylammonium and ammonium. *Molecular and General Genetics* **121**, 239–45.

ARST, H. N., MACDONALD, D. W. & COVE, D. J. (1970). Molybdate metabolism in *Aspergillus nidulans.* I. Mutations affecting nitrate reductase and/or xanthine dehydrogenase. *Molecular and General Genetics* **108**, 129–45.

ASPEN, A. J. & MEISTER, A. (1962). Conversion of α-aminoadipic acid to L-pipecolic acid by *Aspergillus nidulans. Biochemistry* **1**, 606–12.

BACH, D. (1928). Les conditions d'action de l'asparaginase de *l'Aspergillus niger. Comptes rendus* **187**, 955–6.

BAINBRIDGE, B. W., DALTON, H. & WALPOLE, J. H. (1966). Identification of the arginosuccinase gene. *Aspergillus Newsletter* **7**, 18.

BAKER, T. I. (1966). Tryptophan: a feedback activator for chorismate mutase from *Neurospora. Biochemistry* **5**, 2654–7.

BAKER, T. I. (1968). Phenylalanine-tyrosine biosynthesis in *Neurospora crassa. Genetics* **58**, 351–9.

BARRATT, R. W. & OGATA, W. M. (1972). *Neurospora* stock list. *Neurospora Newsletter* **19**, 34–105.

BARTHELMESS, I. B., CURTIS, C. F. & KACSER, H. (1974). Control of the flux of arginine in *Neurospora crassa*: derepression of the last three enzymes of the arginine pathway. *Journal of Molecular Biology* **87**, 303–16.

BARTNIK, E. & WEGLENSKI, P. (1973). Ammonium and glucose repression of the arginine catabolic enzymes in *Aspergillus nidulans*. *Molecular and General Genetics* **126**, 75–84.

BARTNIK, E. & WEGLENSKI, P. (1974). Regulation of arginine catabolism in *Aspergillus nidulans*. *Nature* **250**, 590–2.

BARTNIK, E., GUZEWSKA, J. & WEGLENSKI, P. (1973). Mutations simultaneously affecting ammonium and glucose repression of the arginine catabolic enzymes in *Aspergillus nidulans*. *Molecular and General Genetics* **126**, 85–92.

BAUERLE, R. H. & GARNER, H. R. (1964). The assimilation of arginine and lysine in canavanine resistant and sensitive strains of *Neurospora crassa*. *Biochimica et Biophysica Acta* **93**, 316–22.

BENDER, A. E. & KREBS, H. A. (1950). The oxidation of various synthetic α-amino-acids by mammalian D-amino-acid oxidase of *Cobra venom* and the L- and D-amino-acid oxidases of *Neurospora crassa*. *Biochemical Journal* **46**, 210–9.

BENKO, P. V., WOOD, T. C. & SEGEL, I. H. (1967). Specificity and regulation of methionine transport in filamentous fungi. *Archives of Biochemistry and Biophysics* **122**, 783–804.

BENKO, P. V., WOOD, T. C. & SEGEL, I. H. (1969). Multiplicity and regulation of amino acid transport in *Penicillium chrysogenum*. *Archives of Biochemistry and Biophysics* **129**, 498–508.

BERGQUIST, A., LABRIE, D. A. & WAGNER, R. P. 1969). Amino acid synthesis by the mitochondria of *Neurospora crassa*. I. Dependence on respiration of mitochondria. *Archives of Biochemistry and Biophysics* **134**, 401–7.

BERLYN, M. (1967). Gene-enzyme relationships in histidine biosynthesis in *Aspergillus nidulans*. *Genetics* **57**, 561–70.

BLAKELEY, R. M. & SRB, A. M. (1962). Studies of the genetics and physiology of a nitrate non-utilizing strain of *Neurospora*. *Neurospora Newsletter* **2**, 5–6.

BOUVIER, J. (1973). La biosynthèse de la méthionine et le développement sexuel du *Penicillium baarnense Van Beyma*. *Physiologie Végetale* **11**, 83–94.

BRADFIELD, G., SOMERFIELD, P., MEYN, T., HOLBY, M., BABCOCK, D., BRADLEY, D. & SEGEL, I. H. (1970). Regulation of sulphate transport in filamentous fungi. *Plant Physiology* **46**, 720–7.

BRINK, N. G. (1972). Tryptophan transport in *Neurospora crassa* by various types of *mtr* revertants. *Journal of General Microbiology* **73**, 153–60.

BROOKS, C. J. & DEBUSK, A. G. (1973). Cellular compartmentation of aromatic amino acids in *Neurospora crassa*. I. Occupation of a protein synthesis pool by phenylalanine in *tyr-1* mutants. *Biochemical Genetics* **10**, 91–103.

BROQUIST, H. P. (1971). Lysine biosynthesis. In *Methods in enzymology*, vol. XVII, pp. 112–28. Edited by H. Tabor and C. W. Tabor. New York and London: Academic Press.

BROQUIST, H. P. & STIFFEY, A. V. (1959). Saccharopine dehydrogenase. A marker of the aminoadipic acid (AAA) pathway in lysine biosynthesis. *Federation Proceedings* **18**, 198–205.

BURGOYNE, L., CASE, M. E. & GILES, N. H. (1969). Purification and properties of the aromatic (arom) synthetic enzyme aggregate of *Neurospora crassa*. *Biochimica et Biophysica Acta* **191**, 452–61.

BURKE, R. & PATEMAN, J. A. (1962). Glutamic and alanine dehydrogenase determined by one gene in *Neurospora crassa*. *Nature, London* **196**, 450–1.

BURTON, K. (1952). The L-amino-acid oxidase of *Neurospora*. *Biochemical Journal* **50**, 258–68.

CALVO, J. M. & FINK, G. R. (1971). Regulation of biosynthetic pathways in bacteria and fungi. *Annual Review of Biochemistry* **40**, 943–67.

CARSIOTES, M., JONES, R. F., LACY, A. M., CLEARY, T. J. & FRANKHAUSER, D. B. (1970). Histidine-mediated control of tryptophan biosynthetic enzymes in *Neurospora*. *Journal of Bacteriology* **104**, 98–106.

CASE, M. E. & GILES, N. H. (1968). Evidence for nonsense mutations in the *arom* cluster of *Neurospora crassa*. *Genetics* **60**, 49–58.

CASE, M. E. & GILES, N. H. (1971). Partial enzyme aggregate formed by pleiotropic mutants in the *arom* gene cluster of *Neurospora crassa*. *Proceedings of the National Academy of Sciences, U.S.A.* **68**, 58–62.

CASSELTON, P. J. (1969). Concurrent regulation of two enzymes in fungi. *Science Progress* **57**, 207–27.

CASTANEDA, M., MARTUSIELLI, J. & MORA, J. (1967). The catabolism of L-arginine by *Neurospora crassa*. *Biochimica et Biophysica Acta* **141**, 276–86.

CATCHESIDE, D. E. A. (1969). Prephenate dehydrogenase from *Neurospora*: feedback activation by phenylalanine.

Biochemical and Biophysical Research Communications **36**, 651–6.

CATCHESIDE, D. G. (1960). Complementation among histidine mutants of *Neurospora crassa. Proceedings of the Royal Society* **153**, 179–94.

CHOKE, H. C. (1969). Mutants of *Neurospora crassa* permeable to histidinol. *Genetics* **62**, 725–33.

CLUTTERBUCK, A. J. (1974). *Aspergillus nidulans* genetics. In *Handbook of genetics*, vol. 1, Edited by R. C. King. New York: Plenum Press.

COHEN, B. B. & BISHOP, J. O. (1966). Purification of arginosuccinase from *Neurospora* and comparison of some properties of the wild type enzyme and an enzyme formed by inter-allelic complementation. *Genetical Research* **8**, 243–52.

COHEN, B. L. (1972). Ammonium repression of extracellular protease in *Aspergillus nidulans. Journal of General Microbiology* **71**, 293–9.

COHEN, B. L. (1973*a*). The neutral and alkaline proteases of *Aspergillus nidulans. Journal of General Microbiology* **77**, 521–8.

COHEN, B. L. (1973*b*). Regulation of intracellular and extracellular neutral and alkaline proteases in *Aspergillus nidulans. Journal of General Microbiology* **79**, 311–20.

COHEN, B. L., MORRIS, J. E. & DRUCKER, H. (1975). Regulation of two extracellular proteases of *Neurospora crassa* by induction and by carbon-, nitrogen- and sulphur-metabolite repression. *Archives of Biochemistry and Biophysics* **169**, 324–30.

COLBURN, R. W. & TATUM, E. L. (1965). Studies of a phenylalanine-tyrosine requiring mutant of *Neurospora crassa. Biochimica et Biophysica Acta* **97**, 442–8.

COMBÉPINE, G. & TURIAN, G. (1965). Recherches sur la biosynthese de la glycine chez *Neurospora crassa*, type sauvage et mutants. *Pathological Microbiology* **28**, 1018–30.

COVE, D. J. (1966). The induction and repression of nitrate reductase in the fungus *Aspergillus nidulans. Biochimica et Biophysica Acta* **113**, 51–6.

COVE, D. J. & PATEMAN, J. A. (1963). Independently segregating genetic loci concerned with nitrate reductase activity in *Aspergillus nidulans. Nature, London* **198**, 262–3.

COVE, D. J. & PATEMAN, J. A. (1969). Autoregulation of the synthesis of nitrate reductase in *Aspergillus nidulans. Journal of Bacteriology* **97**, 1374–8.

CREASER, E. H. & GARNER, M. (1969). Proteins immunologically related to *Neurospora* histidinol dehydrogenase. *Journal of General Microbiology* **55**, 417–23.

CREASER, E. H., BENNETT, R. B., & BENNETT, D. J. (1967). The purification and properties of histidinol dehydrogenase from *Neurospora crassa. Biochemical Journal* **103**, 36–41.

CREASER, E. H., BENNETT, D. J. & DRYSDALE, R. B. (1965). Studies on biosynthetic enzymes. Mutant forms of histidinol dehydrogenase from *Neurospora crassa. Canadian Journal of Biochemistry and Physiology*, **43**, 993–1000.

CYBIS, J. & WEGLENSKI, P. (1969). Effects of lysine on arginine uptake and metabolism in *Aspergillus nidulans. Molecular and General Genetics* **104**, 282–7.

CYBIS, J. & WEGLENSKI, P. (1972). Arginase induction in *A. nidulans.* The appearance and decay of the coding capacity of messenger. *European Journal of Biochemistry* **30**, 262–8.

CYBIS, J., PIOTROWSKA, M. & WEGLENSKI, P. (1970). Control of ornithine-transcarbamylase formation in *Aspergillus nidulans. Bulletin of the Academy of Polish Science and Biology* **18**, 669–72.

CYBIS, J., PIOTROWSKA, M. & WEGLENSKI, P. (1972*a*). Genetic control of the arginine pathways in *Aspergillus nidulans.* Common regulation of anabolism and catabolism. *Molecular and General Genetics* **118**, 273–7.

CYBIS, J., PIOTROWSKA, M. & WEGLENSKI, P. (1972*b*). The genetic control of the arginine pathway in *Aspergillus nidulans.* Mutants blocked in arginine biosynthesis. *Acta Microbiologica Poland* **4**, 163–9.

DARLINGTON, A. J. & SCAZZOCCHIO, C. (1967). Use of analogues and the substrate-sensitivity of mutants in the analysis of purine uptake and breakdown in *Aspergillus nidulans. Journal of Bacteriology* **93**, 937–40.

DAVIS, R. H. (1962). A mutant form of ornithine transcarbamylase found in a strain of *Neurospora* carrying a pyrimidine-proline suppressor gene. *Archives of Biochemistry and Biophysics* **97**, 185–91.

DAVIS, R. H. (1970). Sources of urea in *Neurospora crassa. Biochimica et Biophysica Acta* **215**, 412–4.

DAVIS, R. H. & MORA, J. (1968). Mutants of *Neurospora crassa* deficient in ornithine-γ-transaminase. *Journal of Bacteriology* **96**, 383–8.

DAVIS, R. H. & THWAITES, W. M. (1963). Structural gene for ornithine transcarbamylase in *Neurospora. Genetics* **48**, 1551–8.

DAVIS, R. H. & ZIMMERMAN, J. D. (1965). A mutation of *Neurospora* affecting the assimilation of exogenous metabolites. *Genetics* **52**, 439.

DAVIS, R. H., LAWLESS, M. B. & PORT, L. A. (1971). Arginaseless *Neurospora* genetics, physiology and polyamine synthesis. *Journal of Bacteriology* **102**, 299–305.

DE-ANGELI, L. C., POCCHIARI, F., RUSSI, S., TONOLO, A., ZURITA, V. E., CIARANFI, E. & PERKIN, A. (1970). Effect of L-asparaginase from *Aspergillus terreus* on ascites sarcoma in the rat. *Nature, London* **225**, 549–50.

DEBUSK, B. G. & DEBUSK, A. G. (1965). Molecular transport in *Neurospora crassa*. I. Biochemical properties of a phenylalanine permease. *Biochimica et Biophysica Acta* **104**, 139–50.

DEBUSK, B. G. & DEBUSK, A. G. (1967). The free amino-acid pool of *Neurospora. Neurospora Newsletter* **11**, 3.

DEDEKEN, R. H. (1963). Biosynthese de l'arginine chez la levure. I. Le sort de la N α-acetylornithine. *Biochimica et Biophysica Acta* **78**, 606–16.

DEKKER, E. E. (1960). Enzymic formation of glyoxylic acid from γ-hydroxyglutamic acid. *Biochimica et Biophysica Acta* **40**, 174–5.

DEMOSS, J. A. (1962). Studies on the mechanism of the tryptophan synthetase reaction. *Biochimica et Biophysica Acta* **62**, 279–93.

DEMOSS, J. A. (1965*a*). The conversion of shikimic to anthranilic acid by extracts of *Neurospora crassa. Journal of Biological Chemistry* **240**, 1231–5.

DEMOSS, J. A. (1965*b*). Biochemical diversity in the tryptophan pathway. *Biochemical and Biophysical Research Communications* **18**, 850–7.

DEMOSS, J. A. & BONNER, D. M. (1959). Studies on normal and genetically altered tryptophan synthetase from *Neurospora crassa. Proceedings of the National Academy of Sciences, U.S.A.* **45**, 1405–12.

DEMOSS, J. A. & WEGMAN, J. (1965). An enzyme aggregate in the tryptophan pathway of *Neurospora crassa. Proceedings of the National Academy of Sciences, U.S.A.* **54**, 241–7.

DEMOSS, J. A., JACKSON, R. W. & CHALMERS, J. H. (1967). Genetic control of the structure and activity of an enzyme aggregation in the tryptophan pathway of *Neurospora crassa. Genetics* **56**, 413–24.

DOX, A. W. (1909). The intracellular enzymes of lower fungi, especially those of *Pencillium camemberti. Journal of Biological Chemistry* **6**, 461–7.

DOY, C. H. (1968). Control of aromatic biosynthesis particularly with regard to the common pathway and the allosteric enzyme 3 deoxy-D-arabino-heptulsonate-7-phosphate synthase. *Reviews of Pure and Applied Chemistry* **18**, 41–78.

DRUCKER, H. (1972). Regulation of exocellular proteases in *Neurospora crassa*: induction and repression of enzyme synthesis. *Journal of Bacteriology* **110**, 1041–9.

DRUCKER, H. (1973). Regulation of exocellular proteases in *Neurospora crassa*: role of *Neurospora* proteases in induction. *Journal of Bacteriology* **116**, 593–9.

DULAVIER-KLUTCHKO, C. & FLAVIN, M. (1965). Enzymatic syntheses and cleavage of cystathionine in fungi and bacteria. *Journal of Biological Chemistry* **240**, 2537–49.

DUNN, E. & PATEMAN, J. A. (1972). Urea and thiourea uptake in *Aspergillus nidulans. Heredity* **29**, 129.

EDWARDS, J. M. & JACKMAN, L. M. (1965). A branch point intermediate in aromatic biosynthesis. *Australian Journal of Chemistry* **18**, 1227–39.

EL-ERYANI, A. A. (1969). Genetic control of phenylalanine and tyrosine biosynthesis in *Neurospora crassa. Genetics* **62**, 711–23.

EMERSON, R. L., PUZISS, M. & KNIGHT, S. G. (1950). The D-amino acid oxidase of molds. *Archives of Biochemistry* **25**, 299–308.

ENGLESBERG, E., IRR, J., POWER, J. & LEE, N. (1965). Positive control of enzyme synthesis by *Gene C* in the L-arabinose system. *Journal of Bacteriology* **90**, 946–59.

ESSER, K. & KUENEN, R. (1967). Biosynthetic pathways. In *Genetics of fungi*, pp. 356–410. Berlin: Springer-Verlag Press.

EYLAR, O. R. & SCHMIDT, E. L. (1959). A survey of heterotrophic micro-organisms from soil for ability to form nitrite and nitrate. *Journal of General Microbiology* **20**, 473–81.

FAWOLE, M. O. & CASSELTON, P. J. (1972). Observations on the regulation of glutamate dehydrogenase activity in *Coprinus lagopus*. *Journal of Experimental Botany* **23**, 530–51.

FINCHAM, J. R. S. (1953). Ornithine transaminase in *Neurospora crassa* and its relation to the biosynthesis of proline. *Biochemical Journal* **53**, 313–20.

FINCHAM, J. R. S. (1962). Genetically determined multiple forms of glutamic dehydrogenase in *Neurospora crassa*. *Journal of Molecular Biology* **4**, 257–74.

FINCHAM, J. R. S. & BOULTER, A. B. (1956). Effects of amino acid on transaminase production in *Neurospora crassa*: evidence for four different enzymes. *Biochemical Journal* **62**, 72–7.

FINCHAM, J. R. S. & BOYLEN, J. B. (1957). A block in arginine synthesis in *Neurospora crassa* due to gene mutation. *Biochemical Journal* **61**, XXIII-XXIV.

FINK, G. R. (1971). Gene clusters and the regulation of biosynthetic pathways in fungi. In *Metabolic pathways*, vol. V. pp. 199–223. Edited by H. J. Vogel. New York and London: Academic Press.

FLAVIN, M. & SEGEL, A. (1964). Purification and properties of the cystathionine γ-cleavage enzyme of *Neurospora*. *Journal of Biological Chemistry* **239**, 2220–7.

FLAVIN, M. & SLAUGHTER, C. (1960*a*). Purification and properties of threonine synthetase of *Neurospora*. *Journal of Biological Chemistry* **235**, 1103–11.

FLAVIN, M. & SLAUGHTER, C. (1960*b*). Threonine synthetase mechanism: studies with isotopic hydrogen. *Journal of Biological Chemistry* **235**, 1112–8.

FLAVIN, M. & SLAUGHTER, C. (1964). Cystathionine cleavage enzymes of *Neurospora*. *Journal of Biological Chemistry* **239**, 2212–9.

FLAVIN, M. & SLAUGHTER, C. (1965). Synthesis of the succinic ester of homoserine, a new intermediate in the bacterial biosynthesis of methionine. *Biochemistry* **4**, 1370–5.

FLAVIN, M. & SLAUGHTER, C. (1967*a*). Enzymatic synthesis of homocysteine or methionine directly from *O*-succinyl-homoserine. *Biochimica et Biophysica Acta* **132**, 400–5.

FLAVIN, M. & SLAUGHTER, C. (1967*b*). The depression and function of enzymes of reverse *trans*-sulfuration in *Neurospora*. *Biochimica et Biophysica Acta* **132**, 406–11.

FRANK, W. & JILGE, G. (1961). Enzymatic transformations of C2 acids by microorganisms. II. Glyoxylate-α-ketoglutarate condensing enzyme from *Aspergillus niger*. *Archives of Microbiology* **39**, 88–94.

GAERTNER, F. H. & COLE, K. W. (1973). Properties of chorismate synthase in *Neurospora crassa*. *Journal of Biological Chemistry* **248**, 4602–9.

GALBRAITH, J. C. & SMITH, J. E. (1969). Changes in activity of certain enzymes of the tricarboxylic acid cycle and the glyoxylate cycle during the initiation of conidiation of *Aspergillus niger*. *Canadian Journal of Microbiology* **15**, 1207–12.

GILES, N. H., CASE, M. E., PARTRIDGE, C. W. H. & AHMED, S. I. (1967*a*). A gene cluster in *Neurospora crassa* coding for an aggregate of five aromatic synthetic enzymes. *Proceedings of the National Academy of Sciences, Washington* **58**, 1453–60.

GILES, N. H., PARTRIDGE, C. W. H., AHMED, S. I. & CASE, M. E. (1967*b*). The occurrence of two dehydrogenases in *Neurospora crassa*, one constitutive and one inducible. *Proceedings of the National Academy of Sciences, Washington* **58**, 1930–7.

GOLDIN, B. R. & FRIEDEN, C. (1971). L-glutamate dehydrogenases. *Current Topics in Cellular Regulation* **4**, 77–117.

GOLDSMITH, J., LIVONI, J. P., NORBERG, C. L. & SEGEL, I. H. (1973). Regulation of nitrate uptake in *Penicillium chrysogenum* by ammonium ion. *Plant Physiology* **52**, 362–7.

GORDON, M., HASKINS, F. A. & MITCHELL, H. K. (1950). The growth promoting properties of quinic acid. *Proceedings of the National Academy of Sciences, Washington* **36**, 427–30.

GREENBERG, D. M. (1969*a*). Biosynthesis of amino acids and related compounds. In *Metabolic pathways*, vol. III, pp. 237–315. Edited by D. M. Greenberg. New York and London: Academic Press.

GREENBERG, D. M. (1969*b*). Carbon catabolism of amino acids. In *Metabolic pathways*, vol. III, pp. 96–179. Edited by D. M. Greenberg, New York and London: Academic Press.

GROSS, S. R. (1962). On the mechanism of complementation at the *leu-2* locus of *Neurospora*. *Proceedings of the National Academy of Sciences, Washington* **48**, 922–30.

GROSS, S. R. (1965). The regulation of synthesis of leucine biosynthesis in

Neurospora. Proceedings of the National Academy of Sciences, Washington **54**, 1538–46.

GROSS, S. R. (1969). Genetic regulatory mechanisms in the fungi. *Annual Review of Genetics* **3**, 395–424.

GROVER, A. K. & KAPOOR, M. (1973). Studies on the regulation, subunit structure, and some properties of NAD-specific glutamate dehydrogenase of *Neurospora. Journal of Experimental Botany* **2**, 847–60.

GUPTA, R. K. & PRAMER, D. (1970). Amino acid transport by the filamentous fungus *Arthrobotrys conoides. Journal of Bacteriology* **103**, 120–30.

HACKETTE, S. L., SKYE, G. E., BURTON, C. E. & SEGEL, H. (1970). Characterisation of an ammonium transport system in filamentous fungi with methylammonium ^{14}C as the substrate. *Journal of Biological Chemistry* **245**, 4241–50.

HALSALL, D. M. & CATCHESIDE, D. E. A. (1971). Structural genes for DAHP synthase isoenzymes in *Neurospora crassa. Genetics* **67**, 183–8.

HALSALL, D. M. & DOY, C. H. (1969). Studies concerning the biochemical genetics and physiology of activity and allosteric inhibition mutants of *Neurospora crassa* 3-deoxy-D-arabino-heptulsonate-7-phosphate synthase. *Biochimica et Biophysica Acta* **185**, 432–46.

HALSALL, D. M., CATCHESIDE, D. E. A. & DOY, C. H. (1971). Some properties of the 3-deoxy-D-arabino-pepulosonate-7-phosphate synthase isoenzymes from mutant strains of *Neurospora crassa. Biochimica et Biophysica Acta* **227**, 464–72.

HAYAISHI, O. (1962). *Oxygenases.* New York: Academic Press.

HAYSMAN, P. & BRANCH HOIRE, H. (1971). Some genetic and physiological characteristics of urease-defective strains of *Neurospora crassa. Canadian Journal of Genetics and Cytology* **13**, 256–69.

HEPPEL, L. A. (1971). In *Structure and function of biological membranes,* pp. 223–247. Edited by L. I. Rothfield New York and London: Academic Press.

HOGG, R. W. & BROQUIST, H. P. (1968). Homocitrate formation in *Neurospora crassa.* Relation to lysine biosynthesis. *Journal of Biological Chemistry* **243**, 1839–45.

HOLDEN, J. T. (1962). The composition of microbial amino acid pools. In *Amino acid pools,* pp. 73–108. Edited by J. T. Holden. Amsterdam: Elsevier Publishing Company.

HOROWITZ, N. H. (1944). The *d*-amino acid oxidase of *Neurospora. Journal of Biological Chemistry* **154**, 141–9.

HUNTER, D. R. & SEGEL, I. H. (1971). Acidic and basic amino acid transport systems of *Penicillium chrysogenum. Archives of Biochemistry and Biophysics* **144**, 168–83.

HUTTER, R. & DEMOSS, J. A. (1967). Organisation of the tryptophan pathway: a phylogenetic study of the fungi. *Journal of Bacteriology* **94**, 1896–907.

HYNES, M. J. (1970). Induction and repression of amidase enzymes in *Aspergillus nidulans. Journal of Bacteriology* **103**, 482–7.

HYNES, M. J. (1972). Mutants with altered glucose repression of amidase enzymes in *Aspergillus nidulans. Journal of Bacteriology* **111**, 717–22.

HYNES, M. J. (1973*a*). The effect of the carbon source on nitrate-reductase activity in *Aspergillus nidulans. Journal of General Microbiology* **79**, 155–9.

HYNES, M. J. (1973*b*). Alteration in the control of glutamate uptake in mutants of *Aspergillus nidulans. Biochemical and Biophysical Research Communications* **54**, 685–9.

HYNES, M. J. (1974*a*). The effects of the carbon source on glutamate dehydrogenase activities in *Aspergillus nidulans. Journal of General Microbiology* **81**, 165–70.

HYNES, M. J. (1974*b*). The effects of ammonium, L-glutamate and L-glutamine on nitrogen catabolism in *Aspergillus nidulans. Journal of Bacteriology* **120**, 116–23.

HYNES, M. J. & PATEMAN, J. A. (1970*a*). The genetic analysis of regulation of amidase synthesis in *Aspergillus nidulans.* I. Mutants unable to utilise acrylamide. *Molecular and General Genetics* **108**, 97–106.

HYNES, M. J. & PATEMAN, J. A. (1970*b*). The genetic analysis of regulation of amidase synthesis in *Aspergillus nidulans.* II. Mutants resistant to fluoroacetamide. *Molecular and General Genetics* **108**, 107–16.

HYNES, M. J. & PATEMAN, J. A. (1970*c*). The use of amides as nitrogen sources by *Aspergillus nidulans. Journal of General Microbiology* **63**, 317–24.

IMADA, A., IGARASI, S., NAKAMAMA, K. & ISONO, M. (1972). Asparaginase and glutaminase activities in micro-organisms. *Journal of General Microbiology* **76**, 85–99.

INGLE, J., JOY, K. W. & HAGEMAN, R. H. (1966). The regulation of activity of the enzymes involved in the assimilation of nitrate by higher plants. *Biochemical Journal* **100**, 577–88.

JACOBSON, E. S. & METZENBERG, R. L. (1968). A new gene which affects uptake of neutral and acidic amino acids in *Neurospora crassa. Biochimica et Biophysica Acta* **156**, 140–7.

JACOBSON, J. W., HART, B. A., DOY, C. W. & GILES, N. H. (1972). Purification and stability of the multi-enzyme complex encoded in the *arom* gene cluster in *Neurospora crassa. Biochimica et Biophysica Acta* **289**, 1–12.

JACOBY, G. A. (1964). The induction and repression of amino acid oxidation in *Pseudomonas fluorescens. Biochemical Journal* **92**, 1–8.

JACOBY, W. B. (1954). An interrelationship between tryptophan, tyrosine and phenylalanine in *Neurospora.* In *Amino acid metabolism*, pp. 909–13. Edited by W. D. McElroy and B. Glass. New York: Baltimore Press.

JACOBY, W. B. & BONNER, D. M. (1953). Kynureninase from *Neurospora:* purification and properties. *Journal of Biological Chemistry* **205**, 699–707.

JACOBY, W. B. & BONNER, D. M. (1956). Kynurenine transaminase from *Neurospora. Journal of Biological Chemistry* **221**, 689–95.

JENKINS, M. B. & WOODWARD, V. W. (1970). Purification and properties of homoserine dehydrogenase from *Neurospora crassa. Biochimica et Biophysica Acta* **212**, 21–31.

JONES, O. T. G. (1963). The accumulation of amino acids, with particular reference to the plant parasitic fungus *Botrytis fabae. Journal of Experimental Botany* **14**, 399–411.

JONES, O. T. G. & WATSON, W. A. (1962). Uptake of amino acids by the fungus *Botrytis fabae. Nature, London* **194**, 947–8.

JOY, K. W. & HAGEMAN, R. H. (1966). The purification and properties of nitrite reductase from higher plants and its dependence on ferrodoxin. *Biochemical Journal* **100**, 263–73.

JUNGWIRTH, C., GROSS, S. R., MARGOLIN, P. & UMBARGER, H. E. (1963). The biosynthesis of leucine. I. The accumulation of β-carboxy β-hydroxy isocaproate by leucine auxotrophs of *Salmonella typhimurium* and *Neurospora crassa. Biochemistry* **2**, 1–6.

KABACK, H. R. (1970). Transport. *Annual Review of Biochemistry* **39**, 561–93.

KAPLAN, M. M. & FLAVIN, M. (1965*a*). Enzymatic synthesis of L-cystathionine from the succinic ester of L-homoserine. *Biochimica et Biophysica Acta* **104**, 390–6.

KAPLAN, M. M. & FLAVIN, M. (1965*b*). On the methionine pathway in the fungi and bacteria and the mechanism of the isomerization reaction. *Journal of Biological Chemistry* **240**, 3928–33.

KAPOOR, M. & BRAY, D. F. (1968). Feedback inhibition of glutamine synthetase of *Neurospora crassa* by nicotinamide-adenine dinucleotide. *Biochemistry* **7**, 3583–8.

KAPOOR, M. & GROVER, A. K. (1970*a*). Catabolite-controlled regulation of glutamate dehydrogenases of *Neurospora crassa. Canadian Journal of Microbiology* **16**, 33–40.

KAPOOR, M. & GROVER, A. K. (1970*b*). Allosteric interaction of D-glutamate with the glutamate dehydrogenases of *Neurospora crassa. Canadian Journal of Microbiology* **16**, 1341–9.

KAPOOR, M., BRAY, D. F. & WARD, G. W. (1969). Glutamine synthetase of *Neurospora crassa.* Inactivation by urea and protection by some substrates and allosteric effectors. *Archives of Biochemistry and Biophysics* **134**, 423–33.

KAPPY, M. S. & METZENBERG, R. L. (1965). Studies on the basis of ethionine-resistance in *Neurospora. Biochimica et Biophysics Acta* **107**, 425–33.

KAPPY, M. S. & METZENBERG, R. L. (1967). Multiple alterations in metabolite uptake in a mutant of *Neurospora crassa. Journal of Bacteriology* **94**, 1629–35.

KASHMIRI, S. V. S. & GROSS, S. R. (1970). Mutations affecting the regulation of the production of the enzymes of leucine biosynthesis in *Neurospora. Genetics* **64**, 423–40.

KENNY, F. T. (1967). Turnover of a rat liver tyrosine transaminase. Stabilisation after inhibition of protein synthesis. *Science* **156**, 525–8.

KERR, D. S. (1971). *O*-acetylhomoserine sylfhydrylase from *Neurospora.* Purification and consideration of its function in homocysteine and methionine synthesis. *Journal of Biological Chemistry* **246**, 95–102.

KERR, D. S. & FLAVIN, M. (1970). The

regulation of methionine synthesis and the nature of cystathionine-γ-synthase in *Neurospora. Journal of Biological Chemistry* **245**, 1842–55.

KINGHORN, J. R. & PATEMAN, J. A. (1972). L-glutamate and L-aspartate transport in *Aspergillus nidulans. Heredity* **29**, 128.

KINGHORN, J. R. & PATEMAN, J. A. (1973*a*). NAD and NADP L-glutamate dehydrogenase activity and ammonium regulation in *Aspergillus nidulans. Journal of General Microbiology* **78**, 39–46.

KINGHORN, J. R. & PATEMAN, J. A. (1973*b*). A new class of mutants affecting ammonium regulation and methylammonium resistance in *Aspergillus nidulans.* Heredity **31**, 427–8.

KINGHORN, J. R. & PATEMAN, J. A. (1974*a*), The effect of the carbon source on ammonium regulation in *Aspergillus nidulans. Molecular and General Genetics* **128**, 95–8.

KINGHORN, J. R. & PATEMAN, J. A. (1974*b*). The regulation of NAD L-glutamate dehydrogenase in *Aspergillus nidulans. Genetical Research* **23**, 119–24.

KINGHORN, J. R. & PATEMAN, J. A. (1975*a*). The structural gene for NADP L-glutamate dehydrogenase in *Aspergillus nidulans. Journal of General Microbiology* **86**, 290–300.

KINGHORN, J. R. & PATEMAN, J. A. (1975*b*). Mutations which affect amino acid transport in *Aspergillus nidulans. Journal of General Microbiology* **86**, 174–84.

KINGHORN, J. R. & PATEMAN, J. A. (1975*c*). Mutants of *Aspergillus nidulans* lacking nicotinamide adenine dinucleotide-specific glutamate dehydrogenase. *Journal of Bacteriology* (in press).

KINGHORN, J. R. & PATEMAN, J. A. (1975*d*). Studies of partially repressed mutants at the *tamA* and *areA* loci in *Aspergillus nidulans. Molecular and General Genetics* (in press).

KINSEY, J. A. & STADLER, D. R. (1969). Interaction between analogue resistance and amino acid auxotrophy in *Neurospora. Journal of Bacteriology* **97**, 1114–7.

KINSRY, S. C. & MCELROY, W. D. (1958). *Neurospora* nitrate reductase: the role of phosphate flavine and cytochrome-c reductase. *Archives of Biochemistry and Biophysics* **73**, 466–83.

KIRITANI, K., NARISE, S., BERGQUIST, A. & WAGNER, R. P. (1965). The overall *in vitro* synthesis of valine from pyruvate by *Neurospora* homogenates. *Biochimica et Biophysica Acta* **100**, 432–43.

KITTO, G. B., KOTTKE, M. E., BERTLAND, L. H., MURPHEY, W. H. & KAPLAN, N. O. (1967). Studies on malate dehydrogenase and aspartate aminotransferases from *Neurospora crassa. Archives of Biochemistry and Biophysics* **121**, 224–32.

KNIGHT, S. G. (1948). The L-amino acid oxidase of molds. *Journal of Bacteriology* **55**, 401–7.

KOLMARK, H. G. (1969*a*). Genetic studies of urease mutants in *Neurospora crassa. Mutation Research* **8**, 51–63.

KOLMARK, H. G. (1969*b*). Urease defective mutants in *Neurospora crassa. Molecular and General Genetics* **104**, 219–34.

KORNBERG, H. L. (1958). The metabolism of C_2 compounds by micro-organisms. *Biochemical Journal* **68**, 535–42.

KORNBERG, H. L. & ELSEN, S. R. (1961). The metabolism of two compounds by micro-organisms. *Advances in Enzymology* **23**, 401–7.

LABRIE, D. A., LEITER, E. H., BERGQUIST, A. & WAGNER, R. P. (1972). Synthesis of amino acids from succinate by the mitochondria of *Neurospora crassa. Canadian Journal of Biochemistry* **50**, 982–7.

LÉJOHN, H. B. (1971). Enzyme regulation, lysine pathways and cell wall structures as indicators of major lines of evolution in fungi. *Nature, London* **231**, 164–8.

LESTER, G. (1963). Regulation of early reactions in the biosynthesis of tryptophan in *Neurospora crassa. Journal of Bacteriology* **85**, 468–75.

LESTER, G. (1966). Genetic control of amino acid permeability in *Neurospora crassa. Journal of Bacteriology* **91**, 677–84.

LESTER, G. (1968). *In vivo* regulation of intermediate reactions in the pathway of tryptophan biosynthesis in *Neurospora crassa. Journal of Bacteriology* **96**, 1768–73.

LESTER, G. (1971). Regulation of tryptophan biosynthetic enzymes in *Neurospora crassa. Journal of Bacteriology* **107**, 193–202.

LEWIS, C. M. & FINCHAM, J. R. S. (1970*a*). Regulation of nitrate reductase in the Basidiomycete *Ustilago maydis. Journal of Bacteriology* **103**, 55–61.

LEWIS, C. M. & FINCHAM, J. R. S. (1970*b*). Genetics of nitrate reductase in *Ustilago maydis. Genetical Research* **16**, 151–63.

LIN, E. C. C. (1970). The genetics of bacterial transport systems. *Annual Review of Genetics* **4**, 225–62.

LINGENS, F., GOEBEL, W. & VESSELER, H. (1967). Regulation der biosynthese der aromatischen aminosäuren in *Claviceps paspali. European Journal of Biochemistry* **2**, 442–7.

MACDONALD, D. W., COVE, D. J. & CODDINGTON, A. (1974). Cytochrome-c reductases from wild-type and mutant strains of *Aspergillus nidulans. Molecular and General Genetics* **128**, 187–99.

MACDONALD, D. W., ARST, H. N. & COVE, D. J. (1974). The threonine dehydratase structural gene in *Aspergillus nidulans. Biochimica et Biophysica Acta* (In press).

MACMILLAN, A. (1956). The entry of ammonia into fungal cells. *Journal of Experimental Botany* **7**, 113–26.

MAGILL, C. W., SWEENEY, H. & WOODWARD, V. W. (1972). Histidine uptake in strains of *Neurospora crassa* with normal and mutant transport systems. *Journal of Bacteriology* **110**, 313–20.

MATSUBARA, H. & FEDER, J. (1971). Other bacterial, mold and yeast proteases. In *The enzymes*, 3rd edition, pp. 721–95. Edited by P. D. Boyer. New York and London: Academic Press.

MCCURDY, M. D. & CANTINO, E. C. (1960). Isocitrase, glycine–alanine transaminase and development in *Blastocladiella emersonii. Plant Physiology* **35**, 463–76.

MEDINA, A. & NICHOLAS, D. J. D. (1957*a*). Hyponitrite reductase in *Neurospora. Nature, London* **179**, 533–4.

MEDINA, A. & NICHOLAS, D. J. D. (1957*b*). Metallo enzymes in the reduction of nitrite to ammonia in *Neurospora. Biochimica et Biophysica Acta* **25**, 138–41.

MEISTER, A. (1953). Preparation and enzymatic reactions of the keto analogues with asparagine and glutamine. *Journal of Biological Cemistry* **200**, 571–89

MEISTER, A. (1965). *Biochemistry of the amino acids* (2 vols.) New York: Academic Press.

MESSENGUY, F. & WIAME, J. M. (1969). The control of ornithine transcarbamylase activity by arginase in *Saccharomyces cerevisiae. Federation of Europeam Biochemistry Society Letters* **3**, 47–50.

METZENBERG, R. L. (1972). Genetic regulatory systems in *Neurospora. Annual Review of Genetics* **6**, 111–33.

METZENBERG, R. L. & MITCHELL, H. K. (1956). Isolation of prephenic acid from *Neurospora. Archives of Biochemistry* **64**, 51–6.

MINSON, A. C. & GREASER, E. H. (1969). Purification of a trifunctional enzyme catalysing three steps of the histidine pathway from *Neurospora crassa. Biochemical Journal* **114**, 49–56.

MITCHELL, H. K. & HOULAHAN, M. B. (1948). An intermediate in the biosynthesis of lysine in *Neurospora. Journal of Biological Chemistry* **174**, 883–7.

MOHLER, W. C. & SUSKIND, S. R. (1960). The similar properties of tryptophan synthetase and a mutationally altered enzyme in *Neurospora crassa. Biochimica et Biophysica Acta* **43**, 228–99.

MONDER, C. & MEISTER, A. (1959). α-Ketoglutaramic acid as a product of enzymic transamination of glutamine in *Neurospora. Biochimica et Biophysica Acta*, **28**, 202–3.

MORA, J., SALCEDA, R. & SANCHEZ, S. (1972). Regulation of arginase activity by intermediates of the arginine biosynthetic pathway in *Neurospora crassa. Journal of Bacteriology* **110**, 870–7.

MORGAN, D. H. (1965). Acetylornithine transaminase in *Neurospora. Neurospora Newsletter* **8**, 8.

MORGAN, D. H. (1966). Regulation of arginine biosynthesis in *Neurospora crassa. Heredity* **21**, 342.

MORGAN, D. H. (1970*a*). Selection and characterisation of mutants lacking arginase in *Neurospora crassa. Molecular and General Genetics* **108**, 291–302.

MORGAN, D. H. (1970*b*). Arginase mutants of *Neurospora crassa. Heredity* **26**, 154.

MORTIMORE, G. E., WOODSIDE, K. H. & HENRY, J. E. (1972). Compartmentation of free valine and its relation to protein turnover in perfused rat liver. *Journal of Biological Chemistry* **247**, 2776–84.

MORGAN, A. G. & MACMILLAN, A. (1954). The assimilation of nitrogen from ammonium salts and nitrate by fungi. *Journal of Experimental Botany* **5**, 232–52.

MUNKRES, K. D. (1965*a*). Simultaneous genetic alteration of *Neurospora* malate dehydrogenase and aspartate transaminase. *Archives of Biochemistry and Biophysics* **112**, 340–6.

MUNKRES, K. D. (1965*b*). Physicochemical identity of *Neurospora* malate dehydrogenase and aspartate aminotransferase. *Archives of Biochemistry and Biophysics* **112**, 347–54.

MUNKRES, K. D., GILES, N. H. & CASE, M. E. (1965). Genetic control of *Neurospora* malate dehydrogenase and aspartate aminotransferase. I. Mutant selection, linkage and complementation studies. *Archives of Biochemistry and Biophysics* **109**, 397–403.

NAGAI, S. & FLAVIN, M. (1966). Acetylhomoserine and methionine biosynthesis in *Neurospora. Journal of Biological Chemistry* **241**, 3861–3.

NASON, A. & EVANS, H. J. (1953). Triphosphate pyridine nucleotide nitrate reductase in *Neurospora. Journal of Biological Chemistry* **202**, 655–73.

NASON, A., ABRAHAM, R. G. & AUERBACH, B. C. (1954). Enzymatic reduction of nitrite to ammonia by reduced pyridine nucleotides. *Biochimica et Biophysica Acta* **15**, 159–61.

NEWMEYER, D. (1962). Gene influencing the conversion of citrulline to arginosuccinate in *Neurospora crassa. Journal of Microbiology* **28**, 215–30.

NEWMEYER, D. (1967). Arginine synthesis in *Neurospora crassa;* genetic studies. *Journal of General Micriobiology* **16**, 449–62.

NICHOLAS, D. J. D. (1959*a*). Metalloenzymes in nitrate assimilation of plants with special reference to microorganisms. *Symposia of the Society for Experimental Biology XIII*, 1–23.

NICHOLAS, D. J. D. (1959*b*). IVth International Congress of Biochemistry, Vienna, 1958. XIII (Colloquia) pp. 307–31. London: MacMillan (Pergamon) Press.

NICHOLAS, D. J. D. (1963). The metabolism of inorganic nitrogen and its compounds in micro-organisms. *Biological Reviews (Cambridge Philosophical Society)* **38**, 530–68.

NICHOLAS, D. J. D. (1965). Utilisation of inorganic nitrogen and amino acids by fungi. In *The fungi*, vol. 1, pp. 349–76. Edited by G. C. Ainsworth and A. S. Sussman. New York and London: Academic Press.

NICHOLAS, D. J. D. & NASON, A. (1954). Molybdenum and nitrate reductase. III. Molybdenum as a constituent of nitrate reductase. *Journal of Biological Chemistry* **207**, 352–60.

NICHOLAS, D. J. D. & WILSON, P. W. (1964). A dissimilatory nitrate reductase from *Neurospora crassa. Biochimica et Biophysica Acta* **86**, 466–76.

NICHOLAS, D. J. D., MEDINA, A. & JONES, O. T. G. (1960). A nitrite reductase from *Neurospora crassa. Biochimica et Biophysica Acta* **37**, 468–76.

NORD, F. F. & MULL, R. P. (1945). Recent progress in the biochemistry of *Fusaria. Advances in Enzymology* **5**, 165–205.

OAKS, S. & BIDWELL, R. G. S. (1970). Compartmentation of intermediary metabolites. *Annual Review of Plant Physiology* **21**, 43–66.

OHNISHI, E., MACLEOD, H. & HOROWITZ, M. H. (1962). Mutants of *Neurospora* deficient in D-amino oxidase. *Journal of Biological Chemistry* **237**, 138–42.

OLSHAN, A. R. & GROSS, S. R. (1974). Role of the *leu-3* cistron in the regulation of the synthesis of isoleucine and valine biosynthetic enzymes of *Neurospora. Journal of Bacteriology* **118**, 374–84.

PALL, M. L. (1969). Amino acid transport in *Neurospora crassa.* I. Properties of two amino acid transport systems. *Biochimica et Biophysica Acta* **173**, 113–27.

PALL, M. L.(1970*a*). Amino acid transport in *Neurospora crassa.* II. Properties of a basic amino acid transport system. *Biochimica et Biophysica Acta* **203**, 139–49.

PALL, M. L. (1970*b*). Amino acid transport in *Neurospora crassa.* III. Acidic amino acid transport. *Biochimica et Biophysica Acta* **211**, 513–20.

PATEMAN, J. A. (1969). Regulation of synthesis of glutamate dehydrogenase and glutamine synthetase in microorganisms. *Biochemical Journal* **115**, 768–75.

PATEMAN, J. A. & COVE, D. J. (1967). Regulation of nitrate reduction in *Aspergillus nidulans. Nature, London* **215**, 1234–9.

PATEMAN, J. A. & FINCHAM, J. R. S. (1964). Complementation and enzyme studies of revertants induced in an *am* mutant of *N. crassa. Genetical Research* **6**, 419–32.

PATEMAN, J. A., KINGHORN, J. R. & DUNN, E. (1974). Regulatory aspects of L-glutamate transport in *Aspergillus nidulans. Journal of Bacteriology* **119**, 534–42.

PATEMAN, J. A., REVER, B. M. & COVE, D. J. (1967). Genetical and biochemical studies of nitrate reduction in *Aspergillus nidulans. Biochemical Journal* **104**, 103–11.

PATEMAN, J. A., COVE, D. J., REVER, B. M. & ROBERTS, D. B. (1964). A common co-factor for nitrate reductase and xanthine dehydrogenase which also regulates the synthesis of nitrate reductase. *Nature, London* **201**, 58–60.

PATEMAN, J. A., DUNN, E., KINGHORN, J. R. & FORBES, E. (1974). Ammonium and methylammonium transport in wild-type and mutant cells of *Aspergillus nidulans. Molecular and General Genetics* **133**, 225–36.

PATEMAN, J. A., KINGHORN, J. R., DUNN, E. & FORBES, E. (1973). Ammonium regulation in *Aspergillus nidulans. Journal of Bacteriology* **114**, 943–50.

PEES, E. (1966). Lysine, histidine and isoleucine mutants. *Aspergillus Newsletter* **7**, 11–2.

PIENIAZEK, N. J., BÓL, J., BALBIN, E. & STEPIEN, P. P. (1974). An *Aspergillus nidulans* mutant lacking serine transacetylase: Evidence for two pathways of cysteine biosynthesis. *Molecular and General Genetics* **132**, 363–6.

PIOTROWSKA, M., SAWACKI, M. & WEGLENSKI, P. (1969). Mutants of the arginine-proline pathway in *Aspergillus nidulans. Journal of General Microbiology* **55**, 301–5.

RAO, E. & DeBUSK, A. G. (1975). A mutant of *Neurospora* deficient in the general (PmG) amino acid transport system. *Neurospora Newsletter* **22**, 12–13.

RAVEL, J. M., NORTON, S. J., HUMPHREYS, J. S. & SHIRE, W. (1962). Asparagine biosynthesis in *Lactobacillus arabinosus* and its control by asparagine through enzyme inhibition and repression. *Journal of Biological Chemistry* **237**, 2845–9.

RINES, H. W., CASE, M. E. & GILES, N. H. (1969). Mutants in the *arom* gene cluster of *Neurospora crassa* specific for biosynthetic dehydroquinase. *Genetics* 789–800.

ROBERTS, C. F. (1967). Complementation analysis of the tryptophan pathway of *Aspergillus nidulans. Genetics* **55**, 233–9.

ROBINSON, J. H., ANTHONY, C. & DRABBLE, W. T. (1973*a*). The acidic amino acid permease of *Aspergillus nidulans. Journal of General Microbiology* **79**, 53–65.

ROBINSON, J. H., ANTHONY, C. & DRABBLE, W. T. (1973*b*). Regulation of the acidic amino acid permease of *Aspergillus nidulans. Journal of General Microbiology* **79**, 65–89.

RODWELL, V. W. (1969*a*). Biosynthesis of amino acids and related compounds. In *Metabolic pathways*, vol. III pp. 317–73. Edited by D. M. Greenberg. New York and London: Academic Press.

RODWELL, V. W. (1969*b*). Carbon catabolism of amino acids. In *Metabolic pathways*, vol. III, pp. 191–235. Edited by D. M. Greenberg. New York and London: Academic Press.

ROESS, W. B. & DEBUSK, A. G. (1968). Properties of a basic amino acid permease in *Neurospora crassa. Journal of General Microbiology* **52**, 421–32.

SAKAMI, W. (1955). The biochemical relationship between glycine and serine. In *Amino-acid metabolism*, pp. 658–83. Edited by W. D. McElroy and B. Glass. New York: Baltimore Press.

SANCHEZ, S., MARTINEZ, L. & MORA, J. (1972). Interactions between amino acid transport systems in *Neurospora crassa. Journal of Bacteriology* **112**, 276–84.

SANDERS, P. P. & BROQUIST, H. P. (1966). Saccharopine, an intermediate of the aminoadipic acid pathway of lysine biosynthesis. *Journal of Biological Chemistry* **241**, 3435–40.

SARIS, N. E. & VIRTANEN, A. I. (1957*a*). On hydroxylamine compounds in *Azotobacter* cultures. I. Formation of hydroxylamine compounds. *Acta Chemica Scandinavica* **11**, 1438-40.

SARIS, N. E. & VIRTANEN, A. I. (1957*b*). On hydroxylamine compounds in *Azotobacter* cultures. II. On the chemical nature of the bound hydroxylamine fraction in *Azotobacter* cultures. *Acta Chemica Scandinavica* **11**, 1440–2.

SCAZZOCCHIO, C. & DARLINGTON, A. J. (1968). The induction and repression of the enzymes of purine breakdown in *Aspergillus nidulans. Biochimica et Biophysica Acta* **166**, 557–68.

SCHEETZ, R. W., WHELAN, H. A. & WRISTON, J. C. (1971). Purification and properties of an L-asparaginase from *Fusarium tricinctum. Archives of Biochemistry and Biophysics* **142**, 184–9.

SCHLOEMER, R. H. & GARRETT, R. H. (1974*a*). Nitrate transport system in *Neurospora crassa. Journal of Bacteriology* **118**, 259–69.

SCHLOEMER, R. H. & GARRETT, R. H. (1974*b*). Uptake of nitrite by *Neurospora crassa. Journal of Bacteriology* **118**, 270–4.

SCHMALFUSS, K. & MOTHES, K. (1930). Über die fermentative desauridierung durch *Aspergillus niger. Biochemische Zeitschrift* **221**, 134–53.

SCHWEET, R. S., HOLDEN, J. T. & LOWRY, P. H. (1954*a*). Lysine metabolism in *Neurospora. Federation Proceedings* **13**, 293.

SCHWEET, R. S., HOLDEN, J. R. & LOWRY, P. H. (1954*b*). The metabolism of lysine in *Neurospora. Journal of Biological Chemistry* **211**, 517–28.

SEECOF. K. L. & WAGNER, R. P. (1959). Transaminase activity in *Neurospora crassa.* I. Purification and substrate specificity of a phenylpyruvate transaminase. *Journal of Biological Chemistry* **234**, 2689–93.

SHAPIRO, B. M. & STADTMAN, E. R. (1970). The regulation of glutamine synthesis in micro-organisms. *Annual Review of Microbiology* **24**, 501–25.

SILVER, W. S. & MCELROY, W. D. (1954). Enzyme studies on nitrate and nitrite mutants of *Neurospora. Archives of Biochemistry and Biophysics* **51**, 379–94.

SINHA, U. (1967). Aromatic amino acid biosynthesis and *p*-fluorophenylalanine resistance in *Aspergillus nidulans. Genetical Research* **16**, 261–72.

SINHA, U. (1969). Genetic control of the uptake of amino acids in *Aspergillus nidulans. Genetics* **62**, 495–505.

SKYE, G. E. & SEGEL, I. H. (1970). Independent regulation of cysteine and cystine transport in *Penicillium chrysogenum. Archives of Biochemistry and Biophysics* **138**, 306–18.

SLAYMAN, C. W. (1973). The genetic control fo membrane transport. In *Current topics in membranes and transport*, vol. 4. pp. 1–174. Edited by F. Bronner and A. Kleinzeller. New York and London: Academic Press.

SOJKA, G. A. & GARNER, H. R. (1967). The serine biosynthetic pathway in *Neurospora crassa. Biochimica et Biophysica Acta* **148**, 92–7.

SORGER, G. J. (1965). Simultaneous induction and repression of nitrate reductase and TPNH-cytochrome-c reductase in *Neurospora crassa, Biochimica et Biophysica Acta* **99**, 234–5.

SORGER, G. J. & GILES, N. H. (1965). Genetic control of nitrate reductase in *Neurospora crassa. Genetics* **52**, 777–88.

SRB, A. M. & HOROWITZ, N. H. (1944). The ornithine cycle in *Neurospora* and its genetic control. *Journal of Biological Chemistry* **154**, 129-39.

STADLER, D. R. (1966). Genetic control of the uptake of amino acids in *Neurospora. Genetics* **54**, 677-85.

STADLER, D. R. (1967). Suppressors of amino acid uptake of *Neurospora. Genetics* **57**, 935–42.

STEWART, G. R. & MOORE, D. (1974). The activities of glutamate dehydrogenases during mycelial growth and sporophore development in *Coprinus lagopus* (sensu Lewis). *Journal of General Microbiology* **83**, 73–81.

ST LAWRENCE, P., MALING, B. D., ALTWERGER, L. & RACHMELER, M. (1964). Mutational alteration of permeability in *Neurospora*: effects on growth and the uptake of certain amino acids and related compounds. *Genetics* **50**, 1383–402.

STRICKLAND, W. N. (1971). Regulation of glutamate dehydrogenase in *Neurospora crassa* as a response to carbohydrate and amino acids in the media. *Australian Journal of Biological Sciences* **24**, 905–15.

STUART, W. D. & DEBUSK, A. G. (1971). Molecular transport. I. *In vitro* studies of isolated glycoprotein subunits of the amino acid transport system of *Neurospora crassa* conidia. *Archives of Biochemistry and Biophysics* **144**, 512–8.

SUBRAMANIAN, K. M. & SORGER, G. J. (1972*a*). Regulation of nitrate reductase in *Neurospora crassa.* Regulation of transcription and translation. *Journal of Bacteriology* **110**, 547–53.

SUBRAMANIAN, K. M. & SORGER, G. J.(1972*b*). Regulation of nitrate reductase in *Neurospora crassa:* stability *in vivo. Journal of Bacteriology* **110**, 538–46.

SUBRAMANIAN, K. M., WEISS, R. L. & DAVIS, R. H. (1973). Use of external biosynthetic and organellar arginine by *Neurospora. Journal of Bacteriology* **115**, 284–90.

TAMENBAUN, S. W., GARNJOBST, L. & TATUM, E. L. (1954). A mutant of *Neurospora* requiring asparagine for growth. *Journal of Bacteriology* **41**, 484–95.

TATUM, E. L. & PERKINS, D. D. (1950). Genetics of micro-organisms. *Annual Review of Microbiology* **4**, 129–50.

TATUM, E. L., EHRENSVÄRD, G. & GARNJOBST, L. (1954). Synthesis of aromatic compounds by *Neurospora. Proceedings of the National Academy of Sciences* **40**, 271–6.

TEAS, H. J., HOROWITZ, N. H. & FLING, M. (1948). Homoserine as a precursor of threonine and methionine in *Neurospora. Journal of Biological Chemistry* **172**, 651–8.

THAYER, P. S. & HOROWITZ, N. H. (1951). The L-amino acid oxidase of *Neurospora. Journal of Biological Chemistry* **192**, 755–67.

THIAMIN, K. V. (1963). *The life of bacteria*, 2nd edition. London: MacMillan Press.

THWAITES, W. M. & PENDYALA, L. (1969). Regulation of amino acid assimilation in a strain of *Neurospora crassa* lacking basic amino acid transport activity. *Biochimica et Biophysica Acta* **192**, 455–61.

TISDALE, J. H. & DEBUSK, A. G. (1970). Developmental regulation of amino acid transport in *Neurospora crassa. Journal of Bacteriology* **104**, 689–97.

TRUPIN, J. S. & BROQUIST, H. P. (1965). Saccharopine, an intermediate of the aminoadipic acid pathway of lysine biosynthesis. I. Studies in *Neurospora crassa. Journal of Biological Chemistry* **240**, 2524–30.

TSAI, H., TSAI, J. H. J. & YU, P. H. (1973). Effects of yeast proteinase and its inhibitor on the inactivation of tryptophan synthase from *Saccharomyces cerevisiae* and *Neurospora crassa. European Journal of Biochemistry* **40**, 225–32.

TURIAN, H. (1961). Cycle glyoxylique, transaminase alanine-glyoxylate et differentiation sexuelle chez *Allomyces* et *Neurospora. Pathological Microbiology* **24**, 819–39.

UMBARGER, E. A. & DAVIS, B. D. (1962). Pathways of amino acid biosynthesis. In *The bacteria*, vol. 3, pp. 167–251. Edited by I. C. Gunsalus and R. Y. Stainer. New York and London: Academic Press.

VOGEL, H. J. (1960). Two modes of lysine synthesis among lower fungi: evolutionary significance. *Biochimica et Biophysica Acta* **41**, 172–3.

VOGEL, H. J. (1961). Lysine synthesis and phylogeny of lower fungi: some chytrids *versus Hyphochytrium. Nature, London* **189**, 1026–7.

VOGEL, H. J. & BONNER, D. M. (1954). On the glutamate-proline-ornithine interrelation in *Neurospora crassa. Proceedings of the National Academy of Sciences, U.S.A.* **40**, 688–94.

VOGEL, H. J. & DAVIS, B. D. (1952). Glutamate gamma-semialdehyde and delta-1-pyrroline-5-carboxylic acid, intermediates in the biosynthesis of proline. *Journal of American Chemical Society* **74**, 109–12.

VOGEL, R. H. & KOPAC, M. J. (1959). Glutamic-γ-semialdehyde in arginine and proline synthesis of *Neurospora*: a mutant-tracer analysis. *Biochimica et Biophysica Acta* **36**, 505–10.

VOGEL, R. H. & VOGEL, H. J. (1963). Evidence for acetylated intermediates of arginine synthesis in *Neurospora crassa. Genetics* **48**, 914.

WAGNER, P. R., BERGQUIST, A. & BARBEE, T. (1965). The synthesis *in vitro* of valine and isoleucine from pyruvate and α-ketobutyrate in *Neurospora. Biochimica et Biophysica Acta* **100**, 444–50.

WAGNER, P. R., BERGQUIST, A., BARBEE, T. & KIRITANI, K. (1965). Genetic blocks in the isoleucine-valine pathway of *Neurospora crassa. Genetics* **49**, 865–82.

WALKER, G. C. & NICHOLAS, D. J. D. (1961). An iron requirement for a dissimilatory nitrate reductase in *Neurospora crassa. Nature, London* **189**, 141–2.

WAMPLER, D. E. & FAIRLEY, J. L. (1967). Argininosuccinate synthetase of *Neurospora crassa. Archives of Biochemistry and Biophysics* **121**, 580–6.

WEBBER, B. B. & CASE, M. E.(1960). Genetical and biochemical studies of histidine-requiring mutants of *Neurospora crassa.* I. Classification of mutants and characterization of mutant groups. *Genetics* **45**, 1605–15.

WEBSTER, R. E. & GROSS, S. R. (1965). The α-isopropylmalate synthetase of *Neurospora.* I. The kinetics and end-product control of α-isopropylmalate synthetase function. *Biochemistry* **4**, 2309–18.

WEBSTER, R. E., NELSON, C. A. & GROSS, S. R. (1965). The α-isopropylmalate synthetase of *Neurospora.* II. The relation between structure and function and complementation interactions. *Biochemistry* **4**, 2319–27.

WEGLENSKI, P. (1966). Genetical analysis of proline mutants and their suppressors in *Aspergillus nidulans. Genetical Research* **8**, 311–21.

WEGMAN, J. & DEMOSS, J. A. (1965). The enzymatic conversion of anthranilate to indolylglycerol phosphate in *Neurospora crassa. Journal of Biological Chemistry* **240**, 3781–8.

WEISS, R. L. (1973). Intracellular localisation of ornithine and arginine pools in *Neurospora. Journal of Biological Chemistry* **248**, 5409–13.

WEISS, R. L. & DAVIS, R. H. (1973). Intracellular localization of enzymes of arginine metabolism in *Neurospora. Journal of Biological Chemistry* **248**, 5403–8.

WHITAKER, A. & MORTON, A. G. (1971). Amino acid transport in *Penicillium griseofulvum. Transactions of the British Mycological Society* **56**, 353–69.

WIEBERS, J. L. & GARNER, H. R. (1964). Use of S-methylcysteine and cystathionine by methioneless *Neurospora* mutants. *Journal of Bacteriology* **88**, 1798–804.

WIEBERS, J. L. & GARNER, H. R. (1967*a*). Homocysteine and cysteine synthetases of *Neurospora crassa.* Purification, properties and feedback control of activity. *Journal of Biological Chemistry* **242**, 12–23.

WIEBERS, J. L. & GARNER, H. R. (1967*b*). Acyl derivatives of homoserine as substrate for homocysteine in *Neurospora crassa*, yeast and *Escherichia coli. Journal of Biological Chemistry* **242**, 5644–9.

WILEY, W. R. (1970). Tryptophan transport in *Neurospora crassa*: a tryptophan-binding protein released by cold shock. *Journal of Bacteriology* **103**, 656–62.

WILEY, W. R. & MATCHETT, W. H. (1966). Tryptophan transport in *Neurospora crassa.* I. Specificity and kinetics. *Journal of Bacteriology* **92**, 1698–705.

WILLETTS, A. J. (1972). Metabolism of threonine by Penicillia: growth on threonine as a sole carbon and nitrogen source. *Journal of General Microbiology* **73**, 71–83.

WOLFINBARGER, L. & DEBUSK, A. G. (1971). Molecular transport. II. *In vivo* studies of transport mutants of *Neurospora crassa* with altered amino acid competition patterns. *Archives of Biochemistry and Biophysics* **144**, 503–11.

WOLFINBARGER, L. & DEBUSK, A. G. (1972). The kinetics of L-aspartate transport in *Neurospora crassa* conidia. *Biochimica et Biophysica Acta* **290**, 355–67.

WOLFINBARGER, L. & KAY, W. W. (1973*a*). Acidic amino acid transport in *Neurospora crassa* mycelia. *Biochimica et Biophysica Acta* **330**, 335–43.

WOLFINBARGER, L. & KAY, W. W. (1973*b*). Transport of C_4-dicarboxylic acids in *Neurospora crassa. Biochimica et Biophysica Acta* **307**, 243–57.

WOLFINBARGER, L., JERVIS, H. H. & DEBUSK, A. G. (1971). Active transport of L-aspartic acid in *Neurospora crassa. Biochimica et Biophysica Acta* **249**, 63–8.

WOODIN, T. S. & NISHIOKA, L. (1973). Chorismate mutase isozyme patterns in three fungi. *Biochimica et Biophysica Acta* **309**, 224–31.

WOODWARD, C. K., READ, C. P. & WOODWARD, V. W. (1967). *Neurospora crassa* mutants defective in the transport of amino acids. *Genetics* **56**, 598–610.

WOOTTON, J. C. (1967). Short-term changes in the amino acid pool of *Neurospora* mycelium. *Neurospora Newsletter* **12**, 3.

WOOTTON, J. C., CHAMBERS, G. K., TAYLOR, J. G. & FINCHAM, J. R. S. (1973). Amino-acid sequence homologies between the NADP-dependent glutamate dehydrogenase of *Neurospora* and the bovine enzyme. *Nature, New Biology* **241**, 42–3.

YANOFSKY, C. & BONNER, D. M. (1950*a*). Studies on the conversion of 3-hydroxyanthranilic acid to niacin in *Neurospora. Journal of Biological Chemistry* **190**, 211–8.

YANOFSKY, C. & BONNER, D. M. (1950*b*). Evidence for the participation of kynurenine as a normal intermediate in the biosynthesis of niacin in *Neurospora. Proceedings of the National Academy of Sciences, U.S.A.* **36**, 167–76.

YANOFSKY, C. & REISSIG, J. L. (1953). L-serine dehydrase of *Neurospora. Journal of Biological Chemistry* **202**, 567–77.

YU, P. H., KULA, M-R, & TSAI, H. (1973). Studies on the apparent instability of *Neurospora* tryptophan synthase. Evidence for protease. *European Journal of Biochemistry* **32**, 129–35.

YURA, T. (1959). Genetic alteration of pyrroline-5-carboxylate reductase in *Neurospora crassa. Proceedings of the National Academy of Sciences, U.S.A.* **45**, 197–209.

YURA, T. & VOGEL, H. J. (1955). On the biosynthesis of proline in *Neurospora crassa*: enzymatic reduction of Δ'-pyrroline-5-carboxylate. *Biochimica et Biophysica Acta* **17**, 582–9.

YURA, T. & VOGEL, H. J. (1959). Pyrroline-5-carboxylate reductase of *Neurospora crassa*: partial purification and some properties. *Journal of Biological Chemistry* **234**, 335–8.

CHAPTER 8

Nucleic Acid and Protein Synthesis in Filamentous Fungi

D. R. BERRY and E. A. BERRY

8.1 Introduction

In this review we have attempted to cover the whole field of nucleic acid and protein synthesis in the filamentous fungi. In addition to the major section on nucleocytoplasmic systems, we have also included sections on mitochondrial and viral systems. Since the enzymes, nucleic acids and protein components isolated from crude extracts may belong to any of the systems, it is our view that workers in the field should be aware of the developments in each of these areas when designing experiments and interpreting data. Recent studies on the DNA-dependent RNA polymerases give ample justification to this approach and with the recent discovery of so many viruses in fungi, it is apparent that viral systems also have an increasing role to play in our understanding of nucleic acid and protein synthesis in fungi.

Not all aspects of nucleic acid and protein synthesis have been studied in the same detail in the filamentous fungi. Whereas the mitochondrial systems of *Neurospora crassa* have been studied as intensively as any mitochondrial system and have been included in several reviews in recent years (p. 262), the nucleocytoplasmic system is less well understood and has been reviewed only once (Van Etten, 1969). Direct information relevant to virus directed nucleic acid and protein synthesis is almost non existent.

8.2 Nucleocytoplasmic Systems

DNA and DNA Replication

CHARACTERIZATION The field of DNA synthesis in the filamentous fungi is less well studied than other areas of fungal cell biology. The only aspect which can be considered to have been investigated in any detail is that of base composition. Much data has been accumulated for studies of a taxonomic nature (Storck & Alexopoulos, 1970), much of which is not relevant to the study of DNA synthesis. Considerable variation in the guanine and cytosine (G+C) content of DNA has been observed; however, the variation with species, genera and families was found to be less than in phyla and other higher taxa, as might be expected if the taxonomic system had a natural validity. The average G+C content for the fungi studied was 50.5%.

The fungi resemble other eukaryotes in having at least two types of DNA; nuclear and mitochondrial (p.263). In some species, cytoplasmic DNAs have also been recognized. In *Blastocladiella emersonii*, four bands of DNA were recognized in caesium chloride gradients (Myers & Cantino, 1971); nuclear DNA of density 1.731 g ml^{-1} contributed the bulk of the extract and two other bands at 1.715 and 1.705 were tentatively attributed to nucleolar and mitochondrial DNA (but see p. 240). A fourth peak of density 1.687 g ml^{-1} appeared to be associated with complex organelles found in the motile spores referred to as γ particles. There may be a similarity between this and the associations of DNA particles with the parabasal bodies of flagella in certain protozoa.

The four components can also be identified by their unique T_m values on a temperature melting curve. The 1.715 g ml^{-1} peak was attributed to nucleolar DNA because its G + C content (56%) most closely resembled that of ribosomal RNA of *Blastocladiella emersonii* (50.3%). If this is so, the high concentration of this fraction (12.5%) indicates an exceptionally high tandem duplication of rRNA genes. A similar cytoplasmic DNA component has been recognized in *Allomyces* (Ojha & Turian, 1971). In *Coprinus lagopus*, two DNA fractions were recognized by their melting curves; 90% of the DNA had a T_m of 91.5 and a G + C content of 52.3% while 10% had a T_m of 82.5 and a G + C content of 32% (Dutta, Penn & Knight, 1972). Two major peaks of DNA were reported when DNA preparations from *Histoplasma capsulatum* and *Blastomyces dermatitis* were centrifuged in caesium chloride gradients. The larger peak was attributed to nuclear DNA and the minor one to mitochondrial DNA. However, the large size of the minor peak leads one to suspect that at least part of this fraction may be of nuclear or nucleolar origin.

GENOME SIZE AND HETEROGENEITY Results obtained using standard chemical techniques to estimate the size of the fungal genome have not proved very reliable (Storck & Alexopoulos, 1970). A figure of 2.6×10^{-8} μg DNA/nucleus (approximately 1.6×10^{8} D) was obtained for conidia of *Aspergillus nidulans* (Bainbridge, 1971) but the difficulties of obtaining a reliable value for vegetative mycelium are well illustrated in this paper. In the absence of reliable values for the relative lengths of the G1, S and G2 phases of the cell cycle, a combination of DNA amounts and numbers of nuclei per sample are not sufficient to provide an accurate value for a haploid nucleus. Fortunately, a less direct, but more precise method of estimating genome size has become available. In the absence of redundant DNA, the genome size of a DNA sample is directly proportional to its heterogeneity. This heterogeneity can now be measured using renaturation kinetics. By comparing the *Cot* value at which 50% of the single stranded molecules reassociate for an unknown sample with that of a known sample such as the *Escherichia coli* genome, an estimate of the genome size can be obtained (Britten & Kohne, 1968). The genome size of *Mucor bacilliformis*, *Neurospora crassa*, and *Coprinus lagopus* were estimated by this method to be 2.0, 2.2 and 2.5×10^{10}D respectively (Dutta & Ojha, 1972). In this paper the degree of homology between the DNA of *N. crassa* and several other species was estimated by hybridization. A value of 97% was obtained with homologous DNA, 76% and 78% with *N. intermedia* and *N. sitophila* and 10% and 14% with *Mucor azygospora* and *Coprinus lagopus.*

The nuclear DNA of fungi is not homogeneous. Brooks & Huang (1972) calculated that 20% of the DNA of *Neurospora crassa* is redundant. This preparation was free of mitochondrial DNA so the redundancy could not be attributed to multiple copies of this fraction. The redundant DNA was calculated to have a length of 10^5 base pairs and a repetition frequency of 60. After purification, it was tested for its homology to rRNA and tRNA using a membrane filter technique and the rRNA and tRNA cistrons were found to constitute only 2.3% and 1.2% of the redundant DNA respectively. In similar experiments on *Neurospora crassa* carried out by Dutta (1973), a value of 10% redundant DNA was obtained. The rRNA genes of *N. crassa* were isolated by Chattopadhyay, Kohne & Dutta (1972). Ribosomal RNA was hybridized to *N. crassa* DNA, and the hybrid molecules were separated out using a hydroxyapatite column. By a repeated cycling of this procedure, almost pure DNA/RNA duplex was obtained from which the ribosomal cistron could be released. Ribosomal RNA cistrons were found to contribute 0.95% of the genome. From the known genome size and the length of the rRNA molecules, the number of cistrons was estimated to be approximately 100. A similar repetition frequency for rRNA genes was also obtained by comparing the renaturation frequency of the isolated genes with that of the whole genome.

Using similar techniques, it was estimated that if there are approximately 60 different tRNA genes in *Neurospora crassa*, then each is reiterated 44 times (Ray & Dutta, 1972; Dutta & Ray, 1973). The low reiteration values attributed to rRNA and tRNA cistrons in *Neurospora crassa* are not consistent with high levels of 'nucleolar' DNA such as have been reported in *Blastocladiella emersonii* and named on the basis of a similar G + C content to that of rRNA.

CHROMATIN The DNA of the fungi is associated in the nucleus with nuclear proteins. However, the nature of these chromosomal proteins appears to vary in different groups of fungi. No histones were found in *Allomyces arbuscula* by Stumm & Van Went (1968). The absence of histones was also reported in *Neurospora crassa* (Dwivedi, Dutta & Block, 1969) and in *N. tetrasperma, Microsporum gypseum* and *Phycomyces blakesleanus* (Leighton, Dill, Stock & Phillips, 1971). However, histone-like proteins have been isolated from *Achlya bisexualis* using carboxymethylcellulose chromatography (Horgen, Nagao, Chia & Key, 1973). The chromatin preparations used in these experiments had a low level of RNA, indicating a low level of contamination with ribosomes and ribosomal proteins, and the distribution of the putative nuclear basic proteins on acrylamide gels did not correspond to those obtained from ribosomes. However, the same workers using the same techniques were not able to detect histone in *Blastocladiella emersonii.* The presence of histones in *Neurospora crassa* has been reported recently by Hsiang & Cole (1973). They describe two fractions, one slightly lysine-rich, resembling F2b of calf thymus, and the other lysine-rich, resembling F2a of calf thymus. They found that the ratio of DNA/RNA/histone/non-histone protein was 1.0: 0.14: 0.24: 0.6 in their chromatin preparations. They suggest that the failure of other workers to detect histone can be attributed to low yields of chromatin, proteolytic degradation and a difficulty in packing fungal chromatin by centrifugation.

DNA REPLICATION The complexities of the chromosome structure make it inevitable that the mechanism of DNA and genome replication in the fungi will be more complex than in the prokaryotic cell.

Both guanine-8-^{14}C and ^{32}P phosphate were incorporated into nuclear and mitochondrial DNA in *Botryodiplodia theobromae* during spore germination (Dunkle & Van Etten, 1972). Incorporation commenced after a 2-3 h lag. No evidence was obtained in the experiment for amplification of genes during spore formation. In germinating spores of *Phycomyces blakesleeanus*, nuclear division occurred at 4–5 h after germination, whereas DNA synthesis did not commence until 8 h, indicating that the spore nuclei were in the G_2 phase of the cell cycle (Van Assche & Carlier, 1973). Studies using isotopes have unfortunately been limited by the absence of thymidine kinase in many fungi (Grivell & Jackson, 1968). It has been shown that DNA was not specifically labelled when labelled thymidine, thymidine monophosphate and thymidine triphosphate were fed. The incorporation of label into RNA as well as DNA suggested that the thymidine was being converted first into uracil and then into other pyrimidine nucleotides. Kessel & Rosenberger (1968) circumvented this difficulty in demonstrating the synchronous division of nuclei in *Aspergillus nidulans* by using an autoradiographic technique in which they fed labelled adenine, then removed labelled RNA by treatment with RNAase.

A DNA polymerase enzyme has been purified from *Rhizopus stolonifer* (Gong, Dunkle & Van Etten, 1973). This enzyme, which was purified over 500-fold was found to require Mg^{2+}, DNA, a reducing agent, dGTP, dCTP and dATP in addition to the labelled dTTP. The authors postulate that the enzyme is nuclear in origin since attempts to isolate a DNA polymerase from mitochondria were unsuccessful. In addition, the molecular weight of the enzyme (70–75 000 D) is much higher than that found for mitochondrial DNA polymerase in yeast.

The enzyme was purified from both spores and germinated spores and found to be the same in both preparations. However, it was observed that whereas enzyme from germinated spores did not require added DNA, that from ungerminated spores did. It is possible that the enzyme is bound more tightly to native DNA in germinated spores. When the DNA from several sources was tested, the enzyme was found to be more active with denatured DNA than native DNA.

Temperature-sensitive mutants were isolated from *Ustilago maydis* in an attempt to obtain strains with modified DNA polymerase activities. From 400 temperature-sensitive mutants tested, 5 were shown to be blocked in DNA synthesis, at the restrictive temperature of 32°C. The DNA polymerase activity of these mutants was tested, and in one case it was found that the activity of cells grown at the restrictive temperature for 5 h was only 10–25% that of the wild type. The enzyme was partially purified by DNA cellulose chromatography and found to be heat sensitive. The enzyme does not appear to be a repair enzyme, since the mutant strain is only marginally more sensitive to ionizing radiation at the restrictive temperature. Furthermore, the enzyme is evidently not the only DNA polymerase in this strain since considerable DNA polymerase activity remains even after extended incubation at 32°C (Jeggo, Unrau, Banks & Holliday, 1973).

The mechanism by which DNA damaged by ionising radiation is repaired, involves DNA breakdown and resynthesis. The enzymes involved however, are unlikely to be the same as those involved in DNA replication. A repair mechanism which involves the excision of cytosine dimers has been described in *Neurospora crassa* (Worthy & Eppler, 1972). In this organism, the number of cytosine dimers produced was found to increase linearly with the ultraviolet dose and to decrease during photoreactivation. In a subsequent paper, Worthy & Eppler (1973) estimated the number of breaks by studying the size distribution of DNA obtained, using a standardized isolation procedure. Ultraviolet irradiation was found to decrease the molecular weight of the DNA obtained, presumably by increasing the number of breaks in the DNA. During the period when the repair mechanism was operative, the number of apparent breaks decreased. Using a combination of D_2O, ^{15}N and ^{32}P labelling, they were able to demonstrate that ^{32}P was incorporated into existing DNA during the period when repair would be taking place, and that the amount of ^{32}P incorporated, increased with the dose of ultraviolet radiation. It has recently been reported that the repair mechanism in *Ustilago maydis* is dependent upon adequate levels of pyrimidines in general, and probably thymidine triphosphate in particular (Moore, 1974).

RNA and Transcription

CHARACTERIZATION OF RNA RNA synthesis has been studied in several fungi in an attempt to obtain a clearer understanding of fungal gene expression and its control. In the fungi, as in other organisms, ribosomal RNA (rRNA) and transfer RNA (tRNA) have been readily extracted using the phenol method and isolated by acrylamide gel electrophoresis (Loening, 1968), methylated albumen kieselguhr (MAK) column chromatography (Gottlieb, Rao & Shaw, 1968), and by sucrose gradient centrifugation (Mirkes, 1974; Murphy & Lovett, 1966). Gottlieb, Rao & Shaw (1968) experienced difficulty in isolating RNA from *Ustilago maydis*. The RNA appeared to be tightly bound to a protein factor and was not released by phenol extraction. It was, however, readily separated by prior treatment with NaCl. This is included in the extraction medium of Loening (1968), and Lovett & Haselby (1971).

Loening (1968) established that within limits, the rate of movement of RNA in gels was inversely proportional to the log of the molecular weight. Using this technique he determined the molecular weights of the rRNA in the following fungi: 1.30 and 0.73×10^6 for *Aspergillus*; 1.30 and 0.68×10^6 for *Botrytis*; 1.30 and 0.71×10^6 for *Chaetomium*; and 1.28 and 0.72×10^6 for *Rhizopus*. A more detailed analysis of the rRNA of fungi was carried out by Lovett & Haselby (1971). The RNA of the smaller ribosomal subunit was found to have a molecular weight of between 0.71 and 0.75×10^6 in all the groups of fungi studied. However, the values obtained for heavy rRNA fell into distinct groups (Table 8.1). The high values obtained for the Myxo-

Table 8.1. Molecular weights of ribosomal RNA in fungi (from Lovett & Haselby, 1971)

Organisms	***Molecular weight* × *10⁶***		***Number of determinations***
	Large	***Small***	
Acrasiales			
Dictyostelium discoideum	1.42 ± 0.014	0.73 ± 0.01	4
Myxomycetes			
Physarum polycephalum	1.45 ± 0.007	0.75	3
Hyphochytridiomycetes			
Rhizidiomyces apophysatus	1.36 ± 0.016	0.73 ± 0.007	3
Oomycetes			
Achlya ambisexualis	1.40	0.72 ± 0.01	2
Phytophthora capsici	1.43 ± 0.007	0.73 ± 0.007	3
Chytridiomycetes			
Entophlyctis sp.	1.36 ± 0.01	0.73	2
Rhizophlyctis rosea	1.34	0.73	2
Allomyces arbuscula	1.34 ± 0.01	0.74	2
Blastocladiella emersonii	1.32	0.73	
Zygomycetes			
Mucor racemosus	1.33 ± 0.007	0.71 ± 0.01	3
Phycomyces blakesleanus	1.34 ± 0.005	0.72 ± 0.008	4
Ascomycetes			
Saccharomyces cerevisiae	1.31 ± 0.012	0.72 ± 0.007	3
Sordaria fimicola	1.30 ± 0.022	0.72	2
Basidiomycetes			
Ustilago maydis	1.32 ± 0.01	0.74 ± 0.01	2
Coprinus lagopus	1.30	0.73	2
Schizophyllum commune	1.32 ± 0.01	0.73 ± 0.002	2
Higher plants			
Glycine max	1.32 ± 0.022	0.69	3

mycetes, Acrasiales and Oömycetes suggested a closer affinity to the animal kingdom than the fungi as represented by the Ascomycetes and the Basidiomycetes. Using acrylamide gel electrophoresis, the 28 S and 18 S cytoplasmic rRNA can readily be distinguished from the 25 S and 16 S mitochondrial rRNA.

Lovett & Leaver (1969) claimed that some sequence homology between the RNA of the large and small ribosomal subunits was indicated by the formation of high molecular weight aggregates of heavy and light ribosomal RNA in *Blastocladiella emersonii*, produced by heating to 40–60°C in a *p*-amino-salicylate/triisopropylnaphthalene sulphonate/phenol extraction medium. Aggregates of molecular weight equivalent to the sum of the two subunits have also been obtained in *Mucor*, *Coprinus* and *Rhizidiomyces* (Lovett & Haselby, 1971). Sequence homology between the RNA of the heavy and light ribosomal subunits has not been demonstrated directly by hybridization in filamentous fungi, as it has in yeast (Retel & Planta, 1970).

Peaks that are heavier than mature rRNAs and run before them on gels have been described by several workers (Lin, Davies, Tripathi, Raghn &

Gottlieb, 1971; Gamow & Prescott, 1972; Holloman, 1970). These may well be ribosomal precursor RNA, but in view of the reported formation of high molecular weight aggregates, this should not be assumed in the absence of confirmatory experiments. In *Phycomyces blakesleanus* however, two peaks were found to have S values of 29 S and 32 S (Gamow & Prescott, 1972). It was demonstrated that after 5 min pulse labelling with ^{32}P, the isotope passed primarily into the 29 S and 32 S peaks. After 15 min however, all the rRNA and tRNA fractions were labelled, and the activity of the 29 S and 32 S peaks had decreased.

Transfer RNA is not such a discrete fraction on gels, sucrose gradients and MAK columns, and has an S value of between 4 and 5. Messenger RNA is a more difficult fraction to recognize and isolate. Perhaps the most elegant demonstration of messenger activity in fungi was presented by Ramakrishnan and Staples (1970). RNA was extracted from *Uromyces phaseoli* using the phenol method, cleaned up on a Sephadex G50 column then fractionated on a 5–20% sucrose gradient. Fractions from this gradient were then tested for messenger activity by adding them to an *in vitro* protein-synthesizing system prepared from *Escherichia coli.* It was found that, although the region of the gradient containing from 4 S to 19 S RNA did not contain the bulk of the RNA, most of the messenger activity was found in this fraction. This messenger activity could be destroyed by treatment with RNAase. RNA from this region of the gradient has been found to be rapidly labelled in *Aspergillus niger* (Aitkhazhin, Azimuratoua, Kim & Darkanbaev, 1972), and to have a high specific activity when ^{3}H-adenine was fed to germinating cultures of *Neurospora crassa* conidia (Mirkes, 1974).

MAK column chromatography was used by Mehadevan to isolate mRNA from *Neurospora crassa.* The identification of the mRNA fraction was confirmed using a hybridization technique (Bagwat & Mehadevan, 1970, 1973). The percentage of DNA which could hybridize with RNA was determined at 8, 16 and 24 h after the induction of germination. The value obtained at 8 h (6%) was higher than that recorded at 16 and 24 h (3and 4% respectively). This suggested that more genes were being expressed at this time. The possibility that rRNA and tRNA were making a major contribution to these values was eliminated by saturating the DNA with cold rRNA and tRNA prior to the addition of the labelled mRNA fraction. Differences between the mRNA fractions obtained at different times were indicated by competition experiments in which, for example, cold mRNA from the 8 h stage was bound to the DNA prior to the addition of labelled mRNA from the 16 h stage. If identical messages were being produced at both stages, then the cold mRNA would have saturated the corresponding genes and prevented the binding of the identical labelled mRNA. The reduction in the percentage of labelled mRNA band gives an indication of the degree of homology of the sample. Approximately 40% homology was found between 8 and 16 h samples and 60% homology between 16 and 24 h samples. Conversely, the failure of the cold sample to completely inhibit the binding of the labelled mRNA fraction indicates the diversity of the two fractions.

In an attempt to determine more accurately the percentage of the genome involved in RNA production at any one time, Dutta (1973) first separated out repeated from non-repeated DNA on a hydroxyapatite column. RNA was obtained using a hot phenol extraction. RNA and DNA were then hydridized and the single-stranded DNA separated from DNA/RNA hybrids by elution through a hydroxyapatite column. The results indicated that 35% of the non-reiterated DNA was transcribed in mid log phase mycelium.

Recently, a more specific technique has become available for recognising mRNA (Darnell, Wall & Tushinski, 1971). Most eukaryotic mRNAs have been found to have a sequence of polyadenylic acid at the 3′ end of the molecule. Such mRNAs can be separated out by chromatography on a poly-dT cellulose column (Aviv & Leder, 1972). Silver & Horgen (1974) have isolated an RNA fraction containing 35.5% AMP residues from *Achlya ambisexualis.* This is a similar value to that obtained from messengers of eukaryotic cells. Actinomycin D caused a rapid reduction in the specific activity of this fraction in cells which were being fed ^{3}H-uridine. ^{3}H-Adenine was rapidly incorporated into the poly-A-rich RNA and its incorporation was inhibited by cordycepin, a drug which has been shown to inhibit the post-transcriptional addition of poly-A. This poly-A rich fraction has been isolated from whole homogenates, from purified nuclei, and from polyribosomes. If it is mRNA, it might be expected to possess a broad size distribution. Electrophoresis of the poly-A fraction on acrylamide gels showed that its size varied from 4–30 S. The poly-A component, which is resistant to T_1 and pancreatic RNAase was found by electrophoresis to have a molecular weight of 16 000–18 000 D which corresponds to a chain length of 60 nucleotides. This is smaller than the poly-A sequences isolated from animal messengers but similar to the values obtained in yeast (McLaughlin, Warner, Edmonds, Nakazato & Vaughan, 1973). Treatment of *Achlya* with antheridiol stimulated a 3- to 4-gold increase in the incorporation of ^{3}H-uridine into a poly-A rich fraction after a lag of 2–3 h at a time when the initials of the male sex organs were beginning to appear (Silver & Horgen, 1974). A putative messenger RNA fraction has been isolated from *Rhizopus stolonifer* during sporangiospore germination (Roheim, Knight & Van Etten, 1974).

This fraction was isolated on a poly-dT cellulose column and found to constitute 4.2% of the total RNA. In spores labelled with ^{3}H-adenine, 5.5% of the radioactivity was found in the poly-A-rich fraction. When material was centrifuged through a sucrose density gradient, some material was found sedimenting in the region of rRNA but the majority was lighter than the 26 S rRNA subunit and had a heterogeneous size distribution. Other lines of evidence in support of this fraction being mRNA were that it could be isolated from polyribosomes, and that if the spores were supplied with ^{3}H-methyl-methionine only 0.5% of the label in the total RNA was present in this fraction. This compares with 7%, if uracil was supplied. Finally, it was demonstrated that whereas rRNA hybridized with only 1.25% of the total DNA, this mRNA fraction hybridized with 7.9%. Furthermore, complete saturation of the DNA with unlabelled rRNA caused less than a 5% decrease in the amount of mRNA hybridizing to DNA.

METHYLATION OF RNA Ribosomal and transfer RNA have been shown to contain methylated bases such as methylguanine in those organisms which have been studied. Such bases are methylated after transcription in reactions which involve methylmethionine as a methyl donor. The ^{3}H-methyl group from ^{3}H-methylmethionine was found to be incorporated into rRNA (24 S and 17 S), precursor rRNA (29 S and 32 S), and tRNA (4 S) in *Phycomyces blakesleanus* within 5 m (Gamow & Prescott, 1972). After 10 m, all the label had passed into the rRNA and tRNA peaks, indicating that the methylation reaction is very rapid. It has also been demonstrated in *Rhizopus stolonifer* (Roheim, Knight & Van Etten, 1974) that virtually all of the ^{3}H-labelled methylmethionine is incorporated into rRNA and tRNA.

In *Neurospora crassa*, cycloheximide has been shown to inhibit the production of rRNA. During cycloheximide inhibition, a high molecular weight fraction appears. This probably represents a precursor molecule, since if ^{14}C-uracil is supplied it becomes rapidly labelled. If the cycloheximide is then removed, the label passes into rRNA. Cycloheximide has been shown to inhibit the methylation of the high molecular weight RNA fraction. It does not, however, have any effect on pseudouridine formation (Viau & Davis, 1970). A tRNA methylase activity has also been demonstrated in *Neurospora crassa* (Wong, Scarborough & Borek, 1971). The methylase activity was found to be much higher in ungerminated conidia than germinated conidia and a difference in the specificity of the extracts from ungerminated and germinated conidia was observed.

BIOSYNTHESIS OF RNA Many combinations of techniques have been used to study the biosynthesis of RNA and its role in growth and development. A model approach to such studies was demonstrated by Murphy & Lovett (1966) in their studies on zoospore formation in *Blastocladiella emersonii*. In this organism, the vegetative phase grows as a coenocyte. Zoospore formation was induced at 15.5 h by transfer into a dilute salts solution, and was completed by 19.0–19.5 h. The exponential increase in nucleic acid, protein and dry weight during the growth phase was terminated by the transfer to dilute salts medium such that total RNA reached its highest value at 16.5 h. The overall rates of apparent RNA and protein synthesis were estimated by observing the incorporation of ^{14}C-uracil and ^{14}C-leucine into cold trichloroacetic acid precipitates. After transfer, a short burst of uracil incorporation was followed by a rapid fall-off in the rate of incorporation to almost zero at 17 h. In contrast, leucine incorporation was stimulated by transfer and remained high until 18 h when it also began to fall off. In an attempt to establish that the changes in the rate of incorporation observed reflected changes in RNA and protein synthesis, the pool sizes of uracil and leucine were studied together with the rates of uptake of the labelled precursors into the pools. The pool size of leucine showed a 2-fold increase at the period of maximum incorporation so the value observed must represent a conservative estimate of the true rate of protein synthesis. Thus, although leucine was evidently taken up, it was probably diluted out by endogenous leucine. Uracil was also readily taken up by the plants up to 16 h. However, after this time the rate of uptake dropped and studies on the pool size of uracil showed that after 17 h it also began to fall and that uracil was excreted into the medium. Two points referred to here deserve special emphasis. Firstly, the capacity of a cell to take up a precursor may differ at different stages of its development and secondly, any restrictions upon uptake may not necessarily apply to all the precursors being fed. Analysis of uracil-labelled RNA on sucrose gradients showed that at 16 h label was incorporated into rRNA and tRNA and into a heavy peak, which was presumably precursor rRNA. Double labelling techniques were used to demonstrate that little or no turnover of rRNA occurred during the formation of the ribosomes of the nuclear cap.

Such results indicate that experiments in which rates of RNA synthesis are calculated from rates of incorporation of a labelled precursor into RNA should be treated with caution. Experiments in which rates of incorporation into different RNA fractions are compared are probably less open to such problems. During the germination of conidia of *Aspergillus oryzae*, it has been reported that rRNA is synthesized first, then tRNA and finally mRNA (Ono, Kimura & Yanagita, 1966; Tanaka, Ono & Yanagita, 1966). More recent studies by Roheim, Knight & Van Etten (1974) indicate that all three fractions are synthesized within 8 min in *Rhizopus stolonifer*. The problems of cell permeability have also been emphasised by studies with inhibitors (Tisdale & De Busk, 1972). Conidia of *Neurospora crassa* were found to be impermeable to actinomycin D and α-amanitin, although the vegetative mycelium was not. The resistance of conidial development to these antibiotics is probably therefore a reflection of this impermeability rather than the existence of stable messenger RNA. Ethylenediaminetetraacetic acid was also found to inhibit the uptake of labelled uridine and orotic acid into trichloroacetic acid insoluble material by *Histoplasma capsulatum*, giving the impression of an inhibition of RNA synthesis (Cheung, Kobayashi, Schlessinger & Medoff, 1974). The possibility that changes in permeability to inhibitors may occur at different stages in the life cycle of an organism should not be overlooked. The resistance of spore induction, germination and mitosporangium formation to actinomycin D in *Allomyces arbuscula* (Burke, Seale & McCarthy, 1972) may be attributable to such changes in permeability.

Different bases have been used as labelled precursors for studies on RNA synthesis, *e.g.* guanine in *Histoplasma capsulatum* (Cheung *et al.*, 1974), adenine in *Neurospora crassa* (Mirkes, 1974), and in several experiments labelled CO_2 has been used as a precursor for RNA (Tanaka, Ono & Yanagita, 1966; Stallknecht & Mirocha, 1971). Evidence has been presented that CO_2 is incorporated more readily into purines than into pyrimidines, which could lead to incorrect interpretations of results in single-stranded RNA such as mRNA, in which purines and pyrimidines are not necessarily present in the same amounts. The discovery of terminal poly-A sequences in mRNA (p. 245) emphasizes this, and indicates a possible complication of using adenine as a labelled precursor. The uptake of bases has been facilitated by the use of auxotrophic strains which require an exogenous supply of the appropriate base (Kritskii & Chernysheva, 1968; Arst & Scazzocchio, 1972).

Information on the cellular control of RNA synthesis in fungi is lacking. Studies in which *Achlya bisexualis* was grown in continuous culture have shown that whereas the DNA and protein content of the mycelium is more or less independent of specific growth rate, the RNA content increases linearly with this function (Griffin, Timberlake & Cheney, 1974). In studies on surface culture of *Neurospora crassa*, a diurnal rhythm in the level of RNA in the mycelium and in the rate of incorporation of ^{3}H-uridine into RNA was observed (Martens & Sargent, 1974). This rhythm in the production of RNA coincides with a similar rhythm of conidium formation; the peak of RNA synthesis occurring slightly before the peak of conidium formation and of RNA content. Conidia have been demonstrated to contain higher levels of RNA than vegetative mycelium. The relationship between RNA and protein synthesis was studied by Arst & Scazzocchio (1972). They followed the incorporation of uracil and leucine into strains which were auxotrophic for uracil, leucine, and other amino acids. They showed that under conditions of amino acid starvation, RNA synthesis was inhibited and they concluded by analogy with bacterial systems that the control of RNA synthesis was stringent.

ENZYMES OF RNA SYNTHESIS Although DNA-dependent RNA polymerase activities have been detected in several species during the last few years, the study of the enzymology of fungal RNA synthesis is still in its infancy. An RNA polymerase fraction was isolated from *Blastocladiella emersonii* by Horgen (1971), using DEAE cellulose chromatography. He obtained three distinct components (I, II and III), two of which (I and II) were found to occur in the nucleus (Horgen & Griffin, 1971*a*), and the third (III) was located in the mitochondrial fraction. The three enzymes were found to be sensitive to different inhibitors. Enzyme III, the mitochondrial enzyme, was inhibited by rifampin at $150\,mg\,ml^{-1}$, whereas the other two enzymes remained unaffected. Enzyme II was found to be inhibited by α-amanitin and enzyme I by cycloheximide. Studies in other eukaryotic organisms have indicated that polymerase I is a nucleolar enzyme which is involved in the biosynthesis of rRNA (Roeder & Rutter, 1970*a*, *b*), and polymerase II is responsible for the synthesis of DNA-like RNA.

Three polymerase enzymes have been separated by DEAE cellulose chromatography from *Achlya bisexualis* (Timberlake, McDowell & Griffin, 1972), *Allomyces arbuscula* (Cain & Nester, 1973), *Rhizopus stolonifer* (Gong & Van Etten, 1972), and *Histoplasma capsulatum* (Boguslawski, Schlessinger, Medoff & Kobayashi, 1974). Difficulty was experienced in obtaining reproducible patterns in *H. capsulatum* using DEAE cellulose columns and better results were obtained with the cation exchanger phospho-cellulose. The results of inhibitor studies in these fungi have been less consistent. Polymerase I from *A. bisexualis* was found to be inhibited by cycloheximide, but polymerase I in *R. stolonifer* and *A. arbuscula* were unaffected by this inhibitor. The results with α-amanitin were equally varied; 80% inhibition of an unspecified enzyme fraction was obtained in *H. capsulatum.* Polymerase II was inhibited by α-amanitin in *A. arbuscula* but only 30% inhibition of polymerase II was observed in *R. stolonifer.* Rifampin was not found to inhibit any of the polymerases in *R. stolonifer* or *A. arbuscula.* These results suggest that the three inhibitors referred to cannot be used in all systems to distinguish the three polymerases.

It has been demonstrated that the inhibitory effect of cycloheximide on RNA synthesis is separable from its effect on protein synthesis (Timberlake & Griffin, 1974). A series of chemical derivatives of cycloheximide were obtained and their effect on the incorporation of labelled uracil and leucine into a trichloroacetic acid precipitate was studied. Whereas several compounds, including cycloheximide, inhibited the incorporation of both compounds, streptovitacin A and streptimidine inhibited leucine uptake but not uracil and two others, cycloheximide acetate and dehydrocycloheximide actually stimulated uracil incorporation whilst inhibiting protein synthesis. Evidence has been presented that cycloheximide also inhibits the methylation of RNA (p. 246).

The properties of RNA polymerase enzymes have been studied to a limited extent. All three were found to be more active in *Allomyces arbuscula* and *Rhizopus stolonifer* when supplied with denatured DNA than with native DNA (Cain & Nester, 1973; Gong & Van Etten, 1972) and to require divalent cations such as Mg^{2+} and Mn^{2+}. The three enzymes of *Histoplasma capsulatum* and *R. stolonifer* were found to have different optima of ionic strength (Boguslawski *et al.*, 1974; Gong & Van Etten, 1972). Polymerase I and II have been further purified by ammonium sulphate precipitation, DEAE cellulose chromatography and glycerol gradient centrifugation (Gong & Van Etten, 1974). RNA polymerase I was found to have three major subunits of molecular weight 188 000, 104 000, and 32 000 D; while polymerase II was found to have subunits of 215 000, 114 000, 37 000 and 27 000 D.

When cultures of the pathogenic *Histoplasma capsulatum* were transferred from growth at 23°C to growth at 37°C, RNA synthesis was halted for several hours. The effects of temperature on the isolated RNA polymerases were studied. The α-amanitin sensitive enzyme was inhibited at 37°C whereas the other two enzymes were more active at this temperature, at least in the early stages of the assay. The inhibition of the former fraction could not be attributed to denaturation of the enzyme, since if enzyme preparations kept at 37°C for 30 m were transferred to incubations at 23°C the rate of RNA synthesis was stimulated. Temperature was found to be a critical parameter in studies on the inhibition of RNA polymerase by actinomycin D in *Neurospora crassa* (Totten & Hove, 1971). 5 μg of actinomycin D were found to inhibit by 28%, 42% and 92% at 25, 30 and 35°C respectively. Actinomycin D was found to inhibit all three polymerases by 70% in *Rhizopus stolonifer* at 30°C (Gong & Van Etten, 1974).

No change in the RNA polymerase content of *Histoplasma capsulatum* and *Allomyces arbuscula* was observed at different stages of their life cycles (Cain & Nester, 1973; Boguslawski *et al.*, 1974). However, in *Rhizopus stolonifer*, polymerase I and III were present in both germinated and ungerminated spores, whereas polymerase II was absent in ungerminated spores and only appeared at 3 h after germination. In an attempt to explain the lack of polymerase II activity, more rigorous extraction techniques in case the enzyme was more tightly bound, were adopted without success, and experiments to test for the presence of an endogenous inhibitor gave negative results (Gong & Van Etten, 1972).

Although there is no direct evidence that cyclic AMP is involved in the control of RNA synthesis in filamentous fungi, it would be surprising if it is not at least indirectly involved. There is increasing evidence that cyclic AMP is a key regulatory compound in the fungi as well as in other organisms (Flavia & Torres, 1972*a*, *b*).

Protein Synthesis

TRANSFER RNA AND THE CHARGING REACTION The existence of transfer RNA has been demonstrated directly or indirectly in several fungi. A soluble RNA fraction, which had a G + C content of 61% was isolated by Henney & Storck (1963*a*) from *Neurospora crassa* using sucrose gradient centrifugation. This differed from the ribosomal fraction with a G + C content of 50% and was present both in growing mycelium and conidia. Although the presence of functional tRNA and the corresponding synthetase enzymes have been indicated by the preparation of a soluble enzyme fraction from fungal extracts for use in *in vitro* protein synthesizing systems (p. 257), only in a few instances have specific tRNAs and synthetase enzymes been demonstrated by the addition of individually labelled amino acids. One such instance was demonstrated in *N. crassa* by Brown & Novelli (1968), in which individually labelled amino acid tRNAs were separated out using reverse phase chromatography on hydrophobic diatomateous earth on which dimethyldiamylammonium chloride in isoamyl acetate was immobilized. Both cytoplasmic and mitochondrial tRNAs were distinguished in this study.

Aminoacyl acceptor activity for all twenty amino acids commonly found in proteins, has been found in both conidia and germinating conidia of *Botryodiplodia theobromae* and *Rhizopus stolonifer* (Van Etten & Brambl, 1968; Van Etten, Koski & El-Olemy, 1969). In these studies it was demonstrated that tRNA existed *in vivo* in both the charged and uncharged state. The presence of transfer RNA and aminoacyl tRNA synthetases has also been demonstrated in *Fusarium* (Merlo, Roker & Van Etten, 1972), *Neurospora crassa* (Shearn & Horowitz, 1969*a*; Emmet, Williams, Frederick & Williams, 1972), *Aspergillus oryzae* (Horikoshi, Ohtaka & Ikeda, 1969), and *Blastocladiella emersonii* (Schmoyer & Lovett, 1969).

It has recently been suggested that, in addition to mitochondrial and cytoplasmic species of arginyl tRNA, a separate nuclear species may also exist (Emmet *et al.*, 1972).

It has been established in *Escherichia coli* that the terminal-CCA of tRNA is labile. Transfer RNA nucleotidyl transferase, the enzyme involved in the attachment of the terminal-CCA to tRNA, has been isolated from *Neurospora crassa*:

$$\text{tRNA} + 2\text{CTP} + \text{ATP} \xrightleftharpoons{\text{Mg}^{2+}} \text{tRNA-pCpCpA} + 3\text{PP}_i$$

The *Neurospora crassa* enzyme has been found to be several hundredfold more active *in vitro* than the corresponding enzymes from *E. coli* and rat liver, incorporating 24 100 nM AMP per hour (Hill & Nazario, 1973). In addition to being a substrate, ATP was found to be a competitive inhibitor of CTP incorporation and CTP of ATP incorporation. The observation that the K_m and K_i values for ATP were similar, suggested a common site of action. CTP also appears to act at a single site both as a substrate and as an inhibitor; however, the site is different from the ATP binding site.

The conditions for the attachment of amino acids to tRNA in *Neurospora crassa* have been studied in detail by Shearn & Horowitz (1969*a*).

$$\text{tRNA}_1 + \text{aa}_1 + \text{ATP} \rightarrow \text{tRNA}_1\ \text{aa}_1 + \text{AMP} + \text{PP}_i$$

Of the 20 amino acids studied, 14 were found to give optimal rates of charging in a common reaction mixture; however 6: alanine, glutamate, glutamine, methionine, glycine and serine required different conditions. The critical variables were found to be ATP, tRNA, Mg^{2+} concentrations, pH and the presence of sulphydryl reagents. Around 90% of tRNA molecules were found to have acceptor activity. Analogues were shown to be bound at the same site as the corresponding amino acid but with reduced efficiency.

Studies on the interspecific acylation of tRNAs (Vanderhoff, Travis, Murray & Key, 1972) indicate that although synthetase enzymes from *Neurospora crassa* were active with tRNAs from pea root, soybean hypocotyl, carrot, corn, yeast, *Neurospora* and *Escherichia coli*, maximal rates of charging were only obtained when both the tRNAs and the synthetase enzymes were derived from *Neurospora*. This suggests some degree of species specificity. However, experiments with phenylalanine tRNA synthetase II from *N. crassa* suggest that the specificity of the charging reaction may not be so precise. This enzyme was found to be active with alanine tRNA and valine tRNA from *Escherichia coli*. The reaction with the *E. coli* tRNAs was, however, different from the normal charging reaction in that it was inhibited by sodium chloride, whereas the reaction with *N. crassa* tRNA was stimulated by the same treatment (Holten & Jacobsen, 1969). It is not clear whether sodium chloride was acting upon the synthetase enzyme or the tRNA.

The attachment of arginine to the corresponding tRNA in *Neurospora crassa* has been shown to be inhibited by arginosuccinate, a precursor of arginine (Nazario, 1967). Arginosuccinate accumulates in *N. crassa* strains carrying the *Arg 10* mutation when it is grown in the presence of growth limiting amounts of arginine or in non-limiting arginine plus citrulline. The level of charged arginine tRNA was determined by assaying the accepting capacity of arginine tRNA in extracts incubated at pH 8.9, which is known to remove amino acids from aminoacyl tRNAs, and in unincubated extracts. The difference between the two values gives the proportion of tRNA which was charged *in vivo*. Low values for charged arginine tRNA were obtained from *Arg 10* cultures grown in conditions in which arginosuccinate accumulates. In confirmation of these results, arginine tRNA synthetase was purified and shown to be inhibited by arginosuccinate *in vitro*.

Transfer RNA acceptor activity has been found to be influenced by the addition of ethionine. Addition of ethionine to cultures of *Neurospora crassa*, results in a decrease in protein synthesis and in the inhibition of tyrosinase (Shearn & Horowitz, 1969*b*). It was observed that the level of tRNA acceptor activity dropped rapidly in such cultures. Using methylated albumin kieselguhr columns, Shearn & Horowitz (1969*b*) went on to study the effect of ethionine on the acceptor activity of individual tRNAs. They found that not all transfer RNAs were affected in the same way. Glutamate, histidine, lysine, proline, tryptophan and tyrosine tRNAs decreased in activity whereas serine, valine and methionine tRNAs increased in activity. The decrease in tRNA acceptor activity appears to be attributable to an increase in ribonuclease activity and a decrease in RNA synthesis. It is considered that the tRNAs which were not reduced by ethionine treatment were probably more resistant to ribonuclease attack. The relative amounts of different iso-accepting species of glutamyl and seryl tRNA were found to change but no new species of tRNAs were observed in ethionine-treated cultures.

The total level of tRNA acceptor activity has been shown to vary in different stages of development (Van Etten, Koski & El-Olemy, 1969). The values obtained for spores and germinating spores in *Botryodiplodia theobromae* and *Rhizopus stolonifer* were 25% and 37% for *Botryodiplodia* and 59% and 69% for *Rhizopus*, respectively. Amino acid composition however, does not appear to be controlled by the relative abundance of different species of tRNA. The distribution of tRNA acceptor activities in *Neurospora crassa* was not found to be correlated with the composition of the amino acid pool or the amino acid content of the soluble protein of the organism (Shearn & Horowitz, 1969*a*).

Evidence has been presented that aminoacyl tRNAs may be involved in the regulation of amino acid biosynthesis. A mutant *trp 5* which is deficient in L-tryptophan tRNA synthetase has been isolated from *Neurospora crassa*. The mutation appears to be in the structural gene, since it affects both the K_m and the specific activity of the enzyme. Strains carrying this mutation have a reduced ability to repress the enzymes involved in tryptophan biosynthesis (Nazario, Kinsey & Ahmad, 1971). Interest in the possible importance of isoaccepting species in the control of translation (Sueoka & Kano-Sueoka, 1970) has stimulated a series of studies on the occurrence and relative abundance of such tRNAs in filamentous fungi.

Transfer RNAs from germinated and non-germinated spores of *Rhizopus stolonifer* were analysed by cochromatography on benzoylated DEAE cellulose (BDC) columns by Merlo, Roker & Van Etten (1972). Two peaks were observed for the tRNAs of alanine, lysine, methionine, phenylalanine and valine. However, whereas differences in the relative abundance of two isoaccepting species of valine and lysine between germinated and non germinated spores were observed, no differences were observed in the other amino acid tRNAs. Experiments in which tRNA from both germinated and non germinated spores were charged using synthetase enzymes from both types of spore, indicated that the differences in tRNA profiles observed could not be attributed to differences in synthetase activity. An unusual result was obtained with isoleucine tRNA. Although only a single peak was observed in both germinated and ungerminated spores, the positions of the peaks on the column were not identical. Evidence was presented that differences in the synthetase enzymes may be involved in this difference in the properties of isoleucine tRNA.

The tRNA composition of shaken and static cultures of *Neurospora crassa* has also been studied (Nazario, 1972). Two peaks of arginine tRNA were detected on a BDC column, one of which was strongly bound to poly-AG and bound to a lesser extent to poly-CGA, and the other which bound to poly-CGA but not to poly-AG. The relative abundance of those two species differed in shaken and static cultures. It was postulated that the difference may be attributable to the absence of the terminal AMP residue, a tRNA lacking this being inactive as an acceptor molecule. However, treatment with ATP and nucleotidyl transferase did not produce any significant increase in acceptor activity.

The appearance of an extra peak of leucyl tRNA in vegetative cells of *Aspergillus oryzae* compared with conidia has been reported, together with differences in the relative frequency of two peaks of methionyl tRNA in conidial and vegetative preparations (Horikoshi, Ohtaka & Ikeda, 1969).

The role of methionyl tRNA in initiation is discussed in a subsequent section.

Unfortunately, most of the techniques used to study isoaccepting species of tRNA do not separate out cytoplasmic from mitochondrial tRNA in the initial extraction. For isoaccepting species to be functional in regulation, they must be present in the same pool. This lack of separation may not be a problem, if the synthetase enzymes are specific to mitochondrial or cytoplasmic tRNA molecules (Küntzel, 1969*a*) (see also p. 267). The absence of multiple peaks when aminoacyl tRNAs are analysed (Merlo, Roker & Van Etten, 1972) suggests that mitochondrial tRNAs may not be present in sufficient amounts to produce a significant contamination of whole cell extracts.

RIBOSOMES The conventional techniques used for isolating ribosomes in other organisms have been found to be successful in fungi. Ribosomes and polyribosomes have been detected using the analytical ultracentrifuge in, for example, *Neurospora crassa* (Henney & Storck, 1964) and *Uromyces phaseoli* (Staples, Ramakrishnan & Yanif, 1970). However, more frequently, ribosomes have been isolated by the centrifugation of a 10 000 g supernatant through a 10–40% linear sucrose gradient. For the isolation of ribosomal subunits, a less concentrated gradient is used, as for example the 5–30% gradient used for studies in *N. crassa* by Mirkes (1974). Ribosomes dissociate into subunits upon incubation at low concentrations of magnesium ion. Some values which have been obtained for the sedimentation coefficients of fungal ribosomes and ribosomal subunits are shown in Table 8.2. *Aspergillus niger* ribosomes were reported to contain 53% RNA, 47% protein (Moyer & Storck, 1964), while values of 67%, 55% and 58% RNA were obtained in ascospores, conidia and hyphae of *N. crassa* respectively (Henney & Storck, 1963*b*). The 75% value obtained in *Erysiphe graminis* (Leary & Ellingboe, 1971) is probably a high value resulting from caesium chloride used during analysis causing dissociation of rRNA and protein. Ribosomal RNA is discussed in more detail on p. 242.

Although the dissociation of ribosomes into subunits is a reversible process, evidence presented by Van Etten (1971) showed that some damage does occur and that this is primarily associated with the 60S subunit.

Little is known in detail of the structure and biosynthesis of ribosomes in filamentous fungi. Labelled 28S and 18S rRNA was incorporated into ribosomes during a 5 min pulse of ^{3}H adenine in germinating *N. crassa* conidia (Mirkes, 1974), indicating a potentially rapid turnover of ribosomes. The incorporation of leucine into ribosomal protein has been studied during zoosporangium development in *Blastocladiella emersonii,* (Adelman & Lovett, 1972). Ribosomes were extracted from exponentially growing cells, stripped of nascent peptide by treatment with puromycin and washed in ammonium sulphate. The ribosomal proteins were then extracted and analysed by sodium lauryl sulphate (SLS)/polyacrylamide gel electrophoresis. The gels were sliced into 1 mm sections and counted for radioactivity. A constant pattern of incorporation was obtained throughout the period of exponential growth when *de novo* synthesis was still in progress. After zoospore induction however, when ribosome synthesis is

Table 8.2 Sedimentation values reported for ribosomes, ribosomal subunits and polyribosomes in some fungi

Species	Ribosome	Large subunit	Small subunit	Reference
Aspergillus niger	80	60	40	Moyer & Storck, 1964
Botryodiplodia theobromae	77	60	36	Van Etten, 1971
Blastocladiella emersonii	80	63	41	Lovett, 1963
Erysiphe graminis	80			Leary & Ellingboe, 1971
Neurospora crassa	77	60	39	Alberghina & Suskind, 1967
Neurospora crassa	80			Henney & Storck, 1963*b*
Neurospora crassa	78			Sargent, 1973
Schizophyllum commune	80	60	40	Leary, Morris & Ellingboe, 1969
Uromyces phaseoli	82	55	38	Staples, Bedigian & Williams, 1968

	Polysomes						
Species	1	2	3	4	5	6	Reference
Neurospora crassa	80	125	158	190	210	240	Henney & Storck, 1964
Neurospora crassa		115	149	176			Alberghina & Suskind, 1967
Uromyces phaseoli	82	121	146	189	204		Staples, Bedigian & Williams, 1968

known to be severely repressed, it was found that leucine was still being incorporated into ribosomal protein.

Further analysis of these results indicated that although no new species of ribosomal proteins were being synthesized during the differentiation of zoospores, the banding patterns obtained in differentiating cells differed quantitatively from that in exponentially growing cells.

In *Neurospora crassa*, no differences in the banding pattern on acrylamide gels between ribosomal protein derived from vegetative and conidial stages of development were observed (Rothschild, Hirkgura & Suskind, 1967). The protein content of *N. crassa* ribosomes was studied by Alberghina & Suskind (1967) using both biochemical and antigenic techniques. The amino acid content of the ribosomal protein fraction was determined. It was found to contain between 1 and 2 half-cystine molecules per molecule of protein, and the ratio of basic amino acids (arginine, lysine and histidine) to acidic amino acids (aspartate and glutamate) was found to be 0.84. Although a large number of bands were obtained by gel electrophoresis, only a few precipitin bands were resolved by immunodiffusion tests. When aggregation of subunits was prevented by the alkylation of -SH groups, a molecular weight of 20 000 was obtained for the ribosomal protein fraction. It is not obvious how a single subunit molecular weight or the limited antigenic diversity observed can be reconciled with the multiple bands obtained by gel electrophoresis and the evidence of multiple ribosomal proteins obtained in other organisms.

In an *in vivo* experiment using colonies of *Trichoderma viride*, it was demonstrated that labelled uridine incorporated into RNA in the actively growing peripheral zone of the colony did not move with the advancing hyphae when colonies were transferred to unlabelled substrates. It seems unlikely, in view of this data, that ribosomes migrate during growth (Stavy, Stavy & Galun, 1970).

Protein synthesis in vivo

Extensive studies in many organisms have shown that active protein synthesis is associated with polysome formation (Lengyel & Soll, 1969; Haselkorn & Rothman-Denes, 1973). Using an analytical ultracentrifuge, polysomes were demonstrated in growing hyphae and germinating conidia of *Neurospora crassa* by Henney & Storck (1964). They did not, however, find any polysomal aggregates in non-germinated conidia. In contrast, Staples, Bedigian & Williams (1968), found that uredospores of the bean rust fungus, *Uromyces phaseoli*, contained polysomes which decreased in number during spore germination. The sedimentation values for polysomes from a selected number of organisms are shown in Table 8.2. The high S values observed for what are apparently monomer peaks probably indicates a non-specific binding of monosomes and ribosomal subunits.

In *Uromyces phaseoli*, 57% of the ribosomal material was found to be aggregated into polysomes (Staples, Ramakrishnan & Yanif, 1970). However, such figures are notoriously unreliable, since breakdown during extraction can produce a rapid decrease in the values observed. A critical study of the conditions required for the extraction of polyribosomes from *Neurospora crassa* has recently been carried out by Sargent (1973). Preparations containing 72% of polysomal material have been obtained using liquid nitrogen to freeze the material during homogenization, and a high level (0.33 M) of potassium chloride. A direct correlation between the level of potassium chloride and the level of polysomes in the preparation was indicated. The stabilizing effect of potassium chloride can probably be attributed to its effect on ribonuclease, since ribonuclease activity in the extract decreased with increasing potassium chloride concentration. This effect is considered to be one of ionic strength rather than a specific effect of potassium chloride. The fragility of the polysomes in *Fusarium solani* may be attributable to the low level of potassium chloride in the extraction medium (Cochrane, Rado & Cochrane, 1971). Functional ribosomes and polysomes have also been isolated from lyophilized mycelia of *Schizophyllun commune* (Leary, Morris & Ellingboe, 1969). Monosomes, however, were predominant in this preparation so it is possible that some breakdown occurred during the extraction.

In *Uromyces phaseoli*, it was found that the majority of the protein synthesizing activity was associated with a microsome fraction and that the label could be released from this fraction by treatment with sodium deoxycholate. The relative amounts of ribosomal RNA material in the microsomal and soluble fractions were studied and compared with the value obtained for total RNA using a phenol extraction technique. The majority of the ribosomes were found to be membrane bound and only 37% could be extracted directly (Yanif & Staples, 1969).

A high level of membrane bound polyribosomes has also been reported in stationary phase mycelium of *Aspergillus niger* (Moyer & Storck, 1964). In *Neurospora crassa*, both free and membrane-bound ribosomes have been detected by electron microscopy and by isolation (Grinwich & Trevithick, 1973). The amount of membrane-bound rRNA was lower than that reported in *Uromyces phaseoli* (below 25%), and electron micrographs indicated that the ribosomes were less frequent and more irregularly spaced than in rough endoplasmic reticulum from plant and animal cells.

In germinating spores of *Peronospora tabacina*, protein synthesizing activity was associated with both the 20 000 **g** pellet and the 117 000 **g** pellet, *i.e.* microsomes and polysomes. The activities of both fractions increased during germination. However, whereas the increase in the 117 000 **g** pellet was inhibited when RNA synthesis was inhibited, the increase in the 20 000 **g** pellet was not. Thus the protein synthesis associated with the 20 000 **g** microsomal fraction does not appear to be dependent upon the synthesis of new messengers (Holloman, 1971).

The turnover of polyribosomes *in vivo* has been studied by Mirkes (1974) in an elegant series of experiments using germinating *Neurospora crassa* conidia. In this system, the polysome population was found to increase from 3% to 70% of the ribosomal material during the first hour after germination. The initial rise in polysome population was very rapid and if the spores were harvested by washing the colonies in water followed by filtration of the conidia, rather than harvesting them in the dry state, an initial value of 30% polysomes was obtained. Polysome formation at this stage is evidently very rapid.

When the conidia were fed with labelled leucine, label was incorporated into the polysome fractions but not into monosomes. As the percentage of polysomes increased between 15 and 60 min after the onset of germination, so the amount of label incorporated into the polysome fractions increased. The failure to find label in the monosomic fraction indicates that it is not a product of RNAase degradation during the extraction, since treatment of leucine-labelled polysomes with RNAase converts them to monosomal material which is labelled because the labelled nascent peptide is not released by the completion of translation but remains bound to the ribosomes. These results may indicate that the monosome is a natural component of the protein synthesizing system. However, it should be noted that the release of nascent peptide by treatment with puromycin results in the dissociation of the ribosomes into subunits. The functional relationship between monosomes and subunits in bacteria and eukaryotes has been shown to be complex (Lengyel & Soll, 1969); Haselkorn & Rothman-Denes, 1973). These results may indicate that this is also the case in fungi.

Holloman (1973) has obtained results in *Peronospora tabacina* which suggests that the large ribosomal subunit is more tightly bound to membranes than is the smaller subunit and that the small subunit is released from the membranes of freshly harvested conidia when the germination inhibitor is removed, by washing at 0°C. During germination, the smaller ribosomal subunit appears to reassociate with the membrane fraction as the level of protein synthesis increases. Treatment with cycloheximide prevents this reassociation. Holloman (1973) has presented a model of the initiation of protein synthesis, in which the small ribosomal subunit separates from the membrane-bound large ribosomal subunit complex before binding to mRNA to form an initiation complex. This initiation complex then reassociates with the large ribosomal subunit (Fig. 8.1).

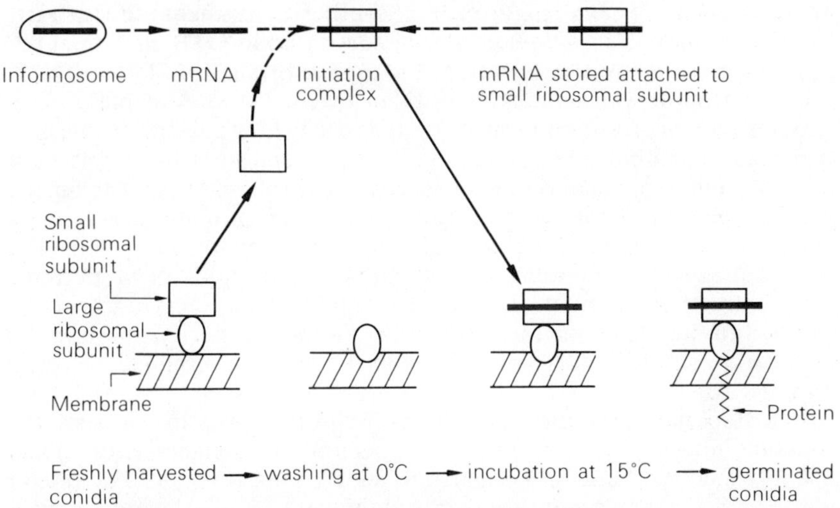

Fig. 8.1 Diagrammatic representation of the proposed changes in membrane-bound ribosomes during germination (from Holloman, 1973).

If the spores of *Neurospora crassa* are fed with ^{14}C uracil during germination, the polysomal material is rapidly labelled. Phenol extraction of the RNA followed by sucrose gradient centrifugation indicates that label is distributed between rRNA, tRNA and a polydisperse fraction of between 6 and 15S which probably represents mRNA. An increase in the amount of label incorporated into the polydisperse fraction during the first 30 min of spore germination may indicate a stimulation of mRNA synthesis at this stage, when the frequency of polysomes is also shown to be increasing (Mirkes, 1974).

In Vitro *Protein-Synthesizing Systems*

The characteristics required of an *in vitro* protein-synthesizing system were described for *Penicillium cyclopium* by Van Etten, Parisi & Ciferri (1966). The experiments involved the use of polyuridylic acid as a messenger, soluble RNA from yeast and a 105 000 **g** supernatant and ribosomes from *P. cyclopium.* Incorporation of labelled phenylalanine into nascent peptide was shown to be dependent upon each of the above components. In addition, an ATP-generating system consisting of ATP, phosphoenolpyruvate and pyruvate kinase was required. A dependence upon GTP was also demonstrated, but this did not appear to be absolute. It seems probable that some GTP was present in the soluble fraction. The incubation was inhibited by puromycin and RNAase but was only slightly inhibited by streptomycin and DNAase.

The incubation using extracts from *Penicillium chrysogenum* described by Hindle, Kornfield & Knight (1966) which was reported to be independent of exogenously supplied ATP, GTP and magnesium ions and to be insensitive to RNAase treatment but sensitive to streptomycin and chloramphenicol, appears to have been contaminated with bacteria.

In a critical study of an *in vitro* protein-synthesizing system using extracts from *Botryodiplodia theobromae*, a dependence upon GTP, and an ATP-generating system was demonstrated, and the importance of Mg^{2+} and NH_4^+ ion concentration was demonstrated (Van Etten, 1968). Maximal rate of phenylalanine incorporation was restricted to a narrow band of magnesium concentration around 20 mM. The optimal concentration for ammonium ions was found to be 80 mM (and not higher than 160 mM), ammonium ion however was replaceable by potassium ion. A pH optimum of 8.0 and a temperature optimum of 20°C were also demonstrated for this system. The activity of the system was found to be stimulated by the addition of spermine to the extraction medium. The level of endogenous messenger in the ribosomal preparation must have been low in these experiments since the level of incorporation of phenylalanine in the absence of added synthetic messenger was low.

When studying the optimal conditions for *in vitro* protein synthesis, the distinction between systems in which endogenous messenger is present and those in which synthetic messengers are added must be made. The role of polysomes as the active unit in *in vitro* protein synthesis was demonstrated in *Uromyces phaseoli* by Staples, Bedigian & Williams (1968).

In *Fusarium solani*, ribosomes from ungerminated and germinated conidia were found to be equally active in promoting the formation of poly-phenylalanine using poly-U as the messenger in an *in vitro* protein-synthesizing system, but the ribosomes from germinated conidia were found to be ten times more active at synthesizing peptide from endogenous mRNA (Rado & Cochrane, 1971). Germinated conidia were shown to have a higher polysome content (Cochrane, Rado & Cochrane, 1971).

In cell-free systems obtained from the peripheral and central zones of surface colonies of *Trichoderma viride*, the incorporation of ^{14}C leucine was three times as high in the outer zone than in the central zone when the system was dependent upon endogenous messenger, but only 10% higher when poly-U was supplied as an exogenous messenger (Stavy, Stavy & Galun, 1970). The use of endogenous messengers is probably a more accurate guide to the *in vivo* situation. However, this does not preclude the use of synthetic messengers in more specific situations.

It has been shown in *Neurospora crassa*, that ribosomes derived from vigorously growing mycelium (E) are ten times more active than ribosomes from stationary phase mycelium (S) in supporting *in vitro* protein synthesis, in the absence of a synthetic messenger. When poly-U was used as an exogenous messenger, ribosomes from E were most active at low concentrations of Mg^{2+} ions. However, at higher concentrations, ribosomes from S were most active (Alberghina, Sturani & Ursino, 1969). The level of polysomes in stationary phase mycelium is probably low; so *in vitro* protein synthesis using ribosomes from this phase will be dependent upon an exogenous messenger. A high level of Mg^{2+} ions has been shown to be essential for the non-specific initiation of translation of synthetic messengers in prokaryotes. Polysomes from *Schizophyllum commune* were shown to be stable at low concentrations of Mg^{2+} ions in an *in vitro* protein-synthesizing system using endogenous mRNA (Leary, Morris and Ellingboe, 1969).

The information available on the enzymology of translation in filamentous fungi is limited. Yanif & Staples (1969) obtained results which suggested that sodium deoxycholate (DOC) removed an essential factor for translation from preparations of ribosomes from *Uromyces phaseoli.* An aminoacyl tRNA binding enzyme was isolated from the same species (Yanif & Staples, 1971). In *Peronospora tabacina,* transfer factors required for peptide elongation were removed from a microsome fraction by washing with 1 M KCl and 1% Triton-X (Holloman, 1973). The activity of the microsomes could be restored by adding back this fraction. *In vitro* protein-synthesizing systems using these transfer factors and aminoacyl tRNAs as a source of amino acids were inhibited by cycloheximide, anisomycin, fusidic acid and sparsomycin. These inhibitors have been shown in other organisms to act at the level of peptide chain elongation (Pestka, 1971) so the results indicate that the enzymes involved in translation in fungi resemble those in other organisms.

Indirect evidence as to the nature of these enzymes involved in translation in filamentous fungi has been obtained by substituting ribosomal and supernatant fractions from *in vitro* protein-synthesizing systems from fungi with corresponding fractions from other organisms. The ribosomal and supernatant fractions of *Penicillium cyclopium* have been shown to be interchangeable with the homologous fractions prepared from castor bean, and *Saccharomyces cerevisiae* but not with those from *Escherichia coli* (Van Etten, Parisi & Ciferri, 1966). In addition, the cytoplasmic supernatant enzymes of *Neurospora crassa* cannot replace the mitochondrial fraction from the same organism or the homologous fraction in *E. coli* (Küntzel, 1969*a*). The filamentous fungi appear to have a typical eukaryotic cytoplasmic system of protein synthesis.

It is known from studies on other organisms that puromycin binds to nascent peptide, and releases it from ribosomes. The property has been used to estimate the rate of synthesis of nascent peptide in *Neurospora crassa.* After feeding ^3H-labelled puromycin, the number of nascent peptides can be related to the amount of label incorporated into 5% trichloroacetic acid precipitable material. Using such a technique, it has been demonstrated that the higher the growth rate, the higher the number of nascent peptides per unit weight of RNA (Alberghina, Sturani & Schiaffonati, 1971). This presumably indicates that a higher percentage of the ribosomal material is associated with polysomes. These workers also calculated the rate of protein synthesis *in vitro.* From observations of incorporation rates of 4500 pmoles phenylalanine per mg RNA in 30 min, a rate of protein synthesis of 1 molecule of phenylalanine per ribosome per min was obtained for *in vitro* experiments. These values compare with similar values obtained in mammals, but are two or three orders of magnitude lower than the rates observed *in vivo.*

INITIATION AND TERMINATION The most common N-terminal amino acids in the soluble proteins of *Neurospora crassa* are glycine, alanine and serine which account for 75% of the terminal residues. Glycine alone accounts for 35% of them, although it only constitutes 10% of the total amino acid composition. Unlike *Escherichia coli,* methionine does not frequently occur in the N-terminal position. Although absent in the terminal position in vegetative cells, phenylalanine has been found in this position in conidia (Rho & De Busk, 1971*a*). The dinitrophenyl method used in the above experiments, was also used to determine the N-terminal amino acids of nascent peptide. To obtain a high yield of nascent peptide, stationary phase mycelium was transferred from a spent medium to a fresh minimal medium, then after 15 min, labelled amino acids were added. The incubation was then terminated with cycloheximide after a further 5 min. Treatment with puromycin was also used as a method to obtain nascent peptide. In both preparations, methionine was the most common N-terminal amino acid. Similar results were obtained in *in vitro* systems (Rho & De Busk, 1971*b*). Analysis of methionyl tRNA on benzoylated DEAE cellulose (BDC) columns showed the presence of two peaks of acceptor activity. One of these two peaks, methionyl tRNAf, was formylable with a soluble fraction from *Escherichia coli* whilst the other, methionyl tRNAm, was not. Methionyl tRNAf was found to be bound to an AUG ribosome complex more efficiently than the methionine tRNAm, as would be expected if AUG were the initiator codon and if methionyl tRNAf the transfer RNA involved in the formation of the initiation complex. However, as in other eukaryotes no evidence that N-formyl methionine is produced in fungi, or involved in initiation has been presented (Lovett, 1974). Analysis of nascent peptide produced in an *in vitro* synthesis supplied with methionyl tRNAf and methionyl tRNAm gave 80% of the methionyl tRNAf in the terminal position and 20% internal compared with 29% terminal and 71% internal for methionyl tRNAm (Rho & De Busk, 1971*c*). The high level of both these acceptor activities makes it unlikely that either can be attributed to a mitochondrial contaminant.

Experiments which may be relevant to the *in vivo* mechanism of termination have been carried out by Mehadevan & Bhagwat (1969). They showed that 8-azaguanine inhibited protein synthesis by up to 50% in *Neurospora crassa*. This inhibitor appears to act by preventing the completion of polypeptide chains. It has been proposed that nonsense triplets are produced in the mRNA which cause incomplete translation (Bhagwat & Mehadevan, 1972).

BREAKDOWN OF POLYSOMES In addition to the unfortunate destruction of polysomes and ribosomes by RNAase during their extraction, it seems likely that rRNA, mRNA and hence polysomes, are also broken down naturally at varying rates. It has been reported that the ribosomes from vegetative hyphae of *Aspergillus oryzae* are much more resistant to autolysis at 43°C than are conidial ribosomes (Horikoshi, Ohtaka & Ikeda, 1969). They observed that the level of RNAase in the microsomal fraction of conidia was four times higher than the corresponding fraction from vegetative hyphae. The supernatant fraction was six times higher in RNAase activity (Horikoshi & Ikeda, 1969). In addition to this difference, it was also reported that an *in vitro* protein-synthesizing system containing vegetative ribosomes and poly-U was more resistant to the RNAase activity present in the supernatant than was a similar system containing conidial ribosomes.

Work on *Neurospora crassa* also indicates that ribosomes contain considerable RNAase activity. The ribosome preparation used was capable of giving a linear rate of amino acid incorporation for up to 1 h. So it appears that the RNAase activity of these ribosomes does not have an uncontrolled destructive effect on the protein-synthesizing system. The average RNAase activity of the ribosomal preparations as OD 260 units/mg ribosomal protein/h was found to be around 12. However, treatment with urea resulted in a 200% increase, and with NH_4^+ and K^+ at 0.5 M a 600% and 430% increase respectively. The effect of ammonium and potassium ions is somewhat surprising in view of the requirement for these in *in vitro* protein-synthesizing systems. The release of four times the original RNAase activity observed in the pellet by incubation in 4 M urea suggests that RNAase is intimately involved in the structure of the ribosome (Ursino, Sturani & Alberghina, 1969).

The activity of ribosome bound RNAase has been shown to vary with different stages of the life cycle of *Blastocladiella emersonii* (Adelman & Lovett, 1974*a*). However, no correlation was observed in these experiments between ribosome-bound RNAase activity and the protein-synthesizing activity of the ribosomes. However, the variation in the amino acid incorporating activity of the ribosomes was found to be correlated with the breakdown of rRNA during ribosome extraction which is presumably carried out by other nucleases normally physically separated from ribosomes (Adelman & Lovett, 1974*a*).

ENDOGENOUS INHIBITORS OF PROTEIN SYNTHESIS AND STABLE MESSENGERS Considerable interest has been shown in the control of translation, particularly with respect to the initiation of protein synthesis in germinating spores (Van Etten, 1969; Lovett, 1975). Protein synthesis is normally low or absent in spores and increases drastically upon germination. In several species, the onset of protein synthesis has been associated with hydration of the spores and the available evidence indicates that inhibitors present in the spore are removed upon hydration. Two germination inhibitors, methyl-3,4-dimethoxycinnamate and methylferulate have been isolated from uredospores of *Uromyces phaseoli* (Macko, Staples, Allen & Renwick, 1971). Removal of these inhibitors by washing, stimulated amino acid incorporation by endogenous messenger RNA (Yanif & Staples, 1969). An unidentified compound extracted from conidia of *Glomerella cingulata* was found to inhibit the incorporation of amino acids into germinating conidia (Lingappa, Lingappa & Bell, 1973). In *Peronospora tabacina*, a stimulation of the *in vitro* amino acid incorporating activity of conidial ribosomes was obtained by washing with water at 0°C (Holloman, 1971). Recently, Mirkes (1974) has shown that an increase in the frequency of polysomes from 3% to 30% occurred when conidia of *Neurospora crassa* were harvested in distilled water as opposed to a dry harvest.

The most intensively studied example of an inhibitor of protein synthesis associated with spore germination is that of *Blastocladiella emersonii*. The zoospores of *B. emersonii* contain only 80S monosomes which are restricted to a membrane-bound nuclear cap (Murphy & Lovett, 1966). These ribosomes are inactive in a protein-synthesizing system in which polyuridylic acid is supplied as a messenger, although ribosomes from the growth phase are fully active in such a system. If the ribosomes isolated from the nuclear cap are washed with 0.5 M KCl, then they become activated for polyphenylalanine synthesis in the *in vitro* system. The KCl wash contains a compound which inhibits the system if added back (Schmoyer & Lovett, 1969). The inhibitor has been found both tightly bound to the ribosomes and also free in the cytoplasmic supernatant. Its identity has not yet been established although it has been shown to be a low molecular weight compound, possibly a nucleotide (Adelman & Lovett, 1974*b*). The inhibitor appears to be produced at the same time as the induction of zoosporangia and zoospores. A self inhibitor of *Dictyostelium discoideum* spores, which blocks protein synthesis *in vivo*, has been shown to be N.N.dimethyl-guanosine (Bacon, Sussman & Paul, 1973).

Several early reports suggested that fungal spores lacked polysomes and hence probably also mRNA (Van Etten, 1969). However, Staples, Bedigian & Williams (1968) found that uredospores of *Uromyces phaseoli* did contain polysomes. Polysomes have also been shown to be present in the spores of *Fusarium solani* (Rado & Cochrane, 1971) and *Puccinia graminis* (Dunkle, Maheshwari & Allen, 1969).

Holloman (1969) reported in *Peronospora tabacina* that whereas cycloheximide inhibited protein synthesis during spore germination, two inhibitors of RNA synthesis, ethidium bromide and 5-fluorouracil (each of which was shown to inhibit RNA synthesis by at least 98%) did not affect germination or protein synthesis *in vivo*. It was concluded that the initial phase of protein synthesis after germination utilized mRNA which was endogenous within the spores.

The existence of mRNA in spores which contain polysomes is to be expected. However, experiments which indicate the presence of messenger RNA in spores which do not contain polysomes are more surprising and perhaps more exciting.

Blastocladiella emersonii zoospores contain only monosomes (Lovett, 1963), but it has been reported that the initial phase of protein synthesis in zoospore germination is unaffected by actinomycin D at levels which have been shown to inhibit the incorporation of uracil into RNA (Soll & Sonneborn, 1971; Lovett, 1968). Lovett (1975) has recently studied the level of poly-A and poly-A associated RNA in zoospores of *Blastocladiella emersonii*. The reasons for believing that such a fraction represents mRNA were discussed above (p. 245). Approximately 2.5% of total RNA was found to be associated with poly-A, and of this, 80% was localized in the nuclear cap. This fraction however did not appear to be bound to ribosomes.

The rapid increase in polysomes when spores of *Neurospora crassa* are suspended in distilled water for 10 min may be attributable to the attachment of ribosomes to existing mRNA (Mirkes, 1974).

8.3 Mitochondrial Systems

Introduction

The mitochondria of filamentous fungi, like those of other organisms, are semi-autonomous units within the cytoplasm, whose growth and reproduction depend not only upon the nucleocytoplasmic system but also upon their own mitochondrial system. There is an extensive literature on mitochondrial biogenesis, and there are excellent general reviews available (Borst & Kroon, 1969; Rabinowitz & Swift, 1970; Borst & Grivell, 1971; Borst, 1972; Schatz & Mason, 1974).

Studies on this system in the filamentous fungi are in the forefront of the field, and have made a considerable contribution to an understanding of the nature and role of mitochondrial genetic systems in all organisms.

Early indications of the semi-autonomy of mitochondria in the filamentous fungi were given by Luck (1963), who showed that a pulse of labelled choline became randomly distributed amongst the lipid fractions of all mitochondria in *Neurospora*, after growth on unlabelled medium. This suggested that the mitochondria had increased in number by division of pre-existing structures.

Mitochondrial DNA and Replication

CHARACTERIZATION OF MITOCHONDRIAL DNA Presumptive evidence that mitochondria of filamentous fungi contained their own genetic material in order to direct at least part of the process of mitochondrial biogenesis was available as early as 1952, when Mitchell & Mitchell showed that the *poky* mutants of *Neurospora*, which displayed gross mitochondrial disfunction, were extrachromosomally inherited. This concept was more generally accepted, when Luck & Reich (1964) demonstrated, using buoyant density determinations on DNA from *Neurospora*, that mitochondria contained a unique species of DNA, which differed in base content from nuclear DNA.

Table 8.3 Nucleotide composition of mitochondrial DNA of fungi (from Villa & Storck, 1968). Nuclear DNA values are also given for comparison

Species	G + C content (moles %)	
	Mt-DNA	**nuclear-DNA**
Zygomycetes		
Cunninghamella echinulata	34	34
Mucor fragilis	38	39
M. rouxii	38	37
Ascomycetes		
Ceratocystis ulmi	40	56
Gelasinospora autosteria	41	54
G. calospora	41	55
G. cerealis	42	55
G. tetrasperma	44	—
Neurospora crassa	42	54
N. sitophila	43	55
Sordaria macrospora	42	54
Chaetomium globosum	34	58
Basidiomycetes		
Daedalia confragosa	31	57
Schizophyllum commune	28	61

Subsequently, several of the filamentous fungi have been shown to contain their own mitochondrial DNA, using CsCl buoyant density determinations (see Table 8.3).

From such studies, some generalizations can be made:

1. There is no relationship between the base content of mitochondrial DNA (Mt-DNA) and the homologous nuclear DNA.

2. Mt-DNAs of closely related species tend to have similar base contents.
3. All Mt-DNA so far examined has been double-stranded and does not seem to be associated with any histone-like proteins.
4. Mt-DNA of most organisms is homogeneous in base content. However, *Neurospora crassa*, with two distinct species, and *Neurospora sitophila* with three species of different buoyant density, are striking exceptions to this generalization (Reich & Luck, 1966).

Most of the work on the size and structure of Mt-DNA in filamentous fungi has been carried out in *Neurospora crassa*. Early studies obtained heterogeneous populations of linear molecules up to 25 μm in length. Although Schäfer, Bugge, Grandi & Küntzel (1971) obtained a homogeneous population of linear molecules, 25 μm in length, more recent studies by Clayton & Brambl (1972), and by Agsteribbe, Kroon & Van Bruggen (1972) have separately demonstrated a preponderance of 20 μm circles. Clark-Walker & Gleason (1973) have also obtained 14 μm circles from a total DNA extract of the water mould *Saprolegnia*, which they suggest is mitochondrial in origin.

Although, in the case of most organisms, the molecular weight of Mt-DNA calculated from renaturation kinetics, and that calculated from length measurements in the electron microscope agree closely, thus indicating that Mt-DNA molecules are neither heterogeneous, nor contain any repeated sequences, *Neurospora crassa* seems to be the exception again. From studies of renaturation kinetics in this organism, Wood & Luck (1969) believe that there is some indication of repeated sequences in *Neurospora* Mt-DNA. The presence of repeated sequences may possibly cause the heterogeneous buoyant densities so far found also only in *Neurospora* spp.

REPLICATION OF MITOCHONDRIAL DNA Observations on Mt-DNA replication are fairly sparse in filamentous fungi. Reich & Luck (1966) have carried out a Meselson-Stahl experiment, transferring mycelium of *Neurospora crassa* grown on ^{15}N medium, to ^{14}N medium and examining the distribution of heavy nitrogen in nuclear and Mt-DNA after several replication cycles. The authors concluded that both species of DNA showed a semi-conservative type of replication, although the results in mitochondria were complicated by the large ^{15}N pool size.

As for the timing of Mt-DNA synthesis, Hawley & Wagner (1967) have shown that mitochondria divide synchronously in one area of mycelium. This may indicate a concomitant synchrony of Mt-DNA synthesis. Mt-DNA synthesis has been shown to occur approximately 3 h after germination of *Botryodiplodia theobromae* spores (Dunkle, Van Etten & Brambl, 1972), and shortly after germination of uredospores of *Uromyces phaseoli* (Staples, 1974).

MITOCHONDRIAL DNA MUTATIONS The extrachromosomally inherited '*poky*' mutations in *Neurospora crassa* have been extensively investigated. *Poky* phenotypes display slow growth, deficiency of cytochromes a, a_3 and b, excessive production of cytoplasmic cytochrome c and a reduced capacity for mitochondrial protein synthesis (see below). No gross change in Mt-DNA can be detected (Kuriyama & Luck, 1974). Woodward & Munkres (1966) claimed originally that the *poky mi-1* mutant was a point mutation affecting mitochondrial 'structural protein'. This is, however, now considered to be untenable on several grounds (Borst, 1972), and has more recently been shown to be a result of a deficiency of small ribosomal subunits (Rifkin & Luck, 1971; see below for further discussion).

Rowlands & Turner (1973) have recently isolated a mitochondrially inherited oligomycin-resistant mutant in *Aspergillus nidulans*. This is of interest, in view of the extensive volume of research on extrachromosomally inherited antibiotic resistance mutants in yeast (see several articles in Kroon & Saccone, 1974).

Mitochondrial RNA and Transcription

The incorporation of labelled nucleoside triphosphate into high molecular weight RNA within mitochondria has been shown to occur in many species, *e.g.* in *Neurospora* (Reich & Luck, 1966), and *Agaricus campestris* (Vogel & Kemper, 1967). However, much of the investigations on mitochondrial RNAs (Mt-RNA) have been primarily concerned with the transcriptional origin of such RNAs, since the presence of an RNA species in mitochondria does not *a priori* indicate that it has been synthesised there. Two main methods have been used to indicate the transcriptional origin of a Mt-RNA species from Mt-DNA; either hybridisation to Mt-DNA, or *in vitro* incorporation experiments, using labelled nucleoside triphosphates.

RIBOSOMAL RNA It can be seen from Table 8.4 that, in general, the sedimentation coefficients and the base contents are quite different in mitochondrial and cytoplasmic rRNAs, a generalization which is also true in other organisms. The anomalous data in *Agaricus* may well be caused by contamination of mitochondrial rRNA with cytoplasmic rRNA. Low G + C content is also characteristic of Mt-ribosomal RNA. Edelman, Verma & Littauer (1970) showed from thermal denaturation experiments in *Aspergillus*, that the G + C content of the ordered regions in the Mt-ribosomal RNA chains were also low, being 27 and 32% for heavy and light Mt-ribosomal

Table 8.4 Comparison of size and base content of ribosomal RNA in filamentous fungi. The cytoplasmic data are given in brackets for comparison

Species	Sedimentation coefficient of: large subunit	small subunit	Base content (moles %)	Reference
Neurospora crassa	23 S (26 S)	16 S (17 S)	37	Küntzel & Noll, 1967
	or 25 S (28 S)	19 S (18 S)		Lizardi & Luck, 1971
Aspergillus nidulans	23.5 S (26.5 S)	15.5 S (17.0 S)	32 (51)	Edelman, Verma & Littauer, 1970
			or 31 (53)	Verma *et al.*, 1971
Ustilago maydis	23 S (25 S)	16 S (18 S)		Lin *et al.*, 1971
Agaricus campestris	28 S (28 S)	18 S (18 S)	52.3 (51.7)	Pollard, Stemler & Blaydes, 1966

RNA components, and 55.1 and 51% for cytoplasmic-ribosomal RNA components. This low G + C content, leading to an unusual secondary structure, may explain the lack of agreement between the size of Mt-ribosomal RNA molecules estimated from sedimentation values, and that estimated from polyacrylamide gel electrophoresis or length measurements in the electron microscope. Estimates from these latter techniques give a molecular weight for fungal Mt-ribosomal RNA lying somewhere between that for *Escherichia coli* and fungal cytoplasmic-ribosomal RNA (Verma, Edelman, Herzberg & Littauer, 1970; Borst & Grivell, 1971).

Kuriyama & Luck (1973) have investigated the path of synthesis of Mt-ribosomal RNA in *Neurospora crassa.* Using pulse chase and hybridization competition techniques, they have detected a short lived 32 S Mt-ribosomal RNA together with 2 molecules, each slightly larger than the mature rRNA molecules and have constructed the following hypothetical sequence:

		$(1.6 \times 10^6$ D)		$(1.28 \times 10^6$ D)
32 S precursor	→	25 S precursor	→	25 S
$(2.4 \times 10^6$ D)	→	19 S precursor	→	19 S
		$(0.9 \times 10^6$ D)		$(0.72 \times 10^6$ D)

Approximately 22% of the precursor molecule is lost during maturation.

Kuriyama & Luck (1974) have also studied the methylation during maturation of Mt-ribosomal RNA in *Neurospora.* By studying incorporation of labelled methionine into methionine-requiring strains, they have shown that the 32 S precursor molecule is the major site for methylation, and that the amount of methylation in mature Mt-ribosomal RNA is 1.4 methyl groups/100 nucleotides. Because the *poky* phenotype is characterized by a deficiency in small Mt-ribosomal subunits (Rifkin & Luck, 1971), they also investigated methylation during maturation of rRNA in *poky* mitochondria. They showed that these rRNA molecules were under-methylated, being 70% and 55% of control values for the large and small molecules respectively, and that most of the immature 19 S rRNA is degraded before incorporation into the small ribosome subunit, possibly as a consequence of this.

Although both bacterial and cytoplasmic eukaryotic ribosomes contain a 5 S rRNA as part of the large subunit, such an RNA species seems to be absent in mitochondrial ribosomes. Careful experiments by Lizardi & Luck (1971) did not detect any such species in mitochondria of *Neurospora crassa.* It is possible either that 5 S Mt-RNA is slightly smaller and therefore has a similar mobility to the tRNAs on acrylamide gels, or that the mitochondrial equivalent is left attached to the large rRNA molecule (from which, in the bacterial precursor molecule, it is cleaved).

The similarity of fungal Mt-DNA base composition and that of Mt-ribosomal RNA suggested that the latter might be a mitochondrial gene product. Wood & Luck (1969) carried out hybridization experiments, binding mitochondrial and cytoplasmic rRNAs from *Neurospora* to nuclear and Mt-DNA immobilized on nitrocellulose filters. They found that Mt-ribosomal RNA bound to Mt-DNA, and that cytoplasmic rRNA bound to cytoplasmic DNA. They obtained a figure of 9% hybridization of Mt-ribosomal RNA to total Mt-DNA. This value has since been shown by Schäfer & Küntzel (1972) to be too high, and these workers obtained a figure of 2.5% hybridization, giving an estimate of 1 gene for each rRNA molecule per mitochondrial genome.

TRANSFER RNA *Neurospora* mitochondria have been shown to contain a full complement of tRNAs (Barnett & Brown, 1967). At least 15 of these species do not occur in the cytoplasmic system as shown by their behaviour in counter-current distribution and reversed phase chromatography, and by their acylation characteristics (Epler, 1969). One at least (leu-tRNA) was shown to have different anticodon recognition properties from its cytoplasmic counterpart (Epler & Barnett, 1967).

In addition, mitochondria of *Neurospora* also contain a full complement of aminoacyl-tRNA synthetases, at least 3 of which are found exclusively in the mitochondria (Barnett, Brown & Epler, 1967).

A striking indication of the uniqueness of the mitochondrial system is the presence of N-formyl methionyl-tRNA (fMet-tRNA) in *Neurospora* (Küntzel & Sala, 1969). Mitochondrial extracts also contain a formylase which reacts only with mitochondrial fMet-tRNA (Epler, Shugart & Barnett, 1970).

An alteration in the behaviour of several Mt-transfer RNA species on benzoylated DEAE-cellulose columns was found by Brambl & Woodward (1972) in the *poky mi-1* mutant of *Neurospora.* They attributed this to an under-methylation of many transfer RNA species in the mutant, a process which probably takes place in the mitochondrion. This is interesting in view of the under-methylation of rRNA also shown to occur in this mutant (see above).

There has been very little work carried out on the origin of the Mt-transfer RNAs in the filamentous fungi. However, it is very likely, by analogy with other systems, that most, if not all the Mt-transfer RNAs are transcribed from Mt-DNA, especially in view of Blossey & Küntzel's (1972) undocumented claim that the mitochondrial genome of *Neurospora crassa* contains 40 genes for tRNA.

MESSENGER RNA There is very little information on the characteristics of mitochondrial mRNA. Studies on hybridization of Mt-RNA to Mt-DNA in *Neurospora*, have indicated that in addition to rRNA, at least 10% (Schäfer & Küntzel, 1972) to 20% (Blossey & Kuntzel, 1972) of the genome is saturated by minor RNA species, which are probably mRNA. In addition, *in vitro* experiments on isolated mitochondria indicate that the transcription products of Mt-DNA contain mRNA that may be subsequently translated into protein (see below).

RNA POLYMERASE AND TRANSCRIPTION Schäfer *et al.* (1971) showed that Mt-DNA from *Neurospora* could be transcribed by *Escherichia coli* RNA polymerase, and that sigma factor was required to give rifampicin resistance to the system. Thus Mt-DNA contains bacterial types of promoter regions. The presence of an RNA polymerase within *Neurospora* mitochondria was indicated by the actinomycin D-inhibited ability of mitochondria to incorporate nucleoside triphosphates into high molecular weight RNA (Reich & Luck, 1966).

Using DEAE cellulose chromatography, Tellez de Iñon, Leoni & Torres (1974), and Horgen & Griffin (1971*b*) have purified multiple forms of RNA polymerase from *Neurospora crassa*, and from *Blastocladiella emersonii* respectively, one of which was mitochondrial in origin, in each case. Küntzel & Schäfer (1971) have also characterized the Mt-RNA polymerase from *Neurospora*, demonstrating it to be α-amanitin resistant and rifampicin sensitive, and to consist of a single polypeptide chain of molecular weight 64 000, thus indicating a resemblance with T7 RNA polymerase in its simplicity of structure. It has a strong preference for native (*Neurospora* Mt-DNA) template.

IMPORT AND EXPORT OF MITOCHONDRIAL RNA It has been shown, in filamentous fungi, that Mt-ribosomal RNA is transcribed from Mt-DNA, and that cytoplasmic rRNA is transcribed from nuclear DNA. This is possibly also true for transfer RNA species. Thus there is probably no import or export of stable RNA species in mitochondria. The situation is not so clear in the case of mRNA. However, in view of Küntzel & Blossey's (1974) recent demonstration in *Neurospora*, that the proteins produced *in vivo* using endogeneous mRNA and those produced *in vitro* using Mt-DNA, *Escherichia coli* RNA polymerase and an *E. coli* translation system were identical, it would seem likely that mRNA is also not imported or exported into or from mitochondria.

Mitochondrial Protein Synthesis

RIBOSOMES The ability of a ribosome fraction extracted from mitochondria of *Neurospora crassa* to incorporate labelled phenylalanine in poly-U-directed poly-Phe synthesis was demonstrated by Küntzel (1969*a*), although the activity obtained was of the order of 300-fold less than that of *Escherichia coli* ribosomes (Borst & Grivell, 1971). Before this, appropriate staining techniques had demonstrated the presence of ribosome-like particles in *Neurospora* mitochondria under the electron microscope (Luck, 1964).

Ribosomes with a reported sedimentation coefficient of 73 S containing subunits of approximately 50–52 S and 37–39 S have been extracted from *Neurospora* by several workers (Küntzel & Noll, 1967; Borst & Grivell, 1971), and although the equivalent S values for *Aspergillus* were reported as 67, 50 and 32 S (Edelman, Verma & Littauer, 1970), these figures are certainly too low, and *Aspergillus* ribosomes are probably identical with those of Neurospora (Borst & Grivell, 1971).

Until recently, therefore, a figure of about 73 S was accepted as being the approximate sedimentation coefficient for fungal mitochondrial ribosomes. However, Datema, Agsteribbe & Kroon (1974) have recently thrown doubt upon this assumption by isolating an 80 S ribosome from *Neurospora* mitochondria. These 80 S ribosomes were only found when extraction was carried out in the presence of Mg^{2+} and heparin as an RNAase inhibitor. Omission of one or other of the above, ageing of the preparation, or incubation with a post-ribosomal mitochondrial supernatant gave 73 S ribosomes. Since mobility in acrylamide gels was identical for both 73 S and 80 S ribosomes, and the subunits from each had identical S values (52 and 39 S), it was suggested that the native 80 S mitochondrial ribosome underwent a conformational change under certain conditions, producing a 73 S ribosome.

Elegant work by Lizardi & Luck (1972), using polyacrylamide gel electrophoresis and isoelectric focussing showed that each of the 53 ribosomal proteins found on mitochondrial ribosomes of *Neurospora* were different from those on the corresponding cytoplasmic ribosomes. This was confirmed by Hallermayer & Neupert (1974) using immunological techniques.

Küntzel & Noll (1967), and Küntzel (1969*b*) have demonstrated the existence of fractions in *Neurospora* mitochondria with the characteristics of polysomes, from sucrose gradients at S values of 73 S, 103 S, 134 S, 160 S, and 186 S, from the monomer to the pentamer. However, these polysomes are uncharacteristically resistant to RNAase and Michel & Neupert (1973) have suggested that they are, in fact, aggregates of monosomes caused by hydrophobic interaction of their nascent polypeptides. Such a phenomenon has been shown to occur in HeLa cells (Ojala & Attardi, 1972).

Mitochondrial ribosomes in *Neurospora* seem to be located near or perhaps even attached to the inner membrane of the mitochondrion (Michel & Neupert, 1973). It has been suggested (Forrester, Nagley & Linnane, 1970) that the phenomenon is of general occurrence.

INITIATION OF PROTEIN SYNTHESIS *Neurospora* mitochondria contain fMet-tRNA, fMet-tRNA synthetase and transformylase activity (see above). Sala & Kuntzel (1970) have shown that mitochondrial ribosomes from *Neurospora* can bind fMet-tRNA and synthesize fMet-puromycin in presence of 5 mM Mg^{2+}. This reaction is stimulated by the presence of the triplet ApUpG, and GTP. After washing ribosomes with 1 M NH_4Cl, a treatment which removes initiation factors from bacterial ribosomes, this binding capacity is lost, but it can be restored by replacement with bacterial initiation factors. Cytoplasmic ribosomes are inactive in these reactions. Mitochondrial ribosomes thus resemble those of bacteria in the initiation of protein synthesis, and probably contain bacterial-type initiation factors.

PEPTIDE CHAIN ELONGATION Küntzel (1969*a*) showed that cell-free systems from *Neurospora* cytoplasm and mitochondria could not cooperate in poly-U-dependent chain elongation, although those from Neurospora mitochondria and *Escherichia coli* could. He suggested that this was due to the incompatibility of the chain elongation factors in the two *Neurospora* systems. Grandi & Küntzel (1970) showed that this was, in fact, the case. They isolated two complementary peptide chain elongation factors, G and T, from a 100 000 **g** mitochondrial supernatant of *Neurospora*. Both these factors were 'bacterial' in type, as they could be replaced by the bacterial G and T factors in a cell-free system. Mitochondrial T factors seem to be more labile than those of bacteria.

SITE OF SYNTHESIS OF MITOCHONDRIAL PROTEINS Several techniques have been used in order to ascertain which mitochondrial proteins are translated on cytoplasmic ribosomes, and which on mitochondrial ribosomes. These are:

1. *Pulse labelling with amino acids in the presence of specific inhibitors.* Mitochondrial and cytoplasmic systems are affected by different inhibitors.

Mitochondrial protein synthesis is inhibited by bacterial inhibitors *i.e.* by chloramphenicol and also by tetracycline, lincomycin, erythromycin, and neomycin. Cytoplasmic protein synthesis is inhibited by compounds which do not affect bacterial systems *e.g.* cycloheximide. Emetin and anisomycin have also been used (Borst & Grivell 1971). (The action of fusidic acid here is odd, in that it inhibits both cytoplasmic and bacterial systems but does not affect mitochondrial systems in *Neurospora* (Grandi, Helms & Küntzel, 1971).)

2. In vitro *synthesis in isolated mitochondria.* This is a method only satisfactory for investigating mitochondrially synthesized proteins (Ibrahim, Burke & Beattie, 1973).

3. *The effect of mitochondrial gene mutation upon mitochondrial proteins.* Results from each of these methods need cautious interpretation, as each method has its own pitfalls.

Using method 1, Hawley & Greenawalt (1970) have given a general estimate of 15% of proteins of the mitochondrion synthesized on mitochondrial ribosomes. There is a voluminous amount of literature pertaining to the identity of this elusive 15% of proteins, and consequently also to the 85% of mitochondrial proteins which are synthesized on cytoplasmic ribosomes. Table 8.5 indicates the mitochondrial proteins which are definitely not synthesized on mitochondrial ribosomes and are therefore imported into the mitochondrion.

Table 8.5 Mitochondrial proteins not synthesized on mitochondrial ribosomes of filamentous fungi

Species	Proteins	Method used (see text)	Reference
Neurospora crassa	of outer membrane	1, 2	Neupert & Ludwig, 1971
Pythium ultimum	cytochrome c	CAP inhibition *in vivo*	Marchant & Smith, 1968
Neurospora crassa	cytochrome c	1, 2	Schatz & Mason, 1974
N. crassa	malate dehydrogenase	3	Wagner, 1969
N. crassa	Mt-DNA polymerase	2	Schatz & Mason, 1974
N. crassa	Mt-RNA polymerase	1	Barath & Küntzel, 1972
N. crassa	leu-tRNA synthetase	3	Schatz & Mason, 1974
N. crassa	ribosomal proteins	1, 2	Lizardi & Luck, 1972
N. crassa	cytochrome oxidase—4 smaller subunits	1	Sebald, Weiss & Jackl, 1972

N.B. To complete the list, from studies of petites in yeast (method 3), it has been shown that citric acid cycle enzymes, ferrochelatase, cytochrome b_2, mitochondrial polypeptide elongation factors, some subunits of ATPase and others, are synthesized on cytoplasmic ribosomes (Schatz & Mason, 1974).

The story in the case of the mitochondrially synthesized mitochondrial proteins is not so straightforward, however, owing to the fact that most, if not all, of these proteins are hydrophobic and firmly bound to the inner membrane. They are therefore difficult to extract and characterise (Borst, 1972). Results from different groups, hence, vary quite widely. Table 8.6 lists the molecular weight and number of proteins identified by different

Table 8.6 Mitochondrial proteins synthesized on mitochondrial ribosomes of filamentous fungi

Species	Protein species: No. of products	Protein species: Mol. Wt × 10^3	Protein species: Function	Method used (see text)	Reference
Neurospora crassa	7+	33, 33, 28, 25 21, 17, 11 <2, >200 ?	—	1.	Swank, Sheir & Munkres, 1971
Aspergillus nidulans	4	40, 27, 18, 13	—	1.	Turner, 1973
Neurospora crassa	±5	>300, 180, 58, 51, 11	—	2; MtDNA +*E. coli* polymerase	Blossey & Küntzel, 1972
N. crassa	Many	15–250	—	2; endogenous mRNA *and* 1.	Blossey & Küntzel, 1972
N. crassa	1	—	—	1	Birkmayer, 1971
N. crassa	Many	<10, 11	—	1	Michel & Neupert, 1973
N. crassa	Many	10, 12	—	2 (endogenous mRNA) 2 (Mt DNA + *E. coli* polymerase)	Küntzel & Blossey, 1974
N. crassa	4	41, 29, 21, 10	3 heavy subunits of cytochrome oxidase	1	Lansman, Rowe & Woodward, 1974
N. crassa	3	41, 29, 21	3 heavy subunits of cytochrome oxidase	1	Sebald, Machleidt & Otto, 1973
N. crassa	1	30	cytochrome b apoprotein	1	Weiss & Ziganke, 1974
So far found only in yeast	4	—	ATPase 4/10 mitochondrially coded	—	Schatz & Mason 1974

groups as being synthesized on mitochondrial ribosomes, together with any identification as to the function of such proteins. Many groups obtaining low molecular weight products suggest that higher molecular weight proteins are aggregates (Küntzel & Blossey, 1974; Schatz & Mason, 1974). On the other hand, groups obtaining higher molecular weight products suggest that the low molecular weight polypeptides are the result of degradation (Lansman, Rowe & Woodward, 1974).

From Table 8.6 the proteins which have an identified function are therefore cytochrome b, cytochrome oxidase and oligomycin sensitive ATPase. Two of these are multisubunit enzymes, of which some subunits are synthesized in the mitochondria and some in the cytoplasm. It seems that the 'mitochondrial' subunits form the hydrophobic 'binding' part of the molecule, while the 'cytoplasmic' subunits form the hydrophilic 'enzymic' part of the molecule. Rowe, Lansman & Woodward (1974) even claim that these 3 or 4 hydrophobic subunits are 'common' to many of the mitochondrial enzymes, and are, in fact basic elements of the inner membrane. Molecular weight studies on the relevant subunits do not, however, entirely bear out this neat hypothesis.

Other possible candidates for synthesis on mitochondrial ribosomes are:

1. *A protein directly or indirectly concerned with methylating activity.* It has been shown above that *poky mi-1* mutants in *Neurospora* display undermethylated Mt-ribosomal and transfer RNA. In addition, they possess undermethylated cytochrome c (Kuriyama & Luck, 1974). In normal strains, this methylation takes place on attachment to the inner mitochondrial membrane. Possibly either the methylating enzyme itself, or a part of the inner membrane protein which binds the (cytoplasmically synthesized) methylating enzyme, is synthesized by the mitochondrial system.

2. *Since the mitochondrial and cytoplasmic systems are so closely coordinated, both systems probably synthesize regulator proteins.* Barath & Küntzel (1972) have shown that when the mitochondrial system in *Neurospora* is inhibited by chloramphenicol, then the cytoplasmically synthesized Mt-RNA polymerase accumulates in the cytoplasm to such an extent that it crystallizes. This has also been shown to occur in the case of cytochrome c (Borst, 1972). The authors suggest that this is caused by the inhibition of a mitochondrial regulator protein, which normally controls the synthesis of Mt–RNA polymerase. They also postulate the presence of a regulator protein coded for by the nuclear genome, which acts on mitochondrial transcription.

3. *The demonstrated requirement for a functional mitochondrial protein-synthesizing system in order to continue mitochondrial replication (Borst, 1972) may possibly be due to a mitochondrially synthesized protein of the inner membrane which binds the mitochondrial RNA polymerase.* In this event, the polymerase would become non-functional, when not bound. Alternatively, it may be due to the lack of synthesis of a regulator protein on the Barath and Küntzel model.

The products so far known to be transcribed from Mt-DNA are summarized in Table 8.7. It can be seen that these products account for 6.84×10^6 Daltons of the Mt-DNA. As the average molecular weight of Mt-DNA in filamentous fungi is $3–5 \times 10^7$ Daltons, a lot of information on Mt-DNA remains to be accounted for.

Table 8.7 Products coded for by mitochondrial DNA (Schatz & Mason, 1974)

Product	Molecular weight	ds DNA required (Mol. weight)
Large rRNA	0.56×10^6	1.12×10^6
Small rRNA	0.33×10^6	0.66×10^6
± 20 tRNA spp.	0.56×10^6	1.12×10^6
Cytochrome oxidase subunits	4×10^4	0.8×10^6
	3×10^4	0.6×10^6
	2×10^4	0.4×10^6
ATPase subunits	3×10^4	0.6×10^6
	2×10^4	0.4×10^6
	2×10^4	0.4×10^6
	7×10^3	0.14×10^6
Cytochrome b subunit	3×10^4	0.6×10^6
	Total	6.84×10^6

8.4 Viral Systems

Introduction

A discussion of fungal viruses is pertinent to this review, not only because the production of progeny virus particles involves nucleic acid and protein synthesis but also because the presence of virus might be expected to affect the metabolism of host protein and nucleic acid to a greater or lesser extent.

The presence of a virus in fungi was first suggested by Sinden & Hauser (1957), who postulated a virus as the etiological agent for dieback disease in *Agaricus bisporus.* This was subsequently confirmed by Hollings (1962). Viruses were independently discovered in the *Penicillium* group, in the search for the identity of the antiviral substances 'statolon', and 'helenine', present in mycelial extracts from *Penicillium stoloniferum* and *Penicillium funiculosum* (Ellis & Kleinschmidt, 1967; Lampson, Tytell, Field, Nemes & Hillemann, 1967). Their occurrence has subsequently been reported in more than 60 species from 50 genera of fungi (Lemke & Nash, 1974; see Table 8.8).

Evidence for the presence of viruses listed in Table 8.8 may consist simply of the presence of virus-like particles in the electron microscope, often combined with pathological symptoms in the host fungus. As these conditions cannot be said to answer Koch's postulates, many authors have, in fact, avoided the use of the term 'virus'. Koch's postulates have only been fulfilled in the case of the *Penicillium chrysogenum* virus Pc-V (Lemke, Nash & Pieper, 1973; Lemke & Nash, 1974), and the *Agaricus bisporus* viruses (Hollings & Stone, 1971).

Characterization of Fungal Viruses

Appropriate data for viruses which have been well characterized is included in Table 8.9. The *Agaricus bisporus* viruses are included in here, despite the lack of data, because of their economic importance. In addition, the three

Table 8.8 Reported occurrence of viruses and virus-like particles in fungi (Lemke & Nash, 1974)

Basidiomycetes	Fungi Imperfecti	*P. claviforme*
Agaricus bisporus	*Alternaria tenuis*	*P. cyaneofulvum*
A. campestris	*Arthrobotrys* sp.	*P. funiculosum*
Boletus sp.	*Aspergillus flavus*	*P. multicolor*
Coprinus lagopus	*A.foetidus*	*P. stoloniferum*
Hypholoma sp.	*A. glaucus*	*P. variabile*
Laccaria laccata	*A. niger*	*P. notatum*
Lentinus edodes	*Botrytis* sp.	*P. citrinum*
Polyporus sp.	*Candida tropicalis*	*Periconia circinata*
Puccinia graminis	*C. utilis*	*Piricularia oryzae*
Schizophyllum commune	*Cephalosporium chrysogenum*	*Rhodotorula glutinis*
Thanatephorus cucumeris	*Chromelosporium* sp.	*Sclerotium cepivorum*
Tilletiopsis sp.	*Chrysosporium* sp.	*Scopulariopsis* sp.
Ustilago maydis	*Colletotrichum lindemuthianum*	*Spicaria* sp.
Ascomycetes	*Fusarium moniliforme*	*Stemphylium botryosum*
Daldinia sp.	*Gliocladium* sp.	*Trichothecium* sp.
Diplocarpon rosae	*Gliomastic* sp.	*Verticillium* sp.
Hypoxylon sp.	*Gonatobotrys* sp.	Phycomycetes
Neurospora crassa	*Helminthosporium maydis*	*Aphelidium* sp.
Gaeumannomyces graminis	*H. oryzae*	*Choanephora* sp.
Peziza ostracoderma	*H. victoriae*	*Mucor* sp.
Saccharomyces carlsbergensis	*Kloeckera* sp.	*Paramoebidium arcuatum*
S. cerevisiae	*Mycogone perniciosa*	*Plasmodiophora brassicae*
Saccharomycodes ludwigii	*Paecilomyces* sp.	*Rhizopus* sp.
	Penicillium brevi-compactum	*Schizochytrium aggregatum*
	P. chrysogenum	*Syncephalastrum* sp.
		Thraustochytrium sp.

examples listed at the end of the table differ from the rest, principally because they probably do not contain double-stranded RNA genomes.

The virus-like particles found in *Neurospora crassa* were obtained from the respiratory deficient cytoplasmic mutant '*Abnormal-1*', and resemble some members of the myxo-, arbo-, rhabdo-virus groups. They are polymorphic vesicles, with an electron-dense nucleoid of 120–170 nm diameter, surrounded by one or two membrane envelopes. The genome seems to be single-stranded RNA, as determined by base content and sensitivity to ribonuclease. The virion also contains lipoprotein, glycoprotein and phospholipid (Küntzel *et al.*, 1973). They seem to be associated with the mitochondria, and further investigation is awaited with interest.

The viruses that have been extracted from *Thraustochytrium* sp. are thought to be of the Herpes group, because of their morphology and life cycle. So far they have only been studied in the electron microscope, and their nucleic acid has been assumed to be double-stranded DNA by analogy with this group (Kazama & Schornstein, 1973).

Tikchonenko *et al.* (1974) claim to have isolated viruses from various *Penicillium* species which are phage-like, both in their morphology and in their possession of a double-stranded DNA genome. Extracts from *Penicillium* species are added to bacteria and plaques are produced which contain these phage-like viruses. However, on the data given, the possibility that the phages are derived from bacterial contaminants in the extracts cannot be eliminated. In addition, it is not clear whether the phages which are characterized are obtained from the mycelial extract, or from the plaques on the bacterial plates. In view of this, it is possible that some components of the fungal extract, possibly the fungal double-stranded RNA viruses themselves, induce lysogenic phages in the bacteria to undergo the lytic cycle. It is of interest that this effect is only obtained with mycelium containing double-stranded RNA viruses.

Apart from the above examples, the other viruses so far characterized in the fungi have been shown to possess double-stranded (ds) RNA genomes. They also possess polyhedral capsids from 35-40 nm in diameter (except for the *Agaricus* viruses 1, 2, 3 and 5), and where examined, have more than one RNA segment per genome.

Several double-stranded RNA viruses are now known to exist in other than fungal hosts. They include viruses with a vertebrate host *e.g.* Reovirus, Bluetongue virus, African horse sickness virus, Colorado tick fever virus and Cytoplasmic polyhedrosis virus. The double-stranded RNA viruses with plant hosts also multiply in their insect vectors *e.g.* Wound tumour virus, Rice dwarf virus, Maize rough dwarf virus and Fiji disease virus. There is also a double-stranded RNA phage, Ø6, (Wood, 1973). All these ds-RNA viruses are larger than those occurring in fungi (80-60 nm capsid diameter), and those so far investigated contain a number (6-15) of RNA molecules, one each of which is represented in each capsid (Wood, 1973).

The fungal viruses so far investigated, although they possess a segmented genome (2-6 pieces), seem to encapsidate each piece separately—they are multiparticle systems. For example, Wood & Bozarth (1972) showed that for Pc-V from *Penicillium chrysogenum*, the average molecular weight of a virion (calculated from the S value) was 13×10^6 D. The RNA content per virion (calculated from buoyant density studies, u/v spectra and chemical analysis) was 11-15%. This gives a value of 2×10^6 D for the average molecular weight of the total genome in one virion. Since there are three segments of dsRNA, of molecular weight 1.89, 1.99 and 2.18×10^{-6}, each virion can only contain one RNA segment. In addition, partially degraded virions show no more than one protruding RNA molecule. Similar calculations have been carried out on AfV-s and AfV-f from *Aspergillus foetidus* (Ratti & Buck, 1972) and on PsV-s from *Penicillium stoloniferum* (Buck & Kempson-Jones, 1973).

It can be seen from these few paragraphs that the study of fungal viruses is still a very young but rapidly expanding field. However, even at this stage, it is evident that any studies on biosynthesis and regulation of nucleic acid and protein synthesis in the fungi must take account of any possible contribution to host metabolism made by the presence of a virus.

Replication of Fungal Viruses

There have been few studies on replication of fungal viruses, partly because of the unusual virus-host relationship. Many fungal viruses are latent and

Table 8.9 Characterization of fungal viruses

Fungus	Virus	Virion: Capsid morphology +diameter (nm)	Virion: No. of components[A]	Virion: Sedimentation coefficient[B]	Virion: Buoyant density (g/ml)[C]	Genome: Type of nucleic acid[D]	Genome: Sedimentation coefficient	Genome: Polyacrylamide gel electrophoresis (Mol. wt × 10^6)	Genome: No. of genome segments	References
Agaricus bisporus	1	Polyhedral, 25								
	2	Polyhedral, 29								
	3	Bacilliform, 19 × 50	—	53 S, 74 S, 77 S, 130 S	—	—	—	—	—	(1)
	4	Polydedral, 35								
	5	Polyhedral, 50								
Aspergillus foetidus	AfV-f	Polyhedral 33–37	2	4 components 145–158 S	4 components 1.351–1.380	dsRNA	13.5 S	2.31, 1.87, 1.70, 1.44	4	(2), (3)
	AfV-s	Polyhedral, 33–37		2 components 146 S, 172 S	2 components 1.396, 1.435	dsRNA	13.5 S	2.76, 2.24	2	
Penicillium brevicompactum	Pb-V[E]	Polyhedral, 36–40	1	128 S, 147 S	—	dsRNA	—	2.18, 1.99, 1.89	3	(4)
Penicillium chrysogenum	Pc-V[E]	Polyhedral, 40	1	150 S (218 S minor)	1.354 (+ minor components)	dsRNA	13.0 S	2.18, 1.99, 1.89	3	(5), (6)
Penicillium cyaneofulvum	Pcy-V	Polyhedral, 32.5	1	157 S	1.39	dsRNA	12.5 S	—	—	(7)
Penicillium stoloniferum	PsV-f	Polyhedral, 34	2	61 S, 83 S, 104 S	several components 1.299–1.376	dsRNA	10–12 S	0.99, 0.89, 0.23 + ssRNA	3	(8), (9), (10)
	PsV-s	Polyhedral, 34		E 66 S	1.297	EMPTY →				
				M 87 S	1.332	ssRNA	16.4 S	{0.47, 0.56}	2	
				L 101 S	{1.358, 1.362}	dsRNA	11.7 S	{0.94, 1.11}	(0.94, 1.11)	
				H 113 S	{1.384, 1.390}	dsRNA+ssRNA	11.7, 15.3 S	{0.94 + ss component, 1.11 + ss component}		

(Table 8.9 continued)

Fungus	Virus	Virion: Capsid morphology +diameter (nm)	No. of components[A]	Sedimentation coefficient[B]	Buoyant density (g/ml)[C]	Genome: Type of nucleic acid[D]	Sedimentation coefficient	Polyacrylamide gel electrophoresis (Mol. wt. × 10^6)	No. of genome segments	References
Periconia circinata	Pci-V	Polyhedral, 32	1	66 S, 140 S, 150 S	—	dsRNA	11.5 S, 13.5 S	1.75, 1.40, 1.25, 1.10, 0.48, 0.42	6	(11)
Ustilago maydis	Um-V	Spherical, 40–41	1	5 components, 110–160 S	—	dsRNA	—	2.87–0.06	5	(12)
Saccharomyces cerevisiae	'killer' virus	Polyhedral, 39	1	150 S	—	dsRNA	±15.5 S, 13 S	2.5, 1.4	2	(13), (14)
Neurospora crassa		Central core + Envelope, 120–170	—	—	1.13–1.2	ssRNA	33 S	—	—	(15)
Penicillium brevicompactum, *P. chrysogenum*, *P. stoloniferum* }		Binal, Phage-Type	3	—	1.48–1.51	dsDNA?	—	—	—	(16)
Thraustochytrium sp.	—	Polyhedral + envelope	—	—	—	dsDNA?	—	—	—	(17)

[A] Number of distinct, probably unrelated viruses determined by one or more of following: (a) serological specificity; (b) electrophoresis in polyacrylamide gel, in sucrose density gradients or in agarose gel; (c) DEAE cellulose chromatography.

[B] Sedimentation coefficients ($S_{20,w}$) derived either from analytical ultracentrifrigation or from sucrose density gradients.

[C] Buoyant densities determined in CsCl.

[D] ds = double stranded, ss = single stranded: strandedness of RNA determined by melting behaviour, by resistance to RNAase, by chromatography of RNA hydrolysate and by effect on interferon production.

[E] Viruses serologically related.

Modified from Lemke & Nash, (1974). Yeast-like fungi have been included here for the sake of completeness, but will not be further discussed.

1. Hollings & Stone, 1971.
2. Ratti & Buck, 1972.
3. Banks *et al.*, 1970.
4. Wood, Bozarth & Mislivec, 1971.
5. Wood & Bozarth, 1972.
6. Nash *et al.*, 1973.
7. Banks *et al.*, 1969.
8. Bozarth, Wood & Mandelbrot, 1971.
9. Buck & Kempson-Jones, 1973.
10. Van Frank, Ellis & Kleinschmidt, 1971.
11. Dunkle, 1974.
12. Wood & Bozarth, 1973.
13. Herring & Bevan, 1974.
14. Bevan, Herring & Mitchell, 1973.
15. Küntzel *et al.*, 1973.
16. Tikchonenko *et al.*, 1974.
17. Kazama & Schornstein, 1973.

lysis of mycelium is unpredictable, when it occurs at all. In addition, methods of infection other than heterokaryosis are generally ineffective.

Electron microscope studies in *Penicillium* spp, show that virus particles appear extensively only in older hyphae. They become aggregated into crystalline structures enclosed within vesicles (Border *et al.*, 1972). Virus particles have been shown to be present in ascospores, conidia (Sansing, Detroy, Freer & Hesseltine, 1973) and basidiospores (Dieleman-Van Zaayen, 1972). Detroy, Lillehoj & Hesseltine (1974) have studied the increase in amount of the 5 double-stranded RNA molecules of PsV-f and PsV-s during submerged fermentation of *Penicillium stoloniferum.* They showed that dsRNA was first detectable at 36 h, and increased in amount until 72 h, when mycelial autolysis and loss of viral RNA commenced.

RNA REPLICASE ACTIVITY Because dsRNA viruses require a transcription process in order to produce mRNA, the virions themselves might be expected to contain an RNA replicase exhibiting such a capacity (Baltimore, 1971). There are two studies which indicate that fungal viruses may carry such an enzyme within their capsids. Lapierre, Astier-Manifacier & Cornuet (1971) showed that purified PsV-f and PsV-s particles contained an activity which could synthesize ssRNA, using dsRNA within the virions as template. The assay mixture contained the 4 nucleotides (UTP or CTP labelled), Mg^{2+}, EDTA, Dithiothreitol, and phosphoenol-pyruvate + pyruvate kinase in Tris-HCl buffer. The characteristics of the activity are shown in Table 8.10. From this, it can be seen that the enzyme requires the presence of all 4 nucleotides. It was not inhibited by actinomycin D, rifampicin or DNAase. It was inhibited by ethidium bromide and by RNAase, (although only partially). The authors suggest from the results that the enzyme uses a dsRNA template. It was insensitive to added synthetic polynucleotides, either single- or double-stranded, and also to ssRNA from Turnip yellow mosaic virus.

Nash *et al.* (1973) have demonstrated an RNA replicase activity in *Penicillium chrysogenum* virions, using a somewhat similar assay system. They also showed that the activity required the presence of the 4 nucleotides, and Mg^{2+}. They found in addition that incorporation of Triton-X 100 in the assay system doubled the enzyme activity. It would be of interest to know whether RNAase inhibition was more effective in the presence of Triton-X (see below).

Table 8.10 RNA replicase assay in *Penicillium stoloniferum* viruses (Lapierre et al., 1971)

Assay conditions	c.p.m. ^{32}P	Assay conditions	c.p.m. ^{3}H
COMPLETE	4626	COMPLETE	1467
− ATP, − GTP, − UTP	282	− GTP	442
+ Actinomycin D	4665	+ Ethidium bromide	208
+ DNAase	4600	+ DNAase	1211
+ RNAase	2300	+ Rifampicin	1287
+ NaCl (0.2 M)	5500	+ Actinomycin D	1200

Morgan & Chater (1974) have investigated the virion proteins from PsV-s and PsV-f, using gel electrophoresis. They postulate that one of the minor protein bands which they obtained from each virus, is likely to be the RNA replicase (molecular weights 81 000 for PsV-f and 79 000 for PsV-s). They suggest that the replicase in each virus is coded for by one of the two 1×10^6D dsRNA segments present in each virus. Similar speculations have been made in the case of Reovirus (Wood, 1973). It is interesting to note that this leaves very little viral genetic information available for any protein other than the coat protein.

Effects of Virus on Fungal Host Metabolism

The effect of fungal viruses on their hosts varies. Many are apparently latent—for example, many of the *Penicillium* and *Aspergillus* viruses. Despite the presence of virus in the order of 0.1% by weight of the mycelium (Banks, Buck & Fleming, 1971), the infection seems to have no adverse effect on the growth rate compared with virus-free strains, at least in submerged fermentation experiments on *Penicillium stoloniferum* (Detroy, Lillehoj & Hesseltine, 1974). The presence of the *Agaricus* viruses is more marked, and they can produce drastic reductions in crop productivity (Hollings & Stone, 1971). Viruses can occasionally lyse host mycelium, exhibiting plaque-like phenomena for example in *Schizophyllum commune* (Koltin, Berick, Stamberg & Ben-Shaul, 1973) and in *Penicillium chrysogenum* (Borré, Morgantini, Ortali & Tonolo, 1971). Some viruses are lethal to the host fungus, *e.g.* the 'killer' phenomenon in *Ustilago maydis* (Day, Anagnostakis, Wood & Bozarth, 1972) and in *Saccharomyces cerevisiae* (Berry & Bevan, 1972; Herring & Bevan, 1974).

It has been suggested that fungal viruses might influence the production of secondary metabolites in fungi (Lemke & Ness, 1970). There have been few studies on this and results are as yet equivocal. Production of penicillin from *Penicillium chrysogenum* seems to be independent of virus titre, although industrial strains do seem to contain virus particles (Lemke, Nash & Pieper, 1973). However, *P. notatum* can produce penicillin in the absence of virus (Volkoff, Walters & Dejardin, 1972). In *P. brevicompactum* and *P. stoloniferum*, an inverse relationship has been shown to exist between the presence of virus, and the production of mycophenolic acid (Detroy, Freer & Fennell, 1973). Again, in aflatoxin producing strains of *Aspergillus flavus* and *A. parasiticus*, there is some indication of a link between aflatoxin production and the absence of virus (Wood, Bozarth, Adler & Mackenzie, 1974).

The possibility that viruses might also influence the virulence of plant pathogenic fungi towards their host has also been considered. There is a little evidence for this, in a few cases. In *Helminthosporium maydis* for example, virus is present in strains causing severe disease in maize. Milder strains did not contain virus (Bozarth, Wood & Nelson, 1972; see also review by Lemke & Nash, 1974, for further examples).

It is quite likely that any effects which viruses may have on pathogenicity, secondary metabolite production, *etc.* of the host are mediated by a virus-induced alteration in host gene expression, and are hence having an effect on the host, via regulation of nucleic acid and protein biosynthesis.

Conclusions

To summarize, it seems that the fungal viruses characterized so far have segmented genomes, which are probably encapsidated in separate particles. Investigations carried out on enzyme activity associated with virions themselves seem to indicate that an RNA replicase activity is associated with the virion.

It may be useful at this point to examine the state of play in the more extensively studied Reovirus. It seems that the 10 dsRNA segments of the Reovirus genome remain inside the inner capsid (subviral particle) of the infecting virus, throughout the life cycle. An RNA replicase, identified as one of the three protein subunits of the inner capsid, transcribes single-stranded RNA messenger molecules from each dsRNA template. These find their way on to host cytoplasmic ribosomes, where they are translated. mRNA molecules then act as templates for the production of new progeny dsRNA genomes. However, this must be carried out within newly formed subviral particles, as no free dsRNA is detectable in the cytoplasm (Silverstein, Astell, Levin, Schönberg & Acs, 1972; Schönberg, Silverstein, Levin & Acs, 1971; Acs *et al.*, 1971). Kaempfer & Kaufman (1973) reported that dsRNA molecules are extremely inhibitory to protein synthesis in rabbit reticulocyte lysates, as they bind specifically to IF-3, an initiation factor required in the recycling of ribosomes.

As so many of the fungal viruses (and dsRNA viruses in general) seem to have little or no adverse effects on their host metabolism, it would appear that the Reovirus system of surrounding dsRNA at all times by capsid protein might indeed also be the most likely system in operation in many, if not all the dsRNA fungal viruses. It is possible that the H particles of PsV-s, which contain single stranded (ss) RNA as well as dsRNA are particles which are transcribing mRNA from the dsRNA segment within the capsid (Buck & Kempson-Jones, 1973). It may be that Lapierre *et al.* (1971) are picking up such an activity with their assay system, in which case, if it is occurring within the viral capsid, it might be expected to show the partial resistance to RNAase which they report. Alternatively, the ssRNA molecules may be included in certain particles (*i.e.* the H particles) adventitiously, in which case the enzyme activity detected may be converting these ssRNA molecules to 'genome' dsRNA. In this case, not only would the product be protected within the capsid, but it would also be double-stranded and therefore perhaps even more likely to be relatively RNAase resistant. These two alternatives might be resolved by isolating the H particles and assaying them separately.

Evidence from molecular weight determination of virus components has suggested that the fungal dsRNA viruses are multiparticles. Confirmation of this by more classical methods, for example by showing that infectivity is only obtained when all particles are present, or by combining different strains or mutants of the various components, may require the use of fungal protoplasts, in view of the difficulty of carrying out such infectivity experiments on fungal mycelium.

Multiparticle viruses are common amongst RNA viruses of plants (Matthews, 1970), and have the obvious advantage of ensuring that recombination of the viral genome will take place (in a manner analogous to independent assortment of chromosomes during meiosis in higher organisms) whenever virion components of different strains are mixed, without requiring the use of host recombination enzymes, which are presumably specific for DNA genomes. Such mixing can take place outside the host cell. The disadvantage of such a multicomponent or multiparticle system is that it requires a high multiplicity of infection to ensure that each of the components enters the host cell. In the fungi however, infection probably occurs 'horizontally' via heterokaryosis (although it is interesting to note that mitochondria may *not* be exchanged in this process—see Casselton & Condit, 1972) and 'vertically' via the spores. These methods of infection are likely to ensure that the new host receives its full complement of viral components, and hence that it can support the replication of the viral genome and can produce progeny virions.

It is probable, therefore, that fungal dsRNA viruses enjoy the pros of a multiparticle system without suffering its cons. Indeed it is possible that the unique genetic system of fungi, that of selection of nuclei, has allowed the evolution of a unique genetic system of its viruses, that of selection of multiparticle components.

8.5 References

ACS, G., KLETT, H., SCHÖNBERG, M., CHRISTMAN, J., LEVIN, D. H. & SILVERSTEIN, S. C. (1971). Mechanism of Reovirus double-stranded ribonucleic acid synthesis *in vivo* and *in vitro*. *Journal of Virology* **8**, 684–9.

ADELMAN, T. G., & LOVETT, J. S. (1972). Synthesis of ribosomal protein without *de novo* ribosome production during differentiation in *Blastocladiella emersonii*. *Biochemical and Biophysical Research Communications* **49**, 1174–82.

ADELMAN, T. G. & LOVETT, J. S. (1974*a*). Ribosome function *in vitro* and *in vivo* during the life cycle of *Blastocladiella emersonii*. *Biochimica et Biophysica Acta* **349**, 240–9.

ADELMAN, T. G. & LOVETT, J. S. (1974*b*). Evidence for a ribosome associated translation inhibitor during differentiation of *Blastocladiella emersonii*. *Biochimica et Biophysica Acta* **335**, 236–45.

AGSTERIBBE, E., KROON, A. M. & VAN BRUGGEN, E. F. J. (1972). Circular DNA from mitochondria of *Neurospora crassa*. *Biochimica et Biophysica Acta* **269**, 299–303.

AITKHAZHIN, M. A., AZIMURATOUA, R. Z., KIM, T. N. & DARKANBAEV, T. B. (1972). Isolation and characterisation of a rapidly labelled cytoplasmic RNA fraction of *Aspergillus niger*. *Biochimiya* **37**, 1276–81.

ALBERGHINA, F. A. M., STURANI, E. & SCHIAFFONATI, L. (1971). The peptidyl puromycin reaction with *Neurospora crassa* ribosomes. *Archiv für Mikrobiologie* **80**, 166–75.

ALBERGHINA, F. A., STURANI, E. & URSINO, D. J. (1969). Ribosomal changes related to changes in the rate of growth in *Neurospora crassa* mycelia. *Biochimica et Biophysica Acta* **195**, 576–8.

ALBERGHINA, F. A. M. & SUSKIND, S. R. (1967). Ribosomes and ribosomal proteins from *Neurospora crassa*. I. Physical chemical and immunochemical properties. *Journal of Bacteriology* **94**, 630–49.

ARST, H. N. & SCAZZOCCHIO, C. (1972). Control of nucleic acid synthesis in *Aspergillus nidulans*. *Biochemical Journal* **127**, 188.

AVIV, H. & LEDER, P. (1972). Purification of biologically active globin messenger RNA by chromatography on oligothymidylic acid cellulose. *Proceedings of the National Academy of Sciences, U.S.* **69**, 1408–12.

BACON, C. W., SUSSMAN, A. S. & PAUL, A. G. (1973). Identification of a self inhibitor from spores of *Dictyostelium discoideum*. *Journal of Bacteriology* **113**, 1061–3.

BAINBRIDGE, B. W. (1971). Macromolecular composition and nuclear division during spore germination in *Aspergillus nidulans*. *Journal of General Microbiology* **66**, 319–25.

BALTIMORE, D. (1971). Expression of animal virus genomes. *Bacteriological Reviews* **35**, 235–41.

BANKS, G. T., BUCK, K. W., CHAIN, E. B., DARBYSHIRE, J. E. & HIMMELWEIT, F.

(1969). *Penicillium cyaneo-fulvum* virus and interferon stimulation. *Nature, London* **223**, 155–8.

BANKS, G. T., BUCK, K. W., CHAIN, E. B., DARBYSHIRE, J. E., HIMMELWEIT, F., RATTI, G., SHARPE, T. J. & PLANTEROSE, D. N. (1970). Antiviral activity of double stranded RNA from a virus isolated from *Aspergillus foetidus*. *Nature, London* **227**, 505–7.

BANKS, G. T., BUCK, K. W. & FLEMING, A. (1971). The isolation of viruses and viral ribonucleic acid from filamentous fungi on a pilot plant scale. *Chemical Engineering* **251**, 259–61.

BARATH, Z. & KÜNTZEL, H. (1972). Induction of mitochondrial RNA polymerase in *Neurospora crassa*. *Nature, New Biology* **240**, 195–7.

BARNETT, W. E. & BROWN, D. H. (1967). Mitochondrial transfer RNAs. *Proceedings of the National Academy of Sciences, U.S.* **57**, 452–8.

BARNETT, W. E., BROWN, D. H. & EPLER, J. L. (1967). Mitochondrial-specific aminoacyl-RNA synthetases. *Proceedings of the National Academy of Sciences, U.S.* **57**, 1775–81.

BERRY, E. A. & BEVAN, E. A. (1972). A new species of double-stranded RNA from yeast. *Nature, London* **239**, 279–80.

BEVAN, E. A., HERRING, A. J. & MITCHELL, D. J. (1973). Preliminary characterisation of two species of DS-RNA in yeast and their relationship to the 'killer' character. *Nature, London* **245**, 81–6.

BHAGWAT, A. S. & MEHADEVAN, P. R. (1970). Conserved mRNA from conidia of *Neurospora crassa*. *Molecular and General Genetics* **109**, 142–51.

BHAGWAT, A. S. & MEHADEVAN, P. R. (1972). Effect of 8-azaguanine on polyribosomes of *Neurospora crassa*. *Indian Journal of Biochemistry and Biophysics* **9**, 111–5.

BHAGWAT, A. S. & MEHADEVAN, P. R. (1973). Differential gene action in *Neurospora crassa*. *Journal of Bacteriology* **113**, 572–5.

BIRKMAYER, G. D. (1971). Isolierung und partiellecharakterisierung eines mitochondrial synthetisierten proteins aus *Neurospora crassa*. *Hoppe-Seyler's Zeitschrift für Physiologische Chemie* **352**, 761–3.

BLOSSEY, H. C. & KÜNTZEL, H. (1972), *In vitro* translation of mitochondrial DNA from *Neurospora crassa*. *FEBS Letters* **24**, 335–8.

BOGUSLAWSKI, G., SCHLESSINGER, D., MEDOFF, G. & KOBAYASHI, G. (1974). Ribonucleic acid polymerases of the yeast phase of *Histoplasma capsulatum*. *Journal of Bacteriology* **118**, 480–5.

BORDER, D. J., BUCK, K. W., CHAIN, E. B., KEMPSON-JONES, G. F., LHOAS, P. & RATTI, G. (1972). Viruses of *Penicillium* and *Aspergillus* species. *Biochemical Journal* **127**, 4P–6P.

BORRÉ, E., MORGANTINI, L. E., ORTALI, V. & TONOLO, A. (1971). Production of lytic plaques of viral origin in *Penicillium*. *Nature, London* **229**, 568–9.

BORST, P. (1972). Mitochondrial nucleic acids. *Annual Review of Biochemistry* **41**, 333–76.

BORST, P. & GRIVELL, L. A. (1971). Mitochondrial ribosomes. *FEBS Letters* **13**, 73–88.

BORST, P. & KROON, A. M. (1969). Mitochondrial DNA: physicochemical properties, replication and genetic function. *International Review of Cytology* **26**, 107-90.

BOZARTH, R. F., WOOD, H. A. & MANDELBROT, A. (1971). The *Penicillium stoloniferum* virus complex: two similar double-stranded RNA virus-like particles in a single cell. *Virology* **45**, 516–23.

BOZARTH, R. F., WOOD, H. A. & NELSON, R. R. (1972). Virus-like particles in virulent strains of *Helminthosporium maydis*. *Phytopathology* **62**, 748.

BRAMBL, R. M. & WOODWARD, D. O. (1972). Altered species of mitochondrial transfer RNA associated with the *mi-1* cytoplasmic mutation in *Neuropora crassa*. *Nature, New Biology* **238**, 198–200.

BRITTEN, R. J. & KOHNE, D. E. (1968). Repeated sequences in DNA. *Science* **161**, 529–40.

BROOKS, R. R. & HUANG, P. C. (1972). Redundant DNA of *Neurospora crassa*. *Biochemical Genetics* **6**, 41–9.

BROWN, D. H. & NOVELLI, G. D. (1968). Chromatographic differences between cytoplasmic and mitochondrial tRNAs of *Neurospora crassa*. *Biochemical and Biophysical Research Communications* **31**, 262–6.

BUCK, K. W. & KEMPSON-JONES, G. F. (1973). Biophysical properties of *Penicillium stoloniferum* virus. *Journal of General Virology* **18**, 223–35.

BURKE, D. J., SEALE, T. W. & McCARTHY, S. J. (1972). Protein and RNA synthesis during the diploid life cycle of *Allomyces arbuscula*. *Journal of Bacteriology* **110**, 1065–72.

CAIN, A. K. & NESTER, E. W. (1973). RNA polymerase in *Allomyces arbuscula*. *Journal of Bacteriology* **115**, 769–76.

CASSELTON, L. A. & CONDIT, A. (1972). A mitochondrial mutant of *Coprinus lagopus*. *Journal of General Microbiology* **72**, 521–7.

CHATTOPADHYAY, S. K., KOHNE, D. E. & DUTTA, S. K. (1972). Ribosomal RNA genes of *Neurospora crassa*. Isolation and characterisation. *Proceedings of the National Academy of Sciences U.S.* **69**, 3256–9.

CHEUNG, S. C., KOBAYASHI, G. S., SCHLESSINGER, D. & MEDOFF, G. (1974). RNA metabolism during morphogenesis in *Histoplasma capsulatum*. *Journal of General Microbiology* **82**, 301–7.

CLARK-WALKER, G. D. & GLEASON, F. H. (1973). Circular DNA from the water mould *Saprolegnia*. *Archiv für Mikrobiologie* **92**, 209–16.

CLAYTON, D. A. & BRAMBL, R. M. (1972). Detection of circular DNA from mitochondria of *Neurospora crassa*. *Biochemical and Biophysical Research Communications* **46**, 1477–82.

COCHRANE, J. C., RADO, T. A. & COCHRANE, V. W. (1971). Synthesis of macromolecules and polyribosome formation in early stages of spore germination in *Fusarium solani*. *Journal of General Microbiology* **65**, 45–55.

DARNELL, J. E., WALL, R. & TUSHINSKI, R. J. (1971). An adenylic acid rich sequence in mRNA of HeLa cells and its possible relationship to reiterated sites in DNA. *Proceedings of the National Academy of Sciences, U.S.* **68**, 1321–5.

DATEMA, R., AGSTERIBBE, E. & KROON, A. N. (1974). The mitochondrial ribosomes of *Neurospora crassa*. 1. On the occurrence of 80 S ribosomes. *Biochimica et Biophysica Acta* **335**, 386–95.

DAY, P. R., ANAGNOSTAKIS, S. L., WOOD, H. A. & BOZARTH, R. F. (1972). Positive correlation between extracellular inheritance and mycovirus transmission in *Ustilago maydis*. *Phytopathology* **62**, 753.

DETROY, R. W., FREER, S. N. & FENNELL, D. I. (1973). Relationship between the biosynthesis of virus-like particles and mycophenolic acid in *Penicillium stoloniferum* and *Penicillium brevicompactum*. *Canadian Journal of Microbiology* **19**, 1459–62.

DETROY, R. W., LILLEHOJ, E. B. & HESSELTINE, C. W. (1974). Replication of virus-like particles in *Penicillium stoloniferum* mycelia. *Canadian Journal of Microbiology* **20**, 113–7.

DIELEMAN-VAN ZAAYEN, A. (1972). Intracellular appearance of mushroom virus in fruiting-bodies and basidiospores of *Agaricus bisporus*. *Virology* **47**, 94–104.

DUNKLE, L. D. (1974). Double-stranded RNA mycovirus in *Periconia circinata*. *Physiological Plant Pathology* **4**, 107–16.

DUNKLE, L. D. & VAN ETTEN, J. L. (1972). Characteristics and synthesis of deoxyribonucleic acid during fungal spore germination. In *Spores V* Ed. H. O. Halvorson, R. Hanson and L. L. Campbell, Washington, American Society for Microbiology, 283–9.

DUNKLE, L. D., MAHESHWARI, R. & ALLEN, P. J. (1969). Infection structures from rust uredospores: effect of RNA and protein synthesis inhibitors. *Science* **163**, 481–2.

DUNKLE, L. D., VAN ETTEN, J. L. & BRAMBL, R. M. (1972). Mitochondrial DNA synthesis during fungal spore germination. *Archiv für Mikrobiologie* **85**, 225–32.

DUTTA, S. K. (1973). Transcription of nonrepeated DNA in *Neurospora crassa*. *Biochimica et Biophysica Acta* **324**, 482–7.

DUTTA, S. K. & OJHA, M. (1972). Relatedness between major taxonomic groups of fungi based on the measurement of DNA nucleotide sequence homology. *Molecular and General Genetics* **114**, 232–40.

DUTTA, S. K. & RAY, R. (1973). Partial characterisation of tRNA genes isolated from *Neurospora crassa*. *Molecular and General Genetics* **125**, 295–300.

DUTTA, S. K., PENN, S. R. & KNIGHT, A. (1972). Characterisation of DNA from *Coprinus lagopus* and *Mucor azygospora*. *Experientia* **28**, 582–4.

DWIVEDI, R. S., DUTTA, S. K. & BLOCK, D. P. (1969). Isolation and characterisation of chromatin from *Neurospora crassa*. *Journal of Cell Biology* **43**, 51–8.

EDELMAN, M., VERMA, I. M. & LITTAUER, U. Z. (1970). Mitochondrial ribosomal RNA from *Aspergillus nidulans*: characterisation of a novel molecular species. *Journal of Molecular Biology* **49**, 67–83.

ELLIS, L. F. & KLEINSCHMIDT, W. J. (1967). Virus-like particles of a fraction of statolon, a mould product. *Nature, London* **215**, 649–50.

EMMET, N., WILLIAMS, C. M., FREDERICK, L. & WILLIAMS, L. S. (1972). Arginyl tRNA and synthetase of *Neurospora crassa*. *Mycologia* **64**, 499–509.

EPLER, J. L. (1969). The mitochondrial and cytoplasmic transfer RNAs of *Neurospora crassa*. *Biochemistry* **8**, 2285–90.

EPLER, J. L. & BARNETT, W. E. (1967). Coding properties of *Neurospora* mitochondrial and cytoplasmic leucine tRNAs. *Biochemical and Biophysical Research Communications* **28**, 328–33.

EPLER, J. L., SHUGART, L. R. & BARNETT, W. E. (1970). N-formylmethionyl-tRNA in mitochondria of *Neurospora*. *Biochemistry* **9**, 3375–9.

FLAVIA, M. M. & TORRES, H. N. (1972*a*). Adenylate cyclase activity in lubrol treated membranes from *Neurospora crassa*. *Biochimica et Biophysica Acta* **289**, 428–32.

FLAVIA, M. M. & TORRES, H. N. (1972*b*). Activation of membrane bound adenylate cyclase by glucagon in *Neurospora crassa*. *Proceedings of the National Academy of Sciences U.S.* **69**, 2870–2.

FORRESTER, I. T., NAGLEY, P. & LINNANE, A. W. (1970). Yeast mitochondrial ribosomal RNA: a new extraction procedure and unusual physical properties. *FEBS Letters* **11**, 59–61.

GAMOW, E. & PRESCOTT, D. M. (1972). Characterisation of RNA synthesised by *Phycomyces blakesleanus*. *Biochimica et Biophysica Acta* **259**, 223–7.

GONG, C., DUNKLE, L. D. & VAN ETTEN, J. L. (1973). Characteristics of deoxyribonucleic acid polymerase isolated from spores of *Rhizopus stolonifer*. *Journal of Bacteriology* **115**, 762–8.

GONG, C. & VAN ETTEN, J. L. (1972). Changes in soluble ribonucleic acid polymerases associated with the germination of *Rhizopus stolonifer* spores. *Biochimica et Biophysica Acta* **272**, 44–52.

GONG, C. & VAN ETTEN, J. L. (1974). Purification and properties of RNA polymerases I and II from germinated spores of *Rhizopus stolonifer*. *Canadian Journal of Microbiology*—in Press.

GOTTLIEB, D., RAO, M. V. & SHAW, P. D. (1968). Changes in RNA during germination of teliospores of *Ustilago maydis*. *Phytopathology* **58**, 1593–7.

GRANDI, M., HELMS, A. & KÜNTZEL, H. (1971). Fusidic acid resistance of mitochondrial G factor from *Neurospora crassa*. *Biochemical and Biophysical Research Communications* **44**, 864–71.

GRANDI, M. & KÜNTZEL, H. (1970). Mitochondrial peptide chain elongation factors from *Neurospora crassa*. *FEBS Letters* **10**, 25–8.

GRIFFIN, D. H., TIMBERLAKE, W. E. & CHENEY, J. C. (1974). Regulation of macromolecular synthesis, colony development and specific growth rate of *Achyla bisexualis* during balanced growth. *Journal of General Microbiology* **80**, 381–8.

GRINWICH, K. D. & TREVITHICK, J. R. (1973). Evidence for the existence of membrane-bound ribosomes in *Neurospora crassa*. *Journal of General Microbiology* **79**, 173–9.

GRIVELL, A. R. & JACKSON, J. F. (1968). Thymidine kinase: evidence for its absence from *Neurospora crassa* and some other micro-organisms and the relevance of this to the specific labelling of DNA. *Journal of General Microbiology* **54**, 307–17.

HALLERMAYER, G. & NEUPERT, W. (1974). Immunological difference of mitochondrial and cytoplasmic ribosomes of *Neurospora crassa*. *FEBS Letters* **41**, 264–8.

HASELKORN, R. & ROTHMAN-DENES, L. B. (1973). Protein synthesis. *Annual Review of Biochemistry* **42**, 397–438.

HAWLEY, E. S. & GREENAWALT, J. W. (1970). An assessment of *in vivo* mitochondrial protein synthesis in *Neurospora crassa*. *Journal of Biological Chemistry* **245**, 3574–83.

HAWLEY, E. S. & WAGNER, R. P. (1967). Synchronous mitochondrial division in *Neurospora crassa*. *Journal of Cell Biology* **35**, 489–99.

HENNEY, H. & STORCK, R., (1963*a*). Nucleotide composition of RNA from *Neurospora crassa*. *Journal of Bacteriology* **85**, 812–26.

HENNEY, H. & STORCK, R. (1963*b*). Ribosomes and RNA in three morphological states of *Neurospora crassa*. *Science* **142**, 1675–6.

HENNEY, H. R. & STORCK, R. (1964). Polyribosomes and morphology in *Neurospora crassa*. *Proceedings of the National Academy of Sciences, U.S.* **51**, 1050–5.

HERRING, A. J. & BEVAN, E. A. (1974). Virus-like particles associated with the double-stranded RNA species found in killer and sensitive strains of the yeast *Saccharomyces cerevisiae*. *Journal of General Virology* **22**, 387–94.

HILL, R. & NAZARIO, M. (1973). Purification and kinetic properties of *Neurospora* transfer RNA nucleotidyltransferase. *Biochemistry* **12**, 483–5.

HINDLE, C. W., KORNFIELD, J. M. & KNIGHT, S. G. (1966). Incorporation of amino acids into protein by cell-free extracts of *Penicillium chrysogenun*. *Archiv für Mikrobiologie* **53**, 41–9.

HOLLINGS, M. (1962). Viruses associated with a die-back disease of cultivated mushroom. *Nature. London* **196**, 962–5.

HOLLINGS, M. & STONE, O. M. (1971). Viruses that infect fungi. *Annual Review of Phytopathology* **9**, 93–118.

HOLLOMAN, D. W. (1969). Biochemistry of germination in *Peronospora tabacina* conidia: evidence for the existence of stable mRNA. *Journal of General Microbiology* **55**, 267–74.

HOLLOMAN, D. W. (1970). RNA synthesis during fungal spore germination. *Journal of General Microbiology* **62**, 75–87.

HOLLOMAN, D. W. (1971). Protein synthesis during germination of *Peronospora tabacina* conidia. *Archives of Biochemistry and Biophysics* **145**, 643–9.

HOLLOMAN, D. W. (1973). Protein synthesis during germination of *Peronospora tabacina* conidia, an examination of the events involved in the initiation of germination. *Journal of General Microbiology* **78**, 1–13.

HOLTEN, V. Z., & JACOBSEN, K. B. (1969). Studies on the aminoacylation of valine and alanine specific tRNA of *E. coli* by amino acyl tRNA synthetase from *N. crassa* and *E. coli*. *Archives of Biochemistry and Biophysics* **129**, 283–9.

HORGEN, P. A. (1971). *In vitro* RNA synthesis in the zoospores of the aquatic fungus *Blastocladiella emersonii*. *Journal of Bacteriology* **106**, 281–2.

HORGEN, P. A. & GRIFFIN, D. H. (1971*a*). RNA polymerase III of *Blastocladiella emersonii* is mitochondrial. *Nature, New Biology* **234**, 17–8.

HORGEN, P. A. & GRIFFIN, D. H. (1971*b*). Specific inhibitors of the 3 RNA polymerases from the aquatic fungus *Blastocladiella emersonii*. *Proceedings of the National Academy of Sciences, U.S.* **68**, 338–41.

HORGEN, P. A., NAGAO, R. T., CHIA, L. S. Y. & KEY, J. L. (1973). Basic nuclear proteins in the Oömycete fungus *Achlya bisexualis*. *Archiv für Mikrobiologie* **94**, 249–58.

HORIKOSHI, K. & IKEDA, Y. (1969). Protein synthesising activity of dormant conidia. *Biochimica et Biophysica Acta* **190**, 187–92.

HORIKOSHI, K., OHTAKA, Y. & IKEDA, Y. (1969). Properties of ribosomes and tRNA in dormant conidia of *Aspergillus oryzae*. In *Spores* IV. Ed. L. L. Campbell, Washington. American Society for Microbiology 175–9.

HSIANG, M. W. & COLE, R. D. (1973). The isolation of histone from *Neurospora crassa*. *Journal of Biological Chemistry* **248**, 2007–13.

IBRAHIM, N. G., BURKE, J. P. & BEATTIE, D. S. (1973). Mitochondrial protein synthesis *in vitro* is not an artifact. *FEBS Letters* **29**, 73–6.

JEGGO, P. A., UNRAU, P., BANKS, G. R. & HOLLIDAY, R. (1973). A temperature sensitive DNA polymerase mutant of *Ustilago maydis*. *Nature, New Biology* **242**, 14–5.

KAEMPFER, R. & KAUFMAN, J. (1973). Inhibition of cellular protein synthesis by double-stranded RNA: inactivation of an initiation factor. *Proceedings of the National Academy of Sciences, U.S.* **70**, 1222–6.

KAZAMA, F. Y. & SCHORNSTEIN, K. L. (1973). Ultrastructure of a fungus herpes-type virus. *Virology* **52**, 478–87.

KESSEL, M. & ROSENBERGER, R. E. (1968). Regulation and timing of DNA synthesis in hyphae of *Aspergillus nidulans*. *Journal of Bacteriology* **95**, 2275–81.

KOLTIN, Y., BERICK, R., STAMBERG, J. & BEN-SHAUL, Y. (1973). Virus-like particles and cytoplasmic inheritance of plaques in a higher fungus. *Nature, New Biology* **241**, 108–9.

KRITSKII, M. S. & CHERNYSHEVA, E. K. (1968). Inclusion of adenine $8^{14}C$ into ribosomes and polysomes of germinating conidia and mycelium of *Neurospora crassa ad* strains deficient in adenine. *Doklady Akademii Nauk. SSSR* **183**, 703–6.

KROON, A. M. & SACCONE, C. (1974). *The biogenesis of mitochondria*. New York and London: Academic Press.

KÜNTZEL, H. (1969*a*). Specificity of mitochondrial and cytoplasmic ribosomes from *Neurospora crassa* in poly-U dependent cell-free systems. *FEBS Letters* **4**, 140–2.

KÜNTZEL, H. (1969*b*). Mitochondrial and cytoplasmic ribosomes from *Neurospora crassa*: characterisation of their subunits. *Journal of Molecular Biology* **40**, 315–20.

KÜNTZEL, H. & BLOSSEY, H. C. (1974). Translation products *in vitro* of mitochondrial messenger RNA from *Neurospora crassa*. *European Journal of Biochemistry* **47**, 165–71.

KÜNTZEL, H. & NOLL, H. (1967). Mitochondrial and cytoplasmic polysomes from *Neurospora crassa. Nature, London* **215**, 1340–5.

KÜNTZEL, H. & SALA, F. (1969). Kettenaufangsmechanismus der mitochondriellen Proteinbiosynthese. *Hoppe-Seyler's Zeitschrift für Physiologische Chemie* **350**, 1158.

KÜNTZEL, H. & SCHÄFER, K. P. (1971). Mitochondrial RNA polymerase from *Neurospora crassa. Nature, New Biology* **231**, 265–9.

KÜNTZEL, H., BARATH, Z., ALI, I., KIND, J. & ALTHAUS, H. H. (1973). Virus-like particles in an extranuclear mutant of *Neurospora crassa. Proceedings of the National Academy of Sciences, U.S.* **70**, 1574–8.

KURIYAMA, Y. & LUCK, D. J. L. (1973). Ribosomal RNA synthesis in mitochondria of *Neurospora crassa. Journal of Molecular Biology* **73**, 425–37.

KURIYAMA, Y. & LUCK, D. J. L. (1974). Methylation and processing of mitochondrial ribosomal RNA's in *poky* and wild type *Neurospora crassa. Journal of Molecular Biology* **83**, 253–66.

LAMPSON, G. P., TYTELL, A. A., FIELD, A. K., NEMES, M. M. & HILLEMANN, M. R. (1967). Inducers of interferon and host resistance. I. Double-stranded RNA from extracts of *Penicillium funiculosum. Proceedings of the National Academy of Sciences, U.S.* **58**, 782–9.

LANSMAN, R. A., ROWE, M. J. & WOODWARD, D. O. (1974). Pulse recovery studies on cycloheximide-insensitive protein synthesis in *Neurospora.* Association of products with cytochrome oxidase. *European Journal of Biochemistry* **41**, 15–23.

LAPIERRE, H., ASTIER-MANIFACIER, S. & CORNUET, P. (1971). Activité RNA polymérase associée aux preparations purifiées de virus du *Penicillium stoloniferum. Comptes Rendus des Séances de L'Académie des Sciences, Série D.* **273**, 992–4.

LEARY, J. O. & ELLINGBOE, A. H. (1971). Isolation and characterisation of ribosomes from non-germinated conidia of *Erysiphe graminis f. sp. tritici. Phytopathology* **61**, 1030–1.

LEARY, J. O., MORRIS, A. J. & ELLINGBOE, A. H. (1969). Isolation of functional ribosomes and polysomes from lyophylised fungi. *Biochimica et Biophysica Acta* **182**, 113–20.

LEIGHTON, T. J., DILL, B. C., STOCK, J. J. & PHILLIPS, C. (1971). Absence of histones from the chromosomal proteins of fungi. *Proceedings of the National Academy of Sciences, U.S.* **68**, 677–80.

LEMKE, P. A. & NASH, C. H. (1974). Fungal viruses. *Bacteriological Reviews* **38**, 29–56.

LEMKE, P. A. & NESS, T. M. (1970). Isolation and characterisation of a double-stranded ribonucleic acid from *Penicillium chrysogenum.* Journal of Virology **6**, 813–9.

LEMKE, P. A. & NASH, C. H. & PIEPER, S. W. (1973). Lytic plaque formation and variation in virus titre among strains of *Penicillium chrysogenum. Journal of General Microbiology* **76**, 265–75.

LENGYEL, P. & SOLL, D. (1969). Mechanism of protein biosynthesis. *Bacteriological Reviews* **33**, 264–301.

LIN, F. K., DAVIES, F. L., TRIPATHI, R. K., RAGHN, K. & GOTTLIEB, D. (1971). RNA in spore germination of *Ustilago maydis. Phytopathology* **61**, 645–8.

LINGAPPA, B. T., LINGAPPA, Y. & BELL, E. (1973). A self inhibitor of protein synthesis in the conidia of *Glomerella cingulata. Archiv für Mikrobiologie* **94**, 97–107.

LIZARDI, P. M. & LUCK, D. J. L. (1971). Absence of a 5S RNA component in the mitochondrial ribosomes of *Neurospora crassa. Nature, New Biology* **229**, 140–2.

LIZARDI, P. M. & LUCK, D. J. L. (1972). The intracellular site of synthesis of mitochondrial ribosomal proteins in *Neurospora crassa. Journal of Cell Biology* **54**, 56–74.

LOENING, U. E. (1968). Molecular weight of ribosomal RNA in relation to evolution. *Journal of Molecular Biology* **38**, 355–65.

LOVETT, J. S. (1963). Chemical and physical characteristics of 'nuclear caps' isolated from *Blastocladiella* zoospores. *Journal of Bacteriology* **85**, 1235–46.

LOVETT, J. S. (1968). Reactivation of RNA and protein synthesis during germination of *Blastocladiella* zoospores, and the role of the nuclear cap. *Journal of Bacteriology* **96**, 962–9.

LOVETT, J. S. (1975). Regulation of protein metabolism during spore germination. In *The Fungal Spore* ed, D. Weber and B. Hess, New York, Wiley and Sons, in Press.

LOVETT, J. S. & HASELBY, J. A. (1971). Molecular weights of the ribosomal RNA of fungi. *Archiv für Mikrobiologie* **80**, 191–204.

LOVETT, J. S. & LEAVER, C. J. (1969). High molecular weight artefacts in RNA extracted from *Blastocladiella* at elevated temperatures. *Biochimica et Biophysica Acta* **195**, 319–27.

LUCK, D. J. L. (1963). Formation of mitochondria in *Neurospora crassa. Journal of Cell Biology* **16**, 483–99.

LUCK, D. J. L. (1964). The influence of precursor pool size on mitochondrial composition in *Neurospora crassa. Journal of Cell Biology* **24**, 445–60.

LUCK, D. J. L. & REICH, E. (1964). DNA in mitochondria of *Neurospora crassa. Proceedings of the National Academy of Sciences, U.S.* **52**, 931–8.

MACKO, V., STAPLES, R. C., ALLEN, P. J. & RENWICK, J. A. A. (1971). Identification of the germination self-inhibitor from wheat stem rust uredospores. *Science* **173**, 835–6.

MARCHANT, R. & SMITH, D. G. (1968). The effect of chloramphenicol on growth and mitochondrial structure of *Pythium ultimum. Journal of General Microbiology* **50**, 391–7.

MARTENS, C. L. & SARGENT, M. L. (1974). Circadian rhythms of nucleic acid metabolism in *Neurospora crassa. Journal of Bacteriology* **117**, 1210–5.

MATTHEWS, R. E. F. (1970). *Plant virology.* Academic Press.

MEHADEVAN, P. R. & BHAGWAT, A. S. (1969). Inhibition of protein synthesis in *Neurospora crassa* by 8-azaguanine. *Indian Journal of Biochemistry* **6**, 169–74.

MERLO, D. J., ROKER, H. & VAN ETTEN, J. L. (1972). Protein synthesis during fungal spore germination. VI. Analysis of transfer RNA from germinated and ungerminated spores of *Rhizopus stolonifer. Canadian Journal of Microbiology* **18**, 949–56.

MICHEL, R. & NEUPERT, W. (1973). Mitochondrial translation products before and after integration into the mitochondrial membrane in *Neurospora crassa. European Journal of Biochemistry* **36**, 53–67.

MIRKES, P. E. (1974). Polysomes, ribonucleic acid and protein synthesis during germination of *Neurospora crassa* conidia. *Journal of Bacteriology* **117**, 196–202.

MITCHELL, M. B. & MITCHELL, H. K. (1952). A case of 'maternal' inheritance in *Neurospora crassa. Proceedings of the National Academy of Sciences, U.S.* **38**, 442–9.

McLAUGHLIN, C. S., WARNER, J. R., EDMONDS, M., NAKAZATO, H. & VAUGHAN, M. H. (1973). Polyadenylic acid sequences in yeast messenger RNA. *Journal of Biological Chemistry* **248**, 1466–71.

MOORE, P. D. (1974). Evidence for an inducible repair mechanism in pyrimidine mutants of *Ustilago maydis. Heredity* **33**, 136.

MORGAN, D. H. & CHATER, K. F. (1974). Gene-protein relationships in the viruses of *Penicillium stoloniferum. Heredity* **33**, 133.

MOYER, R. C. & STORCK, R. (1964). Properties of ribosomes and RNA from *Aspergillus niger. Archives of Biochemistry and Biophysics* **104**, 193–201.

MURPHY, M. N. & LOVETT, J. S. (1966). RNA and protein synthesis during zoospore differentiation in synchronised cultures of *Blastocladiella. Developmental Biology* **14**, 68–95.

MYERS, R. B. & CANTINO, E. D. (1971). DNA profile of the spores of *Blastocladiella emersonii.* Evidence for γ particle DNA. *Archiv für Mikrobiologie* **78**, 252–67.

NASH, C. H., DOUTHART, R. J., ELLIS, R. F., VAN FRANK, R. M., BURNETT, J. P. & LEMKE, P. A. (1973). On the mycophage of *Penicillium chrysogenum. Canadian Journal of Microbiology* **19**, 97–103.

NAZARIO, M., (1967). The accumulation of arginosuccinate in *Neurospora crassa.* II. Inhibition of arginyl tRNA synthesis by arginosuccinate. *Biochimica et Biophysica Acta* **145**, 146–52.

NAZARIO, M. (1972). Different arginine tRNA species prevalent in shaken and unshaken cultures of *Neurospora. Journal of Bacteriology* **112**, 1072–82.

NAZARIO, M., KINSEY, J. A. & AHMAD, M. (1971). *Neurospora* mutant deficient in tryptophanyl tRNA synthetase activity. *Journal of Bacteriology* **105**, 121–6.

NEUPERT, W. & LUDWIG, G. D. (1971). Sites of biosynthesis of outer and inner membrane proteins of *Neurospora crassa* mitochondria. *European Journal of Biochemistry* **19**, 523–32.

OJALA, D. & ATTARDI, G. (1972). Expression of the mitochondrial genome in HeLa cells. X. Properties of mitochondrial polysomes. *Journal of Molecular Biology* **65**, 273–89.

OJHA, M. & TURIAN, G. (1971). Interspecific transformations and DNA characteristics in *Allomyces. Molecular and General Genetics* **112,** 49–59.

ONO, T., KIMURA, K. & YANAGITA, T. (1966). Sequential synthesis of various molecular species of RNA in the early phase of conidia germination in *Aspergillus oryzae. Journal of General and Applied Microbiology* **12**, 13–26.

PESTKA, S. (1971). Inhibition of ribosome functions. *Annual Review of Microbiology* **25**, 487–562.

POLLARD, C. J., STEMLER, A. & BLAYDES, D. F. (1966). Ribosomal ribonucleic acids of chloroplastic and mitochondrial preparations. *Plant Physiology* **41**, 1323–9.

RABINOWITZ, M. & SWIFT, H. (1970). Mitochondrial nucleic acids and their relation to the biogenesis of mitochondria. *Physiological Reviews* **50**, 376–427.

RADO, T. A. & COCHRANE, V. W. (1971). Ribosomal competence and spore germination in *Fusarium solani. Journal of Bacteriology* **106**, 301–4.

RAMAKRISHNAN, L. & STAPLES, R. C. (1970). Evidence for a template RNA in resting uredospores of the bean rust fungus. *Contributions from the Boyce Thompson Institute* **24**, 197–202.

RATTI, G. & BUCK, K. W. (1972). Virus particles in *Aspergillus foetidus*: a multicomponent system. *Journal of General Virology* **14**, 165–75.

RAY, R. & DUTTA, S. K. (1972). Isolation of tRNA genes from *Neurospora crassa. Biochemical and Biophysical Research Communications* **47**, 1458–63.

REICH, E. & LUCK, D. J. L. (1966). Replication and inheritance of mitochondrial DNA. *Proceedings of the National Academy of Sciences, U.S.* **55**, 1600–8.

RETEL, J. & PLANTA, R. J. (1970). On the mechanisms of the biosynthesis of ribosomal RNA in yeast. *Biochimica et Biophysica Acta* **224**, 458–69.

RHO, H. M. & DeBUSK, A. G. (1971*a*). N-terminal residues of *Neurospora crassa* proteins. *Journal of Bacteriology* **107**, 840–5.

RHO, H. M. & DeBUSK, A. G. (1971*b*). NH_2 terminal methionine in nascent peptides from *Neurospora crassa. Biochemical and Biophysical Research Communications* **42**, 319–25.

RHO, H. M. & DeBUSK, A. G. (1971*c*). Protein chain initiation by methionyl tRNA in the cytoplasm of *Neurospora crassa. Journal of Biological Chemistry* **246**, 6566–9.

RIFKIN, M. R. & LUCK, D. J. L. (1971). Defective production of mitochondrial ribosomes in the *poky* mutant of *Neurospora crassa. Proceedings of the National Academy of Sciences, U.S.* **68**, 287–90.

ROEDER, R. G. & RUTTER, W. J. (1970*a*). Multiple RNA polymerases and RNA synthesis during sea urchin development. *Biochemistry* **9**, 2543–53.

ROEDER, R. G. & RUTTER, W. J. (1970*b*). Specific nucleolar and nucleoplasmic RNA polymerases. *Proceedings of the National Academy of Sciences, U.S.* **65**, 675–82.

ROHEIM, J. R., KNIGHT, R. H. & VAN ETTEN, J. L. (1974). Synthesis of ribonucleic acids during the germination of *Rhizopus stolonifer* sporangiospores. *Developmental Biology* in press.

ROTHSCHILD, H., HIRKGURA, H. & SUSKIND, S. R. (1967). Ribosomes and ribosomal proteins from *Neurospora crassa.* II,. Ribosomal proteins in different wild type strains and during various stages of development. *Journal of Bacteriology* **94**, 1800–1.

ROWE, M. J., LANSMAN, R. A. & WOODWARD, D. O. (1974). A comparison of mitochondrially synthesised proteins from whole mitochondria and cytochrome oxidase in *Neurospora. European Journal of Biochemistry* **41**, 25–30.

ROWLANDS, R. T. & TURNER, G. (1973). Nuclear and extranuclear inheritance of olibomycin resistance in *Aspergillus nidulans. Molecular and General Genetics* **126**, 201–16.

SALA, F. & KÜNTZEL, H. (1970). Peptide chain initiation in homologous and heterologous systems from mitochondria and bacteria. *European Journal of Biochemistry* **15**, 280–6.

SANSING, G. A., DETROY, S. N., FREER, S. N. & HESSELTINE, W. (1973). Virus particles from conidia of *Penicillium* species. *Applied Microbiology* **26**, 914–8.

SARGENT, M. L. (1973). Use of liquid nitrogen and high ionic strength for the isolation of functional polyribosomes from *Neurospora crassa. Biochimica et Biophysica Acta* **324**, 267–74.

SCHÄFER, K. P. & KÜNTZEL, H. (1972). Mitochondrial genes in *Neurospora*: a single cistron for ribosomal RNA. *Biochemical and Biophysical Research Communications* **46**, 1312–9.

SCHÄFER, K. P., BUGGE, G., GRANDI, M. & KÜNTZEL, H. (1971). Transcription of mitochondrial DNA *in vitro* from *Neurospora crassa. European Journal of Biochemistry* **21**, 478–88.

SCHATZ, G. & MASON, T. L. (1974). Biosynthesis of mitochondrial proteins. *Annual Review of Biochemistry* **43**, 51–87.

SCHÖNBERG, M., SILVERSTEIN, S. C., LEVIN, D. H. & ACS, G. (1971). Asynchronous synthesis of the complementary strands of the Reovirus genome. *Proceedings of the National Academy of Sciences, U.S.* **68**, 505–8.

SCHMOYER, I. R. & LOVETT, J. S. (1969). Regulation of protein synthesis in zoospores of *Blastocladiella emersonii. Journal of Bacteriology* **100**, 854–64.

SEBALD, W., MACHLEIDT, W. & OTTO, J. (1973). Products of mitochondrial protein synthesis in *Neurospora crassa.* Determination of equimolar amounts of three products in cytochrome oxidase on the basic of amino-acid analysis. *European Journal of Biochemistry* **38**, 311–24.

SEBALD, W., WEISS, H. & JACKL, G. (1972). Inhibition of assembly of cytochrome oxidase in *Neurospora crassa* by chloramphenicol. *European Journal of Biochemistry* **30**, 413–7.

SHEARN, A. & HOROWITZ, N. H. (1969*a*). A study of transfer RNA in *Neurospora.* I. Attachment of amino acids and amino acid analogues. *Biochemistry* **8**, 295–343.

SHEARN, A. & HOROWITZ, N. H. (1969*b*). A study of RNA in *Neurospora.* II. Failure to detect transfer RNA alteration in tyrosinase derepressed cultures. *Biochemistry* **8**, 304–22.

SILVER, J.C. & HORGEN, P. A. (1974). Hormonal regulation of presumptive mRNA in the fungus *Achlya ambisexualis Nature, London* **249**, 252–4.

SILVERSTEIN, S. C., ASTELL, C., LEVIN, D. H., SCHÖNBERG, M. & ACS, G. (1972). The mechanisms of Reovirus uncoating and gene activiation *in vivo. Virology* **47**, 797–806.

SINDEN, J. W. & HAUSER, E. (1957). It is 'La France'. *Mushroom Growers Association Bulletin* **95**, 407–9.

SOLL, D. R. & SONNEBORN, D. R. (1971). Zoospore germination in *Blastocladiella emersonii.* III. Structural changes in relation to protein and RNA synthesis. *Journal of Cell Science* **9**, 679–99.

STALLKNECHT, G. F. & MIROCHA, C. J. (1971). Fixation and incorporation of CO_2 in RNA by germinating uredospores of *Uromyces phaseoli. Phytopathology* **61**, 400–5.

STAPLES, R. C. (1974). Synthesis of DNA during differentiation of bean rust uredospores *Physiological Plant Pathology* **4**, 415–24.

STAPLES, R. C., BEDIGIAN, D. & WILLIAMS, P. H. (1968). Evidence for polysomes in extracts of bean rust uredospores. *Phytopathology* **58**, 151–4.

STAPLES, R. C., RAMAKRISHNAN, L. & YANIF, L. (1970). Membrane bound ribosomes in germinated uredospores *Phytopathology* **60**, 58–62.

STAVY, R., STAVY, L. & GALUN, E. (1970). Protein synthesis in aged and young zones of *Trichoderma* colonies. *Biochimica et Biophysica Acta* **217**, 468–76.

STORCK, R. & ALEXOPOULOS, C. J. (1970). DNA of fungi. *Bacteriological Reviews* **34**, 126–56.

STUMM, C. & VAN WENT, J. (1968). Histone bei dem Phycomyceten *Allomyces arbuscula. Experientia* **24**, 1112–3.

SUEOKA, N. & KANO-SUEOKA, T. (1970). Transfer RNA and cell differentiation. *Progress in Nucleic Acid Research and Molecular Biology* **10**, 23–55.

SWANK, R. T., SHEIR, G. I. & MUNKRES, K. D. (1971). *In vivo* synthesis, molecular weights and proportions of mitochondrial proteins in *Neurospora crassa. Biochemistry* **10**, 3924–31.

TANAKA, K., ONO, T. & YANAGITA, T. (1966). Carbon dioxide incorporation into RNA in the early phase of conidia germination in *Aspergillus oryzae* with special reference to tRNA synthesis. *Journal of General and Applied Microbiology* **12**, 329–36.

TELLEZ DE IÑON, M. T., LEONI, P. D. & TORRES, H. N. (1974). RNA polymerase activities in *Neurospora crassa. FEBS Letters* **39**, 91–5.

TIKCHONENKO, T. I., VELIKODVORSKAYA, G. A., BOBKOVA, A. F., BARTOSHEVICH, YU.E., LEBED, E. P., CHAPLYGINA, N. M. & MAKSIMOVA, T. S. (1974). New fungal viruses capable of reproducing in bacteria. *Nature, London* **249**, 454–6.

TIMBERLAKE, W. E. & GRIFFIN, D. H. (1974). Differential effects of analogues of cycloheximide on protein and RNA synthesis in *Achlya. Biochimica et Biophysica Acta* **349**, 39–46.

TIMBERLAKE, W. E., McDOWELL, L. & GRIFFIN, D. H. (1972). Cycloheximide inhibition of the DNA dependent RNA polymerase of *Achlya bisexualis. Biochemical and Biophysical Research Communications* **46**, 942–7.

TISDALE, J. H. & DeBUSK, A. G. (1972). Permeability problems encountered when treating conidia of *Neurospora crassa* with RNA synthesis inhibitors. *Biochemical and Biophysical Research Communications* **48**, 816–22.

TOTTEN, R. E. & HOVE, H. B. (1971). Temperature dependent actinomycin D; effect on RNA synthesis during synchronous development in *Neurospora crassa. Biochemical Genetics* **5**, 521–32.

TURNER, G. (1973). Cycloheximide resistant synthesis of mitochondrial-membrane components in *Aspergillus nidulans, European Journal of Biochemistry* **40**, 201–6.

URSINO, D. J., STURANI, E. & ALBERGHINA, F. A. M. (1969). RNA degrading enzymes associated with purified ribosomes from *Neurospora crassa* mycelia. *Biochimica et Biophysica Acta* **179**, 500–2.

VAN ASSCHE, J. A. & CARLIER, A. R. (1973). The pattern of protein and nucleic acid synthesis in germinating spores of *Phycomyces blakesleanus. Archiv für Mikrobiologie* **93**, 129–36.

VAN ETTEN, J. L. (1968). Protein synthesis during fungal spore germination. I. Characteristics of an *in vitro* phenylalanine incorporating system prepared from germinating spores of *Botryodiplodia theobromae. Archives of Biochemistry and Biophysics* **125**, 13–21.

VAN ETTEN, J. L. (1969). Protein synthesis during fungal spore germination. *Phytopathology* **59**, 1060–4.

VAN ETTEN, J. L. (1971). Preparation of biologically active ribosomal subunits from fungal spores. *Journal of Bacteriology* **106**, 704–6.

VAN ETTEN, J. L. & BRAMBL, R. M. (1968). Protein synthesis during fungal spore germination. II. Amino acyl soluble RNA synthetase activities during germination of *Botryodiplodia theobromae* spores. *Journal of Bacteriology* **96**, 1042–8.

VAN ETTEN, J. L., KOSKI, R. K. & EL-OLEMY, M. M. (1969). Protein synthesis during fungal spore germination. IV. Transfer ribonucleic acid from germinated and ungerminated spores. *Journal of Bacteriology* **100**, 1182–6.

VAN ETTEN, J., PARISI, B. & CIFERRI, O. (1966). Interchangeability of ribosomes and protein synthesising enzymes from *Penicillium cyclopium* and those prepared from other organisms. *Nature, London* **212**, 932–3.

VAN FRANK, R. M., ELLIS, L. F. & KLEINSCHMIDT, W. J. (1971). Purification and physical properties of mycophage PS1. *Journal of General Virology* **12**, 33–42.

VANDERHOFF, L. N., TRAVIS, R. C., MURRAY, M. G. & KEY, J. L. (1972). Interspecies aminoacylation of tRNA from several higher plants, *Neurospora*, yeast and *E. coli. Biochimica et Biophysica Acta* **269**, 413–8.

VERMA, I. M., EDELMAN, M. & LITTAUER, U. Z. (1971). A comparison of nucleotide sequences from mitochondrial and cytoplasmic ribosomal RNA of *Aspergillus nidulans. European Journal of Biochemistry* **19**, 124–9.

VERMA, I. M., EDELMAN, M., HERZBERG, M. & LITTAUER, U. Z. (1970). Size determination of mitochondrial ribosomal RNA from *Aspergillus nidulans* by electron microscopy. *Journal of Molecular Biology* **52**, 37–40.

VIAU, J. P. & DAVIS, F. F. (1970). Effect of cycloheximide on the synthesis and modification of ribosomal RNA in *Neurospora crassa. Biochimica et Biophysica Acta* **209**, 190–5.

VILLA, V. D. & STORCK, R. (1968). Nucleotide composition of nuclear and mitochondrial deoxyribonucleic acid of fungi. *Journal of Bacteriology* **96**, 184–90.

VOGEL, F. S. & KEMPER, L. (1967). Intrinsic RNA synthesis in mitochondria of *Agaricus campestris*, underscoring the probability of an extra-nuclear genetic system. *Experimental Cell Research* **47**, 209–21.

VOLKOFF, O., WALTERS, T. & DEJARDIN, R. A. (1972). An examination of *Penicillium notatum* for the presence of *Penicillium chrysogenum* type virus particles. *Canadian Journal of Microbiology* **18**, 1352–3.

WAGNER, R. P. (1969). Genetics and phenogenetics of mitochondria. *Science* **163**, 1026–31.

WEISS, H. & ZIGANKE, B. (1974). Cytochrome b in *Neurospora crassa* mitochondria. Site of translation of the heme protein. *European Journal of Biochemistry* **41**, 63–71.

WONG, R. S. L., SCARBOROUGH, G. A. & BOREK, E. (1971). Transfer RNA methylases during the germination of *Neurospora crassa. Journal of Bacteriology* **108**, 446–50.

WOOD, H. A. (1973). Viruses with double-stranded RNA genomes. *Journal of General Virology* **20**, 61–85.

WOOD, H. A. & BOZARTH, R. F. (1972). Properties of virus-like particles of *Penicillium chrysogenum*: one double-stranded RNA molecule per particle. *Virology* **47**, 604–9.

WOOD, H. A. & BOZARTH, R. F. (1973). Heterokaryon transfer of virus-like particles associated with a cytoplasmically inherited determinant in *Ustilago maydis. Phytopathology* **63**, 1019–21.

WOOD, H. A., BOZARTH, R. F., ADLER, J. & MACKENZIE, D. W. (1974). Proteinaceous virus-like particles from an isolate of

Aspergillus flavus. Journal of Virology **13**, 532–4.

WOOD, H. A., BOZARTH, R. F. & MISLIVEC, P. B. (1971). Virus-like particles associated with an isolate of *Penicillium brevi-compactum. Virology* **44**, 592–8.

WOOD, D. D. & LUCK, D. J. L. (1969). Hybridisation of mitochondrial ribosomal RNA. *Journal of Molecular Biology* **41**, 211–24.

WOODWARD, D. O. & MUNKRES, K. D. (1966). Alterations of a maternally inherited mitochondrial structural protein in respiratory-deficient strains of *Neurospora. Proceedings of the National Academy of Sciences, U.S.* **55**, 872–80.

WORTHY, T. E. & EPLER, J. L. (1972). Repair of light-induced UV damage to DNA of *Neurospora crassa. Journal of Bacteriology* **110**, 1010–6.

WORTHY, T. E. & EPLER, J. L. (1973). Characterisation of excision repair in *Neurospora crassa. Journal of Bacteriology* **115**, 498–505.

YANIF, Z. & STAPLES, R. C. (1969). Transfer activity of ribosomes from germinating uredospores. *Contributions from the Boyce Thompson Institute* **24**, 157–63.

YANIF, Z., & STAPLES, R. C. (1971). The purification and properties of amino acyl tRNA binding enzyme from bean rust uredospores. *Biochimica et Biophysica Acta* **232**, 717–25.

CHAPTER 9

Reserve Carbohydrates in Fungi

H. J. BLUMENTHAL

9.1 Introduction

When fungal cells are placed in a balanced medium containing a sugar such as glucose or sucrose as their major source of carbon and energy, the cells grow and during this relatively luxurious state of carbohydrate metabolism they usually accumulate what are referred to in this chapter as reserve carbohydrates. Since low molecular weight, soluble carbohydrates such as the polyols (= sugar alcohols) and the disaccharide trehalose appear to play important roles as carbon and energy reserves amongst the filamentous fungi, this chapter is not entitled reserve macromolecules.

Lipid reserves are used, along with carbohydrates, during the germination of many fungal spores (Hashimoto, Wu & Blumenthal, 1972; Sussman & Douthit, 1973). A similar situation probably exists in the mycelia of many filamentous fungi (Foster, 1949). For example, in some of the dermatophytes, 60–74% of the mycelial weight is lipid, which is regarded as the primary storage material rather than carbohydrate (*cf.* Grappel, Bishop & Blank, 1974). However, no specific section will be devoted to discussing the status of reserve lipids in this chapter.

This chapter is restricted primarily to considerations of the biosynthesis, degradation and function of the polyols, trehalose, glycogen and certain other intracellular and extracellular polysaccharides of filamentous fungi, although appropriate examples from other microbes will also be considered. This is a selective rather than a comprehensive review and a conscious attempt has been made to use the most recent citations.

9.2 Polyols and Trehalose

Lewis & Smith (1967) have reviewed the distribution and the metabolism of polyols in fungi and green plants. In a later review, Smith, Muscatine & Lewis (1969) stated that 'It is now generally accepted that the commonest and most abundant soluble carbohydrates in mycelia of most fungi (with the exception of Phycomycetes) are trehalose and polyols, especially mannitol and arabitol.' In a typical carbon balance with an *Aspergillus* sp. growing in a simple glucose and urea salts medium, 35% of the glucose carbon was converted to mannitol, along with smaller amounts of glycerol (10%), erythritol (2.5%) and glycogen (1%) (Lee, 1967*a*). Lee calculated that 34% of glucose carbon was used for growth and 53% was used for polyol formation.

In all of the fermentations reported in the literature, no more than about a 50% yield of mannitol has been observed. Under these conditions, such large amounts of this very soluble substance cannot be retained by the mycelium and mannitol may be found in culture filtrates.

Only trehalose and D-mannitol were invariably present in the mycelial neutral, water-soluble carbohydrates of *Claviceps purpurea*, which varied in composition with changes in the nutrient solution on which the fungus was grown and with the strain used. Other polyols sometimes found were glycerol, D-arabitol, threitol, erythritol and xylitol, while galactitol was found only when the cells were grown on galactose. These investigators suggested that the polyols '. . . may serve as a form of stored reducing power as well as a reserve of carbon nutrients . . .' although they recognized that other studies were needed to decide if they were useful reserves or metabolic by-products (Vining & Taber, 1964).

Earlier studies on the formation of mannitol from sugar in *Aspergillus niger* showed an increase to 10% of the mycelial dry weight in 48 h. The mannitol was then rapidly utilized when the sugar in the medium was exhausted, decreasing to 3.8% by 72 h (*cf.* Foster, 1949).

In more recent studies, it has been possible to show differences in the formation and utilization sequence of mannitol and a pentitol. Corina & Munday (1971*b*) found that mannitol accumulated while *Aspergillus clavatus* utilized glucose, and then the fungus used the mannitol after the glucose was metabolized. In turn, ribitol, a pentitol, accumulated as the mannitol disappeared. From the $^{3}H/^{14}C$ isotope ratio patterns, they concluded that the predominant function of mannitol was that of a storage compound, probably in preparation for conidiation, whereas ribitol appeared to be involved primarily in hydrogen acceptor mechanisms occurring during the utilization of storage materials such as fatty acids.

In an important series of studies with the marine Hyphomycete *Dendryphiella salina*, Holligan & Jennings (1972*a, b, c, d*) also found that there were differences in the timing of glucose conversion to mannitol and a pentitol, in this case the more commonly found arabitol, and they suggested that mannitol was a precursor for arabitol synthesis. From these and the preceding studies, the general sequence appeared to be: glucose → mannitol → pentitol (arabitol or ribitol).

Growth of filamentous fungi, such as *Aspergillus niger*, on acetate as the sole carbon source also leads to the production of the acyclic polyols mannitol, arabitol and erythritol, as well as the disaccharides trehalose and maltose (Barker, Gómez-Sánchez & Stacey, 1958). The conversion of acetate to mannitol is widespread in fungi and in this case the acetate probably is assimilated via the glyoxylate cycle (Cochrane, 1958; Holligan & Jennings, 1972*b*).

Although mannitol was isolated from *Aspergillus niger* conidia almost 50 years ago, and suggested as a possible substrate for the endogenous metabolism of germinating fungal conidia 25 years ago, it remained for

Horikoshi, Iida & Ikeda (1965) to show that the large amounts of mannitol in *Aspergillus oryzae* conidia were rapidly consumed in the early stages of germination, providing evidence that the mannitol was a major substrate for endogenous respiration after conversion to fructose by mannitol dehydrogenase.

Perhaps it is appropriate now to consider the enzymes concerned with the biosynthesis and catabolism of mannitol. At the present time, unfortunately, the differentiation between the anabolic and catabolic enzymes involved in mannitol metabolism is nor clear. One important principle of metabolic regulation is that exactly the same sequence of enzymatic steps will not be used for catabolic and anabolic pathways in the living cell. Consequently, we can be sure that a single enzyme will not perform both functions, although this does not rule out multiple forms of the same enzyme, *e.g.* separate enzymes requiring NAD and NADP.

Mannitol may be formed either by reduction of fructose directly (eqn. *1*) and/or by reduction of fructose-6-P (eqn. *2*) followed by hydrolysis of the resulting mannitol-1-P by a specific phosphatase (eqn. *3*). Thus far there have been no reports in fungi of the direct reduction of mannose to mannitol, of mannose-1-P reduction to mannitol-1-P, or of epimerization of other hexitols to mannitol.

$$\text{Fructose} + \text{NADH} \xleftrightarrow[\text{dehydrogenase}]{\text{Mannitol (NAD)}} \text{Mannitol} + \text{NAD} \qquad (1a)$$

$$\text{Fructose} + \text{NADPH} \xleftrightarrow[\text{dehydrogenase}]{\text{Mannitol (NADP)}} \text{Mannitol} + \text{NADP} \qquad (1b)$$

$$\text{Fructose-6-P} + \text{NADPH} \xleftrightarrow[\text{dehydrogenase}]{\text{Mannitol-1-P (NADP)}} \text{Mannitol-1-P} + \text{NADP} \qquad (2a)$$

$$\text{Fructose-6-P} + \text{NADH} \xleftrightarrow[\text{dehydrogenase}]{\text{Mannitol-1-P (NAD)}} \text{Mannitol-1-P} + \text{NAD} \qquad (2b)$$

$$\text{Mannitol-1-P} + H_2O \xrightarrow[\text{phosphatase}]{\text{Mannitol-1-}} \text{Mannitol} + H_3PO_4 \qquad (3)$$

$$\text{Mannitol} + \text{Phosphoenol} - \text{pyruvate (PEP)} \xleftrightarrow[\text{phosphotransferase}]{\text{Mannitol-PEP}} \text{Mannitol-1-P} + \text{Pyruvate} \qquad (4)$$

$$\text{Mannitol} + \text{Acetyl phosphate} \xleftrightarrow[\text{phosphotransferase}]{\text{Mannitol-Acetyl-P}} \text{Mannitol-1-P} + \text{Acetate} \qquad (5)$$

$$\text{Mannitol} + \text{ATP} \xrightarrow{\text{Mannitol kinase}} \text{Mannitol-1-P} + \text{ADP} \qquad (6)$$

Since the enzyme reactions indicated in equations *1a, b* and *2a, b* are reversible *in vitro*, these enzyme reactions can also be considered to be involved in the catabolism of mannitol. However for equations *2a, b* to be operative, free mannitol would first have to be phosphorylated by either PEP (eqn. *4*; Leighton, Stock & Kelln, 1970), acetyl phosphate (eqn. *5*; Lee, 1967*b*) or ATP (eqn. *6*; Leighton, Stock & Kelln, 1970). To further complicate the picture, both mannitol dehydrogenase (eqn. *1*) and mannitol-1-P dehydrogenase (eqn. *2.*) may each exist as separate NAD- and NADP-linked enzymes, sometimes together in the same cell.

Since free mannitol has both reflective and rotational symmetry, one cannot distinguish the 'top' from the 'bottom' of the molecule. Thus mannitol-1-P and mannitol-6-P are the same molecules. In such cases the rules of carbohydrate nomenclature require the use of the lower number, hence the use of mannitol-1-P. The problems that this symmetry of mannitol produces in attempts to estimate glucose catabolic pathways in fungi (Blumenthal, 1965) which form mannitol from glucose have recently been discussed (Holligan & Jennings, 1972*c*, *d*).

A study by Yamada, Okamoto, Kodama & Tanaka (1959) clearly suggested that mannitol biosynthesis in *Piricularia oryzae* occurred by sequential reduction of fructose-6-P by a NAD-linked mannitol-1-P dehydrogenase (eqn. *2b*), followed by cleavage of the resulting mannitol-1-P by a specific mannitol-1-phosphatase (eqn. *3*). Since these investigators had not indicated the presence of any mannitol dehydrogenase activity in their enzyme preparations, there appeared to be no question that this was *the* enzymatic mechanism for fungal mannitol biosynthesis. Lee (1967*b*) also thought that these reactions would be the biosynthetic ones for mannitol formation in *Aspergillus*, particularly because the mannitol-1-phosphatase reaction was irreversible and thus would drive the paired reactions towards mannitol accumulation. Corina & Munday (1971*b*) also demonstrated the presence of the NAD-linked mannitol-1-P dehydrogenase in *Aspergillus clavatus* and measured the capacity of mycelial extracts to synthesize ^{3}H-mannitol from 6-^{3}H-glucose-6-P daily over a 6-day period. The capacity was relatively constant during the first 4 days of growth and then decreased to about one-half the initial rate by the sixth day. With phosphatase(s) and glucose-P isomerase almost surely present in the crude extracts, these results do not rule out the possible participation of a NAD-linked mannitol dehydrogenase acting on ^{3}H-fructose.

There has been evidence that only mannitol dehydrogenase is involved in mannitol catabolism. For example, the conidia of *Aspergillus oryzae* contained separate NAD- and NADP-linked mannitol dehydrogenases (Horikoshi, Iida & Ikeda, 1965) although the *in vitro* activity of the NAD-linked enzyme (eqn. *1a*) was very low, amounting to only about 5 to 10% of the NADP-linked enzyme (eqn. *1b*). In this case, since the conidial mannitol was being broken down to provide energy for conidial germination, the NADP enzyme was considered to be operating to convert mannitol to fructose.

The assumption of separate catabolic and anabolic mannitol enzyme systems was not fully supported by the results of Strandberg (1969), who studied the pathways of both mannitol biosynthesis and utilization in *Aspergillus candidus* extracts. He determined the enzyme levels in both glucose- and mannitol-grown cells, expecting that higher concentrations of the mannitol catabolic enzymes would be found in the mycelium which had been grown with mannitol as the sole carbon source than in glucose-grown cells. In addition to the NAD-linked mannitol-1-P dehydrogenase and a mannitol-1-P phosphatase, he also found a NADP-linked mannitol dehydrogenase in extracts of both glucose- and mannitol-grown mycelia. However, the levels of mannitol dehydrogenase were lower in the mannitol-grown mycelia than in the glucose-grown mycelia, just the opposite effect he had expected if the mannitol dehydrogenase was involved primarily in mannitol catabolism.

The mannitol metabolism of the germinating macroconidia of *Microsporum gypseum* was also examined by Leighton, Stock & Kelln (1970) and spore extracts were found to contain a NAD-linked mannitol-1-P dehydrogenase (eqn. *2b*), a non-specific phosphatase that could hydrolyse mannitol-1-P as well as a number of other sugar phosphates, and a NADP-linked mannitol dehydrogenase (eqn. *1b*). From the results of experiments measuring enzyme levels during a 24 h period of germination, as well as measurements of the enzyme levels in spores using either glucose or mannitol as the carbohydrate source for both sporulation and spore germination, there was only slight suggestive evidence that mannitol dehydrogenase was more directly involved with mannitol utilization than the combined mannitol-1-P dehydrogenase plus phosphatase system. Holligan & Jennings (1972*b*) found dehydrogenase activity in crude extracts of *Dendryphiella salina* with mannitol and NAD or NADP, but not with mannitol-1-P. On that basis they concluded that free fructose was the substrate for mannitol formation with mannitol dehydrogenase.

The formation of mannitol-1-P by the mannitol-1-P dehydrogenase reaction probably serves a useful physiological function. In glucose-grown *Escherichia coli*, which has not been reported to accumulate free mannitol as reserve product, the intact bacterial suspensions accumulate mannitol-1-P during anaerobic utilization of glucose. It was suggested by Helle & Klungsoeyr (1962) that the mannitol-1-P (NAD) dehydrogenase served as a means to reoxidize NADH, whilst reducing fructose-6-P to mannitol-1-P, when other means for reoxidation of NADH were limiting in the absence of oxygen.

In filamentous fungi there is evidence that the hexosemonophosphate (=pentose phosphate) pathway is involved in polyol metabolism. The addition of nitrate to a glucose-containing medium significantly increased the intracellular content of arabitol, but not mannitol, in the starved *Dendryphiella salina* mycelium. Holligan & Jennings (1972*b*) concluded that '. . . the quantity of arabitol synthesized is related to the activity of the pentose-phosphate pathway which in turn is influenced by the form of nitrogen supplied to the fungus'. Their studies with 1-^{14}C- or 6-^{14}C-glucose confirmed that the activity of the hexosemonophosphate pathway was highest with nitrate as the nitrogen source (Holligan & Jennings, 1972*c*). Furthermore, they were able to demonstrate that mannitol was synthesized partly by a 'direct' route and partly from hexose-P reformed after cycling *via* the hexosemonophosphate pathway. Also, the mannitol pool appeared to be turning over more rapidly than the arabitol pool.

Additional indications of some type of linkage between the hexosemonophosphate pathway and polyol metabolism in filamentous fungi have come from genetic studies with mutants of *Aspergillus nidulans*. Hankinson (1974) noted that four of the hexosemonophosphate pathway enzymes and mannitol-1-P dehydrogenase were each subject to at least two independent regulatory mechanisms and suggested that glucose-6-P, or a closely related intermediate, might be an inducer of these enzymes. In *Azotobacter agilis* it has been demonstrated that ribitol is an inducer for the bacterial mannitol dehydrogenase, although ribitol itself is not a substrate for the enzyme (Marcus & Marr, 1961). Apparently this interesting finding has not been applied in any fungal system.

All of the studies using enzyme assays were performed on fungal extracts *in vitro* and it is well known that the actual *in vivo* conditions are not always reflected in the results of *in vitro* experiments. Reactions reversible *in vitro* may not be reversible *in vivo*. Consequently, the identification of specific biosynthetic and catabolic mannitol enzymes is still unresolved.

Recently, Jennings & Austin (1973) described a most unusual relationship between the polyols and cellular polysaccharides in the marine fungus *Dendryphiella salina*. They found a stimulatory effect, following the absorption of the nonmetabolizable sugar 3-*O*-methyl glucose, on the conversion of mannitol and arabitol to polysaccharides and other insoluble compounds. These investigators reported that the total soluble carbohydrate concentration in the cytoplasm remained constant during the uptake of the sugar analogue and suggested that this maintenance of a constant concentration while absorbing soluble carbohydrates from the medium was a means of regulating the hyphal osmotic pressure to prevent hyphal bursting or changes in hyphal structure.

Trehalose is an α,α-diglucoside, the most widely distributed disaccharide in fungi, and known to be found along with the polyols in fungi since the studies of Bourquelot in the late nineteenth century. In the case of trehalose, as opposed to mannitol, the anabolic and catabolic enzymes are clearly differentiated. The biosynthesis of trehalose occurs via a sugar nucleotide in a two-step process:

$$\text{UDP-Glucose} + \text{D-Glucose-6-P} \xrightleftharpoons[\text{synthase (UDP-forming)}]{\text{Trehalose-P}} \text{UDP} + \text{Trehalose-6-P} \qquad (7)$$

$$\text{Trehalose-6-P} + H_2O \xrightarrow[\text{phosphatase}]{\text{Trehalose}} \text{Trehalose} + H_3PO_4 \qquad (8)$$

The catabolism of trehalose is effected by the action of the single carbohydrase, trehalase:

$$\text{Trehalose} + H_2O \xrightarrow{\text{Trehalase}} 2\ \text{D-Glucose} \qquad (9)$$

Sussman (1966) has reviewed the role of trehalose in *Neurospora*. Conidia contain about 5–10% trehalose, which is rapidly consumed during germination. There is then a resynthesis of the trehalose by the mycelium, reaching levels of about 2–4% as soon as conidiation begins. In aconidial mutants, trehalose continues to accumulate until growth ceases. Sussman therefore suggested that trehalose was used during both conidial germination and formation. In dormant ascospores, lipid is used as the endogenous substrate and the trehalose, present in concentrations up to 14%, is unused, even after storage for many years. However, immediately following ascospore germination, the trehalose is rapidly degraded by the trehalase.

The fact that the large amounts of the reserve carbohydrate trehalose in the cytoplasm are not used in ungerminated *Neurospora* ascospores, in spite of the fact that it coexists with active forms of the enzyme trehalase, suggests that there is some form of separation of the substrate and enzyme. A current explanation for these findings is that activation of a variety of fungal spores by heat, solvents, *etc.*, lessens a permeability barrier separating the enzyme and substrate. Although there is more than one trehalase isozyme in *Neurospora*, and the genetic control of these multiple forms has been studied (Yu, Garrett & Sussman, 1971), their function is not fully understood. The differences in the trehalose-trehalase relationships between the asexual and sexual spores of the same *Neurospora* species is exemplified by the finding that trehalose is easily and almost completely extracted from intact ungerminated conidia while very little of the sugar can be extracted from intact ungerminated ascospores (Sussman, 1966).

Trehalose also apparently serves as a carbon and energy source in *Dictyostelium discoideum* during germination (*cf.* Smith & Galbraith, 1971). In this organism, some enzymes like trehalose-P synthase may not manifest their activity until special methods are applied to remove inhibitors (*cf.* Sussman & Douthit, 1973). This emphasizes the dangers of negative results in enzyme studies. A good example of the complexity that inhibitors and/or activators may introduce into enzyme studies comes from the recent report of Ulane & Cabib (1974). They purified the protein inhibitor of the yeast protease which, in turn, activated the inactive, or 'zymogen', form of the enzyme chitin synthetase.

The role of glycolipids, such as the acylated trehalose found in *Pullularia*, is largely unknown although it has been suggested that such compounds may serve as an endogenous reserve in some fungal spores and dormancy breaking may involve their metabolism.

The formation and utilization of trehalose and polyols do not always follow parallel courses. Vining & Taber (1964) found that with an inorganic nitrogen source, *Claviceps purpurea* formed increasing amounts of trehalose while polyol accumulation decreased. In *Schizophyllum commune* spores or germlings 1-^{14}C-arabitol was converted primarily into trehalose while this was not the case for 1-^{14}C-mannitol (Aitken & Niederpruem, 1973).

Holligan & Jennings (1972*b*) have noted that '. . . in fungi containing both mannitol and arabitol, growth on inorganic nitrogen leads to arabitol accumulation, whereas fungi which store mannitol and trehalose show increased levels of trehalose.'

Two new microbial enzymes have recently been reported for the catabolism of trehalose, although neither has yet been found in fungi. A trehalose phosphorylase was described in *Euglena gracilis* which converted trehalose plus inorganic phosphate into glucose-1-P and glucose, while trehalose-6-P was cleaved by a phosphotrehalase in the bacterium *Bacillus popilliae* with the formation of glucose and glucose-6-P (*cf.* Bhumiratana, Anderson & Costilow, 1974).

9.3 Glycogen and Starch

Both glycogen and starch, the recognized reserve polysaccharides of animals and plants, respectively, were first reported as fungal products about 90 years ago. Cramer first extracted a 'spore starch' from *Penicillium glaucum* spores towards the end of the last century, which constituted 17% of the dry spore weight, and gave the typical blue colouration of plant starch with iodine. Glycogen, which was detected at about the same time within many Basidiomycetes and a yeast, was recognized on the basis of the red-brown colour it produced on addition of iodine (Foster, 1949; Gorin & Spencer, 1968).

It was soon recognized that starch was not found in all fungi. In a study of 45 *Penicillium* species, only 22 produced mycelial starch. Furthermore, the starch formers elaborated smaller amounts of organic acids than the non-starch formers. In these studies, both the starch and the organic acids were considered to be acting as reserve substrates (*cf.* Foster, 1949).

In more recent years, studies with filamentous fungi have concentrated on fungal glycogen. Glycogens, from both animal and microbial sources, consist of multiply branched molecules containing numerous chains of $\alpha(1 \rightarrow 4)$-linked D-glucose residues with resulting molecular weights of about 10 million. The chains average about 10–12 glucose residues and there are $\alpha(1 \rightarrow 6)$ branch linkages attached. The structure of glycogen has recently been reviewed by Manners (1971).

Fungal glycogen is often found in amounts approximating 5% of the mycelial or spore dry weight (Cochrane, 1958), although it has been reported to be present in levels as high as 37% in *Phymatotrichum omnivorum* sclerotia (Ergle, 1947). Low CO_2 concentrations enhance glycogen utilization during the vegetative growth of this organism (Gunasekaran, 1972). In *Neurospora crassa*, glycogen accumulates in the subapical and generally older hyphal zones, the absence of the reserve material in the hyphal tip being in keeping with the importance of glycolysis in that zone (*cf.* Turian & Bianchi, 1972).

The mechanism for the biosynthesis of glycogen occurs, like that of trehalose, by means of sugar nucleotides. The enzyme glycogen synthetase, or UDPG:glycogen α-4 glucosyl transferase, was discovered only 17 years ago but the many studies on the regulation of this enzyme attest to its importance as a primary reserve in diverse types of living cells, including the filamentous fungi.

$$\text{UDP-glucose} + (1,4\text{-}\alpha\text{-D-Glucosyl})_n \xleftrightarrow[\text{synthetase}]{\text{Glycogen}} \text{UDP} + (1,4\text{-}\alpha\text{-D-Glucosyl})_{n+1} \qquad (10)$$

Glycogen phosphorylase is the enzyme responsible for the catabolism of glycogen.

$$(1,4\text{-}\alpha\text{-D-Glucosyl})_n + H_3PO_4 \xleftrightarrow[\text{phosphorylase}]{\text{Glycogen}} (1,4\text{-}\alpha\text{-D-Glucosyl})_{n-1} + \alpha\text{-D-Glucose-1-P} \qquad (11)$$

Since phosphorylase is reversible and can synthesize glycogen *in vitro*, it was thought for many years to be the enzyme responsible for both the synthesis and catabolism of glycogen *in vivo*. However, it is now known that the phosphorylase functions only as a catabolic enzyme *in vivo*. In fact, it was the study of glycogen biosynthesis and breakdown that led to the recognition of the generalization that pathways for the biosynthesis and catabolism of a substance must differ to an extent that will allow for the separate regulation of the two processes.

There are two different modes of regulation of glycogen synthetase. The first is 'metabolite control' by allosteric effectors, such as glucose-6-P and adenine nucleotides, which are *not* substrates for the enzyme (eqn. *10*), and the second is interconversion of the glycogen synthetase between two separable forms, a glucose-6-P-independent (I) form and a glucose-6-P-dependent (D) form (Cabib, Rothman-Denes & Huang, 1973; Téllez-Iñón, Terenzi & Torres, 1969).

The metabolite control mechanism has been demonstrated in a number of filamentous fungi, following its discovery in yeast. For example, in the water mould *Blastocladiella emersonii*, in which a glycogen-like polysaccharide constitutes 4 to 8% of the dry weight of the organism, Plessmann Camargo, Meuser & Sonneborn (1969) have shown that glucose-6-P stimulates the glycogen synthetase in subcellular fractions of the organism to varying degrees in different phases of the fungal life cycle. For zoospores, stimulation of the enzyme by glucose-6-P was as high as 90-fold, while for growing cells it was only around 4-fold. The glycogen synthetase reaction was also inhibited by nucleotides and nucleotide derivatives such as ATP, ADP and UDP, but glucose-6-P could completely reverse the inhibition by ATP. As a result of these findings it was concluded that glycogen synthesis was controlled in a major way, but not necessarily the only way, by the intracellular concentration of glucose-6-P.

The antagonistic effects of glucose-6-P and ATP have also been observed in the glycogen synthetase of *Dictyostelium.* From studies with yeast, this type of control apparently predominates in resting cells, while the conversion of the D to the I form appears to be the main factor in the accumulation during the later growth phases. Since the D form is more sensitive to inhibition, this form has been suggested to be inactive in living yeast (Cabib, Rothman-Denes & Huang, 1973).

The interconvertible D and I forms of glycogen synthetase have been reported to occur in *Neurospora crassa* (Téllez-Iñón, Terenzi & Torres, 1969). They found that the conversion of the I to D form *in vitro* required ATP-Mg^{2+} while the D to I conversion required only Mg^{2+}. There is also some evidence that these *in vitro* enzyme interconversions in *N. crassa* extracts may also occur *in vivo.* It has previously been reported (Trinci & Collinge, 1973) that *N. crassa* mycelia grown with L-sorbose as the carbon source, instead of sucrose, led to major morphological and physiological changes in the fungus. Téllez-Iñón, Terenzi & Torres (1969) observed that sorbose-grown mycelia accumulated much less glycogen than mycelia grown in sucrose. Furthermore, extracts from sorbose-grown mycelium showed lower levels of glycogen synthetase than sucrose-grown mycelium, and the glycogen synthetase from sucrose-grown mycelium also exhibited a 3- to 6-fold greater stimulation by glucose-6-P than did the sorbose-grown enzyme. The physiological significance of these observations is enhanced when the results of recent studies with *Aspergillus clavatus* growing on glucose are considered (Corina & Munday, 1971*b*). They found that the levels of a sugar tentatively identified as sorbose were shown to accumulate in the mycelia along with ribitol after the utilization of the mannitol which in turn had accumulated during glucose metabolism. Roughly, the sequence appeared to be: glucose → mannitol → ribitol + sorbose. Although these investigators did not report the cellular levels of glycogen or of glycogen synthetase, one can predict that the least active form of the enzyme will be predominant at the time that the sorbose is accumulating.

Téllez-Iñón & Torres (1970) have reported that glycogen phosphorylase, the enzyme responsible for glycogen utilization in *Neurospora crassa* mycelia, also exists in two interconvertible forms, one active (*a* form) and the other inactive (*b* form) in the absence of 5′-AMP. Furthermore, they found that the conversion of the *b* to the *a* form required ATP-Mg^{2+} and that the conversion proceeded at a faster rate in the presence of cyclic AMP. The enzyme adenyl cyclase, which forms cyclic AMP from ATP, was also detected in the mycelium. Thus, the interconversions of the phosphorylated and dephosphorylated forms of glycogen synthetase and glycogen phosphorylase are important features of glycogen metabolism in filamentous fungi, as they are in mammalian cells.

9.4 Other Polysaccharides

Although all of the di- or polysaccharide reserve carbohydrates considered thus far, trehalose, glycogen and starch, contain only glucose with α-linkages, not all such compounds serve as reserve substances. One example is the polysaccharide nigeran (mycodextran), which is a linear polymer of D-glucose found in aspergilli and penicillia and which consists predominantly of alternate α-D$(1 \rightarrow 3)$- and α-D$(1 \rightarrow 4)$-linked glucopyranosyl residues (Gorin & Spencer, 1968). Early evidence showed that this carbohydrate was not reused in mycelia, and that it increased with age, even in autolysed cultures. These findings have been confirmed more recently by Gold, Mitzel & Segel (1973). Similarly, the extracellular α-glucan, pullulan, elaborated by *Pullularia pullulans* in its yeast form but not its mycelial form, is not normally utilizeable by the fungus during periods of glucose limitation (Catley, 1973).

The cell walls of budding yeasts are rich in mannan-protein complexes which are interspersed with a glucan (Ballou & Raschke, 1974), while the walls of mycelial fungi contain only small quantities of mannose polymers (Bartnicki-Garcia, 1968). Even in yeast there is no evidence that the structural mannan-protein complexes, which are the principal immunogens of the yeast cell, or the mannan-enzymes play any significant role as reserve carbohydrates.

The status of the strictly structural polysaccharide components of filamentous fungi, such as chitin, also fall into this class. Corina & Munday (1971*a*) recently re-emphasized this point experimentally by measuring the metabolic stability of the hyphal cell wall carbohydrates of *Aspergillus clavatus*. They measured the mycelial isotope distribution after growth on $U^{14}C$- and 6-^{3}H-glucose and the $^{3}H/^{14}C$ ratios of re-isolated monosaccharides. Their results suggested that glucose and glucosamine became metabolically inert once they had been incorporated into the wall polymers.

In spite of these examples of polysaccharides that do not appear to serve as endogenous reserves, there have been some suggestions that certain glucans may play a role as reserve material. Bull & Chesters (1966), while reviewing the occurrence of laminarins, which are $\beta(1 \rightarrow 3)$ glucans, mentioned the possibility that $\beta(1 \rightarrow 3)$ glucans may also occur as reserve material in some filamentous fungi. They cited the sclerotan from *Sclerotinia libertiana* and pachyman from *Portia cocos* as examples. More recently, Tokunaga & Bartnicki-Garcia (1971*a*, *b*) reported that the walls from cysts, hyphae and sporangia of *Phytophthora palmivora* were chiefly β-glucans with $1 \rightarrow 3$-, $1 \rightarrow 4$- and $1 \rightarrow 6$-links. The zoospores were found to depend upon internal reserves for synthesizing their cyst wall. During synchronous encystment, in the absence of exogenous nutrients, the largely water-soluble glucan carbohydrate fraction decreased markedly while there was a concomitant increase in the insoluble cyst wall glucan. Clearly in this case, since there were no exogenous nutrients, only the soluble glucan could be considered to be a major reserve carbohydrate.

Recent studies with the insoluble cell wall glucans from a number of yeasts have demonstrated that the preparations were heterogeneous and contained 15% or more of a soluble, branched $\beta(1\rightarrow6)$-glucan, which could be extracted with dilute acetic acid, in addition to the remaining insoluble branched $\beta(1\rightarrow3)$-glucan. Of interest here is the comment of Manners, Masson & Patterson (1974) that 'The function of the $\beta(1\rightarrow6)$-glucan is not yet known. Since it is soluble and of low molecular weight it might be a reserve rather than a structural material, although this function would duplicate that generally assigned to glycogen.'

There are many examples of fungal polysaccharides that are released as extracellular products, some of which are later re-utilized. For example, Freeman & Macpherson (1949) described two extracellular polysaccharides produced by strains of *Penicillium luteum*, a neutral polygalactose and an acidic polyglucose. They found that the latter polysaccharide supported about the same amount of growth as media containing lactose or glycerol as the sole sources of carbon, whereas the neutral polygalactose was not utilized to any significant extent. They agreed with the generally accepted assumption at that time, that extracellular fungal polysaccharides might act as reserve carbohydrates.

Martin & Adams (1956) surveyed the carbohydrate constituents of intracellular and extracellular polysaccharides of 31 species of fungi and noted that this information could be used to classify the organisms within different groups. Martin (1958) subsequently studied the changes in the amounts of individual sugars within the crude extracellular polysaccharide materials present during the growth of *Mucor racemosus*. He found that after 3 days, the ratios of the sugars in the polysaccharides were glucose : fucose : mannose : galactose 1.0 : 0.9 : 0.5 : 0.2, whereas at 10 days those values were 1.0 : 8.3 : 8.0 : 2.6. Noting the dramatic decrease in the glucose content between the third and tenth days, he suggested that there was "... a gradual shift in the type of polysaccharide produced, perhaps with the concomitant utilization of the glucans already formed.'

A more recent example comes from studies with malonogalactan, an acidic polysaccharide of *Penicillium citrinum* consisting of D-galactose and malonic acid in a 3 : 1 molar ratio (Kohama, Fujimoto, Kuninaka & Yoshino, 1974). These investigators had previously reported that the carboxylesterase(s) and glycosidase(s) produced simultaneously by the fungus could release malonic acid and galactose. Re-utilization of the sugar would then be a simple matter for the mould. Another example of the release of galactose from an extracellular polysaccharide was reported by Gander (1974). He described a rapid decrease in the percentage of galactose in the extracellular peptidophosphogalactomannans produced by *Penicillium charlesii* that was coincident with the appearance of extracellular galactofuranosidase.

Finally, a $\beta(1\rightarrow3)$-glucan is produced extracellularly by *Claviceps fusiformis* during growth on glucose or sucrose which is subsequently degraded in some strains by the production of a β-$(1\rightarrow3)$-glucanase. As a consequence of the release of the glucose from the glucan, there is secondary growth of the organism (Banks, Mantle & Szczyrbak, 1974). It is now known that the synthesis of the truly extracellular $\beta(1\rightarrow3)$-glucanase produced by a thermophilic *Streptomyces* species is semiconstitutive and subject to catabolite repression by metabolizable carbon sources. Of interest is the finding that the enzyme is induced by gentiobiose, a molecule structurally unlike its substrate (Lilley & Bull, 1974). Other aspects of the interesting problem of synthesis and excretion of extracellular polysaccharides and polysaccharases are discussed by Bull (1972).

9.5 Endogenous Metabolism

Foster (1949) rejected the use of the concept of reserve storage materials since he considered that there was '... no more reason for considering fat and polysaccharide depositions as reserve or storage products, than the carbohydrates and organic acids formed and which accumulate outside the cells.' Since these extracellular accumulations also tend to be utilized by the mould after exhaustion of the initial carbohydrate source supplied, he could not accept the idea of extracellular materials as storage or reserve products and preferred to think of all accumulations of such products as shunt, or overflow, products.

Whether one chooses to use the terms shunt products or reserve products is really a problem of semantics. As long as the terms used are clearly defined, there is no problem. In general, most fungal physiologists prefer to think of the reserve products as those intracellular products most easily used by the fungus when the original substrate has been consumed. Manners (1971), in a review on the storage carbohydrates of yeast, considered only trehalose and glycogen, while Sols, Gancedo & Delafuente (1971) considered lipids, as well as trehalose and glycogen, while reviewing the energy reserves of yeast.

As mentioned, the endogenous reserves are usually considered to be those metabolized soon *after* the initial carbon source has been entirely utilized. Actually, the problem is much more complex, since there is evidence that a dynamic state exists in the fungi with significant fractions of the exogenous substrates continually converted to 'endogenous' reserves. These 'new' reserves are then, in turn, metabolized even while the exogenous substrate is still present. The amounts of endogenous reserves that accumulate depend upon the cellular regulatory factors controlling the rate of synthesis and/or the rate of degradation of these substances.

Most cells utilize cellular materials in the absence of a metabolizable exogenous substrate resulting in endogenous respiration. In the filamentous fungi, however, the rate of this endogenous respiration is unusually high as compared to the respiration in the presence of exogenous substrate. The results of an investigation on the endogenous respiration of *Neurospora crassa*, employing a manometric-isotopic technique with non-growing unlabelled mycelium and uniformly-^{14}C-labelled substrates, indicated that the degree of inhibition of endogenous respiration during oxidation of added substrates was influenced by a number of factors, such as the nature and concentration of the exogenous substrates, the age of the cells, the growth temperature, the carbon source used for growth of the cells and whether or not they had been starved (Blumenthal, 1963). When *N. crassa* was grown on glucose as the sole carbon source, the oxidation of added glucose caused a 15–40% inhibition of the utilization of endogenous stores while the oxidation of exogenous acetate caused a 50–100% inhibition. The concentration of acetate, but not glucose, influenced the degree of inhibition. When both glucose and acetate were added as cosubstrates, the resulting inhibition of the oxidation of endogenous materials was not equivalent to the sum of the inhibitions caused by each substrate separately. These results were interpreted to mean that glucose and acetate inhibited the utilization of different endogenous substrates and that there was competition in the cell between the endogenous, as well as exogenous, substrates.

In *Penicillium chrysogenum*, the endogenous respiration of glucose-grown cells was not inhibited by the concurrent oxidation of either glucose or acetate as substrates. However, the endogenous respiration of acetate-grown cells, while not affected by glucose, was very markedly suppressed during acetate oxidation (Blumenthal, Koffler & Heath, 1957). These results, along with those utilizing *Neurospora crassa*, demonstrate that the previous history of an organism, perhaps by regulating the quantitative distribution of enzymes, effector molecules and/or endogenous reserves within the cell, does affect the manner in which the endogenous respiration behaves subsequently in the presence of an added substrate. Furthermore, it mitigates against the use of terms such as primary or secondary endogenous reserves, since under different growth or physiological conditions a given reserve product may or may not be used preferentially. As discussed earlier, trehalose might be considered to be a primary reserve product in *Neurospora* conidia, but fats are the primary reserve product in ungerminated *Neurospora* ascospores and trehalose is utilized only after germination (Sussman, 1966).

Unfortunately, the actual endogenous substrates were not identified in the preceding studies. With the currently available technology, it should be possible to isolate and measure the turnover rates of the various endogenous substrates so that a truer picture of the dynamic interrelationships between the endogenous reserves and the exogenous substrates can be ascertained.

Studies by Mizunuma (1963, 1966) with *Aspergillus sojae* mycelium also provided support for the concept that the endogenous substrate(s) varied depending upon the nature of the growth medium. Media with a high C : N ratio yielded mycelia in which carbohydrate or lipid was the first endogenous substrate utilized, followed by the utilization of nitrogenous substances. When the growth medium contained a low C : N ratio, however, the endogenous utilization of carbohydrate and fat was low while the main endogenous sources were derived from the pool amino acids, proteins and nucleic acids.

Taber & Tertzakian (1965) compared the levels of lipid, trehalose and polyols over a period of 6–7 days in a rich medium, in which penicillin production was low, and in a more sparse medium, which favoured penicillin production. The levels of polyols, particularly mannitol, lipid and trehalose were higher in the rich glucose and yeast extract medium than in the less-rich lactose and glucose penicillin-production medium. They considered the polyols, trehalose and lipids to be 'primary shunt' products, which can be translated to mean endogenous reserves, and penicillin to be a secondary shunt product, which can be translated to mean largely non-re-utilizable, secondary metabolic products. The production of secondary metabolites is discussed in Vol. 1 of this series and in the last section of this book.

Trinci & Righelato (1970) studied some of the changes in cytology and metabolism of glucose-grown *Penicillium chrysogenum* under conditions in which the mycelia were starved for glucose and only the endogenous material provided the substrates for cell maintenance or growth. They found an early loss of protein, RNA and particularly DNA, which decreased *ca.* 75% in 5 h. Surprisingly, it was reported that the total mycelial macromolecular carbohydrate content, as determined by an anthrone method, decreased only after the first two days of starvation. The fungal wall thickness remained unchanged even when the cells were autolysing, again reaffirming the idea that structural polysaccharides, such as chitin, are not utilized as reserve carbohydrates.

The results with the mycelial carbohydrates may be misleading, however, since the anthrone procedure will not detect mannitol, and colour development for the pentoses of DNA and RNA are very low when the anthrone procedure is used under conditions for glucose or glucose polymer determinations. Consequently, the true status of the carbohydrate reserves or of components such as the pentoses of DNA and RNA were probably not revealed by the analytical method selected. Holligan & Jennings (1972*a*) reported that mannitol in the mycelium of *Dendryphiella salina* decreased by about 50% 2 days after glucose had disappeared from the medium.

Reviewing the rate of increase of our knowledge on this complex subject during the past two decades, one can confidently predict that during this next decade answers will become available for many of the questions still remaining about reserve carbohydrates.

9.6 References

AITKEN, W. B. & NIEDERPRUEM, D. J. (1973). Isotopic studies of carbohydrate metabolism during basidiospore germination in *Schizophyllum commune.* II. Changes in specifically labelled glucose and sugar alcohol utilization. *Archiv. für Mikrobiologie,* **88**, 331–44.

BALLOU, C. E. & RASCHKE, W. (1974). Polymorphism of the somatic antigen of yeast. *Science* **184**, 127–34.

BANKS, G. T., MANTLE, P. G. & SZCZYRBAK, C. A. (1974). Large-scale production of clavine alkaloids by *Claviceps fusiformis. Journal of General Microbiology* **82**, 345–61.

BARKER, S. A., GÓMEZ-SÁNCHEZ, A. & STACEY, M. (1958). Studies of *Aspergillus niger.* X. Polyol and disaccharide production from acetate. *Journal of the Chemical Society* **1958**, 2583–6.

BARTNICKI-GARCIA, S. (1961). Cell wall chemistry, morphogenesis, and taxonomy of fungi. *Annual Review of Microbiology* **22**, 87–108.

BHUMIRATANA, A., ANDERSON, R. L. & COSTILOW, R. N. (1974). Trehalose metabolism by *Bacillus popilliae. Journal of Bacteriology* **119**, 484–93.

BLUMENTHAL, H. J. (1963). Endogenous metabolism of filamentous fungi. *Annals of the New York Academy of Science* **102**, 688–706.

BLUMENTHAL, H. J. (1965). Carbohydrate metabolism. 1. Glycolysis. In *The fungi,* Vol. I, 229–68. Edited by G. C. Ainsworth and A. S. Sussman. New York: Academic Press.

BLUMENTHAL, H. J., KOFFLER, H. & HEATH, E. C. (1957). Biochemistry of filamentous fungi. V. Endogenous respiration during concurrent metabolism of endogenous substrates. *Journal of Cellular and Comparative Physiology* **50**, 471–97.

BULL, A. T. (1972). Environmental factors influencing the synthesis and excretion of exocellular macromolecules. *Journal of Applied Chemistry and Biotechnology* **22**, 261–92.

BULL, A. T. & CHESTERS, C. G. C. (1966). The biochemistry of laminarin and the nature of laminarinase. *Advances in Enzymology* **28**, 325–64.

CABIB, E., ROTHMAN-DENES, L. B. & HUANG, K. (1973). The regulation of glycogen synthesis in yeast. *Annals of the*

New York Academy of Sciences **210**, 192–206.

CATLEY, B. J. (1973). The rate of elaboration of the extracellular polysaccharide, pullulan, during growth of *Pullularia pullulans. Journal of General Microbiology* **78**, 33–8.

COCHRANE, V. W. (1958). *Physiology of fungi*, New York: John Wiley & Sons.

CORINA, D. L. & MUNDAY, K. A. (1971*a*). The metabolic stability of carbohydrates in walls of hyphae of *Aspergillus clavatus. Journal of General Microbiology* **65**, 253–7.

CORINA, D. L. & MUNDAY, K. A. (1971*b*). Studies on polyol function in *Aspergillus clavatus*: a role for mannitol and ribitol. *Journal of General Microbiology* **69**, 221–7.

ERGLE, D. R. (1947). The glycogen content of *Phymatotrichum* sclerotia. *Journal of the American Chemical Society* **69**, 2061–2.

FOSTER, J. W. (1949). *Chemical activities of fungi.* New York: Academic Press.

FREEMAN, G. G. & MacPHERSON, C. S. (1949). Studies on metabolic products of the *Penicillium luteum* series. *Biochemical Journal* **45**, 179–89.

GANDER, J. E. (1974). Fungal cell wall glycoproteins and peptidopolysaccharides. *Annual Review of Microbiology* **28**, 103–19.

GOLD, M. H., MITZEL, D. L. & SEGEL, I. H. (1973). Regulation of nigeran accumulation by *Aspergillus aculeatus. Journal of Bacteriology* **113**, 856–62.

GORIN, P. A. J. & SPENCER, J. F. T. (1968). Structural chemistry of fungal polysaccharides. *Advances in Carbohydrate Chemistry* **28**, 367–417.

GRAPPEL, S. F., BISHOP, C. T. & BLANK, F. (1974). Immunology of dermatophytes and dermatophytosis. *Bacteriological Reviews* **38**, 222–50.

GUNASEKARAN, M. (1972). Physiological studies on *Phymatotrichum omnivorum.* I. Pathways of glucose catabolism. *Archiv für Mikrobiologie* **83**, 328–31.

HANKINSON, O. (1974). Mutants of the pentose phosphate pathway in *Aspergillus nidulans. Journal of Bacteriology* **117**, 1121–30.

HASHIMOTO, T., WU, C. D. R. & BLUMENTHAL, H. J. (1972). Characterization of L-leucine-induced germination of *Trichophyton mentagrophytes* microconidia. *Journal of Bacteriology* **112**, 967–76.

HELLE, K. B. & KLUNGSOEYR, L. (1962). Mannitol 1-phosphate formation in *Escherichia coli* during glucose utilization. *Biochimica et Biophysica Acta* **65**, 461–71.

HOLLIGAN, P. M. & JENNINGS, D. H. (1972*a*). Carbohydrate metabolism in the fungus *Dendryphiella salina.* I. Changes in the levels of soluble carbohydrates during growth. *New Phytologist* **71**, 569–82.

HOLLIGAN, P. M. & JENNINGS, D. H. (1972*b*). Carbohydrate metabolism in the fungus *Dendryphiella salina.* II. The influence of different carbon and nitrogen sources on the accumulation of mannitol and arabitol. *New Phytologist* **71**, 583–94.

HOLLIGAN, P. M. & JENNINGS, D. H. (1972*c*). Carbohydrate metabolism in the fungus *Dendryphiella salina.* III. The effect of the nitrogen source on the metabolism of [1-^{14}C]- and [6-^{14}C]-glucose. *New Phytologist* **71**, 1119–33.

HOLLIGAN, P. M. & JENNINGS, D. H. (1972*d*). Unexpected labelling patterns from radioactive sugars fed to plants containing mannitol. *Phytochemistry* **11**, 3447–51.

HORIKOSHI, K., IIDA, S. & IKEDA, Y. (1965). Mannitol and mannitol dehydrogenase in conidia of *Aspergillus oryzae. Journal of Bacteriology* **80**, 326–30.

JENNINGS, D. H. & AUSTIN, S. (1973). The stimulatory effect of the non-metabolized sugar 3-*O*-methyl glucose on the conversion of mannitol and arabitol to polysaccharide and other insoluble compounds in the fungus *Dendryphiella salina. Journal of General Microbiology* **75**, 287–294.

KOHAMA, T., FUJIMOTO, M., KUNINAKA, A. & YOSHINO, H. (1974). Structure of malonogalactan, an acidic polysaccharide of *Penicillium citrinum. Agricultural and Biological Chemistry (Japan)* **38**, 127–34.

LEE, W. H. (1967*a*). Carbon balance of a mannitol fermentation and the biosynthetic pathway. *Applied Microbiology* **15**, 1206–10.

LEE, W. H. (1967*b*). Mannitol acetyl phosphate phosphotransferase of *Aspergillus. Biochemical and Biophysical Research Communications* **29**, 337–42.

LEIGHTON, T. J., STOCK, J. J. & KELLN, R. A. (1970). Macroconidial germination in *Microsporum gypseum. Journal of Bacteriology* **103**, 439–46.

LEWIS, D. H. & SMITH, D. C. (1967). Sugar alcohols (polyols) in fungi and green plants. I. Distribution, physiology and metabolism. *New Phytologist* **66**, 143–84.

LILLEY, G. & BULL, A. T. (1974). The production of β-1,3 glucanase by a thermophilic species of *Streptomyces. Journal of General Microbiology* **83**, 123–33.

MANNERS, D. J. (1971). The structure and biosynthesis of storage carbohydrates in yeast. In *The yeasts,* Vol. 2, 419–39. Edited by A. H. Rose and J. S. Harrison. London: Academic Press.

MANNERS, D. J., MASSON, A. J. & PATTERSON, J. C. (1974). The heterogeneity of glucan preparations from the walls of various yeasts. *Journal of General Microbiology* **80**, 411–7.

MARCUS, L. & MARR, A. G. (1961). Polyol dehydrogenases of *Azotobacter agilis. Journal of Bacteriology* **82**, 224–32.

MARTIN, S. M. (1958). Production of extracellular polysaccharides by *Mucor racemosus. Canadian Journal of Microbiology,* **4**, 317–9.

MARTIN, S. M. & ADAMS, G. A. (1956). A survey of fungal polysaccharides. *Canadian Journal of Microbiology* **34**, 715–21.

MIZUNUMA, T. (1963). Studies on the metabolism of *Aspergilli.* Part III. Substrates of endogenous respiration in various strains of *Aspergilli. Agricultural and Biological Chemistry (Japan),* **27**, 853–7.

MIZUNUMA, T. (1966). Studies on the metabolism of *Aspergilli* Part VIII. Relationship between endogenous formation of ammonia and levels of mycelial components. *Agricultural and Biological Chemistry* (Japan) **30**, 742–9.

PLESSMAN CAMARGO, E., MEUSER, R. & SONNEBORN, D. (1969). Kinetic analyses of the regulation of glycogen synthetase activity in zoospores and growing cells of the water mold, *Blastocladiella emersonii. Journal of Biological Chemistry* **244**, 5910–9.

SMITH, D., MUSCATINE, L. & LEWIS, D. (1969). Carbohydrate movement from autotrophs to heterotrophs in parasitic and mutualistic symbiosis. *Biological Reviews* **44**, 17–90.

SMITH, J. E. & GALBRAITH, J. C. (1971). Biochemical and physiological aspects of differentiation in the fungi. *Advances in Microbial Physiology* **5**, 45–134.

SOLS, A., GANCEDO, C. & DELAFUENTE, G. (1971). Energy-yielding metabolism in yeasts. In *The yeasts,* Vol. 2, 271–305. Edited by A. H. Rose and J. S. Harrison. London: Academic Press.

STRANDBERG, G. W. (1969). D-Mannitol metabolism by *Aspergillus candidus. Journal of Bacteriology* **97**, 1305–9.

SUSSMAN, A. S. (1966). Types of dormancy as represented by condidia and ascospores of *Neurospora.* In *The fungus spore* 235–56. Edited by M. F. Madelin. London: Butterworths.

SUSSMAN, A. S. & DOUTHIT, H. A. (1973). Dormancy in microbial spores. *Annual Review of Plant Physiology* **24**, 311–52.

TABER, W. A. & TERTZAKIAN, G. (1965). Sequential primary and secondary shunt metabolism in *Penicillium chrysogenum. Applied Microbiology* **13**, 590–4.

TÉLLEZ-IÑÓN, M. T., TERENZI, H. & TORRES, H. N. (1969). Interconvertible forms of glycogen synthetase in *Neurospora crassa. Biochimica et Biophysica Acta* **191**, 765–8.

TÉLLEZ-IÑÓN, M. T. & TORRES, H. N. (1970). Interconvertible forms of glycogen phosphorylase in *Neurospora crassa. Proceedings of the National Academy of Sciences* (*U.S.A.*) **66**, 459–63.

TOKUNAGA, J. & BARTNICKI-GARCIA, S. (1971*a*). Cyst wall formation and endogenous carbohydrate utilization during synchronous encystment of *Phytophthora palmivora* zoospores. *Archiv für Mikrobiologie* **79**, 283–92.

TOKUNAGA, J. &. BARTNICKI-GARCIA, S. (1971*b*). Structure and differentiation of the cell wall of *Phytophthora palmivora*: cysts, hyphae and sporangia. *Archiv für Mikrobiologie* **79**, 293–310.

TRINCI, A. P. J. & COLLINGE, A. (1973). Influence of L-sorbose on the growth and morphology of *Neurospora crassa. Journal of General Microbiology* **78**, 179–92.

TRINCI, A. P. J. & RIGHELATO, R. C. (1970). Changes in constituents and ultrastructure of hyphal compartments during autolysis of glucose-starved *Penicillium chrysogenum. Journal of General Microbiology* **60**, 239–49.

TURIAN, G. & BIANCHI, D. E. (1972). Conidiation in *Neurospora. The Botanical Review* **38**, 119–54.

ULANE, R. E. & CABIB, E. (1974). The activating system of chitin synthetase from *Saccharomyces cerevisiae.* Purification and properties of an inhibitor of the activating factor. *Journal of Biological Chemistry* **249**, 3418–22.

VINING, L. C. & TABER, W. A. (1964). Analysis of the endogenous sugars and polyols of *Claviceps purpurea* (Fr.) Tul. by chromatography on ion exchange resins. *Canadian Journal of Microbiology* **10**, 647–57.

YAMADA, H., OKAMOTO, K., KODAMA, K. & TANAKA, S. (1959). Mannitol formation by *Piricularia oryzae. Biochimica et Biophysica Acta* **33**, 271–3.

YU, S., GARRETT, M. K. & SUSSMAN, A. S. (1971). Genetic control of multiple forms of trehalase in *Neurospora crassa. Genetics* **68**, 473–81.

CHAPTER 10

Chemical Nature of Membrane Components

A. H. ROSE

10.1 Introduction

The beautiful and elegant electron micrographs of thin sections through hyphae of filamentous fungi, a series of which have recently been collated by Beckett, Heath & McCaughlin (1975), show clearly that, in common with higher eukaryotic organisms, fungal hyphae contain a complex membraneous system which includes plasma, nuclear, mitochondrial and vesicular membranes, interconnected to some extent by an endoplasmic reticulum. Although the chemical anatomy or molecular architecture of some subcellular organelles in fungi—notably the cell wall—has been the subject of extensive study, regrettably this activity does not extend to membranes. At the time of writing, no one class of membrane from filamentous fungi has been isolated and purified to the point at which extensive chemical analyses are justified. Almost all membranes are made up of a mixture of protein and lipid. Little has been reported on proteins which are thought to be located in membranes in filamentous fungi, but there exists a wealth of data on fungal lipids as shown in the reviews of Brennan, Griffin, Lösel & Tyrrell (1974) and Mangnall & Getz (1973) and in the book of Weete (1974). Almost all of these data on fungal lipids have come from analyses of extracts of intact mycelia. The situation becomes even more frustrating when one realizes that arguably the better of these analytical data are for single-celled fungi or yeasts, and in particular for strains of *Saccharomyces cerevisiae* (Brennan *et al.*, 1974). The polarization of published data becomes even more noticeable when one peruses the published data on biosynthesis of membrane lipids, which are almost exclusively for *S. cerevisiae.*

In writing this review, I have had, perforce, to sift very carefully the data available on fungal lipids so that I can concentrate on those lipids that are most likely to be true membrane components. Likewise, I have necessarily had to borrow extensively from the literature on lipids and membranes, and biogenesis of these compounds and structures, in *Saccharomyces cerevisiae.* Finally, I should stress that this review deals mainly, in so far as it refers to membranes, with the plasma membrane. With the exception of mitochondrial membranes, almost nothing is known about other membranes, even in single-celled fungi. Mitochondria in filamentous fungi are dealt with in Chapter 4 of this volume. Verily, verily can it be stated that, when it comes to membranes in filamentous fungi, ignorance truly prevails.

10.2 Chemical Nature of Membrane Components

Proteins

A substantial body of data on one class of plasma membrane proteins, namely the transport proteins, in bacteria has come from experiments that use a special technique on intact cells rather than isolated plasma membranes. The technique has been termed *cold osmotic shock* and involves incubating bacteria in a hypermolar solution of an inert solute (such as sucrose) containing EDTA followed by centrifugation and resuspension of the cells in ice-cold water containing 10 mM Mg^{2+}. It was developed in large part by Heppel (1969). Bacteria that have been subjected to this treatment are found to have lost some and probably all of the enzymes that lie in the periplasm, the interface between the inside of the wall and the outside of the plasma membrane. In addition, they were found to have lost the ability to transport certain solutes into the cell (Kundig, Kundig, Anderson & Roseman, 1966). It was subsequently shown (Anraku, 1968) that, after being subjected to cold osmotic shock, bacteria release transport proteins, and the technique is now extensively used to obtain these proteins from bacteria (Simoni, 1972).

Cold osmotic shock has been applied to strains of *Neurospora crassa* and *Saccharomyces cerevisiae.* Wiley (1970) showed that the ability of *N. crassa* to transport tryptophan decreased to about 90% of the original value following imposition of cold osmotic shock to germinated conidia of the fungus. He lists the evidence which allowed him to conclude that the protein released is concerned in tryptophan transport into the fungus. A strain of *Saccharomyces cerevisiae* that had been subjected to cold osmotic shock showed a diminished rate of accumulation of two non-metabolizable solutes, namely glucosamine hydrochloride and 2-aminoisobutyrate (Patching & Rose, 1971). However, these workers were unable to prove conclusively that proteins involved in transport of these solutes had been released from the yeast.

Other classes of plasma membrane protein in bacteria, including those that catalyse reactions leading to synthesis of cell wall components, are more firmly held in the membrane than transport proteins. Whether a similar situation prevails in the fungal and yeast plasma membrane is not as yet known, and an answer to this question will only come when more extensive studies have been made on isolated plasma membranes.

Lipids

Analyses of isolated mammalian membranes, particularly the erythrocyte membrane, have shown conclusively that the lipids in these eukaryotic membranes are a mixture of polar lipids and sterols. The principal polar lipids are glycerophospholipids—usually referred to simply as *phospholipids* and *glycolipids.* It is therefore usually assumed—and with some justification—that phospholipids and glycolipids that are extracted from intact cells and mycelia of fungi are true membrane components. This section of the chapter describes, briefly, the phospholipids and glycolipids that have been extracted from filamentous fungi and certain yeasts.

PHOSPHOLIPIDS These lipids can be incorporated into biological membranes because of their amphipathic character, a property which they share with glycolipids. Phospholipids have the general formula shown here

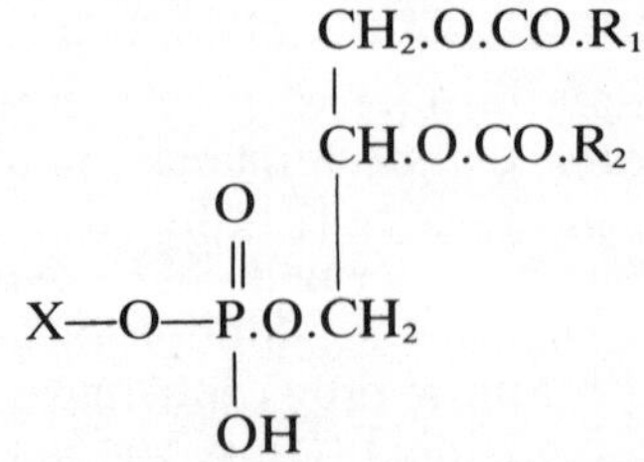

where R_1 and R_2 indicate fatty-acyl residues and where X may be any one of several hydrophilic residues including choline, ethanolamine and inositol. There are two types of structural variation in phospholipid molecules, centering around the nature of the fatty-acyl residues and of the hydrophilic group attached to the phosphodiester bond. Two very fine reviews of the phospholipid composition of yeasts and fungi have recently been published, one very comprehensive by Mangnall & Getz (1973) and the other shorter but nevertheless thorough and engagingly critical by Brennan *et al.* (1974). Either or both of these sources should be consulted for a more detailed treatment of the subject than can be accommodated in this chapter.

Table 10.1 lists some examples of the different types of phospholipid—based on differences in the structure attached to the phosphodiester linkage—found in some filamentous fungi and yeasts. The amount of data available for the species listed in very disparate. Not surprisingly, it is most extensive for the yeasts *Saccharomyces cerevisiae* and *Candida utilis.* The principal phospholipids in the organisms listed in Table 10.1 are phosphatidylethanolamine and phosphatidylcholine, along with smaller amounts of, among others, phosphatidylinositol, phosphatidylserine and phosphatidylglycerol. The reason for the predominance of the two neutral or Zwitterion phospholipids is not known. Indeed it is true to say that we have no explanation for the occurrence of any one phospholipid in yeasts and filamentous fungi.

It is important to realize that the phospholipid composition of yeasts and filamentous fungi varies, to some extent, from strain to strain, within a particular species, and also with the conditions under which the organisms are grown. Again, information on this aspect of phospholipid composition is most extensive for *Saccharomyces cerevisiae.* When grown under aerobic conditions, the major phospholipid of the NCYC 366 strain of this yeast is phosphatidylethanolamine (Hunter & Rose, 1972) but, when the yeast is grown under strictly anaerobic conditions, with nutritional supplements of a sterol and oleic acid, then phosphatidylcholine is the principal phospholipid, there being twice as much of this phospholipid in cell extracts as of phosphatidylethanolamine (Hossack & Rose, 1976). The proportions of phosphatidylcholine and phosphatidylethanolamine in lipids of *S. cerevisiae* NCYC 366, and of other strains of this yeast, can be raised by

Table 10.1 Some examples of phospholipids detected in filamentous fungi and yeasts

Fungus or Yeast	Principal Phospholipids Detected					References
Phycomycetes:						
Phycomyces blakesleeanus	PC[a]	PE[b]	PS[c]			Jack (1966)
Pythium ultimum	PC	PE	PG	DPG		Bowman & Mumma (1967)
Rhizopus nigricans	PC	PE	PE			Jack & Laredo (1968)
Ascomycetes:						
Alternaria oleracea	PC	PE	PS			Jack & Laredo (1968)
Glomerella cingulata	PC	PE	PS			Jack (1964); Jack & Laredo (1968)
Hansenula anomola	PC	PE	PE	PI[d]		Letters (1968)
Lipomyces lipofer	PC	PE	PS			Jack (1966)
Neurospora sitophila	PC	PE	PS			Jack & Laredo (1968)
Saccharomyces cerevisiae	PC	PE	PI	PS	DPG[e]	Hunter & Rose (1971, 1972); Hossack & Rose (1976)
Basidiomycetes:						
Agaricus bisporus (mycelium)	PC	PE	PS			Griffin, Brennan & Lösel (1970); Holz & Schisler (1971)
Clitocybe illudens	PC	PE	PS			Bentley, Lavate & Sweeley (1964)
Coprinus comatus	PC	PE	PS			Jack (1966)
Deuteromycetes:						
Candida lipolytica	PC	PE	PI	PS		Kates & Baxter (1962); Letters (1968)
Humicola grisea var *thermoidea*	PC	PE	PI	DPG		Mumma, Sekura & Fergus (1971)
Rhodotorula graminis	PC	PE	PI	PS		Letters (1968)

[a] PC indicates phosphatidylcholine; [b] PE, phosphatidylethanolamine; [c] PS, phosphatidylserine; [d] PI, phosphatidylinositol; [e] DPG, diphosphatidylglycerol or cardiolipin.

including either choline or ethanolamine in the growth medium (Waechter, Steiner & Lester, 1969; Waechter & Lester, 1971; Ratcliffe, Hossack, Wheeler & Rose, 1973). The significance of this effect of choline and ethanolamine on the phospholipid composition of *S. cerevisiae* will be discussed again in the section of this chapter that deals with phospholipid synthesis in yeasts and filamentous fungi.

Few if any studies have been carried out on the fatty-acyl composition of individual fungal phospholipids. But there is an extensive body of data on the fatty-acyl composition of groups of fungal lipids, or more frequently of the total lipid extracted from filamentous fungi. These data, while being far removed from the ideal, provide some insight into the variations that occur in the fatty-acyl composition of fungal lipids. The account that follows presents from the literature data that, while not specifically in all instances having been obtained from membrane lipids, in all likelihood represent a fair commentary on the fatty-acyl composition of fungal membrane lipids.

Higher fungi, in common with most eukaryotic organisms, contain predominantly $C_{16:0}$, $C_{16:1}$, $C_{18:1}$ and $C_{18:2}$ fatty-acyl residues in their lipids. Lower fungi, especially the Phycomycetes, have a fatty-acyl composition in their lipids which is somewhat different. In *Conidiobolus (Entomophthora) coronatus*, for example, a sizeable proportion of the fatty-acyl residues in the cellular lipids have a chain length less than 16 carbon atoms (Mumma & Bruszewski, 1970). The vast majority of fatty-acyl residues in fungal lipids contain an even number of carbon atoms. But fatty-acyl residues with an odd number of carbon atoms do occasionally occur, and again these are commoner among species in the Phycomycetes. Indeed, Tyrell (1967, 1971) reported that $C_{13:0}$ and $C_{15:0}$ acids can account for as much as 5% of the total fatty-acyl residues in *Conidiobolus heterosporus*. Branched-chain fatty acids, characteristic of several genera of bacteria, are also found in the lipids of some fungi; in species of *Conidiobolus* they can account for as much as 50% of the total fatty-acyl residues in the fungus (Tyrrell, 1968, 1971). A word of caution, here, because many of these branched-chain acids are found in the triacylglycerols of the fungus, and so are not membrane components. Nevertheless, some are in phospholipids, and so must influence membrane properties. Saturated or mono-unsaturated fatty-acyl residues predominate in species in the Ascomycetaceae, whereas the lower fungi contain species that synthesize polyunsaturated fatty-acyl residues. Members of the Zygomycetes, for example, often contain residues of γ-linolenic acid. Moreover, *Blastocladiella emersonii* contains, in addition to α- and γ-linolenic acids, both $C_{20:3}$ and $C_{20:4}$ fatty-acyl residues (Sumner, 1970).

Studies on mutants of *Neurospora crassa* and *Saccharomyces cerevisiae* that are auxotrophic for an unsaturated fatty acid (Keith, Wisnieski, Henry & Williams, 1973) have shown that a fairly even proportion of saturated and mono-unsaturated fatty-acyl residues seemingly needs to be present in cellular lipids. While unsaturated fatty-acyl residues are believed to contribute to fluidity in membranes, and hence to facilitate movement of proteins in membranes, nothing is known of the role of branched-chain or polyunsaturated fatty-acyl residues in membrane phospholipids. The fungi in which the rarer fatty-acyl residues have been detected could prove to be valuable organisms with which to study the physiological role of these fatty-acyl residues.

GLYCOLIPIDS Whereas the chemistry of microbial phospholipids has been studied for many years, knowledge about the composition of the wide range of glycolipids that are found in micro-organisms is of much more recent vintage (Shaw, 1970, 1975). Because of their extremely amphipathic character, it is almost certain that most of the glycolipids extracted from filamentous fungi and yeasts are located in membranes, although this has yet to be proved by analysis of isolated fungal and yeast membranes. Nevertheless, caution must be exercized in discussing fungal glycolipids, since most fungal lipids in this class that have been studied to any great extent are extracellular products (Brennan *et al.*, 1974).

The only truly intracellular glycolipids from filamentous fungi that have been at all well characterized are a relatively small number of acylated sugars, formally not dissimilar from those that occur in bacteria (Shaw, 1975). Laine, Griffin, Sweeley & Brennan (1972) reported on a monoglucosyloxy-octadecenoic acid from *Aspergillus niger.* Evidence for a membrane role for this lipid was adduced by these workers when they showed that the proportion of the total cell lipid accounted for by this glycolipid rose 3-fold when the mould cultures were starved of a source of phosphate, under which conditions synthesis of phospholipids was restricted. More recently, Brennan's laboratory has reported evidence for similar glucosyloxy fatty acids in a *Fusarium* sp., *Penicillium janthcinellum* and *Rhizopus stolonifer* (Byrne & Brennan, 1975). Earlier, an acylated glucose had been reported in *Saccharomyces cerevisiae* (Brennan, Flynn & Griffin, 1970). In addition, acylated trehaloses had been extracted from *Pullularia pullulans* (Merdinger, Kohn & McClain, 1968) and from *Claviceps purpurea* (Cooke & Mitchell, 1969). The assumption that these glycolipids are endogenous reserve lipids merits re-examination now that more is known about the role of glycolipids in membranes.

Sphingolipids differ from glycerophospholipids in that they are based not on glycerol phosphate but on sphingosine derivatives. Sphingolipids are widely found in extracts of filamentous fungi and yeasts, and some of them are glycosylated. As yet, little has been reported on the structure of these glycosylated sphingolipids, although Brennan and his associates recently established that a glycosyl ceramide occurs in *Aspergillus niger* (Brennan, Roe, Byrne & Tighe, 1975); cerebrins are a class of sphingolipids (Brennan *et al.*, 1974). Lester and his colleagues have detected a ceramide textrahexoside in *Neurospora crassa* (Lester, Smith, Wells, Rees & Angus, 1974). Appreciable amounts of a mannosylated inositol ceramide have been detected in the envelope of *Saccharomyces cerevisiae* by Työfinoja, Nurminen & Suomalainen (1974).

NON-GLYCOSYLATED SPHINGOLIPIDS Sphingolipids which do not have a glycosyl residue attached to the molecule are far more common in fungi than their glycosylated counterparts. Many of these lipids are ceramides—that is they are based on cerebrin, which has the structure shown here (Carter, Celmer, Lands, Mueller & Tomizawa, 1954).

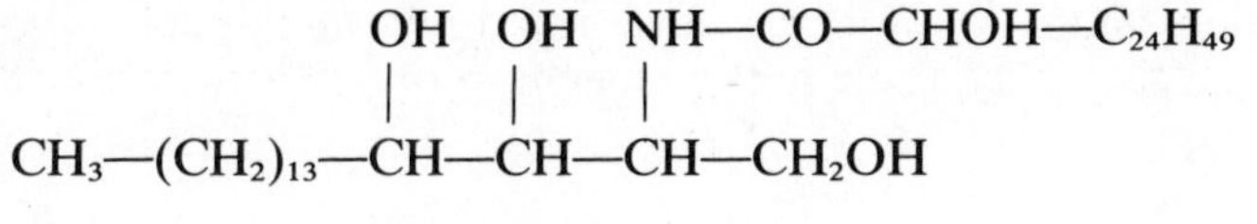

Ceramides have been extracted from a wide range of fungi, including representatives of all of the major taxonomic classes (Brennan *et al.*, 1974), as well as from yeasts (Hunter & Rose, 1971). Organisms for which more detailed reports on ceramide content are available include *Amanita muscaria* (Weiss & Stiller, 1972), *Aspergillus niger* (Wagner & Fiegert, 1969), *Candida utilis* (Stanacev & Kates, 1963), *Fusarium lini* (Weiss, Stiller & Jack, 1973), *Neurospora crassa* (Lester *et al.*, 1974), *Phycomyces blakesleeanus* (Weiss, Stiller & Jack, 1973) and *Saccharomyces cerevisiae* (Sweeley, 1959; Smith & Lester, 1974).

Unfortunately, at present, these lipids are in many ways chemical curiosities, for little if anything is known of the way in which they might be involved in membranes. Moreover, there are reasons for believing that ceramides may not occur in fungal and yeast cells in the form in which their structures are currently viewed by mycological chemists. Steiner, Smith, Waechter & Lester (1969) reported on the effects of mild alkaline treatment on sphingolipids from *Saccharomyces cerevisiae*, with evidence for the operation of degradative processes that may cause other workers to revise their views on ceramide structure.

STEROLS It has been known for many years that sterols occur in filamentous fungi and yeasts, and the standard chemical texts on these compounds (Fieser & Fieser, 1959; Shoppee, 1964) include many early references to sterols that have been detected in fungi. However, the reader would do well to view these early reports with a dollop of healthy scepticism. Techniques for separating sterols from extracts and for identifying minute amounts of these compounds have undergone extensive development in recent years, and it is as well to rely on more recently acquired data, with respect to the sterling efforts of the pioneer microbial biochemists.

Goodwin (1973) has published a detailed list of sterols that have been detected in filamentous fungi and yeasts. It includes data on fungi from all of the principal taxonomic classes, and illustrates the very wide range of sterols that are synthesized by fungi. While the supposedly typical fungal sterols, such as ergosterol and zymosterol, certainly occur with regularity, so does cholesterol which has often been considered a typical animal sterol.

It can safely be predicted that not all of the sterol that occurs in a fungus is in membranes. This is because a reasonably large proportion of the cellular sterol is esterified at C-3 with a long fatty-acyl chain. Sterol esters are not amphipathic, and cannot therefore be incorporated into membranes. As shown in a later section of this chapter, however, it is likely that sterol esters have a function in biogenesis of membranes.

Biosynthesis of terpenoid compounds, including sterols, by fungi is discussed more fully in Chapter 13 of this volume, and the reader should turn to the chapter for a fuller account of fungal sterols.

ACYLGLYCEROLS Mono-, di- and triacylglycerols are synthesized, often in very large amounts, by almost all filamentous fungi and yeasts (Brennan *et al.*, 1974). However, these compounds, like sterol esters, are not amphipathic, and so cannot be incorporated into membranes. Again, it is thought that they play a role in the biogenesis of fungal membranes (see p. 322).

10.3 Isolation and Composition of Fungal Plasma Membranes

The structure and composition of the plasma membrane that surrounds cells is currently the subject of intense investigation in many laboratories. These studies have been extended to filamentous fungi and yeasts, with at the moment by far the greater amount of data being available on the yeast plasma membrane.

Isolation methods

Two types of method have been used to prepare intact plasma membranes from filamentous fungi and yeasts. The first of these involves the conversion of mycelium or cells to protoplasts (or sphaeroplasts, depending upon the terminology used by the workers concerned), lysing these sphaeroplasts, and isolating plasma membranes from the lysate by differential or isopycnic centrifugation. The second method entails separation of plasma membranes or membrane-rich fractions, by centrifugation, from a preparation of disrupted mycelium or cells obtained by subjecting the organisms to mechanical disruption. Both methods have their adherents, and they give rise to plasma membrane preparations that differ somewhat in composition.

Boulton & Eddy (1962) and Boulton (1965) pioneered the use of sphaeroplasts to prepare plasma membranes from *Saccharomyces cerevisiae* using the digestive juice of *Helix pomatia* (the Roman snail) to remove the yeast wall (Eddy & Williamson, 1957). They subjected osmotically sensitive sphaeroplasts to osmotic shock by mixing with ice-cold 25mM tris buffer (pH 7.2) containing mM-magnesium chloride, and isolated plasma membranes by centrifugation. A similar technique was used by Garcia-Mendoza & Villanueva (1967) to prepare plasma membranes from *Candida utilis*, and a year later by Longley, Rose & Knights (1968) to obtain plasma membranes from a strain of *S. cerevisiae* that is particularly susceptible to enzymic removal of the cell wall. My laboratory has since abandoned the use of snail juice for preparing sphaeroplasts, following the discovery (K. Hunter and A. H. Rose, unpubl. results) that the juice contains a phospholipase activity. Use is now made of an extracellular Basidiomycete glucanase (Reese & Mandels, 1959; Huotari, Nelson, Smith & Kirkwood, 1958), which is free from detectable lipase, phospholipase and, if the cultures are harvested at a suitably early phase, from protease activity. Kidby and his colleagues (Schibeci, Rattray & Kidby, 1973) are responsible for the most sophisticated application, so far, of this method for isolating yeast plasma membranes. They prepared sphaeroplasts of *S. cerevisiae*, albeit with snail juice, but used the technique of tagging the sphaeroplast membrane in order to detect plasma membrane material on the gradient used for separating membranes from other structures in the sphaeroplast lysate. These workers explored the use of several reagents for tagging the sphaeroplast membrane, including N-ethylmaleimide, dansyl chloride, fluoro-2,4-dinitrobenzene and sodium iodide, all suitably labelled with radioactive atoms. Iodination of the sphaeroplasts proved to be the most suitable technique, one which has since been adopted by Marriott (1975) and Hossack & Rose (1976) for isolating, respectively, plasma membranes from *C. albicans* and *S. cerevisiae*.

Numerous reports have been published on techniques for preparing protoplasts or sphaeroplasts from filamentous fungi, often using lytic enzymes obtained from streptomycetes (Villanueva, Gacto & Sierra, 1973; Horikoshi, 1973). There are, however, few reports of the isolation of plasma membranes from lysates of sphaeroplasts from filamentous fungi. Some years ago, Garcia-Acha, Aguirre, Lopez-Belmonte, Uruburu & Villanueva (1966) prepared protoplasts from *Fusarium culmorum*, and proceeded to isolate plasma membranes from these structures. Unfortunately, this study failed to encourage other mycologists to use this technique to obtain plasma membranes from other filamentous fungi, and it is only very recently that the technique was extended by Marriott (1975) to the mycelial form of *Candida albicans*. Marriott (1975) used a lytic enzyme from *Streptomyces violaceus* to prepare protoplasts from his mycelia and yeast cells.

Advocacy for the second method of preparing fungal plasma membranes, involving preparations of disrupted cells and mycelia, has come principally from the laboratories of Matile and his colleagues at the Swiss Federal Institute of Technology in Zurich and of Suomalainen and his associates at the laboratories of the State Alcohol Monopoly in Helsinki, Finland. These workers believe that plasma membranes obtained from protoplasts or sphaeroplasts may contain molecular species that are not components of *in vivo* plasma membranes. They prefer to obtain cell-envelope preparations, containing plasma membranes and some associated cell wall material, and usually to remove the wall material by digestion with suitable enzymes. The first report of this technique was by Matile, Moor & Mühlethaler (1967) who homogenized *Saccharomyces cerevisiae* in 0.05 M tris buffer (pH 7.2) containing 0.25 M sucrose and mM EDTA by shaking with Ballotini glass beads. After the beads had been removed, the extract was submitted to differential centrifugation. Following the removal of the mitochondrial fraction, the microsomal fraction was layered on a density gradient of Urografin. On centrifugation, the plasma membrane-rich fraction banded at a density of 1.165–1.70 g cm^{-2}. Unlike Matile and his colleagues, Suomalainen, Nurminen & Oura (1967) treated their isolated envelope fractions to digestion with snail-gut juice to remove wall material. Analytical data reported by these two groups of workers is referred to on page 318.

A recent publication by Nombela, Uruburu & Villanueva (1974) describes the production of plasma membranes from mechanically prepared extracts of *Fusarium culmorum*. The sediment obtained by centrifuging the cell-free extract at 40 000 **g** was fractionated in a discontinuous gradient of sucrose solutions. These workers measured the ATPase and fumarase activities of the bands that separated on the gradient, and assumed that the band with the highest ATPase/fumarase ratio was that rich in plasma membranes, an assumption which was confirmed by electron microscopy of the fraction. The results of their analysis of the plasma membrane-rich fraction are reported on page 318.

Isolation of fungal plasma membranes by using mechanical disruption and isolation of cell-envelope fractions has come under critical scrutiny by several workers in the field. The principal reservation voiced by the critics of this technique is that fragmented membrane fractions obtained by imposition of a mechanical stress can reseal to form vesicles in which may be trapped cellular components that are not true plasma membrane components. Christensen & Cirillo (1972) have reported that yeast plasma membrane fractions can indeed reseal to form vesicles, an observation confirmed by Dubé, Setterfield, Kiss & Lusena (1973) who made a more detailed criticism of the method.

Composition of isolated membranes

Data which have been reported on the chemical composition of plasma membranes isolated from yeasts and filamentous fungi are summarized in Table 10.2. These data are very difficult to compare, firstly because of the different methods used to isolate the plasma membranes, secondly because a wide variety of different analytical methods have been employed, and last but not least because some of the groups of workers such as Matile, Moor & Mühlethaler (1967) do not report (or have never acquired) detailed analyses of the lipid fractions. Nevertheless, certain generalizations can be drawn from these data. All of the preparations examined contain high proportions of protein and lipid, which agrees with the generally held view of the basically protein-lipid nature of all biological membranes (Singer & Nicholson, 1972). Sizable differences have been reported in the carbohydrate content of the isolated plasma membranes, with values ranging from as low as 3% of the membrane dry weight to as high as 30%. All workers who have employed a mechanical disintegration method for releasing plasma membranes from cells or mycelium, and have refrained from using an enzymic digestion of the isolated envelopes, report high values for the carbohydrate content of their organelles. These include Matile *et al.* (1967) with a strain of *Saccharomyces cerevisiae*, and Nombela *et al.* (1974) with *Fusarium culmorum*. The presence of carbohydrate in isolated membranes is usually taken to indicate the existence of undigested wall material on the membrane. This view received support from the data reported by Longley, Rose & Knight (1968) who showed that the small amount of carbohydrate on the plasma membranes which they isolated from strains of *S. cerevisiae* contained about equal amounts of glucose and mannose residues, monomers which make up the bulk (in about equal amounts) of the cell wall polymers in this yeast. Explanations for the high content of carbohydrate in membranes obtained without the use of enzyme digestion are several in number. One possibility is that certain parts of the cell or hyphal wall are covalently bound to the underlying plasma membrane, and so remain attached to the membrane. That these parts of the cell wall do not remain with the plasma membrane when these structures are obtained with methods that involve enzyme digestion could be due to their enzyme-catalysed removal as a result of the action of enzymes present in the various preparations that are employed.

The main variations in the published analyses of isolated plasma membranes concern the contents of different classes of lipid. The need for the presence of amphipathic molecules in membranes implies that triacylglycerols and sterol esters should not be detected as membrane constituents. Nevertheless, Marriott (1975), who employed enzyme digestion with both the yeast and mycelial forms of *Candida albicans*, reported the presence in isolated plasma membranes of both classes of neutral lipids, as did Nurminen & Suomalainen (1973) when reporting on isolated plasma membranes from *Saccharomyces cerevisiae* although these workers did not use enzyme

Table 10.2 Composition of isolated plasma membranes from yeasts and filamentous fungi

Organism	Reference	Content (% dry weight) of			
		Protein	**Lipid**	**Carbohydrate**	**Nucleic acid**
Yeasts					
Candida albicans (yeast form)	Marriott (1975)	52.0	43.0	9.0	0.3
C. utilis	Garcia-Mendoza & Villanueva (1967)	38.5	40.4	5.2	1.1
Saccharomyces cerevisiae	Boulton (1965)	46.5	41.5	3.2	7.5
S. cerevisiae	Christensen & Cirillo (1972)	65.0	29.0	3.0	4.0
S. cerevisiae	Longley, Rose & Knights (1968)	49.3	39.1	5.0	7.0
S. cerevisiae	Matile, Moor & Mühlethaler (1967)	26.6	45.5	30.8	—
S. cerevisiae	Schibeci, Rattray & Kidby (1973)	49.3	2.3	5.0	7.0
S. cerevisiae	Suomalainen, Nurminen & Oura (1967)	35.0	35.0	25.0	—
Filamentous Fungi					
C. albicans (mycelial form)	Marriott (1975)	45.0	31.0	25.0	0.5
Fusarium culmorum	Nombela, Uruburu & Villanueva (1974)	25.0	40.0	30.0	—

digestion. These reports of the presence of neutral lipids are not easy to explain.

10.4 Biogenesis of Fungal Plasma Membranes

In view of the paucity of published data on the chemical composition of isolated fungal plasma membranes, it is hardly surprising to discover that precious little is known about the way in which these membranes are made by the cell. Fortunately, in recent years, a number of workers have addressed themselves to this problem, which is particularly fascinating since it involves both scalar and vectorial metabolism in the cell. An additional justification for these studies is that, once some understanding has emerged about the way in which fungal plasma membranes grow, it may suggest ways in which this can be controlled, and so provide a rational basis for the design of antifungal agents.

The scalar component in the biogenesis of fungal plasma membranes is the synthesis at various loci in the cell of those molecules—proteins, phospholipids, glycolipids and sterols—that make up the membrane. The vectorial component, which must be closely linked to the scalar component, entails the insertion of these molecules into the growing membrane. It is often not appreciated that this latter process must be very tightly coupled to cell wall synthesis. A moment's reflection shows that, if plasma membrane were synthesized without concomitant synthesis of the overlying cell or hyphal wall, extensive infolding of the plasma membrane must take place, a phenomenon which examination of thin sections through cells or hyphae in the electron miscroscope shows rarely if ever to occur. Moreover, if new cell wall were formed without, at the same time, synthesis of underlying plasma membrane, lysis of the cell or hypha would rapidly follow. Mycologists who study biogenesis of the fungal plasma membrane must therefore pay special attention to this tight coupling of membrane and wall biogenesis.

Biosynthesis of membrane components

PROTEIN While much has been reported on the mechanism of protein synthesis in filamentous fungi and yeasts (de Robichon-Sjulmajster & Surdin-Kerjan, 1971)—indeed *Saccharomyces cerevisiae* has been much favoured as a model organism for the study of protein synthesis in eukaryotes—nothing is known of the way in which individual plasma membrane-bound proteins are synthesized in these organisms. Research on this subject will need to await a full characterization of individual membrane-bound proteins. In principle this can be attempted even now, by studying the induced synthesis of individual classes of transport protein.

PHOSPHOLIPIDS The pathways which are generally believed to be followed during synthesis of fungal plasma membrane phospholipids are shown in Fig. 10.1. However, it has yet to be shown that all of these reactions operate in filamentous fungi and yeasts, and again the scanty information that is available is largely confined to *Saccharomyces cerevisiae* and *Neurospora crassa*. Synthesis of coenzyme-A esters of fatty acids by filamentous fungi is described in Chapter 6.

The two-step acylation of glycerol 3-phosphate to yield phosphatidic acid (Fig. 10.1) was reported by Kuhn & Lynen (1965) using extracts of

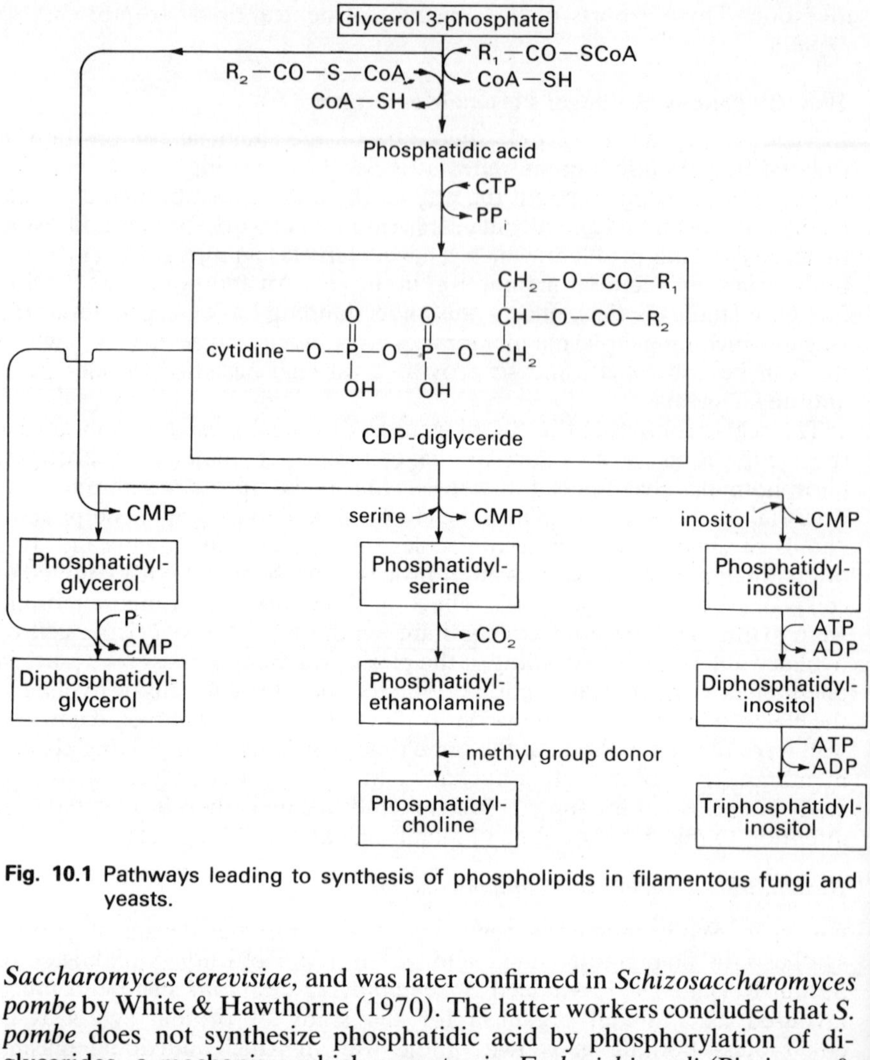

Fig. 10.1 Pathways leading to synthesis of phospholipids in filamentous fungi and yeasts.

Saccharomyces cerevisiae, and was later confirmed in *Schizosaccharomyces pombe* by White & Hawthorne (1970). The latter workers concluded that *S. pombe* does not synthesize phosphatidic acid by phosphorylation of diglycerides, a mechanism which operates in *Escherichia coli* (Pieringer & Kunnes, 1965).

Evidence for operation in *Saccharomyces cerevisiae* of the next step in phospholipid biosynthesis, namely the production of CDP-diglyceride from phosphatidic acid, was obtained by Hutchison & Cronan (1968). These workers reported that the enzyme is associated with a particulate fraction in the yeast, and that its activity is stimulated several-fold by the non-ionic detergent Triton X–100. Interestingly, Steiner & Lester (1972) obtained data which they interpreted to show that CDP-diglyceride may be produced from an endogenous phospholipid other than phosphatidic acid. Because CDP-diglyceride is at a crucial position on the pathway for phospholipid synthesis, further work is urgently required to resolve the origin, or origins, or CDP-diglyceride, particularly if the regulatory mechanisms that operate on these pathways are to be understood.

Two separate pathways are known to lead to synthesis of the two major phospholipids in filamentous fungi and yeasts, namely phosphatidylethanolamine and phosphatidylcholine. Fig. 10.1 shows the reactions that take place on the *methylation pathway*; these involve synthesis of phosphatidylserine from CDP-diglyceride, decarboxylation of phosphatidylserine to yield phosphatidylethanolamine, and stepwise methylation of this phospholipid to give phosphatidylcholine. Evidence for the operation of the first two of these reactions has been reported for *Neurospora crassa* (Sherr & Byk, 1971) and *Saccharomyces cerevisiae* (Steiner & Lester, 1972). Enzymes that catalyse methylation of phosphatidylethanolamine, using S-adenosylmethionine as a methyl-group donor, have been detected in both *Neurospora crassa* (Scarborough & Nyc, 1967*a, b*) and *S. cerevisiae* (Waechter, Steiner & Lester, 1969; Steiner & Lester, 1970). Activity in both organisms was associated with a particulate fraction.

There are reasons for believing that *Neurospora crassa* and *Saccharomyces cerevisiae* can synthesize phosphatidylethanolamine and phosphatidylcholine in another way, using reactions in which CDP-ethanolamine or CDP-choline reacts with a diglyceride. The finding that provision of exogenous ethanolamine or choline in the growth medium enhances synthesis of phosphatidylethanolamine and phosphatidylcholine, respectively, provided the first evidence for operation of this alternative or *cytidine nucleotide pathway*. For *N. crassa*, this evidence was provided by Crocken & Nyc (1964) and more recently by Sherr & Byk (1971), and for *S. cerevisiae* by Waechter, Steiner & Lester (1969), Waechter & Lester (1971) and Ratcliffe *et al.* (1973). Unfortunately, the enzymic basis for these reactions is not well established. Thus, Sherr & Byk (1971) were unable to demonstrate any reaction between CDP-choline and diglyceride with extracts of their strain of *N. crassa*. The situation is further complicated by the suggestion that the cytidine nucleotide pathway may be confined to mitochondria (Magnall & Getz, 1973). That this may not be so is suggested by data obtained in my laboratory by Victoria Sharpe (Victoria Sharpe, J. A. Hossack & A. H. Rose, unpub. observations) which reveal that there are differences in the stretching ability of the plasma membrane in sphaeroplasts obtained from *S. cerevisiae* enriched in phosphatidylcholine as compared with phosphatidylethanolamine. Nevertheless, the role of the cytidine nucleotide pathway for synthesis of phosphatidylethanolamine and phosphatidylcholine will not be resolved until the enzymic basis of the reactions involved has been clarified.

The CDP-diglyceride-dependent incorporation of inositol into phosphatidylinositol (Fig. 10.1) has been reported by Steiner & Lester (1972) using cell-free extracts of *Saccharomyces cerevisiae*. Earlier White & Hawthorne (1970) had failed to obtain evidence for synthesis of phosphatidylinositol by this reaction with cell-free extracts of *Schizosaccharomyces pombe*. Again, the discrepant nature of the scanty evidence available will only be resolved by further research. Diphosphatidylinositol and triphosphatidylinositol occur in *S. cerevisiae* in small amounts (much smaller than the amount of phosphatidylinositol present; Lester & Steiner, 1968). Wheeler, Michell & Rose (1972) reported evidence for the plasma membrane location of the enzyme that catalyses synthesis of diphosphatidylinositol, namely phosphatidylinositol kinase. It is worth noting that there is a rapid turnover of diphosphatidylinositol in *S. cerevisiae*, and that synthesis and subsequent dephosphorylation of this phospholipid constitute an ATPase activity. It would be very informative to know whether the cycle operates as an ATPase in the plasma membrane of *S. cerevisiae*.

GLYCOLIPIDS Almost nothing has been reported on the biosynthesis of those fungal glycolipids that are thought to be truly located in membranes. Since these are, as far as we know, mainly acylated sugars (see p. 313), it can only be presumed that they are synthesized by mechanisms similar to those which operate in bacteria, and which involve transfer of sugar residues from a nucleotide diphosphate sugar to a diglyceride (Shaw, 1975).

STEROLS The biosynthesis in fungal sterols is dealt with in Chapter 13 of this volume.

Assembly of the plasma membrane

Examination of filamentous fungi and yeasts in the light microscope, and more recently in thin sections in the electron microscope, has revealed the accumulation at regions where new plasma membrane and cell wall are being laid down—the hyphal tip in filamentous fungi and the neck in the budding yeast cell—of small electron-transparent vesicles, and this has led to the widespread belief that these vesicles are concerned with the growth process. Unfortunately, the description and nomenclature of these structures is in a most confusing state. In filamentous fungi, they are often referred to as *wall vesicles*, on the not as yet proven assumption that they are involved only in synthesis of new cell wall. Excellent examples of micrographs showing these structures in filamentous fungi appear in the publications by Grove & Bracker (1970) and Heath, Gay & Greenwood (1971). Another group of electron-transparent structures frequently found near the hyphal tip in fungi are known as *lomasomes*. They too have been implicated in synthesis of wall material, but whether they differ fundamentally from wall vesicles is far from being understood (Heath & Greenwood, 1970).

Several groups of workers have reported accumulation of electron-transparent vesicles at the region of cell wall growth in budding yeasts (see Matile, Moor & Robinow, 1969, for a review) none more elegantly than Sentandreu & Northcote (1969). Earlier examinations in the light microscope had revealed that the vesicles in yeast stain with lipid-soluble dyes which is why they came to be known as *lipid droplets*; the term sphaerosome is also used to describe these structures in yeasts.

A more detailed examination of certain of the electron micrographs showing these cell envelope-associated vesicles shows that they are able to fuse with the plasma membrane, a preliminary event presumably in the role of these structures in formation of new plasma membrane and cell wall. Examples of these micrographs showing fusion with the membrane can be found in the references already cited. Micrographs showing fusion of similar structures with the developing ascospores of *Saccharomyces cerevisiae* were published by Beckett, Illingworth & Rose (1973).

In many filamentous fungi, these structures are thought to originate in *Golgi bodies* or equivalent structures that lie a little further back from the hyphal tip. The morphological form of the originating organelle differs in various fungi, but basically always consists of flattened membrane-bound sacs. However, there is no convincing evidence for the existence of a Golgi body in *Saccharomyces cerevisiae* the organism with which much of the physiological research on the smaller vesicles has been conducted. Instead, the vesicles in *S. cerevisiae* are thought to originate from the subcellular organelle which is usually referred to as the vacuole (Sentandreu & Northcote, 1969).

As yet, only somewhat desultory attempts have been made to study the biochemical basis of the growth of the fungal and yeast cell envelope. However, certain aspects of the process have recently been receiving some attention from fungal physiologists particularly with work on the vacuole. Indge (1968) was the first to isolate vacuoles from *Saccharomyces cerevisiae*, and he described certain properties of these subcellular organelles. This finding was followed by a series of papers from Matile and his colleagues at the Swiss Federal Institute in Zurich, Switzerland. A most valuable re-orientation in our thinking of the role of the vacuole in *S. cerevisiae* came in a paper by Matile & Wiemken (1967) in which they suggested that this subcellular organelle may act as a lysosome because of its content of a number of lytic enzyme activities, including esterases, proteases and nucleases. Subsequently, both Matile (1971), with *Neurospora crassa*, and Beteta & Gascon (1971) with *S. cerevisiae* demonstrated that intracellular invertase was almost exclusively located in vacuoles isolated from lysed sphaeroplasts of these organisms by flotation through Ficoll gradients. This claim has since been disputed for *S. cerevisiae* by Holley & Kidby (1973). These workers showed that invertase activity is associated with vacuoles in *S. cerevisiae* only when glucose-repressed cells are derepressed. Otherwise, this enzyme activity is found in smaller subcellular structures, which are very likely the smaller vesicles that are seen to assemble near points of cell wall and membrane growth in *S. cerevisiae*. Indeed, it seems very likely that most workers who have isolated vacuoles from *N. crassa* or *S. cerevisiae* have dealt with preparations that contain in addition the smaller vesicles. This was certainly true of preparations obtained in my laboratory from preparations of lysed sphaeroplasts of *S. cerevisiae* NCYC 366. Hossack, Wheeler & Rose (1973) reported on the large content of triacylglycerols and sterol esters in a mixed, low-density preparation, a report which was followed by one from Cartledge & Rose (1973) which described the various lytic activities of the mixed preparation.

More recently, Cartledge & Rose (1976) have separated the small and large vesicles from a mixed preparation of sphaeroplasts of *Saccharomyces cerevisiae*, and have reported on their lipid content and enzyme activities. It was also shown that, whereas the larger or vacuolar fraction is osmotically sensitive, the smaller vesicles, which measure about 0.2 μm in diameter, are not. Clausen, Christiansen, Jensen & Behnke (1974) reported on a class of lipid particles which they isolated from *S. cerevisiae* (commercial baker's yeast). From the description of these particles, their lipid composition (predominantly sterol esters and triacylglycerols), and size, it would seem that they are very similar indeed to the smaller vesicles isolated by Cartledge & Rose (1976).

Now that preparations of the small and large vesicles from *Saccharomyces cerevisiae* are being obtained in various laboratories, the stage is set for an examination of the role of the smaller vesicles in growth of the plasma membrane, and for a study of the manner in which the smaller vesicles originate from the larger vacuolar vesicles. The next two-three years should therefore witness exciting developments in our understanding of envelope growth on filamentous fungi and yeasts.

10.5 References

ANRAKU, Y. (1968). Transport of sugars and amino acids in bacteria. I. Purification and specificity of the galactose and leucine-binding proteins. *Journal of Biological Chemistry* **243**, 3116–22.

BECKETT, A., HEATH, I. B. & McCAUGHLIN, D. J. (1975). *An atlas of fungal ultrastructure.* London: Longman Group.

BECKETT, A., ILLINGWORTH, R. F. & ROSE, A. H. (1973). Ascospore wall development in *Saccharomyces cerevisiae. Journal of Bacteriology* **113**, 1054–7.

BENTLEY, R., LAVATE, W. V. & SWEELEY, C. C. (1964). Lipid components of two Basidiomycetes—*Calvatia gigantea* and *Clitocybe illudens. Comparative Biochemistry and Physiology* **11**, 263–8.

BETETA, P. & GASCON, S. (1971). Localization of invertase in yeast vacuoles. *Federation of European Biochemical Societies Letters* **13**, 297–300.

BOULTON, A. A. (1965). Some observations on the chemistry and morphology of the membranes released from yeast protoplasts by osmotic shock. *Experimental Cell Research* **37**, 434–59.

BOULTON, A. A. & EDDY, A. A. (1962). The properties of certain particles isolated from yeast protoplasts disrupted by osmotic shock. *Biochemical Journal* **82**, 16P–17P.

BOWMAN, R. D. & MUMMA, R. O. (1967). The lipids of *Pythium ultimum. Biochimica et Biophysica Acta* **144**, 501–10.

BRENNAN, P. J., FLYNN, M. P. & GRIFFIN, P. F. S. (1970). Acylglucoses in *Escherichia coli, Saccharomyces cerevisiae* and *Agaricus bisporus. Federation of European Biochemical Societies Letters* **8**, 322–4.

BRENNAN, P. J., GRIFFIN, P. F. S., LÖSEL, D. M. & TYRRELL, D. (1974). In *Progress in the chemistry of fats and other lipids*, Vol. 14, pp. 51–89. Edited by R. T. Holman. Oxford: Pergamon Press.

BRENNAN, P. J., ROE, J., BYRNE, P. F. S. & TIGHE, J. J. (1975). The glycolipids of *Agaricus bisporus* sporophore. *Proceedings of the Society for General Microbiology II*, 16.

BYRNE, P. F. S., BRENNAN, P. J. (1975). Distribution and formation of glucosyloxy fatty acids of fungi. *Biochemical Society Transactions* **2**, 1346–8.

CARTER, H. E., CELMER, W. D., LANDS, W. E., MUELLER, K. L. & TOMIZAWA, H. H. (1954). Biochemistry of the sphingolipids. VIII. Occurrence of a long chain base in plant phosphatides. *Journal of Biological Chemistry* **206**, 613–23.

CARTLEDGE, T. G. & ROSE, A. H. (1973). Properties of low-density vesicles from *Saccharomyces cerevisiae. Proceedings of the Third International Specialised Symposium on Yeasts*, pp. 251–259. Helsinki.

CARTLEDGE, T. G. & ROSE, A. H. (1976). Isolation and properties of two classes of low-density subcellular vesicles from *Saccharomyces cerevisiae. Journal of Bacteriology* (in preparation).

CHRISTENSEN, M. S. & CIRILLO, V. P. (1972). Yeast membrane vesicles; isolation and general characteristics. *Journal of Bacteriology* **110**, 1190–205.

CLAUSEN, M. K., CHRISTIANSEN, K., JENSEN, P. K. & BEHNKE, O. (1974). Isolation of lipid particles from baker's yeast. *Federation of European Biochemical Societies Letters* **43**, 176–9.

COOKE, R. C. & MITCHELL, D. T. (1969). Sugars and polyols in sclerotia of *Claviceps purpurea, C. nigrificans* and *Sclerotinia curreyana* during germination. *Transactions of the British Mycological Society* **52**, 365–72.

CROCKEN, B. J. & NYC, J. F. (1964). Phospholipid variations in mutant strains of *Neurospora crassa. Journal of Biological Chemistry* **239**, 1727–30.

DUBÉ, J., SETTERFIELD, G., KISS, G. & LUSENA, C. V. (1973). Fate of the plasma membrane of *Saccharomyces cerevisiae* during cell rupture. *Canadian Journal of Microbiology* **19**, 285–90.

EDDY, A. A. & WILLIAMSON, D. H. (1957). A method of isolating protoplasts from yeast *Nature, London* **179**, 1252–3.

FIESER, L. F. & FIESER, M. (1959). *Steroids.* New York: Reinhold Publishing Co.

GARCIA-ACHA, I., AGUIRRE, M. J. R., LOPEZ-BELMONTE, F., URUBURU, F. & VILLANEUVA, J. R. (1966). Isolation of cytoplasmic membranes from *Fusarium culmorum* protoplasts. *Transactions of the British Mycological Society* **49**, 603–9.

GARCIA-MENDOZA, C. & VILLANEUVA, J. R. (1967). Preparation and composition of the protoplast membrane of *Candida utilis. Biochimica et Biophysica Acta* **135**, 189–95.

GOODWIN, T. W. (1973). Comparative biochemistry of sterols in eukaryotic micro-organisms. In *Lipids and biomembranes of eukaryotic micro-organisms*, pp. 1–40. Edited by J. A. Erwin, New York: Academic Press.

GRIFFIN, P. F. S., BRENNAN, P. J. & LÖSEL, D. M. (1970). Free lipids and carbohydrates of *Agricus bisporus* mycelium. *Biochemical Journal* **119**, 11P–12P.

GROVE, S. N. & BRACKER, C. E. (1970). Protoplasmic organization of hyphal tips among fungi: vesicles and Spitzenkörper. *Journal of Bacteriology* **104**, 989–1009.

HEATH, I. B. & GREENWOOD, A. D. (1970). The structure and formation of lomasomes. *Journal of General Microbiology* **62**, 129–37.

HEATH, I. B., GAY, J. L. & GREENWOOD, A. D. (1971). Cell wall formation in the Saprolegniales; cytoplasmic vesicles underlying developing walls. *Journal of General Microbiology* **65**, 225–32.

HEPPEL, L. (1969). The effect of osmotic shock on release of bacterial proteins and on active transport. *Journal of General Physiology* **54**, 95S–113S.

HOLLEY, R. A. & KIDBY, D. K. (1973). Role of vacuoles and vesicles in extracellular enzyme secretion by yeast. *Canadian Journal of Microbiology* **19**, 113–17.

HOLZ, R. B. & SCHISLER, L. C. (1971). Lipid metabolism of *Agaricus bisporus* (Large) Sing. 1. Analysis of sporophore and mycelial lipids. *Lipids* **6**, 176–80.

HORIKOSHI, K. (1973). Comparative studies on β-1, 3-glucanases of micro-organisms. *Proceedings of the Third International Symposium on Yeast Protoplasts*, pp. 25–32. London: Academic Press.

HOSSACK, J. A. & ROSE, A. H. (1976). Fragility of plasma membranes in *Saccharomyces cerevisiae* enriched with different sterols. *Biochimica et Biophysica Acta* in press.

HOSSACK, J. A., WHEELER, G. E. & ROSE, A. H. (1973). Environmentally induced changes in the lipid composition of cells and membranes. *Proceedings of the Third International Symposium on Yeast Protoplasts* pp. 211–27. London: Academic Press.

HUNTER, K. & ROSE, A. H. (1971). Yeast lipids and membranes. In *The yeasts*, Vol. 2, pp. 211–70. Edited by A. H. Rose and J. S. Harrison, London: Academic Press.

HUNTER, K. & ROSE, A. H. (1972). Lipid composition of *Saccharomyces cerevisiae* as influenced by growth temperature. *Biochimica et Biophysica Acta* **260**, 639–53.

HUOTARI, F. I., NELSON, T. E., SMITH, F. & KIRKWOOD, S. (1958). Purification of an exo-β-(1-3)-glucanase from Basidiomycete species QM806. *Journal of Biological Chemistry* **243**, 952–6.

HUTCHISON, H. T. & CRONAN, J. E. JR (1968). The synthesis of cytidine diphosphate diglyceride by cell-free extracts of yeast. *Biochimica et Biophysics Acta* **164**, 606–8.

INDGE, K. J. (1968). The isolation and properties of the yeast cell vacuole. *Journal of General Microbiology* **51**, 441–6.

JACK, R. C. M. (1964). Lipid metabolism of fungi, I. Lipids of the conidia of *Glomerella cingulata. Contributions of the Boyce Thompson Institute* **22**, 311–33.

JACK, R. C. M. (1966). Lipid patterns in the main classes of fungi. *Journal of Bacteriology* **91**, 2101–2.

JACK, R. C. M. & LAREDO, J. A. (1968). Fungal spore phospholipids and the accumulation of selected chemicals. *Lipids* **3**, 459–60.

KATES, M. & BAXTER, R. M. (1962). Lipid composition of mesophilic and psychrophilic yeasts (*Candida* species) as influenced by environmental temperature. *Canadian Journal of Biochemistry and Physiology* **40**, 1213-27.

KEITH, A. D., WISNIESKI, B. J., HENRY, S. & WILLIAMS, J. C. (1973). Membranes of yeast and *Neurospora*: lipid mutants and physical studies. In *Lipids and biomembranes of eukaryotic micro-organisms*, pp. 259–321. Edited by J. A. Erwin, New York: Academic Press.

KUHN, N. J. & LYNEN, F. (1965). Phosphatidic acid synthesis in yeast. *Biochemical Journal* **94**, 240–6.

KUNDIG, W., KUNDIG, F. D., ANDERSON, B. & ROSEMAN, S. (1966). Restoration of active transport of glycosides in *Escherichia coli* by a component of a phosphotransferase system. *Journal of Biological Chemistry* **241**, 3243–6.

LAINE, R. A., GRIFFIN, P. F. S., SWEELEY, C. C. & BRENNAN, P. J. (1972). Monoglucosyl-oxooctadecenoic acid—a glycolipid from *Aspergillus niger. Biochemical Journal* **11**, 2267–71.

LESTER, R. L. & STEINER, M. R. (1968). The occurrence of diphosphoinositide and triphosphoinositide in *Saccharomyces cerevisiae. Journal of Biological Chemistry* **243**, 4889–93.

LESTER, R. L., SMITH, S. W., WELLS, G. B., REES, D. C. & ANGUS, W. W. (1974). The isolation and partial characterization of two novel sphingolipids from *Neurospora crassa*: di (inositolphosphoryl) ceramide and [(gal)$_3$glu] ceramide. *Journal of Biological Chemistry* **249**, 3388–94.

LETTERS, R. (1968). The breakdown of yeast phospholipids in relation to membrane function. *Bulletin de la Societe de Chimie Biologique*, **50**, 1385–93.

LONGLEY, R. P., ROSE, A. H. & KNIGHTS, B. A. (1968). Composition of the protoplast membrane from *Saccharomyces cerevisiae. Biochemical Journal* **108**, 401–12.

MAGNALL, D. & GETZ, R. S. (1973). Phospholipids. In *Lipids and biomembranes of eukaryotic micro-organisms*, pp. 145–95. Edited by J. A. Erwin, New York: Academic Press.

MARRIOTT, M. S. (1975). Isolation and chemical characterization of plasma membranes from the yeast and mycelial forms of *Candida albicans. Journal of General Microbiology* **86**, 115–32.

MATILE, PH. (1971). Vacuoles, lysosomes of *Neurospora crassa. Cytobiologie* **3**, 324–30.

MATILE, PH. & WIEMKEN, A. (1967). The vacuole as the lysosome of the yeast cell. *Archiv für Mikrobiologie* **56**, 148–55.

MATILE, PH, MOOR, H. & MÜHLETHALER, K. (1967). Isolation and properties of plasmalemma in yeast. *Archiv für Mikrobiologie* **58**, 201–11.

MATILE, PH., MOOR, H. & ROBINOW, C. F. (1969). Yeast cytology. In *The yeasts*, Vol. 1, pp. 219–302. Edited by A. H. Rose and J. S. Harrison. London: Academic Press.

MERDINGER, E., KOHN, P. & McCLAIN, R. C. (1968). Composition of lipids in extracts of *Pullularia pullulans. Canadian Journal of Microbiology* **14**, 1021–7.

MUMMA, R. O. & BRUSZEWSKI, T. E. (1970). The fatty acids of *Entomophthora coronata. Lipids* **5**, 915–20.

MUMMA, R. O., SEKURA, T. D. & FERGUS, C. L. (1971). Thermophilic fungi: III. The lipids of *Humicola grisea* var. *thermoides. Lipids* **6**, 589–94.

NOMBELA, C., URUBURU, F. & VILLANUEVA, J. R. (1974). Studies on membranes isolated from extracts of *Fusarium culmorum. Journal of General Microbiology* **81**, 247–54.

NURMINEN, T. & SUOMALAINEN, H. (1973). On the enzymes and lipid composition of cell envelope fractions from *Saccharomyces cerevisiae. Proceedings of the Third International Specialized Symposium on Yeasts*, pp. 169–89. Edited by H. Suomalainen and C. Waller, Helsinki.

PATCHING, J. W. & ROSE, A. H. (1971). Cold osmotic shock in *Saccharomyces cerevisiae. Journal of Bacteriology* **108**, 451–8.

PIERINGER, R. A. & KUNNES, R. S. (1965). The biosynthesis of phosphatidic acid and lysophosphatidic acid by glyceride phosphokinase pathways in *Escherichia coli. Journal of Biological Chemistry* **240**, 2833–8.

RATCLIFFE, S. J., HOSSACK, J. A., WHEELER, G. E. & ROSE, A. H. (1973). Modifications to the phospholipid composition of *Saccharomyces cerevisiae* induced by exogenous ethanolamine. *Journal of General Microbiology* **76**, 445–9.

REESE, E. T. & MANDELS, M. (1959). Glucanases in fungi. *Canadian Journal of Microbiology* **5**, 173–85.

DE ROBICHON-SJULMAJSTER, H. & SURDIN-KERJAN, Y. (1971). Nucleic acid and protein synthesis in yeasts: regulation of synthesis and activity. In *The yeasts*, Vol. 2, pp. 335–418. Edited by A. H. Rose and J. S. Harrison, London: Academic Press.

SCARBOROUGH, G. A. & NYC, J. F. (1967*a*). Methylation of ethanolamine phosphatides by microsomes from normal and mutant strains of *Neurospora crassa. Journal of Biological Chemistry* **242**, 238–42.

SCARBOROUGH, G. A. & NYC, J. F. (1967*b*). Properties of a phosphatidyl monomethyl-ethanolamine N-methyl transferase from *Neurospora crassa. Biochimica et Biophysica Acta* **146**, 111–9.

SCHIBECI, A., RATTRAY, J. B. M. & KIDBY, D. K. (1973). Isolation and identification of yeast plasma membrane. *Biochimica et Biophysica Acta* **311**, 15–25.

SENTANDREU, R. & NORTHCOTE, D. H. (1969). The formation of buds in yeast. *Journal of General Microbiology* **55**, 393–8.

SHAW, N. (1970). Bacterial glycolipids. *Bacteriological Reviews* **34**, 365–77.

SHAW, N. (1975). Bacterial glycolipids and glycophospholipids. *Advances in Microbial Physiology* **12**, 141–67.

SHERR, S. I. & BYK, C. (1971): Choline and serine incorporation into phospholipids of *Neurosopora crassa. Biochimica et Biophysica Acta* **239**, 243–7.

SHOPPEE, C. W. (1964). *Chemistry of the steroids.* London: Butterworths.

SIMONI, R. D. (1972). In *Membrane molecular biology,* pp. 281–322. Edited by C. F. Fox and A. Keith. Stanford: Sinauer Associates Inc.

SINGER, S. J. & NICOLSON, G. L. (1972). The fluid mosaic model of the structure of cell membranes. *Science,* **175**, 720–30.

SMITH, S. W. & LESTER, R. L. (1974). Inositol phosphoryl ceramide, a novel substance and the chief member of a major group of yeast sphingolipids containing a single inositol residue. *Journal of Biological Chemistry* **249**, 3395–405.

STANACEV, B. Z. & KATES, M. (1963). Constitution of cerebrin from the yeast *Torulopsis utilis. Canadian Journal of Biochemistry and Physiology* **41**, 1330–4.

STEINER, M. R. & LESTER, R. L. (1970). *In vitro* study of the methylation pathway of phosphatidylcholine synthesis and the regulation of this pathway in *Saccharomyces cerevisiae. Biochemistry, New York* **9**, 63–9.

STEINER, M. R. & LESTER, R. L. (1972). *In vitro* studies of phospholipid biosynthesis in *Saccharomyces cerevisiae. Biochimica et Biophysica Acta* **260**, 222–43.

STEINER, S., SMITH, S., WAECHTER, C. J. & LESTER, R. L. (1969). Isolation and partial characterization of a major inositol-containing lipid in baker's yeast, mannosyl-diinositol diphosphoryl ceramide. *Proceedings of the National Academy of Sciences of the United States of America* **64**, 1042–8.

SUMNER, J. L. (1970). The fatty acid composition of *Blastocladiella emersonii. Canadian Journal of Microbiology* **16**, 1161–4.

SUOMALAINEN, H., NURMINEN, T. & OURA, E. (1967). Isolation of the plasma membrane of yeast. *Acta Chemical Fenniae* **B40**, 323–6.

SWEELEY, C. C. (1959). A gas chromatographic method for phytosphingosine assay. *Biochimica et Biophysica Acta* **36**, 268–71.

TYÖFINOJA, K., NURMINEN, T. & SUOMALAINEN, H. (1974). The cell-envelope glycolipids of baker's yeast. *Biochemical Journal* **141**, 133–9.

TYRRELL, D. (1967). The fatty acid compositions of 17 *Entomophthora* isolates. *Canadian Journal of Microbiology* **13**, 755–60.

TYRRELL, D. (1968). The fatty acid composition of Entomophthoraceae. II. The occurrence of branched-chain fatty acids in *Conidiobolus denaesporus* Drechsl. *Lipids* **3**, 368–72.

TYRRELL, D. (1971). The fatty acid composition of some Entomophthoraceae. III. *Canadian Journal of Microbiology* **17**, 1115–8.

VILLANUEVA, J. R., GACTO, M. & SIERRA, J. M. (1973). Enzymic composition of a lytic system from *Micromonospora chalcea* AS. *Proceedings of the Third International Symposium on Yeast Protoplasts.* pp. 3–24. London: Academic Press.

WAECHTER, C. J. & LESTER, R. L. (1971). Regulation of phosphatidylcholine biosynthesis in *Saccharomyces cerevisiae. Journal of Bacteriology* **105**, 837–43.

WAECHTER, C. J., STEINER, M. R. & LESTER, R. L. (1969). Regulation of phosphatidylcholine biosynthesis by the methylation pathway in *Saccharomyces cerevisiae. Journal of Biological Chemistry* **244**, 3419–22.

WAGNER, H. & FIEGERT, E. (1969). Spingolipide und Glykolipide von Pilzen und höheren Pflanzen. III. Mi H: Isolierung eines cerebrosids aus *Aspergillus niger. Zeitschrift für Naturforschung* **B24**, 359–69.

WEETE, J. D. (1974). *Fungal lipid biochemistry.* New York: Plenum Press.

WEISS, B. & STILLER, R. L. (1972). Sphingolipids of mushrooms. *Biochemistry, New York* **11**, 4552–7.

WEISS, B., STILLER, R. L. & JACK, R. C. M. (1973). Sphingolipids of the fungi *Phycomyces blakesleeanus* and *Fusarium lini. Lipids* **8**, 25–30.

WHEELER, G. E., MICHELL, R. H. & ROSE, A. H. (1972). Phosphatidylinositol kinase activity in *Saccharomyces cerevisiae. Biochemical Journal* **127**, 64P.

WHITE, G. L. & HAWTHORNE, J. N. (1970). Phosphatidic acid and phosphatidylinositol metabolism in *Schizosaccharomyces pombe. Biochemical Journal* **117**, 203–13.

WILEY, W. R. (1970). Tryptophan transport in *Neurospora crassa:* a tryptophan-binding protein released by cold osmotic shock. *Journal of Bacteriology* **103**, 656–62.

CHAPTER 11

The Cell Wall

R. F. ROSENBERGER

11.1 Introduction

The cell wall of fungi is the rigid structure which both protects the fragile protoplast from damage and maintains the characteristic shape of the cell. This can be readily demonstrated by treating mycelia with enzymes which hydrolyse their walls (Villanueva, 1966). As the wall is digested, the membrane bursts or, with added osmotic protection, protoplasts are liberated which have a spherical shape quite different from that of the original hypha.

Work on the composition and structure of fungal walls has a long history. Van Wisselingh (1898) and others (Wettstein, 1921; Nabel, 1939) showed that fungal walls contain polysaccharides and in particular that chitin, an aminopolysaccharide mainly found in invertebrates, was present in many species. The early investigations, however, relied on qualitative tests performed on whole cells and these as such could only provide very limited information. Modern work on wall structure began with the development of methods for separating the walls from other cell constituents (Aronson & Machlis, 1959; Salton, 1961; Crook & Johnston, 1962). Chemical analysis of such purified walls showed them to be complex mixtures of polymers. Further, the kinds of polymers present could vary in species from different taxa and in a single organism at different stages of its life cycle (Bartnicki-Garcia & Nickerson, 1962; Bartnicki-Garcia, 1968).

While wall fractionations were revealing a complex chemical composition, electron microscope studies were demonstrating that the physical structure was equally complicated (*e.g.* Hunsley & Burnett, 1970). Fungal walls contain a network of fibrils with the spaces in the net filled by matrix polymers and in this they resemble such man-made composites as glass fibre—reinforced—plastic and reinforced concrete. This model of walls as fibre-matrix composites, in which different polymers can fulfill the functions of matrix and fibres and whose mechanical properties can be modified by altering parameters such as fibre concentration, is one on which much present-day work is based (Aronson, 1965; Northcote, 1972). There are still many gaps in our knowledge of the chemical composition of walls and in our understanding of what the physical organization implies. The walls of relatively few species have been studied and in these it has almost always been the wall of that readily available form, the vegetative hypha. The parameters likely to affect the mechanical characters of a composite are only beginning to be studied. However, the available findings do form a basis from which the ultimate goal of cell wall research, the relation between wall metabolism and the morphology and development of the fungal organism, can be faintly glimpsed in the far distance.

11.2 Chemical Composition

General considerations

Wall composition will be fully understood when the answers to three sets of questions, differing in their complexity, have been obtained. These questions are: (a) what are the monomers present; (b) how are the constituent monomers linked in each polymer; and (c) how are the various polymers held together in the cell wall as a whole?

After hydrolysis, the monomers can usually be identified with relative ease and the answers to this question tend to be the most complete. The information has also provided workers in the field with a perhaps not unexpected bonus. As in bacteria (Ghuysen, 1968), variations in overall wall composition tend to follow phylogenetic lines and the data has been of help in classification (Bartnicki-Garcia, 1968). The next stage, determining the linkage of monomers in the various polymers, presents considerably more difficulty, particularly when a polymer is present in small amounts. Thus, while the structure of most of the quantitatively important polymers has been established, this is not true for the minor constituents. The final problem, bonds between the polymers, is the most complex but probably the most important in understanding biosynthesis and the alterations of structure during growth and development. Very little is as yet known about this aspect.

Methods

For a comprehensive list of references on the preparation and analysis of fungal walls, the reader is referred to the review by Taylor & Cameron (1973).

ISOLATION OF WALLS The mycelium is disrupted either mechanically using pressure cells (Novaes-Ledieu, Jimenez-Martinez & Villanueva, 1967; Wessels, Kreger, Marchant, Regensburg & De Vries, 1972), homogenizers (Hamilton & Knight, 1962; Zonneveld, 1971), ultrasonic probes (Bull, 1970*a*; Mahadevan & Mahadkar, 1970), or by shaking with detergent solutions (Mahadevan & Tatum, 1965; Bull, 1970*a*). The wall fraction is then separated from soluble constituents and particulate debris by differential centrifugation. The walls may be further extracted with salt solutions (Gancedo, Gancedo & Asensio, 1966; Mahadevan & Mahadkar, 1970) or lipid solvents (Wessels, 1965) to remove contaminating material.

The above procedures yield walls which are free from cytoplasm and membranes when examined under the electron microscope and which, on hydrolysis, show no typically cytoplasmic components such as ribose. As in the purification of any complex biological structure, there is no easy way of eliminating the possibilities that some genuine wall component has been removed during purification or that cell disruption has allowed some cytoplasmic material to become strongly adsorbed.

FRACTIONATION OF PURIFIED WALLS Since polysaccharides usually make up more than 75% of fungal walls, fractionation is mainly an excercise in polysaccharide separation. Many of the commonly occurring polymers differ in their solubility in acid and alkaline solutions and these properties have been the most widely exploited in fractionation. As an illustration of these techniques, the frequently-used procedure devised by Mahadevan & Tatum (1965) is shown in Fig. 11.1.

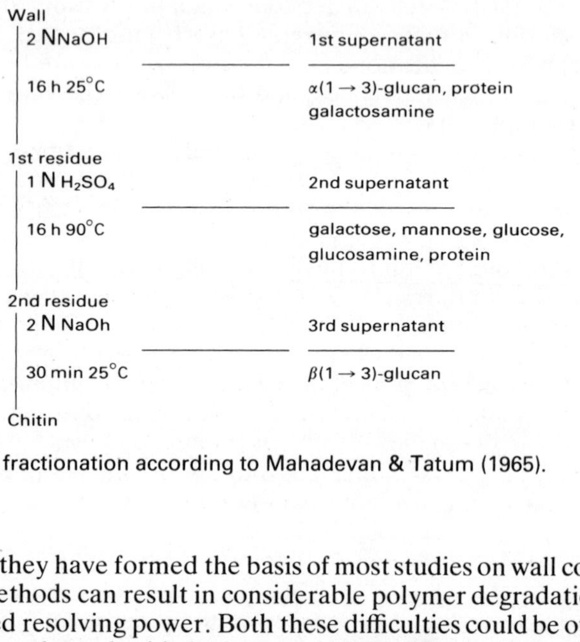

Fig. 11.1 Wall fractionation according to Mahadevan & Tatum (1965).

Although they have formed the basis of most studies on wall composition, the above methods can result in considerable polymer degradation and they have a limited resolving power. Both these difficulties could be overcome by using specific polysaccharide hydrolases for the controlled digestion of walls (*e.g.* Troy & Koffler, 1969). Purified hydrolases with the required specificities have so far not been commercially available and this has limited their use. As such enzymes become more easily obtainable, however, their contribution to studies of fungal walls will no doubt be as great as that of bacteriolytic enzymes in the elucidation of bacterial wall structure (Ghuysen, 1968).

CHARACTERIZATION OF POLYMERS The kind of monomers present in whole walls or in fractions separated from whole walls can be readily determined by hydrolysis and chromatography of the hydrolysates. When polysaccharides are hydrolysed by prolonged heating with mineral acids, however, some destruction of sugars is bound to occur. This may interfere with accurate quantitative determinations of monomers and with the detection of particularly labile compounds such as uronic acids (Gancedo, Gancedo & Asensio, 1966; Bartnicki-Garcia & Reyes, 1968).

If a polysaccharide can be obtained in sufficient purity and yield, the classical methods of carbohydrate chemistry are available for its complete characterization. In mixtures of polysaccharides some of the linkages present can usually be determined by the susceptibility to purified enzymes of known specificity and by the detection of particular disaccharides after partial hydrolysis (*e.g.* Novaes-Ledieu, Jimenez-Martinez & Villanueva, 1967). Among the physical methods available, X-ray powder diagrams have been extensively used to detect chitin and cellulose which tend to be more crystalline than most other wall polymers (Fuller & Barshad, 1960; Aronson, 1965). X-ray diffraction can also be used to characterize glucans other than cellulose though in at least one case this has not proved particularly reliable (Kreger, 1954; Mitchell & Sabar, 1966; Aronson, Cooper & Fuller, 1967; Bartnicki-Garcia, 1968). Infra-red spectroscopy has been employed by several groups to strengthen the evidence for the type of linkage postulated (Aronson, Cooper & Fuller, 1967; Michell & Scurfield, 1967, 1970*a*, *b*; Bull, 1970*a*).

Survey of wall constituents

Table 11.1 lists the most frequently occurring and quantitatively most important polymers in fungal walls while Table 11.2 illustrates some typical wall analyses. The best studied of these compounds are discussed in greater detail below. Regarding this survey, it should be remembered that the walls of only a limited number of species have so far been examined and that these have usually been walls of vegetative hyphae. Future work, which will no doubt both extend the range of species studied and examine different stages of the life cycle, seems almost certain to add new compounds.

$\beta(1 \rightarrow 3)$-GLUCAN, R-GLUCAN This is probably the most widely distributed of all fungal wall polysaccharides. While absent from hyphae of

Table 11.1 Polymers found in fungal walls

Polymer	Constituent monomers	Linkage	Distribution
R-glucan	Glucose	$\beta(1 \rightarrow 3)$ + $\beta(1 \rightarrow 6)$	Most groups except Mucorales
S-glucan	Glucose	$\alpha(1 \rightarrow 3)$	Ascomycetes and Basidiomycetes
Cellulose	Glucose	$\beta(1 \rightarrow 4)$	Phycomycetes
Chitin	N–acetyl-glucosamine	$\beta(1 \rightarrow 4)$	Most group except Phycomycetes and Mucorales
Chitosan	Glucosamine	$\beta(1 \rightarrow 4)$	Mucorales
Aminopolysaccharide	Galactosamine	Unknown	Ascomycetes, Hyphomycetes
Polyuronide	Glucuronic acid	Unknown	Mucorales
Heteropolymers	Mannose, galactose, fucose, xylose, glucose, glucuronic acid	Unknown	Zygomycetes, Ascomycetes Basidiomycetes
Protein	Common amino acids—at times hydroxyproline		All groups, hydroxyproline in Phycomycetes

Zygomycetes (Bartnicki-Garcia & Reyes, 1968; Bartnicki-Garcia, 1968) and of *Agaricus* (Kreger, 1954), it appears to be a wall component in the other species of filamentous fungi examined and in many yeasts (Mahadevan & Tatum, 1965; Wessels, 1965; Novaes-Ledieu, Jimenez-Martinez & Villanueva, 1967; Bartnicki-Garcia, 1968; Bull, 1970*a*). When present it comprises 15–30% of the total wall polysaccharide.

The fungal R-glucan is not a straight-chain $\beta(1 \rightarrow 3)$-glucose polymer like *Euglena* paramylon or plant callose, but contains some $\beta(1 \rightarrow 6)$ branches (Bull & Chesters, 1966; Wessels, 1969*a*; Novaes-Ledieu, Jimenez-Martinez & Villanueva 1967; Bull, 1970*a*; Zonneveld, 1971). This has been most clearly demonstrated by the detection of gentiobiose, the $\beta(1 \rightarrow 6)$-disaccharide, after digestion with purified $\beta(1 \rightarrow 3)$-glucanases. Nothing

Table 11.2 Typical analyses of fungal walls

	Aspergillus nidulans **(Zonneveld, 1971)**	***Phytophthora heveae*** **(Novaes-Ledieu *et al.*, 1967)**
Monomers	**% of wall dry weight**	**% of wall dry weight**
Glucose	53	85–90
Galactose	2.7	0.5–1.0
Mannose	2.0	0.5–1.0
Glucosamine	19.1	2.3
Lipid	4.6	2.5
Polymers		
$\alpha(1 \rightarrow 3)$-glucan	22	—
$\beta(1 \rightarrow 3)$-glucan	21	54
Chitin	19.1	—
Cellulose	—	36
Protein	10.5	4.6–6.7

definite, however, can be said about the frequency of the $\beta(1 \rightarrow 6)$ branch points, about the average length of branches or if in fact the $\beta(1 \rightarrow 3)$-glucan fraction could be a mixture of closely related polymers. Part of this uncertainty is due to the solubility properties of the R glucan. Before it can be extracted from walls, these have to be subjected to mild acid hydrolysis (Mahadevan & Tatum, 1965). The acid treatment breaks some of the bonds (Kreger, 1954; Wessels *et al.*, 1972) and allows the polymers to be extracted but the preliminary hydrolysis prevents any firm conclusions being drawn about the structure of the native polymer. It may be of interest in this connection that several fungi produce an extracellular mucilage which is $\beta(1 \rightarrow 3)$-glucan with $\beta(1 \rightarrow 6)$ branches containing a single glucose residue (Buck, Chen, Dickerson & Chain, 1968; Wessels *et al.*, 1972).

R glucans usually give diffuse X-ray diffraction patterns, presumably because the side chains do not allow easy packing in crystals (Kreger & Meeuse, 1952; Kreger, 1954; Wessels *et al.*, 1972). On mild acid hydrolysis and with the progressive removal of the side branches, the X-ray diagrams become much sharper and like those of straight-chain $\beta(1 \rightarrow 3)$-glucans. Such X-ray patterns have been used to identify R-glucan, as has susceptibility to purified $\beta(1 \rightarrow 3)$-glucanase (Mahadevan & Tatum, 1965; Bull,

1970*a*; Zonneveld, 1971), the appearance of laminaribiose, the $\beta(1 \rightarrow 3)$-disaccharide, in partial hydrolysates (*e.g.* Novaes-Ledieu, Jimenez-Martinez & Villanueva, 1967) and infra-red spectroscopy (Michell & Scurfield, 1967; 1970*a*).

The branched nature of the R glucan and its wide-spread occurrence have raised the question if this polymer could be fulfilling a special function in the wall. In bacteria, the crosslinking of polysaccharides by peptide bridges is a crucial factor in wall construction (Ghuysen, Strominger & Tipper, 1968). Crosslinks between polysaccharide chains are also believed to be of importance in providing strength and rigidity in plant cell walls (Lamport, 1970). It is thus tempting to speculate that crosslinks, whose nature is not yet clear, may also occur between fungal wall polymers. If this were indeed the case, the branched R glucan, whose detachment from the wall requires mild acid hydrolysis, would be a possible candidate for this role.

$\beta(1 \rightarrow 4)$-GLUCAN, CELLULOSE In contrast to its ubiquitous occurrence in plant cell walls, cellulose is found in the walls of only a limited number of fungi, mainly species of Oömycetes (Bartnicki-Garcia, 1968). When present it occurs as a major component of the wall and accounts for 30–45% of the wall polysaccharide (Novaes-Ledieu, Jiminez-Martinez & Villanueva, 1967).

In walls, cellulose occurs in the form of microfibrils (Hunsley & Burnett, 1970). Its marked insolubility and resistance to hydrolysis allow cellulose to be purified from most other wall constituents except chitin and this simplifies its identification. X-ray diffraction (Aronson, 1965), susceptibility to cellulase digestion, the presence of cellobiose in partial hydrolysates and its solubility in Schweitzers reagent (Reeves, 1951; Michell & Scurfield, 1970*b*) have all been used to identify this polymer.

$\alpha(1 \rightarrow 3)$-GLUCAN, S-GLUCAN This straight-chain polysaccharide is a major component in the walls of Ascomycetes and Basidiomycetes (Kreger, 1954; Bacon, Jones, Farmer & Webley, 1968; Bull, 1970*a*; Wessels *et al.*, 1972). It comprises 15–25% of the wall polysaccharides and in at least one species it forms one of the outermost layers of the wall (Hunsley & Burnett, 1970; Wessels *et al.*, 1972). S glucan can be extracted from walls with alkali at room temperature and in such extracts it is usually the major component. It has been identified after precipitation from alkaline extracts by the products of partial hydrolysis, X-ray diffraction and infra-red spectroscopy.

An unexpected function of the S glucan in *Aspergillus nidulans* has been demonstrated by Zonneveld (1972*a*, *b*, 1973, 1974). While the evidence for other organisms suggests that the S glucan is a stable wall polymer (Wessels *et al.*, 1972), this is not the case with *A. nidulans.* Here the S glucan is degraded during fruit-body formation and provides the energy and carbon to drive this morphogenetic process. In keeping with its function as a reserve storage compound, the S glucan content of vegetative *A. nidulans* hyphae can be varied by changing the glucose concentration in the medium.

CHITIN This is poly-*N*-acetylglucosamine with the monomers linked $\beta(1 \rightarrow 4)$ in straight chains. It is one of the most frequently occurring polymers in fungal walls, only $\beta(1 \rightarrow 3)$-glucans having a wider distribution (Bartnicki-Garcia, 1968). Chitin has not been found in the walls of

Oömycetes and Zygomycetes, but occurs in other groups of filamentous fungi as microfibrils (Mahadevan & Tatum, 1965; Aronson, 1965; Hunsley & Burnett, 1970). In the Oömycetes, its place as the fibrillar phase is taken by cellulose while Zygomycetes contain the closely related chitosan (Bartnicki-Garcia & Reyes, 1968).

The chitin content of walls is quite variable, from 5% of the wall dry weight in *Schizophyllum* (Wessels, 1965) to 60% in *Sclerotium* (Bloomfield & Alexander, 1967). Because of its great resistance to hydrolysis chitin can be purified from other wall constituents by heating with strong acids and alkali. Its crystallinity makes it suitable for X-ray diffraction analysis and this and susceptibility to chitinase have been used for identification (Aronson, 1965; Troy & Koffler, 1969).

CHITOSAN This is the deacylated analogue of chitin, *i.e.* $\beta(1 \rightarrow 4)$-polyglucosamine with no or very few acetyl groups. It has so far been described as a major component only in *Mucor* and *Phycomyces* (Bartnicki-Garcia, 1968).

GALACTOSAMINE POLYMERS Galactosamine has been identified in the walls of several Ascomycetes and usually accounts for only a few percent of the total monomers (Mahadevan & Tatum, 1965; Applegarth & Bozoian, 1969; Bull, 1970*a*; Katz & Rosenberger, 1970). It should be remembered, however, that the detection of galactosamine is easy only in those wall fractions which do not contain glucosamine. Where chitin and glucosamine are also present, this compound could be underestimated or missed. In at least two fungi galactosamine has been shown to be a major component of the wall, although nothing is known about the polymer in which it occurs (Applegarth & Bozoian, 1969; Bull, 1970*a*).

POLYURONIDES These have been shown to occur as major components in *Mucor*, where both a homopolymer of glucuronic acid and a heteropolymer of glucuronic acid were found (Bartnicki-Garcia & Reyes, 1968). In hydrolysates of other fungi, either small amounts (Gancedo, Gancedo & Asensio, 1966) or no uronic acids have been detected. Uronic acids are, however, known to be very labile to acid hydrolysis and they could easily have been underestimated or missed entirely in many instances.

OTHER POLYSACCHARIDES Four sugars have been identified as minor components in hydrolysates of walls from different fungal species. These are mannose and galactose in Ascomycetes, mannose, fucose and xylose in Basidiomycetes and mannose, fucose and galactose in Zygomycetes (Bartnicki-Garcia, 1968). There is only very limited information about the polymers in which these sugars occur. In *Mucor rouxii*, mannose, galactose and fucose were found in a heteropolymer with glucuronic acid (Bartnicki-Garcia & Reyes, 1968) and in Ascomycetes it has been suggested that galactose and mannose occur in a heteropolymer with glucose (Ruiz-Herrera, 1967; Zonneveld, 1971).

PROTEINS AND LIPIDS Walls purified by extensive washing and centrifugations usually contain 10–15% by weight of proteins and 5–10% lipid (Johnston, 1965; Ruiz-Herrera, 1967; Mitchell & Taylor, 1969; Mahadevan & Tatum, 1965; Bull, 1970*a*). These findings raised the question if the protein and lipid associated with wall preparations are true wall components or parts of attached membrane fragments (*e.g.* Aronson, 1965; Sentandreu & Northcote, 1968). In the case of the proteins at least it now appears clear that they are structural components of the wall. Although in some strains detergents will remove most of the protein (Bull, 1970*a*) in others even 8 M urea will not dissociate them from the wall (Mitchell & Taylor, 1969). Marked changes are observed in the ultrastructure of walls after treatment with proteolytic enzymes and these can be readily explained only if the proteins are an integral part of the envelope (Hunsley & Burnett, 1970). Wall proteins contain upward of 14 amino acids and interestingly, hydroxyproline occurs in fungi having cellulose but not in those having chitin in the wall (Bartnicki-Garcia, 1968). The presence of hydroxyproline again supports the view that wall proteins are specialized compounds rather than cytoplasmic contaminants. It further suggests the possibility that, at least in the cellulose-containing species, protein may form crosslinks between polysaccharide chains. This is by analogy with the hydroxyproline-containing protein of plant cell walls, which appears to have such a function (Lamport, 1970).

At present there is only very little information about the nature and function of the lipids associated with the wall.

MELANINS These complex pigments are of common occurrence in fungal walls and may be either indolic or catecholic melanins (Nicolaus, Piatelli & Fattorusso, 1964; Bull, 1970*a*). Because of the complexities of melanin chemistry, it is not clear how similar the pigments from different fungi are to each other. Melanins are secondary metabolites and are produced in large amounts only after active growth has ceased (Rowley & Pirt, 1972). In old mycelium, however, these pigments may account for up to 20% of the wall dry weight.

Melanins are deposited on the exterior surface of hyphae or other fungal structures and greatly increase the resistance of the wall to attack by hydrolytic enzymes (Bloomfield & Alexander, 1967; Kuo & Alexander, 1967; Bull, 1970*b*).

11.3 The Organization of Polymers in the Wall

The fungal wall is by no means a homogeneous mixture of its constituent polymers but appears to be a structured and complex assembly. The spatial relations of the polymers to each other seem certain to determine as many properties of the wall as the nature of the polymers themselves. Such relations would affect both static mechanical characteristics and the dynamic changes the walls have to undergo during growth and morphogenesis. So far two general features of wall organization have been established and these are discussed below.

Microfibrillar and matrix phases

At least one of the polymers in the wall always appears to be present in the form of microfibrils (Hawker, 1965; Mahadevan & Tatum, 1967; Hunsley & Burnett, 1968, 1970). The microfibrils mesh together to form a net and the spaces in the net are filled by a matrix of other polymers. The fibrillar network can be seen clearly only after the matrix polymers have been removed by a suitable chemical or enzymatic treatment (Fig. 11.2). The treatment, particularly the type of enzyme, which will reveal microfibrils can thus be used to identify the polymers which make up the matrix, and similarly the treatments that disrupt the fibrils will indicate their chemical composition. The microfibrils consist of either chitin or cellulose, depending on which of these polymers the wall contains. In Ascomycetes there is a second net of glycoprotein fibrils in addition to the chitin net (Mahedevan & Tatum, 1967; Hunsley & Burnett, 1970). The matrix phase is usually composed of $\beta(1 \rightarrow 3)$-glucan and protein.

The mixture of fibrils and matrix must give fungal walls at least some properties in common with such products of modern technology as glass fibre-reinforced-plastic. The case for this has been strongly argued by Northcote (1972) in regard to plant cell walls, which also contain mixed fibril-matrix phases. Composite materials of this type possess remarkable strength for their weight and this is clearly a desirable feature in a protective cell wall (Holliday, 1966; Kelly, 1966; Mark, 1967; Northcote, 1972). But perhaps even more important than the increase in strength is the range of mechanical properties that can be achieved with no change of materials and by having mixed phases. The parameters which affect the characteristics of man-made composites have been extensively studied (Holliday, 1966; Kelly, 1966). Changes in the length and cross-section of fibres, in the area the fibres occupy, in the orientation of fibres and in the degree of interaction between fibres and matrix all produce marked alterations in the behaviour of the composite. But these are precisely the type of parameters that a cell could control and modify during growth and differentiation. Changes in the

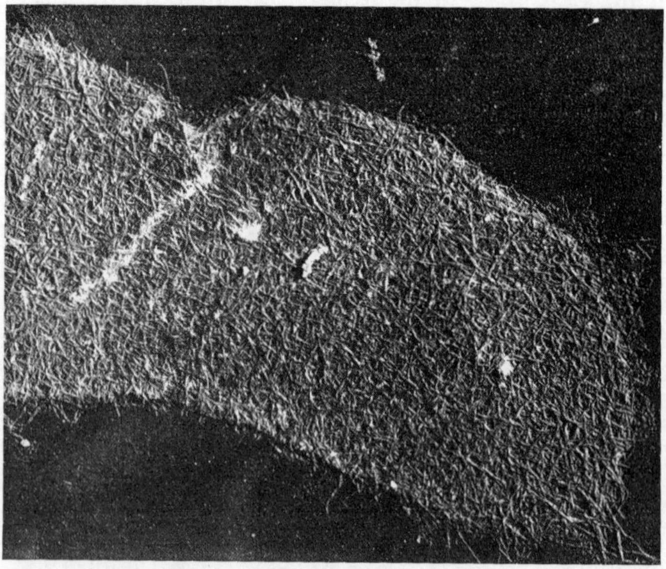

Fig. 11.2 Apex of *Schizophyllum commune* hypha after chemical treatment to reveal microfibrils. (From Hunsley & Burnett (1970).)

relative amounts of fibres and matrix should affect the plasticity of the wall and it appears that such changes do occur in hyphae. As the young apical wall matures, both the diameter and the density of microfibrils increase (Hunsley & Burnett, 1968, 1970) and at the same time the wall becomes more rigid (Robertson, 1959). In theory, at least, modulations of biosynthesis which alter fibre-matrix interactions could provide a range of wall properties sufficient to satisfy the complex requirements of morphogenesis.

Coaxial layers

When ultra-thin sections of hyphae and of spores are examined under the electron microscope, the walls usually contain layers which differ in their electron density (Hawker 1965). More information about the nature of these layers has been obtained by combining electron microscopy with the controlled digestion of walls by specific enzymes or chemical treatments (Mahadevan & Tatum, 1967; Hunsley & Burnett, 1970). The layers consist mainly of a specific polymer or a specific mixture of polymers. The microfibrillar phase, whether chitin or cellulose, and its accompanying matrix material appear to be always on the inner surface of the wall, close to the plasmalemma. The surface layers in *Neurospora* and *Phytophthora* contain the $\beta(1 \rightarrow 3)$-glucan and in *Schizophyllum* $\alpha(1 \rightarrow 3)$-glucan (Wessels *et al.*, 1972). Protective skins such as melanin are deposited as a sheath on the exterior (Rowley & Pirt, 1972). The deposition of wall material in layers may confer some specific properties on the envelope as a whole and the analogy with man-made composites could perhaps be pressed to compare such walls with laminates. However, the occurrence of layers could merely reflect the way the biosynthetic mechanisms operate, particularly when the wall, after its primary deposition, undergoes further thickening (Hunsley & Burnett, 1970). Once a layer has been formed, it may be difficult to intercalate additional material.

Functions of wall polymers in maintaining shape

Hyphae can be treated with specific enzymes or chemicals so that wall polymers are removed consecutively and that finally walls containing a single component are left (Mahadevan & Tatum, 1967; Hunsley & Burnett, 1970). Such experiments have shown that each of the major wall components can, on its own, maintain the characteristic shape of the hyphae. Thus, at least in non-growing mycelium, the maintenance of shape does not reside in any single polymer.

The role of chitin in fungal walls has been investigated by a different approach using a mutant blocked in the synthesis of glucosamine and therefore chitin (Katz & Rosenberger, 1970, 1971*a*). When grown in the presence of an osmotic stabilizer, the hyphae are normal in shape and the walls contain very little chitin. Such hyphae are not osmotically sensitive and do not alter their shape even in distilled water. However, when transferred to a growth medium with no osmotic stabilizer, they swell and lyse. Thus, there does appear to be a stage during wall extension and growth when all the wall components are required to provide the necessary strength and rigidity. Once the wall has been formed, loss of one or even most of the components does not appear to affect its integrity.

11.4 Biosynthesis of the Wall

Sites of wall synthesis

Observations under the light microscope (for review see Robertson, 1965) and autoradiograms (Bartnicki-Garcia & Lippman, 1969; Gooday, 1971; Katz & Rosenberger, 1971*b*) have shown that hyphae possess an active site of wall synthesis at their apex. The apical site appears to be the only one at which material is deposited so as to produce extension and growth. From autoradiograms, it appears that some polymers are also incorporated into the subapical wall and this is generally considered to represent wall thickening and modification. It should be pointed out, however, that while vegetative hyphae grow apically, this is not necessarily true of sporangiophores or ellipsoidal cells (Castle, 1942; Bartnicki-Garcia & Lippman, 1969). Because of the relative ease of experimentation almost all studies have been made on vegetative hyphae and very little is known about wall synthesis in other cell types.

While incorporation of new wall is largely confined to the hyphal tip, there is a growing body of evidence that cytoplasmic vesicles play an important role in wall synthesis (McClure, Park & Robinson, 1968; Girbardt, 1969; Grove & Bracker, 1970; Bracker, 1971). These vesicles arise from endomembrane systems in the hyphae, move towards the apex and can be seen to fuse with the plasmalemma at the tip (Fig. 11.3). The vesicles disappear from the apical region when the hypha ceases to extend, reform at the tip when growth restarts and cytochemical data indicates that they contain polysaccharides. The isolation and characterization of vesicles from

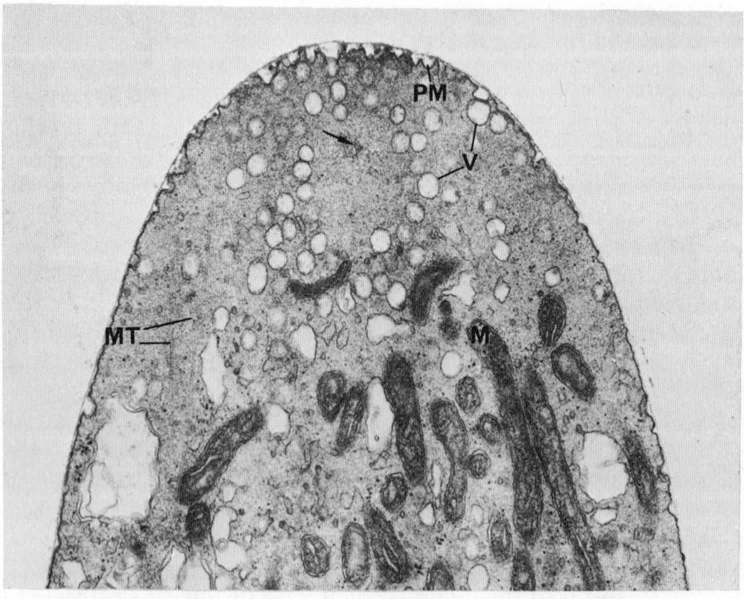

Fig. 11.3 Vesicles (V) in the apical zone of a hypha of *Armillaria mellea*. (From Grove & Bracker (1970).)

filamentous fungi is still in its very early stages and until the nature of their contents have been determined it will not be clear what functions they fulfill. It seems likely that vesicles transport precursors for wall synthesis since long lengths of hyphae supply materials to a single tip (Zalokar, 1959; Trinci, 1971). Their content of polysaccharides indicates that some of the wall polymers may be synthesized inside the vesicles. However, since the enzyme systems making chitin and $\beta(1 \rightarrow 3)$-glucan appear to be attached to the wall (Wang & Bartnicki-Garcia, 1966; McMurrough, Flores-Carreon & Bartnicki-Garcia, 1971; Mishra & Tatum, 1972) it is unlikely that whole wall is preformed inside the vesicles and then inserted as complete units when they fuse with the plasmalemma.

The limitation of wall extension to the hyphal apex indicates the existence of controls which determine the sites of wall incorporation. Little is known about how these controls operate. Inhibition of protein synthesis changes the sites of wall deposition from apical to all round the hypha (Katz & Rosenberger, 1971*b*; Sternlicht, Katz & Rosenberger, 1973). This effect appears analogous to that found in gram-positive bacteria (Rogers, 1970). As new enzymes cannot be formed in the absence of protein synthesis, the findings suggest that the enzymes of wall synthesis are present in all parts of the hypha but that their activity is inhibited except at the apex. It is of interest that both a zymogen form of chitin synthetase and an inhibitor which prevents its activation have been found in yeast (Cabib & Farkas, 1971).

New sites of wall extension are initiated subapically at intervals during growth and these give rise to the side branches. A link must therefore exist between the overall growth process and the controls that determine sites of wall deposition. Some general models have been put forward to describe these relations (Katz, Goldstein & Rosenberger, 1972; Trinci, 1974; Morrison & Righelato, 1974). The frequency with which hyphae initiate new sites of wall deposition, *i.e.* the frequency of branching, can be changed markedly both by adding compounds such as sorbose or deoxycholate to the medium (Tatum, Barrat & Cutter, 1949; Mackintosh & Pritchard, 1963) and by specific mutations (Brody & Tatum, 1967; Scott & Tatum, 1970; Elorza & Arst, 1971). When the frequency of branching is changed, either by chemicals or by mutation, the proportions of wall polymers relative to each other are also altered (Mahadevan & Tatum, 1965; Elorza & Arst, 1971). Further, several of these mutations are in the structural genes specifying such enzymes of intermediary metabolism as phosphoglucomutase and glucose-6-phosphate dehydrogenase. How these alterations in enzyme activity affect the mechanism which controls sites of wall deposition is not clear.

Enzymes of wall biosynthesis

Only a limited number of studies have been carried out on enzyme systems which synthesize fungal wall polymers. Chitin synthetase activity has been demonstrated in several species but the enzymes have not been resolved or purified to any extent (Glaser & Brown, 1957; Camargo, Dietrich, Sonneborn & Strominger, 1967; McMurrough, Flores-Carreon & Bartnicki-Garcia, 1971; McMurrough & Bartnicki-Garcia, 1971). The enzyme system transfers *N*-acetylglucosamine units from uridine-diphospho (UDP) acetylglucosamine to acceptors and is markedly activated by free acetylglucosamine. $\beta(1 \rightarrow 3)$-glucan synthetase from *Phytophthora* (Wang & Bartnicki-Garcia, 1966) and from *Neurospora* (Mishra & Tatum, 1972) also uses UDP glucose to transfer glucose units to unidentified wall acceptors. There is no clear evidence as yet whether lipid intermediates are involved in wall synthesis in the filamentous fungi.

Both the chitin and $\beta(1 \rightarrow 3)$-glucan synthetase are particulate enzymes and most of the activity appears to be associated with the wall fraction. It may be that in fungi the situation is similar to that in plant cells where some wall polymers appear to be made in cytoplasmic vesicles and others by wall-bound enzymes (Northcote, 1972).

Wall hydrolases

Fungi elaborate a variety of polysaccharide hydrolases and frequently these can attack polymers which form part of the organism's own wall (*e.g.* Huotari, Nelson, Smith & Kirkwood, 1968). The hydrolases fall into two quite distinct classes. One type appears to be a typical inducible catabolic exoenzyme whose purpose is the break-down of exogenous polysaccharides for use as carbon and energy sources. The second type of hydrolase is bound quite tightly to the wall and could be active in hydrolysing bonds to allow for extension and morphogenesis (Mahadevan & Mahadkar, 1970; Bartnicki-Garcia, 1973). Although there is as yet no clear evidence connecting wall autolysins with wall extension in filamentous fungi, such a function would be in keeping with current views on wall growth in bacterial and plant cells (Higgins & Shockman, 1971; Lamport, 1970). Wall hydrolases have, on the other hand, been implicated in at least two morphogenetic processes. In *Schizophyllum*, nuclear migration is an essential step in the formation of a dikaryon from sexually compatible monokaryotic strains (Raper, 1966). Nuclei can only migrate when the internal septa have been removed and this is accomplished by changes in activity of wall hydrolases, particularly enzymes acting on the $\beta(1 \rightarrow 3)$-glucan (Wessels, 1966, 1969*a*, *b*; Janszen & Wessels, 1970; Wessels & Koltin, 1972). *Aspergillus nidulans* forms $\alpha(1 \rightarrow 3)$-glucanase at the onset of fruit-body formation and the presence of this enzyme and of its substrate, the wall $\alpha(1 \rightarrow 3)$-glucan, appear essential for the production of perithecia (Zonneveld, 1971, 1972*a*, 1972*b*, 1973, 1974).

11.5 References

APPLEGARTH, D. A. & BOZOIAN, G. (1969). The cell wall of *Helminthosporium sativum*. *Archives of Biochemistry and Biophysics* **134**, 285–9.

ARONSON, J. M. (1965). The cell wall. In *The fungi*, Vol. 1, pp. 49–76. Edited by G. C. Ainsworth and A. S. Sussman. New York: Academic Press.

ARONSON, J. M. & MACHLIS, L. (1959). The chemical composition of the hyphal walls of the fungus *Allomyces*. *American Journal of Botany* **46**, 292–300.

ARONSON, J. M., COOPER, B. A. & FULLER, M. S. (1967). Glucans of Oömycete cell walls. *Science* **155**, 332–5.

BACON, J. S. D., JONES, D., FARMER, V. C. & WEBLEY, D. M. (1968). The occurrence of $\alpha(1 \rightarrow 3)$-glucan in *Cryptococcus, Schizosaccharomyces* and *Polyporus* species and its hydrolysis by a *Streptomyces* culture filtrate lysing cell walls of *Cryptococcus*. *Biochimica et Biophysica Acta* **158**, 313–15.

BARTNICKI-GARCIA, S. (1968). Cell wall chemistry, morphogenesis and taxonomy of fungi. *Annual Review of Microbiology* **22**, 87–105.

BARTNICKI-GARCIA, S. (1973). Hyphal morphogenesis. *Symposium Society General Microbiology* **23**, 245–67.

BARTNICKI-GARCIA, S. & LIPPMAN, E. (1969). Fungal morphogenesis: cell wall construction in *Mucor rouxii*. *Science* **165**, 302–4.

BARTNICKI-GARCIA, S. & NICKERSON, W. S. (1962). Isolation, composition and structure of cell walls of filamentous and yeast-like forms of *Mucor rouxii*. *Biochimica et Biophysica Acta* **58**, 102–19.

BARTNICKI-GARCIA, S. & REYES, E. (1968). Polyuronides in the cell walls of *Mucor rouxii*. *Biochimica et Biophysica Acta* **170**, 54–62.

BRACKER, C. E. (1971). Cytoplasmic vesicles in germinating spores of *Gilbertella persicaria*. *Protoplasma* **72**, 381–97.

BRODY, S. & TATUM, E. L. (1967). The primary biochemical effect of a morphological mutation in *Neurospora crassa*. *Proceedings National Academy of Sciences, U.S.A.* **58**, 923–30.

BUCK, K. W., CHEN, A. W., DICKERSON, A. G. & CHAIN, E. B. (1968). Formation and structure of extracellular glucans produced by *Claviceps* species. *Journal of General Microbiology* **51**, 337–52.

BULL, A. T. (1970*a*). Chemical composition of wild-type and mutant *Aspergillus nidulans* cell walls. The nature of polysaccharide and melanin constituents. *Journal of General Microbiology* **63**, 75–94.

BULL, A. T. (1970*b*). Inhibition of polysaccharases by melanin: enzyme inhibition in relation to mycolysis. *Archives of Biochemistry and Biophysics* **137**, 345–56.

BULL, A. T. & CHESTERS, C. G. C. (1966). The biochemistry of laminarin and the nature of laminarinase. *Advances in Enzymology* **28**, 325–64.

CABIB, E. & FARKAS, V. (1971). The control of morphogenesis: an enzymatic mechanism for the initiation of septum formation in yeast. *Proceedings National Academy of Sciences,* U.S.A. **68**, 2052–56.

CAMARGO, E. P., DIETRICH, C. P., SONNEBORN, D. & STROMINGER, J. L. (1967). Biosynthesis of chitin in spores and growing cells of *Blastocladiella emersonii*. *Journal of Biological Chemistry* **242**, 3121–28.

CASTLE, E. S. (1942). Spiral growth and reversal of spiraling in *Phycomyces* and their bearing on primary wall structure. *American Journal of Botany* **29**, 664–72.

CROOK, E. M. & JOHNSTON, I. R. (1962). The qualitative analysis of the cell walls of selected species of fungi. *Biochemical Journal* **83**, 325–31.

ELORZA, M. V. & ARST, H. N. (1971). Sorbose resistant mutants of *Aspergillus nidulans*. *Molecular and General Genetics* **111**, 185–93.

FULLER, M. S. & BARSHAD, I. (1960). Chitin and cellulose in the cell walls of *Rhizidiomyces* sp. *American Journal of Botany* **47**, 105–9.

GANCEDO, J. M., GANCEDO, C. & ASENSIO, C. (1966). Uronic acids in fungal cell walls. *Biochemische Zeitschrift* **346**, 328–32.

GHUYSEN, J. M. (1968). Use of bacteriolytic enzymes in determination of wall structure and their role in cell metabolism. *Bacteriological Reviews* **32**, 425–64.

GHUYSEN, J. M., STROMINGER, J. L. & TIPPER, D. J. (1968). Bacterial cell walls. *Comparative Biochemistry* **26A**, 53–99.

GIRBARDT, M. (1969). Die Ultrastruktur der Apikalregion von Pilzhyphen. *Protoplasma* **67**, 413–41.

GLASER, L. & BROWN, D. H. (1957). The synthesis of chitin in cell-free extracts of *Neurospora crassa*. *Journal Biological Chemistry* **228**, 729–42.

GOODAY, G. W. (1971). An autoradiographic study of hyphal growth of some fungi. *Journal of General Microbiology* **67**, 125–33.

GROVE, S. N. & BRACKER, C. E. (1970). Protoplasmic organisation of hyphal tips among fungi: vesicles and Spitzenkörper. *Journal of Bacteriology* **104**, 989–1009.

HAMILTON, P. B. & KNIGHT, S. G. (1962). An analysis of the cell walls of *Penicillium chrysogenum*. *Archives of Biochemistry and Biophysics* **99**, 282–7.

HAWKER, L. E. (1965). Fine structure of fungi as revealed by electron microscopy. *Biological Reviews* **40**, 52–92.

HIGGINS, M. L. & SHOCKMAN, G. D. (1971). Prokaryotic cell division with respect to walls and membranes. *Critical Reviews in Microbiology* **1**, 29–72.

HOLLIDAY, L. (1966). *Composite materials* pp. 1–27, Amsterdam: Elsevier.

HUNSLEY, D. & BURNETT, J. H. (1968). Dimensions of microfibrillar elements in fungal walls. *Nature, London* **218**, 462–3.

HUNSLEY, D. & BURNETT, J. H. (1970). The ultrastructural architecture of the walls of some hyphal fungi. *Journal of General Microbiology* **62**, 203–18.

HUOTARI, F. I., NELSON, T. E., SMITH, F. & KIRKWOOD, S. (1968). Purification of exo $\beta(1 \rightarrow 3)$ glucanase from Basidiomycete species QM 806. *Journal of Biological Chemistry* **243**, 952–6.

JANSZEN, F. H. A., & WESSELS, J. G. H. (1970). Enzymic dissolution of hyphal septa in a Basidiomycete. *Antonie van Leeuwenhoek* **36**, 255–7.

JOHNSTON, I. R. (1965). The composition of the cell wall of *Aspergillus niger. Biochemical Journal* **96**, 651–8.

KATZ, D. & ROSENBERGER, R. F. (1970). A mutation in *Aspergillus nidulans* producing hyphal walls which lack chitin. *Biochimica et Biophysica Acta* **208**, 452–60.

KATZ, D. & ROSENBERGER, R. F. (1971*a*). Lysis of an *Aspergillus nidulans* mutant blocked in chitin synthesis and its relation to wall assembly and wall metabolism. *Archiv für Mikrobiologie* **30**, 284–92.

KATZ, D. & ROSENBERGER, R. F. (1971*b*). Hyphal wall synthesis in *Aspergillus nidulans*: effect of protein synthesis inhibition and osmotic shock on chitin insertion and morphogenesis. *Journal of Bacteriology* **108**, 184–90.

KATZ, D., GOLDSTEIN, D. & ROSENBERGER, R. F. (1972). Model for branch initiation in *Aspergillus nidulans* based on measurements of growth parameters. *Journal of Bacteriology* **109**, 1097–100.

KELLY, A. (1966). *Strong solids.* Oxford: Clarendon Press.

KREGER, D. R. (1954). Observations on cell walls of yeasts and some other fungi by X-ray diffraction and solubility tests. *Biochimica et Biophysica Acta* **13**, 1–9.

KREGER, D. R. & MEEUSE, B. J. D. (1952). X ray diagrams of *Euglena* paramylon, of the acid insoluble glucan of yeast cell walls and of laminarin. *Biochimica et Biophysica Acta* **9**, 699–700.

KUO, M. J. & ALEXANDER, M. (1967). Inhibition of the lysis of fungi by melanins, *Journal of Bacteriology* **94**, 624–9.

LAMPORT, D. T. A. (1970). Cell wall metabolism. *Annual Review of Plant Physiology* **21**, 235–70.

McCLURE, W. K., PARK, D. & ROBINSON, P. M. (1968). Apical organisation in the somatic hyphae of fungi. *Journal of General Microbiology* **50**, 177–82.

MacKINTOSH, M. E. & PRITCHARD, R. H. (1963). The production and replica plating of micro-colonies of *Aspergillus nidulans. Genetical Research, Cambridge* **4**, 320–2.

McMURROUGH, I. & BARTNICKI-GARCIA, S. (1971). Properties of a particulate chitin synthetase from *Mucor rouxii. Journal of Biological Chemistry* **246**, 4008–16.

McMURROUGH, I., FLORES-CARREON, A. & BARTNICKI-GARCIA, S. (1971). Pathway of chitin synthesis and cellular localisation of chitin synthetase in *Mucor rouxii. Journal of Biological Chemistry* **246**, 3999–4007.

MAHADEVAN, P. R. & TATUM, E. L. (1965). Relationship of the major constituents of the *Neurospora crassa* cell wall to wild-type and colonial morphology. *Journal of Bacteriology* **90**, 1073–81.

MAHADEVAN, P. R. & TATUM, E. L. (1967). Localisation of structural polymers in the cell wall of *Neurospora crassa. Journal of Cell Biology* **35**, 295–302.

MAHADEVAN, P. R. & MAHADKAR, U. R. (1970). Role of enzymes in growth and morphology of *Neurospora crassa*: cell wall bound enzymes and their possible role in branching. *Journal of Bacteriology* **101**, 941–7.

MARK, R. E. (1967). *Cell wall mechanics of tracheids.* New Haven and London: Yale University Press.

MICHELL, A. J. & SCURFIELD, G. (1967). Composition of extracted fungal walls as indicated by infrared spectroscopy. *Archives of Biochemistry and Biophysics* **120**, 628–37.

MICHELL, A. J. & SCURFIELD, G. (1970*a*). An assessment of infrared spectra as indicators of fungal cell wall composition. *Australian Journal of Biological Science* **23**, 345–60.

MICHELL, A. J. & SCURFIELD, G. (1970*b*). Chitin and cellulose in *Ceratocystis* cell walls. *Transactions of the British Mycological Society* **55**, 488–91.

MISHRA, N. C. & TATUM, E. L. (1972). Effect of L sorbose on polysaccharide synthases of *Neurospora crassa. Proceeding National Academy of Sciences U.S.A.* **69**, 313–7.

MITCHELL, R. & SABAR, N. (1966). Hyphal cell wall structure of two species of *Pythium. Canadian Journal of Microbiology* **12**, 471–5.

MITCHELL, A. & TAYLOR, I. F. (1969). Cell wall proteins in *Aspergillus niger* and *Chaetomium globosum. Journal of General Microbiology* **59**, 103–9.

MORRISON, K. B. & RIGHELATO, R. C. (1974). The relationships between hyphal branching, specific growth rate and colony radial growth rate in *Penicillium chrysogenum. Journal of General Microbiology* **81**, 517–20.

NABEL, K. (1939). Uber die Membran Niederer Pilze, besonders von *Rhizidiomyces bivellatus* nov. spez. *Archiv für Mikrobiologie* **10**, 515–41.

NICOLAUS, R. A., PIATELLI, M. & FATTORUSSO, E. (1964). The structure of melanins and melanogenesis. IV. On some natural melanins. *Tetrahedron* **20**, 1163–72.

NORTHCOTE, D. H. (1972). Chemistry of the plant cell wall. *Annual Review of Plant Physiology* **23**, 113–32.

NOVAES-LEDIEU, M., JIMENEZ-MARTINEZ, A. & VILLANUEVA, J. R. (1967). Chemical composition of hyphal walls of Phycomycetes. *Journal of General Microbiology* **47**, 237–45.

RAPER, J. R. (1966). *Genetics of sexuality in higher fungi.* New York: The Ronald Press Co.

REEVES, R. E. (1951). Cuprammonium-glycoside complexes. *Advances in Carbohydrate Chemistry* **6**, 108–31.

ROBERTSON, N. F. (1959). Experimental control of hyphal branching and branch form in hyphomycetous fungi. *Journal of the Linneaen Society* **56**, 207–11.

ROBERTSON, N. F. (1965). The mechanism of cellular extension and branching. In *The fungi* Vol. 1, pp. 613–23. Edited by G. C. Ainsworth and A. S. Sussman, New York: Academic Press.

ROGERS, H. J. (1970). Bacterial growth and the cell envelope. *Bacteriological Reviews* **34**, 194–214.

ROWLEY, B. I. & PIRT, S. J. (1972). Melanin production by *Aspergillus nidulans* in batch and chemostat cultures. *Journal of General Microbiology* **72**, 553–63.

RUIZ-HERRERA, J. (1967). Chemical components of the cell wall of *Aspergillus nidulans* species. *Archives of Biochemistry and Biophysics* **122**, 118–25.

SALTON, M. R. J. (1961). *Microbial cell walls.* New York: Wiley.

SCOTT, W. A. & TATUM, E. L. (1970). Glucose-6-phosphate dehydrogenase and *Neurospora* morphology. *Proceedings National Academy of Sciences, U.S.A.* **66**, 515–22.

SENTANDREU, K. & NORTHCOTE, D. H. (1968). The structure of a glycopeptide isolated from the yeast cell wall. *Biochemical Journal* **109**, 419–32.

STERNLICHT, E., KATZ, D. & ROSENBERGER, R. F. (1973). Subapical wall synthesis and wall thickening induced by cycloheximide in hyphae of *Aspergillus nidulans. Journal of Bacteriology* **114**, 819–23.

TATUM, E. L., BARRAT, R. W. & CUTTER, V. M. (1949). Chemical induction of colonial paramorphs in *Neurospora* and *Syncephalastrum. Science* **109**, 509–11.

TAYLOR, I. F. P. & CAMERON, D. S. (1973). Preparation and quantitative analysis of fungal cell walls. Strategy and tactics. *Annual Review of Microbiology* **27**, 243–60.

TRINCI, A. P. J. (1971). Influence of the peripheral growth zone on the radial growth rate of fungal colonies. *Journal of General Microbiology* **67**, 325–44.

TRINCI, A. P. J. (1974). A study of the kinetics of hyphal extension and branch initiation of fungal mycelia. *Journal of General Microbiology* **81**, 225–36.

TROY, F. A. & KOFFLER, H. (1969). The chemistry and molecular architecture of the cell walls of *Penicillium chrysogenum. Journal of Biological Chemistry* **244**, 5563–76.

VILLANUEVA, J. R. (1966). Protoplasts of fungi. In *The fungi,* Vol. 2. pp. 2–62. Edited by G. C. Ainsworth and A. S. Sussman. New York: Academic Press.

WANG, M. C. & BARTNICKI-GARCIA, S. (1966). Biosynthesis of $\beta(1 \rightarrow 3)$ and $\beta(1 \rightarrow 6)$ linked glucan by *Phytophthora cinnamoni* hyphal walls. *Biochemical and Biophysical Research Communications* **24**, 832–7.

WESSELS, J. G. H. (1965). Morphogenesis and biochemical processes in *Schizophyllum commune. Wentia* **13**, 1–113.

WESSELS, J. G. H. (1966). Control of cell wall glucan degradation during development in *Schizophyllum commune. Antonie van Leeuwenhoek* **32**, 341–55.

WESSELS, J. G. H. (1969*a*). Biochemistry of sexual morphogenesis in *Schizophyllum commune*: effect of mutations affecting the incompatibility system on cell wall metabolism. *Journal of Bacteriology* **98**, 697–704.

WESSELS, J. G. H. (1969*b*). A $\beta(1 \rightarrow 6)$ glucan glucanohydrolase involved in hydrolysis of cell wall glucan in *Schizophyllum commune. Biochimica et Biophysica Acta* **178**, 191–3.

WESSELS, J. G. H. & KOLTIN, Y. (1972). R-glucanase activity and susceptibility of hyphal walls to degradation in mutants of *Schizophyllum* with disrupted nuclear migration. *Journal of General Microbiology* **71**, 471–5.

WESSELS, J. G. H., KREGER, D. R., MARCHANT, R., REGENSBURG, B. A. & DE VRIES, O. M. H. (1972). Chemical and morphological characterisation of the hyphal wall surface of the Basidiomycete

Schizophyllum commune. Biochimica et Biophysica Acta **273**, 346–58.

WETTSTEIN, F. (1921). Das vorkommen von chitin und seine verwertung als systematisch—phylogenetisches merkmal in pflanzenreich. *Sitzungsberichte der Kaiserlichen Akademie der Wissenschaften in Wien Math. Naturwiss Kl Abt.*1. **30**, 3–20.

VAN WISSELINGH, C. (1898). Mikrochemische untersuchungen uber die zellwande der fungi. *Jahrbuch für Wissenschaftliche Botanik Berlin* **31**, 619–87.

ZALOKAR, M. (1959). Growth and differentiation of *Neurospora* hyphae. *American Journal Botany* **46**, 602–10.

ZONNEVELD, B. J. M. (1971). Biochemical analysis of the cell wall of *Aspergillus nidulans. Biochimica et Biophysica Acta* **249**, 506–14.

ZONNEVELD, B. J. M. (1972*a*). A new type of enzyme, an exo-splitting $\alpha(1 \rightarrow 3)$ glucanase from non-induced cultures of *Aspergillus nidulans. Biochimica et Biophysica Acta* **258**, 541–7.

ZONNEVELD, B. J. M. (1972*b*). Morphogenesis in *Aspergillus nidulans.* The significance of $\alpha(1 \rightarrow 3)$ glucan of the cell wall and $\alpha(1 \rightarrow 3)$ glucanase for cleistothecium development. *Biochimica et Biophysica Acta* **273**, 174–87.

ZONNEVELD, B. J. M. (1973). Inhibitory effect of 2-deoxyglucose on cell wall $\alpha(1 \rightarrow 3)$ glucan synthesis and cleistothecium development in *Aspergillus nidulans. Developmental Biology,* **34**, 1–8.

ZONNEVELD, B. J. M. (1974). $\alpha(1 \rightarrow 3)$ glucan synthesis correlated with $\alpha(1 \rightarrow 3)$ glucanase synthesis, conidiation and fructification in morphogenetic mutants of *Aspergillus nidulans. Journal of General Microbiology* **81**, 445–51.

CHAPTER 12

Hormones in Fungi

J. D. BU'LOCK

12.1 Introduction

In plants and animals alike there are hormone substances each of which serves to link by chemical means a set of causative conditions to a specific complex of effects. Such substances are also important in micro-organisms and no account of the developmental biochemistry of fungi would be complete without a discussion of their hormonal systems.

However, the evidence for *widespread* occurrence of hormone mechanisms in fungi (Machlis, 1966) comes mainly from observational mycology, and the number of systems for which useful chemical and biochemical data exist is very limited indeed. For this reason some comparable data for systems in other types of micro-organism, such as protozoa and algae, provide a helpful background to the fungal topic. Not only do such systems help us to see some general principles, but some of them illustrate very basic mechanisms which also exist in fungi but are less well-characterized. On a very selective basis, therefore, some of these non-fungal systems have been included in this account.

The concept of hormones was developed in vertebrate physiology and brought *via* botany into mycology, so that its use in our field requires quite careful definitions. Slightly modifying some previous statements (Raper (1952) and Machlis (1972), both following Huxley (1935)), but avoiding special terminologies, we define a fungal hormone as a substance

(a) of relatively low molecular weight,
(b) which is produced (or released) by an organism or part of a differentiated organism,
(c) in response to a specific set of circumstances,
(d) and which is transported (actively or passively),
(e) to another similar organism or to another part of the same organism,
(f) where it elicits a specific set of responses.

Each aspect of this itemized definition is worthy of further comment. Thus (a), the macromolecular apparatus of gene expression is not usefully included (otherwise the topic would be almost all-embracing); on the other hand the precedent of some non-fungal systems shows that regulators with quite high molecular weights, such as glycoproteins, may act as diffusible hormones in certain cases. Item (b) is so phrased because we cannot usefully distinguish between hormones which mediate between single undifferentiated cells (*e.g.* in *Saccharomyces*), between differentiated parts of the same thallus (*e.g.* in homothallic Mucorales), and between discrete colonies (*e.g.* in heterothallic Mucorales). Definition of the controls for hormone production or release (c) is an essential objective. Despite (d) above, we shall find it useful at least to look at some cases where actual translocation of a regulator (such as cyclic AMP) is not established, but in general—and partly for simple reasons of scale—the phenomena of hormone translocation and the concept of hormone gradients are particularly important in micro-organisms. Similarly, despite (e) it may be instructive at least to note some interactions between organisms of different or even unrelated species, and any discussion of microbial ecology must eventually include hormone-like effects in mixed-culture situations, though at present these are very little understood. The concept of a specific set of hormone responses, (f), is essential, and it also helps to exclude broader (not necessarily trivial) phenomena such as the effects of pH changes or CO_2 accumulation.

We shall find that in micro-organisms, whose life-cycle is relatively rapid, hormones are more important as regulators of change than of steady states, and since in fungi many of these changes are forms of differentiation associated with means of reproduction this will be the main context of this account. Similarly, the relative prominence of systems in which hormones are translocated through the ambient medium rather than within a cell mass reflects the fact that the fungi have not evolved very far along the road to specialized multicellularity; on the other hand it must also reflect the fact that such systems are rather easier to observe and to explore.

As already noted, the number of fungal systems about which we have substantial information is only a small proportion of those which are known, and a still smaller proportion, surely, of those which must actually exist. For this reason the discussion of general principles has been kept to a minimum, and the systems have simply been dealt with in sequence. It is my own view that we can already see some of the general principles emerging, and hopefully this will also become apparent to the reader.

12.2 Cyclic AMP mediated Differentiation in Amoebae and Slime Moulds

Some notes on differentiation processes in amoebae and in Myxomycetes are included as basic examples of microbial systems mediated by adenine 3′,5′ cyclic monophosphate, cAMP, which is a substance of undoubted but largely unexplored significance in fungi as later sections will show.

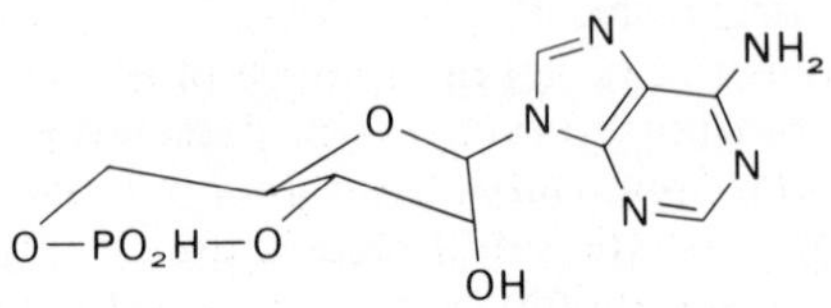

cAMP

Free-living amoebae such as *Acanthamoeba castellani* and *Hartmanella culbertsonii* encyst—that is, they form resistant metabolically-inert cells which survive adverse conditions—when the environment no longer permits them to multiply, for example on nutrient exhaustion or when mitochondrial functions are blocked. Encystment leads to the loss of most functions of the trophozoote stage and an almost complete shut down of metabolism, but the process itself also requires, at least transiently, several new metabolic processes such as the breakdown of reserve glycogen and the synthesis of cellulose and of specific mucopolysaccharides. This is accomplished by a shift in the selection of genes being translated and is dependent upon *de novo* enzyme syntheses, and these changes are effected in a coordinated manner. The link between the environmental signal and the gene-based response is that the environmental changes activate adenyl cyclase in the cell membrane. The cell level of cAMP results from the balanced activities of this cyclase and of the phosphodiesterase which hydrolyses cAMP, and a large transient increase in cAMP level marks the onset of differentiation. A scheme for this relatively simple system, based on the work of Gessat & Jantzen (1974) and Krishna Murti (1973), is set out in Fig. 12.1. This exemplifies what is very probably a general mechanism in cell systems of

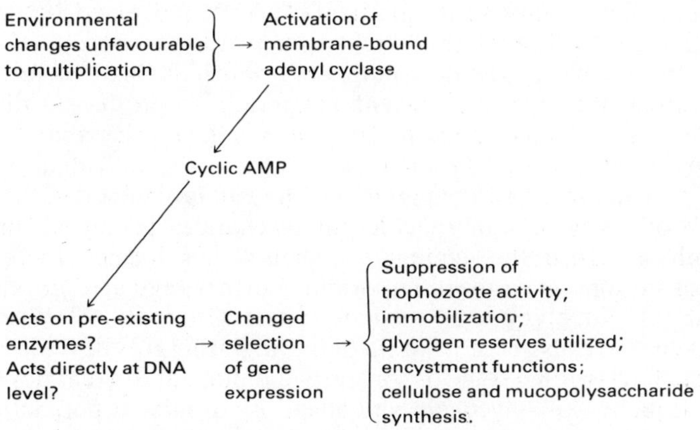

Fig. 12.1 Cyclic AMP as a mediator in the encystment of amoebae.

many kinds, *viz.* the role of cAMP in effecting a switch from vegetative to non-vegetative functions in response to nutritional circumstances. The precise consequences of such a switch cannot be so generally formulated since they must be determined by many different factors in different systems, both genetic and environmental, and it is this which allows the basic mechanism to be used in so many different cases.

In this system there are no data to indicate whether the cAMP affects gene transcription rather directly, as in the control of catabolite repression in bacteria, or through a series of kinase reactions as in many vertebrate systems. Such information is equally lacking for the other cAMP systems to which we shall refer, and so this problem will not be considered further.

In the true amoebae there is no suggestion that the cAMP acts *between* cells as well as within them, and strictly it cannot be termed a hormone. Such a situation does however exist in the cellular slime mould *Dictyostelium discoideum*, the life-cycle of which has received considerable attention in recent years (Sussman & Sussman, 1969; Bonner, 1971; Garrod & Ashworth, 1973). Only the most relevant aspects of this system are summarized here, and the interested reader is referred to the literature cited for a fuller and more detailed account.

Free-living myxamoebae constitute the vegetative phase in *Dictyostelium*; when the nutritional environment becomes inadequate they begin to aggregate, first by individual migration towards centres and then, as they come into mutual contact, in coordinated streaming. The attractant 'acrasin' is cAMP, liberated by the action of a membrane-bound cyclase; it is broken down by a phosphodiesterase which can be either membrane-bound or released into the medium, the activity of which may also be modulated by other substances (Panbacker & Bravard, 1972; Chassy, 1972). The first aggregation centres in a population arise randomly, but Garrod & Ashworth (1973) note that amoebae grown to a high glycogen content form many fewer centres than amoebae without glycogen reserves. Chemotaxy in the migration phase depends upon an amoebae receiving a short 'pulse' of cAMP, subsequently releasing such a pulse itself, and migrating for a finite period in the direction from which the first pulse arrived (Robertson, Drage & Cohen, 1972). By making the gradient of cAMP 'perceived' by an amoeba temporal as well as spatial, this mechanism avoids the conceptual difficulty inherent in many 'gradient hypotheses', which is that of providing a concentration gradient of sufficient magnitude to produce a directional effect, given a low overall concentration of the effector (less than 10^{-8} molar cAMP in this case) and a very small receptor (here, the individual amoeba).

In the 'streaming' phase of aggregation the mutual contacts of the cells are also controlled by immunospecific polysaccharides produced during the earlier phase. Once the aggregate or 'grex' has formed these surface properties become even more important. Further regulated development, which may also involve cAMP-mediated effects (Bonner, 1971) occur in the grex. In general, cells at the front tip of the migrating grex develop into stalk cells in the final fruiting stage or sorocarp and those at the rear develop into spores. Equally low-glycogen amoebae, in a mixed population with glycogen-rich amoebae, predominantly afford tip cells in the grex and stalk cells in the sorocarp (Garrod & Ashworth, 1973). Nevertheless, whatever the proportion of low- and high-glycogen cells initially, the ratio of stalk cells to spores in the sorocarp tends towards a 'normal' pattern.

Thus the event of nutrient exhaustion, and the internal state of the cells at that time, determine the aggregation mechanism and predispose the intermediate and final dispositions of the cells, but once the cells come into close association they are also subject to an overall balancing mechanism within the aggregate as a whole. The first phase of this regulatory system involves cAMP as a true hormone, and the second phase may at least partly depend upon the same mechanism; it also involves specific cell surface components produced during the first phase.

Considerable attention has also been given to the metabolic aspects of this differentiation system, but since the slime moulds are not true fungi these aspects are not dealt with here.

12.3 Hormonal Systems in the Phycomycetes

Chemotaxis in Allomyces

In *Allomyces* [a genus of the Blastocladiales (Chytridomycetes, Mastigomycotina)] sexual reproduction requires fusion between the uniflagellate motile 'male' and 'female' gametes. The chemotaxy which brings about their approach is due to the attractant hormone sirenin. This is a bicyclic sesquiterpene diol which is produced by the female gametangia and by the

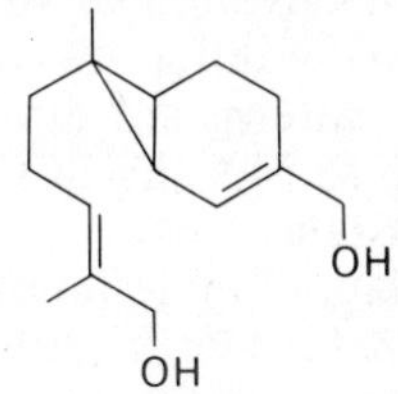

Sirenin

sluggishly-motile female gametes, and released into the surrounding water. Male gametes are attracted to higher concentrations of sirenin; the simplest explanation of the chemotropism is that the male gametes swim in random directions but faster at higher sirenin concentrations, but more subtle explanations are not excluded. Inactivation of the attractant by the male gametes is an essential part of the overall mechanism. Chemotaxis leads to the pairing of gametes and the mechanism which limits interspecific matings is not known. The adhesive contact which leads to plasmogamy is blocked if boric acid is added, which suggests that specific surface polysaccharides may be involved. Sirenin has no effect on the diploid gametes.

This hormone system was first described by Machlis (1958); the structure of sirenin was established by Nutting, Rapoport & Machlis (1968), and Carlile & Machlis (1965) demonstrated the inactivation of sirenin by the male gametes. Sirenin is active at very low concentrations, less than 10^{-10} g ml^{-1}. Neither the chemistry of its formation (presumably *via* farnesyl pyrophosphate), nor the molecular mode of its action (ultimately upon the flagellum) are known, but some genetic aspects of the mating system are instructive. In *Allomyces* the male and female gametangia, usually produced in pairs, are biochemically and spatially distinct; in some species the male gametangia are terminal and the females sub-terminal, and in other species the pattern is reversed; male gametangia are always pigmented, mainly with γ-carotene. Interspecific hybrids forming *almost* exclusively male or female gametangia can be bred (Emerson & Wilson, 1954); nevertheless, from a given haploid mycelium the male and female gametangia and the uninucleate gametes all have the same genotype. This 'phenotypic sexuality' means that in response to some environmentally-determined difference between parts of the coenocytic mycelium—for example, between apical and subapical regions under nutritional stress—different patterns of gene expression occur locally and are stabilized there—in *Allomyces*, by the formation of the separate gametangial compartments. The regulatory mechanism for this differential gene expression seems to involve both DNA transcription and RNA translation (Fahnrich, 1974), and one of its results is the setting-up of the hormone system based on sirenin.

A scheme for the overall situation in *Allomyces* is attempted in Fig. 12.2, showing the segregation of the male and female phenotypes. Genes expressed only in the female phenotype lead to the synthesis of sirenin and the production of female-specific surface substance in the gametes; flagellar activity in the gametes is low and γ-carotene is not formed. Conversely, genes expressed only in the male phenotype code for male-type surface coating, for γ-carotene synthesis, and for high flagellar activity linked to a system for the uptake and inactivation of sirenin. Certainly many other genes must also be subjected to this differential expression. In the predominantly single-sex hybrids such patterns are established in the genotype, but even then each is only expressed locally in the mycelium and under the correct environmental circumstances.

This phenotypic differentiation in *Allomyces* is related to the 'non-functional sexuality' of *Blastocladiella* (Cantino, 1966), and the hormone system at least superficially resembles one suspected to occur in *Synchytrium*

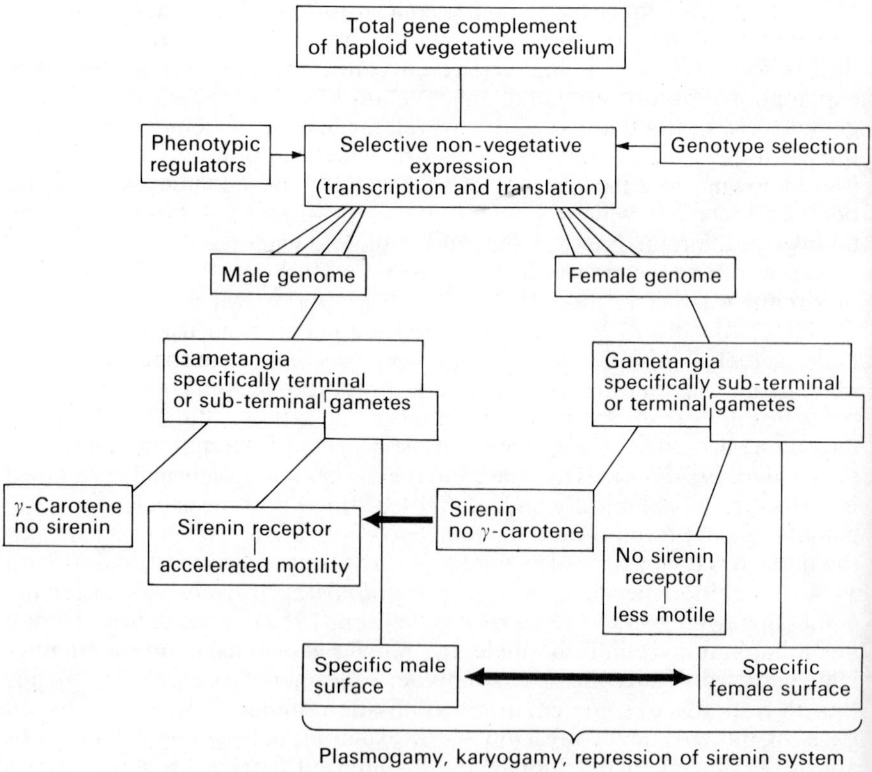

Fig. 12.2 In *Allomyces*, 'male' and 'female' are selective expressions of a common genome. Broad arrows mark the coordinated interaction between them which is secured first by the sirenin system and second by specific surface properties.

(Köhler, 1930). These are all chytrid 'fungi'; in brown algae attractants similarly produced by females and active on male gametes have been characterized in an *Ectocarpus* species (*S*-1-but-1′-enyl-2,5-cycloheptadiene) (Müller, Jaenicke, Donicke & Akintobi, 1971) and in *Fucus* (1,3,5-octatriene) (Müller & Jaenicke, 1973).

Sex hormones in Achyla

One of the most elegantly-explored systems of fungal sex hormones is that discovered in *Achlya* by J. R. Raper (review, 1952; Barksdale, 1969). In this genus of the Saprolegniales (Oömycetes) the vegetative mycelium gives rise to male antheridia, which are specialized branches, or to spherical female oögonia. In most strains of *A. ambisexualis* and *A. bisexualis* predominantly male or female outgrowths are produced, but other strains are homothallic (monoecious) in which both are formed on the same mycelium. Other species are predominantly homothallic, and the heterothallic condition seems merely to represent extremes of imbalance in what is essentially a monoecious system. Though most work has been done on strongly heterothallic strains the same hormone system is common to all (Raper, 1952; Barksdale, 1960).

In classic experiments, Raper showed that both the formation of the sex organs and their subsequent interactions are governed by diffusible hormones, the nature and activities of which have since been considerably clarified by Barksdale and coworkers. An attempt to summarize the system schematically is made in Fig. 12.3.

Vegetative female mycelium secretes the steroid hormone antheridiol. The process is subject to nutritional controls which have not been explored in detail. As in other members of the Saproleginales, the normal sterols of

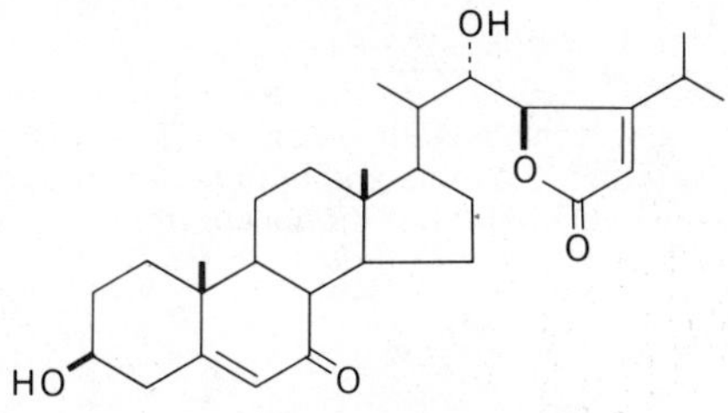

Antheridiol

Achlya are more like those of plants or algae than those of true fungi, probably formed *via* cycloartenol rather than lanosterol (Bu'Lock and Osagie, unpubl. observations) and including typical 24-ethylidene sterols such as fucosterol and its 7-dehydro-derivative (McCorkindale, Hutchinson, Pursey, Scott & Wheeler, 1969). Gooday (1974) notes communicated results of Popplestone and Unrau which confirm that these are the biosynthetic precursors of antheridiol, and a plausible route is that of Fig. 12.4. Antheridiol is detectably active in 10^{-11} M solution and the stereochemistry of the side-chain structure is particularly crucial for the hormone effects (Barksdale *et al.*, 1974).

Diffusion of the antheridiol to male or 'neuter' mycelium initiates a biochemically based morphogenetic sequence which is also subject to

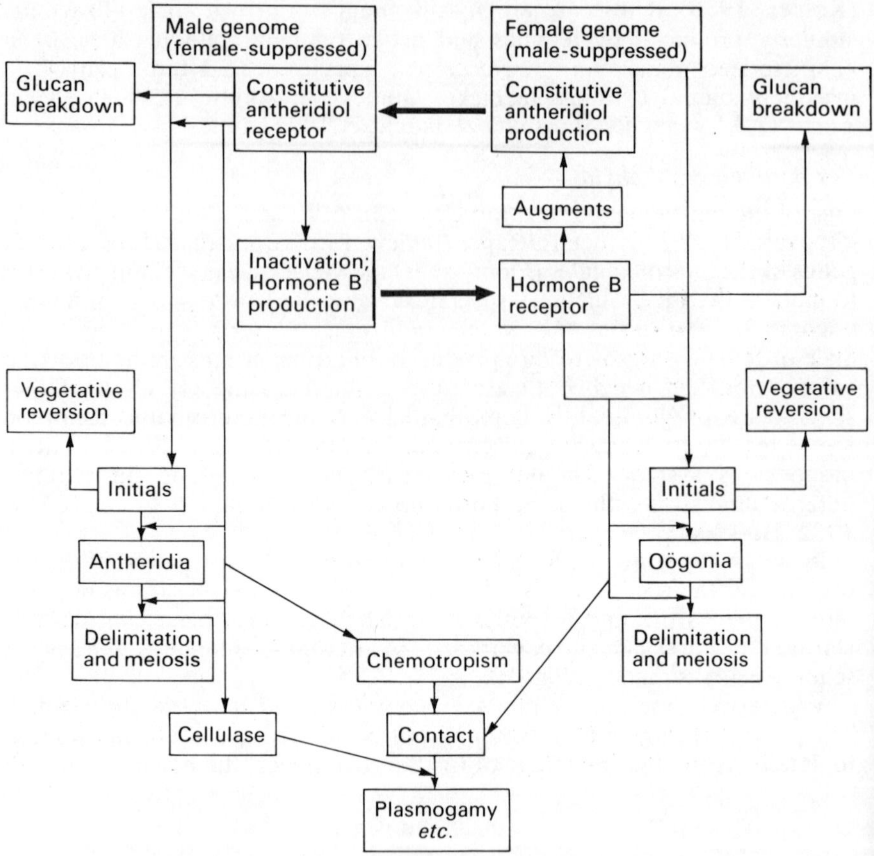

Fig. 12.3 A model for the hormone system in *Achlya*. The scheme does not include the mechanisms by which the male and female genomes are selected (compare Fig. 12.2) and the over-riding nutritional controls.

overriding nutritional controls and which for completion requires that the developing antheridium is exposed to continuous and increasing levels of antheridiol The full process takes 12–14 h and at least for the first 6 h it is continuously dependent upon new transcription of DNA (Horowitz & Russell, 1974) with the formation of new poly(A)-rich messenger RNA (Silver & Horgen, 1974) from which the requisite new enzymes are translated. Further biochemical aspects are noted below. The first morphological effect is the initiation of buds which can still develop as vegetative branches but which in correct nutritional conditions and with continuing antheridiol action elongate rather quickly into antheridia. These develop characteristically and grow towards the source of antheridiol. At 12–14 h the tips swell and the male gametangia are partitioned off from the coenocytic mycelium by a septum; meiosis occurs in the gametangia.

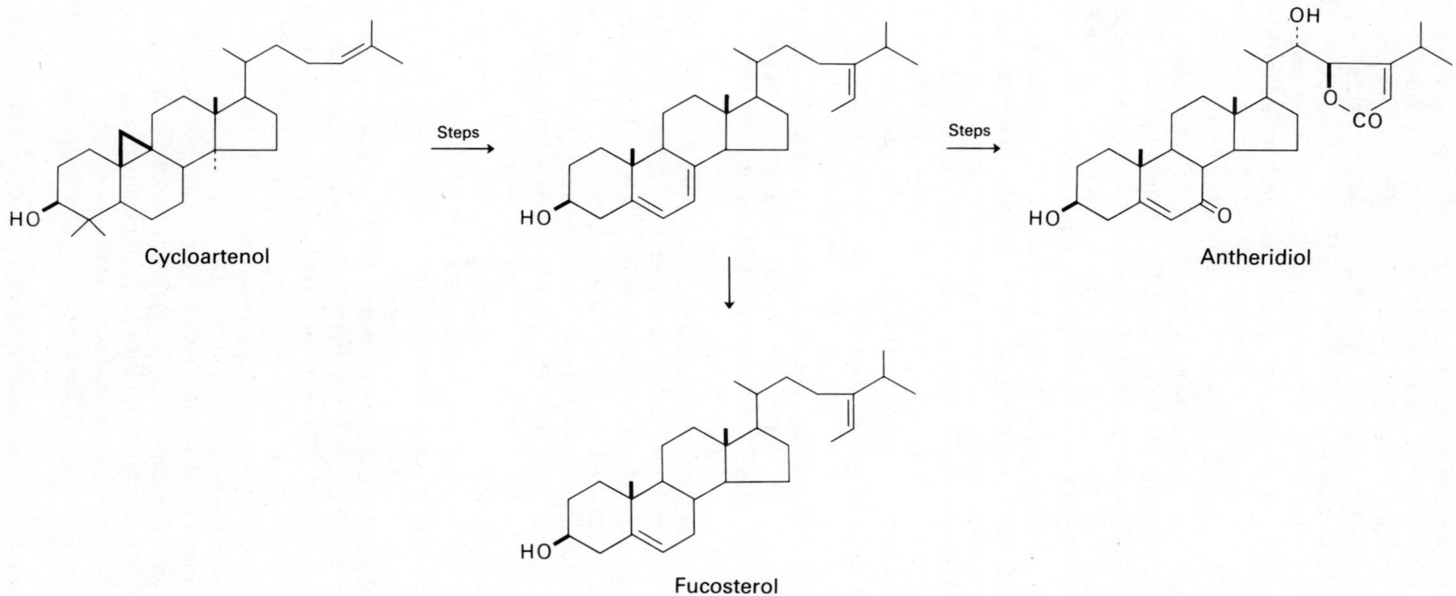

Fig. 12.4 Outline pathway for the biosynthesis of antheridiol from precursors of the major phytosterols of Saprolegniales *etc.*

During development the antheridia take up and inactivate antheridiol and simultaneously they produce and secrete hormone B, a substance of unknown but closely-related structure (see below). This in turn diffuses back towards the female mycelium (or to undifferentiated 'neuter' mycelium) and there induces initials which with continued hormone action develop into oögonia. Meiosis occurs in the oögonia once they are delimited. It is clear from the chemotropism of the antheridia that antheridiol production is intensified as the oögonia develop. When as a result of the chemotropism the two structures make contact, the wall of the oögonium is lysed in the contact region; the antheridium grows through the softened region and its own wall is then dissolved. Nuclear migration and karyogamy follow. Though the genetic situation is by no means clear, most if not all of the sexual process in *Achyla* must be ascribed to phenotypically-determined expression of two different selections from a common gene complement.

So far as the hormone system itself is concerned, the extent of this sexual differentiation has been carefully characterized by Barksdale & Lasure (1973). The female phenotype secretes antheridiol but takes up very little and does not make hormone B; it has a receptor for hormone B, one effect of which is probably to intensify antheridiol production. Conversely, the male phenotype makes little or no antheridiol but is sensitive to it, and under antheridiol stimulation produces hormone B, to which it is insensitive. These phenotypic characteristics appear in the vegetative mycelia of extreme heterothallics and of course are specially emphasized in the differentiated parts of all strains. In the 'neuter' species the situation is less clear-cut and there are some strains which produce very little antheridiol and also secrete hormone B without stimulation by exogenous antheridiol.

It is unlikely, therefore, that hormone B is merely a transformation product from the antheridiol taken up by males. Indeed, Gooday (1974) points out that dilution effects would probably prevent such a system from functioning, and provisional structural data on hormone B [T. C. McMorris and G. P. Arsenault, quoted by Barksdale & Lasure (1973) and Gooday (1974)] are more consistent with the idea that it is an alternative product formed by a 'male' variant of the pathway to antheridiol. A branching pathway, selected phenotypically and giving alternative products each of which intensifies the production of the other, would provide for amplification in the signal system; such a mechanism would be strikingly similar to that in Mucorales (p. 356).

The genetic basis of the difference between the heterothallic and homothallic or neuter conditions is not entirely clear, but the existence of strains which are heterothallic at 25°C but homothallic at 15–20°C is noteworthy. Many of the 'neuter' strains produce no sex organs of either type when grown singly but will react as males to a 'strong' female and as females to a 'strong' male (Raper, 1952; Barksdale, 1960). Thus there is linear hierarchy of sexuality, and a graded imbalance between the capacities to make the two hormones and the capacities to react to them. Since one effect of antheridiol on a neutral mycelium is to make it locally more male, and since hormone B has the opposite effect, it is not surprising that when homothallic strains which are approximately 'balanced' are grown in contact there is little or no reaction in either.

More is known about the mechanism of action of antheridiol than of hormone B but their biochemical effects may be rather similar. Antheridiol, acting on male mycelium and subject to over-riding nutritional constraints, not only activates the production of hormone B but also increases respiration (Warren & Mullins, 1969) and triggers a variety of processes involving polysaccharides. These include breakdown of cytoplasmic glucan reserves (Faro, 1972) and *de novo* synthesis of cellulase (Thomas & Mullins, 1967; Warren & Sells, 1971). The latter is required intracellularly for the localized wall-softening that allows the antheridial initials to develop and a similar process elicited by hormone B must precede development of the oögonia; extracellular cellulase could be functional in the copulation itself. The development of both structures must also involve cell wall synthesis, presumably at the expense of the cytoplasmic glucans already mentioned. The ultrastructural aspects of this hormone effect have recently been described by Mullins & Ellis (1974), who showed that hormone-induced vesicles, presumed to contain the requisite enzymes, aggregate specifically at the sites where initials develop and release their contents behind the softening region of the cell wall.

The way in which the *Achlya* hormones affect gene transcription is not known; in higher eukaryotes, steroid hormones combine with cytoplasmic receptor proteins which then migrate into the nucleus and interact directly with DNA sites.

Other Oömycete systems

In another member of the Saprolegniales, *Dictyuchus monosporus*, Sherwood (1966) showed that a diffusible hormone initiates antheridial development in a manner quite similar to the action of antheridiol in *Achlya*; no corresponding induction of oögonia was demonstrated and the hormone was not characterized. Maximum hormone production was detected in the filtrate from a mixed male/female culture, which could imply some differences from the *Achyla* system. *Sapromyces reinschii* is another Oömycete, but a member of the Leptomitales. Early and painstaking work by Bishop (1940) firmly established the existence of a hormone system in this species, which phenomenologically seems very close to that in *Achyla*; no further work has been reported and the nature of the hormone(s) remains unknown.

Hormone mechanisms in all types of Oömycetes had in fact been postulated from simple observations as long ago as 1881 (de Bary; see Machlis, 1966). There does not seem to have been any cross-testing of antheridiol against species other than *Achlya*, but in this connection the situation with regard to sterols in another Oömycete group, the Peronosporales (Pythiaceae and Peronosporaceae) deserve comment. These species cannot synthesise sterols at all (*e.g.* Elliott, Hendrix, Knights & Partes, 1964; Hendrix, 1966), so systems of endogenous hormones like antheridiol are excluded. However it is also a significant characteristic of the Peronosporales that they are dependent upon exogenous sterols for both sexual and asexual morphogenesis (Haskins, Tulloch & Micetich, 1964; Hendrix, 1965). The requirement is met by a rather broad range of sterols but phytosterols of the 24-ethylidene type are most active (Elliott, 1972; Child & Haskins, 1971) and metabolism to some specific product may occur. The situation is still quite obscure and will doubtless be clarified; meanwhile some older observations of hormone-like effects of plant hosts on these fungi, and of some associations with sterol-producing species such as *Fusarium*, are at least consistent with our present knowledge.

The Trisporic Acid System in the Mucorales

The Mucorales (Zygomycetes) are simple true fungi of filamentous habit, in which sexual reproduction is unambiguously either homothallic or heterothallic; for the heterothallic species there are two fixed mating-types, which are designated *plus* and *minus* since neither can be assigned a distinctively male or female role. The mating-type of heterothallic species can be defined over the entire order because observable interspecific reactions occur. Sexual differentiation is controlled by a hormone system which has been most thoroughly explored in heterothallic species but which functions equally in the homothallics. The subject has been extensively reviewed (van den Ende & Stegwee, 1971; Gooday, 1973, 1974; Bu'Lock, 1975) and although the historical background to the problem is a fascinating chapter in mycology the reader is referred to the cited accounts for these aspects.

There is no connection between mating-type and the morphology of the sexual organs. Heterothallic species are isogametangious and while homothallics are not, a character may be found in the *plus* gametangium of one species and in the *minus* of another. Consequently, it should not be surprising—though mycologists have found it so—that the same hormones govern differentiation of the gametangia in both mating-types. There are four of these hormones, the trisporic acids (TA). All are comparably active

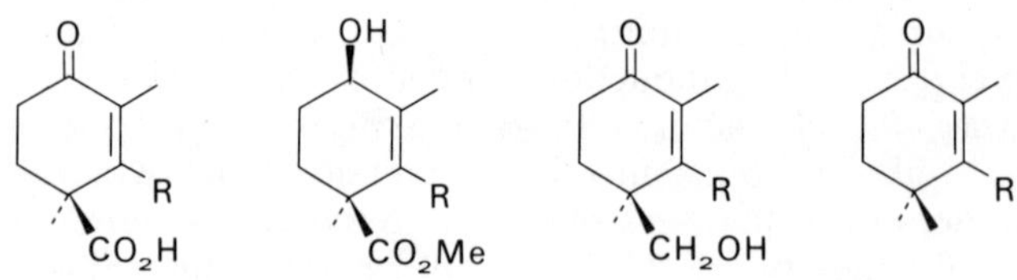

Trisporic acids P^+ prohormones P^- prohormones

$$\left(R = \text{—CH=CH} \cdot \text{CMe=CH} \cdot \text{CH}_2 \cdot \text{CH}_2 \begin{cases} \text{COCH}_3 \\ or \\ \text{CHOH—CH}_3 \end{cases} \right)$$

though 9-*cis*-trisporic acid B is the most active and probably the most important *in vivo*, equivalent to both the 'plus-gamone' and the 'minus-gamone' of earlier workers. The TA are formed from β-carotene by a series of cleavage and oxygenation reactions which is only complete when *plus* and *minus* strains are grown together, that is, in mixed culture, or close together in surface cultures, or separated by a membrane permeable to small molecules. They can be isolated from the culture medium and also from the mycelia of mixed plus/minus cultures.

In heterothallic species there is only one gene locus, MT, which is directly linked to the mating-type, and this has the two functioning alleles MT*plus* and MT*minus*. This is a regulatory locus with several distinct functions, one of which directly concerns hormone synthesis; others are noted later. In vegetative mycelia the pathway from mevalonate via β-carotene to TA is subject to over-riding nutritional control, probably by catabolite repression, and also to two different repression mechanisms (Bu'Lock, Jones, Quarrie & Winskill, 1973; Bu'Lock, 1975). One of these affects early enzymes in the pathway, including those which control the rate of β-carotene synthesis. This mechanism acts similarly in both mating-types, and can be lifted by TA (see below); in single strains of *Blakeslea trispora* these enzymes operate at about 1–10% of the derepressed level. The other repression mechanism, which is virtually complete and is insensitive to derepression by TA, acts differently in the two mating-types; in *plus* strains it prevents synthesis of the enzyme(s) needed to form the 4-keto group of TA, and in *minus* strains it prevents synthesis of the enzyme(s) forming the 1-CO_2H group (Bu'Lock *et al.*, 1973). This second type of repression could be more or less directly due to the gene products from MT*plus* and MT*minus* alleles. Its effect is that

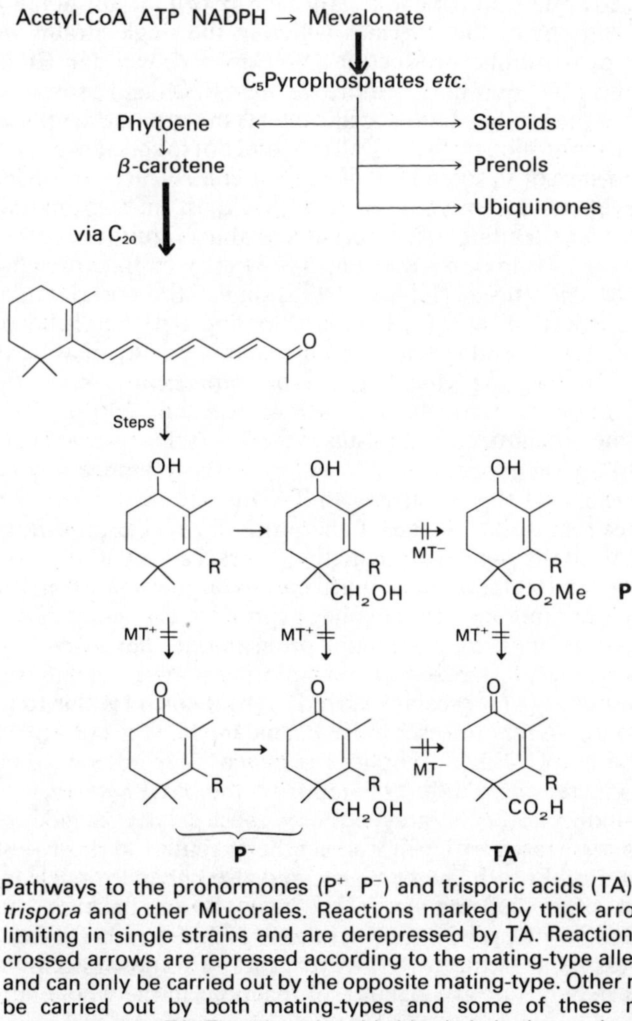

Fig. 12.5 Pathways to the prohormones (P^+, P^-) and trisporic acids (TA) in *Blakeslea trispora* and other Mucorales. Reactions marked by thick arrows are rate-limiting in single strains and are derepressed by TA. Reactions marked by crossed arrows are repressed according to the mating-type allele as shown, and can only be carried out by the opposite mating-type. Other reactions can be carried out by both mating-types and some of these may also be derepressed by TA. The trisporic acid side-chain is shown simply as R. MT^+ and MT^- are the mating-type gene products.

each mating-type produces—at a level which is low because of the first-mentioned mechanism of TA-sensitive MT-insensitive repression—a TA precursor which only the opposite mating-type can convert into TA. These are the P^+ and P^- prohormones, whose structures (Bu'Lock, Jones, & Winskill, 1974) are shown above. The prohormones fulfil the roles ascribed to both progamones and zygotropic hormones in earlier accounts; their formation is summarized in Fig. 12.5.

When heterothallic mycelia of opposite mating-types are adjacent and nutritional conditions are correct, the prohormones diffuse between them either aerially (Mesland, Huisman & van den Ende, 1974) or through the medium, and are converted into TA. Using the existing complement of enzymes, *i.e.* without *de novo* protein synthesis at this stage, P^+ is converted into TA in the *minus* partner and P^- is converted into TA in the *plus.* The TA thus produced initiate morphogenesis in each mycelium (see below) and also derepress synthesis of the enzymes which in the single strains were rate-limiting for prohormone production (Werkman & van den Ende, 1973). This 'amplification' step does require *de novo* RNA and protein synthesis. Evidence that the mating-type-specific effects in hormone synthesis are due to repression controlled by the MT alleles, and not to specific deletions of the 'missing' enzymes or to some kind of positive control mechanism, is that: (a) very low levels of the 'missing' enzymes do exist in each mating-type of *Blakeslea trispora,* leading to a barely-detectable gratuitous synthesis of TA (Sutter, Capage, Harrison & Keen, 1973); (b) in the so-called *laniger* mutants of *Mucor mucedo* (Köhler, 1935) single-strain production of TA is probably constitutive; and (c) balanced mating-type heterokaryons have been observed in several species, and the characteristics of at least some of these are better explained by cross-repression than by cross-complementation or -induction (Esser & Kuenen, 1967). A suggested scheme for the regulatory mechanisms in the TA system is given in Fig. 12.6.

About 10^{-8} g of exogenous TA will detectably induce zygophores in *Mucor mucedo,* and the sensitivity to TA formed *in situ* from exogenous prohormones may well be higher. In heterothallic species, the morphogenic effects of TA are the same in both mating-types, *i.e.* essentially independent of the MT alleles, but their biochemical basis is largely unknown. Zygophore initials grow out randomly from hyphae exposed to exogenous TA, but if the TA is formed *in situ* from exogenous prohormone they grow towards the prohormone source. Hence, for example, as a *plus* zygophore extends towards the *minus* (a) it receives more P^-, (b) it converts this to more TA, and (c) it consequently releases more P^+; meanwhile the converse situation occurs in the *minus.* Each zygophore is increasingly influenced by the TA level and so increasingly differentiated from the basal hyphae.

The TA-induced effects in zygophores must clearly include a range of phenomena associated with cell wall synthesis similar to those described in *Achlya,* but of unknown biochemistry, and also enhanced nuclear division. Carotenoids often accumulate in the zygophores, because the overall TA-derepression of prohormone synthesis includes hitherto rate-limiting steps in β-carotene synthesis. However, the well-known accumulation of carotenoids depends on the balance of reactions in the whole derepressed sequence, and may be very great in some circumstances and quite small in others. This aspect of TA-derepression also causes increases in other

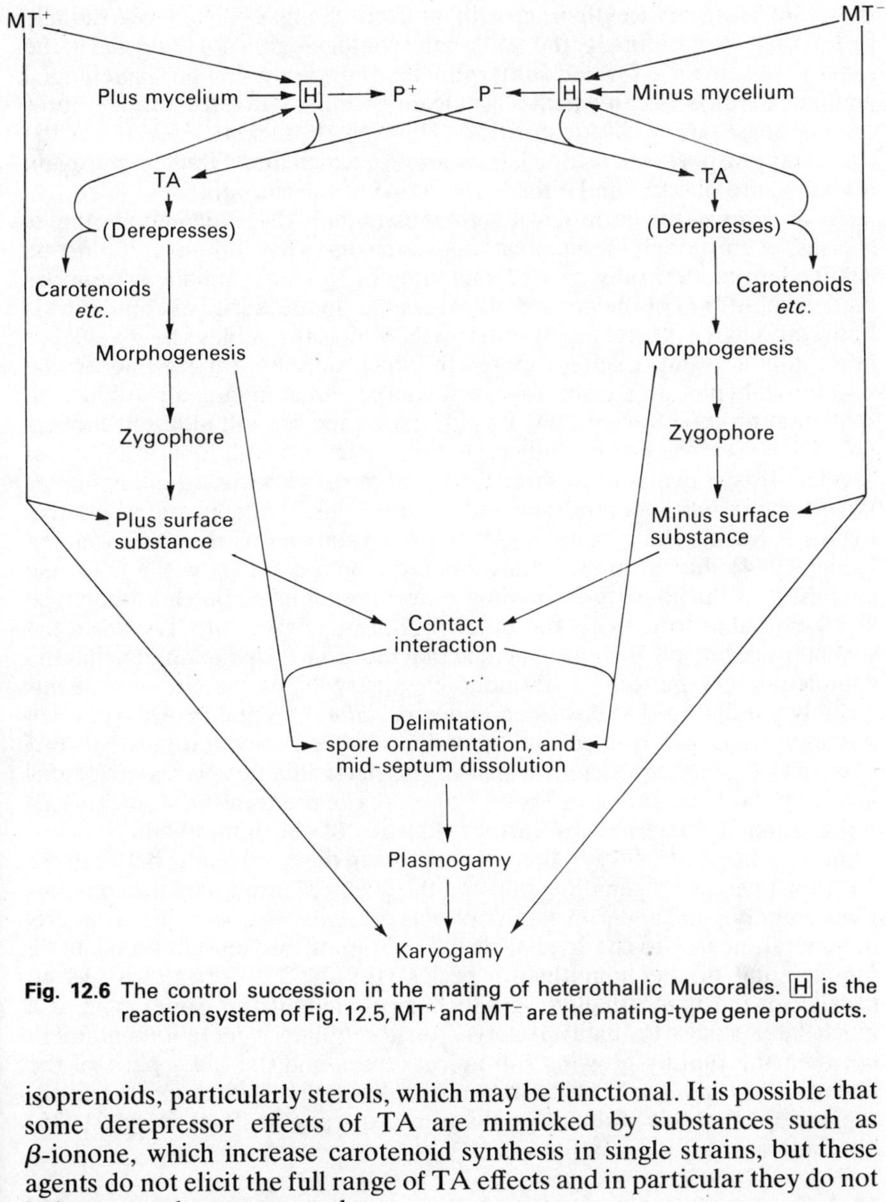

Fig. 12.6 The control succession in the mating of heterothallic Mucorales. H is the reaction system of Fig. 12.5, MT^+ and MT^- are the mating-type gene products.

isoprenoids, particularly sterols, which may be functional. It is possible that some derepressor effects of TA are mimicked by substances such as β-ionone, which increase carotenoid synthesis in single strains, but these agents do not elicit the full range of TA effects and in particular they do not induce zygophore outgrowth.

When *plus* and *minus* zygophores of the same or similar species make contact, as a result of the chemotropic effect already described, they adhere firmly in a manner which does not occur in any other context. This effect is plausibly ascribed to hormonally-stimulated formation of species- and mating-type-specific substances on the surface of the zygophores, and represents the second point in the sexual process at which the action of the MT alleles is apparent. If this type of contact does not occur—as when two disparate species are involved—the zygophores may go on extending until they abort; otherwise their growth pattern changes, they grow radially rather than longitudinally, the walls in the contact region are lysed and at the same time septa are formed subterminally, thus delimiting progametangial compartments between which plasmogamy occurs. Ultrastructural features of this phase are well-documented (Hawker & Beckett, 1971), but the cytogenetic aspects of ensuing karyogamy, in which the MT alleles are again involved, are obscure and outside the scope of this account.

As in *Achlya*, the hormonal system is essentially the same in homothallic species, even though its manifestations are somewhat different; it must be acknowledged that only an understanding of the heterothallic system has made exploration of the homothallics feasible. In these species copulation is between a large gametangial outgrowth, which resembles the zygotes of heterothallics, and a small outgrowth which may be a mere bud on the vegetative hypha. In a given species the large gametangium always has the same mating-type but this may be *plus* in one species and *minus* in another (*cf. Allomyces*), as can be shown by cross-reactions with heterothallic test species. It is convenient to refer to the large outgrowing gametangium as having the *induced* mating-type and the small gametangium and vegetative hypha as having the *constitutive*. It has been shown (Werkman & van den Ende, 1974) that in mass culture homothallic species show the hormone chemistry of the constitutive mating-type, *i.e.* producing one hormone type and being able to convert the other prohormone type into TA; from the sexual process itself we may conclude that the induced gametangium has the complementary pattern of hormone chemistry. Thus the situation seems entirely parallel to that discussed above in *Allomyces* and in *Achlya*, *viz.* a phenotypically-determined and spatially localized selection between two patterns of gene expression, which in the heterothallic species are segregated by the functioning of the MT alleles. The mechanism of phenotypic segregation is suggested by further features of the homothallic species; usually and perhaps always, there is a septum in the basal hypha between the outgrowing induced gametangium and the eventual location of the constitutive gametangium, and irrespective of *plus* or *minus* character the latter are, in general, nearer to the hyphal tips. The role of this septum seems to be fundamental in the homothallic species (Bu'Lock, 1975) and it has no parallel in the heterothallics, in which septa delimit the gametangia at a much later stage. Probably it serves to accentuate a metabolic gradient between the rapidly-growing sub-apical region and the older parts of the hypha, and this gradient controls the phenotypic selection; there are suggestions that this gradient mechanism involves cAMP (Bu'Lock, 1975) but these are as yet unconfirmed.

12.4 Hormonal Systems in the Ascomycetes

Filamentous Ascomycetes

Despite rather numerous observations tending to suggest that there are endogenous factors regulating both growth and asexual differentiation in imperfect fungi, none of these studies has been taken to a stage at which we can profitably discuss the system in terms of hormonal mechanisms. Perhaps the nearest approach is in the work of Robinson, Park & Mclure (1969) on the pigment bikaverin, which is produced by several Fusaria including

Gibberella fujikuroi, in which it is a typical 'secondary metabolite' whose production is controlled by growth-linked 'catabolite repression' (Bu'Lock, Detroy, Hostálék & Munim-al-Shakarchi, 1974). For this reason, the accumulation of this quinone in the mycelium of surface or batch culture is seen simultaneously with other signs of senescence, including vacuolation of the hyphae. However, like many other quinones, bikaverin has a range of cytotoxic effects and it can cause vacuolation in the hyphae of many fungi including at least some of the producing species. Whether such autotoxicity qualifies bikaverin as an endogenous regulator in a useful sense is open to some doubts, but related effects may well underly some of the many observations of 'staling' and similar phenomena. They suggest a mechanism whereby the production of a specific substance has the effect of amplifying the signals of nutritional stress.

The existence of hormone mechanisms in the sexual development of Ascomycetes is more securely based, but none of the molecular aspects has been elucidated; in *Neurospora*—surely one of the most intensively-studied species—even the existence of hormones is a matter of suspicion rather than of strict demonstration. In these fungi the male components, *i.e.* the nuclear donors in the copulation mechanism, may be differentiated antheridia, or specialized microconidia, or conidia or even vegetative male hyphae; the female component is usually a more specialized ascogonium, from which a tubular trichogyne is developed. In some species distinct male and female strains exist; others can form both types of organ, *i.e.* show phenotypic bisexuality, and such strains may be self-fertile or they may be of two genotypic compatibility types. Hormones may control the induction of sexual primordia, their development, and the mechanism of copulation leading to plasmogamy and formation of the ascus; a good general account is that of Machlis (1966).

In *Bombardia lunata*, Zickler (1952) showed that the trichogynes are attracted chemotropically to differentiated male spermatia or indeed to vegetative mycelium of the opposite compatibility type; the attractant could be detected in filtrates and was heat-stable. In *Glomerella cingulata* strongly self-fertile cultures or strongly-mating paired cultures produce diffusing substances, detectable in filtrates, which will promote either self-mating or mutual mating in other cultures (Markert, 1949; Driver & Wheeler, 1955).

A more detailed picture has been given for *Ascobolus stercocarius* (Bistis, 1956, 1957), in which the male elements are oidia which also retain their ability to develop asexually. Oidia exposed to mycelia of the opposite compatability type develop male function and then, apparently through a diffusible mediator, will initiate the formation of ascogonial primordia in the other compatibility type. Further, an oidium promotes the full development of the ascogonium and the trichogyne which then grows chemotropically towards the oidium. During this process both elements develop specific surface properties; the oidial wall softens before contact with the trichogyne, and dissolution of the trichogyne tip, which normally occurs after it penetrates the oidium, can also be induced prior to contact. In this system, therefore, a rather full observational picture has been built up, which is readily understandable in the light of what is now known about *Achlya* and the Mucorales, but no progress towards chemical and biochemical data has since been reported.

Precisely the opposite situation exists for what is said to be a hormonal system controlling perithecium formation in *Gibberella zeae (Fusarium roseum* 'Graminearum') (Wolf & Mirocha, 1973), in which the agent is chemically and biosynthetically identified but its effects are virtually unknown. Zearalenone is a macrocyclic lactone polyketide produced by

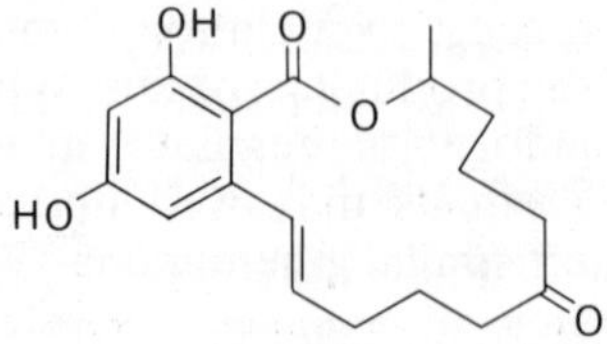

Zearalenone

many strains of *G. zeae*; it was observed that zearalenone production was typical in isolates which formed abundant perithecia (presumably these are strongly self-fertile homothallics). In young cultures (of a producing strain) small quantities of added zearalenone increased the number of perithecia eventually formed, and larger amounts decreased the number; in older cultures the effects were diminished. Unfortunately, no accounts of the morphological phenomena, or indeed of the time course of perithecial appearance, have been published, nor have studies been carried out on, for example, the effects of zearalenone on non-producing or non-perithecial strains. Until such studies appear, the designation of zearalenone as a sexual hormone must be treated with reservation.

Mating Systems in Yeasts

Yeasts such as *Saccharomyces* form haploid single cells which are of two genetically-determined mating-types, *a* and *α*. When two such cells are in proximity, normal budding is inhibited and the cells elongate towards each other (Levi, 1956). In *Kluyveromyces lactis* somewhat different phenomena of directed budding are described (Herman, 1970). In addition there is strong agglutination between the cells, though many yeasts are also self-agglutinative. Cell contact is followed by lysis of the partitioning walls and a diploid cell is formed; diploid cells can multiply vegetatively like the haploids, but on release of catabolite repression they can form asci with four meiospores. The participation of extracellular hormones in the mating-process was first shown by Levi (1956); subsequent studies have added considerably to our knowledge of the system and somewhat to our understanding, and the various reports have yet to be drawn together into a mutually consistent picture. Nevertheless, in view of the importance of the system some attempt to present its main features in a clear manner has been made in the present account.

Sakai & Yanagishima (1972) reported on the mutual agglutinability of *a* and *α* cells, which is abolished by treatment with protease and is ascribed to immunospecific surface proteins (see below). They found that agglutinability was constitutive in some *a*-type strains but in others it was induced by a heat-stable filtrable secretion from *α*-type cells, which they called *α*-substance 1. The induction was blocked by cycloheximide and no corresponding effect of *a*-type on *α* was observed.

Osumi, Shimoda & Yanagishima (1974) presented ultrastructural data on the conjugation process, showing elongation of the cells towards the area of contact. An electron-dense surface material accumulates in the initial contact area and the contacting walls begin to thin, particularly in the median (glucan) layer; the two nuclei migrate towards the dissolving walls at an early stage, and karyogamy occurs soon afterwards. Shimoda & Yanagishima (1972) had earlier shown the importance of lytic processes, particularly lysis of polysaccharides, in the conjugation process.

Yanagishima (1969) also described evidence that both *a* and *α* cells produce hormones which can be extracted from the culture media with methylene chloride (*i.e.* relatively small non-polar molecules) which cause cell expansion. However, the further exploration of these substances (designated *a*-hormone and *α*-hormone) seems to have become confused with a study of rather non-specific swelling effects produced by sterols and other surface-active materials, some of which are naturally present in the cultures (Sakurai *et al.*, 1974), and this part of the picture is consequently most obscure. One activity ascribed to *a*-hormone is to induce *α*-cells to produce a water-soluble non-extractable *α*-substance-II.

Both normal budding and nuclear replication are halted in conjugating cells. Shimoda & Yanagishima (1973) showed that there was a marked reduction in the rate of DNA synthesis, but not of RNA or protein synthesis, in mixed cultures, and that a corresponding effect in *α* cells could be induced by filtrates from *a*-cultures.

Meanwhile Duntze and coworkers, in less contradictory investigations of more limited scope, had successfully isolated from culture filtrates of *α* cells an oligopeptide hormone (molecular weight *ca.* 1400, containing 7–9 aminoacids) with effects on the *a*-cells which seem to include those separately ascribed to at least two substances by the Japanese school, *i.e.* specific elongation and an arrest in the initiation of DNA synthesis (so that the cells are held in G1 phase); the production of this *α*-peptide is genetically linked to the mating-type (Duntze, MacKay & Manney, 1970; Duntze, Stotzler, Bucking-Throm & Kalbitzer, 1973).

As a very tentative reconstruction of the *Saccharomyces* system the following is suggested. Even prior to contact, secretion of the *α*-peptide of Duntze *et al.* (1973) and possibly of a related *a*-product, causes in the opposite mating types the halt in DNA replication, localized lytic wall-softening (leading to elongation) and localized deposits of mating-type specific surface proteins. These effects are accentuated when the cells make contact and the surface proteins agglutinate. In the intermediate stages, lysis may release small molecules such as sterols and *n*-octanoic acid which enhance the swelling effect non-specifically and this process might also be promoted by the initial hormones.

The role of the surface-specific materials is very clearly shown in *Hansenula wingei*, in which the adhesion between opposite mating-types has been shown to be due to immunospecific surface components in each; both are mannan-proteins, one multivalent and of high molecular weight and one univalent and rather smaller (Crandall & Brock, 1968; Crandall, Lawrence & Saunders, 1974). These substances are not diffusible hormones, but their interaction governs later stages of cell fusion and it is worth noting that rather similar glucoproteins do function as sex hormones in the alga *Volvox* (Starr & Jaenicke, 1974; Kochert & Yates, 1974). The involvement of surface interactions in fungal systems has been repeatedly noted in earlier sections of this chapter, but the *Hansenula* system is so far the only one to have been characterized in molecular terms.

The observation that in conjugating *a*-type cells of *Saccharomyces cerevisiae* the cell cycle is suspended at the G-1 phase should be compared with the situation in *Ustilago violacea*, one of the smut fungi (Heterobasidiomycetes), in which the dikaryophase is established following plasmogamy between uninucleate yeast-like haploids or sporidia. The sporidia have two mating-types, a_1 and a_2, and in synchronous cultures the a_1 type will conjugate only during the G1 phase, whereas a_2 cells will enter sexual morphogenesis during most of the cell cycle (Cummins & Day, 1973). Extracellular materials probably mediate pre-copulation stages in this organism too, but they have not been studied so far.

Finally, this admittedly fragmentary account of mating systems in yeasts should include the observation that the transformation of diploid cells into asci with meiospores seems to be regulated by a catabolite repression mechanism which can be lifted by additions of cyclic AMP (Tsuboi & Yanagishima, 1973).

12.5 Hormonal Systems in the Basidiomycetes

In the development of the carpophore or fruit-body of typical ground and coprophilous Agaricaceae there is a generally-recognized sequence (Taber, 1966). Mycelial 'lattices' develop through primordia into a small bud in which differentiation of the cap and gills begins at an early stage. In the cultivated mushroom *Agaricus bisporus* development begins when the 'button' is only 0.2 cm tall and all the cells are differentiated by the time it is 2 cm tall; subsequent 'growth' to the mature carpophore largely consists of cell elongation, particularly in the stipe and most especially in the upper part of the stipe close to the cap.

In this and other agarics several authors have suspected that at least parts of the coordinated morphogenic process are controlled by hormone-like substances, but the evidence for regulatory agents actually diffusing from one cell type to another is all rather indirect. Thus Gruen (1963) by careful measurements of morphogenesis in *Agaricus bisporus* carpophores from which different parts had been excised at various stages concluded that an endogenous factor present in the lamellae partly controlled the early stages of stipe elongation and perhaps cap expansion as well.

Matthews & Niederpruem (1972) noted that additions of cyclic AMP to mycelial preparations of *Coprinus lagopus* markedly accelerated the production of primordia, and recently Uno, Yamaguchi & Ishikawa (1974) have shown that the brief exposure of basal mycelium to light, which is a prerequisite for fruiting in *Coprinus macrorhizus*, is followed by a marked transient rise in mycelial cAMP. Earlier, it had been shown that cAMP would induce fruiting in certain monokaryons of *C. macrorhizus*, and that the ability of dikaryons and of some other monokaryons to fruit spontaneously was associated with higher levels of adenyl cyclase, and lower levels of phosphodiesterase, in these cultures (Uno & Ishikawa, 1973).

Preliminary work with *Coprinus lagopus* carpophores (Bu'Lock and Darbyshire, unpubl. observations) also suggests very strongly that cAMP is involved in the detailed coordination of carpophore development. Transient peaks of cAMP can be measured in just those parts of the developing carpophore where chemical changes—in particular the mobilization of carbohydrate reserves and synthesis of cell wall materials—are proceeding most rapidly. The observations neither prove nor preclude the diffusion of cAMP between cells, and they are only mentioned here, without any account of the very extensive background of metabolic data on fruiting in this species, because they suggest that in the exploration of such broad-spectrum regulators as cAMP, the fungal systems of differentiation may come to play as large a role as any other 'model'. Certainly the case of cAMP, with so many activities in so many different types of cell and at such widely-different levels of evolutionary complexity, emphasizes that in studying regulatory problems the specification of the system being regulated is every bit as important as the identification of a mediator.

ACKNOWLEDGEMENT I am grateful to Mr. Brian Jones for his notable help in compiling the data for this chapter.

12.6 References

BARKSDALE, A. W. (1960). Interthallic sexual reactions in *Achlya*. *American Journal of Botany* **47**, 14–23.

BARKSDALE, A. W. (1969). Sexual hormones of *Achlya* and other fungi. *Science* **166**, 831–7.

BARKSDALE, A. W. & LASURE, L. L. (1973). Induction of gametangial phenotypes in *Achyla*. *Bulletin of the Torrey Botanical Club* **100**, 199–202.

BARKSDALE, A. W., McMORRIS, T. C., SESHADRI, R., ARUNACHALAM, T., EDWARDS, J. A., SUNDEEN, J. & GREEN, D. M. (1974). Response of *Achyla ambisexualis* E87 to hormone antheridiol and certain other steroids. *Journal of General Microbiology* **82,** 295–9.

BISHOP, H. (1940). A study of sexuality in *Sapromyces reinschii*. *Mycologia* **32**, 505–29.

BISTIS, G. (1956). Sexuality in *Ascobolus stercorarius*. I. Morphology of the ascogonium; plasmogamy; evidence for a sexual hormone mechanism. *American Journal of Botany* **43,** 389–94.

BISTIS, G. (1957). Sexuality in *Ascobolus stercorarius*. II. Preliminary experiments on various aspects of the mating process. *American Journal of Botany* **44**, 436–43.

BONNER, J. T. (1971). Aggregation and differentiation in the cellular slime molds. *Annual Review of Microbiology* **25**, 75–92.

BU'LOCK, J. D. (1975). Cascade expression of the mating-type locus in Mucorales. Proceedings of the 1974 Symposium on the Genetics of Industrial Micro-organisms. London and New York: Academic Press.

BU'LOCK, J. D., JONES, B. E. & WINSKILL, N. (1974). Structures of the mating-type specific prohormones of Mucorales. *Journal of Chemical Society: Chemical Communication*, 708–9.

BU'LOCK, J. D., JONES, B. E., QUARRIE, S. A. & WINSKILL, N. (1973). The biochemical basis of sexuality in Mucorales. *Die Naturwissenschaften* **60**, 550–1.

BU'LOCK, J. D., DETROY, R. W., HOSŤÁLÉK, Z. & MUNIM-AL-SHAKARCHI, A. (1974). Regulation of secondary biosynthesis in *Gibberella fujikuroi*. *Transactions of the British Mycological Society* **62**, 367–89.

CANTINO, G. C. (1966). Morphogenesis in aquatic fungi. In *The fungi*, Vol. II, pp. 283–337. Edited by G. C. Ainsworth and A. S. Sussman. London and New York: Academic Press.

CARLILE, M. J. & MACHLIS, L. (1965). The response of male gametes of *Allomyces* to the sexual hormone sirenin. *American Journal of Botany* **52**, 478–83.

CHASSY, B. M. (1972). Cyclic nucleotide phosphodiesterase in *Dictyostelium discoideum*: interconversion of two enzyme forms. *Science* **175**, 1016–8.

CHILD, J. J. & HASKINS, R. H. (1971). Induction of sexuality in heterothallic *Pythium* spp. by cholesterol. *Canadian Journal of Botany* **49**, 329–32.

CRANDALL, M. A. & BROCK, T. D. (1968). Molecular basis of mating in the yeast

Hansenula wingei. *Bacteriological Reviews* **32**, 139–63.

CRANDALL, M. A., LAWRENCE, L. M. & SAUNDERS, R. M. (1974). Molecular complementarity of yeast glycoprotein mating factors. *Proceedings of the National Academy of Sciences, U.S.A.* **71**, 26–9.

CUMMINS, J. E. & DAY, A. W. (1973). Cell cycle regulation of mating type alleles in the smut fungus *Ustilago violacea*. *Nature (London)* **245**, 259–60.

DRIVER, C. H. & WHEELER, H. E. (1955). A sexual hormone in *Glomerella*. *Mycologia* **47**, 311–6.

DUNTZE, W., MACKAY, V., & MANNEY, T. R. (1970). *Saccharomyces cerevisiae*: a diffusible sex factor. *Science* **168**, 1472–3.

DUNTZE, W., STOTZLER, D., BÜCKING-THROM, E. & KALBITZER, S. (1973). Purification and partial characterisation of α-factor, a mating-type-specific inhibitor of cell reproduction from *Saccharomyces cerevesiae*. *European Journal of Biochemistry* **35**, 357–65.

ELLIOTT, C. G. (1972). Sterols and the production of oöspores by *Phytophthora cactorum*. *Journal of General Microbiology* **72**, 321–7.

ELLIOTT, C. G., HENDRIE, M. R., KNIGHTS, B. A. & PARTES, W. (1964). A steroid growth factor requirement for a fungus. *Nature (London)* **203**, 427–8.

EMERSON, R. & WILSON, C. M. (1954). Interspecific hybrids and the cytogenetics and cytotaxonomy of *Euallomyces*. *Mycologia* **46**, 393–434.

ESSER, K. & KUENEN, R. (1967). *Genetics of fungi*. Berlin-Heidelberg: Springer-Verlag, pp. 85–6.

FAHNRICH, P. (1974). Untersuchungen zur Entwicklung des Phycomyceten *Allomyces arbuscula*. II. Einfluss von Inhibitoren der Protein-und Nucleinsauresynthese auf die Gametogenese. *Archives of Microbiology* **99**, 147–53.

FARO, S. (1972). The role of a cytoplasmic glucan during morphogenesis of sex organs in *Achlya*. *American Journal of Botany* **59**, 919–23.

GARROD, D. R. & ASHWORTH, J. M. (1973). Development of the cellular slime mould *Dictyostelium discoideum*. *Symposium of the Society of General Microbiology* **23**, 407–35.

GESSAT, M. & JANTZEN, H. (1974). Die Bedeutung von Adenosin-3′,5′-monophosphat fur die Entwicklung von *Acanthamoeba castellanii*. *Archives of Microbiology* **99**, 155–66.

GOODAY, G. W. (1973). Differentiation in the Mucorales. *Symposium of the Society for General Microbiology* **23**, 269–93.

GOODAY, G. W. (1974). Fungal sex hormones. *Annual Review of Biochemistry* **43**, 35–49.

GRUEN, H. E. (1963). Endogenous growth regulation in carpophores of *Agaricus bisporus*. *Plant Physiology* **38**, 652–66.

HASKINS, R. A., TULLOCH, A. P. & MICETICH, R. G. (1964). Steroids and the stimulation of sexual reproduction of a species of *Pythium*. *Canadian Journal of Microbiology* **10**, 187–93.

HAWKER, L. E. & BECKETT, A. (1971). Fine structure and development of the zygospore of *Rhizopus sexualis* (Smith) Callen. *Philosophical Transactions of the Royal Society of London*, **B 263**, 71–100.

HENDRIX, J. W. (1965). Influence of sterols in growth and reproduction of *Pythium* and *Phytophthora*. *Phytopathology* **55**, 790–805.

HENDRIX, J. W. (1966). Inability of *Pythium aphanidermatum* and *Phytophthora palmivora* to incorporate acetate into digitonin-precipitable sterols. *Mycologia* **58**, 307–15.

HERMAN, A. I. (1970). Interspecies sex specific growth responses in *Kluyveromyces*. *Antonie van Leeuwenhock* **36**, 421–5 (*cf.* also **37**, 379–84).

HOROWITZ, P. K. & RUSSELL, P. J. (1974). Hormone-induced differentiation of antheridial branches in *Achyla ambisexualis*: dependence on ribonucleic acid synthesis. *Canadian Journal of Microbiology* **20**, 977–80.

HUXLEY, J. S. (1935). Chemical regulation and the human concept. *Biological Reviews* **10**, 427–41.

KOCHERT, G. & YATES, I. (1974). Purification and partial characterisation of a glycoprotein sexual inducer from *Volvox carteri*. *Proceeding of the National Acadamy of Sciences, U.S.A.* **71**, 1211–14.

KÖHLER, E. (1930). Beobachtungen an Zoosporen aufschwemmungen von *Synchytrium endobioticum*. *Zentralblatt fur Bakteriologie und Parasitenkunde, Abteilung II* **82**, 1–10.

KÖHLER, F. (1935). Genetische Studien an *Mucor mucedo* Brefeld. *Zeitschrift für induktive Abstammungs- und Verebungslehre* **70**, 1–54.

KRISHNA MURTI, C. R. (1973). Biochemistry of amoebic encystment. *Biochemical Society Transactions* **1**, 1104–5.

LEVI, J. D. (1956). Mating reaction in yeast. *Nature (London)* **117**, 753–4.

MACHLIS, L. (1958). Evidence for a sexual hormone in *Allomyces. Physiologia Plantarum* **11**, 181–92.

MACHLIS, L. (1966). Sex hormones in fungi. In *The fungi*, Vol. II, pp. 415–33. Edited by G. C. Ainsworth and A. S. Sussman. London and New York: Academic Press.

MACHLIS, L. (1972). The coming of age of sex hormones in plants. *Mycologia* **64**, 234–48.

MARKERT, C. L. (1949). Sexuality in the fungus *Glomerella. American Naturalist* **83**, 227–31.

MATTHEWS, T. R. & NIEDERPRUEM, D. J. (1972). Differentiation in *Coprinus lagopus*. I. Control of fruiting and cytology of initial events. *Archiv für Mikrobiologie* **87**, 257–66.

McCORKINDALE, N. J., HUTCHINSON, S. A., PURSEY, B. A., SCOTT, W. T. & WHEELER, R. (1969). A comparison of the types of sterol found in species of Saprolegniales and Leptomitales with those found in some other Phycomycetes. *Phytochemistry* **8,** 861–7.

MESLAND, D. A. M., HUISMAN, J. G. & VAN DEN ENDE, H. (1974). Volatile sexual hormones in *Mucor mucedo. Journal of General Microbiology* **80**, 111–7.

MÜLLER, D. G. & JAENICKE, L. (1973). Fucoserraten, the female sex attractant of *Fucus serratus* L. (Phaeophyta). *FEBS Letters* **30**, 137–9.

MÜLLER, D. G., JAENICKE, L., DONIKE, M. & AKINTOBI, T. (1971). Sex attractants in a brown alga: chemical structures. *Science* **171**, 815–7.

MULLINS, J. T. & ELLIS, E. A. (1974). Sexual morphogenesis in *Achlya*: ultrastructural basis for the hormonal induction of antheridial hyphae. *Proceedings of the National Academy of Sciences, U.S.A.* **71**, 1347–50.

NUTTING, W. H., RAPOPORT, H. & MACHLIS, L. (1968). The structure of sirenin. *Journal of the American Chemical Society* **90**, 6434–8.

OSUMI, M., SHIMODA, C. & YANAGISHIMA, N. (1974). Mating reaction in *Saccharomyces cerevisiae*. V. Changes in the fine structure during the mating reaction. *Archives of Microbiology* **97**, 27–38.

PANBACKER, R. G. & BRAVARD, L. J. (1972). Phosphodiesterase in *Dictyostelium discoideum* and the chemotactic response to cyclic adenosine monophosphate. *Science* **175**, 1014–5.

RAPER, J. R. (1952). Chemical regulation of sexual processes in the Thallophytes. *Botanical Reviews* **18**, 447–545.

ROBERTSON, A., DRAGE, D. J. & COHEN, M. H. (1972). Control of aggregation in *Dictyostelium discoideum* by an external periodic pulse of cyclic adenosine monophosphate. *Science* **175**, 333–5.

ROBINSON, P. M., PARK, D. & McLURE, W. K. (1969). Observations of induced vacuoles in fungi. *Transactions of the British Mycological Society* **52**, 447–50.

SAKAI, K. & YANAGISHIMA, N. (1972). Mating reaction in *Saccharomyces cerevisiae.* II. Hormone regulation of agglutinability of α-type cells. *Archiv für Mikrobiologie* **84**, 191–8.

SAKURAI, A., TAMURA, S., YANAGISHIMA, N., SHIMODA, C., HAGIYA, M. & TAKAO, N. (1974). Isolation and identification of a sexual hormone in yeast. *Agricultural and Biological Chemistry (Japan)* **38**, 231–2.

SHERWOOD, W. A. (1966). Evidence for a sexual hormone in the water mold *Dictyuchus. Mycologia* **58**, 215–20.

SHIMODA, C. & YANAGISHIMA, N. (1972). Mating reaction in *Saccharomyces cerevisiae.* III. Changes in autolytic activity. *Archiv für Mikrobiologie* **85** 310–8.

SHIMODA, C. & YANAGISHIMA, N. (1973). Mating reaction in *Saccharomyces cerevisiae.* III. Retardation of deoxyribonucleic acid synthesis. *Physiologia Plantarum* **29**, 54–9.

SILVER, J. C. & HORGEN, P. A. (1974). Hormonal regulation of presumptive mRNA in the fungus *Achlya ambisexualis. Nature (London)* **249**, 252–4.

STARR, R. C. & JAENICKE, L. (1974). Purification and characterisation of the hormone initiating sexual morphogenesis in *Volvox. Proceedings of the National Academy of Sciences, U.S.A.* **71**, 1050–4.

SUSSMAN, M. & SUSSMAN, R. R. (1969). Patterns of RNA synthesis and of enzyme accumulation and disappearance during cellular slime mould cytodifferentiation. *Symposium of the Society for General Microbiology* **19**, 403–35.

SUTTER, R. P., CAPAGE, D. A., HARRISON, T. L. & KEEN, W. A. (1973). Trisporic acid biosynthesis in separate *plus* and *minus* cultures of *Blakeslea trispora*. Identification by *Mucor* bioassay of two mating-type-specific components. *Journal of Bacteriology* **114**, 1074–82.

TABER, W. A. (1966). Morphogenesis in Basidiomycetes. In *The fungi* vol. II, pp. 381–412. Edited by G. C. Ainsworth and

A. S. Sussman. London and New York: Academic Press.

THOMAS D. DES S. & MULLINS, J. T. (1967). Role of enzymatic wall-softening in plant morphogenesis: hormone induction in *Achlya*. *Science* **156**, 84–5.

TSUBOI, M. & YANAGISHIMA, N. (1973). Effect of cyclic AMP, theophylline, and caffeine on the glucose repression of sporulation in *Saccharomyces cerevisiae*. *Archiv für Mikrobiologie* **93**, 1–12.

UNO, I. & ISHIKAWA, T. (1973). Metabolism of adenosine 3′,5′-cyclic monophosphate and induction of fruiting bodies in *Coprinus macrorhizus*. *Journal of Bacteriology* **113**, 1249–55.

UNO, I., YAMAGUCHI, M. & ISHIKAWA, T. (1974). The effect of light on fruiting body formation and adenosine 3′,5′-cyclic monophosphate metabolism in *Coprinus macrorhizus*. *Proceedings of the National Academy of Sciences, U.S.A.* **71**, 479–83.

VAN DEN ENDE, H. & STEGWEE, D. (1971). Physiology of sex in the *Mucorales*. *Botanical Reviews*, **37**, 22–36.

WARREN, C. O. & MULLINS, J. T. (1969). Respiratory metabolism in *Achlya ambisexualis*. *American Journal of Botany* **57**, 1135–41.

WARREN, C. O. & SELLS, B. H. (1971). Cellulase induction during standardised vegetative growth in *Achlya*. *Journal of General Microbiology* **67**, 367–9.

WERKMAN, B. A., & VAN DEN ENDE, H. (1973). Trisporic acid synthesis in *Blakeslea trispora*. Interaction between *plus* and *minus* mating types. *Archiv für Mikrobiologie* **90**, 365–74.

WERKMAN, B. A. & VAN DEN ENDE, H. (1974). Trisporic acid synthesis in homothallic and heterothallic Mucorales. *Journal of General Microbiology* **82**, 273–8.

WOLF, J. C. & MIROCHA, C. J. (1973). Regulation of sexual reproduction in *Gibberella zeae* (*Fusarium roseum* 'Graminearum') by F-2(Zearelenone). *Canadian Journal of Microbiology* **19**, 725–34.

YANAGISHIMA, N. (1969). Sexual hormones in yeast. *Planta (Berlin)* **87**, 110–8.

ZICKLER, H. (1952). Zur Entwicklungsgeschichte des Askomyceten *Bombardia lunata*. *Archiv für Protistenkunde* **98**, 1–71.

CHAPTER 13

The Biosynthesis of Terpenes and Steroids

N. J. McCORKINDALE

13.1 Biosynthesis of Acyclic Terpenoid Precursors

Fungal products derived wholly or in part from an isoprenoid chain (1.1) range from simple dimethylallyl compounds to complex polycyclic products, many with skeleta unique to fungal metabolism. Current understanding of

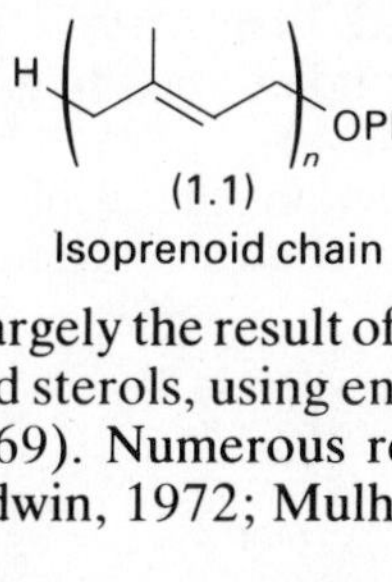

(1.1)
Isoprenoid chain

the isoprenoid pathway is largely the result of intensive investigations on the biosynthesis of squalene and sterols, using enzymes isolated from livers and yeasts (*cf. e.g.* Popjak, 1969). Numerous reviews have appeared on this subject (*e.g.* Goad & Goodwin, 1972; Mulheirn & Ramm, 1972; Packter, 1973).

The isoprenoid allylpyrophosphate chains are derived from mevalonic acid (1.2), usually formed from acetate, although leucine can also act as a source (Fig. 13.1). The enzymic transformation of mevalonic acid into the

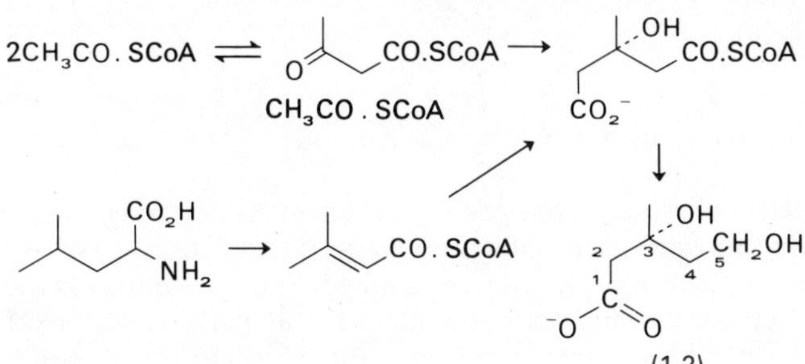

(1.2)

Fig. 13.1 Formation of mevalonate (1.2).

'chain-building unit' isopentenylpyrophosphate (1.3), and thence into the 'chain-initiating unit', dimethylallylpyrophosphate (1.4) are highly stereospecific and have been worked out in considerable mechanistic and stereochemical detail using stereospecifically tritiated and deuterated precursors (Clayton, 1965; Popjak & Cornforth, 1966; Cornforth, 1968, 1969, 1973; Popjak, 1970). The steps are outlined in Fig. 13.2.

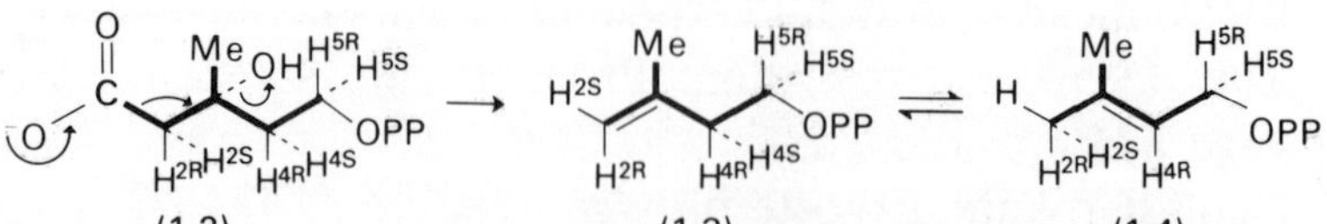

Fig. 13.2 Formation of IPP (1.3) and DMAPP (1.4) from mevalonate (1.2).

Extension of terpenoid allylpyrophosphate chains of the type (1.1) was further demonstrated in these systems to involve successive condensations with further molecules of isopentenylpyrophosphate (IPP) (Fig. 13.3).

Four observations in particular help to elucidate the stereochemistry of the bond-forming processes indicated in Fig. 13.2 and 13.3:

(a) In IPP, the hydrogen atom *trans* to the methyl group is derived from the 2-*pro*-R hydrogen of mevalonate (H^{2R}).

(b) In the allylpyrophosphates, the methyl group or methylene group *trans* to the —CH_2OPP group is derived exclusively from C-2 of mevalonate.

(c) The hydrogen atom of IPP derived from the 4-*pro*-S hydrogen of mevalonate (H^{4S}) is lost and that from 4-*pro*-R (H^{4R}) is retained in isomerization (Fig. 13.2) or condensation to give a higher allylic pyrophosphate (Fig. 13.3).

(d) In the condensation reaction, the displacement of pyrophosphate is accompanied by inversion (at the carbon atom corresponding to C-5 of mevalonate).

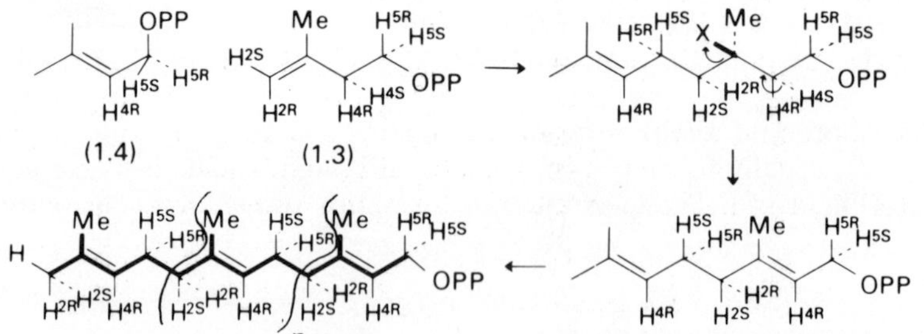

Fig. 13.3 Formation of terpenoid chains, *e.g.* farnesylpyrophosphate (1.5, $n = 1$).

The stereochemical changes observed in the condensation of IPP with allylic pyrophosphates is conveniently accounted for by a two-stage process: attack of a nucleophile X^- (perhaps an enzymic grouping) at the IPP double bond with concomitant *trans* alkylation by an allylic pyrophosphate (*cf.* Fig. 13.3), followed by a *trans anti-coplanar* elimination of XH^{4S} (Cornforth, 1968).

Recently, the stereochemistry of the protonation involved in the isomerization of IPP has been elucidated (Cornforth, 1973). Farnesylpyrophosphate was produced from (2R, 2-^{3}H) or (2S, 2-^{3}H) mevalonate using an enzyme fraction from pig liver prepared in deuterium oxide. The chirality of the (^{2}H, ^{3}H) doubly labelled methyl group so generated in the intermediate dimethylallylpyrophosphate (DMAPP) was preserved in the farnesylpyrophosphate formed. The latter was degraded to acetic acid, the chirality of which, determined by a special assay, confirmed the stereospecificity of the process but was opposite to that expected if isomerization of IPP proceeded by *trans* addition of XH to the double bond followed by *trans* elimination of XH^{4S}, *i.e.* by a process analogous to that involved in chain extension. The results are consistent with a straightforward concerted mechanism involving elimination of H^{4S} and protonation on the opposite side of the molecule.

Since the isomerization of IPP to DMAPP is reversible (Shah, Clelland & Porter, 1965; Holloway & Popjak, 1968) and since the reverse reaction involves loss of hydrogen from the methyl group of DMAPP derived from C-2 of mevalonate, the amount of deuterium or tritium incorporated into terpenoids from mevalonate deuterated or tritiated as well as ^{14}C labelled at C-2, is often somewhat less than would be expected from the ^{14}C incorporation. The extent of the loss of label depends on how quickly the IPP is utilised to initiate or extend the appropriate prenylpyrophosphate chain (*cf.* Cornforth, 1973).

The chain-extension process discussed above, gives, in turn, the all-*trans* acyclic prenylpyrophosphates, namely geranyl (C_{10}), farnesyl (C_{15}), geranylgeranyl (C_{20}) and geranylfarnesyl (C_{25}) pyrophosphate, which are the precursors of terpenes, steroids and carotenoids. Long chain prenylpyrophosphates (C_{35}, C_{40}, C_{45}, C_{50}) formed in the same way, appear to be involved in the biosynthesis of fungal ubiquinones (*e.g.* 1.6) (Stone &

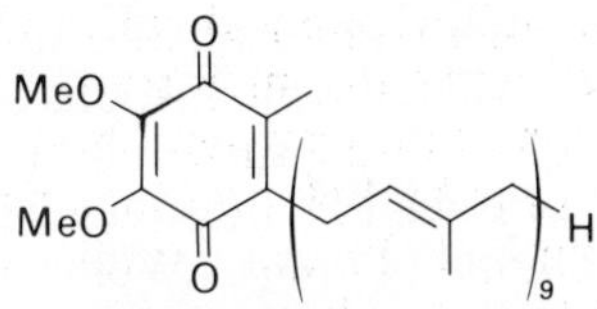

(1.6) Ubiquinone Q-9

Hemming, 1967; Dada, Threlfall & Whistance, 1968). This type of polyprenyl quinone is widely distributed, also occurring in plants, animals, bacteria and algae, and is involved in respiratory chain processes (Threlfall & Whistance, 1971).

Triterpenes and steroids are formed from farnesylpyrophosphate (1.5, $n = 1$) via the C_{30} hydrocarbon squalene (1.7), the two farnesyl units undergoing a head to tail condensation. Hypothetical mechanisms for this process (*cf.* Mulheirn & Ramm, 1972) had to account for the hydrogen atoms at C-1 of the two farnesylpyrophosphate molecules having the fate shown in Fig. 13.4.

At one of these centres, inversion occurs as in the chain extension process. In the other C-1 centre, one of the hydrogen atoms (the *pro*-R hydrogen derived from 5-*pro*-R of mevalonate) retains its configuration while the

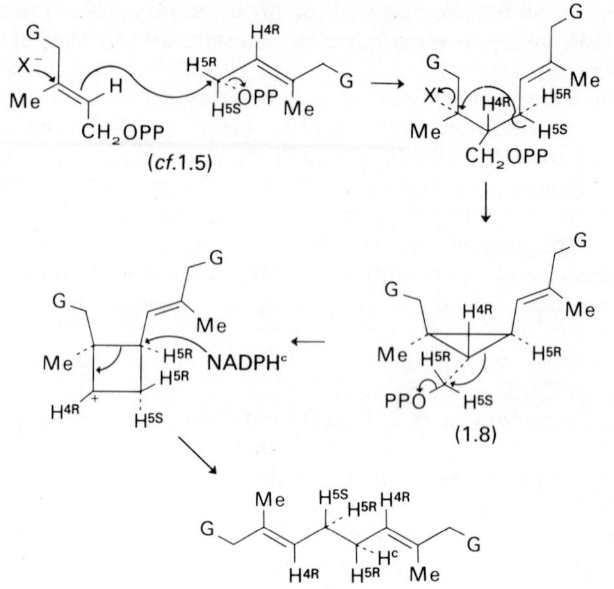

Fig. 13.4 Formation of squalene *via* presqualene pyrophosphate (1.8). G = geranyl.

other hydrogen atom is replaced stereospecifically by one from the co-enzyme NADPH (Popjak & Cornforth, 1966; Cornforth, Cornforth, Donninger & Popjak, 1965*a*; Donninger & Popjak, 1965).

Recently, the mechanism has been clarified with the discovery of an intermediate, presqualene pyrophosphate (1.8). In mammalian liver, the soluble enzyme fraction effects, in the presence of ATP, conversion of mevalonate into isopentenylpyrophosphate, and the isomerization and chain-extension processes, while the microsomal fraction, in the presence of NADPH effects conversion of farnesylpyrophosphate into squalene. If the latter system is deprived of NADPH, presqualene pyrophosphate accumulates (Rilling, 1970; Edmond, Popjak, Wong & Williams, 1971). This intermediate has similarly been obtained from yeast (Epstein & Rilling, 1970) and from bramble tissue culture (Heintz, Benveniste, Robinson & Coates, 1972). In the presence of NADPH, this is readily converted by the microsomal enzymes into squalene, completing the interrupted sequence from farnesyl pyrophosphate to squalene (Altman, Kowerski & Rilling, 1971). Adopting the revised RRR absolute stereochemistry recently proposed for the cyclopropane ring (Cornforth, 1973; *cf*. Altman, Kowerski & Rilling, 1971), a stereochemically satisfying mechanism can be advanced (Fig. 13.4) for the formation of squalene *via* presqualene pyrophosphate (*cf*. Edmond *et al*., 1971). The possibility that presqualene pyrophosphate might not, contrary to appearance be an obligatory intermediate, has however been discussed (Cornforth, 1973).

In the same way, geranylgeranyl pyrophosphate, incubated with a purified microsomal squalene synthetase in the absence of NADPH affords the C_{40} analogue of presqualene pyrophosphate (Qureshi, Barnes & Porter, 1972).

This compound which was called prephytoene pyrophosphate can be converted enzymically to the carotenoid precursor phytoene (Altman *et al.*, 1972).

One notable exception to the stereochemical pattern set out above which gives rise to prenyl chains and in turn various cyclic terpenoids, has been found in a group of polyprenols which have been associated with cell wall biosynthesis (Barr & Hemming, 1972). These have been isolated from plant and animal tissues and also from yeasts and fungi (Hemming, 1970). The dolichols isolated from *Aspergillus fumigatus* (*e.g.* 1.9) are hexahydropolyprenols containing 18–24 isoprenoid units in which the double bonds are mainly *cis*.

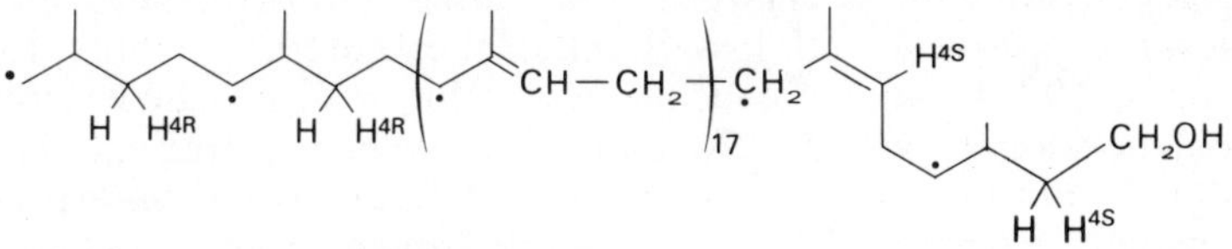

(1.9) Dolichols

The incorporation of (2-^{14}C, 4R-4^{3}H) and (2-^{14}C, 4S-4^{3}H) mevalonate into these polyprenols was studied and a remarkable coincidence found between the number of *cis* and *trans* double bonds and the number of 4-*pro*-S and 4-*pro*-R hydrogens respectively of mevalonate retained in the biosynthesis.

The formation of *cis* double bonds in rubbers with retention of 4-*pro*-S mevalonoid hydrogens (Archer *et al.*, 1965) was suggested (Popjak & Cornforth, 1966) to involve the same mechanism as leads to hydrogen elimination in the chain extension process (see Fig. 13.3) but with the substrate adopting a different orientation of the —CH_2 . CH_2OPP group, suitable for *trans* elimination of X and H^{4R}, thus giving a double bond which is *cis* instead of *trans*.

The biosynthesis of this type of polyprenol can be rationalized as an initial formation of a C_{15} or C_{20} prenyl pyrophosphate with 'biogenetically *trans*' double bonds (*i.e.* 4-*pro*-R mevalonoid hydrogens are retained in their formation) followed by modified chain extension giving a series of *cis* double bonds (Hemming, 1970).

As will be discussed later, the structures of many sesquiterpenes (*e.g.* trichodiene, 1.10) are such that they could plausibly be derived by cyclization of 2-*cis*-6-*trans* farnesyl pyrophosphate rather than all *trans* farnesyl

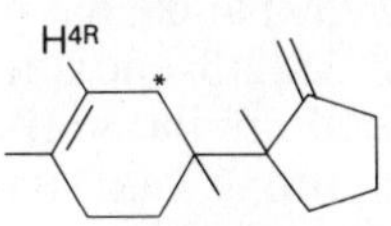

(1.10)
Trichodiene

pyrophosphate (Herout, 1971; Roberts, 1972; Rücker, 1973). Recently, however, the *cis* double bonds in nerol and 2-*cis*-6-*trans* farnesol produced by cell-free extracts from two plants were shown to be biogenetically *trans* since three hydrogen atoms derived from 4-*pro*-R of mevalonate were incorporated (Jedlicki, Jacob, Faini & Cori, 1972). Similar results were found for 2-*cis*-6-*trans* farnesol produced by cell-free extracts of *Andrographis paniculata* tissue culture (Overton & Roberts, 1973) and by cell-free extracts from the fungus *Trichothecium roseum* (Evans, Holton & Hanson, 1973). In the last case, the trichodiene (1.10) produced at the same time, also showed retention of the 4-*pro*-R mevalonoid hydrogen at the position shown corresponding to C-2 of farnesol.

The formation of this particular *cis* double bond can be explained by a mechanism which leaves the 4-*pro*-R mevalonoid hydrogen in question intact. Strong evidence that *trans* to *cis* isomerization of the corresponding unsaturated aldehyde (formed but transiently) is involved, has been provided using the fungus *Helminthosporium sativum* which was able to convert (1-^{2}H) geraniol and (1-^{2}H) *trans, trans*-farnesol to nerol and 2-*cis*-6-*trans*-farnesol respectively, with loss of half of the deuterium content (Suzuki & Marumo, 1972). Similar conclusions to this were reached using (2-^{14}C, 5-$^3H_2$) mevalonate with the *Trichothecium roseum* cell-free extracts (Evans, Holton & Hanson, 1973) and with the *Andrographis paniculata* preparation (Overton & Roberts, 1973, 1974), six hydrogens from C-5 of mevalonate being incorporated into *trans, trans*-farnesol and only five into the *cis,trans*-isomer. The trichodiene (1.10) produced by *T. roseum* in this way, proved to have lost a tritium atom from the starred position corresponding to the 1-position of farnesol (Evans, Holton, & Hanson, 1973). Evidence that this corresponds specifically to the 5-*pro*-S mevalonoid hydrogen atom was provided (Evans, Hanson & Marten, 1974) by the retention of all three 5-*pro*-R mevalonoid hydrogens of farnesyl pyrophosphate during the biosynthesis of trichothecanes, the products of further metabolism of trichodiene. This loss of hydrogen from the C-1 position of farnesol may prove to be a useful probe in deciding whether the formation of a particular sesquiterpene from farnesylpyrophosphate proceeds via the *cis,trans*-isomer. It remains to be seen which particular hydrogen is lost in further cases.

13.2 The Primary Cyclization Products of Squalene Oxide

The biosynthetic pathways leading to the cyclic terpenoids of different organisms seem to follow the same route to the intermediate prenyl pyrophosphates and squalene. Significant differences only become apparent in the processes which follow cyclization. It was demonstrated by ^{18}O labelling that the oxygen atom in lanosterol is derived from oxygen rather than water (Tchen & Bloch, 1957) and it has now been fully established that oxidation of squalene to the 3S-2,3-epoxide precedes cyclization to tetracyclic triterpenes (Dean, de Montellano, Bloch & Corey, 1967; Willett *et al.*, 1967; Stone, Roeske, Clayton & van Tamelen, 1969).

Proton-initiated cyclization of suitably orientated squalene oxide (2.1) as indicated in Fig. 13.5, is thought to give the key intermediate cation (2.2) or an enzyme-stabilized equivalent, from which, by different modifications, the three main prototypes of tetracyclic triterpene, cycloartenol (2.3), lanosterol (2.4) and protosta-17(20)Z,24-dien-3β-ol (2.5), are formed. Cycloartenol is established as the precursor of plant and algal sterols (Hall, Smith, Goad & Goodwin, 1969; Hewlins, Ehrhardt, Hirth & Ourisson, 1969) and is formed from squalene 2,3-oxide by, *inter al.*, (Goad & Goodwin, 1972),

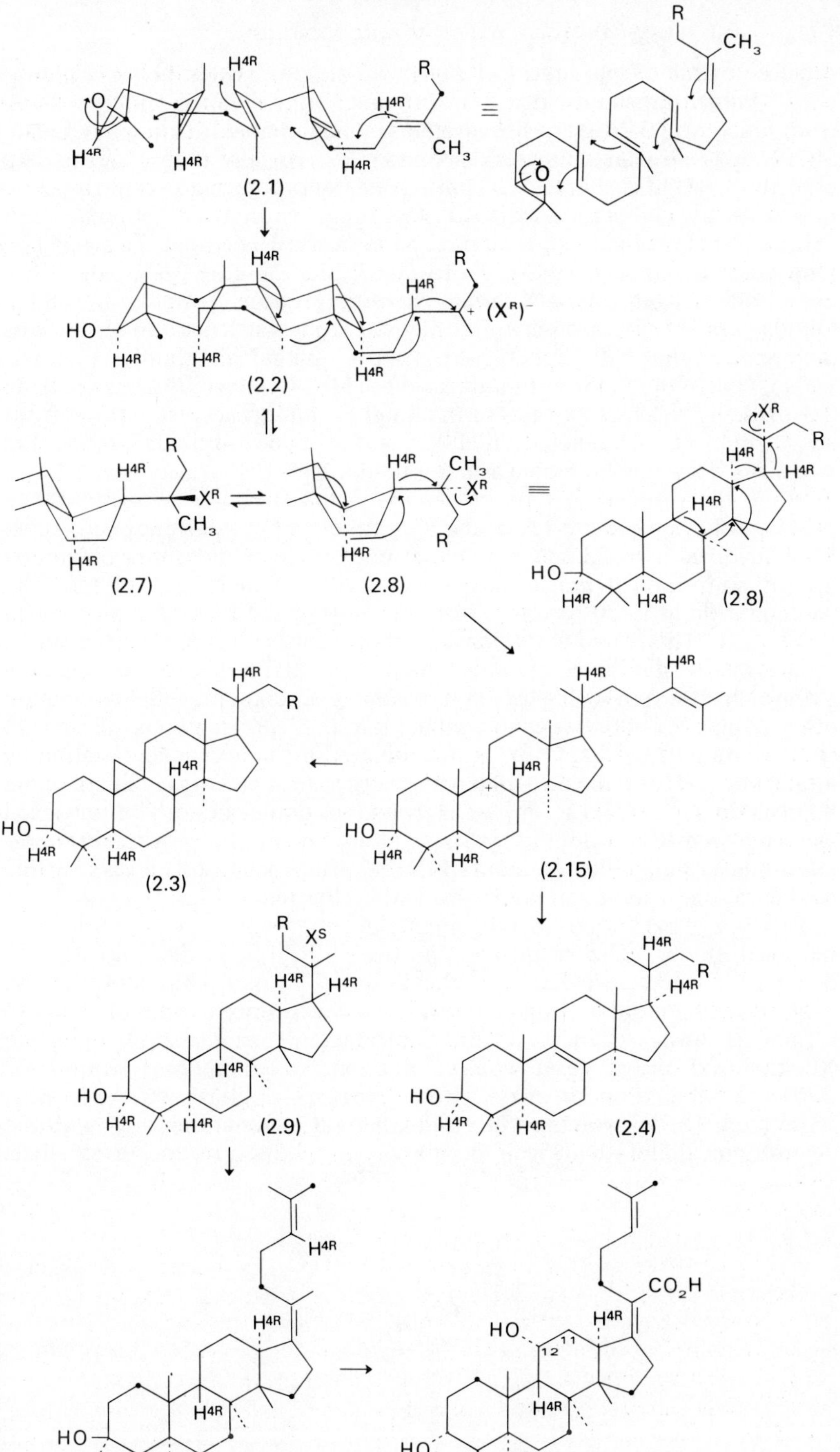

Fig. 13.5 Cyclization of squalene oxide.

cell-free extracts from leaves of *Phaseolus vulgaris* (Rees, Goad & Goodwin, 1968*a*). Lanosterol is the precursor of animal sterols and is formed from squalene 2,3-oxide by liver systems (Yamamoto & Bloch, 1970; Goad, 1970); it is also an obligatory precursor of typically fungal sterols like ergosterol (Akhtar, Parvez & Hunt, 1968) and is formed from squalene 2,3-oxide by cell-free extracts of the fungi *Phycomyces blakesleeanus* (Mercer & Johnson, 1969) and *Cephalosporium caerulens* (Kawaguchi, Kobayashi & Okuda, 1973). Protosta-17(20)Z,24-dien-3β-ol which has been isolated from *Cephalosporium caerulens* (Hattori, Igarashi, Iwasaki & Okuda, 1969) is the probable intermediate from which are derived various members of the $\Delta^{17(20)}$ prostanoid group of fungal metabolites (Turner, 1971). One of these, the antibiotic fusidic acid (2.6) is formed from squalene 2,3-oxide by cell-free extracts of the fungi *Fusidium coccineum* (Godtfredson, Lorck, van Tamelen, Willet & Clayton, 1968) and *Cephalosporium caerulens* (Kawaguchi, Kobayashi & Okuda, 1973).

While the protostane type of triterpene has the same skeleton as the protosterol intermediate (2.2), the biosynthesis of cycloartenol and lanosterol involves a backbone rearrangement by concerted shifts of methyl groups and three 4-*pro*-R mevalonoid hydrogen hydrogen atoms. The configuration at C-20 resulting from the shift of the hydrogen atom from C-17 is R. This can be rationalized if the cation formed by the initial cyclization is stabilized by attachment at C-20 of a grouping X^R (*cf.* Cornforth, 1968) in such a way that, assuming X^R takes precedence over the other groups at C-20, this centre will be R as in (2.7). This would also be the configuration at C-20 if $(X^R)^-$ in fact initiated the concerted cyclization by attack at C-20 from the direction which sees a clockwise arrangement of the main chain, H^{4R}, —CH_2R and —CH_3 round the double bond. The concerted shifts of 4-*pro*-R mevalonoid hydrogen atoms and methyl groups, terminating in nucleophilic displacement of the X^R group as in (2.8) gives the 20R configuration observed in lanosterol and cycloartenol (Fig. 13.5).

This type of backbone rearrangement does not occur in the biosynthesis of the protostanes. The prototype, protosta-17(20)Z,24-dien-3β-ol (2.5) could be formed by collapse of the cation (2.2) with elimination of the 4-*pro*-R mevalonoid hydrogen at C-17. However, since the double bond so formed is always Z in these fungal products (Turner, 1971), the same X^R-stabilized cation (2.8) would not seem to be involved since *trans anti-coplanar* elimination of HX^R would give an E double bond. An indirect route from (2.7) involving *trans anti-coplanar* elimination of X^R with the 2-*pro*-S mevalonoid hydrogen atom at C-22, followed by concerted allylic

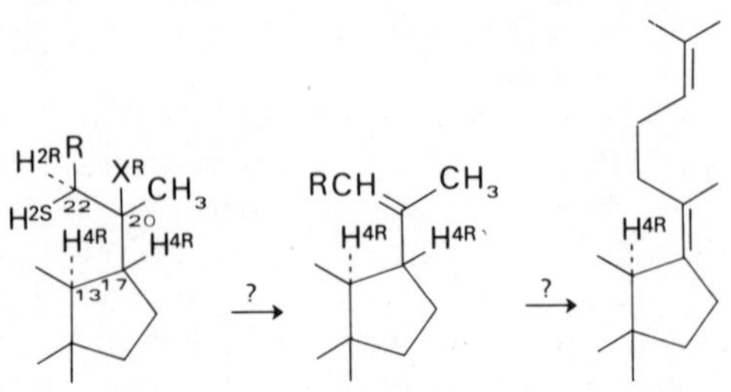

Fig. 13.6 Formation E 17(20) double bonds in protostanes: a disproved postulate.

rearrangement (Fig. 13.6) has been discounted (Caspi, Ebersole, Godtfredson & Vangedal, 1972) since it was demonstrated that no loss or configurational change of the hydrogen atoms at C-22 occurs in the formation of fusidic acid (2.6). The 4-*pro*-R mevalonoid hydrogen at C-13 was shown to be intact in fusidic acid as was that at C-9 (Caspi & Mulheirn, 1970). Nucleophilic displacement of X^R in (2.8) by a second grouping X^S or stabilization of the cation (2.2) by attachment at C-20 of a grouping X^S in a manner such as to give a 20S configuration as in (2.9), followed by *trans anti-coplanar* elimination of HX^S might explain the formation of the Z-17(20) double bond in these compounds (*cf.* Mulheirn & Caspi, 1971).

Fusisterol (2.10), which is formed by *Cephalosporium caerulens* (Hattori *et al.*, 1969) and by *Fusidium coccineum* (Arigoni, 1968), has an R configuration at C-20 like lanosterol and cycloartenol. This could be formed from the X^R-stabilized cation (2.8) by a concerted process (Fig. 13.7*a*) limited to loss of the hydrogen atom at C-13 and migration of the 4-*pro*-R mevalonoid hydrogen atom at C-17 to C-20 with displacement of X^R. An alternative would be enzyme mediated allylic rearrangement of a $\Delta^{17(20)}$ protostane (Fig. 13.7*b*).

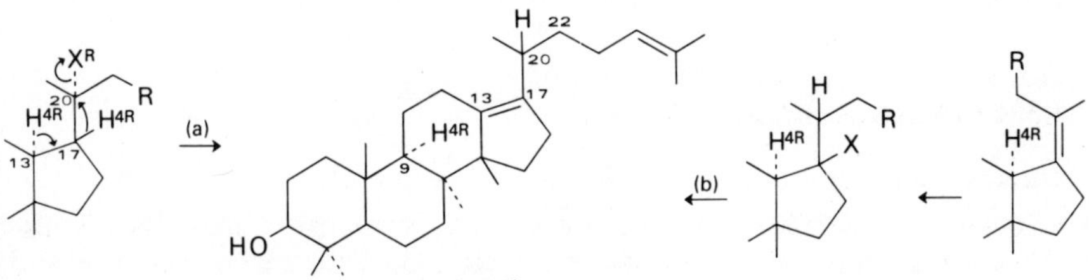

(2.10)

Fig. 13.7 Possible formation of fusisterol (2.10).

With acid, 24,25-dihydrofusisterol acetate smoothly underwent shifts of hydride and methyl groups to afford 24,25-dihydrolanosterol (2.11) (Arigoni, 1968). This type of model for the *in vivo* processes leading to the triterpenes has been provided by non-enzymic and enzymic cyclizations of modified squalene oxides, *e.g.* the oxide (2.12) cyclizes with Lewis acid as shown in Fig. 13.8 to the 24,25-dihydro-$\Delta^{13(17)}$-protosterol (2.13) and the dihydrolanosterol isomer 24,25-dihydroparkeol (2.14), which isomerizes to dihydrolanosterol with mineral acid (van Tamelen & Anderson, 1972). The oxide (2.12) with 2,3-oxido-squalene cyclase afforded 24,25-dihydrolanosterol itself (van Tamelen & Freed, 1970).

The cycloartenol and lanosterol pathways represent different ways in which the backbone-rearranged cation (2.15) or its stabilized equivalent can collapse. In the formation of lanosterol (2.4), loss of the 4-*pro*-R mevalonoid hydrogen from C-8 gives the 8,9 double bond (Cornforth *et al.*, 1965*b*). In the formation of cycloartenol (2.3), loss of a hydrogen atom from the C-19 methyl group gives a cyclopropane ring and leaves the 4-*pro*-R mevalonoid hydrogen at C-8 intact (Rees, Goad & Goodwin, 1968*b*).

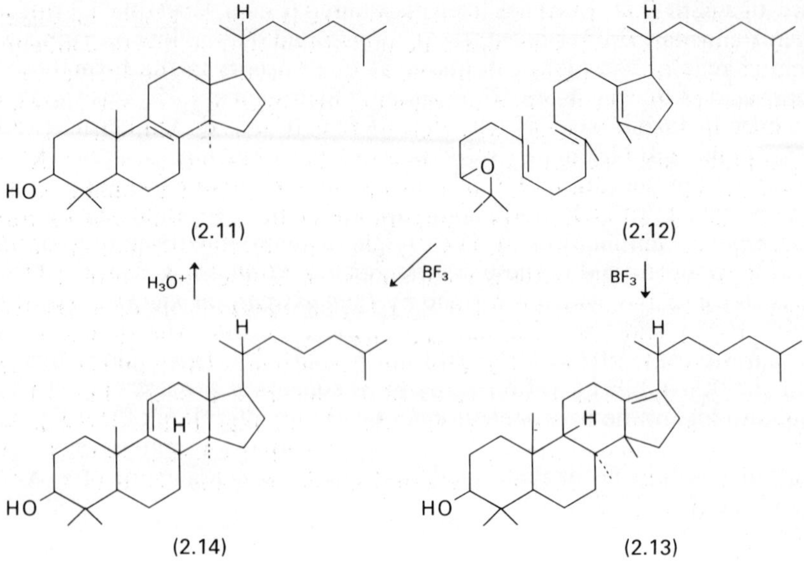

Fig. 13.8 *In vitro* model for cyclizations of squalene oxide.

13.3 Cholesterol and Ergosterol

Pathways Leading from Triterpene Alcohols to Sterols

Cycloartenol and lanosterol give rise to Δ^5-4-desmethyl sterols like cholesterol and ergosterol by several stages affecting both the carbon skeleton and the hydrogenation pattern. The carbon skeleton is altered by oxidative removal of methyl groups at C-4 and C-14 and in some cases by alkylation at C-24. In the case of cycloartenol, opening of the cyclopropane ring to give a Δ^8-19-methylsterol occurs at an early stage. The hydrogenation pattern is altered at several points by saturation or desaturation and one double bond migration occurs from Δ^8 to Δ^7. In addition, 24-alkylation and removal of the methyl groups occur by special mechanisms which affect hydrogens on adjacent carbons.

The criteria for intermediacy of (a) natural occurrence (established in some cases by trapping experiments) and (b) bioconvertibility to the end-product, have been met by large numbers of sterols and considerable effort has gone into attempts to define an order in the transformations involved between the primary cyclization products lanosterol and cycloartenol and the end-products ergosterol and cholesterol (see Goad & Goodwin, 1972; Barton, Corrie, Marshall & Widdowson, 1973). However there has been an increasing awareness of the lack of specificity of the enzymes involved (Fryberg, Oehlschlager & Unrau, 1973; Schroepfer *et al.*, 1972; Goad & Goodwin, 1972). This was highlighted by evidence obtained using four mutants of *Saccharomyces cerevisiae.* Although lacking the ability to effect a specific stage between lanosterol and ergosterol, namely C-24 methylation, Δ^8 to Δ^7 isomerization, 5,6-dehydrogenation or 22,23-dehydrogenation, each mutant was found to accumulate a series of sterols each of which was the result of a different selection and/or order of the permitted transformations, showing that these can occur independently of each other (Barton, Corrie, Widdowson, Bard & Woods, 1974). Similarly, numerous pathways have been shown to exist between lanosterol and cholesterol in animal systems (Schroepfer *et al.*, 1972) and between cycloartenol and phytosterols (Goad & Goodwin, 1972).

One of the possible routes from lanosterol to ergosterol *via* naturally occurring sterols is illustrated in Fig. 13.9. The particular sequence of processes in this route is (a) 24-alkylation; (b) 14α-demethylation; (c) 4-demethylation (loss of C-30); (d) 4-demethylation (loss of C-31); (e) Δ^8 to Δ^7 isomerization; (f) 24(28)-hydrogenation; (g) 5,6-dehydrogenation, and (h) 22,23-dehydrogenation. Table 13.1, however, lists some further

Table 13.1 4-Desmethyl sterols from yeast not included in Fig. 13.9

Ergostenols	Δ^5, Δ^8
Ergostadienols	$\Delta^{5,24(28)}$, $\Delta^{7,22}$, $\Delta^{8,22}$
Ergostatrienols	$\Delta^{5,7,14}$, $\Delta^{5,7,24(28)}$, $\Delta^{8,14,24(28)}$
Ergostatetraenols	$\Delta^{5,7,14,22}$, $\Delta^{5,7,22,24(28)}$
Cholestadienols	$\Delta^{8,24}$
Cholestatrienols	$\Delta^{5,7,24}$, $\Delta^{7,22,24}$, $\Delta^{8,22,24}$

potential intermediates at the 4-desmethylsterol level alone. These have been isolated from yeast and in some cases they can be converted enzymically into ergosterol. Comprehensive schemes to include all the potential precursors of ergosterol (including C_{30} and C_{28} sterols) appear as a network of interlinking pathways (*cf.* Barton *et al.*, 1973; Weete, 1973). The same pattern applies to cholesterol biosynthesis (Schroepfer *et al.*, 1972). The relative importance of different pathways and timing of individual stages are hard to gauge and may well vary with different organisms.

24-Alkylation

Although the enzymes involved are by no means specific (Russell, van Aller & Nes, 1967), 24-alkylation appears to be one of the earliest processes in the formation of 24-substituted sterols, except perhaps in yeast (Barton *et al.*, 1973; Fryberg, Oehlschager & Unrau, 1973). Alkylation at C-24 is known to require a 24,25 double bond (Russell, van Aller & Nes, 1967) and to involve transfer of the S-methyl group of S-adenosyl methionine (Jaureguiberry, Law, McLoskey & Lederer, 1965; Moore & Gaylor, 1969). The formation of 24-methyl groups has been shown to proceed *via* 24-methylene intermediates in yeast (Akhtar, Parvez & Hunt, 1966, 1969). This is reflected in the observed retention of only two deuterium atoms from ($^2H_3$-methyl) methionine in ergosterol produced by various fungi, for example *Polyporus sulphureus* (Jaureguiberry *et al.*, 1965), *Phycomyces blakesleeanus* (Goulston, Goad & Goodwin, 1967), and *Oospora virescens* (Varenne, Polonsky, Bellavita & Ceccherelli, 1971). It was also shown that the 4-*pro*-R mevalonoid hydrogen migrated from C-24 to C-25 during the formation of ergosterol in yeast (Akhtar, Hunt & Parvez, 1967). This may

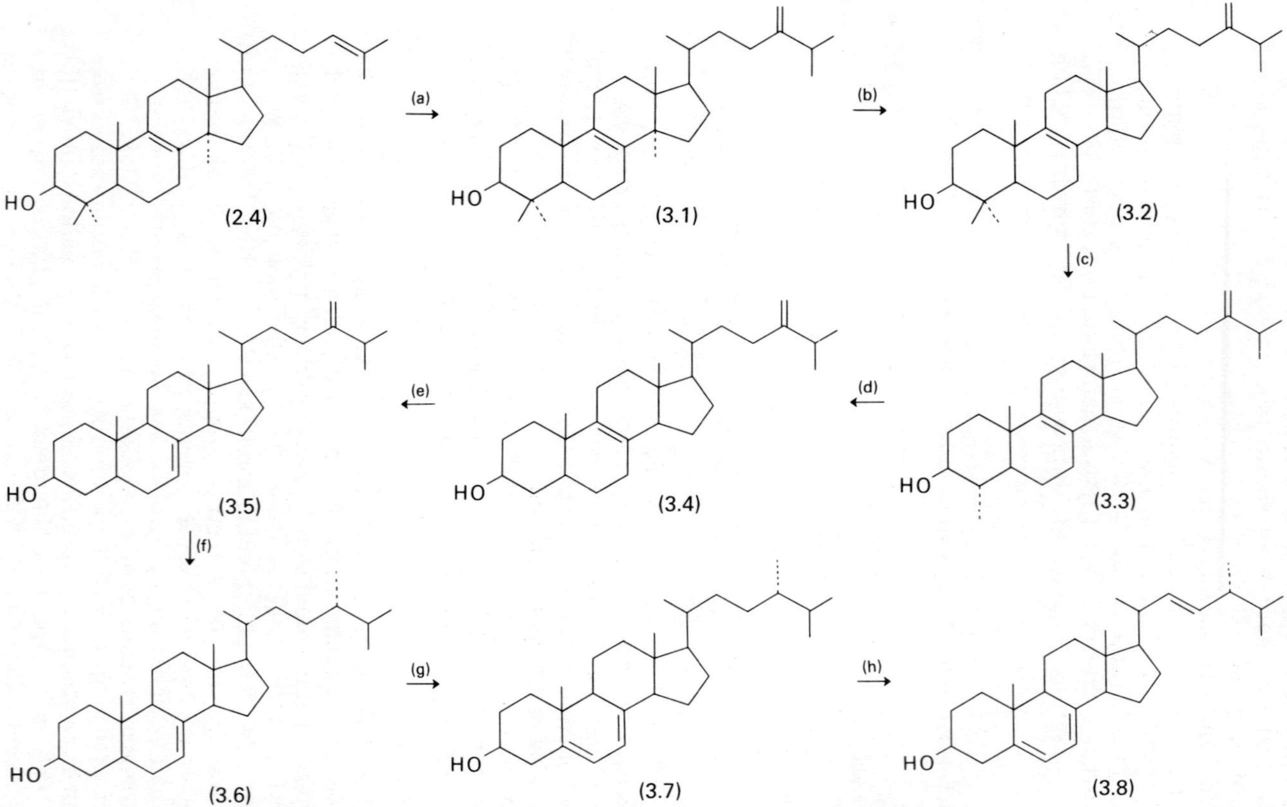

Fig. 13.9 One route between lanosterol (2.4) and ergosterol (3.8) *via* established intermediates.
Sterols: 24-methylenelanosterol (3.1); 4,4-dimethylergosta-$\Delta^{8,24(28)}$-dienol (3.2); 4α-methylergosta-$\Delta^{8,24(28)}$-dienol (3.3); ergosta-$\Delta^{8,24(28)}$-dienol (3.4); ergosta-$\Delta^{7,24(28)}$-dienol (3.5); ergosta-Δ^{7}-enol (3.6); ergosta-$\Delta^{5,7}$-dienol (3.7).
Processes: (a) 24-alkylation; (b) 14α-demethylation; (c) 4-demethylation (loss of C-30); (d) 4-demethylation (loss of C-31); (e) $\Delta^{8} \rightarrow \Delta^{7}$ isomerization; (f) 24(28) hydrogenation; (g) 5,6-dehydrogenation; (h) 22,23-dehydrogenation.

also be inferred to occur during the formation of ergosterol in the fungus *Aspergillus fumigatus* as indicated by the number of 4-*pro*-R mevalonoid hydrogens retained (Stone & Hemming, 1967). These two observations are accommodated in the mechanism shown in Fig. 13.10 (*cf.* Lederer, 1969).

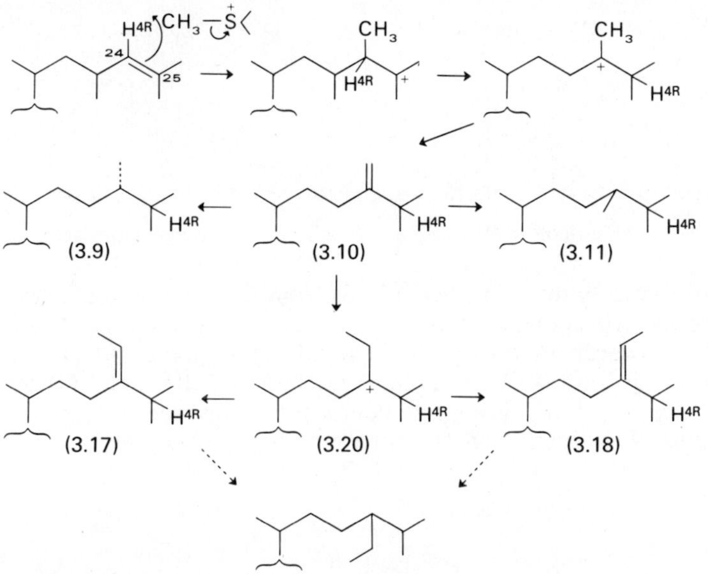

Fig. 13.10 Mechanisms of alkylation of the sterol side chain.
Sterol side chains: ergosta-$\Delta^{5,7}$-dienol (3.9); 24-methylenelanosterol or 24-methylenecholesterol (3.10); campesterol (3.11); fucosterol (3.17); isofucosterol (3.18).

Apart from 24-methylene lanosterol (3.1) which has been isolated from *Phycomyces blakesleeanus* and other fungi (Goulston, Goad & Goodwin, 1967), modified lanostanes occur in a number of Basidiomycetes (*cf.* Turner, 1971). Almost all of the fungal C_{31} triterpenes have a 24-methylene group, many being either 21-oic acids, *e.g.* hydroxymethylglutaryl polyporenic acid A (3.12) from *Piptoporus betulinus* (Bryce, Campbell & McCorkindale, 1967) or 27-oic acids, *e.g.* eburicoic acid (3.13) found in a number of Polypores including *Polyporus sulphureus* (*cf.* Turner, 1971).

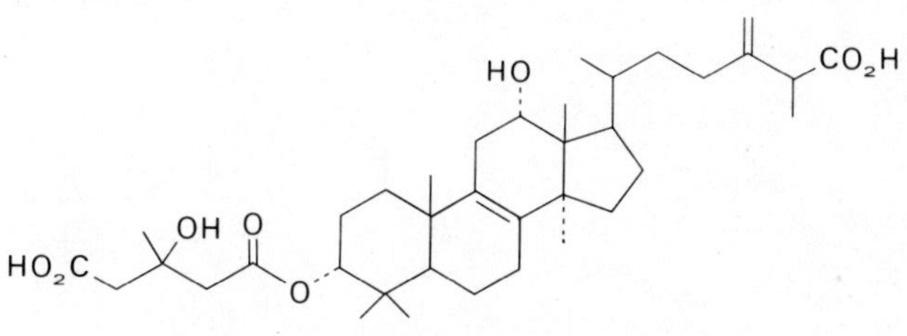

(3.12)
Hydroxymethylglutaryl polyporenic acid A

Cultures of the latter were able to convert both 24-methylenedihydrolanosterol (3.1) (Barton, Harrison, Moss & Widdowson, 1970) and trametanolic acid (3.14) (Devys & Barbier, 1969) into eburicoic acid, demonstrating the lack of specificity of the enzymes involved.

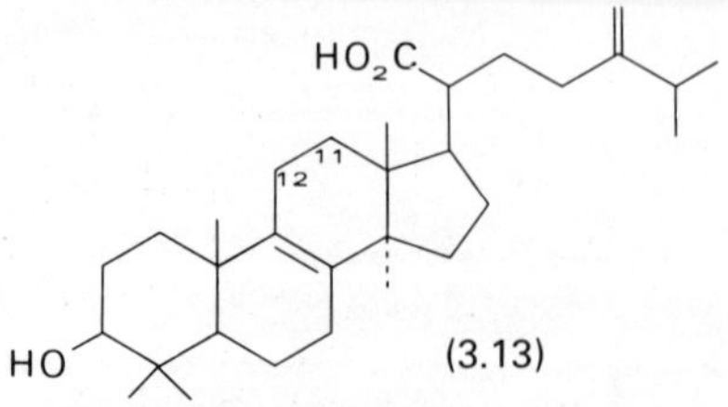

(3.13)

Eburicoic acid

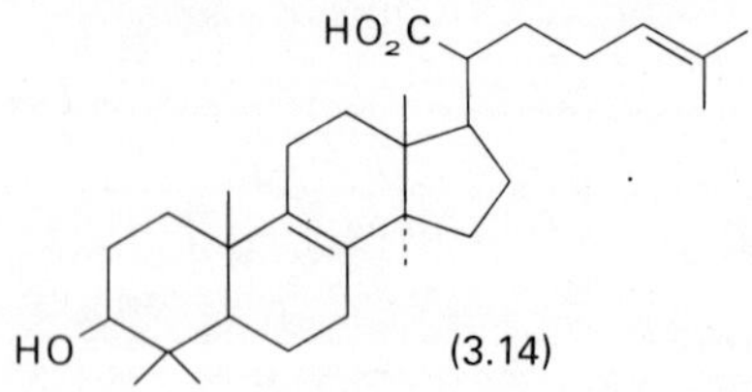

(3.14)

Trametanolic acid

Carboxyacetylquercinic acid (3.15) from *Daedalea quercina* was the first 24-methyl triterpenoid to be isolated from fungal sources (Adam *et al.*, 1967). However, Polonsky and Richroch (*cf.* Lederer, 1969) found that in biosynthesis from (^{14}C,$^{2}H_3$-methyl)-methionine, although ^{14}C was incorporated into the methyl group, no deuterium atoms were retained, equilibration possibly occurring *via* the corresponding butenolide (3.16).

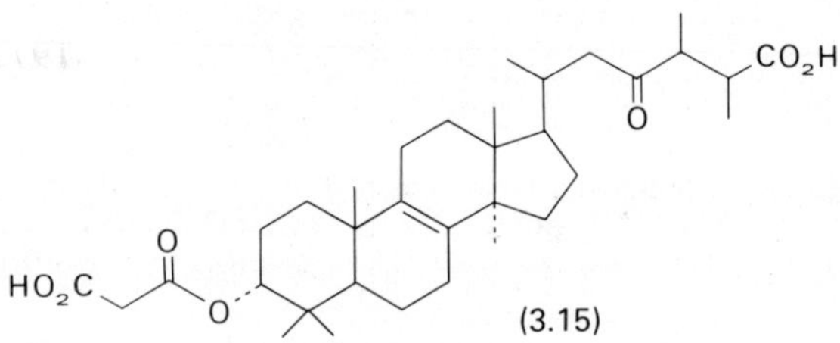

(3.15)

Carboxyacetylquercinic acid

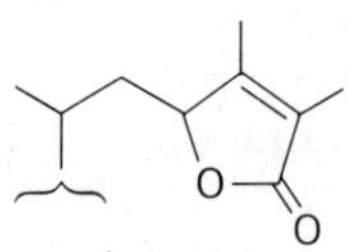

(3.16)

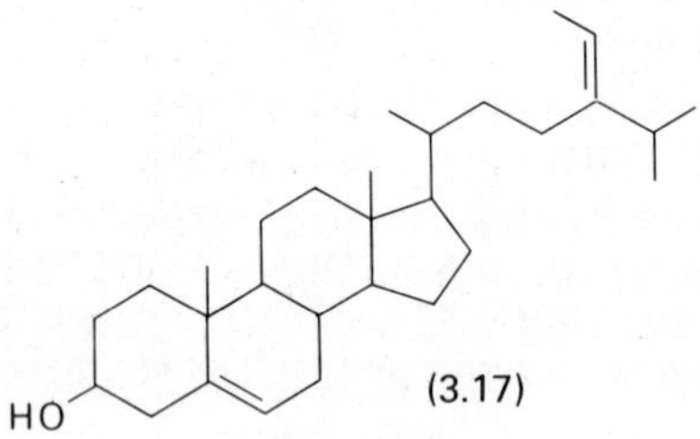

(3.17)

Fucosterol

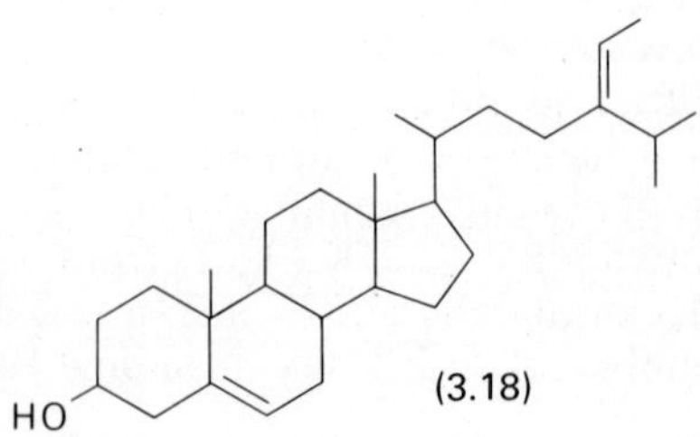

(3.18)

Isofucosterol

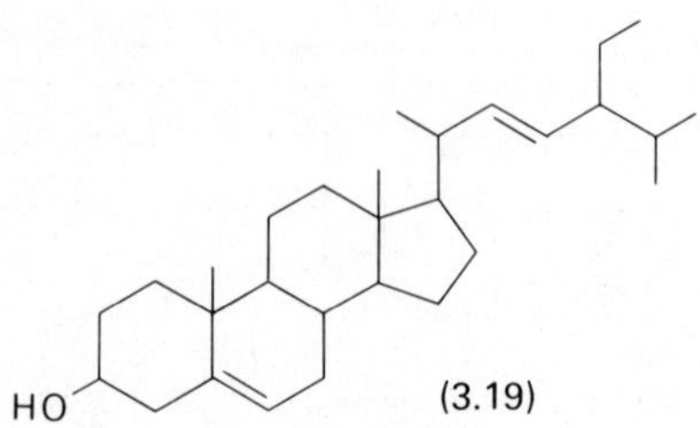

(3.19)

Poriferasterol

No 24-ethylidene triterpenes have been reported from fungal or other sources, but the (E)24-ethylidene sterol fucosterol (3.17), which is common in brown algae, has been found in a number of fungi of the Oömycetes class (McCorkindale, Hutchinson, Pursey, Scott & Wheeler, 1969). The addition of the second carbon is thought to proceed, after removal of one or more nuclear methyl groups, by alkylation of a 24-methylene intermediate (3.10) with *S*-adenosyl-methionine as in Fig. 13.10. In keeping with this, fucosterol produced by the marine brown alga *Fucus spiralis*, was shown to retain three 4-*pro*-R mevalonoid hydrogens, one being located at C-25 (Goad & Goodwin, 1969), and to have the expected 4 : 2 tritium to carbon ratio when biosynthesized from (^{14}C,$^{3}H_3$-methyl)methionine (Goad *et al.*, 1966).

Although fucosterol (3.17) rather than its Z-isomer, isofucosterol (3.18) is accumulated by algae, isofucosterol was more efficiently converted into poriferasterol (3.19) by the alga *Ochromonas malhamensis* (Knapp, Greig, Goad & Goodwin, 1971) and could thus be a dynamic intermediate in the formation of this 24-ethyl sterol. Isofucosterol, which appears to be the isomer produced by plants, was shown to be derived from 24-methylenecholesterol in *Pinus pinea* seeds (van Aller, Chikamatsu, de Souza, John & Nes, 1969), and to have the 'shifted' 4-*pro*-R mevalonoid hydrogen at C-25 (Raab, de Souza & Nes, 1968).

A variant on the mechanisms of methylation shown in Fig. 13.10, is encountered in some algae. Thus all three hydrogens from the methyl group of methionine are retained in the 24-methyl group of ergost-Δ^7-enol produced by the green alga *Chlorella vulgaris* (Tomita, Uomori & Minato, 1970), and in that of ergost-Δ^5-enol produced by a *Trebouxia* species, the algal symbiont of a lichen (Goad, Knapp, Lenton & Goodwin, 1972). Evidently, reduction of the intermediate cation (3.20) occurs, rather than stabilization *via* formation of a 24(28) double bond.

4-Demethylation

The occurrence of 4α-methylsterol intermediates in the biosynthesis of cholesterol and ergosterol has been known for some time (*cf.* Schroepfer *et al.*, 1972). However, in one of the few cases where one terminal process invariably seems to precede another, the 4α-methyl group derived from C-2 of mevalonate in fact is the first to be removed, the sequence of reactions involved being as outlined in Fig. 13.11.

In the rat liver system, hydroxylation of the 4α-methyl group of lanosterol, followed by stepwise oxidation gives a 4α-methyl-4-carboxy-3-sterol (3.21) (Miller & Gaylor, 1970*a*; Hornby & Boyd, 1970). Decarboxylation of the corresponding 4-carboxy-3-ketone (3.22) (Sharpless *et al.*, 1968) is followed by inversion of the methyl group remaining into the α-configuration (Sharpless *et al.*, 1969; Rahman, Sharpless, Spencer & Clayton, 1970). 4α-Demethylation of cycloartenol (2.3) by the fern *Polypodium vulgare* was shown to follow the same course, giving the corresponding 4α-methyl sterol (Ghisalberti, de Sousa, Rees, Goad & Goodwin, 1969).

Removal of the second methyl group from C-4 may well proceed in an analogous manner (Fig. 13.11). Thus oxidation to a 4α-carboxylic acid (3.23) in a rat liver system has been demonstrated (Miller & Gaylor, 1970*b*). Also, in the conversion of (2,2,4-$^{3}H_3$) obtusifoliol (3.24) into poriferasterol

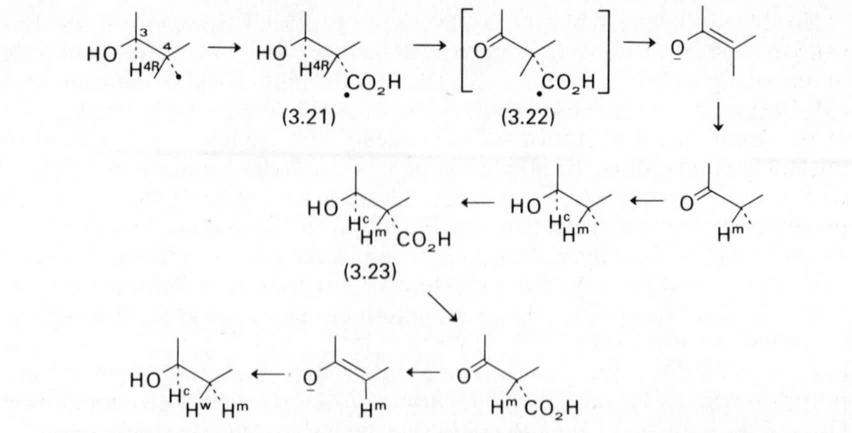

Fig. 13.11 4-Demethylation of sterols. • Labelling from (2-^{14}C)mevalonate; H^{4R}: hydrogen derived from 4-*pro*-R of mevalonate; H^c: hydrogen from NADPH; H^m and H^w: hydrogen from the medium.

(3.25) the 3α-hydrogen is lost, and, following the decarboxylation, the hydrogen entering at C-4 adopts an axial 4β-configuration (Knapp, Goad & Goodwin, 1973).

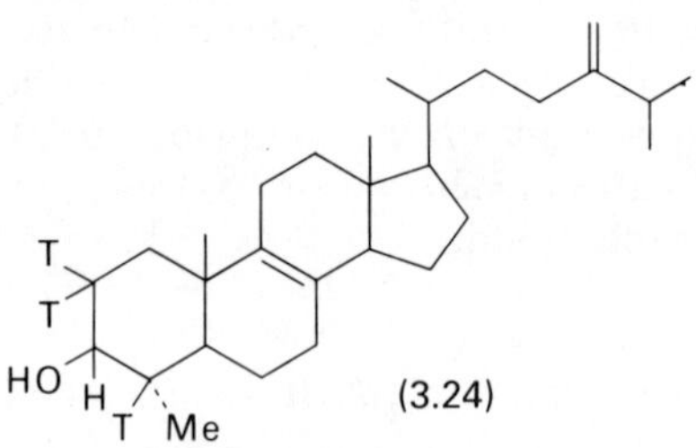

(3.24)

(2,2,4-$^3H_3$) obtusifoliol

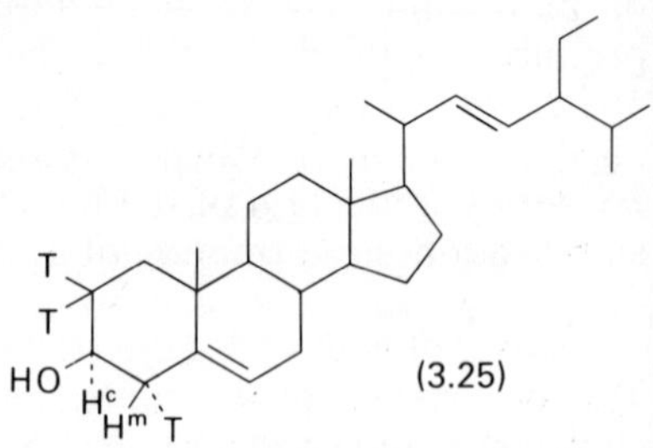

(3.25)

Poriferasterol

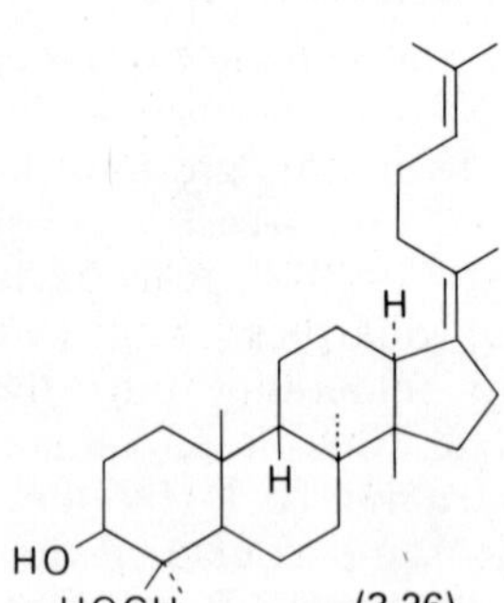

(3.26)

Diol

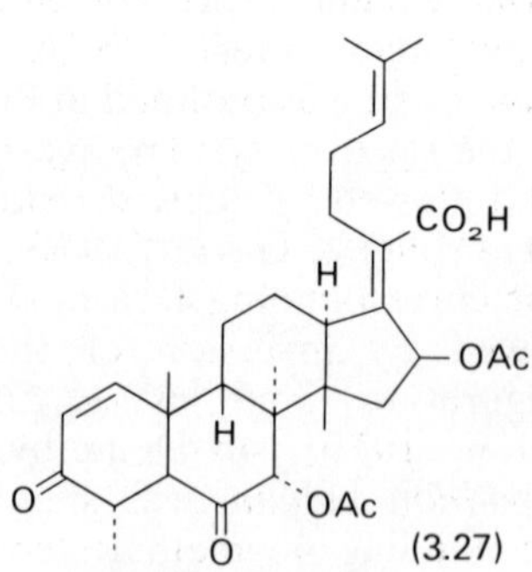

(3.27)

Helvolic acid

By contrast with the 4-demethylation of the backbone-rearranged sterols, the formation of fusidic acid (2.6) from the first cyclized 4,4-dimethylsterol (2.5) involves loss of the 4β-methyl group, the remaining methyl group (which is derived from C-2 of mevalonate) retaining its 4α-configuration (Caspi & Mulheirn, 1970). A probable intermediate in this process, the diol (3.26) was isolated from the fungus *Cephalosporium caerulens* and shown to be converted into helvolic acid (3.27) in this organism (Okuda *et al.*, 1968).

14α-Demethylation

Removal of the 14α-methyl group in the formation of cholesterol by liver microsomes, has been shown to occur *via* introduction of a 14,15 double bond for both Δ^8 and Δ^7 precursors (Alexander, Akhtar, Boar, McGhie & Barton, 1971). The carbon atom at C–14 has been shown to be eliminated as formic acid rather than carbon dioxide as in the case of the carbon atoms at C-4, and a mechanism (Fig. 13.12) has been proposed, involving elimination of the elements of formic acid from a 15α-hydroxy-14α-formyl intermediate (3.28) (Alexander, Akhtar, Boar, McGhie & Barton, 1972). This mechanism was preferred to an alternative involving desaturation of a $\Delta^{8(14)}$ intermediate (Schroepfer *et al.*, 1972) since $\Delta^{8(14)}$ sterols do not appear to have an intermediary role (Alexander *et al.*, 1971).

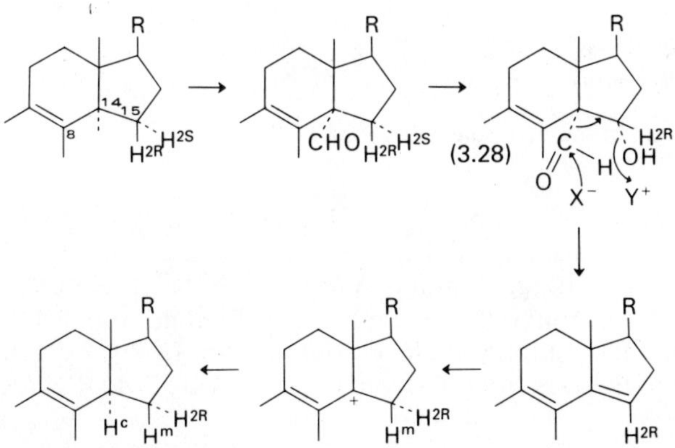

Fig. 13.12 14α-Demethylation of sterols. H^m: hydrogen from the medium; H^c: hydrogen from NADPH.

Evidence has been provided that the 15β (2-*pro*-R mevalonoid) hydrogen of lanosterol is retained and the 15α (2-*pro*-S mevalonoid) hydrogen lost in cholesterol formation by rat liver (Caspi, Ramm & Gain, 1969; Akhtar, Rahimtula, Watkinson, Wilton & Munday, 1969; Fiecchi *et al.*, 1972), and in ergosterol formation by yeast (Caspi & Ramm, 1969). In the liver system at least, the 14,15 double bond undergoes *trans* reduction, there being introduced a hydrogen atom (H^m) from the medium into the 15β-position and another (H^c) from the 4-position of NADPH into the 14α-position of cholesterol (Akhtar, Wilton, Watkinson & Rahimtula, 1972). The 2-*pro*-R mevalonoid hydrogen at C-15 thus undergoes inversion during 14α-demethylation.

Δ⁸ to Δ⁷ isomerization and 7,8 hydrogenation

An interesting stereochemical difference in the biosynthesis of ergosterol and cholesterol (shown in Fig. 13.13) concerns isomerization from Δ^8 to Δ^7.

This has been shown to result in loss of the 2-*pro*-S mevalonoid hydrogen from the 7β-position during the biosynthesis of cholesterol by rat liver (Caspi & Ramm, 1969; Fiecchi *et al.*, 1972) and of poriferasterol (3.19) by the alga *Ochromonas malhamensis* (Smith, Goad & Goodwin, 1968*a*), while loss of the 2-*pro*-R mevalonoid hydrogen from the 7α-position occurs in ergosterol biosynthesis by yeast (Caspi & Ramm, 1969) or by *Aspergillus fumigatus* (*cf.* Goodwin, 1973).

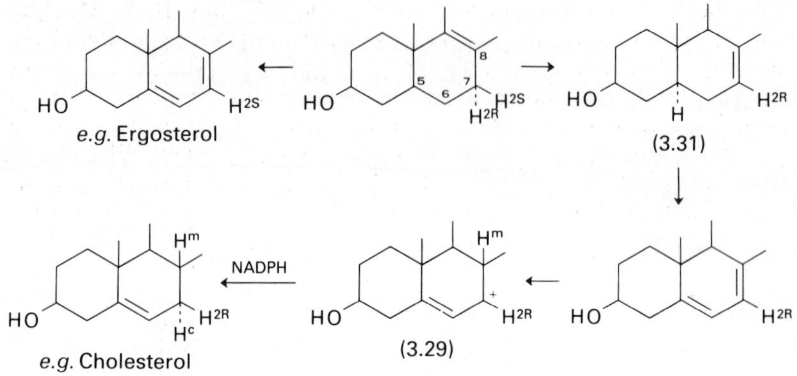

Fig. 13.13 Origin of the hydrogens at C-7 and C-8.

In cholesterol, the retained 2-*pro*-R mevalonoid hydrogen has its configuration inverted during the dehydrogenation-hydrogenation sequence Δ^7 to $\Delta^{5,7}$ to Δ^5 (Caspi, Greig, Ramm & Varma, 1968). The reduction occurs by addition of a proton $(H^m)^+$ from the medium to the 8β-position and reduction of the resulting allylic carbonium ion (3.29) by addition to the 7α-position of hydride ion (H^c) from the coenzyme NADPH (Akhtar *et al.*, 1972).

5,6-Dehydrogenation and 22,23-dehydrogenation

A plausible mechanism for 5,6-dehydrogenation would seem to be the reverse of the *trans* reduction process occurring in the saturation of the 7,8 and 14,15 double bonds. However, it has been shown that this dehydrogenation amounts to an overall *cis* removal of the 5α hydrogen together with the 5-*pro*-S mevalonoid hydrogen situated in the 6α-position whether in rat liver (Akhtar, Parvez & Hunt, 1969), yeast (Akhtar & Marsh, 1967), the alga *Ochromonas malhamensis* (Smith, Goad & Goodwin, 1968*b*) or the fungus *Aspergillus fumigatus* (Bimpson, Goad & Goodwin, 1969). Furthermore, it was shown that in cholesterol biosynthesis, the hydrogen lost from the 6α-position is lost to the water of the medium rather than as hydride to the coenzyme NAD^+ (Aberhart & Caspi, 1971). *Cis* removal of hydrogens by an enzyme-oxygen complex (Fig. 13.14) is therefore postulated (Wilton & Akhtar, 1970). An alternative mechanism in which 5α-hydroxylation is followed by dehydration has also been considered in studies with yeast (Akhtar & Parvez, 1968; *cf.* Topham & Gaylor, 1972). Although ergosta-$\Delta^{7,22}$-dien-3β,5α-diol (*cf.* 3.30) was incorporated into ergosterol, it did so with loss of the 3α-hydrogen atom (Topham & Gaylor, 1972). Since

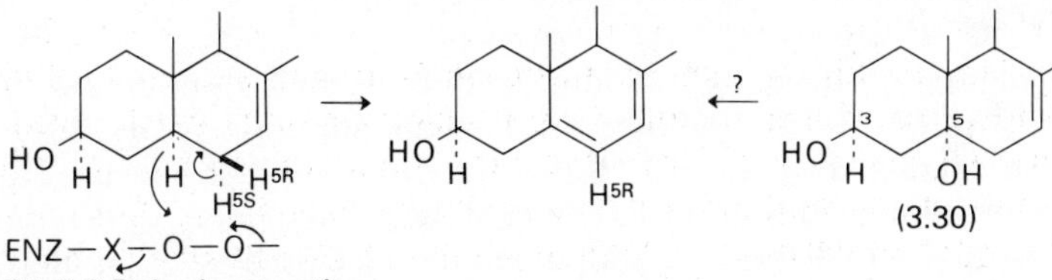

(3.30)

Fig. 13.14 5,6-Dehydrogenation.

ergosta-$\Delta^{7,22}$-dienol (*cf.* 3.31) is incorporated into ergosterol with retention of the 3α-hydrogen atom (Akhtar & Parvez, 1968) the 5α-hydroxy compound may not be an obligatory intermediate. It might, for example, revert to ergosta-$\Delta^{7,22}$-dienol *via* ergosta-$\Delta^{4,7}$-dien-3-one.

The 22,23 double bond present in ergosterol and in certain sterols of plants and algae, is thought to be introduced by a similar mechanism. Which pair of *cis* hydrogens are removed seems to vary from plant to plant and the pair removed in the formation of ergosterol by the fungus *Aspergillus fumigatus,* namely the 2–*pro*-S mevalonoid hydrogen in the 22-*pro*-S position and the 5-*pro*-S mevalonoid hydrogen in the 23-*pro*-S position (Bimpson, Goad & Goodwin, 1969), appears to be the opposite to those removed (22-*pro*-R and 23-*pro*-R) in the formation of algal sterols (Goad & Goodwin, 1972).

Reduction of 24,25 and 24(28) double bonds

In rat liver, reduction of the 24,25 double bond of the lanosterol side chain involves a *cis* addition of hydrogen. The mechanism is, however, somewhat similar to reduction of the 7,8 and 14,15 double bonds. Protonation at C-24 by addition of a 24-*pro*-S hydrogen is followed by reduction of the resulting tertiary carbonium ion by delivery at C-25 of a hydride ion from NADPH to give the configuration shown in (3.32) (Galli-Kienli, Varma, Mulheirn, Yagen & Caspi, 1973; Yagen, O'Gradrick, Caspi & Tamm, 1974). Differentiation of the methyl group derived from C-2 of mevalonate from the other methyl group at C-25 was achieved by its microbiological hydroxylation, X-ray analysis then establishing the configuration at C-25 (Duchamps, Chilester, Wickramasinghe, Caspi, & Yagen, 1971).

Reduction of the 24(28) double bond does not affect the 4-*pro*-R mevalonoid hydrogen at C-25 and probably occurs by a somewhat similar mechanism, *i.e.* protonation at C-28 giving a tertiary carbonium ion, reduction of which could give a 24S methyl group (*cf.* 3.33) as in ergosta-$\Delta^{5,7}$-dienol (equivalent to 24R in ergosterol according to the revised nomenclature rules, I.U.P.A.C./I.U.B., 1970) or 24R methyl group (*cf.* 3.34) as in campesterol.

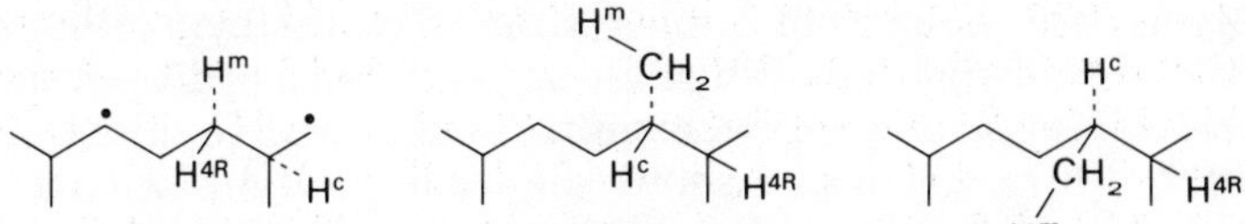

(3.32) (3.33) 24S Methyl group (3.34) 24R Methyl group

24-Ethyl sterols

The sterol mixture from higher plants frequently consists mainly of the ethyl sterol sitosterol (3.35) together with small amounts of its 22-dehydro derivative stigmasterol (3.36). Recently, three fungal species from the Hyphochytridiomycetes and Chytridiomycetes have been reported to contain a 24-ethyl-cholesterol, accompanied in one case by smaller amounts of a 22-dehydro-24-ethylcholesterol (Bean, Patterson & Motta, 1973). Moreover, stigmast-Δ^7-enol (3.37) has been isolated from uredospores of the rust *Puccinia graminis* (Nowak, Kim & Röhringer, 1972). This sterol also occurs, together with its 24-ethylidene analogue, in the uredospores of the rusts *Melampsora lini* (Jackson & Frear, 1968) and *Uromyces phaseoli* (Lin, Langenbache & Knoche, 1972). In the last case evidence has been provided that the uredospores are capable of synthesizing sterols and that these are not merely metabolites of the host sterol (Lin & Knoche, 1974).

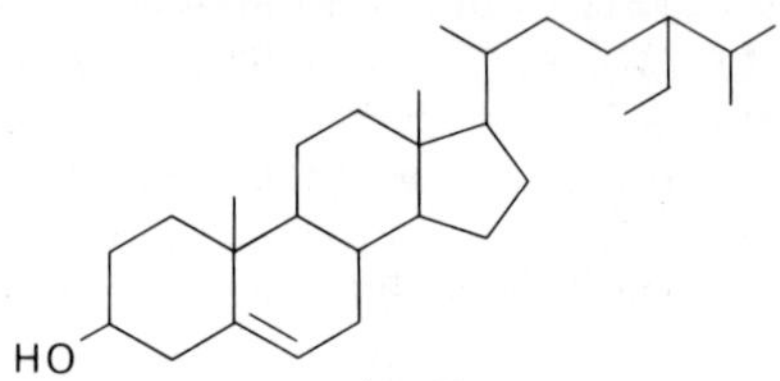

(3.35)
Sitosterol

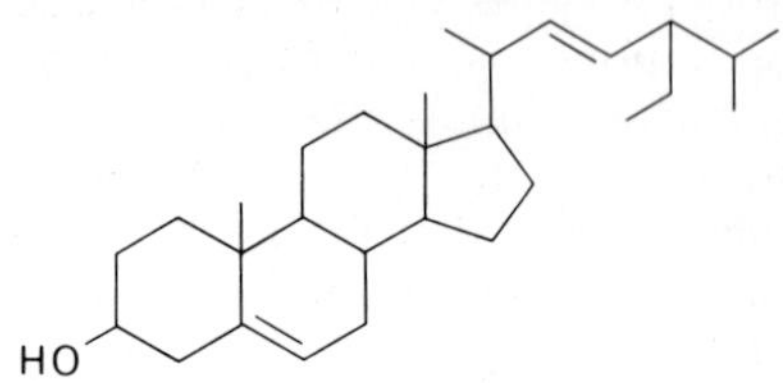

(3.36)
Stigmasterol

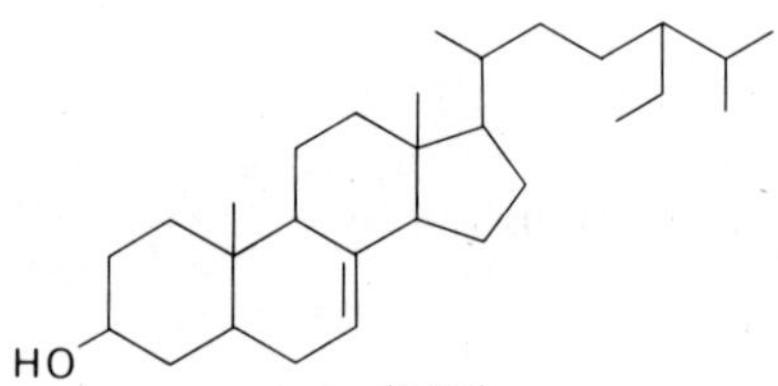

(3.37)
Stigmast-Δ^7-enol

There seems to be considerable variation possible in the mechanism of formation of the side chains of 24-ethylsterols (*cf.* Lederer, 1969). In the alga *Ochromonas malhamensis*, poriferasterol (3.19) appears to be formed *via* an ethylidene intermediate and the 'shifted' 4-*pro*-R mevalonoid hydrogen at C-25 is retained (*cf.* Fig. 13.15*a*) (Smith, Goad & Goodwin, 1972). In a number of plants, however, the loss of this hydrogen suggests that a Δ^{24} compound is formed (Fig. 13.15*b*), followed by 24,25 hydrogenation (Tomita & Uomori, 1973; Armarego, Goad & Goodwin, 1973; Randall, Rees & Goodwin, 1972). A 24-ethylidene intermediate is probable in this case (Goad *et al.*, 1966). Finally, in the Myxomycete *Dictyostelium discoideum*, five deuterium atoms from ($^2H_3$-methyl)methionine are retained in the ethyl group of stigmast-Δ^{22}-enol and a hydrogen shifts from C-23 to C-24, suggesting the mechanism *c* in Fig. 13.15 (Lenfant, Zissmann & Lederer, 1967). Fig. 13.15 shows the location of 4-*pro*-R mevalonoid hydrogens and of hydrogens from the medium (H^m) introduced into the ethyl group by mechanisms *a, b* and *c*. If the reduction processes follow the lines observed in the reduction of Δ^7, Δ^{14} and Δ^{22} in cholesterol (Akhtar *et al.*,

1972; Galli-Kienli *et al.*, 1973), diene reduction occurring *via* an allylic carbonium ion, the respective hydrogens introduced from the medium and from the coenzyme might be predicted to be as shown.

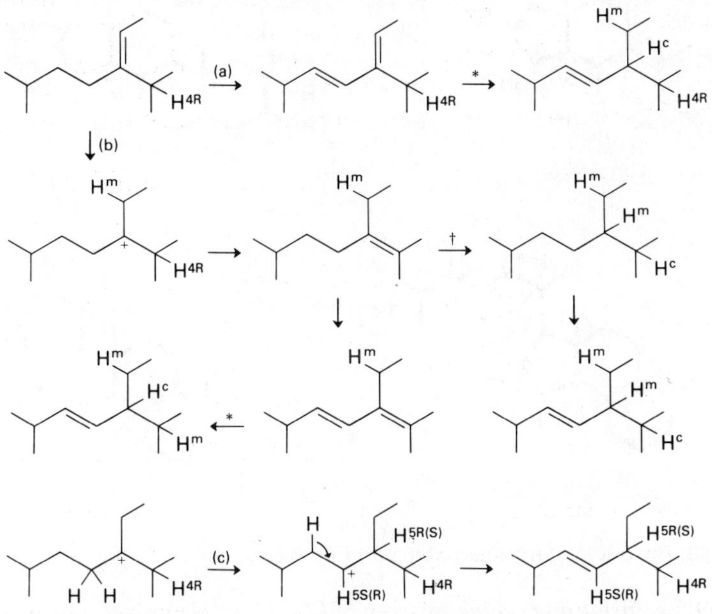

Fig. 13.15 Proposed schemes for formation of 24-ethyl-Δ^{22} side chains in different organisms.
*: Reduction via an allylic carbonium ion; †: Reduction as in cholesterol; H^m: hydrogen from the medium; H^c: hydrogen from coenzyme.

The origin of the hydrogen atoms in sterols

Current understanding of sterol biosynthesis now encompasses not only the way in which the carbon atoms of mevalonate make up the carbon framework of sterols (Fig. 13.2 to 13.5), but also the derivation of the various hydrogen atoms from the various mevalonoid hydrogen atoms, from the medium (H^m) or from NADPH (H^c) (*cf.* Mulhiern & Ramm, 1972). Comparison of the pattern in lanosterol with that in cholesterol or ergosterol (Fig. 13.16) reflects the changes involved in the late stages of the biosynthesis of these 4-demethyl sterols.

As discussed on p. 372, the coupling of two molecules of farnesyl-pyrophosphate to give squalene is an asymmetric process, the 5-*pro*-S mevalonoid hydrogen at C-1 of one residue being stereospecifically replaced by a hydrogen (H^c) from NADPH. In two fungi, at least, the squalene, once formed, is probably liberated into a metabolic pool, allowing oxidation of either end to occur before cyclization to terpenoids. Thus eburicoic acid (3.13), biosynthesized in *Polyporus sulphureus* from (1-$^3H_2$) farnesyl-pyrophosphate was found to have equal distribution of label at C-11 and C-12, corresponding to the C-1 positions of the two farnesyl units (Lawrie, McLean, Pauson & Watson, 1965). Similarly, fusidic acid (2.6) biosyn-

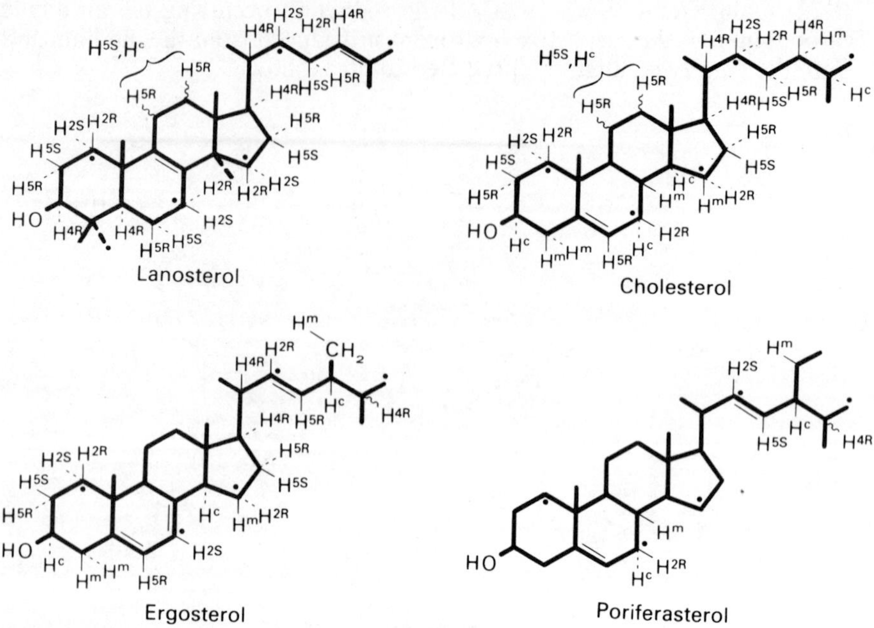

Fig. 13.16 The origin of the hydrogen atoms in sterols.

thesized in *Fusidium coccineum* from 5S(2-^{14}C, 5-^{3}H)mevalonate was found to have tritium content at C-11 and C-12, each equivalent to half a tritium atom (Ebersole, Godtfredsen, Vangedal & Caspi, 1973).

Comparative aspects of sterol biosynthesis

Some of the main variations in the biosynthetic processes used by different organisms are summarized in Table 13.2. The sterols produced by fungi of the orders Saprolegniales and Leptomitales, *e.g.* 24-methylenecholesterol and fucosterol (McCorkindale *et al.*, 1969; Popplestone & Unrau, 1973), are closer akin to those of algae and plants than to those of higher fungi. These aquatic fungi are unusual in having cell walls composed mainly of cellulose-glucan rather than chitin-glucan (Bartnicki-Garcia, 1968; Dietrich, 1973); their pathway of lysine biosynthesis involving diaminopimelic acid is that followed by bacteria, algae and plants and differs from the pathway involving α-aminoadipic acid followed by all other fungal orders (Vogel, 1965); also, the properties of the five enzymes involved in the biosynthesis of tryptophan by these fungi has a characteristic pattern distinct from that of other fungi (Hütter & Demoss, 1967). The details of their biosynthesis of sterols thus should be of some interest.

In this connection, it has been shown that both cycloartenol and lanosterol are incorporated into the sterols of *Saprolegnia ferax* although the former is a more efficient precursor (Bu'lock and Osagie, pers. comm.). A similar result has been achieved with the alga *Ochromonas malhamensis* (Hall *et al.*, 1969) and with various plants, *e.g. Nicotiana tabacum* (Hewlins *et al.*, 1969). In the last two cases, it was subsequently fully established that the natural precursor is cycloartenol (*cf.* Goad & Goodwin, 1972).

Table 13.2 Variations in sterol biosynthesis with type of organism

Higher fungi, yeasts, animals	Lanosterol	Squalene oxide cyclization	Cycloartenol	Higher plants, algae
Higher fungi, yeasts, brown algae	*via* $CH_2=$	24–Alkylation	not *via* $CH_2=$	Myxomycetes, green algae
Higher fungi, yeasts, algae	24S	24-Alkyl group	24R	Higher plants, algae, Myxomycetes
Algae, Oömycetes	E	24–Ethylidene	Z	Higher plants
Higher fungi, yeasts,	Loss of 7α (H^{2R})	Δ^8 to Δ^7 isomerization	Loss of 7β (H^{2S})	Higher plants, algae, animals
Higher fungi, yeast, animals.	Loss of 15α	Formation of Δ^{14}		

Fungal terpenoids derived by further metabolism of sterols

As discussed earlier, ergosterol is the metabolic end-product of numerous pathways. Unlike the tetraenols given in Table 13.1, ergosta-$\Delta^{5,7,9(11),22}$-tetraenol (3.38) from *Mucor rouxii* is not an intermediate in ergosterol biosynthesis but seems to be a product of its further metabolism (Atherton, Duncan & Safe, 1972). This also seems to be true for ergosta-$\Delta^{4,6,8(14),22}$-tetraen-3-one (3.39) produced by *Penicillium rubrum* (White, Perkins & Taylor, 1973). Ergosterol peroxide (3.40) is accepted by *P. rubrum* as a substrate for biosynthesis of this ketone and has also been converted by a yeast enzyme to ergosterol (Topham & Gaylor, 1972) but it is doubtful whether it is a normal intermediate in these systems. The status of ergosterol peroxide as a natural product requires close scrutiny in view of the facility with which it is formed from ergosterol by aerial oxidation, particularly if catalysed by the type of pigment often present in fungal extracts (Adam, Campbell & McCorkindale, 1967).

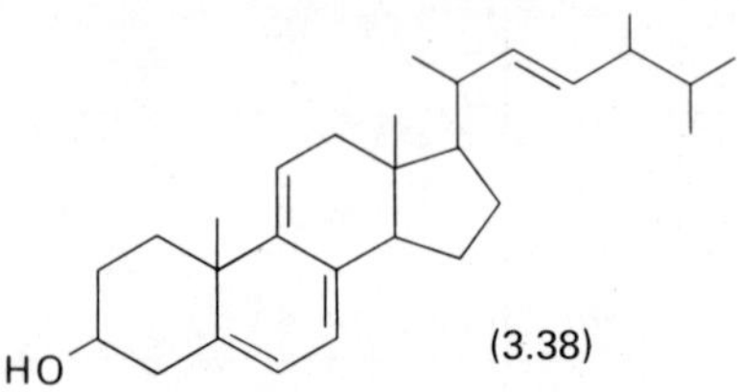

(3.38)

Ergosta-$\Delta^{5,7,9(11),22}$-tetraenol

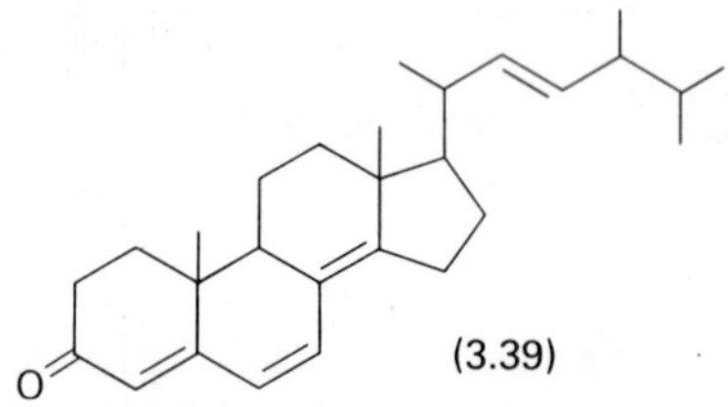

(3.39)

Ergosta-$\Delta^{4,6,8(14),22}$-tetraen-3-one

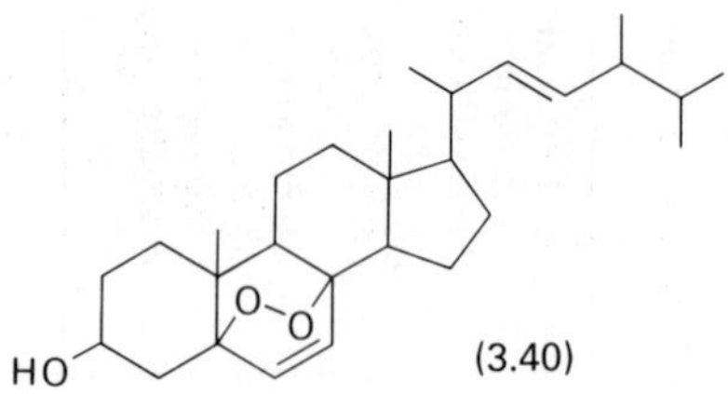

(3.40)

Ergosterol peroxide

Wortmannin (3.41) from *Penicillium wortmannii* (MacMillan, Vanstone & Yeboah, 1968) is derived by fairly extensive modification of lanosterol (MacMillan, Simpson & Yeboah, 1972), and a similar origin can be deduced for viridin (3.42), an antibiotic from *Gliocladium virens* (Brian & McGowan, 1945). Each has lost the sterol side chain and wortmannin retains all the carbons of a 4-methylsterol precursor. Viridin, which has lost,

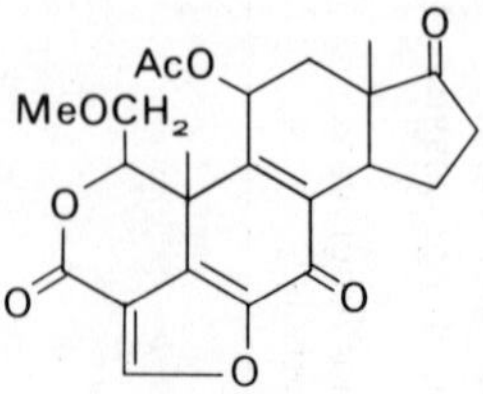

(3.41)
Wortmannin

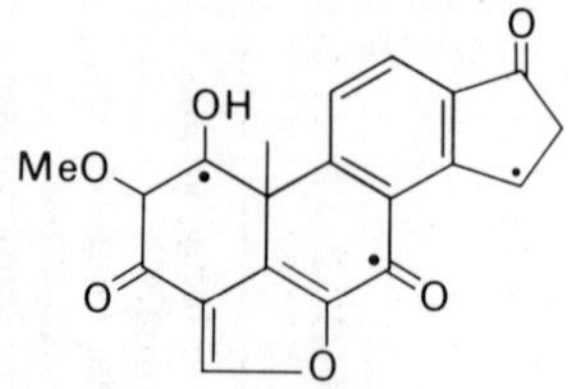

(3.42)
Viridin

in addition, the steroidal C-18 methyl group, was shown to have the labelling pattern indicated, when biosynthesized from (2-^{14}C) mevalonate (Grove, 1969). Since the furan methine carbon was unlabelled, it is evident that as in the first 4-demethylation of lanosterol *en route* to ergosterol, the methyl eliminated from C-4 has been the 4α-methyl group derived from C-2 of mevalonate.

13.4 Sesterterpenes

Fungal sesterterpenes form a small group typified by the phytotoxins ophiobolin C(4.1) from *Cochliobolus miyabeanus* (*cf.* Canonica & Fiecchi, 1970; Cordell, 1974) and ophiobolin F from *Cochliobolus heterostrophus* which has been assigned the structure (4.2) (Nozoe, Morisaki, Fukushima & Okuda, 1968; Nozoe & Morisaki, 1969). Studies with variously labelled mevalonates (Canonica, Fiecchi, Galli-Kienli, Ranzi & Scala 1968; Canonica & Fiecchi, 1970) indicated that the ophiobolins are derived from geranyl-farnesylpyrophosphate (1.5, $n = 3$). Since a cell-free extract of *C. heterostrophus* was shown to incorporate all-*trans* geranylfarnesylpyrophosphate rather than its 2-*cis* isomer (Nozoe & Morisaki, 1969), the formation

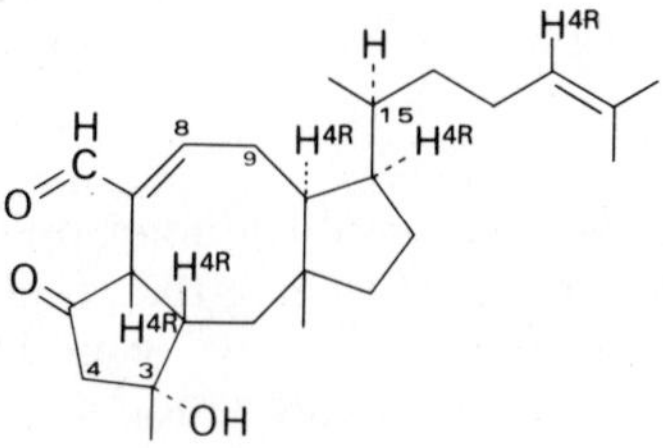

(4.1)
Ophiobolin C

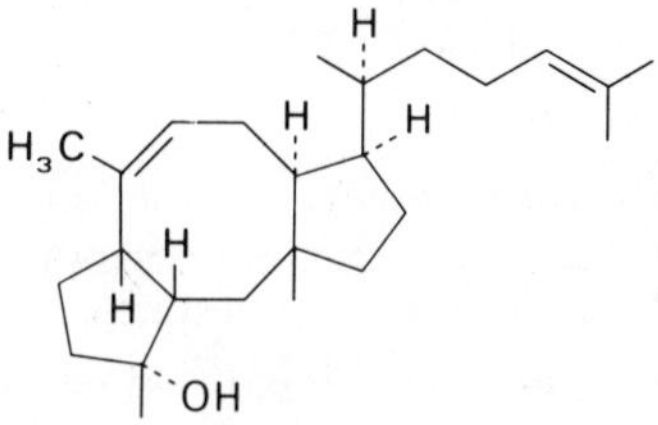

(4.2)
Ophiobolin F

of the ophiobolins can be represented as in Fig. 13.17 (Canonica *et al.*, 1968).

This figure accommodates one of the remarkable features of the biosynthesis, namely that the 2-*pro*-R mevalonoid hydrogen in the 8α-position (ophiobolane numbering) undergoes a 1,5 migration to the 15-position (Canonica *et al.*, 1967*a*, *b*). The first step is thought to be intramolecular alkylation as shown, generating a species (4.3) deficient in electrons at C-15, which may be stabilized by attachment of an enzymic grouping X at C-15 (*cf.* 4.4). In the mechanism shown, the hydrogen migration is part of a concerted process in which the A/B ring junction is formed, the 8,9 double bond is formed and the 8α-hydrogen migrates to C-15, neutralizing the charge (or displacing the enzymic grouping). The positive charge located at C-3 in the product (4.5) could be neutralized by attachment of OH^- in the α-position (Canonica *et al.*, 1968). Alternatively, another enzymic grouping X′ might stabilize the cation as in (4.6), and thus direct the stereochemistry of the hydroxyl addition. This would be effectively the same as if X^- initiated the concerted reaction by nucleophilic attack from the β-side. A necessary feature of this scheme, the derivation of the 3α-hydroxyl group from water rather than from molecular oxygen, has been proved using ^{18}O (Canonica *et al.*, 1967*a*; Nozoe, Morisaki, Tsuda & Okuda, 1967).

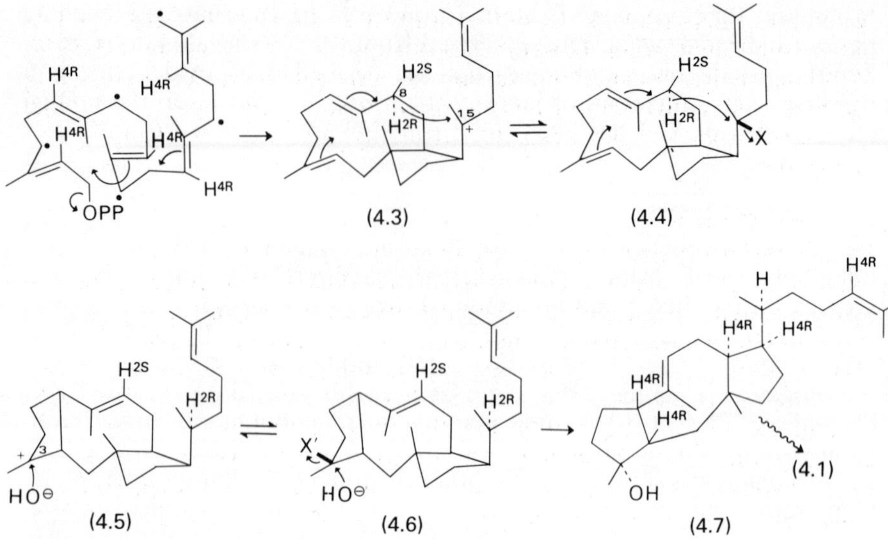

Fig. 13.17 Proposed biogenesis of the ophiobolins from geranylfarnesylpyrophosphate.

This concerted cyclization-migration leads to a *trans* 8,9 double bond as in (4.7). Evidently this isomerizes in subsequent steps to give the *cis* double bond present in the ophiobolins. This isomerization may occur at the $\alpha\beta$-unsaturated aldehyde stage as in the farnesol isomerization sequence discussed earlier. If so, the 8,9 double bond in ophiobolin F (4.2) could not isomerize by such a process and it is interesting that the stereochemistry of its 8,9 double bond remains to be established, since structural proof rests on palladium-charcoal catalysed hydrogenation to the known alcohol (4.8) (Nozoe & Morisaki, 1969).

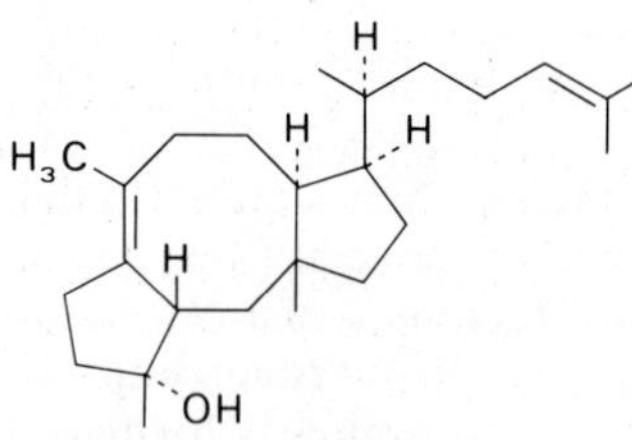

(4.8)

13.5 Diterpenes

The majority of fungal diterpenes are derived in theory from geranylgeranylpyrophosphate (5.1) by proton-initiated cyclization, giving rise, as in Fig. 13.18, to a bicyclic intermediate (5.2) or its enantiomer (5.3). Deprotonation of (5.3) gives the labdadiene (5.4) which can cyclize to the pimarane cation (5.5). Of the relatively few fungal diterpenes which are derived from these without rearrangement of the carbon skeleton, virescenol A (5.6), which occurs as a D-altropyranoside in *Oospora virescens*

(Polonsky, Baskevitch, Bellavita & Ceccherelli, 1970) is obviously related to the cation (5.5). The labelling pattern shown in (5.6) from 2- and 1-^{13}C labelled acetate was established by c.m.r. spectroscopy and was as expected from Fig. 13.18 (Polonsky *et al.*, 1972).

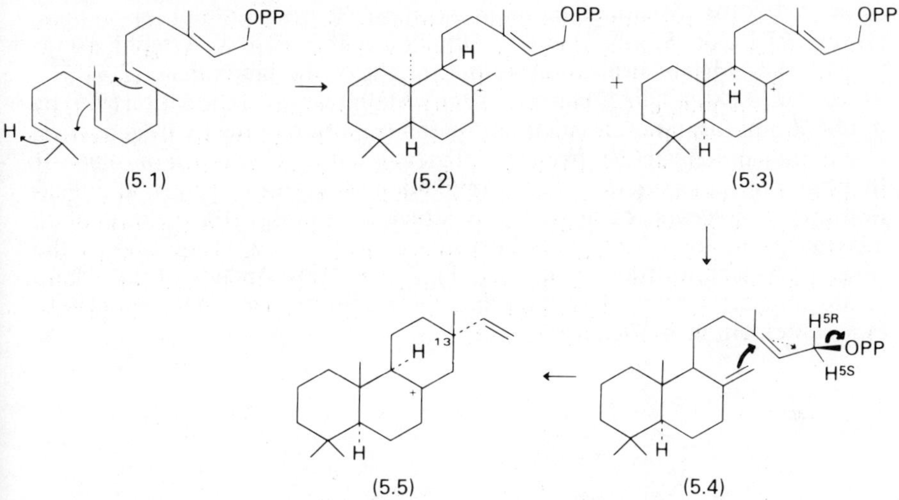

Fig. 13.18 Formation of 13S pimaranes.

The C_{16} diterpenoid lactone (5.7) from an *Acrostalagmus* species (Ellestad, Evans, Kunstmann, Lancaster & Morton, 1970), similarly gave the labelling pattern shown, indicating that the skeleton is probably formed from the labdadiene (5.4) by loss of four carbons of the side chain (Kakisawa, Sato, Ruo & Hayashi, 1973). The co-occurring diacid (5.8.) (Ellestad, Evans & Kuntsmann, 1971) could be an intermediate in the oxidative sequence. If these were extensively degraded triterpenes, C-12 would correspond to a C-5 mevalonoid carbon in lanosterol (*cf.* Fig. 13.16) and thus carry no label.

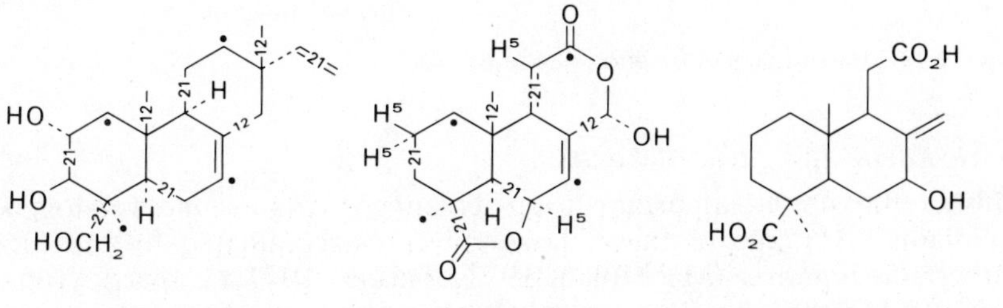

• Carbon atom labelled from 2-labelled mevalonate or 2-labelled acetate.
—21— The bond joining two atoms labelled from 2- and 1-labelled acetate respectively.

Rosenonolactone

The concerted rearrangement/cyclization of the labdadiene (5.4), shown in Fig. 13.19, is thought to give rise to a group of fungal diterpenes typified by rosenonolactone (5.10) from *Trichothecium roseum* (Harris, Robertson & Whalley, 1958). The figure is supported by the labelling pattern, shown to be as in (5.10) for rosenonolactone biosynthesized from (2-^{14}C) mevalonate (Birch, Rickards, Smith, Harris & Whalley, 1959; Britt & Arigoni, 1958) and by the efficient utilization by the fungus of the labdadiene (5.4) as a precursor (Achilladelis & Hanson, 1969). Stabilization of the cation (5.9) by double bond formation is ruled out by the retention of the hydrogens from C-2 of mevalonate at C-1, from C-5 of mevalonate at C-6 and of the 4-*pro*-R hydrogen of mevalonate at C-5 (Achilladelis & Hanson, 1969). Stabilization might, however, be achieved by attachment in the 10α-position of an enzymic grouping X, which is to be displaced by the carboxylate anion in the subsequent lactonization as shown in Fig. 13.19. Introduction of the 7-keto group appears to be a final modification (Achilladelis & Hanson, 1969; Holtzapfel, Birch & Richards, 1969).

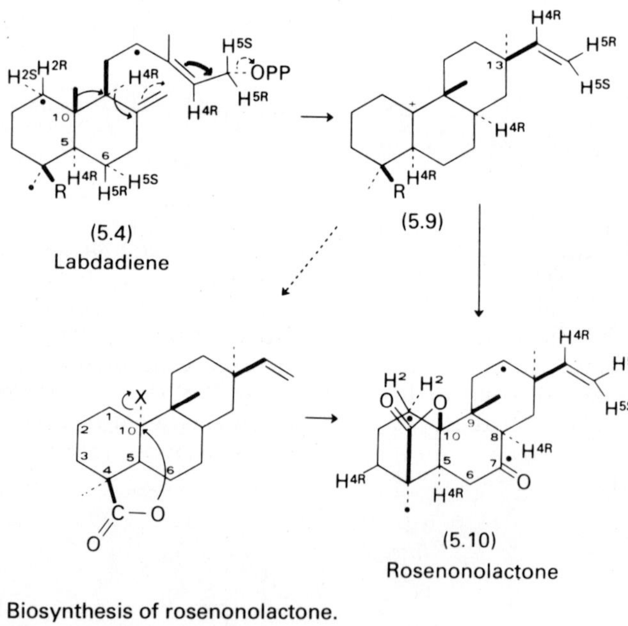

Fig. 13.19 Biosynthesis of rosenonolactone.

(−)Kaurene and gibberellic acid

Gibberellins are an important group of plant growth substances endogenous to plants. A range of these, however, are accumulated by the fungus *Gibberella fujikuroi* (MacMillan, 1971; Turner, 1971). Copalylpyrophosphate (5.11), the labdadiene formed by deprotonation of the 'enantiomeric' cation (5.2), has been shown to be the precursor of the gibberellins in *G. fujikuroi* (Hanson & White, 1969). Three main changes involving the skeleton occur during the conversion of copalylpyrophosphate into gibberellic acid (5.15), namely formation of the tetracyclic (−)kaurene ring system (5.14), contraction of ring B and a lactonization sequence involving expulsion of the ring A/B angular methyl group.

Cyclization of copalylpyrophosphate (5.11) has been studied using enzyme preparations from *Gibberella fujikuroi* (Fall & West, 1971; Evans & Hanson, 1972), and appears to proceed as indicated in Fig. 13.20 *via* the cation (5.12). Although the pimaradiene (5.17) is utilized by *G. fujikuroi* as a substrate for the biosynthesis of gibberellic acid (5.15), the hydrogens on carbons adjacent to C-8 in the latter are all mevalonoid hydrogens as indicated, so that pimaradienes like (5.17) which would be formed by deprotonation of (5.12) are not normal precursors. The fate of different mevalonoid carbons and hydrogens during cyclization of the intermediate cation (5.12) to a bicyclo(3 : 2 : 1) octane system (5.13) and the subsequent rearrangement of this to the bicyclo(3 : 2 : 1)octane system of the gibberellins has been determined and supports Fig. 13.20 (Evans, Hanson & White, 1970; Evans, Hanson & Mulheirn, 1973). In particular, the fate of the 5-*pro*-R and 5-*pro*-S mevalonoid hydrogens corresponding to the 1-position of farnesylpyrophosphate was established by conversion of gibberellic acid

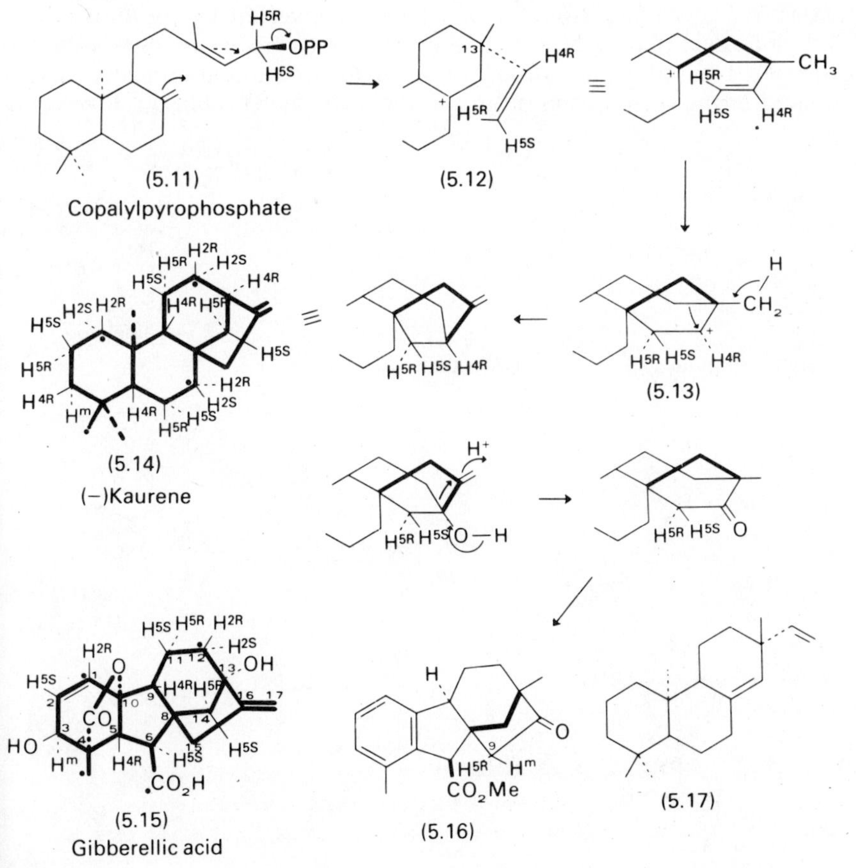

Fig. 13.20 Formation of (−)kaurene.

(5.15) into methyl gibberate (*cf.* 5.16). The more readily exchangeable, *exo* hydrogen at C-14 was shown to be 5-*pro*-S mevalonoid and the *endo* hydrogen 5-*pro*-R mevalonoid (Evans, Hanson & Mulheirn, 1973*b*). This stereochemistry is consistent with the copalylpyrophosphate cyclization being a concerted process giving a vinyl group with a *cis* arrangement of the 4-*pro*-R and 5-*pro*-S mevalonoid hydrogens. These hydrogens would be expected to be *trans* in a 13R diterpene like rosenonolactone (5.10) (Fig. 13.19).

The main pathway to gibberellic acid leads from (−)kaurene *via* the intermediates indicated in Fig. 13.21. 7β-Hydroxykaurenoic acid (5.18) has been shown to be a normal intermediate (Cross, Norton & Stewart, 1968; Lew & West, 1971; Hanson, Hawker & White, 1972) formed by stepwise oxidation of the 4α-methyl group of (−)kaurene to a 4α-carboxyl group followed by 7β-hydroxylation (*cf.* Bearder, Macmillan, Wels, Chaffey & Phinney, 1974). The exact mechanism of ring contraction of 7β-hydroxykaurenoic acid (5.18) to the C_{20} gibberellin skeleton is not clear. Pertinent facts are that the extruded carbon is that at C-7 derived from C-2 of mevalonate (Birch *et al.*, 1959). Ring contraction gives GA_{12} aldehyde (5.19) initially. This has been isolated from cultures of *Gibberella fujikuroi* (Hanson, Hawker & White, 1972) and is converted by the fungus into gibberellic acid (Cross, Norton & Stewart, 1968). The 5-*pro*-S mevalonoid hydrogen from the 6α-position of 7β-hydroxy-kaurenoic acid (5.18) is retained in the 6α-position of gibberellic acid (5.15) (Hanson, Hawker &

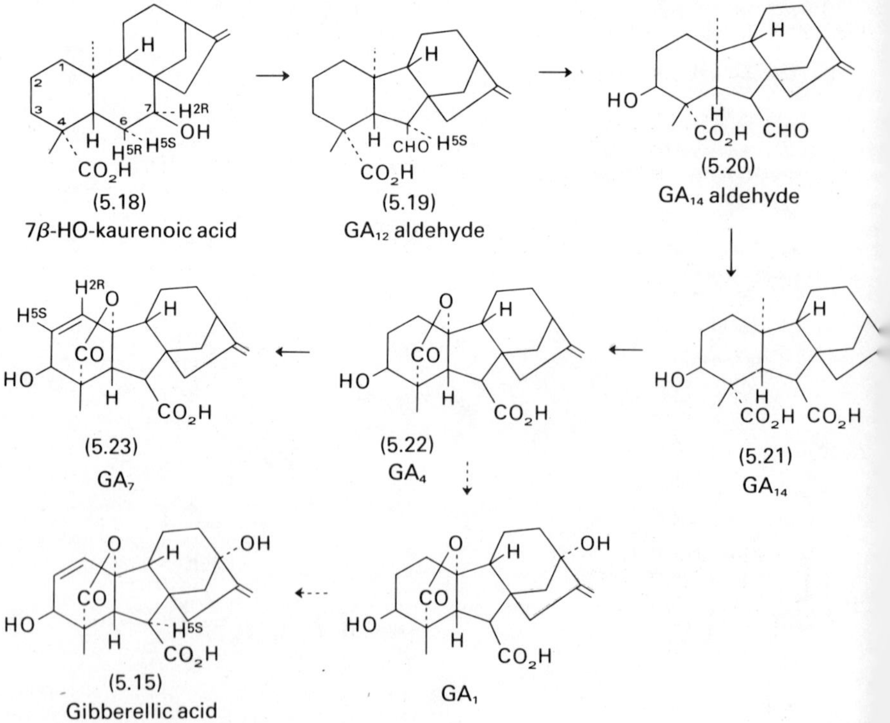

Fig. 13.21 The main pathway to gibberellic acid.

White, 1972; Evans, Hanson & Mulheirn, 1973). Ring contraction of 7β-hydroxykaurenoic acid (5.18) could result as in Fig. 13.22, from loss of the 5-*pro*-R mevalonoid hydrogen from the 6β-position to give a species deficient in electrons at C-6 (either the cation 5.24 or the radical 5.25). In keeping with this, the 6β-tritium atom was lost during the conversion of (^{14}C,6β-^{3}H)-7β-hydroxykaurenoic acid into GA_{12} aldehyde by cell-free extracts of *Cucurbita maxima* (MacMillan, pers. comm.). This conversion was accompanied by 6α-hydroxylation, a process known to be competitive to rather than intermediate in the ring contraction. The first C_{20} gibberellin to be formed on the main pathway to gibberellic acid is GA_{14} (5.21), the order of events being hydroxylation of GA_{12} aldehyde to GA_{14} aldehyde (5.20) followed by oxidation to GA_{14} (Bearder, MacMillan & Phinney, 1973*a*; Hedden, MacMillan & Phinney, 1974).

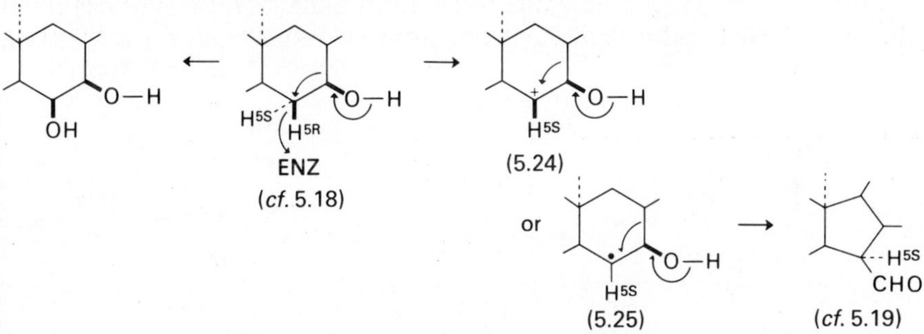

Fig. 13.22 Contraction of ring B in gibberellin biosynthesis.

Unlike the 14α-demethylation sequence in sterol biosynthesis (p. 385), elimination of the methyl group from C-10 in the gibberellins is not attended by loss of hydrogens from any of the adjacent carbon atoms (Hanson & White, 1969). Furthermore, the carbon atom is not eliminated by decarboxylation of the tricarboxylic acid GA_{13} (5.26) since this is not utilized as a precursor of gibberellic acid (Hanson & Hawker, 1972). A Baeyer-Villiger mechanism has also been considered (Hanson, & White, 1969;

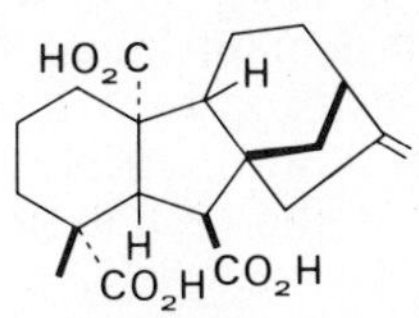

(5.26)
GA_{13}

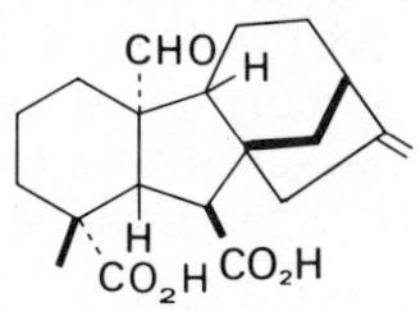

(5.27)
GA_{36}

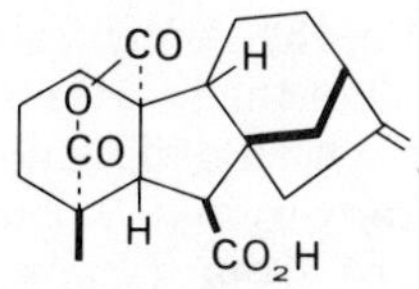

(5.28)

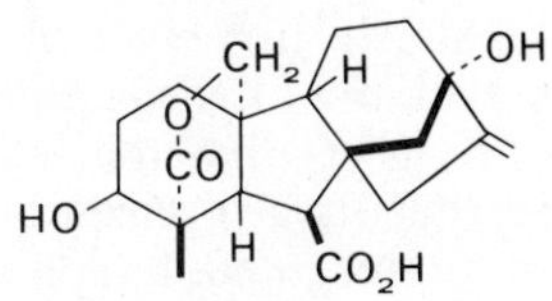

(5.29)
GA_{38}

Durley, Railton & Pharis, 1974) but the corresponding formyl compound, GA_{36} (5.27), which has been isolated from *Gibberella fujikuroi* (Bearder & MacMillan, 1972) is not converted by the fungus to C_{19} gibberellins (MacMillan, pers. comm.). The possible participation of the 4α-carboxylic function in the elimination process has been suggested. With one possible intermediate, namely the anhydride (5.28) of GA_{13}, only a low incorporation was observed, extensive hydrolysis to the triacid GA_{13} (5.26) being a competitive process (Hanson & Hawker, 1972). Other workers have found that hydrolysis of the anhydride can preclude any significant incorporation (MacMillan, pers. comm.). One hypothesis in which a 10-formyl or a 10-carboxy compound would not be involved would require participation of the 4α-carboxylic function as a peracid. If this formed a bridged intermediate as hydride is transferred from the C-10 substituent to the enzymic hydride acceptor (Fig. 13.23), the C-10 substituent (CH_3 or CH_2OH) could then be expelled as formaldehyde or formic acid to give the C_{19} gibberllin, GA_4 (5.22)

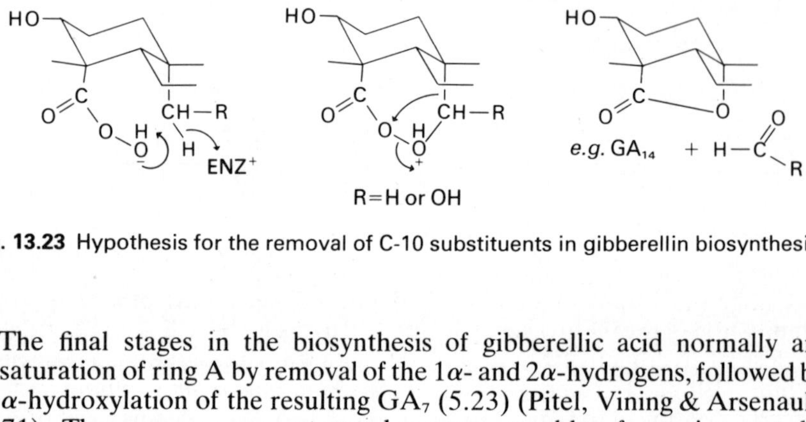

Fig. 13.23 Hypothesis for the removal of C-10 substituents in gibberellin biosynthesis.

The final stages in the biosynthesis of gibberellic acid normally are desaturation of ring A by removal of the 1α- and 2α-hydrogens, followed by 13α-hydroxylation of the resulting GA_7 (5.23) (Pitel, Vining & Arsenault, 1971). The enzymes present are, however, capable of carrying out the processes in the reverse order (*cf.* Bearder, MacMillan & Phinney, 1973*b*).

Variations from the above in the timing of different steps are observed in plants. For example a pathway is operative in seedlings of *Pisum sativum* wherein 3β- and 13α-hydroxylation precede 10-demethylation, and in this series, further metabolism of lactonized 10-hydroxymethyl gibberellin (GA_{38}, 5.29), which accumulates initially, coincides with increased formation of 10-formyl and C_{19} gibberellins (Durley, Railton & Pharis, 1974).

Pleuromutilin and aphidcolin

The diterpenes discussed so far, all have as a precursor, an (8,9-*ante*)labdane cation derived by chair-chair cyclization of geranylgeranylpyrophosphate. In two types of fungal diterpene, the configuration of the 4-*pro*-R mevalonoid hydrogen derived from C-9 of the presumed labdane precursor, suggests that the latter is the (8,9-*syn*)labdane cation (5.31) formed by chair-boat cyclization of geranylgeranylpyrophosphate (5.30). These are represented by pleuromutilin (5.36) from *Pleurotus mutilis* (Arigoni, 1968) and aphidcolin (5.41) from *Cephalosporium aphidcola* (Dalziel, Hesp, Stevenson & Jarvis, 1973).

The biosynthesis of pleuromutilin has been worked out in some detail and is proposed to be as outlined in Fig. 13.24 (*cf.* Arigoni, 1968). Hydride shift of the 4-*pro*-R mevalonoid hydrogen at C-9 to the 8β-position gives a cation (5.32) which could possibly be stabilized by attachment at the 9α-position of an enzymic X grouping. The concerted rearrangement of this cation (5.32), involving one hydride and one methyl shift, gives the ring-contracted

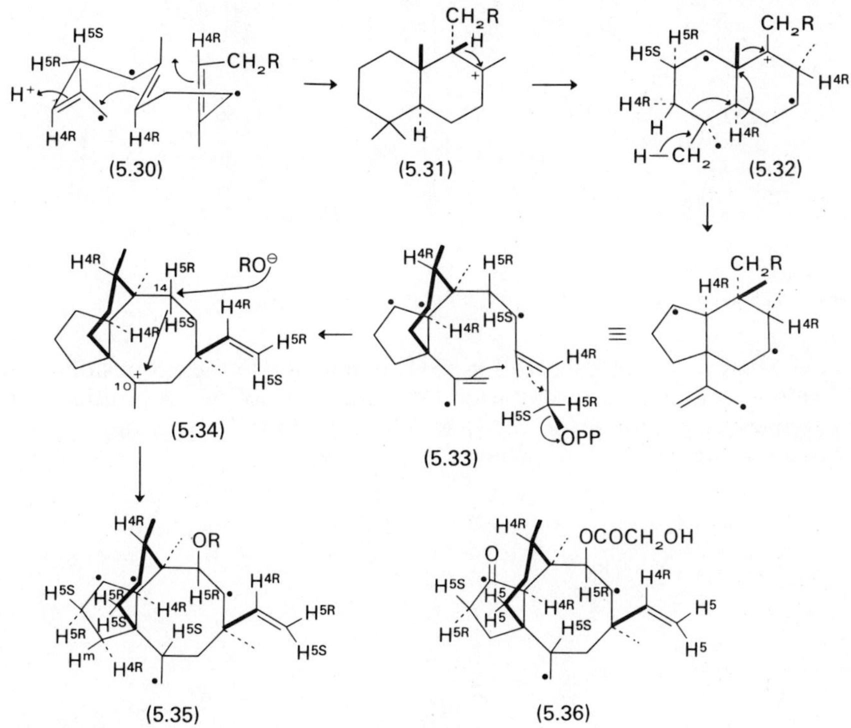

Fig. 13.24 The biogenesis of pleuromutilin (5.36).

intermediate (5.33) which can undergo concerted cyclization to the tricyclic cation (5.34), reminiscent of the labdadiene to pimarene cyclization (Fig. 13.18). Finally, what in effect is nucleophilic displacement of the 5-*pro*-S mevalonoid hydrogen at C-14 in the intermediate (5.34) by RO^- (*i.e.* hydroxyl or glycollate ion) is accompanied by a transannular shift of this hydride ion to the C-10 position. This scheme accounts for the stereochemistry of the pleuromutilin skeleton (5.35) and predicts a fate for the various mevalonoid hydrogens and carbons as indicated. An elegant series of labelling studies and degradations (Arigoni, 1968; Birch, Holtzapfel & Rickards, 1966) rigorously established the origin of most of these carbons and hydrogens in pleuromutilin as shown in (5.36), in agreement with this scheme.

Tracer studies on the biosynthesis of the antibiotic aphidcolin (5.41) have not so far been reported, but it has been suggested (Dalziel *et al.*, 1973) that the skeleton (5.40) could be formed *via* a bicyclo (2 : 2 : 2)octane cation

(5.39) as shown in Fig. 13.25. If H[b] (as well as H[a]) proves to be derived from the 4-*pro*-R hydrogen of mevalonate, this hydrogen is probably in the 9β-position in the appropriate labdane precursor, *i.e.* (5.31) and after

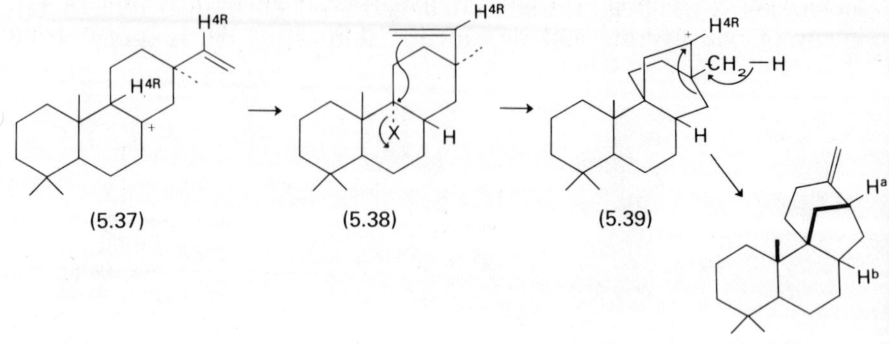

(5.37) (5.38) (5.39) (5.40)

Fig. 13.25 Hypothesis for the formation of aphidcolin (5.41).

cyclization to a pimarane cation (5.37), migrates to the 8β-position. The resulting cation may be stabilized by attachment to the 9α-position of an enzymic X⁻ grouping (*cf.* 5.38) which is displaced in cyclization to the proposed bicyclo(2 : 2 : 2)octane cation (5.39).

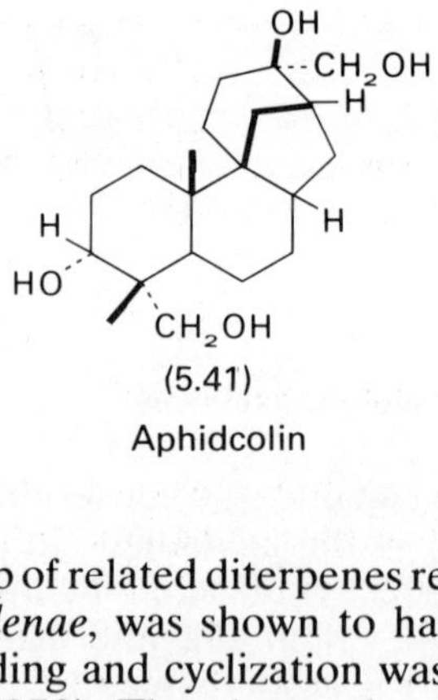

(5.41)
Aphidcolin

Cyathin, A_3

Cyathin A_3, one of a group of related diterpenes recently isolated from bird's nest fungus, *Cyathus helenae*, was shown to have structure (5.42) and a unique type of chain folding and cyclization was suggested for its biosynthesis (Ayer & Taube, 1973). The observed stereochemistry could result

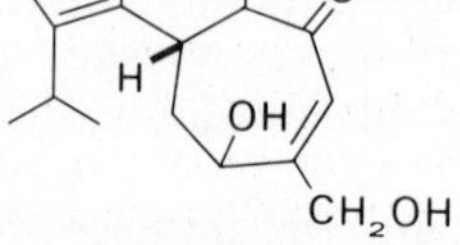

(5.42)
Cyathin A_3

from the sequence outlined in Fig. 13.26 involving formation of the monocyclic, 'stabilized cation' (5.43), further cyclization linked to a carbon-carbon bond shift, deprotonation and finally formation of the tricyclic skeleton (5.44).

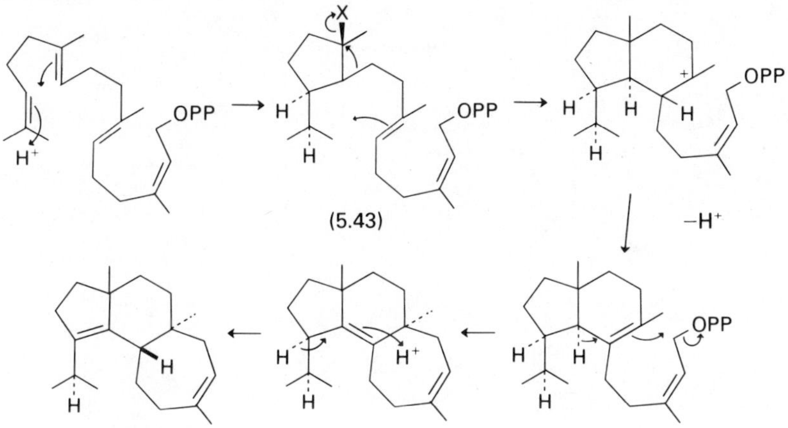

Fig. 13.26 Hypothesis for the formation of the cyathins, *e.g.* (5.42).

13.6 Sesquiterpenes

Sesquiterpenes represent the largest single class of known terpenoids, encompassing many variations in skeleton. A number of the known classes (Herout, 1971; Roberts, 1972; Rücker, 1973) are represented by fungal products (Turner, 1971), often transformed in a characteristic manner as illustrated below.

Illudin M and hirsutic acid C

Current proposals on the biosynthesis of the protoilludane group of fungal sesquiterpenes are outlined in Fig. 13.27. Humulene (6.2), which is produced by cyclization of farnesylpyrophosphate (6.3), is thought to undergo cyclization to the key octadiene shown in two conformations (6.1 and 6.4). The tricyclic cations (6.5) and (6.8) respectively could be formed by cyclization of these, although it is also possible that either ion might normally be formed by rearrangement of the other. Rearrangement of the tricyclic ion (6.5), with loss of a proton as shown, could give an olefin (6.6) from which hirsutic acid C (6.10) and 5-dihydrocoriolin C (6.11) might be formed. Loss of a proton from the other tricyclic ion (6.8) gives a hydrocarbon with the skeleton of illudol (6.12) from *Clitocybe illudens* (McMorris, Nair & Anchel, 1967), while ring contraction could provide the precursors (6.7) and (6.9) of illudin M (6.13) and marasmic acid (6.14) respectively. The lactone (6.15) exemplifies a new group of sesquiterpenes isolated from various species of *Lactarius* (*cf.* Magnusson & Thoren, 1973; Daniewski, Kocor & Zoltowska, 1973) and could be derived from the cyclooctadiene (6.1) either by ring contraction or, more probably, *via* (6.9) with opening of the cyclopropane ring.

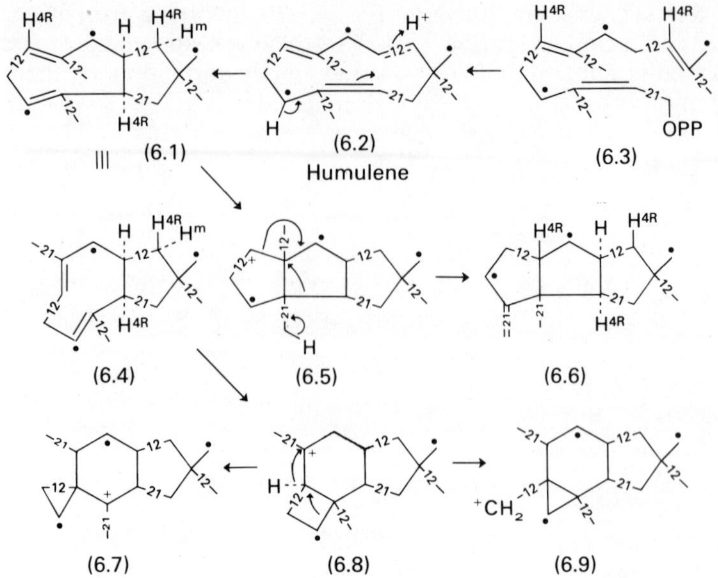

Fig. 13.27 Formation of the protoilludanes. '12' superimposed on a line indicates that the carbon atom at the left and right hand extremities are derived respectively from C-1 and C-2 of one acetate unit. • carbon atom derived from C-2 of mevalonate or C-2 of acetate.

In accordance with Fig. 13.27, illudin M (6.13), biosynthesized by *Clitocybe illudens* was found to retain only one 4-*pro*-R mevalonoid hydrogen. It also retained three hydrogens from C-2 of mevalonate, only one being located in the cyclopropane ring as demanded by the postulated intermediacy of the diene (6.1) (Hanson & Marten, 1973). One third of the label from (2-^{14}C)mevalonate was located at the carbonyl group and one

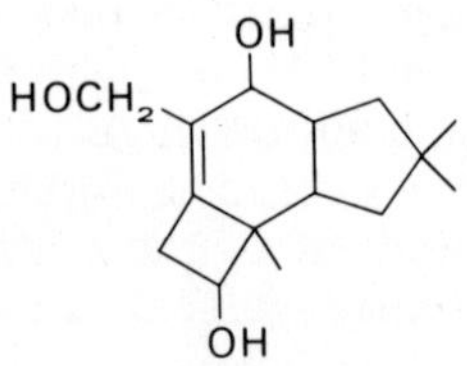

(6.12)
Illudol

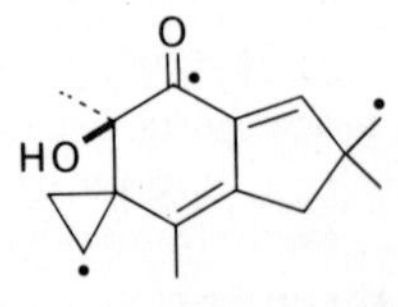

(6.13)
Illudin M

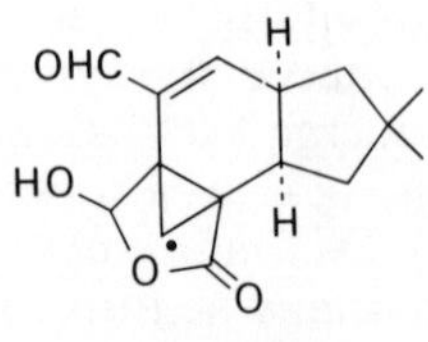

(6.14)
Marasmic acid

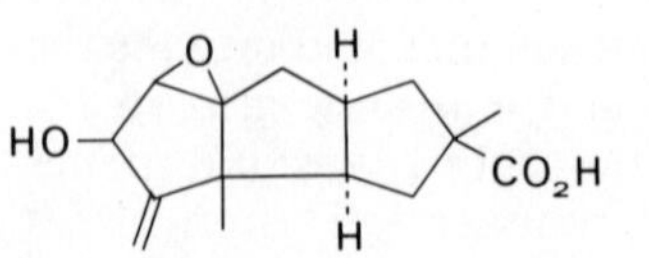

(6.10)
Hirsutic acid

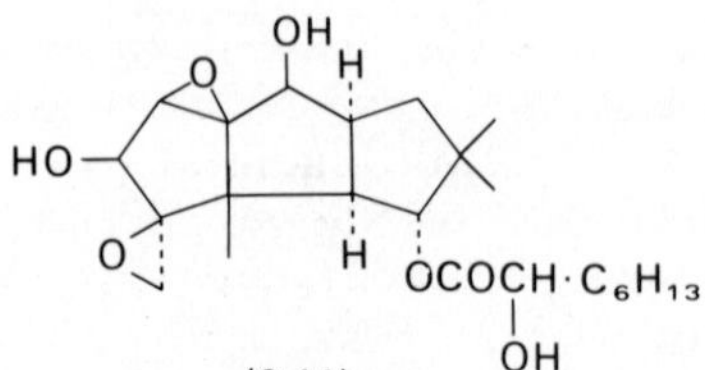

(6.11)
5-Dihydrocoriolin C

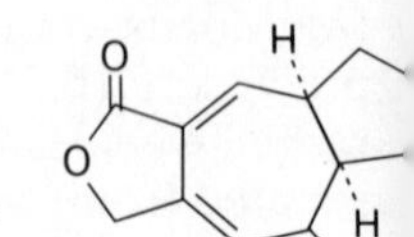

(6.15)

third in the cyclopropyl carbon as shown (Anchel, McMorris & Singh, 1970). Similarly, marasmic acid (6.14), biosynthesized from (2-^{14}C)-mevalonate by *Marasmius congenus*, was shown to have one third of its label in the *gem* dimethyl group and just under a third in the cyclopropane ring (Dugan, de Mayo, Nisbet, Robinson & Anchel, 1966).

Particularly valuable evidence for the validity of this theory resulted from biosynthetic experiments involving the use of ^{13}C n.m.r. spectroscopy. In hirsutic acid C (6.10) from *Stereum hirsutum*, each carbon atom was shown to be derived from C-1 or C-2 of acetic acid in accordance with the predictions of Fig. 13.27 as shown (Feline, Mellows, Jones & Phillips, 1974). Similar results were obtained for 5-dihydrocoriolin C (6.11) from *Coriolus consors* (Tanabe, Suzuki & Janowski, 1974). Perhaps the most striking experiment has been the biosynthesis of 5-dihydrocoriolin C (6.11) from $^{13}CH_3{}^{13}CO_2Na$. The success of this type of experiment depends on the fact that although the statistical chance of adjacent acetate units being labelled as a result of a total incorporation of 1–10% of $^{13}CH_3{}^{13}CO_2Na$ may be small, each acetate unit which is incorporated without C-C cleavage will show mutual coupling of its two carbons. In dimethylallylpyrophosphate, for example, two pairs of carbons will show this enforced co-original coupling while the carbon derived from C-2 of mevalonate will be uncoupled, having lost its partner during decarboxylation of hydroxymethylglutarate (see p. 370). If at any time during a biosynthetic sequence, a pair of ^{13}C labelled, co-originating acetate atoms are separated by rearrangement, each of these atoms will then have a low probability of being located next to a labelled atom and will in the main show no ^{13}C-^{13}C coupling. In 5-dihydrocoriolin C (6.11), mutual coupling was evident within the six pairs of incorporated carbon atoms co-originating from acetate, thus excluding mechanisms in which such pairs are separated by bond shifts at some stage (Tanabe, Suzuki & Jankowski, 1974).

Other fungal sesquiterpenes derived from trans,trans-farnesylpyrophosphate

Fig. 13.28 summarizes the proposed biogenesis of five other groups of fungal sesquiterpene from farnesylpyrophosphate. Concerted cyclization as shown gives the drimane type (*cf* 6.16) exemplified by pebrolide (6.17) from *Penicillium brevi-compactum* (Calzadilla, Ferguson, Hutchinson & McCorkindale, 1968) which retains the expected proportions of tritium from C-2 of mevalonate at C-1, and in the acetoxymethyl group (McCorkindale, Baxter & Calzadilla, unpublished results). The germacrane ion (6.18), formed by 1,10-cyclization of farnesylpyrophosphate, can give rise (as 6.18*a*) to the guiane type represented by a number of products from *Lactarius* species, such as the fulvene (6.19) from *L. deliciosus* (Vokac, Samek, Herout & Sorm, 1970). It can also (as 6.18*b*) afford the eremophilane type represented by phomenone (6.20) from a *Phoma* species (Riche *et al.*, 1974) possibly by way of the concerted methyl and hydride shifts shown. Botrydial (6.24), recently isolated from *Botrytis cinerea* (Tschesche, Felhaber, Geipel, Merker & Welmar, 1974) could, in theory be formed from humulene (6.2) via caryophyllene (6.21). Ring expansion to (6.22), followed by cyclization could give the hypothetical tricyclic precursor (6.23) of botrydial (6.24). Collybolide (6.27) and related sesquiterpenes from *Collybia maculata* (Bui *et al.*, 1974) represent a new class of sesquiterpene based on the monocyclic

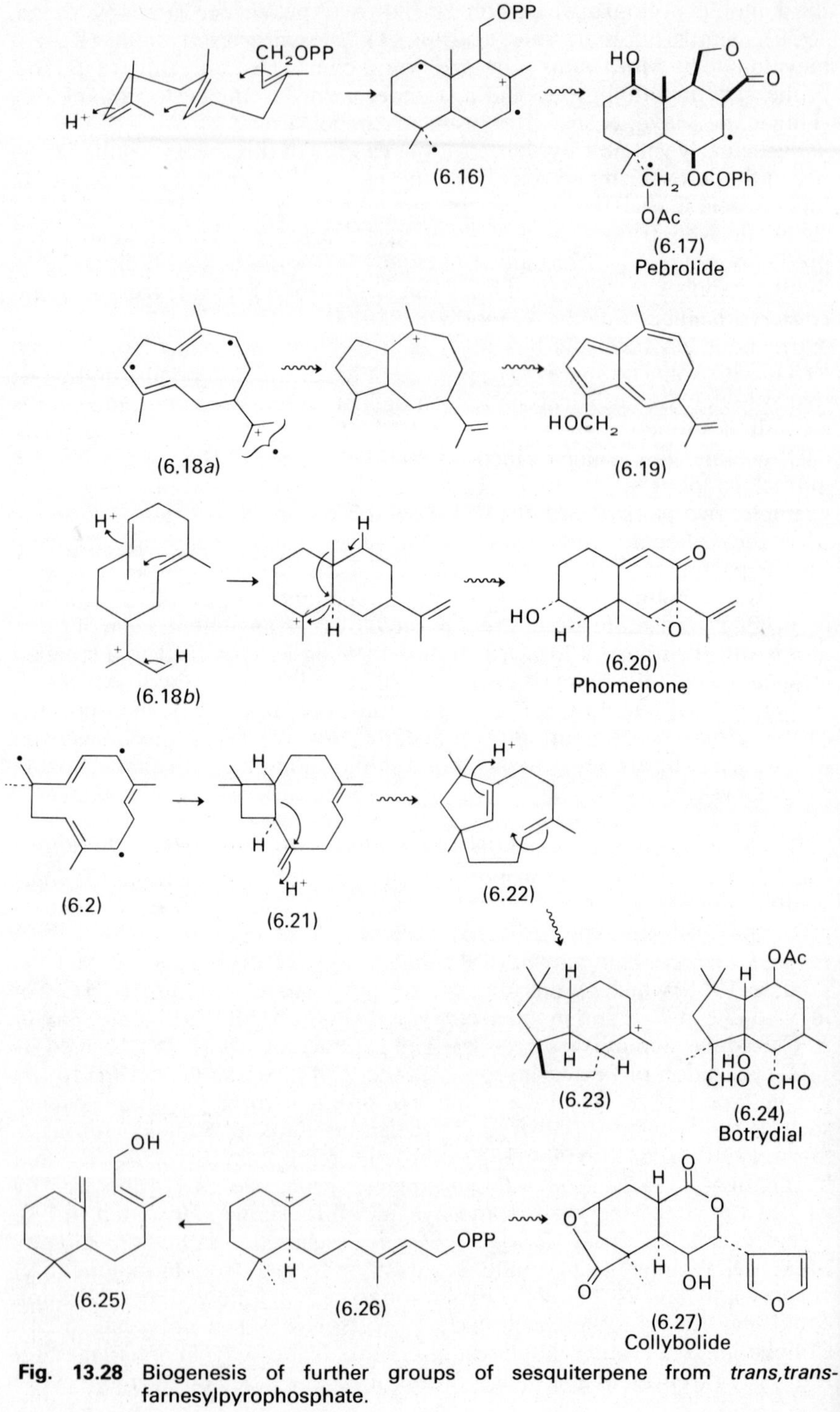

Fig. 13.28 Biogenesis of further groups of sesquiterpene from *trans,trans*-farnesylpyrophosphate.

cation (6.26). *Trans*-γ-monocyclofarnesol (6.25) itself is produced by *Helminthosporium siccans* (Suzuki, Suzuki & Nozoe, 1971).

Helicobasidin and the trichothecanes

As discussed on p. 374, the trichothecane sesquiterpenes produced by a number of fungi (*cf.* Turner, 1971) are derived from farnesylpyrophosphate via its *cis,trans*-isomer, in which the 5-*pro*-S mevalonoid hydrogen from C-1 has been lost. The configuration of the retained 5-*pro*-R mevalonoid hydrogen in the *cis,trans*-farnesylpyrophosphate (6.28) is uncertain at present, but it is retained in the 11α-position in trichothecin, the crotonate ester of trichothecolone (6.34) (Evans, Hanson & Marten, 1974).

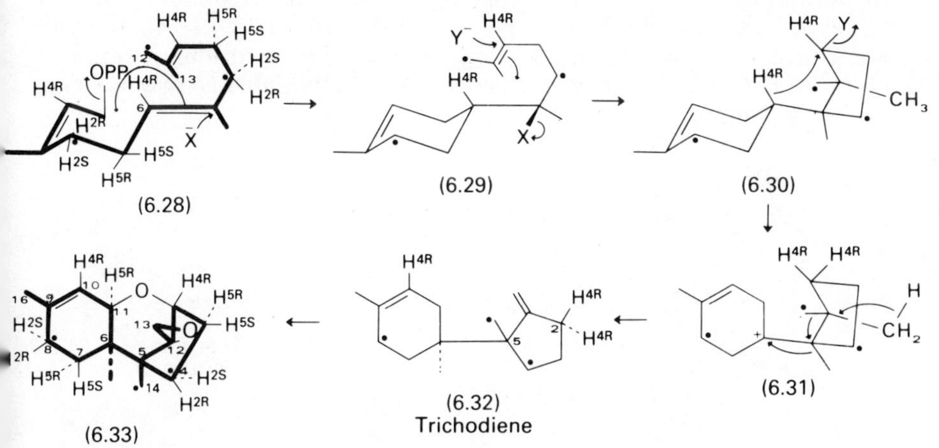

Fig. 13.29 Formation of the trichothecane nucleus.

The biogenesis of the trichothecane skeleton outlined in Fig. 13.29 is based on extensive incorporation/degradation studies on several trichothecanes, establishing the origin of various carbons and hydrogens in relation to those of mevalonic acid (*e.g.* Evans, Hanson & Marten, 1974; Achilladelis, Adams & Hanson, 1972; Achini, Müller & Tamm, 1971; Machida & Nozoe, 1972*a*). Other important factors have been recognition of trichodiene (6.32) as a key intermediate in the biosynthesis of trichothecin by *Trichothecium roseum* (Machida & Nozoe, 1972*b*) and studies on helicobasidin (6.35) from *Helicobasidium mompa* which has the skeleton of the first formed bicyclic intermediate (6.30) (Natori, Inouye & Nishikawa, 1967; Adams & Hanson, 1972).

The latter could be formed in three steps from *cis,trans*-farnesylpyrophosphate (6.28). Addition to the central double bond required to be in an overall *cis* fashion (Arigoni, Cane, Müller & Tamm, 1973). This would be achieved by the first two steps indicated, namely attack at the central double bond by an enzymic nucleophile X^- accompanied by *trans* alkylation by the allylic pyrophosphate system (*cf.* the prenyl chain extension process discussed in Section 13.1). Subsequent attack at the isopropylidene double bond of the product (6.29) by a second enzymic nucleophile Y^-, accompanied by intramolecular *trans* alkylation involving displacement of X^- would give the intermediate (6.30) in which the grouping Y is now appropriately situated for intramolecular displacement by the 4-*pro*-R mevalonoid methine hydrogen. This hydride shift was detected by two 4-*pro*-R mevalonoid hydrogens being found in the five-membered ring of helicobasidin (6.35), presumably at C-2 (Nozoe, Morisaki & Matsumoto, 1970; Adams & Hanson, 1972). From the way in which trichodiene (6.32) is probably formed from the cation (6.31), it can be assumed that trichodiene will also have two 4-*pro*-R mevalonoid hydrogens at C-2, the incoming hydride from the central prenyl unit adopting the β-position. Assuming α-oxygenation at C-2 occurs with retention of configuration, the retained 4-*pro*-R mevalonoid hydrogen at C-2 of a trichothecane will be that originating at the central double bond of farnesylpyrophosphate. In agreement with this, (6-^{3}H;12,13-$^{14}C_2$)farnesol was converted by the fungus *Myrothecium roridum* to roridin, an ester of verrucarol (6.36), in which C-2 carried the tritium label (Arigoni, Cane, Muller & Timm, 1973).

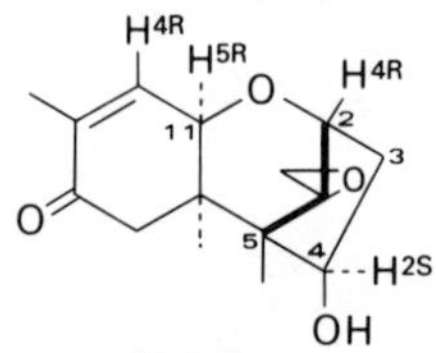

(6.34)
Trichothecolone

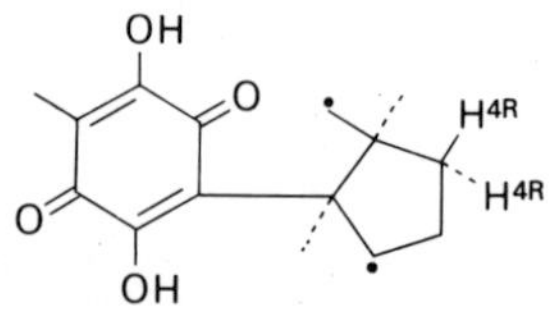

(6.35)

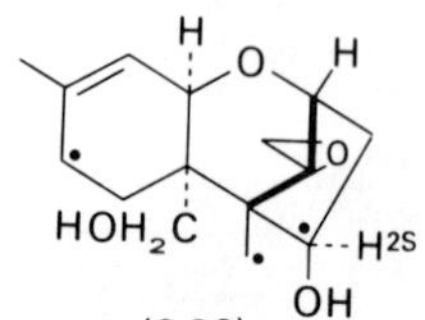

(6.36)
Verrucarol

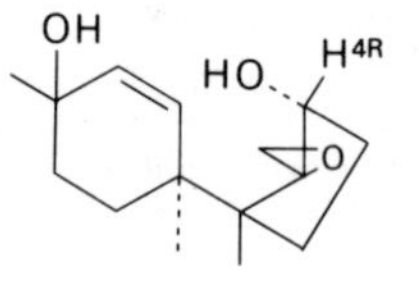

(6.37)

Formation of trichodiene (6.32) from the cation (6.31) involves loss of a proton and shifts of two methyl groups, perhaps in the concerted manner shown in Fig. 13.29. The *gem* dimethyl group migrating from the 12β-position to the 5β-position is that which is derived from C-2 of mevalonate (*cf.* Hanson, Marten & Swerns, 1974). Details of the route from trichodiene to the trichothecanes involving oxygenations and formation of a pyran ring, remain to be determined. A plausible intermediate could be trichodiol (6.37) but it is also possible that this is formed hydrolytically from the co-occurring trichothecene (6.33) (Machida & Nozoe, 1972*a*).

Culmorin and fumagillin

Fig. 13.30 summarizes the proposed biogenesis of two other types of fungal sesquiterpene where derivation from *cis,trans*-farnesyl pyrophosphate can be considered. Culmorin (6.39) from *Fusarium culmorum* (Barton & Werstiuk, 1968) represents the longifoline type of sesquiterpene. A 1,3 hydride shift in the *cis,trans*-unsaturated ion (6.38) could be involved in the biosynthesis.

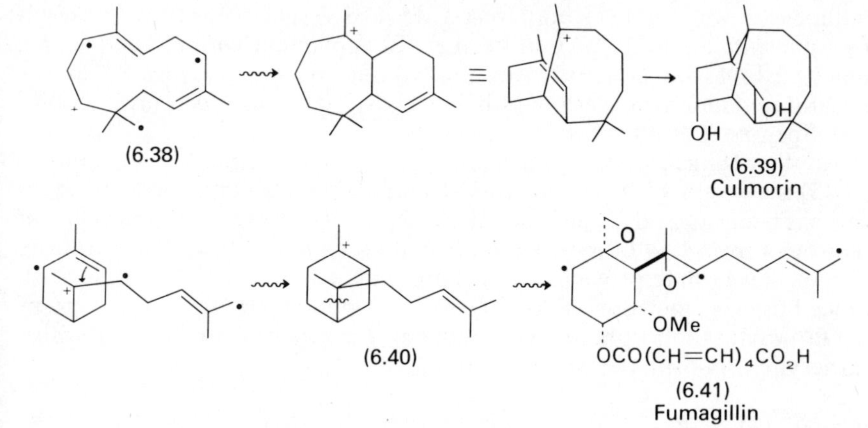

Fig. 13.30 Biogenesis of two other types of sesquiterpene proposed to be formed from *cis,trans*-farnesylpyrophosphate.

Fumagillin (6.41) from *Aspergillus fumigatus* (*cf.* Turner & Tarbell, 1962) could be derived *via* the bergamotane ion (6.40). One third of the activity from (2-^{14}C)mevalonate was shown to be located in the isopropylidene group (Birch & Hussain, 1969).

Helminthosporal and avocetin

Sativene (6.47) and helminthosporal (6.48) from *Helminthosporium sativum* (de Mayo & Williams, 1965; de Mayo, Spenser & White, 1963) probably represent the cadinane type of sesquiterpene (*cf.* the ion 6.45), modified by intramolecular cyclization to 6.46 and a 1,2 bond shift as shown (Fig. 13.31). The candinane ion (6.45) is probably formed *via* germacrene D (6.42) (Yoshihara, Ohta, Sakai & Hirose, 1969; Morikawa & Hirose, 1969), two routes to which seem possible, involving 1,3 hydride shifts in the

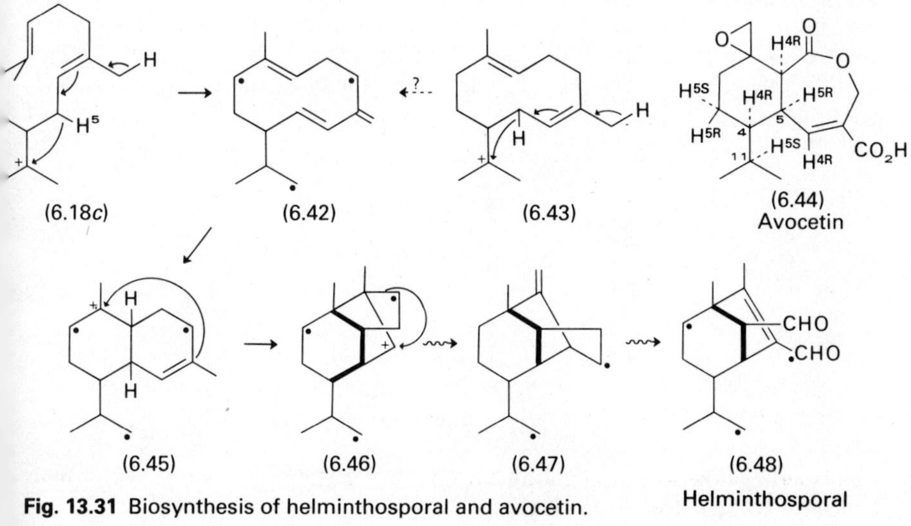

Fig. 13.31 Biosynthesis of helminthosporal and avocetin.

monocyclic ions (6.18c) and (6.43) derived respectively from farnesylpyrophosphate and its *cis,trans*-isomer. In agreement with either scheme, one third of the activity from (2-^{14}C)mevalonate was located in the unsaturated aldehydic carbon of helminthosporal (6.48) (de Mayo, Robinson, Spenser & White, 1962).

Another fungal product which is probably derived from the cadinane ion (6.45) is avocetin (6.44) from *Anthostoma avocetta*, and the biosynthesis of this has been studied in some detail (Arigoni, 1973). The retention at C-4 of hydrogen from C-4 of mevalonate and at C-5 and C-11 of hydrogen from C-5 of mevalonate provides compelling evidence for the 1,3 hydride shift. Since both C-5 mevalonoid hydrogens corresponding to the C-1 hydrogens of farnesylpyrophosphate are retained, the intermediacy of *cis,trans*-farnesylpyrophosphate can be ruled out.

Triprenyl phenolics

Phenolic polyketides are of widespread occurrence among fungi (Turner, 1971) and a number of sesquiterpenoids are derived by triprenylation of such phenols, accompanied by varying degrees of cyclization. Orsellinic acid (6.49) is probably the phenolic precursor of grifolin (6.54) from *Grifola confluens* (Goto, Kakisawa & Hirata, 1963) and also of siccanin (6.56) from *Helminthosporium siccans* (Hirai *et al.*, 1967). Although cyclization of the farnesyl chain may occur by a concerted process in the case of the drimanyl quinone, tauranin (6.55), produced by *Oospora aurantia* (Kawashima, Nakanishi & Nishikawa, 1964), it is now known that a stepwise cyclization sequence is involved in the case of siccanin (*cf.* Fig. 13.32). Incubation of farnesylpyrophosphate and orsellinic acid with a cell-free extract from *H. siccans* gave siccanochromene A (6.53) in quantitative yield (Suzuki & Nozoe, 1972) and this can be converted by whole cells of the fungus into siccanin (Suzuki & Nozoe, 1971). These systems were also used to establish the intermediacy of presiccanochromenic acid (6.51), siccanochromenic acid (6.52) and siccanochromene B (6.57). The fungus also produces the monocyclofarnesol (6.26) (Suzuki, Suzuki & Nozoe, 1971) raising the possibility that partial cyclization of farnesylpyrophosphate may occur before alkylation of orsellinic acid takes place (Suzuki & Nozoe, 1972). This could, however, merely reflect lack of specificity in the cyclase, farnesylpyrophosphate being the normal alkylating agent. *In vitro* synthesis of grifolin (6.54) from orcinol and farnesol has been accomplished, acid

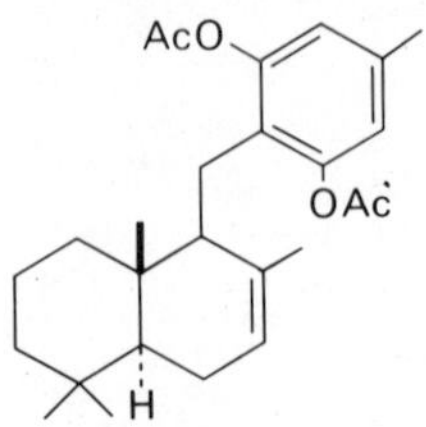

(6.58)

catalysed cyclization then giving a product (6.58) with the same skeleton as tauranin (6.55) (Ima-ye & Kakisawa, 1973).

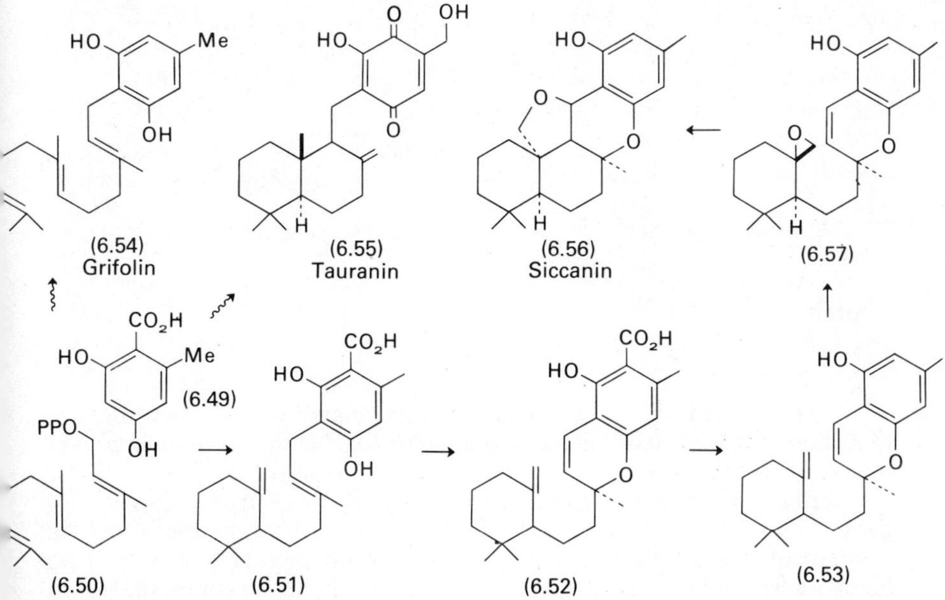

Fig. 13.32 Biosynthesis of siccanin.

It has been suggested that in the biosynthesis of ascochlorin (6.60), epoxidation of the prenyl chain occurs prior to cyclization, and that the resulting cyclohexanol cation (6.59) undergoes concerted shifts of hydride, methyl and hydride to give the terpenoid framework of ascochlorin (Tanabe

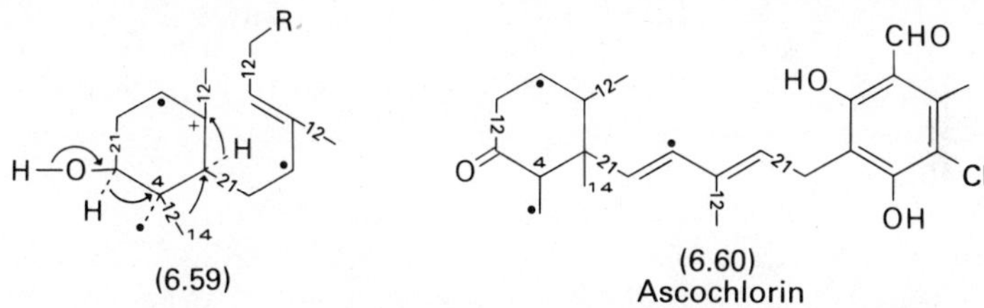

& Suzuki, 1974). This was supported by examination of a sample biosynthesized by *Nectria coccinea* from $^{13}CH_3{}^{13}CO_2$ Na by ^{13}C n.m.r. spectroscopy, a technique discussed earlier in connection with the biosynthesis of 5-dihydrocoriolin C. The carbons corresponding to C-2 of mevalonate appeared as singlets as expected and in the terpenoid moeity, only five of the expected six pairs of co-originating atoms showed ^{13}C–^{13}C coupling. Separation of the C-1, C-14 pair of (6.59) as a result of the methyl shift during its conversion to (6.60) was highlighted by the signals appropriate for these carbons appearing as enriched singlets in the spectrum of the latter.

Triprenylation of a C-methylated polyketide phenolic appears to be involved in the biosynthesis of the cochlioquinones e.g. cochlioquinone A (6.61) from *Cochliobolus miyabeanus* (Canonica *et al.*, 1973). As in the case of tauranin (6.55), a drimanylphenol of the type (6.62) is a likely intermediate.

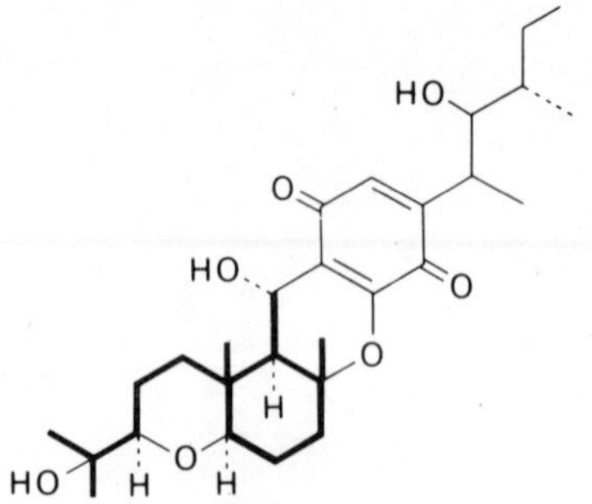

(6.61)
Cochlioquinone A

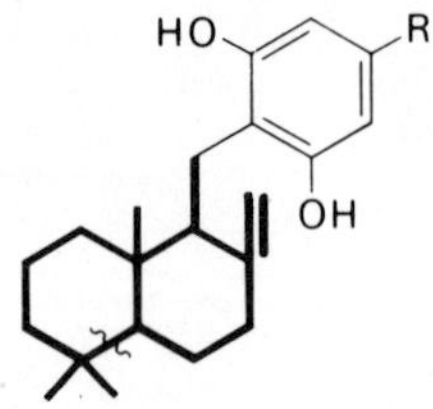

(6.62)

As indicated in Fig. 13.33 the species prenylated in the biosynthesis of mycophenolic acid (6.68) is the phthalide (6.64) derived from methylorsellinic acid (6.63) (Canonica *et al.*, 1972). O-methylation at this stage blocks the prenylation. This contrasts with the lack of specificity exhibited in the final stages of mycophenolic acid biosynthesis. One pathway *via* the C-prenylphthalide (6.65) and O-desmethylmycophenolic acid (6.66) has been supported by efficient incorporation of these compounds into mycophenolic acid (Canonica *et al.*, 1972). However, the O-methyl-C-prenylphthalide (6.67) has been isolated from a strain of *Penicillium brevicompactum* and is converted by the fungus into mycophenolic acid even more efficiently than is the co-occurring C-prenylphthalide (6.65) under parallel conditions (McCorkindale, Baxter, & Turner, unpubl. results).

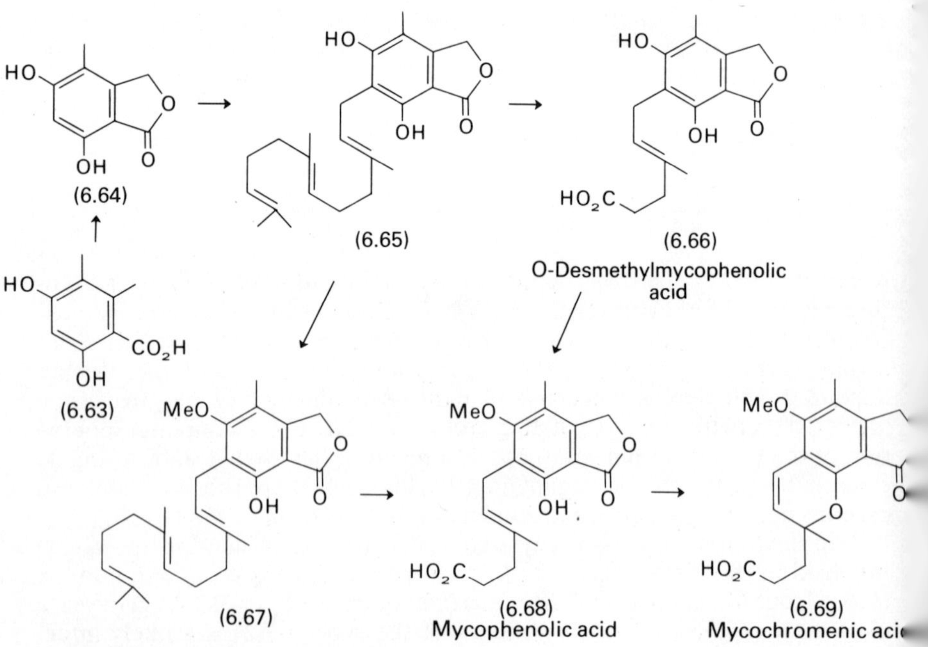

Fig. 13.33 Biosynthesis of mycophenolic acid.

Evidently the enzyme system can effect side chain cleavage and O-methylation in either order.

The chromene rings of mycochromenic acid (6.69) from *Penicillium brevi-compactum* and of siccanochromenic acid (6.52) are probably formed *via* quinone methide intermediates as suggested for the biosynthesis of other naturally occurring chromenes (Turner, 1964). The required dehydrogenation/cyclization of mycophenolic acid can be achieved *in vitro* using dichlorodicyano-benzoquinone (Campbell, Calzadilla & McCorkindale, 1966).

13.7 References

ABERHART, D. J. & CASPI, E. (1971). The fate of the 6α-hydrogen of 5α-cholest-7-en-3β-ol in the conversion to 7-dehydrocholesterol by rat liver microsomes. *Journal of Biological Chemistry* **246**, 1387–92.

ACHILLADELIS, B. A. & HANSON, J. R. (1969). Studies in terpenoid biosynthesis. Part V. Biosynthesis of rosenonolactone. *Journal of the Chemical Society* (C) 2010–14.

ACHILLADELIS, B. A., ADAMS, P. M., & HANSON, J. R. (1972). Studies in terpenoid biosynthesis. Part VIII. The formation of the trichothecane nucleus. *Journal of the Chemical Society, Perkin I* 1425-8.

ACHINI, R., MÜLLER, B. & TAMM, C. (1971). Biosynthesis of verrucarol, the sesquiterpene moeity of the verrucarins and roridins. *Chemical Communications* 404–5.

ADAM, H. K., BRYCE, T. A., CAMPBELL, I. M., McCORKINDALE, N. J., GAUDEMER, A., GMELIN, R. & POLONSKY, J. (1967). Metabolites of the Polyporaceae. II. Carboxyacetylquercinic acid—a novel triterpene conjugate from *Daedalea quercina*. *Tetrahedron Letters* 1461–5.

ADAM, H. K., CAMPBELL, I. M. & McCORKINDALE, N. J. (1967). Ergosterol peroxide: a fungal artefact. *Nature, London* **216**, 397.

ADAMS, P. M. & HANSON, J. R. (1972). Studies in terpenoid biosynthesis. Part VII. The biosynthesis of helicobasidin. *Journal of the Chemical Society, Perkin I* 586–8.

AKHTAR, M. & MARSH, S. (1967). The stereochemistry of the hydrogen elimination in the biological conversion of cholest-7-en-3β-ol into cholesterol. *Biochemical Journal* **102**, 462–7.

AKHTAR, M. & PARVEZ, M. A. (1968). The mechanism of the elaboration of ring B in ergosterol biosynthesis. *Biochemical Journal* **108**, 527–31.

AKHTAR, M., HUNT, P. F. & PARVEZ, M. A. (1967). The transfer of hydrogen from C-24 to C-25 in ergosterol biosynthesis. *Biochemical Journal* **103**, 616–22.

AKHTAR, M., PARVEZ, M. A. & HUNT, P. F. (1966). The synthesis of labelled 24-methylenelanosterol and its conversion into ergosterol. *Biochemical Journal* **100**, 38*C*–40*C*.

AKHTAR, M., PARVEZ, M. A. & HUNT, P. F. (1968). The introduction of the C-22–C-23 ethylenic linkage in ergosterol biosynthesis. *Biochemical Journal* **106**, 623–6.

AKHTAR, M., PARVEZ, M. A. & HUNT, P. F. (1969). Studies on the biosynthesis of the ergosterol side chain. *Biochemical Journal* **113**,727–32.

AKHTAR, M., WILTON, D. C., WATKINSON, I. A. & RAHIMTULA, A. D. (1972). Substrate activation in pyridine nucleotide-linked reactions: illustrations from the steroid field. *Proceedings of the Royal Society, London, Series B* **180**, 167–77.

AKHTAR, M., RAHIMTULA, A. D., WATKINSON, I. A., WILTON, D. C. & MUNDAY, K. A. (1969). The status of C-6, C-7, C-15 and C-16 hydrogen atoms in cholesterol biosynthesis. *European Journal of Biochemistry* **9**, 107–11.

ALEXANDER, K., AKHTAR, M., BOAR, R. B., McGHIE, J. F. & BARTON, D. H. R. (1971). The pathway for the removal of C-32 in cholesterol biosynthesis. *Chemical Communications* 1479–81.

ALEXANDER, K., AKHTAR, M., BOAR, R. B., McGHIE, J. F. & BARTON, D. H. R. (1972). The removal of the 32–carbon atom as formic acid in cholesterol biosynthesis. *Journal of the Chemical Society, Chemical Communications* 383–5.

ALTMAN, L. J., KOWERSKI, R. C. & RILLING, H. C. (1971). Synthesis and conversion of presqualene alcohol to squalene. *Journal of the American Chemical Society* **93**, 1782–3.

ALTMAN, L. J., ASH, L., KOWERSKI, R. C., EPSTEIN, W. W., LARSEN, B. R., RILLING, H. C., MUSCIO, F. & GREGONIS, D. E. (1972). Prephytoene pyrophosphate a new intermediate in the biosynthesis of carotenoids. *Journal of the American Chemical Society* **94**, 3257–9.

ANCHEL, M., McMORRIS, T. C. & SINGH, P. (1970). The biogenesis of illudins S and M in *Clitocybe illudens. Phytochemistry* **9**, 2339–43.

ARCHER, B. L., BARNARD, D., COCKBAIN, E. G., CORNFORTH, J. W., CORNFORTH, R. H. & POPJAK, G. (1965). The stereochemistry of rubber biosynthesis. *Proceedings of the Royal Society, London, Series B* **163**, 519–23.

ARIGONI, D. (1968). Some studies in the biosynthesis of terpenes and related compounds. *Pure and Applied Chemistry* **17**, 331–48.

ARIGONI, D. (1973). The biosynthesis of cyclic terpenes. *Chemical Society Half-day Symposium on the biosynthesis of terpenes*, 4 October, London.

ARIGONI, D., CANE, D. E., MÜLLER, B. & TAMM, C. (1973). Verrucarins and roridins. 26. The mode of incorporation of farnesylpyrophosphate into verrucarol. *Helvetica Chimica Acta* **56**, 2946–9.

ARMAREGO, W. L. F., GOAD, L. J. & GOODWIN, T. W. (1973). Biosynthesis of α-spinasterol from (2-^{14}C, 4R-4-$^{3}H_1$)mevalonic acid by *Spinacea oleracea* and *Medicago sativa. Phytochemistry* **12**, 2181–7.

ATHERTON, L., DUNCAN, J. M. & SAFE, S. (1972). Isolation and biosynthesis of ergost-5,7,9(11),22-tetraen-3β-ol from *Mucor rouxii. Journal of the Chemical Society, Chemical Communications* 882–3.

AYER, W. A. & TAUBE, H. (1973). Metabolites of *Cyathus helenae.* A new class of diterpenoids. *Canadian Journal of Chemistry* **51**, 3842-54.

BARR, R. M. & HEMMING, F. W. (1972). Polyprenol phosphate as an acceptor of mannose from guanosine diphosphate mannose in *Aspergillus niger. Biochemical Journal* **126**, 1203–8.

BARTNICKI-GARCIA, S. (1968). Cell wall chemistry, morphogenesis and taxonomy of fungi. *Annual Review of Microbiology* **22**, 87–108.

BARTON, D. H. R. & WERSTIUK, N. H. (1968). Sesquiterpenoids. Part XIV. The constitution and stereochemistry of culmorin. *Journal of the Chemical Society* (C) 148–55.

BARTON, D. H. R., CORRIE, J. E. T., MARSHALL, P. J. & WIDDOWSON, D. A. (1973). Biosynthesis of terpenes and steroids. VII. Unified scheme for the biosynthesis of ergosterol in *Saccharomyces cerevisiae. Bio-organic Chemistry* **2**, 363–73.

BARTON, D. H. R., HARRISON, D. M., MOSS, G. P. & WIDDOWSON, D. A. (1970). Investigations on the biosynthesis of steroids and terpenoids. Part II. Role of 24-methylene derivatives in the biosynthesis of steroids and terpenoids. *Journal of the Chemical Society (C)* 775–85.

BARTON, D. H. R., CORRIE, J. E. T., WIDDOWSON, D. A., BARD, M. & WOODS, R. A. (1974). Biosynthesis of terpenes and steroids. Part IX. The sterols of some mutant yeasts and their relationship to the biosynthesis of ergosterol. *Journal of the Chemical Society, Perkin I* 1326–33.

BEAN, G. A., PATTERSON, G. W. & MOTTA, J. J. (1973). Sterols and fatty acids of some aquatic Phycomycetes. *Comparative Biochemistry and Physiology* **43B**, 935–9.

BEARDER, J. R. & MACMILLAN, J. (1972). Gibberellin A_{36}, isolation from *Gibberella fujikuroi*, structure and conversion to gibberellin A_{37}. *Agricultural and Biological Chemistry* **36**, 342–4.

BEARDER, J. R., MACMILLAN, J. & PHINNEY, B. O. (1973*a*). 3-Hydroxylation of gibberellin A_{12}-aldehyde in *Gibberella fujikuroi* strain REC-193-A. *Phytochemistry* **12**, 2173–9.

BEARDER, J. R., MACMILLAN, J. & PHINNEY, B. O. (1973*b*). Conversion of gibberellin A_1 into gibberellin A_3 by the mutant R-9 of *Gibberella fujikuroi. Phytochemistry* **12**, 2655–9.

BEARDER, J. R., MACMILLAN, J., WELS, C. M., CHAFFEY, M. B. & PHINNEY, B. O. (1974). Position of the metabolic block for gibberellin biosynthesis in mutant B1-41a of *Gibberella fujikuroi. Phytochemistry* **13**, 911–7.

BIMPSON, T., GOAD, L. J. & GOODWIN, T. W. (1969). The stereochemistry of hydrogen elimination at C-6, C-22 and C-23 during ergosterol biosynthesis by *Aspergillus fumigatus* Fres. *Chemical Communications* 297–8.

BIRCH, A. J. & HUSSAIN, S. F. (1969). Studies in relation to biosynthesis. Part XXXVIII. A preliminary study of fumagillin. *Journal of the Chemical Society (C)* 1473–4.

BIRCH, A. J., HOLTZAPFEL, C. W. & RICKARDS, R. W. (1966). The structure and some aspects of the biosynthesis of pleuromutilin. *Tetrahedron* **22**, *Supplement 8*, 359–87.

BIRCH, A. J., RICKARDS, R. W., SMITH, H., HARRIS, A. & WHALLEY, W. B. (1959). Studies in relation to biosynthesis. XXI. Rosenonolactone and gibberellic acid. *Tetrahedron* **7**, 241–51.

BRIAN, P. W. & McGOWAN, J. C. (1945). Viridin, a highly fungistatic substance produced by *Trichoderma viride*. *Nature, London* **156**, 144.

BRITT, J. J. & ARIGONI, D. (1958). The biogenesis of the diterpene rosenonolactone. *Proceedings of the Chemical Society* 224–5.

BRYCE, T. A., CAMPBELL, I. M. & McCORKINDALE, N. J. (1967). Metabolites of the Polyporaceae. I. Novel conjugates of polyperenic acid A from *Piptoporus betulinus*. *Tetrahedron* **23**, 3427–34.

BUI, A-M., CAVÉ, A., JANOT, M-M., PARELLO, J., POTIER, P. & SCHEIDEGGER, U. (1974). Isolation and structural analysis of collybolide, a new sesquiterpene extract from *Collybia maculata* Alb. et Sch. ex. Fries (Basidiomycetes). *Tetrahedron* **30**, 1327–36.

CALZADILLA, C. H., FERGUSON, G., HUTCHINSON, S. A. & McCORKINDALE, N. J. (1968). Pebrolide, a sesquiterpene benzoate from *Penicillium brevicompactum*. *I.U.P.A.C. 5th International Symposium on Chemistry of Natural Products, Abstracts*, p. 287. Oxford; Alden & Mowbray.

CAMPBELL, I. M., CALZADILLA, C. H. & McCORKINDALE, N. J. (1966). Some new metabolites related to mycophenolic acid. *Tetrahedron Letters* 5107–11.

CANONICA, L. & FIECCHI, A. (1970). Structure and biosynthesis of ophiobolins. *Research Progress in Organic, Biological and Medicinal Chemistry* **2**, 51–93.

CANONICA, L., FIECCHI, A., GALLI-KIENLI, M., RANZI, B. M. & SCALA, A. (1967*b*). The stereochemical course of the 1,5-shift of hydrogen in the biosynthesis of ophiobolins. *Tetrahedron Letters* 4657-9.

CANONICA, L., FIECCHI, A., GALLI-KIENLI, M., RANZI, B. M. & SCALA, A. (1968). The stereochemistry of hydrogen eliminations from 4-C of mevalonate in the biosynthesis of ophiobolins. *Tetrahedron Letters* 275–9.

CANONICA, L., RANZI, B. M., RINDONE, B., SCALA, A. & SCOLASTICO, C. (1973). Biosynthesis of the cochlioquinones. *Journal of the Chemical Society, Chemical Communications* 213–14.

CANONICA, L., FIECCHI, A., GALLI-KIENLI, M., RANZI, B. M., SCALA, A., SALVATORI, T. & PELLA, E. (1967*a*). The biosynthesis of ophiobolins. *Tetrahedron Letters* 3371–6.

CANONICA, L., KROSZCZYNSKI, W., RANZI, B. M., RINDONE, B., SANTANIELLO, E. & SCOLASTICO, C. (1972). Biosynthesis of mycophenolic acid. *Journal of the Chemical Society, Perkin I* 2639–43.

CASPI, E. & MULHEIRN, L. J. (1970). Mechanism of squalene cyclization. Biosynthesis of fusidic acid from (4R)-(2-^{14}C,4-^{3}H)-mevalonic acid. *Journal of the American Chemical Society* **92**, 404–6.

CASPI, E. & RAMM, P. J. (1969). Stereochemical differences in the biosynthesis of C_{27}-Δ^7-steroidal intermediates. *Tetrahedron Letters* 181–5.

CASPI, E., RAMM, P. J. & GAIN, R. E. (1969). Stereochemistry of tritium at carbon 15 in cholesterol derived from (3R,2R)-2T-mevalonic acid in rat livers. *Journal of the American Chemical Society* **91**, 4012–13.

CASPI, E., GREIG, J. B., RAMM, P. J. & VARMA, K. R. (1968). Stereochemistry of tritium at C-1 and C-7 in cholesterol derived from (3R,2R)-2T-mevalonic acid. *Tetrahedron Letters* 3829–32.

CASPI, E., EBERSOLE, R. C., GODTFREDSEN, W. O. & VANGELAL, S. (1972). Mechanism of squalene cyclization: the chiral origin of the C-22 hydrogen atoms in fusidic acid. *Journal of the Chemical Society, Chemical Communications* 1191–33.

CLAYTON, R. B. (1965). Biosynthesis of sterols, steroids and terpenoids. *Quarterly Reviews* **19**, 168–230.

CORDELL, G. A. (1974). The occurrence, structure elucidation and biosynthesis of the sesterterpenes. *Phytochemistry* **13**, 2343–64.

CORNFORTH, J. W. (1968). Olefin alkylation in biosynthesis. *Angewandte Chemie, International Edition* **7**, 903–11.

CORNFORTH, J. W. (1969). Exploration of enzyme mechanisms by asymmetric labelling. *Quarterly Reviews* **23**, 125–40.

CORNFORTH, J. W. (1973). The logic of working with enzymes. *Chemical Society Reviews* **2**, 1–20.

CORNFORTH, J. W., CORNFORTH, R. H., DONNINGER, C. & POPJAK, G. (1965*a*). Studies on the biosynthesis of cholesterol. XIX. Steric course of hydrogen eliminations and of C—C bond formations in squalene biosynthesis. *Proceedings of the*

Royal Society (London), Series B **163**, 492–514.

CORNFORTH, J. W., CORNFORTH, R. H., DONNINGER, C., POPJAK, G., SCHIMIZU, Y., ICHII, S., FORCHIELLI, E. & CASPI, E. (1965*b*). The migration and elimination of hydrogen during biosynthesis of cholesterol from squalene. *Journal of the American Chemical Society* **87**, 3224–8.

CROSS, B. E., NORTON, K. & STEWART, J. C. (1968). The biosynthesis of the gibberellins. Part III. *Journal of the Chemical Society (C)* 1054–63.

DADA, O. A., THRELFALL, D. R. & WHISTANCE, G. R. (1968). Biosynthesis of the polyprenyl side chains of terpenoid quinones and chromanols in maize shoots. *European Journal of Biochemistry* **4**, 329–33.

DALZIEL, W., HESP, B., STEVENSON, K. M. & JARVIS, J. A. J. (1973). The structure and absolute configuration of the antibiotic aphidcolin, a tetracyclic diterpenoid containing a new ring system. *Journal of the Chemical Society, Perkin I* 2841–51.

DANIEWSKI, W. M., KOCOR, M. & ZOLTOWSKA, B. (1973). Constituents of higher fungi. V. Structure of lactarorufin B. *Bulletin de l'Academie Polonaise des Sciences Series des Sciences Chimique* **21**, 785–92.

DEAN, P. D. G., DE MONTELLANO, P. R. O., BLOCH, K. & COREY, E. J. (1967). A soluble 2,3-oxidosqualene sterol cyclase. *Journal of Biological Chemistry* **242**, 3014–15.

DE MAYO, P. & WILLIAMS, R. E. (1965). Sativene, parent of the toxin from *Helminthosporium sativum. Journal of the American Chemical Society* **87**, 3275.

DE MAYO, P., SPENCER, E. Y. & WHITE, R. W. (1963). Terpenoids. IV. The structure and stereochemistry of helminthosporal. *Canadian Journal of Chemistry* **41**, 2996–3004.

DE MAYO, P., ROBINSON, J. R., SPENCER, E. Y. & WHITE, R. W. (1962). The biogenesis of helminthosporal. *Experientia* **18**, 359.

DEVYS, M. & BARBIER, M. (1969). Biosynthesis of eburicoic acid from trametanolic acid. *Bulletin de la Societe de Chimie Biologique* **51**, 925–33.

DIETRICH, S. M. C. (1973). Carbohydrates from the hyphal walls of some Oömycetes. *Biochimica et Biophysica Acta* **313**, 95–8.

DONNINGER, C. & POPJAK, G. (1965). Studies on the biosynthesis of cholesterol. XVIII. The stereospecificity of mevaldate reductase and the biosynthesis of asymmetrically labelled farnesyl pyrophosphate. *Proceedings of the Royal Society (London) Series B* **163**, 465–91.

DUCHAMP, D. J., CHIDESTER, C. G., WICKRAMASINGHE, J. A. F., CASPI, E. & YAGEN, B. (1971). *Cis*-reduction of Δ^{24} of lanosterol in the biosynthesis of cholesterol by rat liver enzymes. A revision. *Journal of the American Chemical Society* **93**, 6283–4.

DUGAN, J. J., DE MAYO, P., NISBET, M., ROBINSON, J. R. & ANCHEL, M. (1966). Terpenoids. XIV. The constitution and biogenesis of marasmic acid. *Journal of the American Chemical Society* **88**, 2838–44.

DURLEY, R. C., RAILTON, I. D. & PHARIS, R. P. (1974). Conversion of gibberellin A_{14} to other gibberellins in seedlings of dwarf *Pisum sativum. Phytochemistry* **13**, 547–51.

EBERSOLE, R. C., GODTFREDSEN, W. O., VANGEDAL, S. & CASPI, E. (1973). Mechanism of oxidative cyclization of squalene. Evidence for cyclization of squalene from either end of the molecule in the *in vivo* biosynthesis of fusidic acid by *Fusidium coccineum. Journal of the American Chemical Society* **95**, 8133–40.

EDMOND, J., POPJAK, G., WONG, S. & WILLIAMS, V. P. (1971). Presqualene alcohol. Further evidence on the structure of a C_{30} precursor of squalene. *Journal of Biological Chemistry* **246**, 6254–71.

ELLESTAD, G. A., EVANS, R. H. & KUNSTMANN, M. P. (1971). LL–Z1271β an additional C_{16} terpenoid metabolite from an *Acrostalagmus* species. *Tetrahedron Letters* 497–500.

ELLESTAD, C. A., EVANS, R. H., KUNSTMANN, M. P., LANCASTER, J. E. & MORTON, G. O. (1970). The structure and chemistry of an antibiotic LL-Z1271α, an antifungal carbon-17 terpene. *Journal of the American Chemical Society* **92**, 5483–9.

EPSTEIN, W. W. & RILLING, H. C. (1970). Studies on the mechanism of squalene biosynthesis. The structure of presqualene pyrophosphate. *Journal of Biological Chemistry* **245**, 4597–605.

EVANS, R. & HANSON, J. R. (1972). The formation of (−)kaurene in a cell-free system from *Gibberella fujikuroi. Journal of the Chemical Society, Perkin I* 2382–5.

EVANS, R., HANSON, J. R. & MARTEN, T. (1974). Studies in terpenoid biosynthesis. Part XI. Stereochemistry of some stages in trichothecane biosynthesis. *Journal of the Chemical Society, Perkin I* 857–60.

EVANS, R., HANSON, J. R. & MULHEIRN, L. J. (1973). Studies in terpenoid biosynthesis. Part X. Incorporation of 5S–(5–$^3H_1$)-mevalonic acid into gibberellic acid. *Journal of the Chemical Society, Perkin I* 753–6.

EVANS, R., HANSON, J. R. & WHITE, A. F. (1970). Studies in terpenoid biosynthesis. Part VI. The stereochemistry of some stages in tetracyclic diterpene biosynthesis. *Journal of the Chemical Society (C)* 2601–3.

EVANS, R., HOLTON, A. M. & HANSON, J. R. (1973). Biosynthesis of 2-*cis*-farnesol. *Journal of the Chemical Society, Chemical Communications* 465.

FALL, R. R. & WEST, C. A. (1971). Purification and properties of kaurene synthetase from *Fusarium moniliforme. Journal of Biological Chemistry* **246**, 6913–28.

FELINE, T. C., MELLOWS, G., JONES, R. B., PHILLIPS, L. (1974). Biosynthesis of hirsutic acid using C-13 nuclear magnetic resonance spectroscopy. *Journal of the Chemical Society, Chemical Communications* 63–4.

FIECCHI, A., GALLI-KIENLI, M., SCALA, A., GALLI, G., GROSSI-PAOLETTI, E., CATTOBENI, F. & PAOLETTI, R. (1972). Hydrogen exchange and double bond formation in cholesterol biosynthesis. *Proceedings of the Royal Society (London) Series B* **180**, 147–65.

FRYBERG, M., OEHLSCHLAGER, A. C. & UNRAU, A. M. (1973). Biosynthesis of ergosterol in yeast. Evidence for multiple pathways. *Journal of the American Chemical Society* **95**, 5747–57.

GALLI-KIENLI, M., VARMA, R. K., MULHEIRN, L. J., YAGEN, B. & CASPI, E. (1973). Reduction of Δ^{24} of lanosterol in the biosynthesis of cholesterol by rat liver enzymes. II. Stereochemistry of addition of the C-25 proton. *Journal of the American Chemical Society* **95**, 1996–2001.

GHISALBERTI, E. L., DE SOUSA, N. J., REES, H. H., GOAD, L. J. & GOODWIN, T. W. (1969). Biological removal of the 4α-methyl group during the conversion of cycloartenol into 31-norcycloartenol in *Polypodium vulgare* Linn. *Chemical Communications* 1403–5.

GOAD, L. J. (1970). Sterol biosynthesis. In *Natural substances formed biologically from mevalonic acid.* ed. Goodwin, T. W. pp. 45–77. London & New York: Academic Press.

GOAD, L. J. & GOODWIN, T. W. (1969). Studies in phytosterol biosynthesis: observations on the biosynthesis of fucosterol in the marine brown alga *Fucus spiralis. European Journal of Biochemistry* **7**, 502–8.

GOAD, L. J. & GOODWIN, T. W. (1972). The biosynthesis of plant sterols. In *Progress in phytochemistry.* ed. Reinhold, L., Liwschitz, Y. **3**, 113. New York: Wiley (Interscience).

GOAD, L. J., HAMMAN, A. S. A., DENNIS, A. & GOODWIN, T. W. (1966). Biosynthesis of the phytosterol side chain. *Nature, London* **210**, 1322–4.

GOAD, L. J., KNAPP, F. F., LENTON, J. L. & GOODWIN, T. W. (1972). Observations on the sterol side chain alkylation mechanism in a *Trebouxia* species. *Biochemistry Journal* **129**, 219–22.

GODTFREDSEN, W. O., LORCK, H., VAN TAMELEN, E. E., WILLETT, J. D. & CLAYTON, R. B. (1968). Biosynthesis of fusidic acid from squalene 2,3-oxide. *Journal of the American Chemical Society* **90**, 208–9.

GOODWIN, T. W. (1973). Comparative biochemistry of sterols in eukaryotic micro-organisms. In *Lipids and biomembranes of eukaryotic micro-organisms.* Edited by Erwin, J. A., pp. 1–40. New York & London: Academic Press.

GOTO, T., KAKISAWA, H. & HIRATA, Y. (1963). The structure of grifolin, an antibiotic from a Basidiomycete. *Tetrahedron* **19**, 2079–83.

GOULSTON, G., GOAD, L. J. & GOODWIN, T. W. (1967). Sterol biosynthesis in fungi. *Biochemical Journal* **102**, 15C–17C.

GROVE, J. F. (1969). Viridin. Part VI. Evidence for a steroidal pathway in the biogenesis of viridin from mevalonic acid. *Journal of the Chemical Society (C)* 549–51.

HALL, J., SMITH, A. R. H., GOAD, L. J. & GOODWIN, T. W. (1969). The conversion of lanosterol, cycloartenol and 24-methylenecycloartenol into poriferasterol by *Ochromonas malhamensis. Biochemistry Journal* **112**, 129–30.

HANSON, J. R. & HAWKER, J. (1972). The formation of the C_{19}-gibberellins from gibberellin A_{13} anhydride. *Tetrahedron Letters* 4299–302.

HANSON, J. R. & MARTEN, T. (1973). Incorporation of (2-$^3H_2$) and (4R-4-^{3}H) mevalonoid hydrogen atoms into the sesquiterpenoid illudin M. *Journal of the Chemical Society, Chemical Communications* 171–2.

HANSON, J. R. & WHITE, A. F. (1969). Studies in terpenoid biosynthesis. Part IV. Biosynthesis of the kaurenolides and gibberllic acid. *Journal of the Chemical Society* (*C*) 981–5.

HANSON, J. R., HAWKER, J. & WHITE, A. F. (1972). Studies in terpenoid biosynthesis. Part IX. The sequence of oxidation of ring B in kaurene—gibberellin biosynthesis. *Journal of the Chemical Society, Perkin I* 1892–5.

HANSON, J. R., MARTEN, T. & SWERNS, M. (1974). Studies in terpenoid biosynthesis. Part XII. Carbon-13 nuclear magnetic resonance spectra of the trichothecanes and the biosynthesis of trichothecolone from 2-^{13}C MVA. *Journal of the Chemical Society, Perkin I* 1033–6.

HARRIS, A., ROBERTSON, A. & WHALLEY, W. B. (1958). The chemistry of the fungi. Part XXXI. The structure of rosenonolactone. *Journal of the Chemical Society* 1799–807,

HATTORI, T., IGARASHI, H., IWASAKI, S. & OKUDA, S. (1969). Isolation of 3β-hydroxy-4β-methylfusida-17(20)[16,21-*cis*], 24-diene(3β-hydroxy-protosta-17 (20)[16,21-*cis*,]24-diene) and a related triterpene alcohol. *Tetrahedron Letters* 1023–6.

HEDDEN, P., MACMILLAN, J. & PHINNEY, B. O. (1974). Fungal products. Part XII. Gibberellin A_{14}-aldehyde, an intermediate in gibberellin biosynthesis in *Gibberella fujikuroi. Journal of the Chemical Society, Perkin I* 587–92.

HEINTZ, R., BENVENISTE, P., ROBINSON, W. H. & COATES, R. M. (1972). Plant sterol metabolism. Demonstration and identification of a biosynthetic intermediate between farnesyl pyrophosphate and squalene in a higher plant. *Biochemical and Biophysical Research Communications* **49**, 1547–53.

HEMMING, F. W. (1970). Polyprenols. In *Natural substances formed biologically from mevalonic acid.* Edited by Goodwin, T. W., 105–17. London & New York: Academic Press.

HEROUT, V. (1971). Biochemistry of sesquiterpenoids. In *Aspects of terpenoid chemistry and biochemistry.* Edited by Goodwin, T. W., pp. 53–94. London & New York: Academic Press.

HEWLINS, M. J. E., EHRHARDT, J. D., HIRTH, L. & OURISSON, G. (1969). The conversion of (^{14}C)cycloartenol and (^{14}C)lanosterol into phytosterols by cultures of *Nicotiana tabacum. European Journal of Biochemistry* **8**, 184–8.

HIRAI, K., NOZOE, S., TSUDA, K., IITAKA, Y., ISHIBASHI, K. & SHIRASAKA, M. (1967). The structure of siccanin. *Tetrahedron Letters* 2177–9.

HOLLOWAY, P. W. & POPJAK, G. (1968). Isopentenylpyrophosphate isomerase from liver. *Biochemical Journal* **106**, 835–40.

HOLTZAPFEL, C. W., BIRCH, A. J., & RICKARDS, R. W. (1969). Oxidation of deoxyrosenonolactone by *Trichothecium roseum. Phytochemistry* **8**, 1009–12.

HORNBY, G. M. & BOYD, G. S. (1970). A carboxylic acid intermediate of lanosterol demethylation. *Biochemical and Biophysical Research Communications* **40**, 1452–4.

HÜTTER, R. & DE MOSS, J. A. (1967). Organization of the tryptophan pathway: a phylogenetic study of the fungi. *Journal of Bacteriology* **94**, 1896–907.

I.U.P.A.C./I.U.B. (1970). Revised tentative rules for steroid nomenclature. In Rodd's *Chemistry of carbon compounds.* Edited by Coffey, S., Vol. IID, 422–54. Amsterdam: Elsevier.

IMA-YE, K. & KAKISAWA, H. (1973). Synthesis of grifolin and dihydrodeoxytauranin. *Journal of the Chemical Society. Perkin I* 2591–5.

JACKSON, L. L. & FREAR, D. S. (1968). Lipids of rust fungi. II. Stigmast-7-enol and stigmasta-7,24(28)-dienol in flax rust uredospores. *Phytochemistry* **7**, 651–4.

JAUREGUIBERRY, G., LAW, J. H., McCLOSKEY, J. & LEDERER, E. (1965). Studies on the mechanism of biological carbon alkylation reactions. *Biochemistry* **4**, 347–53.

JEDLICKI, E., JACOB, G., FAINI, F. & CORI, O. (1972). Stereospecificity of the isopentenylpyrophosphate isomerase and prenyl transferase from *Pinus* and *Citrus. Archives of Biochemistry and Biophysics* **152**, 590–6.

KAKISAWA, H., SATO, M., RUO, T-I. & HAYASHI, T. (1973). Biosynthesis of a C_{16}-terpenoid lactone, a plant growth regulator. *Journal of the Chemical Society Chemical Communications* 802–3.

KAWAGUCHI, A., KOBAYASHI, H. & OKUDA, S. (1973). Cyclization of 2,3-oxidosqualene with microsomal fraction of *Cephalosporium caerulens. Chemical and Pharmaceutical Bulletin* **21**, 577–83.

KAWASHIMA, K., NAKANISHI, K. & NISHIKAWA, H. (1964). Structure of tauranin and a note on the C_{16}-acids obtained from di- and triterpenoids. *Chemical and Pharmaceutical Bulletin* **12**, 796–803.

KNAPP, F. F., GOAD, L. J. & GOODWIN, T. W. (1973). Inversion of the 4β-hydrogen during the conversion of the sterol obtusifoliol into poriferasterol by

Ochromonas malhamensis. Journal of the Chemical Society Chemical Communications 149–50.

KNAPP, F. F., GREIG, J. B., GOAD, L. J. & GOODWIN, T. W. (1971). The conversion of 24-ethylidene-sterols into poriferasterol by *Ochromonas malhamensis. Chemical Communications* 707–9.

LAWRIE, W., McLEAN, J., PAUSON, P. L. & WATSON, J. (1965). The biosynthesis of eburicoic acid. *Chemical Communications* 623–4.

LEDERER, E. (1969). Some problems concerning biological C-alkylation reactions and phytosterol biosynthesis. *Quarterly Reviews* **23**, 453–81.

LENFANT, M., ZISSMANN, E. & LEDERER, E. (1967). Biosynthesis of the ethyl side chain of stigmasterol derivatives by the slime mould *Dictyostelium discoideum. Tetrahedron Letters* 1049–52.

LEW, F. T. & WEST, C. A. (1971). (−)Kaur-16-en-7β-ol-19-oic acid, intermediate in gibberellin biosynthesis. *Phytochemistry* **10**, 2065–76.

LIN, H-K, & KNOCHE, H. W. (1974). Origin of sterols in uredospores of *Uromyces phaseoli. Phytochemistry* **13**, 1795–9.

LIN, H-K., LANGENBACH, R. J. & KNOCHE, H. W. (1972). Sterols of *Uromyces phaseoli* uredospores. *Phytochemistry* **11**, 2319–22.

McCORKINDALE, N. J., HUTCHINSON, S. A., PURSEY, B. A., SCOTT, W. T. & WHEELER, R. (1969). A comparison of the types of sterol found in species of the Saprolegniales and Leptomitales with those found in some other Phycomycetes. *Phytochemistry* **8**, 861–7.

MACMILLAN, J. (1971). Diterpenes—the gibberellins. In *Aspects of terpenoid chemistry and biochemistry*. Edited by Goodwin, T. W. pp. 153–80. New York & London: Academic Press.

MACMILLAN, J., SIMPSON, T. J. & YEBOAH, S. K. (1972). Absolute stereochemistry of the fungal product wortmannin. *Journal of the Chemical Society, Chemical Communications* 1063.

MACMILLAN, J., VANSTONE, A. E. & YEBOAH, S. K. (1968). The structure of wortmannin, a steroidal fungal metabolite. *Chemical Communications* 613–4.

McMORRIS, T. C., NAIR, M. S. R. & ANCHEL, M. (1967). The structure of illudol, a sesquiterpenoid triol from *Clitocybe illudens. Journal of the American Chemical Society* **89**, 4562–3.

MACHIDA, Y. & NOZOE, S. (1972*a*). Biosynthesis of trichothecin and related compounds. *Tetrahedron* **28**, 5113–7.

MACHIDA, Y. & NOZOE, S. (1972*b*). Biosynthesis of trichothecin and related compounds, *Tetrahedron Letters* 1969–71.

MAGNUSSON, G. & THOREN, S. (1973). Fungal extractives. V. The stereostructure of two sesquiterpene lactones from *Lactarius. Acta Chemica Scandinavika* **27**, 2396–8.

MERCER, E. I. & JOHNSON, M. W. (1969). Cyclization of squalene-2,3-oxide to lanosterol in a cell-free system from *Phycomyces blakesleeanus. Phytochemistry* **8**, 2329–31.

MILLER, W. L. & GAYLOR, J. L. (1970*a*). Investigation of the component reactions of oxidative sterol demethylation. Oxidation of a 4,4-dimethyl sterol to a 4β-methyl-4α-carboxylic acid during cholesterol biosynthesis. *Journal of Biological Chemistry* **245**, 5375–81.

MILLER, W. L. & GAYLOR, J. L. (1970*b*). Investigation of the component reactions of oxidative sterol demethylation. Oxidation of a 4α-methyl sterol to a 4α-carboxylic acid during cholesterol biosynthesis. *Journal of Biological Chemistry* **245**, 5369–74.

MOORE, J. T. & GAYLOR, J. L. (1969). Isolation and purification of an S-adenosylmethionine: Δ^{24}-sterol methyltransferase from yeast. *Journal of Biological Chemistry* **244**, 6334–40.

MORIKAWA, K. & HIROSE, Y. (1969). Germacrene-C, precursor of δ-elemene. *Tetrahedron Letters* 1799–801.

MULHEIRN, L. J. & CASPI, E. (1971). Mechanism of squalene cyclization. The biosynthesis of fusidic acid. *Journal of Biological Chemistry* **246**, 2494–501.

MULHEIRN, L. J. & RAMM, P. J. (1972). The biosynthesis of sterols. *Chemical Society Reviews* **11**, 259–91.

NATORI, S., INOUYE, Y. & NISHIKAWA, H. (1967). The structures of mompain and deoxyhelicobasidin and the biosynthesis of helicobasidin, quinonoid metabolites of *Helicobasidium mompa* Tanaka. *Chemical and Pharmaceutical Bulletin* **15**, 380–90.

NOWAK, R., KIM, W. K. & RÖHRINGER, R. (1972). Sterols of healthy and rust infected primary leaves of wheat and of non-germinated and germinated uredospores of wheat stem rust. *Canadian Journal of Botany* **50**, 185–90.

NOZOE, S. & MORISAKI, M. (1969). Enzymic formation of a tricyclic sesterterpene alcohol from mevalonic acid and all-trans-geranylfarnesyl pyrophosphate. *Chemical Communications* 1319–20.

NOZOE, S., MORISAKI, M. & MATSUMOTO, H. (1970). Biosynthesis of helicobasidin and related compounds. *Chemical Communications* 926.

NOZOE, S., MORISAKI, M., FUKUSHIMA, K. & OKUDA, S. (1968). The isolation of an acyclic C_{25}-isoprenoid alcohol, geranylnerolidol and a new ophiobolin. *Tetrahedron Letters* 4457–8.

NOZOE, S., MORISAKI, M., TSUDA, K. & OKUDA, S. (1967). Biogenesis of ophiobolins. The origin of the oxygen atoms in the ophiobolins. *Tetrahedron Letters* 3365–8.

OKUDA, T., SATO, Y., HATTORI, T., IGARASHI, H., TSUCHIYA, T. & WASADA, N. (1968). Isolation of 3β - hydroxy-4β-hydroxy-methylfusida-17(20)[16,21-*cis*],24diene. *Tetrahedron Letters* 4769–72.

OVERTON, K. H. & ROBERTS, F. M. (1973). Biosynthesis of 2-*trans*, 6-*trans*- and 2-*cis*, 6-*trans*.-farnesols by soluble enzymes from tissue cultures of *Andrographis paniculata. Journal of the Chemical Society, Chemical Communications* 378–9.

OVERTON, K. H. & ROBERTS, F. M. (1974). Biosynthesis of *trans,trans*- and *cis,trans*-farnesols by soluble enzymes from tissue cultures of *Andrographis paniculata. Biochemical Journal* in press.

PACKER, N. M. (1973). Biosynthesis of isoprenoid-derived compounds formed *via* farnesylpyrophosphate: the sterols. In *Biosynthesis of acetate-derived compounds*, pp. 143–75. London: John Wiley & Sons.

PITEL, D. W., VINING, L. C. & ARSENAULT, G. P. (1971). Biosynthesis of gibberellins in *Gibberella fujikuroi.* The sequence after gibberellin A_4. *Canadian Journal of Biochemistry* **49**, 194–200.

POLONSKY, J., BASKEVITCH, Z., BELLAVITA, N. C. & CECCHERELLI, P. (1970). Structures des virescenols A et B, métabolites d'*Oospora virescens* (Link) Wallr. *Bulletin de la Société Chimique de France*, 1912–18.

POLONSKY, J., BASKEVITCH, Z., BELLAVITA, N. C., CECCHERELLI, P., BUCKWALTER, B. L. & WENKERT, E. (1972). Carbon-13 nuclear magnetic resonance spectroscopy of naturally occurring substances. XI. Biosynthesis of the virescenosides. *Journal of the American Chemical Society.* **94**, 4369–70.

POPJAK, G. (1969). Enzymes of sterol biosynthesis in liver and intermediates of sterol biosynthesis. In *Methods in enzymology*. ed. Clayton, R. B., Vol. 15, pp. 393–454. New York and London: Academic Press.

POPJAK, G. (1970). Conversion of mevalonic acid into prenyl hydrocarbons as exemplified by the synthesis of squalene. In *Natural Substances Formed Biologically from Mevalonic Acid.* ed. Goodwin, T. W., pp. 17–33. London & New York: Academic Press.

POPJAK, G. & CORNFORTH, J. W. (1966). Substrate stereochemistry in squalene biosynthesis. *Biochemical Journal* **101**, 553–68.

POPPLESTONE, C. R. & UNRAU, A. M. (1973). Major sterols of *Achlya bisexualis. Phytochemistry* **12**, 1131–3.

QURESHI, A. A., BARNES, F. J. & PORTER, J. W. (1972). Lycopersene and prelycopersene pyrophosphate. Intermediates in carotene biosynthesis. *Journal of Biological Chemistry* **247**, 6730–2.

RAAB, K. H., DE SOUZA, N. J. & NES, W. R. (1968). The hydrogen migration in the alkylation of sterols at C-24. *Biochimica et Biophysica Acta* **152**, 742–8.

RAHMAN, R., SHARPLESS, K. B., SPENCER, T. A. & CLAYTON, R. B. (1970). Removal of the 4,4-dimethyl carbons in the enzymic conversion of lanosterol to cholesterol. Initial loss of the 4α-methyl group. *Journal of Biological Chemistry* **245**, 2667–71.

RANDALL, P. J., REES, H. H. & GOODWIN, T. W. (1972). Mechanism of alkylation during sitosterol biosynthesis in *Larix decidua. Journal of the Chemical Society, Chemical Communications* 1295–6.

REES, H. H., GOAD, L. J. & GOODWIN, T. W. (1968*a*). Cyclization of 2,3-oxidosqualene to cycloartenol in a cell-free system from higher plants. *Tetrahedron Letters* 723–5.

REES, H. H., GOAD, L. J. & GOODWIN, T. W. (1968*b*). Studies in phytosterol biosynthesis. Mechanism of biosynthesis of cycloartenol. *Biochemical Journal* **107**, 417–26.

RICHE, C., PASCARL-BILLY, C., DEVYS, M., GAUDEMER, A., BARBIER, M. & BOUSQUET, J-F. (1974). Structure crystalline et moleculaire de la phomenone, phytotoxine produit par le champignon *Phoma exigua* var. *non oxidabilis. Tetrahedron Letters* 2765–6.

RILLING, H. C. (1970). Biosynthesis of presqualene pyrophosphate by liver microsomes. *Journal of Lipid Research* **11**, 480–5.

ROBERTS, J. S. (1972). The sesquiterpenes. In *Chemistry of terpenes and terpenoids.*

Edited by Newman, A. A., pp. 88–154. London & New York: Academic Press.

RÜCKER, G. (1973). Sesquiterpenes. *Angewandte Chemie, International Edition* **12**, 793–806.

RUSSELL, R. T., VAN ALLER, R. T. & NES, W. R. (1967). The mechanism of introduction of alkyl groups at C-24 of sterols. II. The necessity of the Δ^{24}bond. *Journal of Biological Chemistry* **242**, 5802–6.

SCHROEPFER, G. J. JR, LUTSKY, B. N., MARTIN, J. A., HUNTOON, S., FOURCANS, B., LEE, W. H. & VERMILION, J. (1972). Recent investigations on the nature of sterol intermediates in the biosynthesis of cholesterol. *Proceedings of the Royal Society London Series B*, **180**, 125–46.

SHAH, D. H., CLELAND, W. W. & PORTER, J. W. (1965). The partial purification, properties and mechanism of action of pig liver isopentenyl pyrophosphate isomerase. *Journal of Biological Chemistry* **240**, 1946–56.

SHARPLESS, K. B., SNYDER, T. E., SPENCER, T. A., MAHESHWARI, K. K., GUHN, G. & CLAYTON, R. B. (1968). Biological demethylation of 4,4-dimethyl sterols. Initial removal of the 4α-methyl group. *Journal of the American Chemical Society* **90**, 6874–5.

SHARPLESS, K. B., SNYDER, T. E., SPENCER, T. A., MAHESHWARI, K. K., NELSON, J. A. & CLAYTON, R. B. (1969). Biological demethylation of 4,4-dimethyl sterols. Evidence for enzymic epimerization of the 4β-methyl group prior to its oxidative removal. *Journal of the American Chemical Society* **91**, 3394–6.

SMITH, A. R. H., GOAD, L. J. & GOODWIN, T. W. (1968*a*). The stereochemistry of hydrogen elimination at C(7) and C(22) in phytosterol biosynthesis by *Ochromonas malhamensis. Chemical Communications* 926–7.

SMITH, A. R. H., GOAD, L. J. & GOODWIN, T. W. (1968*b*), The stereochemistry of hydrogen elimination at C(6) and C(23) in phytosterol biosynthesis by *Ochromonas malhamensis. Chemical Communications* 1259–60.

SMITH, A. R. H., GOAD, L. J. & GOODWIN, T. W. (1972). Incorporation stereospecifically labelled MVA into poriferasterol by *Ochromonas malhamensis. Phytochemistry* **11**, 2775–81.

STONE, K. J., & HEMMING, F. W. (1967). The stereochemistry of hexahydroprenol, ubiquinone and ergosterol biosynthesis in the mycelium of *Aspergillus fumigatus. Biochemical Journal* **104**, 43–56.

STONE, H. J., ROESKE, W. R., CLAYTON, R. B. & VAN TAMELEN, E. E. (1969). Stereochemistry of sterol biosynthesis: interrelationship of the terminal methyl groups in squalene, the C-1, 1′ methyls in squalene 2,3-oxide, and the C-30, 31 methyls in lanosterol. *Chemical Communications* 530–2.

SUZUKI, Y. & MARUMO, S. (1972). Trans to cis 2,3-double bond isomerization of epoxyfarnesol and farnesol by fungus. *Tetrahedron Letters* 5101–4.

SUZUKI, K. T. & NOZOE, S. (1971). Chromene derivatives as intermediates in the biosynthesis of siccanin. *Chemical Communications* 527–8.

SUZUKI, K. T. & NOZOE, S. (1972). Enzymic formation of siccanochromene-A, a key intermediate in the biosynthesis of siccanin. *Journal of the Chemical Society, Chemical Communications* 1166–7.

SUZUKI, K. T., SUZUKI, N. & NOZOE, S. (1971). Isolation and enzymic formation of trans-γ-monocyclofarnesol. *Chemical Communications* 527.

TANABE, M. & SUZUKI, K. T. (1974). Detection of C—C bond fission during the biosynthesis of the fungal triprenylphenol ascochlorin using (1,2-^{13}C)acetate. *Journal of the Chemical Society, Chemical Communications* 445–6.

TANABE, M., SUZUKI, K. T. & JANKOWSKI, W. C. (1974). Biosynthetic studies with carbon-13. The FT-^{13}C NMR spectra of the sesquiterpenoid coriolins. *Tetrahedron Letters* 2271–4.

TCHEN, T. T. & BLOCH, K. (1957). On the mechanism of enzymatic cyclization of squalene. *Journal of Biological Chemistry* **226**, 931–9.

THRELFALL, D. R. & WHISTANCE, G. R. (1971). Biosynthesis of isoprenoid quinones and chromanols. In *Aspects of terpenoid chemistry and biochemistry.* ed. Goodwin, T. W., pp. 357–404. London & New York: Academic Press.

TOMITA, Y. & UMORI, A. (1973). Biosynthesis of isoprenoids 3. Mechanism of alkylation during the biosynthesis of stigmasterol in tissue cultures of higher plants. *Journal of the Chemical Society, Perkin I* 2656–9.

TOMITA, Y., UOMORI, A. & MINATO, H. (1970). Biosynthesis of the methyl and ethyl group at C-24 of phytosterols in *Chlorella vulgaris. Phytochemistry* **9**, 555–60.

TOPHAM, R. W. & GAYLOR, J. L. (1972). Further characterization of the 5α-hydroxysterol dehydrase of yeast.

Biochemical and Biophysical Research Communications **47**, 180–6.

TSCHESCHE, R., FELHABER, H. W., GEIPEL, R., MERKER, H. J. & WELMAR, K. (1974). Botrydial, ein sesquiterpen-antibiotikum aus der nährlösung des pilzes *Botrytis cinerea*. *Chemische Berichte* **107**, 1720–30.

TURNER, A. B. (1964). Quinone methides. *Quarterly Reviews* **18**, 347–60.

TURNER, W. B. (1971). *Fungal metabolites*. London & New York: Academic Press.

TURNER, J. R. & TARBELL, D. S. (1962). The stereochemistry of fumagillin. *Proceedings of the National Academy of Science, U.S.A.* **48**, 733–5.

VAN ALLER, R. T., CHIKAMATSU, H., DE SOUZA, N. J., JOHN, J. P. & NES, W. R. (1969). The mechanism of introduction of alkyl groups at carbon 24 of sterols. III. The second one-carbon transfer and reduction. *Journal of Biological Chemistry* **244**, 6645–55.

VAN TAMELEN, E. E. & ANDERSON, R. J. (1972). Biogenetic-type total synthesis. 24,25-dihydrolanosterol, 24,25-dihydro-$\Delta^{13(17)}$-protosterol, isoeuphenol, (−)-isotirucallol and parkeol. *Journal of the American Chemical Society* **94**, 8225–8.

VAN TAMELEN, E. E. & FREED, J. H. (1970). Biochemical conversion of partially cyclized squalene 2,3-oxide types to the lanosterol system. Views on the normal enzymic cyclization process. *Journal of the American Chemical Society* **92**, 7206–7.

VARENNE, J., POLONSKY, J., BELLAVITA, N. C. & CECCHERELLI, P. (1971). Sur l'ergosterol produit par l'*Oospora virescens* Link (Wallr.). *Biochimie* **53**, 261–2.

VOGEL, H. J. (1965). Lysine biosynthesis and evolution (fungi, gymnosperms, angiosperms). In *Evolving genes and proteins*. Edited by Bryson, V. and Vogel, H. J., pp. 25–40. New York: Academic Press.

VOKAC, K., SAMEK, Z., HEROUT, V. & SORM, F. (1970). On terpenes CCV. The structure of two native orange substances from *Lactarius deliciosus* L. *Collection of Czechoslovak Chemical Communications* **35**, 1296–301.

WEETE, J. D. (1973). Sterols of the fungi: distribution and biosynthesis. *Phytochemistry* **12**, 1843–64.

WHITE, J. O., PERKINS, D. W. & TAYLOR, S. I. (1973). Biosynthesis of ergosta-4,6,8(14)-tetraen-3-one. A novel oxygenative pathway. *Bioorganic Chemistry* **2**, 163–75.

WILLETT, J. D., SHARPLESS, K. B., LORD, K. E., VAN TAMELEN, E. E. & CLAYTON, R. B. (1967) Squalene-2,3-oxide, an intermediate in the enzymatic conversion of squalene to lanosterol and cholesterol, *Journal of Biological Chemistry* **242**, 4182–91.

WILTON, D. C. & AKHTAR, M. (1970). The stereochemistry of hydrogen elimination during 7,8-double bond formation by *Tetrahymena pyriformis*. *Biochemical Journal* **116**, 337–9.

YAGEN, B., O'GRODNICK. J. S., CASPI, E. & TAMM, C. (1974). Reduction of the 24,25-double bond of lanosterol *in vivo* in the rat. Stereochemistry of the addition of the C-25 proton in the biosynthesis of cholesterol. *Journal of the Chemical Society Perkin I* 1994–2000.

YAMAMOTO, S. & BLOCH, K. (1970). Enzymatic studies on the oxidative cyclizations of squalene. In *Natural substances formed biologically from mevalonic Acid*. Edited by Goodwin, T. W., pp. 35–43. London & New York: Academic Press.

YOSHIHARA, K., OHTA, Y., SAKAI, T. & HIROSE, Y. (1969). Germacrene-D, a key intermediate of cadinene group compounds and bourbonenes. *Tetrahedron Letters* 2263–4.

CHAPTER 14

Carotenoids

T. W. GOODWIN

14.1 Introduction

Nature and nomenclature

Carotenoids, the only naturally occurring tetraterpenoids, are widely distributed throughout nature, but they are synthesized *de novo* only by higher plants, algae, fungi and bacteria. They can be represented formally as consisting of eight isoprenoid residues (ip) and formed by the joining tail to tail of two units each consisting of four isoprenoid residues joined head to tail, thus:

ipipipippipipipi

The three hundred or so naturally occurring carotenoids all represent variations on this basic theme (Isler, 1971). A major sub-division is into carotenes (hydrocarbons) and xanthophylls (oxygen-containing carotenes), and most of the pigments considered in this chapter can be classified according to the nature of their C_9 end groups which are given the prefixes shown.

Acyclic ψ

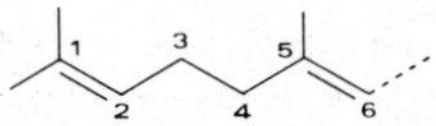

Cyclohexene β

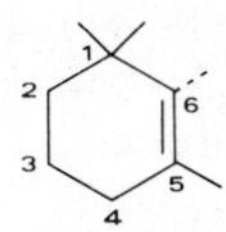

Cyclohexene ε

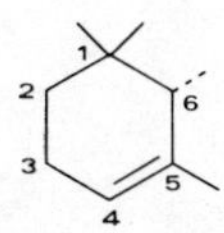

For example, α-carotene (I) is β, ε-carotene; structure (I) also indicates the method of numbering carotenoid molecules: if two end groups are dissimilar then the unprimed numbers are given to the half of the molecule which is associated with the Greek letter cited first in its name. Other basic structures with which we shall be particularly concerned are lycopene (II, ψ,ψ-carotene), β-carotene (III, β,β-carotene) and γ-carotene (IV, β,ψ-carotene).

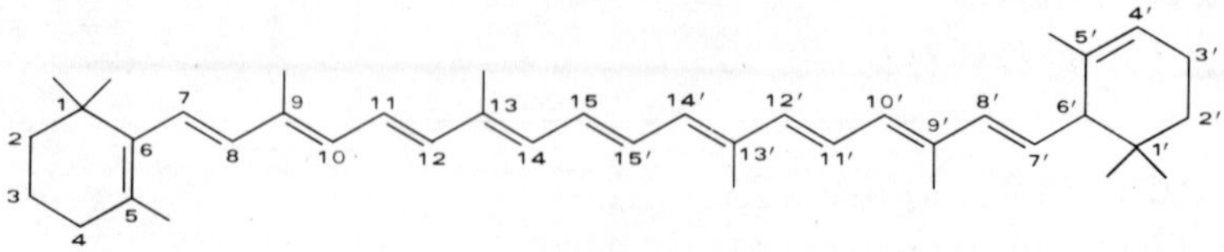

I α-Carotene

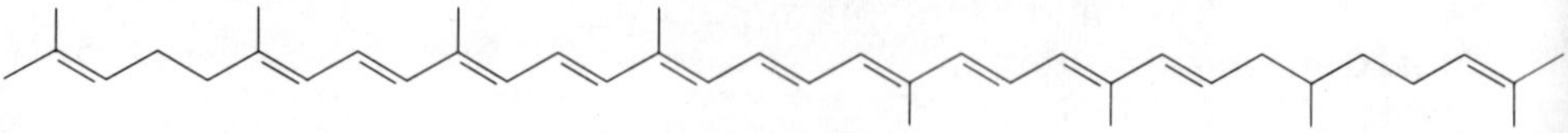

II Lycopene

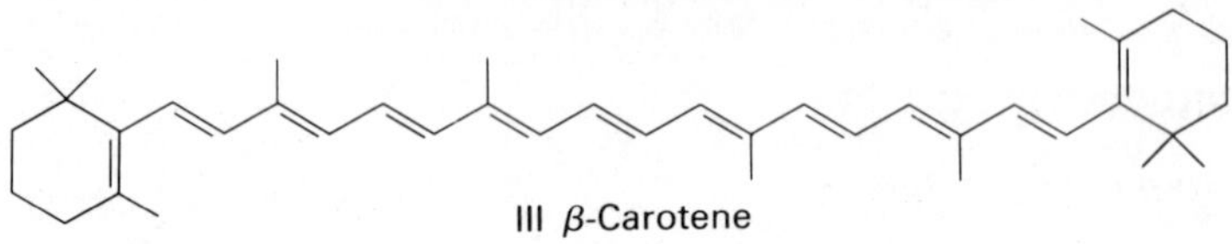

III β-Carotene

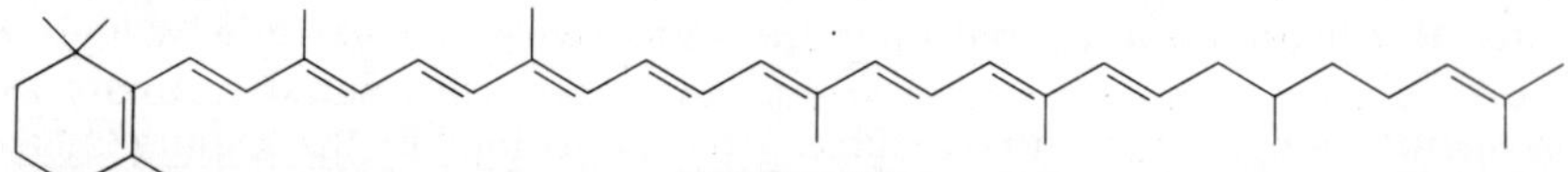

IV γ-Carotene

14.2 Distribution

The distribution of carotenoids in fungi has recently been reviewed (Goodwin, 1972*a*) and all that is necessary here is to outline certain generalizations which emerge when the filamentous fungi are considered. In summary (a) not all the organisms under consideration synthesize carotenoids; (b) many produce only carotenes and this is particularly characteristic of the Phycomycetes; (c) β-carotene and γ-carotene are widely distributed but no carotenes with ε-end groups (α-carotene derivatives) have been unequivocally detected; (d) xanthophylls characteristic of higher plants have rarely, if ever, been detected; (e) acetylenic carotenoids, very characteristic of algae, have not been detected; (f) keto carotenoids, such as canthaxanthin (V), first isolated from *Cantharellus cinnabarinus* are frequently encountered; and (g) unique xanthophylls which appear in the Discomycetes are phillipsiaxanthin (VI) and plectaniaxanthin (VII) and their derivatives; they in some ways (desaturation at 3,4 and hydration at 1,2) represent carotenoids

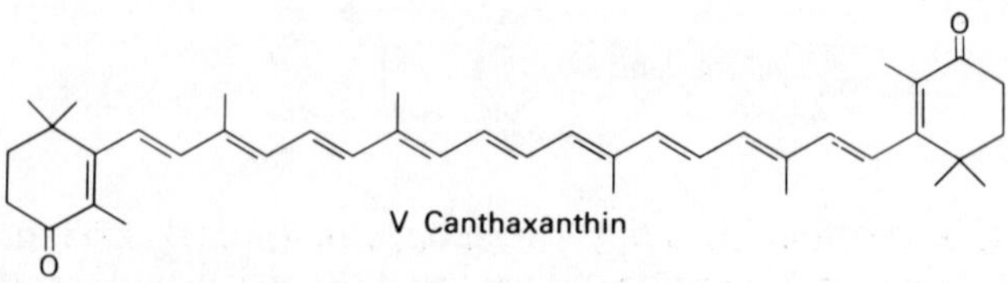

V Canthaxanthin

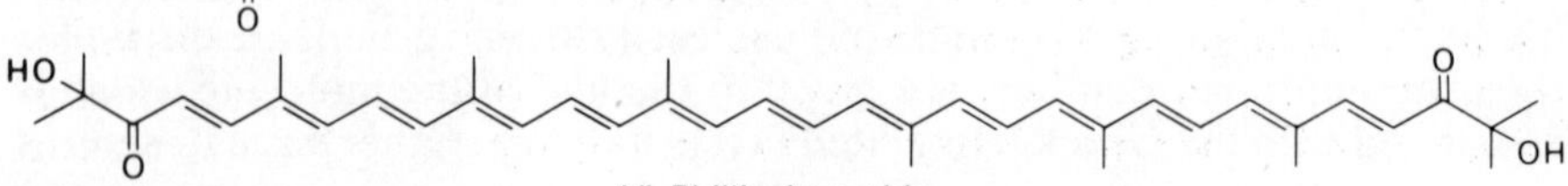

VI Phillipsiaxanthin

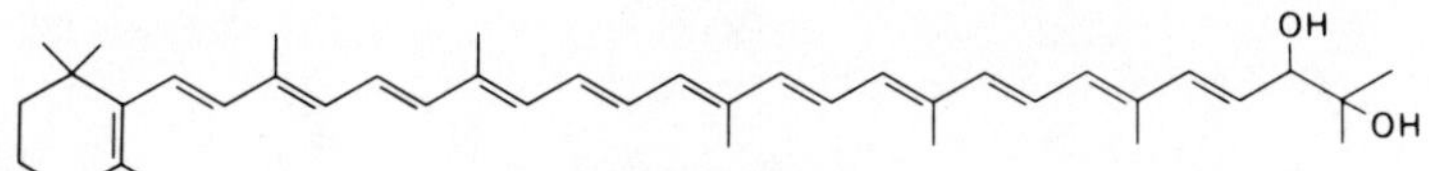

VII Plectaniaxanthin

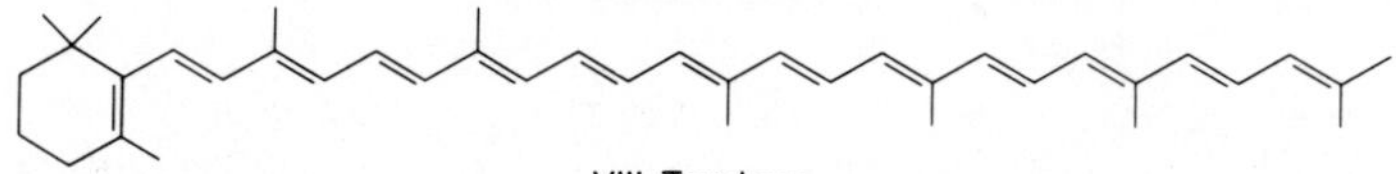

VIII Torulene

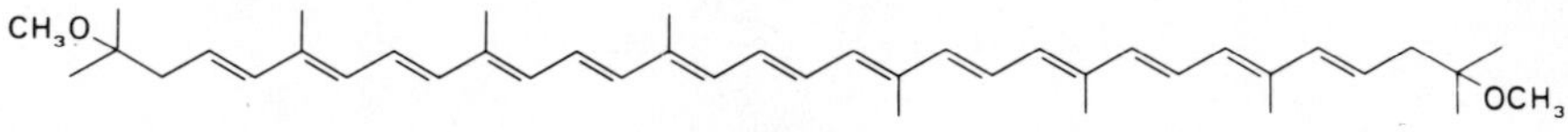

IX Spirilloxanthin

characteristie of the red yeasts (*e.g.* torulene, VIII) and of photosynthetic bacteria (*e.g.* spirilloxanthin, IX).

Sporopollenin the polymer in the wall of the zygospores of *Mucor mucedo* is said to be formed by oxidative polymerization of β-carotene (Gooday, Fawcett, Green & Shaw, 1973).

14.3 Biosynthesis and Metabolism

General pathway to phytoene

The general pathway of biosynthesis of carotenoids from acetyl-CoA is now reasonably well clarified (Goodwin, 1971*a*). For the present context the pathway can be divided into four sections: (a) the conversion of acetyl-CoA into the universal isoprenoid precursor isopentyl pyrophosphate (IPP) via mevalonic acid (MVA) (Fig. 14.1); (b) the conversion of IPP into the first

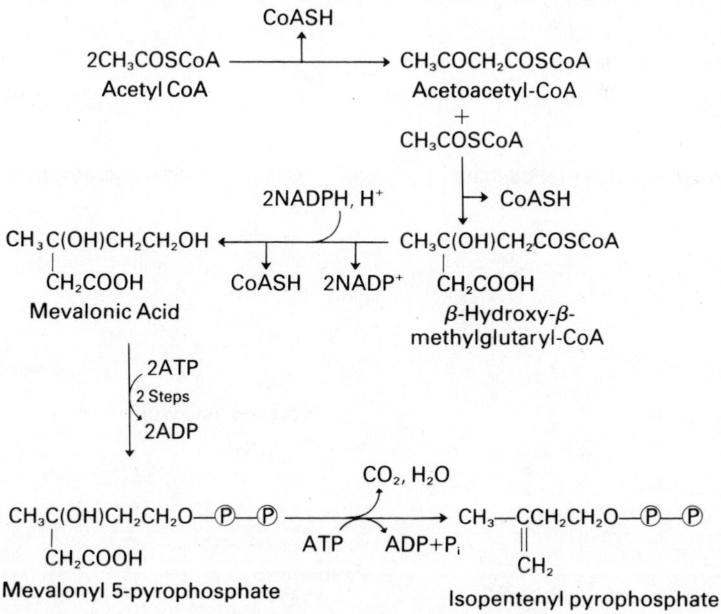

Fig. 14.1 Conversion of acetyl-CoA into isopentenyl pyrophosphate.

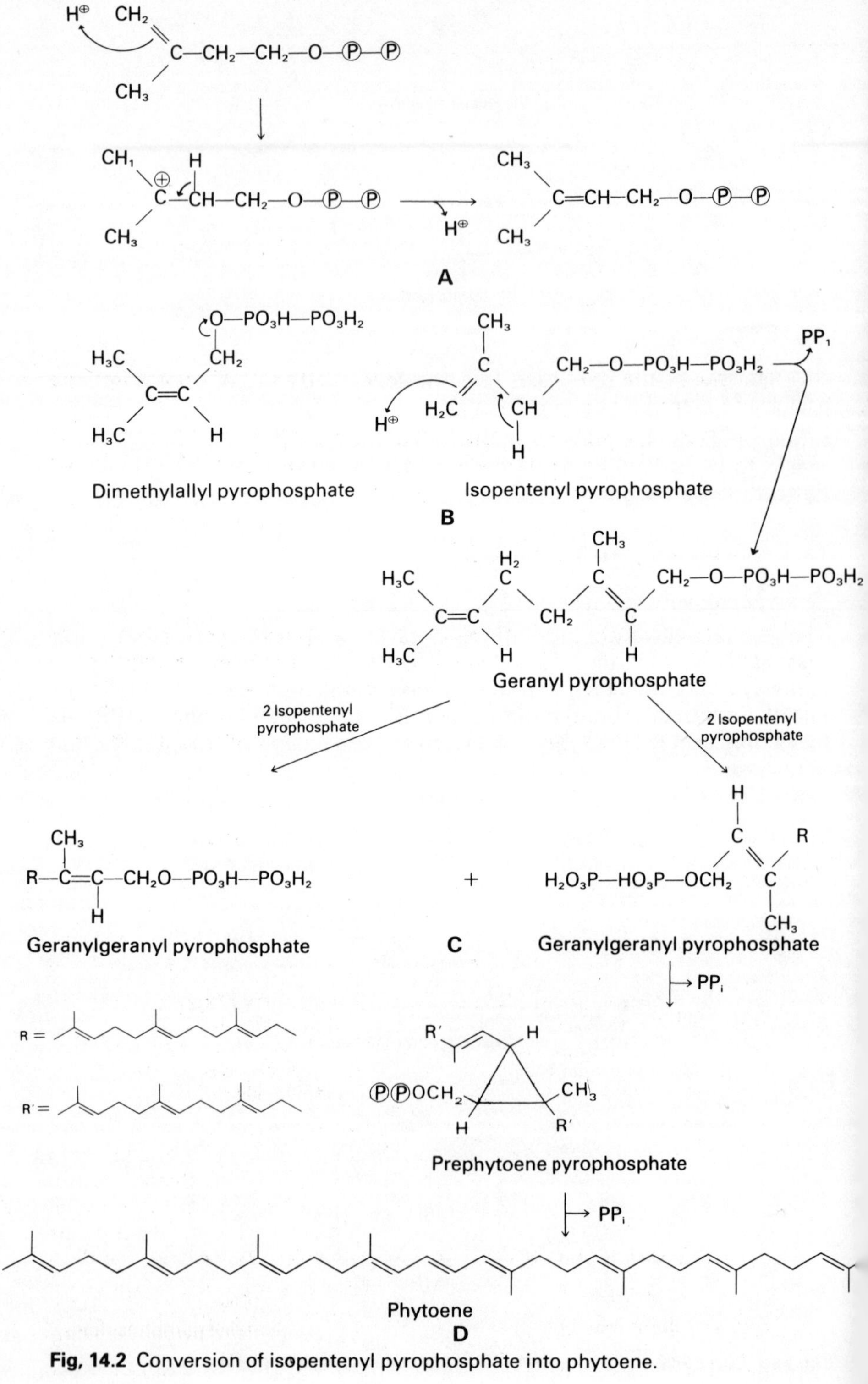

Fig, 14.2 Conversion of isopentenyl pyrophosphate into phytoene.

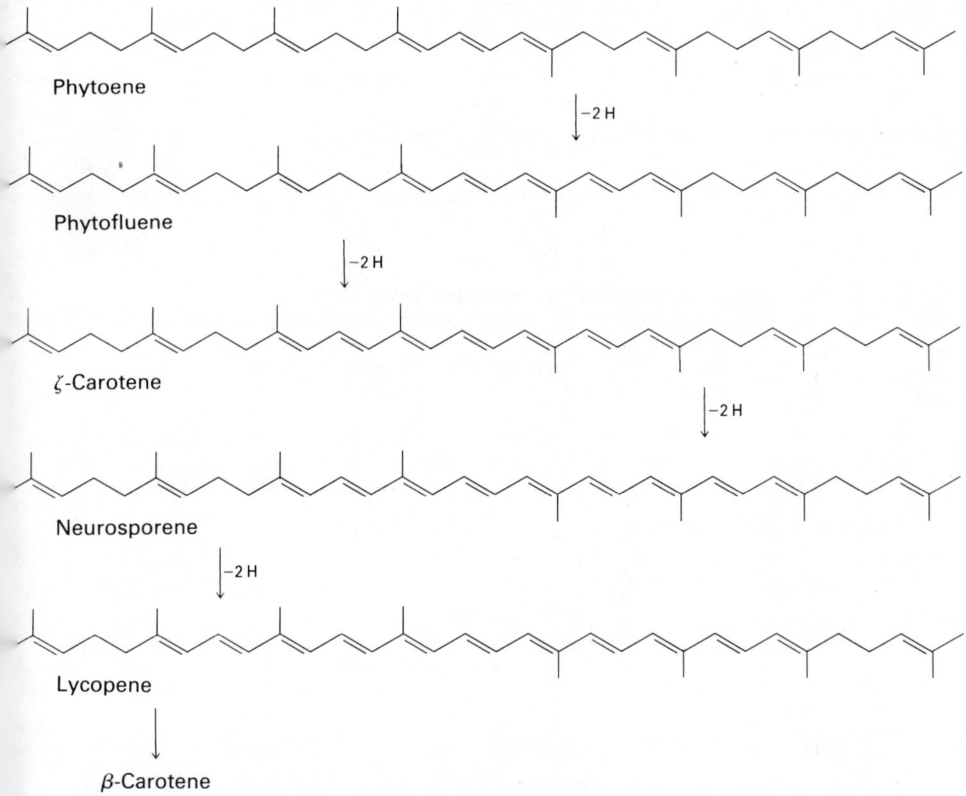

Fig. 14.3 The usual pathway of desaturation of phytoene to lycopene [For ease of presentation phytoene is represented in its all-*trans* form].

C-40 carotenoid, phytoene (Fig. 14.2); (c) the formation of the fully unsaturated acyclic lycopene (Fig. 14.3); and finally (d) the cyclization of lycopene, or other possible polyenes, to β-carotene (see Fig. 14.8). We can now consider how far this pathway has been established in fungi.

The utilization of acetate for the synthesis of β-carotene was first demonstrated in *Phycomyces blakesleeanus* (Schopfer & Grob, 1950, 1952; Schopfer, Grob & Besson, 1952; Schopfer, Grob, Besson & Keller, 1952) and degradation of β-carotene synthesized by *Mucor hiemalis* in the presence of labelled acetate indicates a distribution (Fig. 14.4) consistent with the pathway just outlined (Grob & Bütler, 1955, 1956). Similar results were obtained with *P. blakesleeanus* and the labelling pattern in this organism with [2-^{14}C]-MVA confirms the pathway (Fig. 14.4) (Braithwaite & Goodwin, 1960*a*, *b*, *c*). None of the intermediates between acetyl-CoA and MVA has been shown to be active in fungi although β-hydroxy-β-methyl glutarate (HMG) was active in crude preparation of *P. blakesleeanus* (Chichester, Yokoyama, Nakayama, Lukton & Mackinney, 1959; Yamamoto, Chichester & Nakayama, 1962*a*). However, the observation

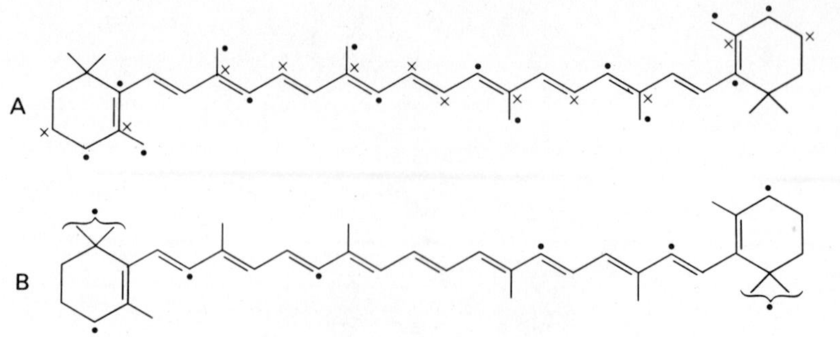

Fig. 14.4 The pattern of labelling in β-carotene formed from:
A. [^{14}C]Acetate (• = carbon atoms arising from C-2 and X = those from C-1 of acetate).
B. [2-^{14}C]Mevalonate (• = carbon atoms arising from C-2 of MVA).

that the amino acids leucine and valine stimulate carotenogenesis in *P. blakesleeanus* (Goodwin & Lijinsky, 1952) led to the elucidation of a pathway from leucine to β-hydroxy-β-methylglutaryl-CoA (HMG-CoA) (Fig. 14.5). Experiments with [1-^{14}C], [2-^{14}C], [3-^{14}C], [4-^{14}C] and [5,5-$^{14}C_2$] leucines showed that only the 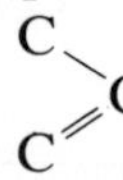 residue, i.e. the end three carbons, was utilized for β-carotene synthesis (Yokoyama, Chichester,

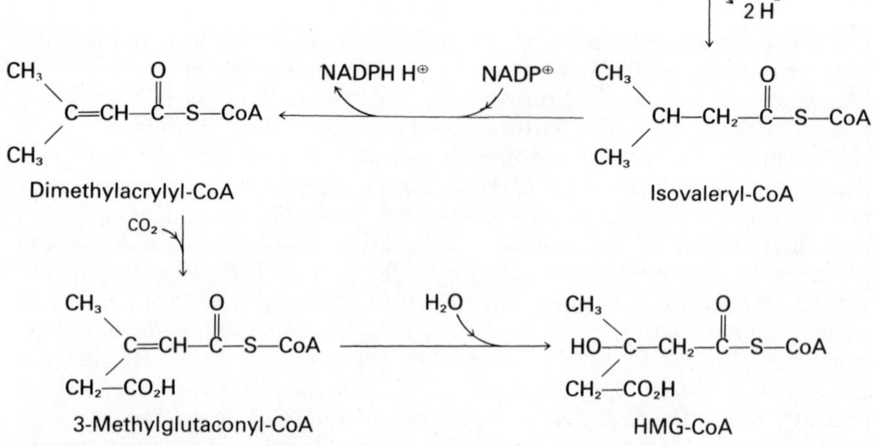

Fig. 14.5 Formation of HMG-CoA from leucine.

Nakayama, Lukton & Mackinney, 1957; Wuhrmann, Yokoyama & Chichester, 1957; Yamamoto, Chichester & Nakayama, 1962*a*, *b*; Yamamoto & Chichester, 1965; Yokoyama, Chichester & Mackinney, 1969). To explain this it must be assumed that HMG-CoA is degraded to acetoacetate and acetyl-CoA (arising from C-2 and C-3 of leucine) and that in the resynthesis of HMG-CoA the pool size of acetyl-CoA is much greater than that of acetoacetyl-CoA and thus the C-2 and C-3 atoms of leucine are diluted out in the final product, β-carotene (Fig. 14.6) (Goodwin, 1971*a*). This proposal also explains the effective fixation of $^{14}CO_2$ into β-carotene by *P. blakesleeanus* only when the fungus is metabolizing leucine (Braithwaite

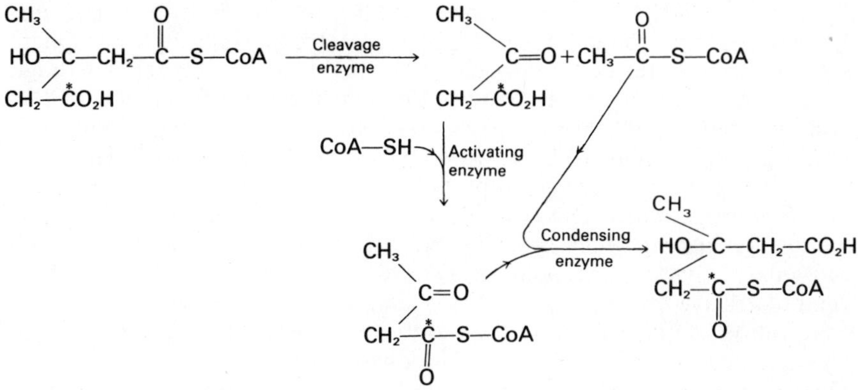

Fig. 14.6 Proposed breakdown and resynthesis of HMG-CoA which would account for incorporation of $^{14}CO_2$ into β-carotene in the presence of unlabelled leucine.

& Goodwin, 1960*a*, *b*, *c*; Chichester *et al.*, 1959; Lowry & Chichester, 1971). From the pathway outlined in Fig. 14.5 the CO_2 fixed in HMG-CoA is located at C-1 which is lost on conversion of HMG-CoA into IPP. Fig. 14.6 indicates how some of this label can be conserved in C-5 of MVA (C-1 of IPP). Furthermore dimethylacrylic acid (not its CoA ester) stimulates carotenogenesis in *P. blakesleeanus* (Reichel & Willis, 1957) and the labelled acid is incorporated into β-carotene in *Blakeslea trispora* (Anderson, Norgard & Porter, 1960). Early experiments had, however, indicated that dimethylacrylic acid was not carotenogenic in *P. blakesleeanus* (Goodwin & Lijinsky, 1952).

As might be expected many experiments have demonstrated the incorporation of [2-^{14}C]MVA into carotenes in fungi; these include *Phycomyces blakesleeanus* (Braithwaite & Goodwin, 1960*a*, *b*, *c*; Chichester *et al.*, 1959; Mackinney, Chandler & Lukton, 1958), *Mucor hiemalis* (Grob, 1959) and *Blakeslea trispora* (Anderson, Norgard & Porter, 1960). More recently crude cell-free systems from native *P. blakesleeanus* and a mutant *CarR 21*(−) converted [2-^{14}C]MVA into β-carotene in good yield (Yokoyama, Nakayama & Chichester, 1962; Lee & Chichester, 1969; Davies, 1973).

The conversion of IPP into dimethylallyl pyrophosphate (DMAPP) (Fig. 14.2A) has been inferred by the observation that iodoacetamide inhibits the conversion of MVA into β-carotene in a cell-free system from *Phycomyces blakesleeanus* (Yokoyama *et al.*, 1962); the IPP → DMAPP isomerase is said to be the only enzyme in the sequence sensitive to iodoacetamide. Geranyl pyrophosphate (C_{10}) is formed by the condensation of one molecule of IPP with one molecule of DMAPP (Fig. 14.2B,); it has not been examined as a carotenoid precursor in fungi but farnesyl pyrophosphate (FPP) (C_{15}) formed by addition of one further IPP molecule (Fig. 14.2) is active in the *Phycomyces* system only in the presence of MVA (Yokoyama *et al.*, 1962), which presumably provides the additional C_5 unit to form geranylgeranyl pyrophosphate (GGPP) (Fig. 14.2C) which, itself, is effectively incorporated into β-carotene (Lee & Chichester, 1969).

The first hydrocarbon precursor formed in *Phycomyces blakesleeanus* is phytoene (Davies, Jones & Goodwin, 1963) presumably formed *via* prephytoene pyrophosphate (Fig. 14.2D) (or prelycopersene pyrophosphate as it has also been termed) although no clear evidence for this exists in fungi. If the dimerization of GGPP followed the same mechanism as that of FPP in forming squalene, the first C-40 product would be lycopersene (15, 15′-dihydrophytoene) and NADPH would be required as a co-factor. However this co-factor is not involved in phytoene biosynthesis in cell-free systems from *P. blakesleeanus* (Davies, 1973). Furthermore, mutants of *P. blakesleeanus* are known which are blocked in β-carotene synthesis but which accumulate phytoene (Meissner & Delbrück, 1968); none is known which accumulates lycopersene. In addition, diphenylamine, which inhibits the desaturation steps in carotenoid synthesis, causes the accumulation of phytoene in *Phycomyces* (Garton, Goodwin & Lijinsky, 1951), *Allomyces* (Turian & Haxo, 1954; Turian, 1957), *Blakeslea trispora* (Thomas & Goodwin, 1967) and *Verticillium albo-atrum* (Valadon & Mummery, 1966, 1969), and, in the case of *Phycomyces*, synthesis of lycopersene was specifically excluded (Davies, Jones & Goodwin, 1963).

Desaturation of phytoene

The pathway outlined in Fig. 14.3 is that generally accepted for conversion of phytoene into lycopene and β-carotene in fungi. The overall reaction has been demonstrated by growing *Phycomyces blakesleeanus* and *Blakeslea trispora* in the presence of diphenylamine (see previous section), washing out the inhibitor and resuspending the organisms in new medium. There is a rapid synthesis of β-carotene at the expense of phytoene. The accumulation in the presence of diphenylamine of very large amounts of phytoene, which are not completely converted into β-carotene can be explained by end-product inhibition. When synthesis of the end-product of the biosynthetic sequence (β-carotene) is inhibited, the metabolite which exerts feedback control at the first specific C-40 step is removed, and thus the first C-40 compound, phytoene, accumulates in excess. The situation may however, be more complex because in the presence of both β-ionone, which stimulates carotenogenesis without itself being incorporated (see p. 439), and diphenylamine, *P. blakesleeanus* synthesizes excessive amounts of both β-carotene and phytoene (Goodwin & Williams, 1965*a*). The mechanism involved in the action of diphenylamine is not known although molecular models show that it fits snugly into the middle of the phytoene molecule when the 15,15′ double bond has the *trans*-configuration (Rilling, 1965).

This emphasizes the importance of defining the exact stereochemistry of the phytoene; in wild-type *P. blakesleeanus* treated with diphenylamine and in the *car-10*(−1) mutant, the 15-*cis* isomer (X) represents over 96% of the polyene which accumulates (Aung Than, Bramley, Davies & Rees, 1972) and this is an active precursor of β-carotene in a cell-free preparation from the wild-type. As phytofluene, the next compound in the sequence also occurs naturally in the 15-*cis* form, but ζ-carotene, the compound following phytofluene does not, then an isomerase would appear to be needed to form all-*trans*-ζ-carotene, the naturally occurring intermediate. No evidence for this yet exists.

A possible alternative pathway from phytoene to lycopene (Fig. 14.7) which bypasses ζ-carotene is suggested by the observation that 7,8,11,12-tetrahydrolycopene is present in mutants *car R21*(−) and *mad-107*(−) of

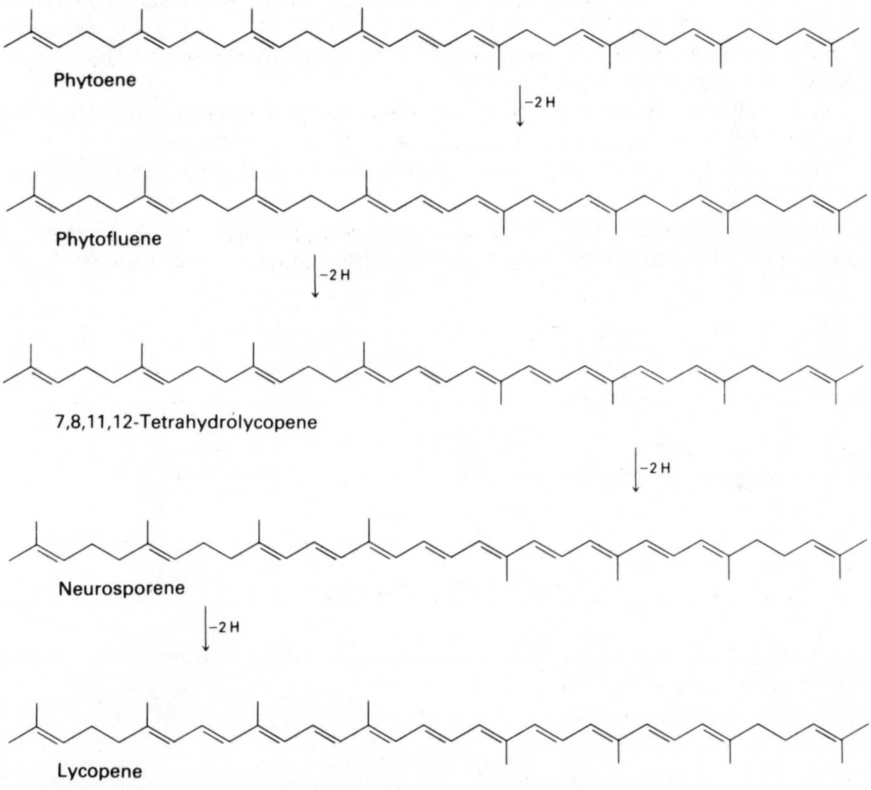

Fig. 14.7 An alternative pathway from phytoene to lycopene [For ease of presentation phytoene is represented in its all-*trans* form).

Phycomyces when grown in the presence of diphenylamine (Davies, 1973). However, comparable cultures from native *Phycomyces* did not accumulate this intermediate (Davies, 1970).

Formation of cyclic carotenes

The immediate question is what is the first stage after phytoene at which cyclization takes place? The most saturated naturally occurring cyclic compound is β-zeacarotene (XI), an isomer of neurosporene. So the possible routes to β-carotene (no derivatives with the ε-end group occur in fungi, see p. 424) are indicated in Fig. 14.8. The evidence in fungi for lycopene as the immediate precursor includes:

(a) mutants of *Phycomyces* have been obtained which accumulate large amounts of lycopene in place of β-carotene (Meissner & Delbrück, 1968);

(b) inhibitors are known which inhibit β-carotene synthesis and cause lycopene to accumulate; these are CPTA [2-(-*p*-chlorophenylthio) triethylamine hydrochloride] with *Blakeslea trispora* and *P. blakesleeanus* (Knypl, 1969: Coggins, Hemming & Yokoyama, 1970) and nicotine with *P. blakesleeanus* (McDermott, Britton & Goodwin, 1973, Davies, 1973). There is no direct evidence in fungi that the lycopene which accumulates in inhibited cultures can be cyclized when the inhibitor is removed, although this has clearly been demonstrated in bacteria (McDermott, Ben-Aziz, Singh, Britton & Goodwin, 1973);

(c) [^{14}C]lycopene is converted into β-carotene in preparations from the *mad-(107)* (−) mutant of *P. blakesleeanus* (Davies, 1973).

The evidence in favour of neurosporene as the first substrate for cyclization includes:

(a) β-zeacarotene (XI), a cyclic isomer of neurosporene, is present in diphenylamine-inhibited cultures of *P. blakesleeanus* and appears to be

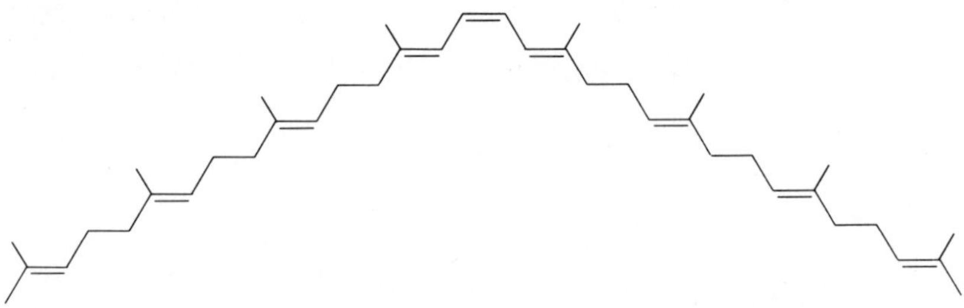

X Phytoene 15-*cis* isomer

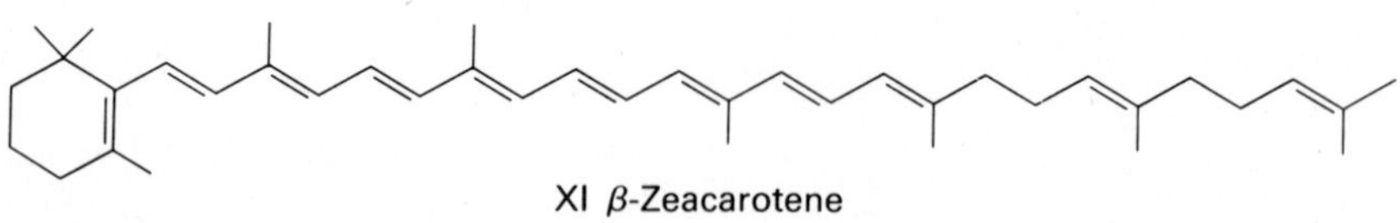

XI β-Zeacarotene

converted into β-carotene on removal of the inhibitor (Williams, Davies & Goodwin, 1965);

(b) [^{14}C]neurosporene is converted into β-carotene in the *P. blakesleeanus* preparation which also cyclizes lycopene (see (b), above) (Davies, 1973). In experiments with organisms other than filamentous fungi the evidence in favour of lycopene is greater (see Goodwin, 1971*a*) but the natural occurrence of β-zeacarotene makes it clear that cyclization can take

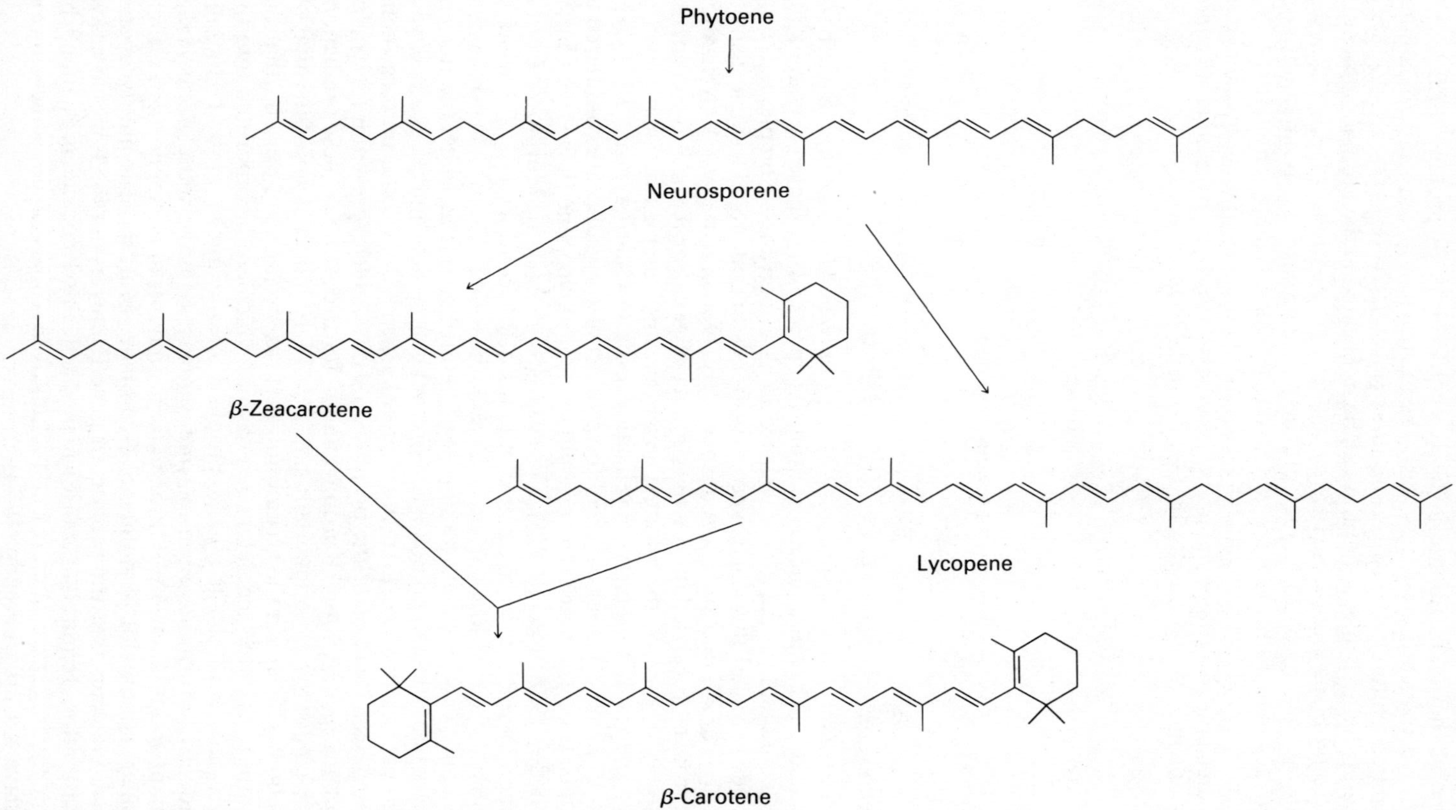

Fig. 14.8 Cyclization pathway for formation of β-cartotene.

place at the neurosporene level. Recently 'cyclic ζ-carotene' (7,8,11,12-tetrahydro-γ-carotene) has also been isolated from diphenylamine-inhibited cultures of *P. blakesleeanus [mad–107*(−1)] (Davies, 1973); it remains to be seen whether this can be implicated in the normal pathway of cyclization (see Goodwin, 1971*a*).

If the nature of the exact precursor for cyclization is in doubt, the mechanism of cyclization seems reasonably clear. Experiments with [2-^{14}C] mevalonic acid stereospecifically tritiated at position 4*R* have revealed that the mechanism outlined in Fig. 14.9 is involved in *Phycomyces blakesleeanus* (Goodwin & Williams, 1965*b*; Goodwin, 1971*b*).

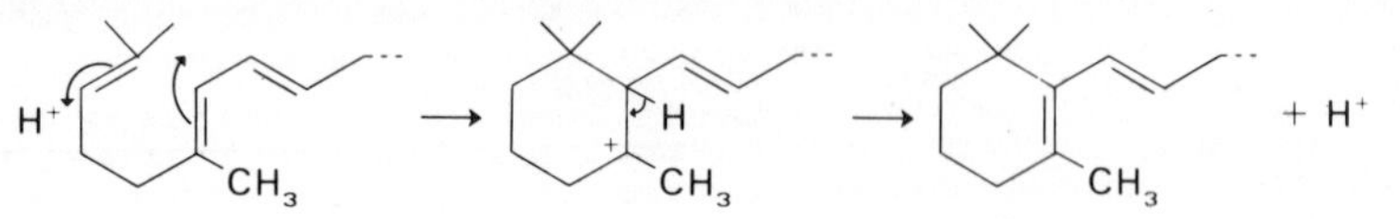

Fig. 14.9 Mechanism for forming the β-ionone ring in cyclic carotenes.

Mutants and carotenogenesis

Three main types of mutants of *Phycomyces blakesleeanus* have been isolated: (a) those accumulating lycopene; (b) those accumulating phytoene; and (c) those synthesizing no C-40 polyenes. Complementation studies revealed that each type corresponds to mutants in a single cistron, termed *car R*, *car B* and *car A*, respectively (Ootaki *et al.*, 1973). Cyclization of lycopene is carried out by the product of gene *car R* and it is considered that two copies of the product in an enzyme complex are concerned in β-carotene formation (two cyclizations) (de la Guardia *et al.*, 1971) (see above). Similarly, four copies of the product of gene *car B* are considered to act in a complex which converts phytoene into lycopene (Eslava & Cerdá-Olmedo, 1974) (see p. 431). De la Guardia, Aragon, Murillo & Cerdá-Olmedo, (1971) have taken two mutants of *Phycomyces*, one, *C2*, which is unable to synthesize carotenoids and the other *C9*, which synthesizes lycopene, and produced a number of heterokaryons in which the nuclear proportions of each homokaryon present can be determined. If an enzyme aggregate is involved then in a heterokaryon *C2* × *C9* the total carotene content as a function of the proportion p of *C2* nuclei present is $(1 - p)$ for lycopene, $p(1 - p)$ for γ-carotene and p^2 for β-carotene. These predictions were confirmed experimentally and Fig. 14.10 indicates how the situation can be visualized on the assumption that two cyclases are present, that the mutant genes produce defective enzymes which are incorporated into the aggregate and that the gene products of the nuclei present combine randomly to form the multi-enzyme complexes. Furthermore, a leaky *carB* mutant accumulates, in addition to large amounts of phytoene, increasingly smaller amounts of phytofluene, ζ-carotene, neurosporene and lycopene, which indicates that the product of gene B, carries out the four successive desaturations required to transform phytoene into lycopene, presumably by the dehydrogenase complex containing four copies of the same enzyme (Eslava & Cerdá-Olmedo, 1974).

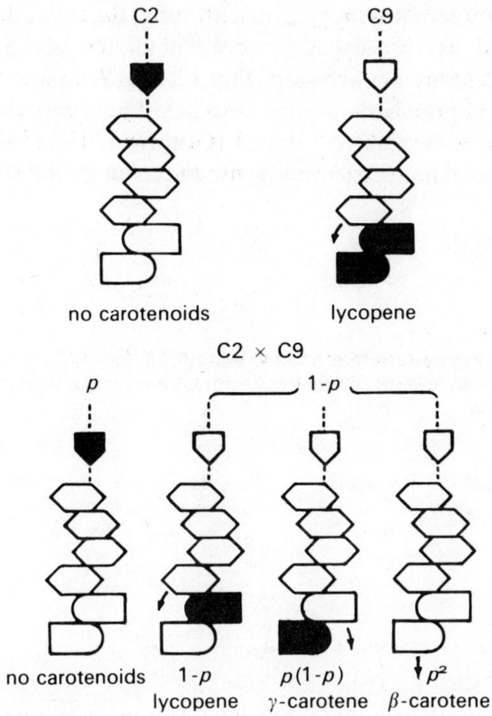

Fig. 14.10 Operations performed by the carotenogenic aggregates of homokaryons C2 and C9 and heterokaryons C2 × C9 indicating proportion expected of each product according to the assumptions indicated on p. 434. *White* symbols = active enzyme; *black* symbols = defective enzymes. Dotted symbols = either active or defective enzymes (de la Guardia *et al.*, 1971) (Reproduced with permission of Professor E. Cerdá-Olmedo).

Stereochemistry of carotene biosynthesis

By use of mevalonate stereospecifically labelled with tritium at C-2, C-4 and C-5 (XII) (Cornforth & Cornforth, 1970) it has been possible to show in *Phycomyces blakesleeanus* that: (a) in the formation of phytoene eight hydrogen atoms, originally in the pro-*S* position at C-4 of eight mevalonate molecules, are lost, and the corresponding 4-*pro-R*-hydrogens are retained (Goodwin & Williams, 1965*a*); (b) at the centre of the phytoene molecule the double bond at C-15 and C-15′ is formed by the loss from each carbon atom of a pro-*R* hydrogen arising from C-5 of mevalonate (Williams, Britton, Charlton & Goodwin, 1967; Buggy, Britton & Goodwin, 1969);

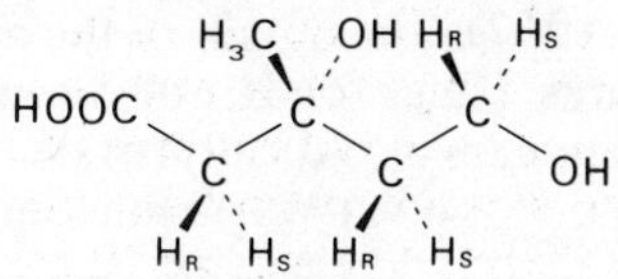

XII Mevalonate

(c) in the desaturation of phytoene to β-carotene, the four additional double bonds are formed by stereospecific removal of the pro-5*R* and pro-2*S* hydrogen of mevalonate at each step (Fig. 14.12) (Williams *et al.*, 1967). In the case of the elimination of the pro-2-*S* hydrogen the results with *Phycomyces* were somewhat equivocal (Goodwin, 1971*b*) but they have been fully established in the carotenogenic *Flavobacterium* 0147 (Goodwin, 1972*b*).

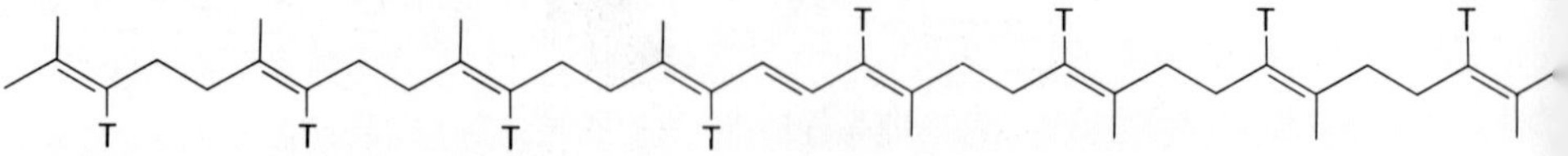

Fig. 14.11 The labelling pattern in phytoene with [4*R*,4-$^3H_1$]MVA as substrate [For ease of presentation phytoene is represented in its all-*trans* form]. T = tritiated.

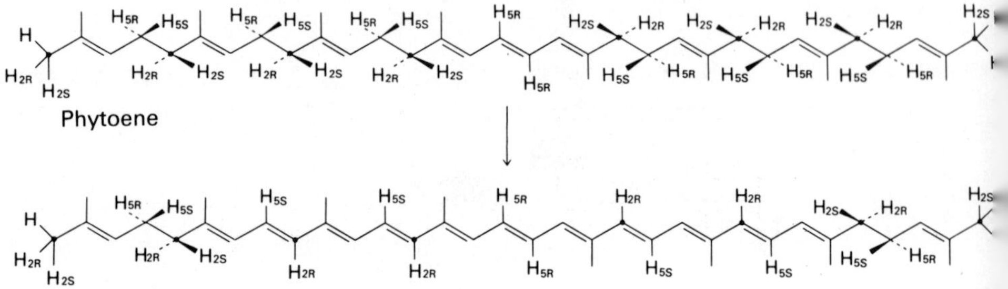

Fig. 14.12 The stereospecific removal of four 5-pro-*R* and four 2-*pro-S* hydrogens of mevalonic acid in the desaturation of phytoene to lycopene.

Control of synthesis

PATTERN OF SYNTHESIS The general pattern of synthesis observed in *Phycomyces blakesleeanus* (Goodwin & Willmer, 1952), *Penicillium sclerotiorum* NRRRL 2074 (Mase, Rabourn & Quackenbush, 1957) and *Epicoccum nigrum* (Gribanovski-Sassu & Foppen, 1967; Foppen & Gribanovski-Sassu, 1967) involves: (a) an initial period of active synthesis leading to maximal concentration: (b) an intermediate stage when the levels persist; and (c) a final stage during which the pigments disappear. In the case of *P. blakesleeanus* the amount of β-carotene synthesized during stage (a) depends on the availability of excess carbohydrate after growth has been completed, because it is only then that carotene synthesis is very active (Fig. 14.13). Thus the final carotene level depends on the C:N ratio in the medium (Goodwin & Willmer, 1952). This also applies to *Sporobolomyces roseus* (Bobkova & Rabotnova, 1967). The details of the metabolic destruction of carotenoids in old cultures (stage c) is not known, but mutatochrome (5,8-epoxy-β-carotene) appears in old cultures of *P. blakesleeanus* (Grob, 1953); 5,6-epoxy-β-carotene is a very probable intermediate in the degradation of β-carotene and this could easily be isomerized to mutatochrome under the strongly acid conditions found in old cultures of *P. blakesleeanus*.

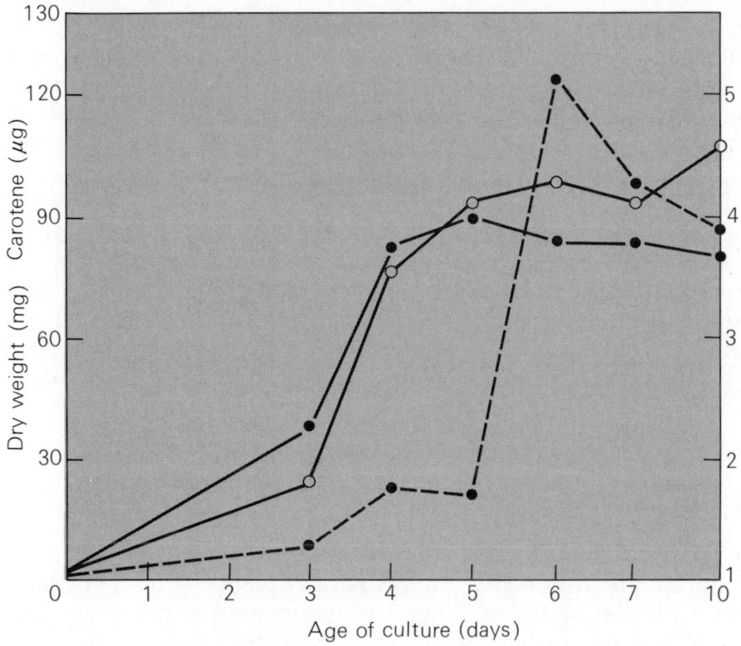

Fig. 14.13 Rate of carotene synthesis in static cultures of *Phycomyces blakesleeanus* compared with growth (Goodwin & Willmer, 1952). ●---●, β-Carotene, ○——○, Dry weight, ●——●, Nitrogen assimilation.

CARBON SOURCES Maltose and glucose are equally effective for β-carotene synthesis in *Phycomyces blakesleeanus* whilst fructose and xylose, although equal to maltose and glucose in supporting growth, are much less carotenogenic; lactose and glycerol, on the other hand, support neither growth nor carotenogenesis (Garton, Goodwin & Lijinsky, 1951). Growth and carotenogenesis are very limited in *P. blakesleeanus* growing on lactate as sole carbon source, but both are stimulated by the addition of acetate which, alone will also support both. The optimal concentration of acetate for carotenogenesis is 0.325%; at higher concentrations pigment production rapidly drops off (Friend, Goodwin & Griffith, 1955).

Wild type *Blastocladiella* is normally non-carotenogenic but in the presence of high levels of bicarbonate it synthesizes γ-carotene. The reason for this is obscure but it has been suggested that bicarbonate inhibits the tricarboxylic acid cycle and thus diverts metabolites normally channelled in this direction into carotenoid synthesis (Cantino & Hyatt, 1953).

NITROGEN SOURCES If asparagine in the normal culture medium is replaced by either valine or leucine then synthesis of β-carotene by *Phycomyces blakesleeanus* is greatly stimulated (Goodwin & Lijinsky, 1952; Chichester, Nakayama, Mackinney & Goodwin, 1955); this observation is explained on p. 429. When asparagine is replaced by $(NH_4)_2SO_4$ growth and carotenogenesis is considerably reduced but addition of any member of the tricarboxylic acid cycle (0.02M) restores growth to normal. However, β-carotene level is normal only in the presence of succinate or acetate

(Friend, Goodwin & Griffiths, 1955; Goodwin, Griffiths & Modi, 1956). [2 − ^{14}C]Glycine is effectively incorporated into β-carotene by *P. blakesleeanus* (Mackinney, Chichester & Nakayama, 1955). As [^{14}C] formate is also effectively incorporated (Reichel & Schreiber, 1968) the pathway from glycine may be:

$$\text{formate} + \text{glycine} \longrightarrow \text{serine} \longrightarrow \text{pyruvate} \longrightarrow \text{acetyl-CoA.}$$

GROWTH FACTORS Low levels of thiamine reduce carotenogenesis slightly in *Phycomyces blakesleeanus* grown on a medium containing glucose as carbon source (Friend & Goodwin, 1954). A slight stimulation of carotene production was observed in *Mucor hiemalis* in the presence of pantothenic acid (Grob, Grindbacher & Schopfer, 1954) whilst riboflavin tended to reduce production in *P. blakesleeanus* (Goodwin, Jamikorn & Willmer, 1953).

TEMPERATURE Carotenoid synthesis in *Phycomyces blakesleeanus* is qualitatively the same in cultures grown over the temperature range 5–25°C (Friend & Goodwin, 1954).

pH On the usual glucose/asparagine medium the starting pH (6.2) drops to 2.6–3.0 during growth of *Phycomyces blakesleeanus*; if the pH change is prevented by buffering the medium, carotenogenesis is almost completely inhibited although growth is unaffected. Similarly, washed mycelia dissimilating glucose will produce β-carotene only in an unbuffered medium (Goodwin & Willmer, 1952).

Bioinduction of carotenogenesis

TRISPORIC ACID AND RELATED COMPOUNDS When (+) and (−) strains of the heterothallic *Choanephora cucurbita* are cultured together the amount of β-carotene produced is twenty times greater than that produced by either strain growing on its own (Barnett, Lilly & Krause, 1956). The same stimulation occurs with mixed cultures of (+) and (−) *Blakeslea trispora* (Ciegler, 1965; Plempel, 1963; Prieto, Spalla, Bianchi & Biffi, 1964; Thomas & Goodwin, 1967). The stimulation is so great that industrial production is feasible under appropriate conditions (Ciegler, 1965). The β-factor responsible for this stimulation consists of a series of acids of which trisporic acid C (XIII) is the major component (Caglioti *et al.*, 1964, 1966; Van den Ende 1968). Trisporic acid also controls gametogenesis in Phycomycetes (see Barksdale, 1969).

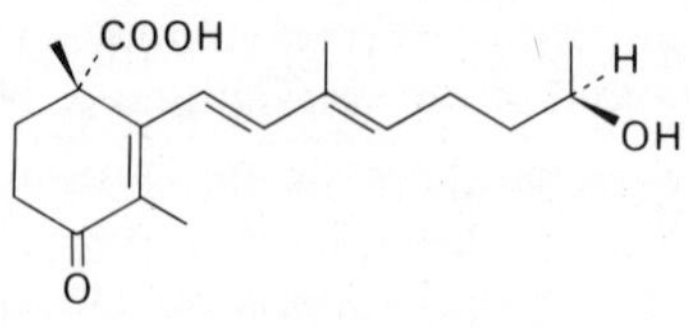

XIII Trisporic acid

Trisporic acid stimulates carotenogenesis in the (−) strain but not in the (+) strain of *Blakeslea trispora* (Thomas & Goodwin, 1967; Sutter & Rafelson, 1968; Yuldashova, Feofilova, Samokhvalov & Bekhtereva,

1972). However, it appears to be the (+) strain which synthesizes the factor because when either the (+) or (−) strain is co-cultured with the homothallic *Zygorhynchus moelleri* trisporic acid is formed only with the (+) strain (Sutter & Rafelson, 1968). Physical contact between the (+) and (−) strains is probably not necessary for trisporic acid production, which is stimulated by a diffusible 'progamone' formed in and released from the (−) strain (Van den Ende, 1968). Trisporic acid is not incorporated into β-carotene and its effect is inhibited by actidione (Thomas, Harris, Kirk & Goodwin, 1967). This means that it probably acts by derepressing synthesis of an enzyme concerned with carotene synthesis. As it stimulates synthesis of enormous amounts of phytoene in the presence of diphenylamine (p. 430) (Thomas & Goodwin, 1967), the enzyme involved must be concerned with the steps of carotenogenesis before phytoene formation.

There is some evidence that trisporic acid is itself formed from β-carotene via retinal (Austin, Bu'Lock & Drake, 1970); if this is so then its action represents a positive feedback system.

β-Ionone has a similar stimulatory effect in *Phycomyces blakesleeanus* (Mackinney, Nakayama, Buss & Chichester, 1952; Mackinney, Chichester & Wong, 1953; Mackinney, Nakayama, Chichester & Buss, 1953; Engel, Würsch & Zimmerman, 1953; Chichester, Wong & Mackinney, 1954) and in heterothallic but not homothallic cultures of *Blakeslea trispora* (Ciegler, Arnold & Anderson, 1959; Reyes, Nakayama & Chichester, 1964). Other terpenes also have a stimulatory effect on carotenogenesis in *B. trispora* (Cederberg & Neujahr, 1969) and vitamin A is particularly active (E. Cerdá-Olmedo, pers. comm.). Synthetic compounds related to β-ionone are stimulatory as are certain amides, lactams, hydrazides and pyridines, in particular succinimide and isonicotinoylhydrazine. The actions of the ionones and the nitrogenous compounds are cumulative (Ninet, Renault & Tissier, 1969).

LIGHT Light stimulates additional carotenoid synthesis in fungi which normally form reasonable amounts in the dark, *e.g. Phycomyces blakesleeanus* (Garton, Goodwin & Lijinsky, 1951; Chichester, Wong & Mackinney, 1954: Bergman, Eslava & Cerdá-Olmedo, 1973) and *Penicillium oxysporum* (Mase, Rabourn & Quakenbush, 1957). However, in other fungi carotene synthesis in the dark occurs to a very limited extent or not at all, but it can be initiated by photoinduction, which involves a short simultaneous exposure to oxygen and light; such fungi include *Fusarium* spp. (Carlile, 1956; Theimer & Rau, 1969), *Neurospora crassa* (Haxo, 1956), *Verticillium* spp. (Valadon & Mummery, 1969, 1971) *Cephalosporium diospyri* (Codner & Platt, 1959; Seviour & Codner, 1973), *Dacrospinax spathularia* (Goldstrohm & Lilly, 1965). *Pyronema confluens* (Carlile & Friend, 1956) and *Syzygites megalocarpus* (Wenger & Lilly, 1966). After photoinduction there is usually a lag period before carotenogenesis begins; in *Fusarium aqueductuum* it is about four hours (Rau, 1967).

Fusarium aqueductuum, which has been examined in detail, reveals the complexity of the situation. In the first place, if after normal photoinduction the fungus is transferred to an oxygen-free environment no carotenoid synthesis occurs. However, access to oxygen within 48 h of photoinduction results in pigment synthesis. Thus the photoinduced-state is reasonably long-lived (Rau, 1969, 1971). Actidione (cycloheximide) when applied before photoinduction inhibits carotenogenesis, but removal at anytime up to 30 h after photoinduction allows pigment synthesis which is proportionally less the longer the period before removal of the inhibitor. This means that some fairly stable 'induction factor' is formed which stimulates synthesis of carotenogenic enzymes (Rau, Lindermann & Rau-Hund, 1968; Rau, 1971). If photo-induced cells are kept for some time under nitrogen and then allowed access to oxygen carotenoid synthesis takes place immediately with no obvious lag phase. This means that the carotenogenic enzymes have been synthesized during the anaerobic period (Lang & Rau, 1972).

In *Verticillium albo-atrum*, cycloheximide, actinomycin D and puromycin inhibited both photoinduced carotenogenesis and protein synthesis, whilst 5-fluorouracil and 5-deoxythymidine inhibited only protein synthesis (Mummery & Valadon, 1973). This suggests that ribosomal RNA is implicated in the photoinduction of carotenogenesis.

Hydrogen peroxide (10^{-1}–10^{-2} M) and *p*-chloromercuribenzoate (5×10^{-5} M) simulate the photoinduction of carotenogenesis in *Fusarium aquaeductuum* in that they induce synthesis in the dark (Rau, Feuser & Rau-Hund, 1967; Theimer & Rau, 1970). The mechanisms involved must however be different from that in photoinduction because (a) the effect of *p*-chloromercuribenzoate and light are additives; (b) inhibition of the *p*-chloromercuribenzoate effect with thiols does not affect photoinduction; and (c) in a mutant in which photoregulation has been repressed and which synthesizes carotenoids in the dark, addition of *p*-chloromercuribenzoate still stimulates additional pigment synthesis (Theimer & Rau, 1969, 1970, 1972). The *p*-chloromercuribenzoate and H_2O_2 effects were not observed in *Verticillium agaricinum* (Valadon & Mummery, 1971).

14.4 References

ANDERSON, D. G., NORGARD, D. W. & PORTER, J. W. (1960). The incorporation of mevalonic acid-2-^{14}C and dimethylacrylic acid-3-^{14}C into carotenes. *Archives Biochemistry and Biophysics* **88**, 68–77.

AUNG THAN, BRAMLEY, P. M., DAVIES, B. H. & REES, A. F. (1972). The stereochemistry of phytoene. *Phytochemistry* **11**, 3187–92.

AUSTIN, D. J., BU'LOCK, J. D. & DRAKE, D. (1970). The biosynthesis of trisporic acid from *β*-carotene via retinol and trisporol. *Experientia* **26**, 348–9.

BARKSDALE, A. W. (1969). The sexual hormones of *Achlya* and other fungi. *Science* **166**, 831–7.

BARNETT, H. L., LILLY, V. G. & KRAUSE, R. F. (1956). Increased production of carotene by mixed (+) and (−) cultures of *Choanophora cucurbitarium*. *Science* **123**, 141.

BERGMAN, K., ESLAVA, A. P. & CERDÁ-OLMEDO, E. (1973). Mutants of *Phycomyces* with abnormal phototropism. *Molecular and General Genetics* **123**, 1–16.

BOBKOVA, T. S. & RABOTNOVA, I. L. (1967). The effect of carbon-nitrogen ratio in the medium on carotenoid biosynthesis by *Sporobolomyces roseus*. *Mikrobiologiya* **36**, 947.

BRAITHWAITE, G. D. & GOODWIN, T. W. (1960*a*). The incorporation of [1-^{14}C] acetate, [2-^{14}C] acetate and $^{14}CO_2$ into lycopene by tomato slices. *Biochemical Journal* **76**, 1–5.

BRAITHWAITE, G. D. & GOODWIN, T. W. (1960*b*). The incorporation of [^{14}C] acetate, [^{14}C] mevalonate and $^{14}CO_2$ into *β*-carotene by the fungus *Phycomyces blakesleeanus*. *Biochemical Journal* **76**, 5–10.

BRAITHWAITE, G. D. & GOODWIN, T. W. (1960*c*). Incorporation of [2-^{14}C] acetate, DL-[2-^{14}C] mevalonate and $^{14}CO_2$ into carrot root preparations. *Biochemical Journal* **76**, 194–7.

BUGGY, M. J., BRITTON, G. & GOODWIN, T. W. (1969). Stereochemistry of phytoene biosynthesis by isolated chloroplasts. *Biochemical Journal* **114**, 641–3.

CAGLIOTI, L., CAINELLI, G., CAMERINO, B., MONDELLI, R., PRIETO, A., QUILICO, A., SALVATORI, T. & SELVA, A. (1964). Sulla costituzione degli acidi trisporici. *Chimica e l'Industria* **46**, 961–6.

CAGLIOTI, L., CAINELLI, G., CAMERINO, B., MONDELLI, R., PRIETO, A., QUILICO, A., SALVATORI, R. & SELVA, A. (1966). The structure of trisporic-C acid. *Tetrahedron Supplements* **7**, 175–87.

CANTINO, E. C. & HYATT, M. T. (1953). Carotenoids and oxidative enzymes in the aquatic Phycomycetes *Blastocladiella* and *Rhizophlyctis. American Journal of Botany* **40**, 688–94.

CARLILE, M. J. (1956). A study of the factors influencing non-genetic variation in a strain of *Fusarium oxysporum. Journal of General Microbiology* **28**, 643–54.

CARLILE, M. J. & FRIEND, J. (1956). Carotenoids and reproduction in *Pyronema confluens. Nature, London* **178**, 369–70.

CEDERBERG, E. & NEUJAHR, H. Y. (1969). Activation of β-carotene synthesis in *Blakeslea trispora* by certain terpenes. *Acta Chemica Scandinavia* **23**, 957.

CHICHESTER, C. O., WONG, P. S. & MACKINNEY, G. (1954). On the biosynthesis of carotenoids. *Plant Physiology* **29**, 238–41.

CHICHESTER, C. O., NAKAYAMA, T., MACKINNEY, G. & GOODWIN, T. W. (1955). On the incorporation of leucine-carbon into carotene by *Phycomyces. Journal of Biological Chemistry* **214**, 515–7.

CHICHESTER, C. O., YOKOYAMA, H., NAKAYAMA, T. O. M., LUKTON, A. & MACKINNEY, G. (1959). Leucine metabolism and carotene biosynthesis. *Journal of Biological Chemistry* **234**, 598–602.

CIEGLER, A. (1965). Microbial carotenogenesis. *Advances in Applied Microbiology* **7**, 1–34.

CIEGLER, A., ARNOLD, M. & ANDERSON, R. F. (1959). Microbiological production of carotenoids. IV. Effect of various grains on production of β-carotene by mated strains of *Blakeslea trispora. Applied Microbiology* **7**, 94–8.

CODNER, R. C. & PLATT, B. C. (1959). Light-induced production of carotenoid pigments by *Cephalosporia. Nature, London* **184**, 471–2.

COGGINS, C. W. JR, HEMMING, G. L. & YOKOYAMA, H. (1970). Lycopene accumulation induced by 2-(4-chlorophenylthio)-triethylamine hydrochloride. *Science* **168**, 1589–90.

CORNFORTH, J. W. & CORNFORTH, R. H. (1970). Chemistry of mevalonic acid. *Biochemical Society Symposium* **29**, 1–15.

DAVIES, B. H. (1970). A novel sequence for phytoene dehydrogenation in *Rhodospirillum rubrum. Biochemical Journal* **116**, 93–9.

DAVIES, B. H. (1973). Carotenoid biosynthesis in fungi. *Pure and Applied Chemistry* **35**, 1–28.

DAVIES, B. H., JONES, D. & GOODWIN, T. W. (1963). The problem of lycopersene in *Neurospora crassa. Biochemical Journal* **87**, 326–9.

DAVIES, B. H., VILLOUTREIX, J., WILLIAMS, R. J. H. & GOODWIN, T. W. (1963). The possible role of β-zeacarotene in carotenoid cyclization. *Biochemical Journal* **89**, 96P.

DE LA GUARDIA, M. D., ARAGON, C. M. G., MURILLO, F. J. & CERDÁ-OLMEDO, E. (1971). A carotenogenic enzyme gate in *Phycomyces*: evidence from quantitative complementation. *Proceedings National Academy of Sciences, U.S.A.* **68**, 2012–15.

ENGEL, B. G., WÜRSCH, J. & ZIMMERMAN, M. (1953). Uber den Einfluss von β-Ionon auf die Bildung von β-carotin durch *Phycomyces blakesleeanus. Helvetica Chimica Acta* **36**, 1771–6.

ESLAVA, A. P. & CERDÁ-OLMEDO, E. (1974). Genetic control of phytoene dehydrogenation in *Phycomyces. Plant Science Letters* **2**, 9–14.

FOPPEN, F. H. & GRIBANOVSKI-SASSU, O. (1967). Lipids produced by *Epicoccum nigrum* in submerged culture. *Biochemical Journal* **106**, 97–100.

FRIEND, J. & GOODWIN, T. W. (1954). The effect of the temperature and thiamine concentration on carotenogenesis by *Phycomyces blakesleeanus. Biochemical Journal* **57**, 434–7.

FRIEND, J., GOODWIN, T. W. & GRIFFITHS, L. A. (1955). The role of carboxylic acids in the biosynthesis of β-carotene by *Phycomyces blakesleeanus. Biochemical Journal* **60**, 649–55.

GARTON, G. A., GOODWIN, T. W. & LIJINSKY, W. (1951). General conditions governing β-carotene synthesis by the fungus *Phycomyces blakesleeanus. Biochemical Journal* **48**, 154–63.

GOLDSTROHM, D. D. & LILLY, V. G. (1965). The effect of light on the survival of pigmented and non-pigmented cells of *Dacyropinax spathularia. Mycologia* **57**, 612–23.

GOODAY, G. W., FAWCETT, P., GREEN, D. & SHAW, E. (1973). The formation of fungal sporopollenin in the zygospore wall of *Mucor mucedo*; a role for the sexual carotenogenesis in the Mucorales. *Journal of General Microbiology* **74**, 233–9.

GOODWIN, T. W. (1971*a*). In *Carotenoids*. Edited by O. Isler, p. 577–636 Basel: Birkhaüser.

GOODWIN, T. W. (1971*b*). Biosynthesis of carotenoids and plant triterpenes *Biochemical Journal* **123**, 293–329.

GOODWIN, T. W. (1972*a*). Carotenoids in fungi and non-photosynthetic bacteria. *Progress in Industrial Microbiology* **11**, 31–88.

GOODWIN, T. W. (1972*b*). Recent development in the biosynthesis of carotenoids. *Biochemical Society Symposium* **35**, 233–44.

GOODWIN, T. W. & LIJINSKY, W. (1952). The effect of different amino acids on carotenogenesis in *Phycomyces blakesleeanus* when used in media containing low concentrations of glucose. *Biochemical Journal* **50**, 268–73.

GOODWIN, T. W. & WILLIAMS, R. J. H. (1965*a*). The stereochemistry of phytoene biosynthesis. *Proceedings Royal Society, Series B*, **163**, 515–8.

GOODWIN, T. W. & WILLIAMS, R. J. H. (1965*b*). A mechanism for the cyclization of an acylic precursor to form β-carotene. *Biochemical Journal* **94**, 5C–7C.

GOODWIN, T. W. & WILLMER, J. S. (1952). Nitrogen metabolism and carotenogenesis in *Phycomyces blakesleeanus*. *Biochemical Journal* **51**, 213–19.

GOODWIN, T. W. GRIFFITHS, L. A. & MODI, V. V. (1956). The action of some antibiotics especially streptomycin on carotenogenesis in *Phycomyces blakesleeanus*. *Biochemical Journal* **62**, 259–68.

GOODWIN, T. W., JAMIKORN, M. & WILLMER, J. S. (1953). The mode of action of diphenylamine in inhibiting carotenogenesis in *Phycomyces blakesleeanus*. *Biochemical Journal* **53**, 531–8.

GRIBANOVSKI-SASSU, O. & FOPPEN, F. H. (1967). The carotenoids of the fungus *Epicoccum nigrum* Link. *Phytochemistry* **6**, 907-9.

GROB, E. C. (1953). Über die Biosynthese des Carotinoide bei *Mucor hiemalis*. *Chimia* **7**, 90–1.

GROB, E. C. (1959). The biosynthesis of carotenoids by micro-organisms. In *Biosynthesis of terpenes and sterols*. Edited by G. E. W. Wolstenholme and M. O'Connor, p. 267–76, London: Churchill.

GROB, E. C. & BÜTLER, R. (1955). Über die Biosynthese des β-carotins bei *Mucor hiemalis*. Die Beteiligung der Essigsäure am Aufbau der Carotinmolekel, insbesondere in den Stellung 14–15 bzw. 14′–15′ und 10–11 bzw 10′–15′ untersucht mit Hilfe von ^{14}C-markierter Essigsäure. *Helvetica Chimica Acta* **38**, 1313–6.

GROB, E. C. & BÜTLER, R. (1956). Über die Biosynthese des β-carotins bei *Mucor hiemalis*. Die Beteiligung der Essigäure am Aufbau der Carotinmolekel, insbesondere in den Stellung 3,4,6 bzw. 3′,4′,6′ untersucht mit Hilfe von ^{14}C-markierter Essigäure. *Helvetica Chimica Acta* **39**, 1975–80.

GROB, E. C., GRUNDBACHER, V. & SCHOPFER, W. H. (1954). Der Einfluss der Pantothensäure, des Pantethins und des phosphorylierten Pantethins auf die Carotinbildung bei *Mucor hiemalis*. *Experientia* **10**, 378.

HAXO, F. (1956). Some biochemical aspects of fungal carotenoids. *Fortschritte der Chemie organischer Naturstoffe* **12**, 169–97.

ISLER, O. (1971). *Carotenoids*. Basel. Birkhauser.

KNYPL, J. S. (1969). Accumulation of lycopene in detached cotyledons of pumpkins treated with (2-chloroethyl) trimethyl ammonium chloride. *Naturwissenschaften* **56**, 572.

LANG, W. & RAU, W. (1972). Untersuchungen über die lichtabhängige Carotinoid-synthese. IX. Zum Induktion mechanismus der Carotinoidbildenden Enzyme bei *Fusarium aquaeductuum*. *Planta* **106**, 345–54.

LEE, T. C. & CHICHESTER, C. O. (1969). Geranylgeranyl pyrophosphate as the condensing unit for the enzymatic synthesis of carotenes. *Phytochemistry* **8**, 603–9.

LOWRY, L. K. & CHICHESTER, C. O. (1971). The role of thiamine, leucine and CO_2 in the biosynthesis of carotenes by the mould *Phycomyces blakesleeanus*. *Phytochemistry* **10**, 323–33.

McDERMOTT, J. C. B., BRITTON, G. & GOODWIN, T. W. (1973). Carotenoid biosynthesis in a *Flavobacterium* sp.: stereochemistry of hydrogen elimination in the desaturation of phytoene to lycopene, rubixanthin and zeaxanthin. *Biochemical Journal* **134**, 1115–7.

McDERMOTT, J. C. B., BEN-AZIZ, A., SINGH, R. K., BRITTON, G. & GOODWIN, T. W. (1973). Recent studies of carotenoid biosynthesis in bacteria. *Pure and Applied Chemistry* **35**, 29–45.

MACKINNEY, G., CHANDLER, B. V. & LUKTON, A. (1958). Carbon sources for carotene and ergosterol: mevalonic and hydroxymethylglutaric acid. *Proceedings 4th International Congress of Biochemistry*, p. 130 London: Pergamon.

MACKINNEY, G., CHICHESTER, C. O. & NAKAYAMA, T. (1955). The incorporation of glycine carbon into β-carotene in *Phycomyces blakesleeanus. Biochemical Journal* **60**, XXXVII.

MACKINNEY, G., CHICHESTER, C. O. & WONG, P. S., (1953). Carotenoids in *Phycomyces. Journal of American Chemical Society* **75**, 5428.

MACKINNEY, G., NAKAYAMA, T., BUSS, C. D. & CHICHESTER, C. O. (1952). Carotenoid production in *Phycomyces. Journal of American Chemical Society* **74**, 3456–7.

MACKINNEY, G., NAKAYAMA, T., CHICHESTER, C. O. & BUSS, C. D. (1953). Biosynthesis of carotene in *Phycomyces. Journal of American Chemical Society* **75**, 236.

MASE, Y., RABOURN, W. J. & QUACKENBUSH, F. W. (1957). Carotene production by *Penicillium sclerotiorum. Archives of Biochemistry and Biophysics* **68**, 150–61.

MEISSNER, G. & DELBRÜCK, M. (1968). Carotenes and retinals in *Phycomyces* mutants. *Plant Physiology*, **43**, 1279–83.

MUMMERY, R. S. & VALADON, L. R. G. (1973). Effect of certain acid and protein inhibitors on carotenogenesis in *Verticillium agaricinium. Physiologia Plantarum* **28**, 254–8.

NINET, L., RENAULT, J. & TISSIER, R. (1969). Activation of the biosynthesis of carotenoids in *Blakeslea trispora. Biotechnology and Bioengineering* **11**, 1195–9.

OOTAKI, T., LIGHTY, A. C., DELBRÜCK, M. & HSU, W. J. (1973). Complementation between mutants of *Phycomyces* deficient with respect to carotenogenesis. *Molecular and General Genetics* **121**, 57–70.

PLEMPEL, M. (1963). Die Chemischen Grudlagen der Sexualreaktion bei Zygomyceten. *Planta* **59**, 492–508.

PRIETO, A., SPALLA, M., BIANCHI, M., BIFFI, G. (1964). Biosynthesis of β-carotene by strains of *Choanephoraceae. Chemistry and Industry* 551.

RAU, W. (1967). Untersuchungen über die lichtabhängige Carotinoidsynthese. IV. Ersatz der Licht. *Planta* **74**, 263–77.

RAU, W. (1969). Untersuchengen über die lichtabhängige Carotinoidsynthese. IV. Die Rolle des Sauerstoffs bei den Lichtinduktion. *Planta* **84**, 30–42.

RAU, W. (1971). Untersuchungen über die lichtabhängige Carotenoidsynthese VII. Reversible Unterbrechnung der Reactionskette durch Cycloheximid und anaerobe Bedingungen. *Planta* **101**, 251–64.

RAU, W., FEUSER, B. & RAU-HUND, A. (1967). Substitution of *p*-chloro- or *p*-hydroxymercuribenzoate for light carotenoid synthesis by *Fusarium aquaeductuum. Biochimica et Biophysica Acta* **136**, 589–90.

RAU, W., LINDERMANN, I. & RAU-HUND, A. (1968). Untersuchungen über Lichtabhängige Carotinoidsyntheses. III. Die Farbstoffe von *Neurospora crassa* in Submerskultur. *Planta* **80**, 309–16.

REICHEL, L. & SCHREIBER, G. (1968). Biosynthesis of β-carotene from 1-carbon units. *Pharmazie* **23**, 594.

REICHEL, L. & WILLIS, M. (1958). Zur Biosynthese des beta-Carotins. *Naturwissenschaften* **45**, 130.

REYES, P., NAKAYAMA, T. O. M. & CHICHESTER, C. O. (1964). The mechanism of β-ionone stimulation of carotenoid and ergosterol production in *Phycomyces blakesleeanus. Biochimica et Biophysica Acta* **90**, 578–92.

RILLING, H. C. (1965). A study of inhibition of carotenoid synthesis. *Archives of Biochemistry and Biophysics* **110**, 39–46.

SCHOPFER, W. H. & GROB, E. C. (1950). Recherches sur la biosynthese des carotenoides chez un microorganisme. *Experientia* **6**, 419.

SCHOPFER, W. H. & GROB, E. C. (1952). Sur la biosynthèse du β-carotene par *Phycomyces* cultive sur un milieu contenant de l'acetate de sodium comme unique source de carbone. *Experientia* **8**, 140.

SCHOPFER, W. H., GROB, E. C. & BESSON, G. (1952*a*). Recherches sur les inhibiteurs de la croissance et de la biogenèse des carotinoides. II. La streptomycine. *Archiv Science (Geneva)* **5**, 198–201.

SCHOPFER, W. H., GROB, E. C., BESSON, G. & KELLER, V. (1952*b*). Recherches sur les inhibiteurs du development et de la biogenèse des carotinoides. I. La streptomycine. *Archiv Science (Geneva)*, **5**, 194–7.

SEVIOUR, L. J. & CODNER, R. C. (1973). Effect of light on carotenoid and riboflavin production by the fungus *Cephalosporium diospyri. Journal of General Microbiology* **77**, 403–15.

SUTTER, R. O. & RAFELSON, M. E. JR. (1968). Separation of β-factor synthesis from stimulated β-carotene synthesis in mated cultures of *Blakeslea trispora*. *Journal of Bacteriology* **95**, 426–32.

THEIMER, R. R. & RAU, W. (1969). Untersuchungen über die lichtabhängige Carotinoidsynthese. *Biochimica et Biophysica Acta* **177**, 180–1.

THEIMER, R. R. & RAU, W. (1970). Untersuchungen über die Lichtabhängige Cartinoidsynthese. *Planta* **92**, 129–37.

THEIMER, R. R. & RAU, W. (1972). Untersuchungen über die Lichtabhängige Carotinoidsynthese. VIII. Die unterschiedlichen Wirkungmechanismen von Licht under Mercuribenzoat. *Planta* **106**, 331–43.

THOMAS, D. M. & GOODWIN, T. W. (1967). Studies on carotenogenesis in *Blakeslea trispora*: the mode of action of trisporic acid. *Phytochemistry* **6**, 355–60.

THOMAS, D. M., HARRIS, R. C., KIRK, J. T. O. & GOODWIN, T. W. (1967). Studies on carotenogenesis in *Blakeslea trispora*. II. The mode of action of trisporic acid. *Phytochemistry* **6**, 361–6.

TURIAN, G. (1957). Recherches sur l'action anticaroténogène de la diphenylamine et ses consequences sur la morphogenèse reproductive chez *Allomyces* et *Neurospora*. *Physiologia Plantarum* **10**, 667.

TURIAN, G. & HAXO, F. (1954). Minor polyene compounds in the sexual phase of *Allomyces javanicus*. *Botanical Gazette* **54**, 254–60.

VALADON, L. R. G. & MUMMERY, R. S. (1966). Inhibition of carotenoid synthesis in a mutant of *Verticillium albo-atrum*. *Journal of General Microbiology* **45**, 531–40.

VALADON, L. R. G. & MUMMERY, R. S. (1969). Biosynthesis of neurosporaxanthin. *Microbios* **1A**, 3.

VALADON, L. R. G. & MUMMERY, R. S. (1971). Effect of light on nucleic acids, protein and carotenoids of *Verticillium agaricinum*. *Microbios* **4**, 227–40.

VAN DEN ENDE, H. (1968). Relationship between sexuality and carotene synthesis in *Blakeslea trispora*. *Journal of Bacteriology* **96**, 1298–303.

WENGER, C. J. & LILLY, V. G. (1966). The effect of light on carotenogenesis, growth and sporulation of *Syzygites megalocarpus*. *Mycologia* **58**, 671–80.

WILLIAMS, R. J. H., DAVIES, B. H. & GOODWIN, T. W. (1965). The presence of β-zeacarotene in cultures of diphenylamine-inhibited *Phycomyces blakesleeanus*. *Phytochemistry* **4**, 759–60.

WILLIAMS, R. J. H., BRITTON, G., CHARLTON, J. M. & GOODWIN, T. W. (1967). The stereospecific biosynthesis of phytoene and polyunsaturated carotenes. *Biochemical Journal* **104**, 767–77.

WUHRMANN, J. J., YOKOYAMA, H. & CHICHESTER, C. O. (1957). Degradation of leucine-derived carotenes. *Journal of American Chemical Society* **50**, 268.

YAMAMOTO, H. Y. & CHICHESTER, C. O. (1965). Dark incorporation of oxygen into antheraxanthin by bean leaf. *Biochimica et Biophysica Acta* **109**, 303.

YAMAMOTO, H. Y., CHICHESTER, C. O. & NAKAYAMA, T. O. M. (1962*a*). Light and dark interconversions of leaf xanthophylls. *Archives of Biochemistry and Biophysics* **96**, 645–9.

YAMAMOTO, H. Y., CHICHESTER, C. O., NAKAYAMA, T. O. M. (1962*b*). Xanthophylls and the Hill reaction. *Photochemistry and Photobiology* **1**, 53–7.

YOKOYAMA, H., CHICHESTER, C. O. & MACKINNEY, G. (1969). Formation of carotene *in vitro*. *Nature, London* **185**, 687–8.

YOKOYAMA, H., NAKAYAMA, T. O. M. & CHICHESTER, C. O. (1962). Biosynthesis of β-carotene by cell-free extracts of *Phycomyces blakesleeanus*. *Journal of Biological Chemistry* **237**, 681–6.

YOKOYAMA, H., CHICHESTER, C. O., NAKAYAMA, T. O. M., LUKTON, A. & MACKINNEY, G. (1957). Carotene, leucine-3-C^{14} and 4-C^{14}. *Journal of American Chemical Society* **79**, 2029–30.

YULDASHEVA, L. S., FEOFILOVA, E. P., SAMOKHVALOV, G. I. & BEKHTEREVA, M. N. (1972). Effects of trisporic acids on formation of ubiquinone-9 and riboflavin by the fungus *Blakeslea trispora*. *Mikrobiologiya* **41**, 430.

CHAPTER 15

Polyketides and Related Metabolites

W. B. TURNER

15.1 Introduction

Polyketide biosynthesis is highly characteristic of the fungi. More fungal secondary metabolites are produced by this route than by any other biosynthetic pathway (Turner, 1971, pp. 74–213) and, moreover, most of the known polyketide-derived natural products have been obtained from the fungi, with a much smaller number having been isolated from bacteria and higher plants. Even in the fungi, polyketide biosynthesis is unevenly distributed and is particularly characteristic of the Fungi Imperfecti. Partly because of this rather limited distribution and partly because the pathway does not lead to any primary metabolites the polyketide route has been largely neglected by biochemists. For this reason little is known about the enzymatic processes involved in polyketide biosynthesis, and the concept was developed largely by organic chemists, particularly by Birch (1967), as a result of an analysis of the structures of natural products. (The analysis revealed several anomolous structures which were subsequently proved incorrect, and the hypothesis has been of assistance in the determination of the structures of natural products.)

In this chapter we shall first consider in simple chemical terms the basic principles of the polyketide hypothesis, principles which have been confirmed by experiments with labelled precursors, usually in fungi. We shall then consider present knowledge of the biochemical mechanisms involved, and then discuss some variations of polyketide biosynthesis and the use of this pathway in combination with other biosynthetic processes. The references are to review articles, to more recent work, and to one or two papers which are considered to be of general importance.

15.2 The Polyketide Hypothesis

Chain assembly and cyclization reactions

Polyketides are formed by condensation of an acetyl unit with three or more malonyl units, with concomitant decarboxylation as in fatty acid biosynthesis (see Chapter 6) but without obligatory reduction of the intermediate

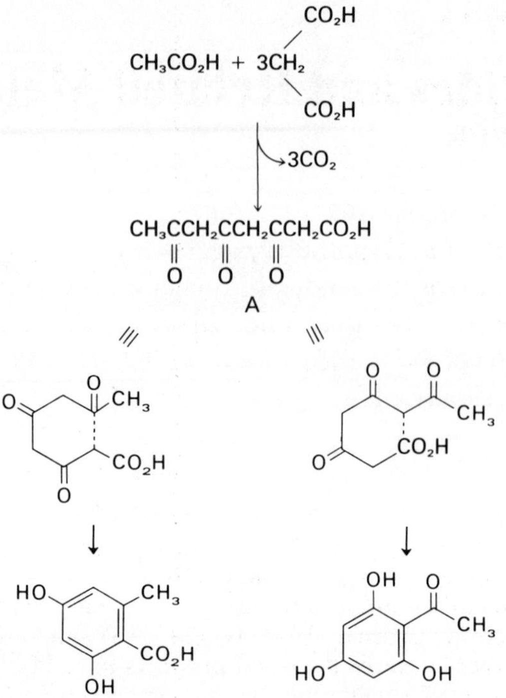

Fig. 15.1 Formation of orsellinic acid and acetylphloroglucinol by alternative cyclizations of the tetraketide intermediate (A).

β-dicarbonyl system. The resulting poly-β-ketomethylene chain (*e.g.* A in Fig. 15.1.) is thus made up of repeating two-carbon units* and the compounds can be classified as triketides, tetraketides, pentaketides, *etc.* according to the number of 'C_2-units' which have contributed to the biosynthesis of the chain; orsellinic acid and acetylphloroglucinol (Fig. 15.1) are thus tetraketides. A survey of the polyketide-derived metabolites of the fungi reveals that tetra-, penta-, hepta-, and octaketides are numerous while tri-, hexa-, nona-, and decaketides are less common. There is a sharp cutout at the decaketide level and siphulin, a lichen product, is the only obviously polyketide-derived compound formed from a chain longer than C_{20}.

The polyketide chains possess reactive methylene groups which can take part in internal aldol-type reactions to give aromatic compounds, a process illustrated in Fig. 15.1 by the formation of orsellinic acid and acetylphloroglucinol by alternative reactions of the intermediate A. Clearly, with longer polyketide chains a greater variety of cyclizations becomes possible but in practice not all the theoretically possible cyclizations of chain length have been observed. In particular the uncyclized residue from the methyl end of a

* Since malonate is formed by carboxylation of acetate with which it is in equilibrium, in whole-cell systems acetate is uniformly incorporated into polyketide compounds which are thus sometimes referred to as 'poly-acetate' compounds, and the repeating C_2-units as 'acetate' units.

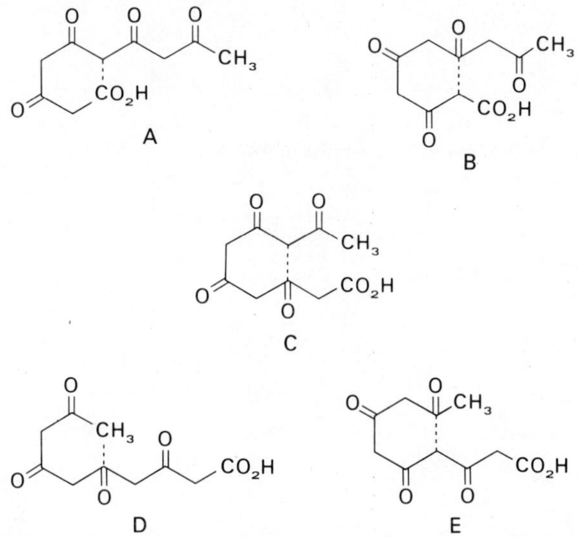

Fig. 15.2 The theoretically possible cyclizations of a pentaketide chain.

polyketide chain is, with one possible exception, never shorter than the residue from the carboxyl end of the chain. This is illustrated in Fig. 15.2 for folding of the (pentaketide) chain derived from five 'C_2-units'; compounds derived from foldings of types A to C are known, but (with the one possible exception noted above) compounds derived from types D and E are not.

Reactions leading to the variety of polyketide structures

Chain formation and cyclization are the basis of polyketide biosynthesis, but the structural variety of the products results from one or more of the reactions discussed below and illustrated in Fig. 15.3. An indication of the variety of structures derivable by the polyketide route is provided by the fact that Fig. 15.3 shows only a fraction of the products derived from just one mode of cyclization (the orsellinic acid/6-methylsalicylic acid type) of the tetraketide intermediate introduced in Fig. 15.1. In fact there are over 60 known metabolites derived from this cyclization, not counting the numerous depsides (*e.g.* lecanoric acid in Fig. 15.3) and depsidones produced by lichens. Moreover there are a further 25 compounds of the triprenyl phenol type illustrated in Fig. 15.3 by grifolin and discussed further below.

(a) Loss of oxygen, possibly by reduction and dehydration (see below). Thus 6-methylsalicylic acid is formed from the tetraketide chain A as illustrated in Fig. 15.3. Although loss of oxygen can be extensive, inspection of the known polyketides reveals a general 'rule'. Compounds which result from a cyclization involving the methylene group α- to the terminal carboxyl group of a polyketide chain (Fig. 15.4) always retain the oxygen atom derived from the β-carbonyl group. This is illustrated in Fig. 15.4 by the formation of the octaketide-derived anthraquinone pachybasin.

(b) Many polyketide-derived aromatic compounds possess 'extra' oxygen atoms not present in the precursor chain, *e.g.* flavipin, aurantiogliocladin, and gentisyl alcohol (Fig. 15.3) or one of the quinonoid oxygens in

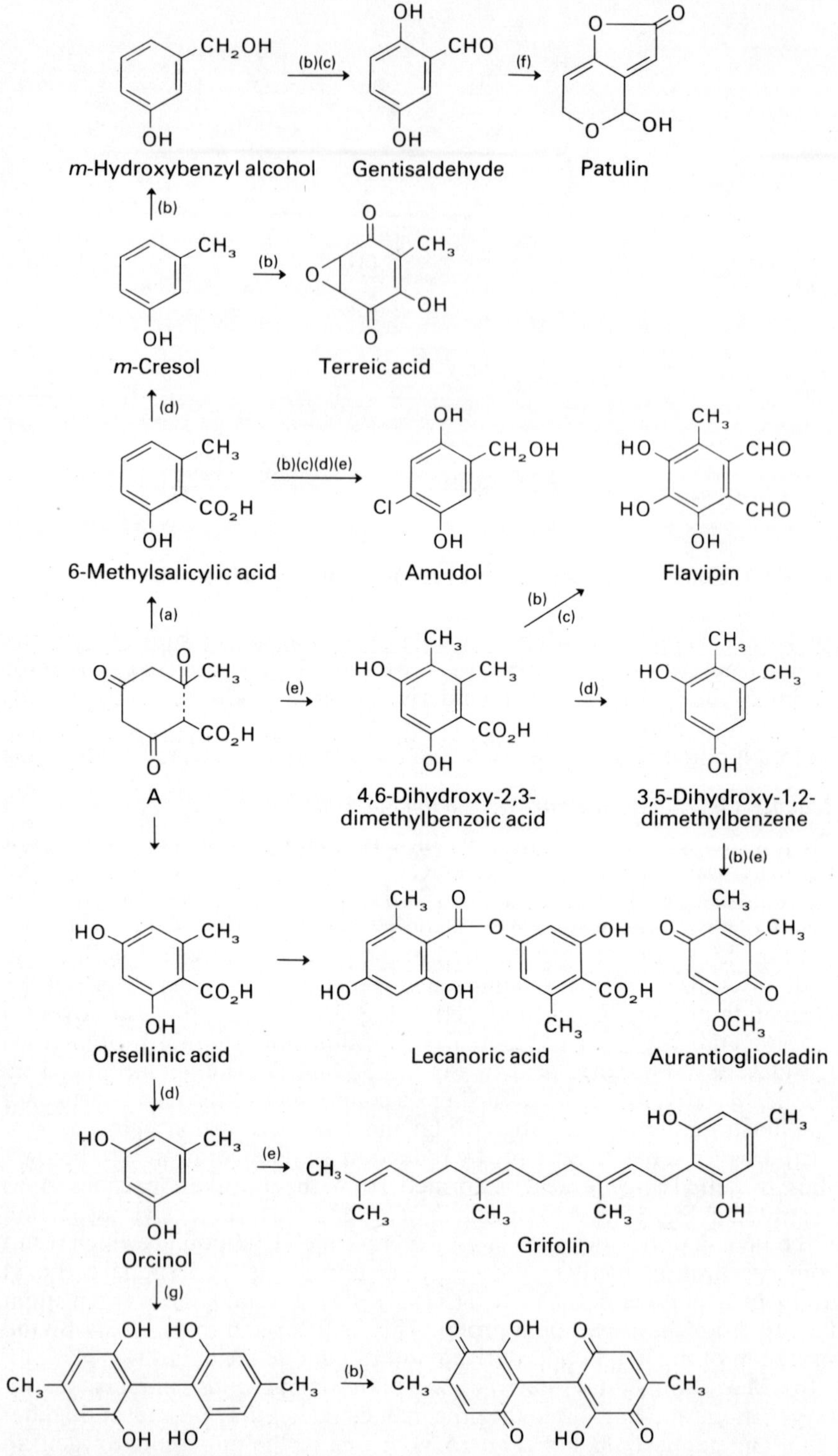

Fig. 15.3 Tetraketide fungal metabolites. The letters on the arrows indicate the type of reaction involved in the formation of each product and correspond to those in the text.

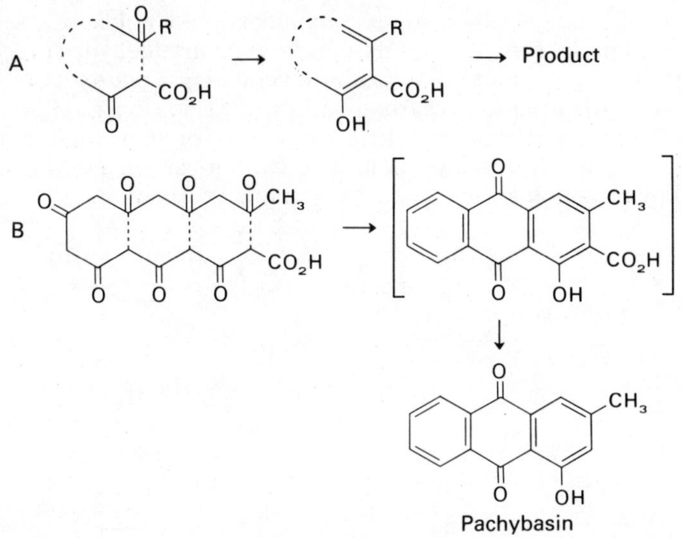

Fig. 15.4 Retention of the β-oxygen atom in polyketide biosynthesis: (A) the general principle; (B) in pachybasin biosynthesis.

pachybasin (Fig. 15.4). These hydroxylation reactions are probably catalysed by mixed function oxygenases.

(c) Methyl or other alkyl groups can be oxidized to hydroxyl or carbonyl functions, *e.g.* flavipin, *m*-hydroxybenzyl alcohol, and gentisaldehyde (Fig. 15.3), and carboxyl groups can be reduced, *e.g.* to form one of the aldehyde groups in flavipin. The hydroxylation of alkyl groups proceeds with retention of configuration (Fig. 15.5).

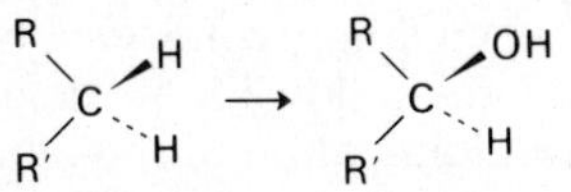

Fig. 15.5 Retention of configuration during biological hydroxylation of alkyl groups.

(d) Decarboxylation can occur, as in the formation of orcinol and *m*-cresol (Fig. 15.3) or pachybasin (Fig. 15.4).

(e) A very common type of reaction involved in the modification of polyketides is electrophilic substitution. The most frequently encountered example is the transfer of the methyl group of *S*-adenosylmethionine either to a carbon atom corresponding to one of the methylene groups of the polyketide chain, *e.g.* in the formation of 4,6-dihydroxy-2,3-dimethylbenzoic acid (Fig. 15.3), or to an oxygen atom *e.g.* in the formation of aurantiogliocladin (Fig. 15.3); the timing of the *C*-methylation reaction is discussed below. Another fairly common alkylation process is the introduction of isoprenoid groups *via* their pyrophosphates; *e.g.* the introduction of a farnesyl chain in the biosynthesis of grifolin (Fig. 15.3); the isoprenoid chains can undergo further modification (see below). A third type of electrophilic process is the introduction of halogen as in amudol (Fig. 15.3).

(f) More drastic (usually oxidative) modification of polyketide skeletons can occur, as in the formation of patulin from gentisaldehyde (Fig. 15.3).

(g) Inter- or intra-molecular oxidative coupling ('one-electron oxidation') can lead to formation of carbon-carbon or carbon-oxygen bonds. Thus phoenicin (Fig. 15.3) is derived from a dimer of orcinol, while usnic acid (Fig. 15.6) results from intermolecular carbon-carbon bond formation followed by intramolecular carbon-oxygen bond formation.

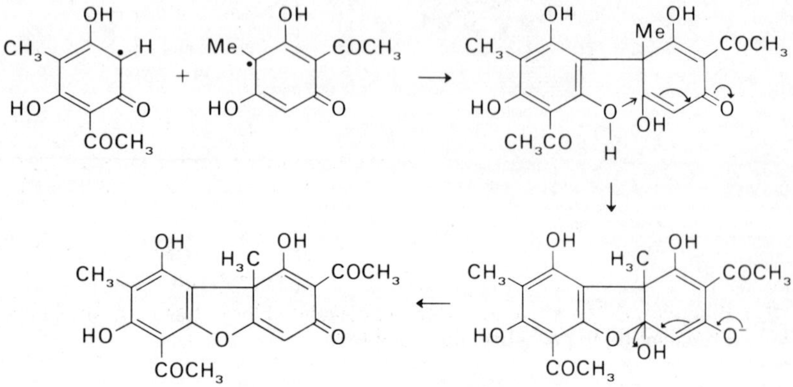

Usnic acid

Fig. 15.6 Oxidative coupling in the formation of usnic acid.

15.3 Techniques for the Study of Polyketide Biosynthesis

As in all biosynthetic studies, there are two basic questions to be answered—what is the nature of the molecular building blocks and what are the intermediate steps between the building blocks and the final product?

The building blocks of polyketide biosynthesis are acetate and malonate, and their involvement has been amply confirmed by numerous experiments using ^{14}C-labelled precursors, followed by degradation of the labelled product to show that no randomization of the label has occurred. The fact that the first two carbon atoms of a polyketide chain are derived from acetate and the remainder from malonate suggests that acetate should be more efficiently and malonate less efficiently incorporated into this position than into the rest of the chain. Such a distinction between the first 'C_2-unit' (the 'starter' group) and the rest would often provide useful information as to the type of chain assembly involved in the biosynthesis of polyketide metabolites. However, although the 'starter effect' has been clearly demonstrated with purified enzyme systems and in a few cases with whole cells, the equilibration of acetate and malonate is normally too rapid in whole cells for a clear effect to be observed. It has therefore not been possible to make as much use of the phenomenon as one might have hoped.

More recently, the use of ^{14}C-labelled acetate for the study of polyketide biosynthesis has been largely replaced by the use of ^{13}C-labelled acetate, using n.m.r. spectroscopy to locate the enriched atoms within the product. This technique has the advantage of making degradation of the labelled product unnecessary and of permitting more complete and precise location of the labelled atoms, but suffers from the disadvantage that high (sometimes unachieveably high) levels of incorporation of the precursor are necessary to give an unambiguous result. The need for a high level of incorporation results from the variability of the intensity of the n.m.r. signals from carbon atoms, and various techniques have been used to minimize this variability (Tanabe, Suzuki & Jankowski, 1973; Cattel, Grove & Shaw, 1973). An important recent development in the study of polyketide biosynthesis is the use of acetate enriched in both carbon atoms with carbon-13 (*e.g.* Seto, Cary & Tanabe, 1973; Seto & Tanabe, 1974). This technique takes advantage of the fact that only those pairs of adjacent carbon atoms which are derived from a single 'C_2-unit' will show carbon-carbon spin-spin coupling, and such atoms are revealed by the presence of satellite bands in the n.m.r. spectrum. The advantages relative to the use of singly-labelled precursors are that lower levels of incorporation can be detected unambiguously and that extra biosynthetic information is obtained.

As well as carbon isotopes, hydrogen isotopes have also been used in the study of polyketide biosynthesis, though not to such effect as the tritium-labelled mevalonates have been used in the study of terpene biosynthesis (Chapter 13). A recent development uses tritium-labelled acetate and n.m.r. spectroscopy, rather than radio-activity, to locate the tritium atoms in the product (Al-rawi, Elvidge, Jaiswal, Jones & Thomas, 1974). Deuterium-labelled acetate, with mass spectroscopic analysis of the deuterated product, has also been used (Scott & Yalpani, 1967).

While the basic steps of polyketide biosynthesis have been established by labelling experiments of the type discussed above, the sequence of steps leading to particular metabolites have usually been less clearly defined and in some cases contradictory results have been reported. The reasons for this uncertainty have been discussed elsewhere (Turner, 1971, pp. 21–2), and several examples of contradictory evidence have been discussed by Turner (1971, *e.g.* pp. 94–7, 102–3). Here it is sufficient to mention a new approach to the problem of elucidating biosynthetic pathways, which suffers from less ambiguity than simple incorporation studies. In this method (Forrester & Gaucher, 1972) labelled putative intermediates are added to cultures and the distribution of radioactivity with time is measured. Kinetic incorporation plots then enable pathway intermediates to be distinguished from metabolic end-products.

The use of purified enzyme systems for biosynthetic studies avoids many of these ambiguities, but in the case of polyketide biosynthesis it has proved difficult to handle the enzymes concerned. Only recently has a pure enzyme, 6-methylsalicylic acid synthetase (see below), been obtained, and experiments with this enzyme have already thrown light on the detailed processes involved. Apart from this only about half a dozen of the enzymes involved in polyketide biosynthesis have been isolated in varying degrees of purity.

15.4 The Mechanism of Polyketide Biosynthesis

As indicated earlier, little is known about the biochemical processes which underlie the polyketide hypothesis, and what follows serves mainly to reveal the questions that remain unanswered.

Relationship of polyketide and fatty acid biosynthesis

The closest analogy for polyketide biosynthesis is provided by fatty acid biosynthesis, a process which has been studied in considerable detail (see Chapter 6). The first point of similarity is the fact that both processes involve the linear condensation of acetate and malonate, activated as their coenzyme A derivatives. Secondly, all the evidence from experiments with cell-free and whole-cell systems suggests that there are no free intermediates between the coenzyme A derivatives and fully assembled and stabilized polyketides, just as there are no free intermediates *en route* to the C_{16} and C_{18} fatty acids. Thirdly, 6-methylsalicylic acid and fatty acids co-produced in the presence of [1-^{14}C, 2-^{3}H]-acetate have distributions of radioactivity consistent with their formation by similar condensation processes. Finally, in the absence of NADPH, which is required to reduce the intermediate carbonyl compounds in fatty acid biosynthesis, highly purified fatty acid synthetase produces triacetic acid lactone (Fig. 15.7) which is a

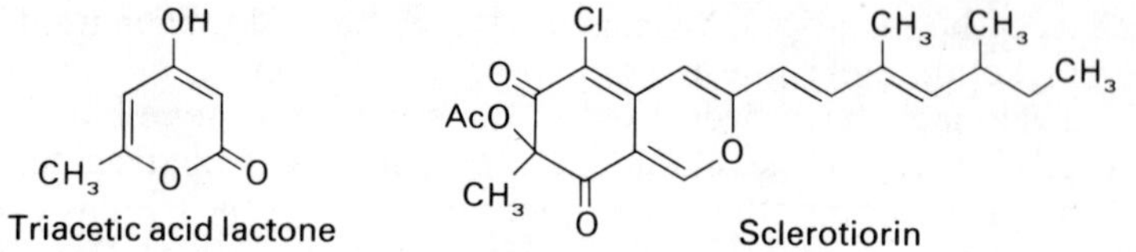

Fig. 15.7 Triacetic acid lactone and sclerotiorin.

stabilized triketide chain. (Triacetic acid lactone is also produced by 6-methylsalicylic acid synthetase in the absence of NADPH, see below.) Although clearly related, fatty acid and polyketide biosyntheses are carried out by separate enzyme systems.

The relationship between polyketide and fatty acid production has been studied in several organisms (*e.g.* Mosbach & Bävertoft, 1971; Koman & Betina, 1973).

Chain assembly

It seems likely that polyketide chains are formed by stepwise condensation of malonyl coenzyme A with a protein-bound chain as in fatty acid biosynthesis, and that the processes of reduction and/or cyclization to give stable products take place on protein-bound intermediates. Granted this, then there is only one possible sequence of events—chain assembly, cyclization, and aromatization—leading to a compound such as orsellinic acid (Fig. 15.1). In such a case, the only problems are the stabilization of the intermediates and the specificity of the cyclization process, and Bu'Lock (1967) has discussed both these problems in terms of metal chelation.

Loss of oxygen

For a compound such as 6-methylsalicylic acid, whose biosynthesis involves the loss of only one oxygen atom relative to orsellinic acid, *six* possible pathways become available (Fig. 15.8) depending on the timing of the reduction, dehydration, and cyclization (aldol) steps. Admittedly, not all of these pathways are equally likely, nor are they mutually exclusive, but none can be ruled out from first principles. Evidence relating to the sequence of

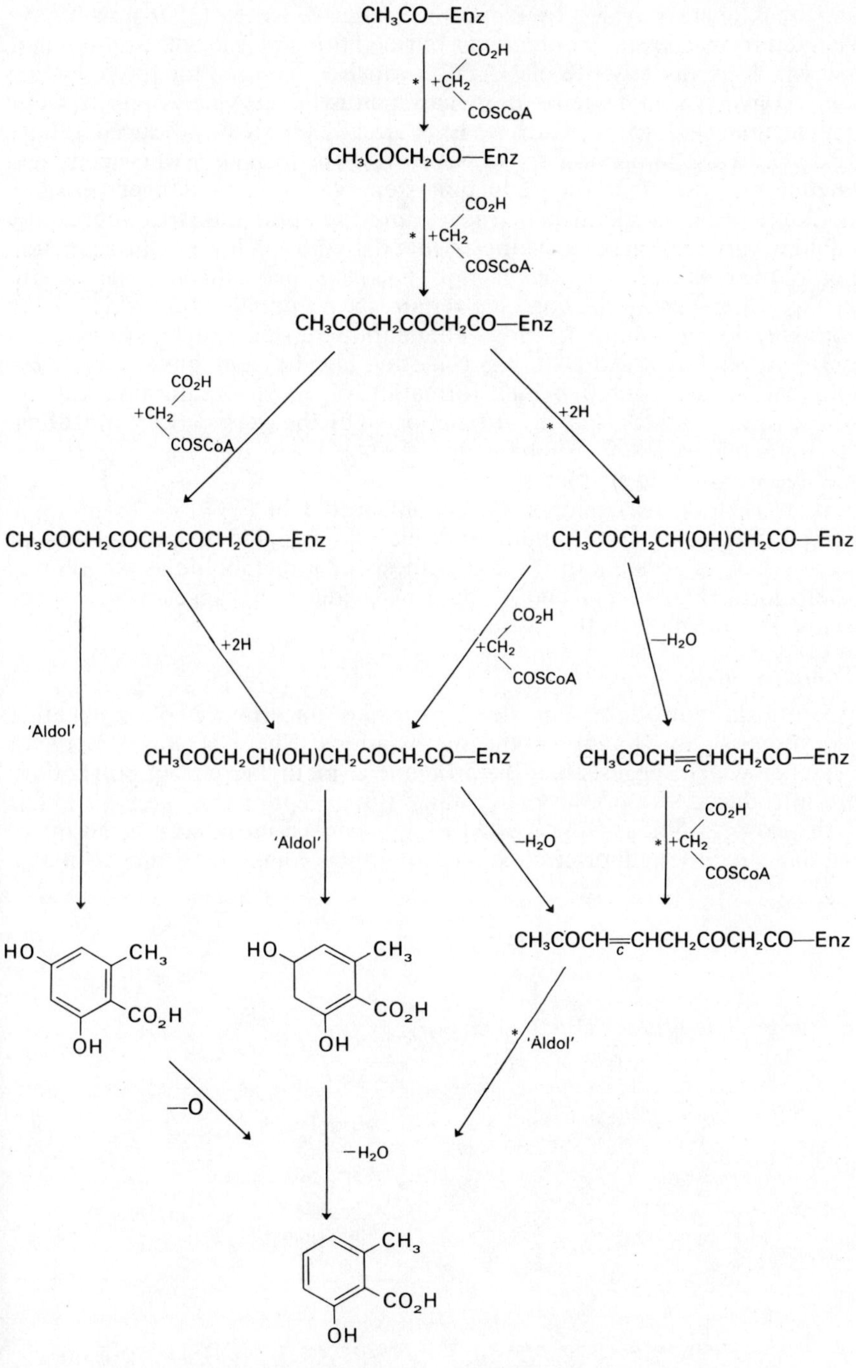

Fig. 15.8 Possible biosynthetic routes to 6-methylsalicylic acid; the most likely sequence is asterisked. (Adapted from Turner, 1971.)

steps has been provided by Dimroth, Walter & Lynen (1970), who have isolated from *Penicillium patulum* a purified 6-methylsalicylic acid synthetase which, in the absence of NADPH which is required for the reduction step, converts acetyl coenzyme A and malonyl coenzyme A into triacetic acid lactone (Fig. 15.7) (which we have already seen to be produced by fatty acid synthetase in the absence of NADPH). The formation of triacetic acid lactone suggests that the reduction step, which is no longer possible, normally occurs at the triketide level, and the authors further suggest, by analogy with fatty acid biosynthesis, that dehydration follows the reduction before the next condensation reaction. This sequence is shown by an asterisk in Fig. 15.8. The cyclization step requires a *cis*-double bond which is the configuration resulting from $\beta\gamma$-elimination during the biosynthesis of unsaturated fatty acids; during 6-methylsalicylic acid biosynthesis, $\beta\gamma$-elimination is favoured by the formation of an $\alpha\beta$-unsaturated ketone. These results and conclusions are supported by the work of Scott, Beadling, Georgopapadakou & Subbarayan (1974), also using a purified 6-methylsalicylic acid synthetase.

If a situation as complex as that illustrated in Fig. 15.8 exists for a compound as simple as 6-methylsalicylic acid, imagine the permutation of steps which is possible in the biosynthesis of a metabolite as complex as sclerotiorin (Fig. 15.7)—and we have no evidence to suggest which of the many possible paths is the favoured one.

Timing of methylations

Sclerotiorin introduces us to a further uncertainty of polyketide biosynthesis—the timing of the methylations. There is now a body of evidence which suggests that the aromatic C-methyl groups of polyketides are introduced at a pre-aromatic stage. But accepting this, we are still left with two possibilities—the methyl groups can be introduced at an intermediate chain-length or after chain assembly is complete; evidence on this

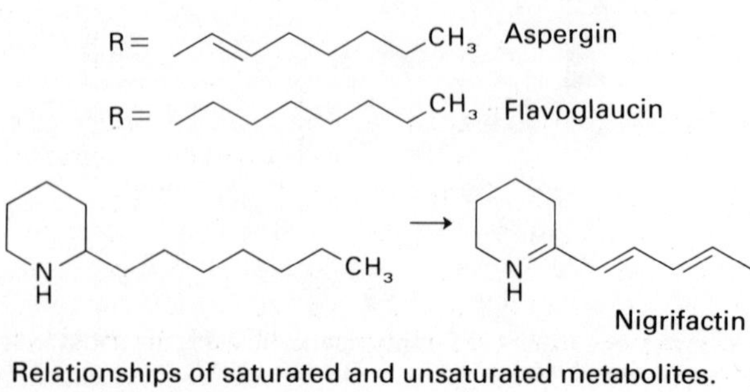

Fig. 15.9 Relationships of saturated and unsaturated metabolites.

point is contradictory. Similar ambiguity exists for the introduction of aliphatic C-methyl groups; for example the two 'extra' methyl groups on the side-chain of sclerotiorin could be introduced at 'active methylene' positions before reduction of the carbonyl groups of the polyketide chain, or at double bonds later in the biosynthetic sequence (*cf.* the introduction of the C-24 methyl group in ergosterol).

Timing of the introduction of double bonds

The side-chain double bonds of sclerotiorin introduce yet another ambiguity—are such double-bonds left behind during formation of the chain, or are they introduced by dehydrogenation of a saturated intermediate? The problem is illustrated by the series auroglaucin, aspergin, and flavoglaucin (Fig. 15.9)—do they form a biosynthetic sequence, and if so in what order? There is no evidence on this point for a fungal metabolite but it has recently been shown (Terashima, Idaka, Kishi & Goto, 1973) that the saturated precursor shown in Fig. 15.9 is efficiently incorporated into nigrifactin by *Streptomyces nigrifaciens.*

Timing of the introduction of oxygen

The evidence concerning the introduction of the 'extra' oxygen atoms possessed by many polyketides leads to two conclusions—the oxygen is introduced late in the biosynthetic sequence, often after the first stabilized products have been formed, and the enzymes (mixed function oxygenases) catalysing the process are relatively non-specific. This latter point leads to the formation of 'metabolic grids' (Bu'Lock, 1965) and has caused confusion in the elucidation of biosynthetic sequences.

15.5 Variations of Polyketide Biosynthesis

In the foregoing discussion we have considered compounds which are derived from a single polyketide chain. There are, however, a few compounds which are formed by condensation of two polyketide chains and many compounds which are formed by condensation of a polyketide chain with a moiety derived by another biosynthetic pathway.

Compounds derived from two polyketide chains

The best-studied examples are citromycetin and mollisin (Fig. 15.10), for which derivations from two chains are firmly established, but there are other compounds which seem likely, by inspection, to be derived from two chains. The arrangement of the two chains shown in Fig. 15.10 for the biosynthesis of mollisin is based on the most recent results with [1,2-^{13}C]-acetate (Seto, Cary & Tanabe, 1973). The two-chain origin of citromycetin was demonstrated with [2-^{14}C]malonate, which revealed the presence of two 'starter' units (see above) but did not distinguish between the two alternatives shown in Fig. 15.10; [1,2-^{13}C]-acetate would permit a distinction between them.

In principle, a different type of 'two-chain' biosynthesis is possible for compounds such as pulvilloric acid (Fig. 15.11) which have saturated normal hydrocarbon side-chains. Such compounds could be formed either by the normal process of polyketide chain-formation, or by the substitution of a fatty acid for acetate as the chain-initiator; the two alternatives are illustrated in Fig. 15.11 for pulvilloric acid. However, hexanoate is not

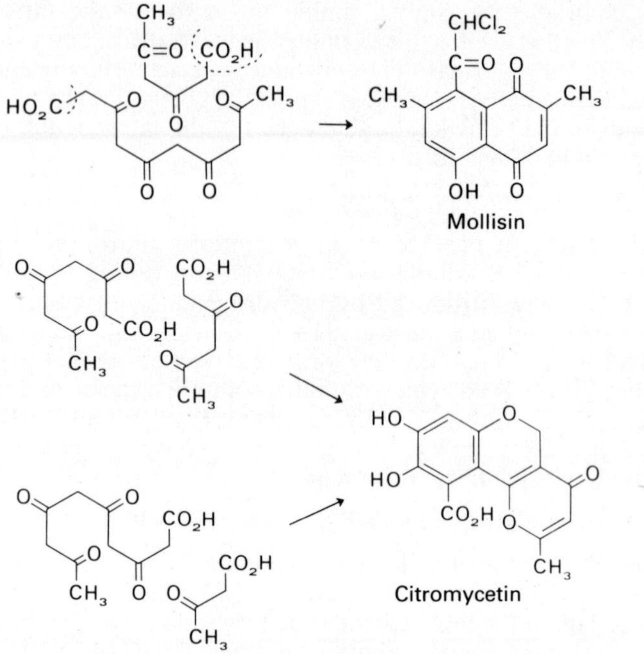

Fig. 15.10 The biosynthesis of mollisin and citromycetin from two polyketide chains.

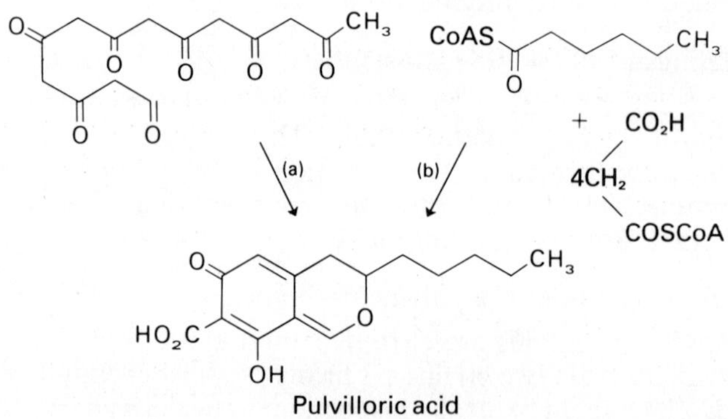

Fig. 15.11 Alternative biosyntheses for pulvilloric acid (for discussion see text).

incorporated into pulvilloric acid, and it is generally assumed that such compounds are formed by the normal process of polyketide biosynthesis.

Compounds of mixed biosynthetic origin

TRIPRENYL PHENOLS There is a group of over twenty fungal metabolites which are formed by introduction of a sesquiterpene side-chain into a tetraketide-derived aromatic ring. We have already seen a simple example in grifolin (Fig. 15.3) and a further relatively simple farnesyl derivative is provided by antibiotic LL-Z1272β (Fig. 15.12), while the related antibiotic

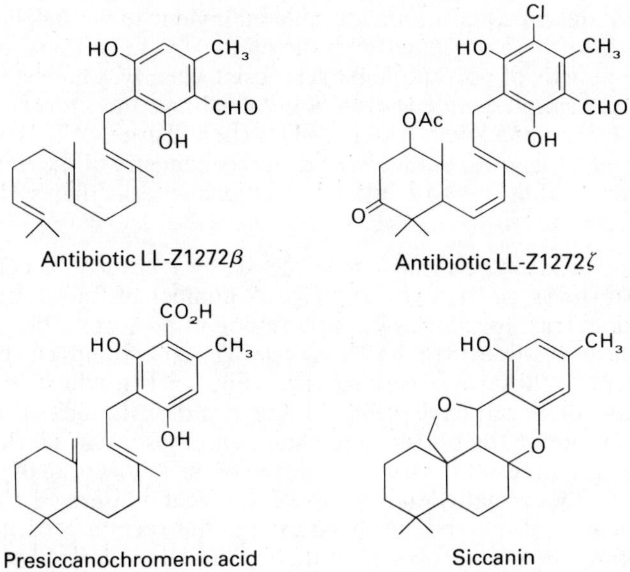

Fig. 15.12 Triprenyl phenols.

LL-Z1272ζ has undergone cyclization of the side-chain with methyl group migration as well as further oxidation and chlorination reactions. Similarly, presiccanochromenic acid (Fig. 15.12) is a precursor of siccanin. In the latter series it seems likely that farnesol undergoes cyclization *before* introduction onto the aromatic ring (Suzuki & Nozoe, 1974).

THE CYTOCHALASANS In these compounds, a polyketide-derived chain has combined with phenylalanine or with tryptophan. The formal process is exemplified in Fig. 15.13 for cytochalasin D, and it will be seen that formation of the bond between C-4 and C-5 involves an unfavourable carbonyl to carbonyl condensation and formation of the bond between C-8

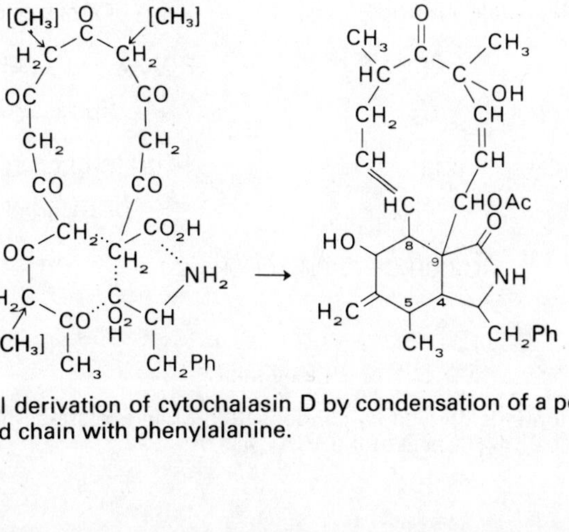

Fig. 15.13 Formal derivation of cytochalasin D by condensation of a polyketide-derived chain with phenylalanine.

and C-9 requires an equally unfavourable methylene to methylene condensation (contrast these processes with the methylene to carbonyl condensations characteristic of polyketide biosynthesis). Clearly, a mechanism more complex than simple condensation is involved and the process has been discussed in terms of a Diels-Alder type reaction (Turner, 1974), which has the advantage of leading to the desired stereochemistry of the ring-system, or of Michael additions to conjugated ketone intermediates (Binder & Tamm, 1973).

COMPOUNDS BIOSYNTHESIZED FROM AN ACETATE-DERIVED CHAIN AND AN INTERMEDIATE OF THE TCA CYCLE A number of fungal metabolites are derived, at least formally, by condensation of an acetate-derived chain with a C_4-dicarboxylic acid of the TCA cycle. The compounds can be divided into two types—those, *e.g.* carlosic acid (Fig. 15.14), which result from condensation of a carboxyl group of the dicarboxylic acid with the α-methylene group of the polyacetate chain, and those, *e.g.* decylcitric acid (Fig. 15.14) which result from condensation of the keto-group of oxaloacetic acid with the α-methylene group of the acetate-derived chain. The description 'acetate-derived' has been used in this section because in some of the compounds, and possibly all of them, the chain is derived from a fatty acid rather than from a polyketide. This is certainly true of decylcitric acid, and an enzyme catalysing its formation from oxaloacetic acid and lauroyl coenzyme A has been isolated from *Penicillium spiculisporum* (Måhlén, 1971). On the other hand, the results of Bloomer, Kappler & Pandey (1972)

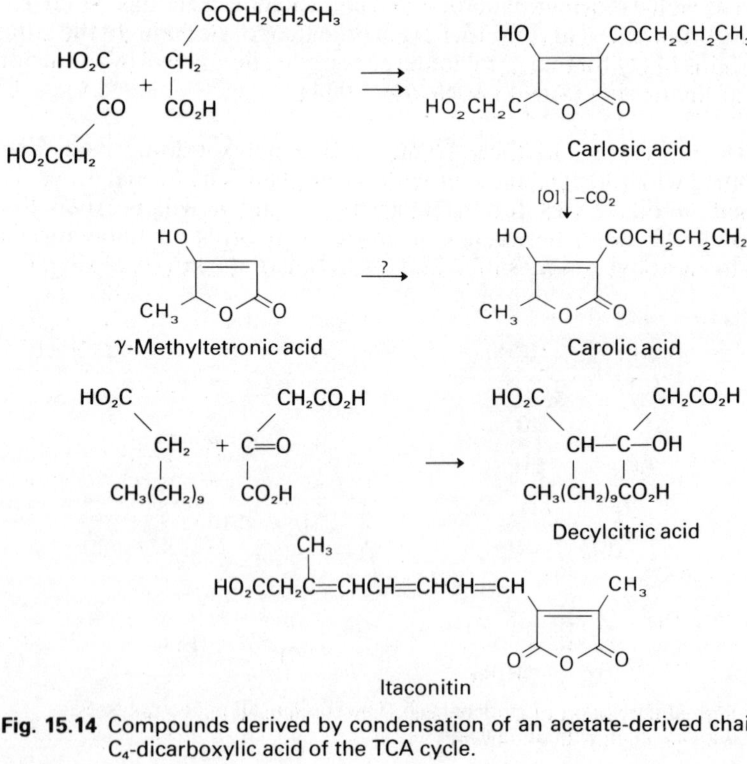

Fig. 15.14 Compounds derived by condensation of an acetate-derived chain with a C_4-dicarboxylic acid of the TCA cycle.

suggest that carolic acid might be formed by acylation of γ-methyltetronic acid (Fig. 15.14), so that the biosynthesis of this and related compounds may not involve condensation with a pre-formed chain, whether polyketide or fatty acid.

The compounds of this general type can undergo further modification *in vivo*, notably by decarboxylation, hydroxylation, and lactone-formation, to give a variety of products. Itaconitin (Fig. 15.14) is one of the more complex examples, and also has the appearance of a polyketide rather than a fatty acid derivation.

15.6 References

AL-RAWI, J. M. A., ELVIDGE, J. A., JAISWAL, D. K., JONES, J. R. & THOMAS, R. (1974). Use of tritium nuclear magnetic resonance for the direct location of ^3H in biosynthetically labelled penicillic acid. *Journal of the Chemical Society, Chemical Communications* 220–1.

BINDER, M. & TAMM, C. (1973). The cytochalasans: a new class of biologically active microbial metabolites. *Angewandte Chemie International Edition* **12**, 370–80.

BIRCH, A. J. (1967). Biosynthesis of polyketides and related compounds. *Science* **156**, 202–6.

BLOOMER, J. L., KAPPLER, F. E. & PANDEY, G. N. (1972). Biosynthesis of carolic acid in *Penicillium charlesii*: the intermediate precursors. *Journal of the Chemical Society. Chemical Communications* 243–4.

BU'LOCK, J. D. (1965). *The biosynthesis of natural products.* pp. 81–3. London: McGraw-Hill.

BU'LOCK, J. D. (1967). *Essays in biosynthesis and microbial development.* pp. 32–40. New York: Wiley.

CATTEL, L., GROVE, J. F. & SHAW, D. (1973). New metabolic products of *Aspergillus flavus.* Part III. Biosynthesis of asperentin. *Journal of the Chemical Society, Perkin Transactions I* 2626–2629.

DIMROTH, P., WALTER, H. & LYNEN, F. (1970). Biosynthese von 6-Methylsalicylsäure. *European Journal of Biochemistry* **13**, 98–110.

FORRESTER, P. I. & GAUCHER, G. M. (1972). Conversion of 6-methylsalicylic acid into patulin by *Penicillium urticae. Biochemistry* **11**, 1102–7.

KOMAN, V. & BETINA, V. (1973). Diphasic production of secondary metabolites by *Penicillium notatum* Westling S-52. II. Biosynthesis of fatty acids and lipids. *Folia Microbiologica* **18**, 133–41.

MÅHLÉN, A. (1971). Properties of 2-decylcitrate synthase from *Penicillium spiculisporum* Lehman. *European Journal of Biochemistry* **22**, 104–14.

MOSBACH, K. & BÄVERTOFT, I. (1971). A comparative study on the biosynthesis of palmitic and orsellinic acids in *Penicillium baarnense. Acta Chemica Scandinavica* **25**, 1931–6.

SCOTT, A. I., BEADLING, L. C., GEORGOPAPADAKOU, N. H. & SUBBARAYAN, C. R. (1974). Biosynthesis of polyketides. Purification and inhibition studies of 6-methylsalicylic acid synthase. *Bioorganic Chemistry* **3**, 238–48.

SCOTT, A. I. & YALPANI, M. (1967). A mass-spectroscopic study of biosynthesis: conversion of deutero-*m*-cresol into patulin. *Chemical Communications* 945–6.

SETO, H. & TANABE, M. (1974). Utilization of ^{13}C-^{13}C coupling in structural and biosynthetic studies. III. Ochrephilone—a new fungal metabolite. *Tetrahedron Letters 1974*, 651–4.

SETO, H., CARY, L. W. & TANABE, M. (1973). Utilization of ^{13}C-^{13}C coupling in structural and biosynthetic studies; the Fourier Transform ^{13}C nuclear magnetic resonance spectrum of mollisin. *Journal of the Chemical Society, Chemical Communications* 867–8.

SUZUKI, K. T. & NOZOE, S. (1974). Biosynthesis of an antibiotic, siccanin. *Bioorganic Chemistry* **3**, 72–80.

TANABE, M., SUZUKI, K. T. & JANKOWSKI, W. C. (1973). Biosynthetic studies with carbon-13: effective use of a paramagnetic ion in the FT-^{13}CNMR spectra of helicobasidin. *Tetrahedron Letters* 4723–6.

TERASHIMA T., IDAKA, E., KISHI, Y. & GOTO, T. (1973). Biosynthesis of nigrifactin. *Journal of the Chemical Society Chemical Communications* 75–6.

TURNER, W. B. (1971). *Fungal metabolites.* London and New York: Academic Press.

TURNER, W. B. (1974). The cytochalasins. *Postepy Higieny I Medycyny Doświadczalnej* **26**, 683–94.

CHAPTER 16

Secondary Metabolites Derived Through the Shikimate-Chorismate Pathway

G. H. N. TOWERS

16.1 The Shikimate-Chorismate Pathway

Shikimic acid (Fig. 16.1, No. 6) was discovered in the seed of a higher plant, *Illicium religiosum* in 1855 but remained a forgotten natural product for many years after. Studies with nutritional mutants of *Escherichia coli* in the 1950s showed it to be an intermediate in the biosynthesis of the aromatic amino acids, phenylalanine, tyrosine and tryptophan (for reviews see Bohm, 1965; Gibson & Pittard, 1968). The metabolic route leading to these amino acids is sometimes referred to as the shikimate pathway. Chorismic acid (Fig. 16.1, No. 9), an intermediate in the pathway, is the precursor of *p*-aminobenzoic, salicylic and 2,3-dihydroxybenzoic acids, the ubiquinones and vitamin K, in addition to the aromatic amino acids. In view of the central role of chorismic acid, as shown in Fig. 16.2, it would perhaps be more realistic to refer to this area of metabolism as the shikimate-chorismate pathway. The biosynthesis of shikimate and chorismate, starting from phosphoenolpyruvate and erythrose-4-phosphate, is shown in Fig. 16.1. The metabolic route leading to the aromatic amino acids occurs in bacteria, fungi and plants. It is not known to be present in animals except for a species of nematode.

Differences in the regulation of the shikimate-chorismate pathway are displayed by bacteria and fungi. In bacteria the enzymes, dehydroquinate synthetase, dehydroquinase, dehydroshikimate reductase, shikimate kinase and 3-enolpyruvylshikimate 5-phosphate synthetase (see Fig. 16.1) do not seem to be associated physically. In a number of fungi, including species of *Rhizopus, Phycomyces, Aspergillus, Coprinus, Ustilago* and *Neurospora*, these five enzymes are physically associated as a multi-enzyme complex (Ahmed & Giles, 1969). In *Neurospora crassa* and *Saccharomyces cereviseae* genetic work shows that this complex is encoded in a cluster of five structural genes (Jacobson, Hart, Doy & Giles, 1972).

Green plants synthesize a large variety of compounds from intermediates of the shikimate-chorismate pathway as well as from phenylalanine, tyrosine

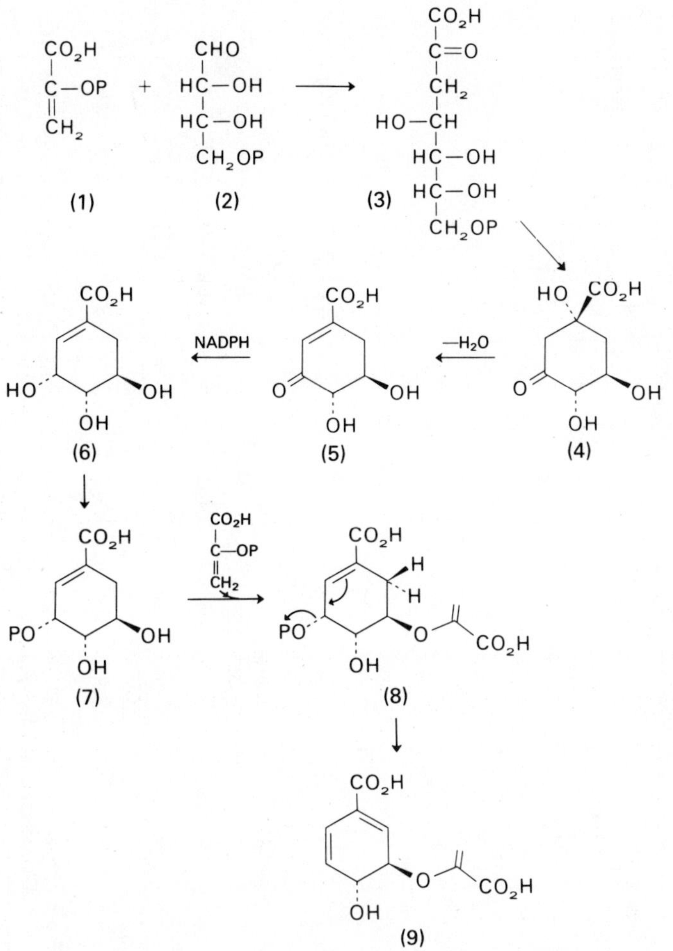

Fig. 16.1 Biosynthesis of shikimic and chorismic acids. 1. Phosphoenolpyruvic acid. 2. D-Erythrose-4-phosphate. 3. 3-Deoxy-D-arabino-heptulosonate 7-phosphate. 4. 5-Dehydroquinic acid. 5. 5-Dehydroshikimic acid. 6. Shikimic acid. 7. Shikimic acid 5-phosphate. 8. 3-Enolpyruvylshikimate 5-phosphate. 9. Chorismic acid.

and tryptophan and derivatives of these amino acids. Most fungi have not been screened for secondary compounds and so far have not been found to synthesize anything like this variety. Nevertheless, there are enough interesting compounds known from fungi to warrant consideration in a separate chapter.

16.2 Compounds Derived from Early Intermediates in the Pathway

Among the compounds to be considered here are *p*-hydroxybenzoic (Fig. 16.2, No. 7) salicylic (Fig. 16.2, No. 5), anthranilic (Fig. 16.2, No. 13) and *p*-aminobenzoic (Fig. 16.2, No. 11) acids and their derivatives.

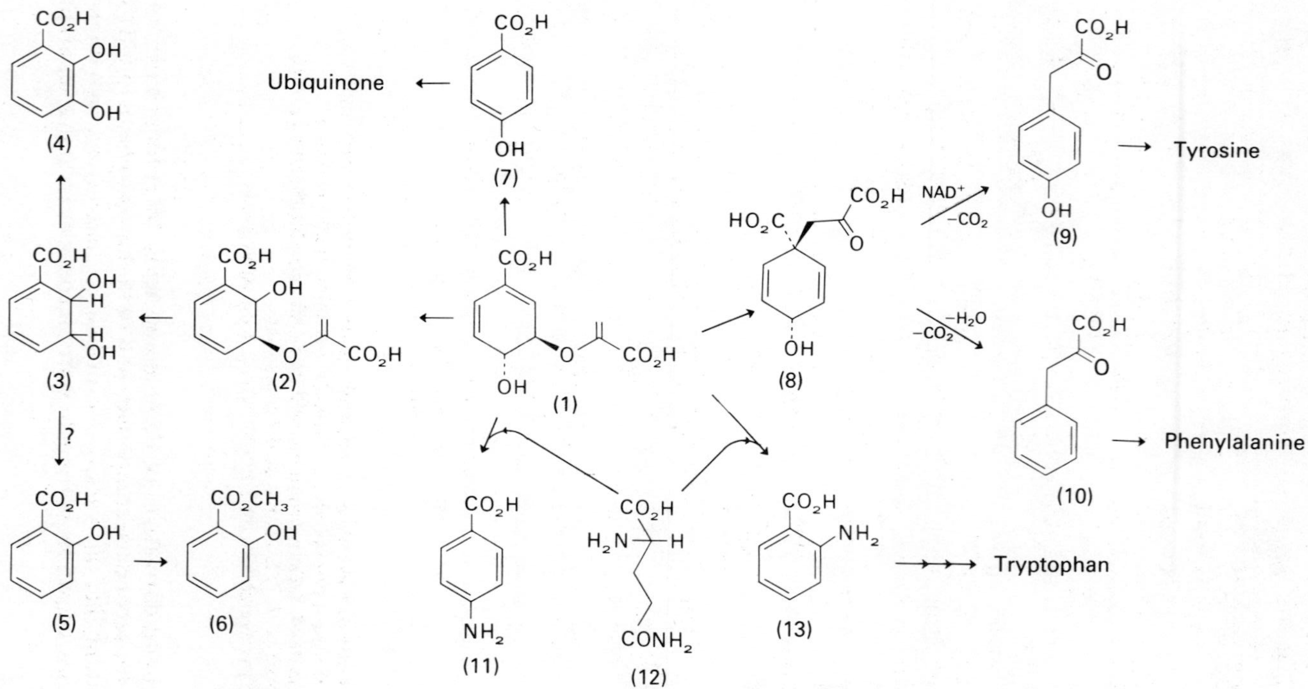

Fig. 16.2 Some compounds derived from chorismic acid (1). (2) Isochorismic acid. (3) 2,3-Dihydro-2,3-dihydroxybenzoic acid. (4) 2,3-Dihydroxybenzoic acid. (5) Salicylic acid. (6) Methylsalicylate. (7) *p*-Hydroxybenzoic acid. (8) Prephenic acid. (9) *p*-Hydroxyphenylpyruvic acid. (10) Phenylpyruvic acid. (11) *p*-Aminobenzoic acid. (12) Glutamine. (13) Anthranilic acid.

p-Hydroxybenzoic acid may arise in more than one way in fungi, *e.g.*, from chorismic acid (Fig. 16.2) or from the oxidation of *p*-coumaric acid (Fig. 16.10, No. 4). Tracer studies with animals show that *p*-hydroxybenzoic acid is the precursor of the quinone ring of the mitochondrial electron transporting substances, the ubiquinones. Ubiquinones are prenylated 1,4-benzoquinones (Fig. 16.3, No. 5), the polyprenyl side chain being derived through the mevalonate pathway and the C-methyl and O-methyl groups being derived from *S*-adenosylmethionine. Odoriferous compounds, such as methylanisate (Fig. 16.3, No. 1) and anisaldehyde (Fig. 16.3, No. 2) which have been identified in some Basidiomycetes, are obviously derived from *p*-hydroxybenzoate.

The closely related hydroxybenzoic acids, protocatechuic (Fig. 16.3, No. 3) and gallic (Fig. 16.3, No. 4) acids have been shown to be formed enzymically from dehydroshikimic acid (Fig. 16.1, No. 5) in *Phycomyces blakesleeanus.* Protocatechuic acid is also formed readily by hydroxylation of *p*-hydroxybenzoic acid. It may undergo enzymatic ring cleavage in fungi to yield aliphatic acids which are oxidized subsequently to CO_2. It is therefore an important intermediate in the turnover of benzenoid compounds.

Another compound derived from chorismate is anthranilic acid (Fig. 16.2, No. 13), an intermediate in the biosynthesis of tryptophan. The amino group of anthranilate is derived from the amide nitrogen of glutamine. There are only a few products of anthranilate metabolism known from fungi. These include *N*-pyruvyl anthranilamide (Fig. 16.3A, No. 6) and cyclic peptide derivatives such as cyclopenin (Fig. 16.3B, No. 3) and viridicatine (Fig. 16.3B, No. 5) from *Penicillium.* The latter two have been shown to be synthesized from anthranilate and phenylalanine (Fig. 16.3B). *p*-Aminobenzoic acid, also the product of a reaction involving chorismate and glutamine (Fig. 16.2), forms part of the tetrahydrofolic acid molecules which are involved in the transfer of one carbon units, *i.e.* C- and O-methylations, in fungi as well as in other organisms. It is also probably the precursor of agaritine (Fig. 16.3A, No. 10) and *p*-hydroxymethyl diazonium ion (Fig. 16.3, No. 7) from *Agaricus bisporus* and *N*-methyl nitrosoaminobenzaldehyde (Fig. 16.3, No. 8), from *Clitocybe suaveolus,* both basidiomycetous fungi.

In certain bacteria derivatives of 2,3-dihydroxybenzoic acid (Fig. 16.2, No. 4) such as 2,3-dihydroxybenzoylglycine and 2,3-dihydroxybenzoylserine are formed in large quantities when the cells are grown in iron-deficient media. The enzymes forming 2,3-dihydroxybenzoate in *Aerobacter aerogenes* are strongly repressed by iron or cobalt ions and it has been suggested that the amino acid conjugates, which can chelate ions, are involved in iron transport (Brot & Goodwin, 1968). 2,3-Dihydroxybenzoic acid has been found in *Claviceps paspali* and in *Aspergillus niger.* Its function if any, is unknown. In these fungi it has been shown to be a degradation product of tryptophan. In bacteria it is biosynthesized, together with salicylic acid, from isochorismic acid and 2,3-dihydro-2,3-dihydroxybenzoic acid (see Fig. 16.2) (Young, Batterham & Gibson, 1968). Methyl salicylate (Fig. 16.2, No. 6), which is responsible for the odour of oil of wintergreen, is produced by species of *Phellinus.*

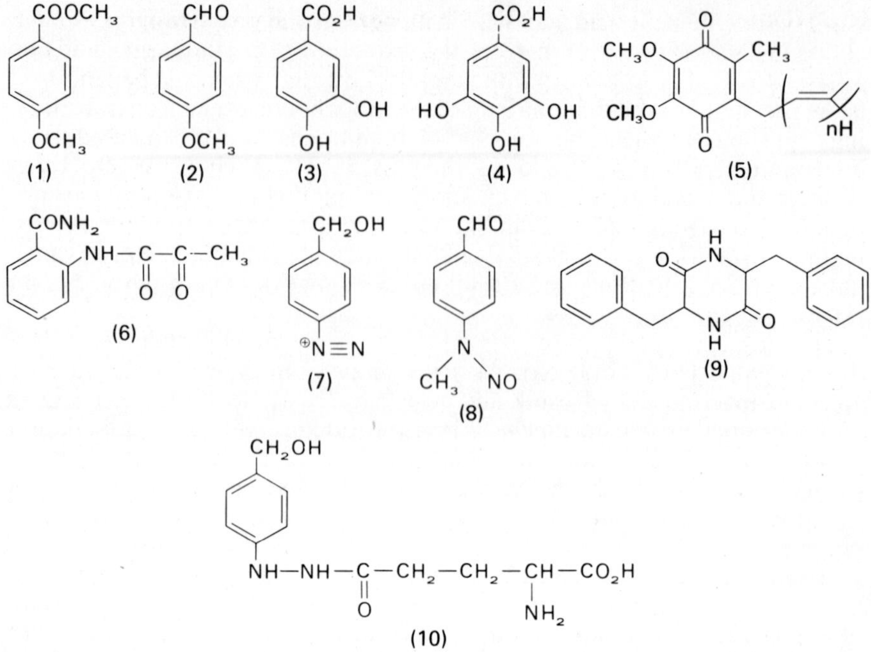

(A)

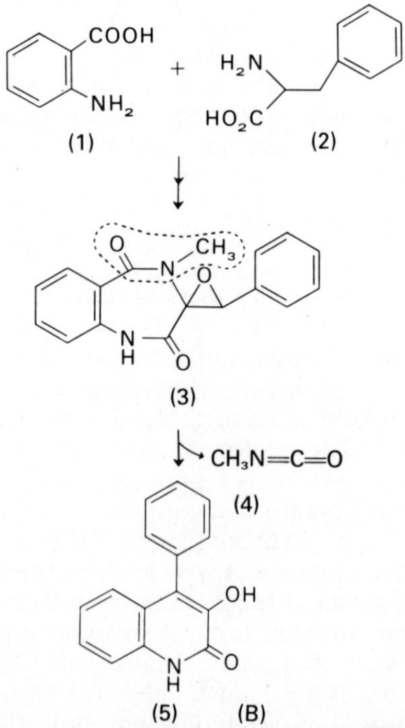

(B)

Fig. 16.3 A. Some metabolites of the shikimate-chorismate pathway. (1) Methylanisate. (2) Anisaldehyde. (3) Protocatechuic acid. (4) Gallic acid. (5) Ubiquinone. (6) N-Pyruvyl anthranilamide. (7) *p*-Hydroxymethyl diazonium ion. (8) N-methyl nitrosoaminobenzaldehyde. (9) L-Phenylalanine anhydride. (10) Agaritine. B. Derivatives of anthranilic acid (1) and phenylalanine (2) found in *Penicillium* spp. (3) Cyclopenin. (4) Methylisocyanate. (5) Visidicatine.

16.3 Compounds Derived from Phenylalanine, Tyrosine, DOPA and Tryptophan

Compounds in which the nitrogen atom is retained

A number of diketopiperazines, which are cyclic anhydrides of amino acids, are known from fungi. An example is L-phenylalanine anhydride (Fig. 16.3A, No. 9) from *Penicillium nigricans.* Condensation products of amino acids also appear as part of the molecule in ergocristine, an alkaloid from *Claviceps purpurea* (Fig. 16.4, No. 4).

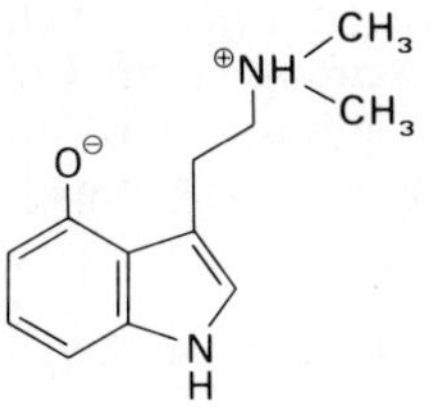

(1)

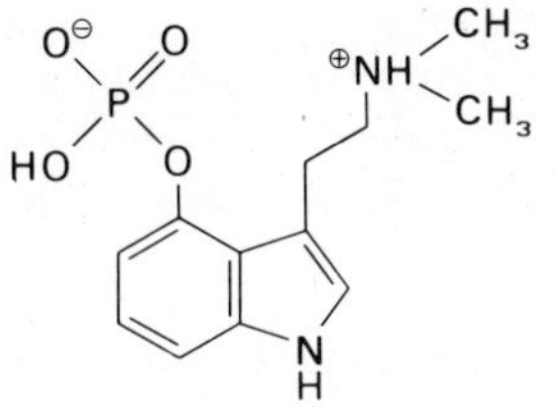

(2)

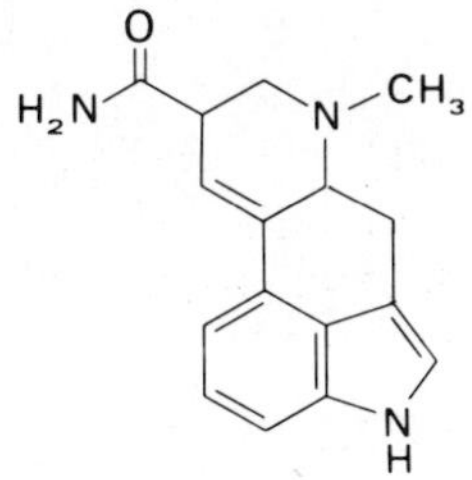

(3)

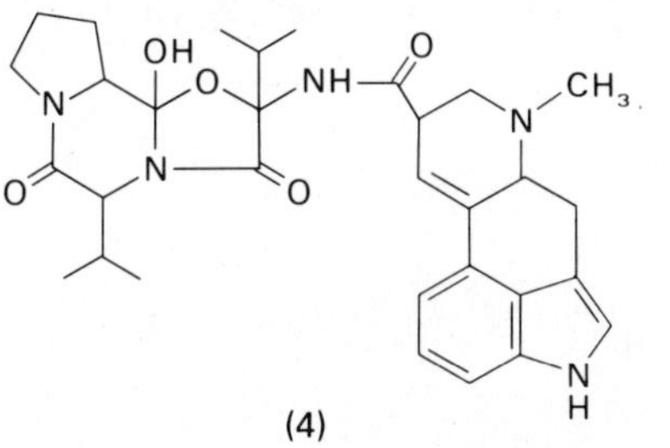

(4)

Fig. 16.4 Some derivatives of tryptophan which occur in fungi. (1) Psilocin. (2) Psilocybin. (3) Lysergic acid amide. (4) Ergocristine.

Ergocristine belongs to a group of alkaloids known as the ergot alkaloids (Willaman & Li, 1970) which are produced in the sclerotia of *Claviceps purpurea*, an Ascomycete which parasitizes the growing kernels in the heads of certain grasses, particularly rye. In the Middle Ages major epidemics of ergotism, known then as *St Anthony's Fire* occurred as the result of including infested rye in grain used for bread. In some epidemics thousands of people died of epileptic-like convulsions and gangrenous loss of limbs. The active principles of ergot are a series of alkaloids whose biosynthesis involves a prenylation of tryptophan, followed by decarboxylation, cyclization and oxidation to produce compounds such as lysergic acid amide (Fig. 16.4, No. 3) which is closely related to the synthetic hallucinogen, LSD or lysergic acid diethylamide. Ergot alkaloids are also produced by other species of *Claviceps* as well as by some species of *Aspergillus* and *Penicillium.*

Other hallucinogenic indole alkaloids occur in species of *Penicillium, Clitocybe* and *Amanita.* Among the better known are psilocybin (Fig. 16.4, No. 2), psilocin (Fig. 16.4, No. 1) and bufotenin. The last compound is also known from the skin of certain species of toad.

Although DOPA (3,4-dihydroxyphenylalanine) has not been reported to be a fungal constituent, three unusual types of compounds, characteristic of certain higher plants, and obviously derived by ring cleavage of DOPA have been found in species of *Amanita* (Chilton, Hsu & Zdybak, 1974). These are the amino acids stizolobinic and stizolobic acids (Fig. 16.5, Nos. 3 and 5) which occur in the leguminous genera *Stizolobium* and *Mucuna* and the pigments known as betalains. Betalains are a series of yellow, red and violet compounds which are characteristic of two large groups of flowering plants, the Centrospermae and the Cactales. In these groups of plants the betalains replace the anthocyanins as flower pigments. The betalamic moiety (Fig. 16.5, No. 6) of the betalains is synthesized by cyclization of the product of 4,5-ring cleavage of DOPA. Interestingly enough, the only enzymes known so far in fungi which catalyse ring cleavage of benzenoid compounds are the 3,4-oxygenases. The fungal syntheses of stizolobic and stizolobinic acids and of betalains indicates that they should also contain 2,3- and 4,5-oxygenases, as is the case with bacteria.

Compounds derived from phenylpyruvic and p-hydroxyphenylpyruvic acids

Simple derivatives of phenylpyruvic (Fig. 16.2, No. 10) and *p*-hydroxyphenylpyruvic acids (Fig. 16.2, No. 9) are commonly found in fungi. A careful investigation in cultures of one species of *Polyporus* revealed the

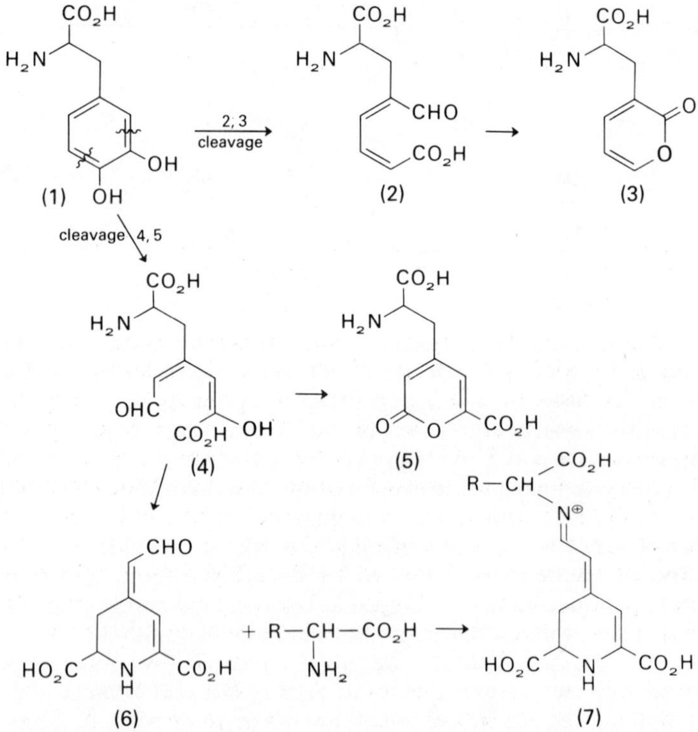

Fig. 16.5 Some derivatives of dihydroxyphenylalanine (DOPA) which have been found in *Amanita* spp. (3) Stizolobinic acid. (5) Stizolobic acid. (6) Betalamic acid. (7) A betalain.

presence of 2-hydroxy, 4-hydroxy- and 3,4-dihydroxyphenylacetic acids as well as three polyhydroxybenzoylformic acids (Crowden, 1967). Among the more unusual compounds is 3-nitro-4-hydroxyphenylacetic acid, found in *Rhizoctonia solani.*

At least thirty terphenylquinones and closely related lactones such as calycin (Fig. 16.6, No. 4) derived from phenylpyruvate or hydroxyphenylpyruvate, are known from lichens and from Basidiomycetes (Thomson 1971). The isolated mycobiont of the lichen, *Candellaria vitellina,* has been shown to synthesize vulpinic acid and calycin and undoubtedly it is the fungal component of the lichens in every case that is responsible for the biosynthesis of these substances. Tracer studies with the lichen *Pseudocyphellaria crocata* indicate that the biogenesis of pulvinic dilactone and calycin is through polyporic acid, an antibacterial and antitumour compound common in *Polyporus* and other Basidiomycetes (see Fig. 16.6) (Maass & Neish, 1967).

The terphenylquinone, atromentin (Fig. 16.7, No. 2) known from *Paxillus atromentosus* and other Basidiomycetes is undoubtedly synthesized from tyrosine or from the corresponding keto acid, *p*-hydroxyphenylpyruvate.

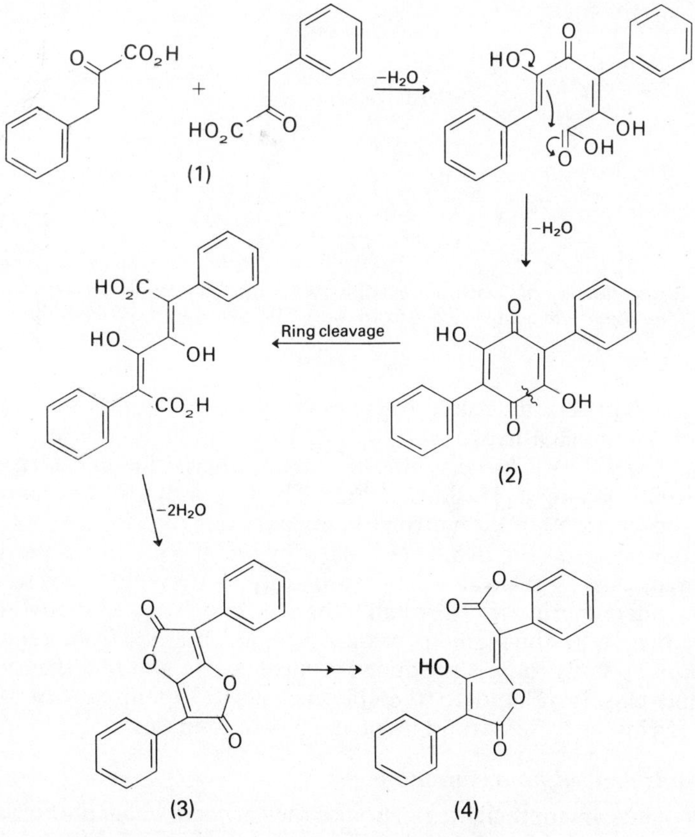

Fig. 16.6 Biosynthesis of pulvinic dilactone (3) and calycin (4) from phenylpyruvic acid (1) with polyporic acid (2) as an intermediate.

The amino- or the keto-acid is also the probable precursor of gyrocyanin (Fig. 16.7, No. 3), a substance which is oxidized to a blue pigment in injured sporophores of the bolete, *Gyroporus cyanescens.* The grevillines (Fig. 16.7, No. 4), a series of pigments in another bolete, are almost certainly yet other products of phenylpyruvate metabolism.

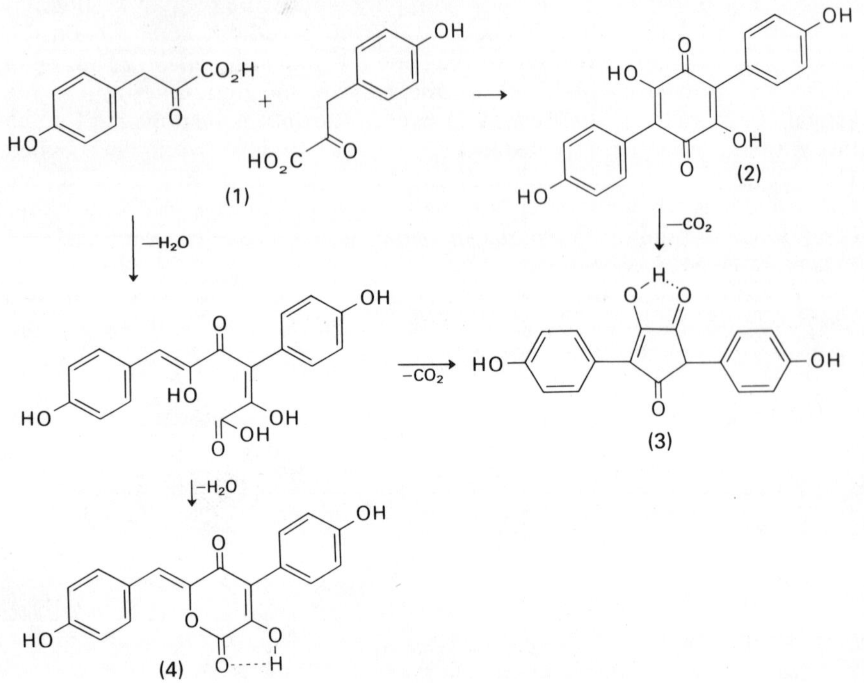

Fig. 16.7 Biosynthesis of various basidiomycetous pigments from *p*-hydroxyphenylpyruvic acid (1). (2) Atromentin. (3) Gyrocyanin. (4) Grevilline B.

Volucrisporin (2,5-di(*m*-hydroxyphenyl)-1,4-benzoquinone), a red pigment with an unusual hydroxylation pattern found in *Volucrispora aurantiaca,* has been shown to be synthesized from *m*-tyrosine in tracer experiments (Read, Vining & Haskins, 1962). The biosynthesis of volucrisporin would proceed through *m*-hydroxy-phenylpyruvate.

Thelephoric acid (Fig. 16.8, No. 4), a pigment found in a number of basidiomycetous genera, is probably derived from 3,4-dihydroxyphenylpyruvate through the hexahydroxyterphenylquinone shown in Fig. 16.8. The pigment, hydnoferrugin, obtained from a species of *Hydnellum* appears to be a product of ring cleavage and recyclization of a compound closely related to thelephoric acid (Gripenberg, 1974). The enzymatic ring-cleavage would involve a 3,4-oxygenase.

Compounds derived from cinnamate

In green plants an appreciable portion of the carbon fixed in photosynthesis is converted ultimately to benzenoid compounds which are formed through cinnamic acid and its derivatives. The benzenoid compounds include the

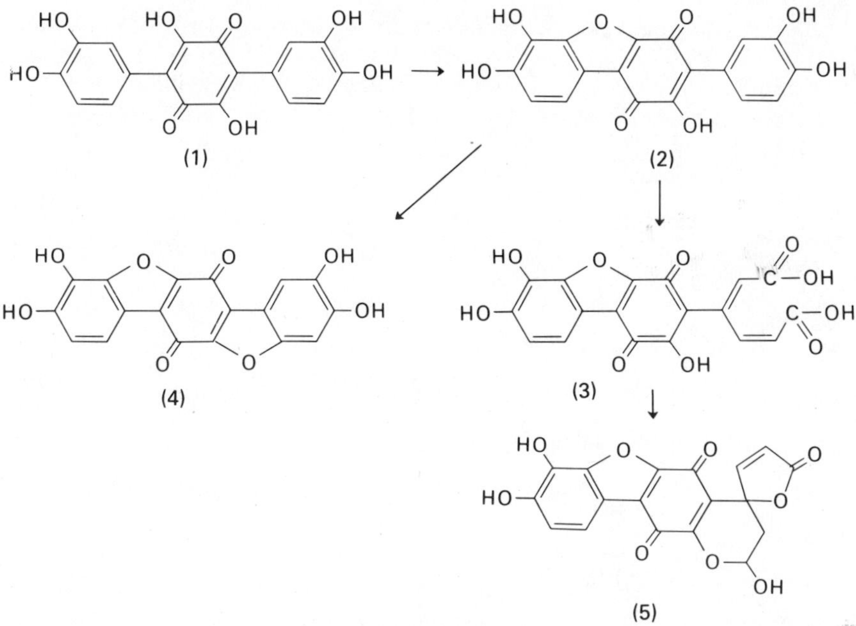

Fig. 16.8 Biosynthesis of hydnoferrugin (5) from ring cleavage of (2), a derivative of hexahydroxyterphenylquinone (1). (4) Thelephoric acid.

cell wall polymers known as lignins, the hydrolysable tannins, the flavonoids and many other compounds such as coumarins, *etc.*

A number of cinnamyl compounds have been isolated from Basidiomycetes (for review see Towers, 1969). Some examples, shown in Fig. 16.9, are cinnamaldehyde from *Stereum subpileatum,* methyl *p*-methoxycinnamate from *Lentinus lepideus,* methyl 3,4-dimethoxy *cis*-cinnamate from uredospores of various rusts, phenylcrotonaldehyde from *Phallus impudicus,* coumarin from *Puccinia graminis* and compounds with an extended side chain such as hispidin from species of *Polyporus* and cortisalin from *Corticium salicinum.*

Tracer studies with ^{14}C-labelled compounds have verified the cinnamate origin of all or part of the carbon skeletons of some of these compounds. The enzyme phenylalanine ammonia lyase (PAL) which catalyses the non-oxidative deaminations of L-phenylalanine to form *trans*-cinnamic acid has been shown to occur in cultures of many basidiomycetous fungi as well as in a few ascomycetous genera. In some fungi, *e.g. Ustilago hordei,* it is specfic for phenylalanine whereas in others, such as *Sporobolomyces roseus* it also catalyses the deamination of tyrosine to give *p*-coumaric acid. *p*-Coumarate may be formed from cinnamate, however, by means of a hydroxylating enzyme which has been shown to occur in *Polyporus* (see Fig. 16.10). *p*-Coumarate, in turn, may be hydroxylated enzymically to give caffeic acid (Fig. 16.10, No. 7). Two derivatives of caffeic acid have been found to be germination self-inhibitors of the uredospores of various rusts. These are methyl 3,4-dimethoxy *cis*-cinnamate (Fig. 16.9, No. 5) which occurs in the

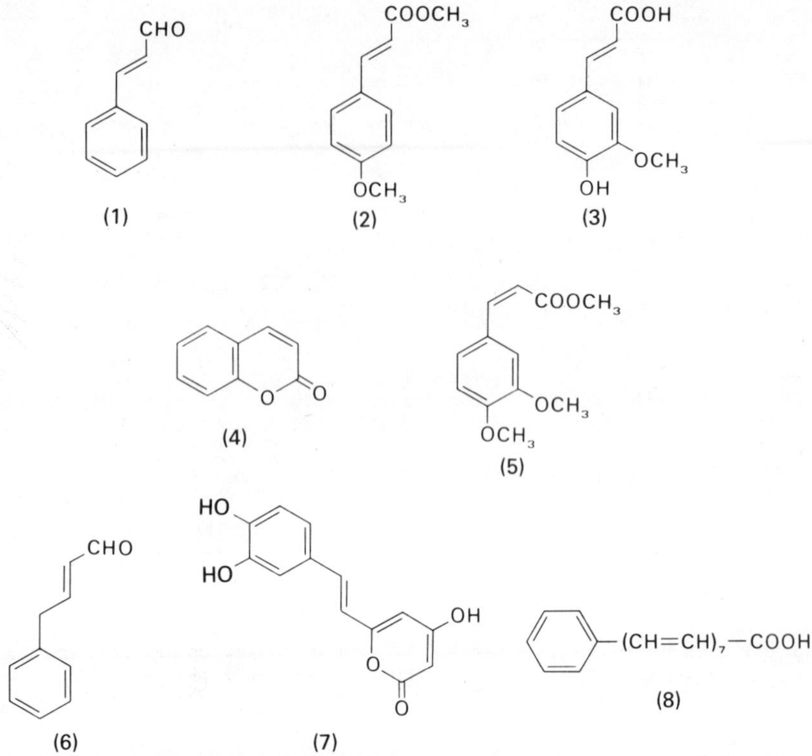

Fig. 16.9 Some derivatives of cinnamic acid which occur in Basidiomycetes. (1) Cinnamaldehyde. (2) Methyl *p*-methoxycinnamic acid. (3) Ferulic acid. (4) Coumarin. (5) Methyl 3,4-dimethoxy *cis*-cinnamic acid. (6) Phenylcrotonaldehyde. (7) Hispidin. (8) Cortisalin.

uredospores of rusts such as sunflower rust (*Puccinia helianthi)* or bean rust (*Uromyces phaseoli)* and methyl *cis*-ferulate which occurs in the uredospores of a race of wheat stem rust (*Puccinia graminis* var. *tritici*) (Macko, Staples, Allen & Renwick, 1971; Allen, 1972). The *trans* isomers of these caffeyl derivatives are inactive. The role of the *cis*-isomers of the hydroxycinnamic acids in controlling germination of spores or any other aspect of growth in other Basidiomycetes has not been examined. The fact that the growth of *Polyporus schweinitzii* in culture is promoted significantly by the addition of small amounts of ferulic acid (Robbins *et al.*, 1963) suggests that this problem bears investigating. Although the interconversion of the *cis* and *trans* isomers of the hydroxycinnamic acids may be effected by the ultraviolet component of sunlight it is possible that an isomerizing enzyme is responsible, in part, for the natural accumulation of the *cis* isomers in rust uredospores.

Germination stimulators have also been detected in rust uredospores. One of these is thought to be coumarin (Fig. 16.9, No. 4) which occurs, along with a very large number of other cinnamyl compounds in the uredospores of wheat stem rust (van Sumere, van Sumere-de Preter, Vining & Ledingham, 1957).

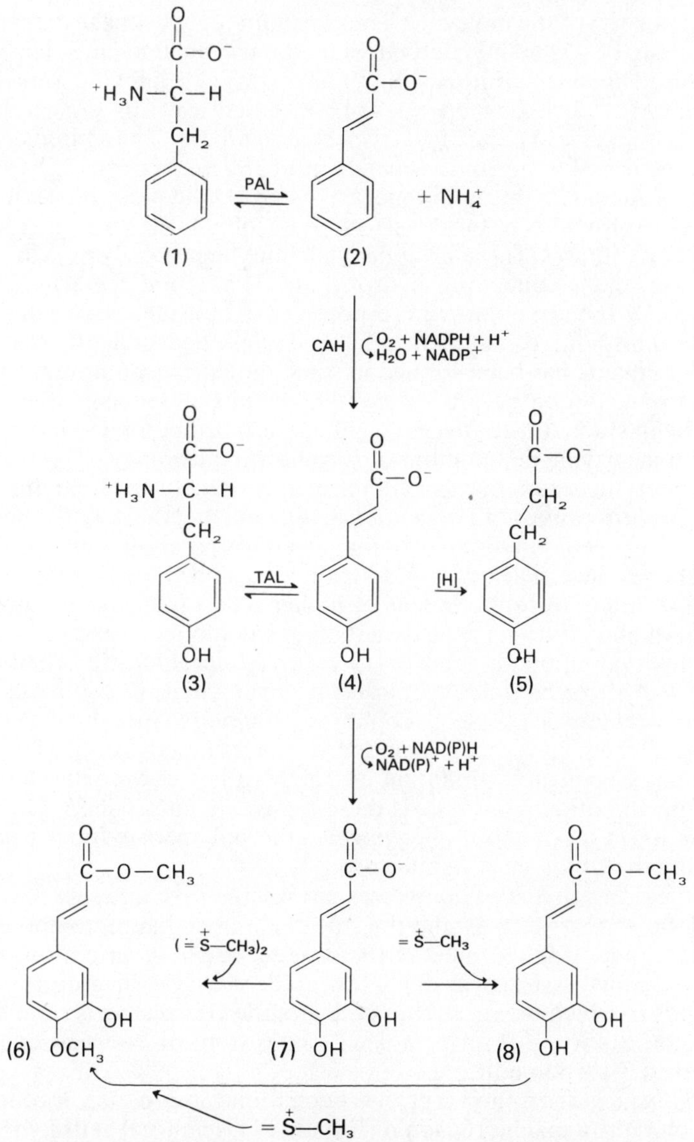

Fig. 16.10 Biosynthesis of cinnamyl compounds from phenylalanine (1) and tyrosine (3) in Basidiomycetes. (2) Cinnamate. (4) *p*-Coumarate. (5) Dihydro *p*-coumarate. (6) Methyl isoferulate. (7) Caffeate. (8) Methyl caffeate. TAL = tyrosine ammonia lyase.

Species of *Polyporus* and *Gymnopilus* produce styrylpyrone pigments such as hispidin which are very similar in structure to compounds which occur in green plants. The biosynthesis of hispidin has been studied in some detail both at the tracer and at the enzymological levels. Radioactive acetate and malonate were shown to be incorporated into the pyrone moiety whereas phenylalanine, *p*-coumarate and caffeate were specifically incorporated into the styryl portion of the molecule. Thus hispidin represents the extension of the side chain of a cinnamyl derivative by the two acetate units. Hispidin is not produced unless laboratory cultures are exposed to light (Towers, Vance & Nambudiri, 1974). Low energies of white light (5×10^4 erg cm^{-2}) were found to be effective at inducing styrylpyrone synthesis. The action spectrum showed a peak at 440 nm and a second one at 380 nm or beyond. The action spectrum is similar to the action spectra of carotenoid biosynthesis in other fungi. Light also affects the levels of some of the enzymes involved in hispidin biosynthesis. These are phenylalanine ammonia lyase, cinnamate hydroxylase and *p*-coumarate hydroxylase. The action spectrum for the appearance of the last enzyme in response to light is identical to the action spectrum of hispidin formation in light. A similar effect of light on flavonoid synthesis in plants has been studied in some detail. The photoregulation of phenylalanine ammonia lyase is a well-known phenomenon in green plants and has been studied extensively. Light has also been found to increase the levels of this enzyme in a number of Basidiomycetes.

It has been suggested that hispidin is oxidatively polymerized during the ripening of sporophores of *Polyporus hispidus* and that the polymer becomes bound to cell wall material effecting chemical cross-linking (Bu'Lock, 1967). The product then serves as a 'lignin' or a kind of toughening polymer although it differs from the lignins of higher plants in having no methoxyl content. Alkaline nitrobenzene oxidation of wood or of extracted lignins of plants yields compounds such as *p*-hydroxybenzaldehyde, vanillin (3-methoxy-4-hydroxybenzaldehyde) and syringaldehyde (3,5-dimethoxy-4-hydroxybenzaldehyde). The sporophores of 'woody' species of *Polyporus* and *Fomes* which have been analysed in this manner do not yield these compounds. Chemical degradation of 'chagi', a hot water extract of *Poria obliqua*, on the other hand, yields these typical lignin products. Obviously there is a lot to learn about phenylpropanoid polymers in fungi especially concerning their biological significance.

It is surprising that Basidiomycetes are unable to synthesize flavonoids although they are able to extend the side chain of cinnamyl compounds by one acetate unit, as in crotonaldehyde, two acetate units as in hispidin, and six acetate units as in cortisalin. Flavonoids are a large group of plant compounds in which the side chain of a cinnamyl derivative is extended by three acetate units to form a molecule consisting of two benzene rings bridged by a 3-carbon unit.

Aspergillus candidus, however, has been found to produce flavonoids in culture; one of these, chlorflavonin (Fig. 16.11) is unusual in its hydroxylation and methylation pattern and in being halogenated (Richards, Bird &

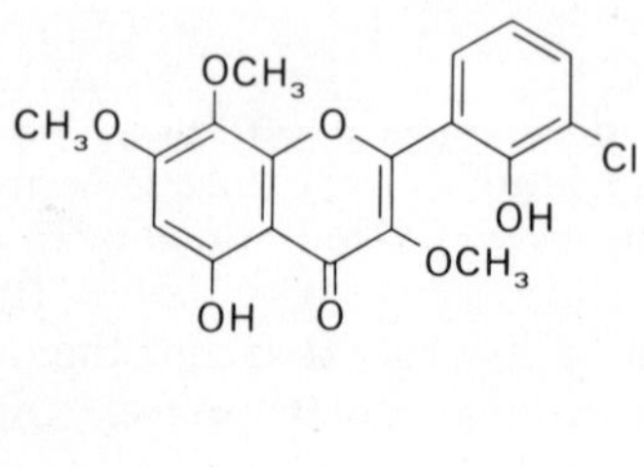

Fig. 16.11 Chlorflavonin.

Munden, 1969). The dechloro analogue is also produced in culture. Tracer studies have shown that chlorflavonin is synthesized from phenylalanine (Marchelli & Vining, 1974). Cinnamic acid is poorly incorporated and *o*-coumaric acid not at all so that the biosynthesis appears to involve a phenylpropanoid compound other than cinnamate such as phenylpyruvate. If this is true then this represents a fungal synthesis of flavonoids distinct from that utilized by plants. Further investigations of the phenolic compounds, derived *via* the shikimate-chorismate pathway, in fungi will undoubtedly lead to more biochemical surprises.

16.4 References

AHMED, S. I. & GILES, N. H. (1969). Organization of enzymes in the common aromatic synthetic pathway: evidence for aggregation in fungi. *Journal of Bacteriology* **99**, 231–7.

ALLEN, P. J. (1972). Specificity of the *cis*-isomers of inhibitors of uredospore germination in the rust fungi. *Proceedings National Academy of Sciences,* U.S.A. **69**, 3497–500.

BOHM, B. A. (1965). Shikimic acid (3,4,5-trihydroxy-1-cyclohexene-1-carboxylic acid). *Chemical Reviews* **65**, 435–66.

BROT, N. & GOODWIN, J. (1968). Regulation of 2,3-dihydroxybenzoylserine synthetase by iron. *Journal of Biological Chemistry* **243**, 510–3.

BU'LOCK, J. D. (1967). *Essays in biosynthesis and microbial development,* pp. 2-7. New York: John Wiley & Sons.

CHILTON, W. S., HSU, C. P. & ZDYBAK, W. T. (1974). Stizolobic and stizologinic acids: L-DOPA oxidation products in *Amanita pantherina. Phytochemistry* **13**, 1179.

CROWDEN, R. K. (1967). Biosynthesis of the polyphenolic acid metabolites of *Polyporus tumulosus* Cooke. *Canadian Journal of Microbiology* **13**, 181-97.

GIBSON, F. & PITTARD, J (1968). Pathways of biosynthesis of aromatic amino acids and vitamins and their control in microorganisms. *Bacteriological Reviews* **32**, 465–92.

GRIPENBERG, J. (1974). Fungus pigments. XXIII. Hydnoferrugin: a novel type of a 2,5-diphenylbenzoquinone-derived pigment. *Tetrahedron Letters,* 619–22.

JACOBSON, J. W., HART, B. A., DOY, C. H. & GILES, N. H. (1972). Purification and stability of the multienzyme complex encoded in the Arom gene cluster of *Neurospora crassa. Biochimica et Biophysica Acta* **289**, 1–12.

MAASS, W. S. G. & NEISH, A. C. (1967). Lichen substances. II. Biosynthesis of calycin and pulvinic dilactone by the lichen, *Pseudocyphellaria crocata. Canadian Journal of Botany* **45**, 59–72.

MACKO, V., STAPLES, R. C., ALLEN, P. J. & RENWICK, J. A. A. (1971). Identification of the germination self-inhibitor from wheat stem rust uredospores. *Science* **173**, 835–6.

MARCHELLI, R. & VINING, L. C. (1974). The biosynthetic origin of chlorflavonin, a flavonoid antibiotic from *Aspergillus candidus. Canadian Journal of Biochemistry* **51**, 1624–9.

READ, G., VINING, L. C. & HASKINS, R. H. (1962). Biogenetic studies on volucrisporin. *Canadian Journal of Chemistry* **40**, 2357–61.

RICHARDS, M., BIRD, A. E. & MUNDEN, J. E. (1969). Chlorflavin, a new antifungal antibiotic. *Journal of Antibiotics* **22**, 388–9.

ROBBINS, W. J., HERVEY, A., PAGE, A. C., GALE, P. H., HOFFMAN, C. H., MOSCATELLI, E.A., KONINSZY, F. R., SMITY, M. C., & FOLKERS, K. (1963). Growth factors for *Polyporus schweinitzii.* Identification of ferulic acid as a new cofactor. *Mycologia* **55**, 742–57.

THOMSON, R. H. (1971). *Naturally occurring quinones,* pp. 153–54. London: Academic Press.

TOWERS, G. H. N. (1969). Metabolism of cinnamic acid and its derivatives in Basidiomycetes. In *Perspectives in biochemistry,* pp. 179–91. Edited by J. B. Harborne and T. Swain. New York: Academic Press.

TOWERS, G. H. N., VANCE, C. P., & NAMBUDIRI, A. M. D. (1974). Photoregulation of phenylpropanoid and styrylpyrone biosynthesis in *Polyporus his-*

pidus. In *Advances in Phytochemistry* **5** (in press). New York: Academic Press.

VAN SUMERE, C. F., VAN SUMERE-DE PRETER, C., VINING, L. C. & LEDINGHAM, G. A. (1957). Coumarins and phenolic acids in the uredospores of wheat stem rust. *Canadian Journal of Microbiology* **3**, 847–62.

WILLAMAN, J. J. & LI, HUI-LIN (1970). Alkaloid-bearing plants and their contained alkaloids. *Lloydia* **33**, 1–286.

YOUNG, I. G., BATTERHAM, T. & GIBSON, F. (1968). Isochorismic acid: a new intermediate in the biosynthesis of 2,3-dihydroxybenzoic acid. *Biochimica et Biophysica Acta* **165**, 567–8.

CHAPTER 17

Secondary Metabolites Derived from Non-Aromatic Amino Acids

J. L. C. WRIGHT and L. C. VINING

17.1 Introduction

Besides fulfilling their primary function as the building units of proteins, amino acids provide biogenetic precursors for a variety of smaller molecules. For example, in fungi as in animals, tryptophan is a progenitor of the nicotinamide moiety of pyridine nucleotides, aspartic acid is the starting point for biosynthesis of the pyrimidines, and glycine is one of the metabolic intermediates from which purines are constructed. However, the range of compounds derived from amino acids includes not only such primary metabolites essential for normal development of most organisms, but also a number of secondary, or accessory metabolites. In the variety of ways in which they are able to transform amino acids into distinctive new molecules, fungi are more versatile than the Eubacteria which, in the main, are limited to combining them as oligopeptides. The fungi synthesize many interesting peptides and, in addition, carry out a range of structural modifications and condensations which convert the amino acids into complex nitrogenous metabolites. These are less numerous and varied than the alkaloids of higher plants, but the difference may reflect only a lack of systematic screening of the fungi for basic secondary products. A high proportion of the substances which are known are antibiotics or toxins and have been discovered because of their biological activity. Fungi not distinguished by some such noteworthy property are usually not examined.

This chapter will deal mainly with secondary metabolites formed from the non-aromatic amino acids. Some substances of mixed biogenetic origin are included, but those derived primarily from intermediates of the shikimate-chorismate pathway have been discussed in the previous chapter. References to the original literature supplement those assembled in a recent book on fungal metabolites (Turner, 1971) which is recommended as a source of additional information.

17.2 Unusual Amino Acids

The pool of soluble amino acids that can be extracted from fungal mycelium invariably contains more than the twenty compounds required for protein synthesis. Some of the extra, such as ornithine, homocysteine, and α-aminoadipic acid, are normal biosynthetic intermediates; others have been formed by decarboxylation of acidic amino acids. γ-Aminobutyric acid is almost universally distributed. It is often accompanied by α-aminobutyric acid, and β-alanine, especially in aging mycelium where cysteic acid and taurine from the breakdown of cysteine may be present as well. In addition some fungi make amino acids of unusual structure.

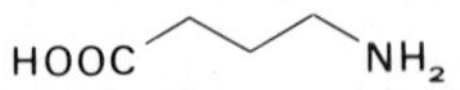

γ-Aminobutyric acid

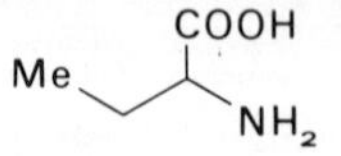

α-Aminobutyric acid

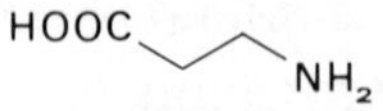

β-Alanine

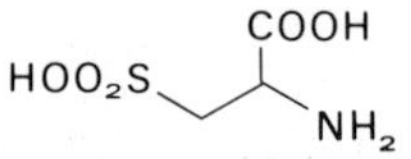

Cysteic acid

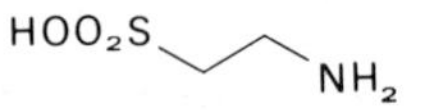

Taurine

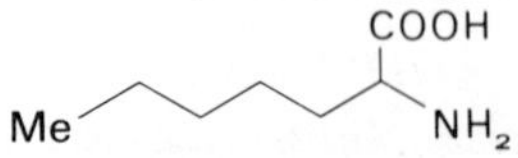

α-Aminoheptanoic acid

Claviceps microcephala produces α-aminoheptanoic acid as well as ergothioneine (Fig. 17.1). The latter was first discovered in 1909 during investigations on the chemistry of ergot, a parasitic sclerotial form of *Claviceps.* Ergothioneine has been detected in animals, plants, and fungi,

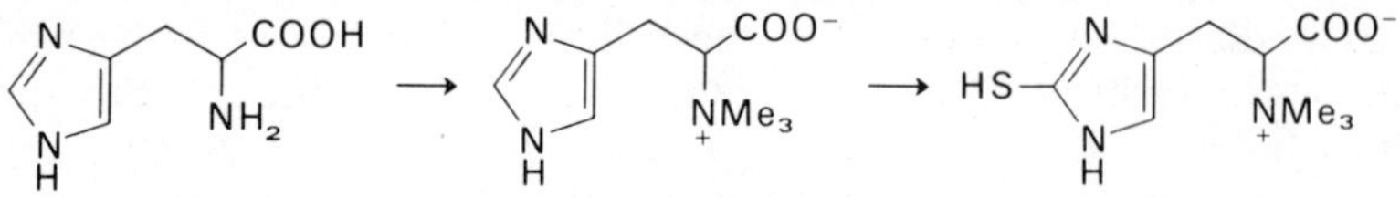

Hercynine Ergothioneine

Fig. 17.1 Biosynthesis of hercynine and ergothioneine.

but its presence in animal tissues is due exclusively to absorption from the gastrointestinal tract. *De novo* synthesis has been proven only in two fungi, *Claviceps purpurea* and *Neurospora crassa*, and in *C. purpurea* ergothioneine is restricted to the conidia (Heath & Wildy, 1957; Askari & Melville, 1962). In *N. crassa* and in *Coprinus comatus* it is accompanied by the closely related hercynine which is an intermediate in its biosynthesis from histidine (Fig. 17.1).

Other fungal betaines are presumably formed by *N*-methylation of the primary amino acids. Several fungi, including *Aspergillus oryzae*, produce stachydrine and the betaine that gives its name to this class of compound. Carnitine occurs in *Neurospora crassa* (Horne & Broquist, 1973) but in animals it has a role in the transport of fatty acids across mitochondrial membranes and thus may not be a true secondary metabolite. In both the fungus and in rats it is biosynthesized from lysine by *N*-methylation, loss of two carbons to give γ-butyrobetaine and β-hydroxylation (Fig. 17.2). *N.*

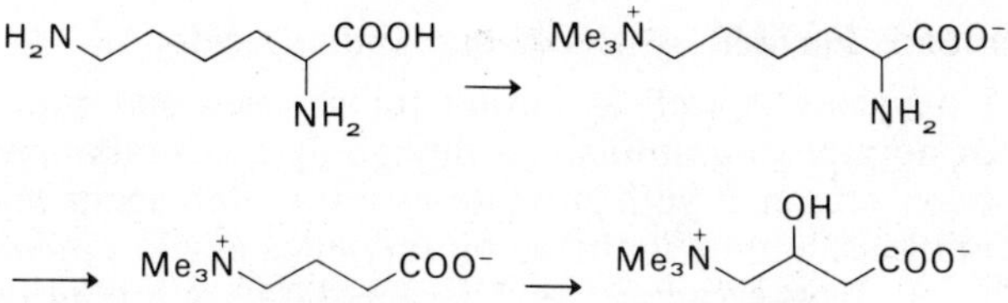

γ-Butyrobetaine Carnitine

Fig. 17.2 Biosynthesis of carnitine.

crassa also produces *S*-methylcysteine but *S*-methylation appears not to proceed beyond this stage. In this respect fungi are unlike marine algae where *S*-dimethylated thetins are abundant. The higher fungi are a source of

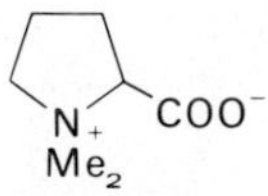

Stachydrine

Betaine

several unique amino acids (Laskin & Lechevalier, 1974). β-Methyllanthionine occurs in the stinkhorn, *Phallus impudicus; Lactarius helvus* contains β-methylene-L-norvaline and 2-methylene-cycloheptene-1,3-diglycine; some edible morels contain *cis*-3-amino-L-proline, α-aminoisobutyric acid and 2,4-diaminobutyric acid. The latter is probably formed by a non-specific amino transferase acting on aspartic semialdehyde, a normal biosynthetic intermediate.

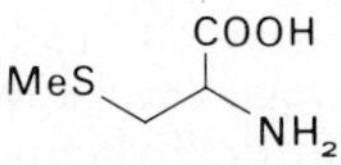

S-Methylcysteine

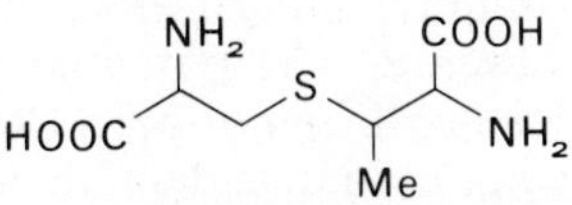

β-Methyllanthionine

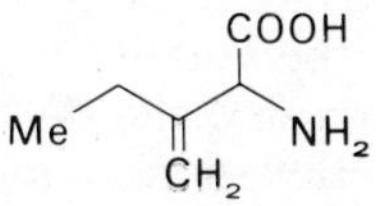

β-Methylene-L-norvaline

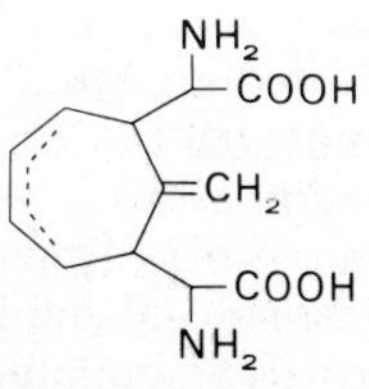

2-Methylenecycloheptene-1,3-diglycine

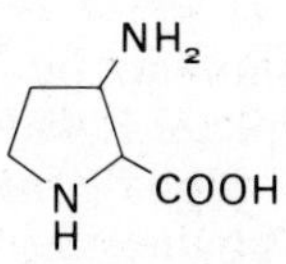

cis-3-Amino-L-proline

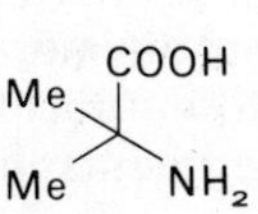

α-Aminoisobutyric acid

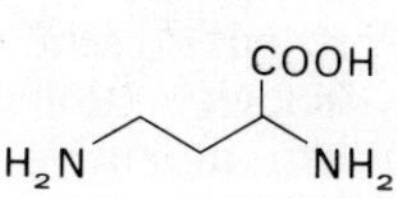

2,4-Diaminobutyric acid

17.3 Compounds Formed by Modifying Amino Acids

Amino acid oxidases as well as amino transferases generate keto acids. These do not normally accumulate, although pyruvic and dimethylpyruvic acids have been obtained with sulphite as a trapping agent in *Aspergillus niger* fermentations, and by inducing biotin deficiency in *Piricularia oryzae* cultures. The products of amino acid decarboxylation are more abundant. Agmatine, putrescine and cadaverine bear an obvious relationship to arginine, ornithine and lysine, respectively, and are common products of microbial decomposition. Volatile amines from the decarboxylation of

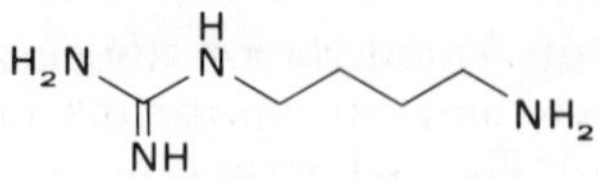

Agmatine

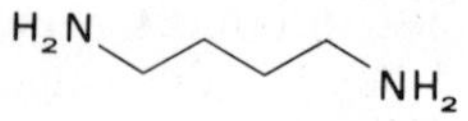

Putresine

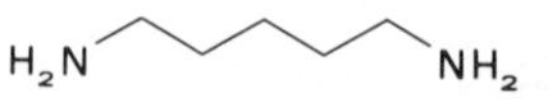

Cadaverine

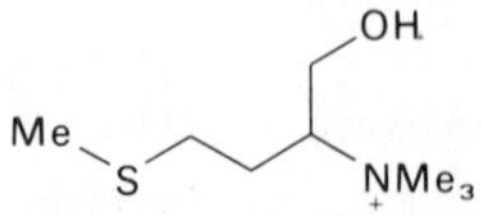

N-Dimethylmethioninol

neutral amino acids are equally common and are associated with the distinctive odours of some higher fungi (Birkinshaw, 1965). Enzymes of broad specificity have been implicated since some fungi have as many as eight amines, and the composition of the mixture correlates closely with their content of free amino acids in the cells. However, decarboxylation may not be the only biosynthetic route, particularly to the lower homologues. *Claviceps microcephala* can aminate acetaldehyde to give ethylamine, and the biosynthesis of di-and trimethylamines, which are common constituents of the volatile fraction, is obscure. The potential precursor amino acids, sarcosine and *N*-dimethylglycine, have not been found in fungi. Successive methylations of methylamine or oxidative demethylation of trimethylamine are alternatives. Trimethylamine is known to be formed in bacteria by metabolism of choline. Whichever mechanisms are used, they are sometimes capable of marked specificity. *Tilletia tritici*, the stinking bunt fungus of wheat, produces trimethylamine in quantity, while in *Claviceps microcephala* isoamylamine and *n*-heptylamine predominate. The distinctive aroma of Camembert cheese depends mainly on the formation of *N*-dimethyl methioninol by *Penicillium camemberti*.

Choline sulphate is widely distributed in micro-organisms and is formed by transfer of sulphate from adenosine-3-phosphate-5′-sulphatophosphate to choline. It is believed to act as a reserve of easily assimilable and activated sulphate. Acetylcholine is responsible for some of the pharmacological effects of ergot, and also of the poisonous red cap mushroom, *Amanita muscaria*. However, the main toxic constituents of *A. muscaria* are the muscarine and muscaridine bases which occur in several isomeric forms. These are also present in some *Clytocybe* species but are more characteristic of the genus *Inocybe*. Biogenesis from a hexose precursor has been suggested, but no definitive biosynthetic experiments have yet been reported for these compounds, or for the unusual 3-butenyltrimethylammonium base found in *A. muscaria*. Muscarine producers are responsible for the majority of mushroom poisonings and can be lethal. More often the victim recovers because the symptoms develop early and the toxin is voided before most of it can be absorbed. The slow-acting

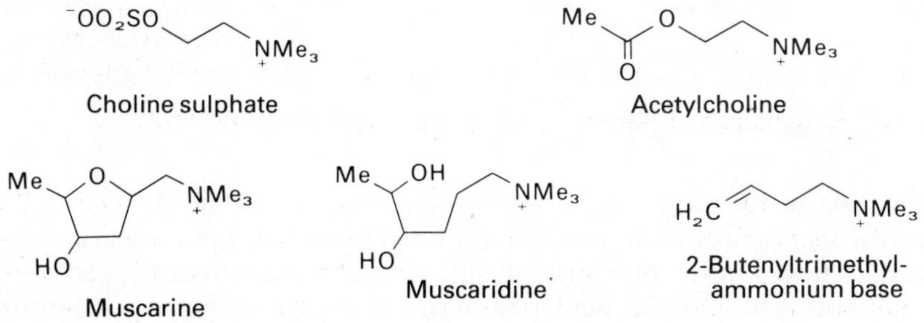

peptide toxins from the *Amanita phalloides* group discussed later are much more dangerous.

Penicillium aurantio-virens and several related species produce the antitumour agent hadacidin which is formed by an initial and apparently rate-limiting *N*-hydroxylation of glycine. This intermediate does not accumulate but is immediately *N*-formylated (Fig. 17.3). *N*-hydroxylation of amino acids is not uncommon in fungi and the derived hydroxamic acids

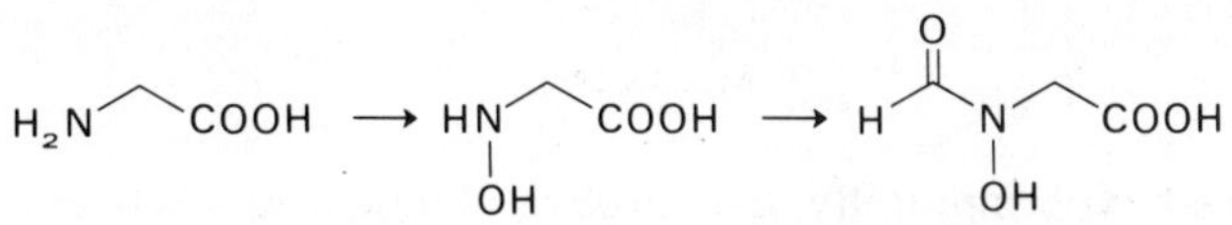

Hadacidin

Fig. 17.3 Biosynthesis of hadacidin.

will be encountered later in a combined form as sideramines. More complete oxidation of the amino group occurs in *Penicillium atrovenetum* and species of *Aspergillus* which make β-nitropropionic acid from L-aspartic acid. β-Alanine is not on the biosynthetic pathway, but β-nitroacrylic acid is reduced to β-nitropropionic acid by *P. atrovenetum*, and may be an intermediate (Fig. 17.4). The formation of nitro groups by oxidation of an appropriate amine seems to be the usual route in fungi as in other micro-organisms, and the first step in biosynthesis of 1-amino-2-nitrocyclopentane carboxylic acid by *Aspergillus wentii* is a reaction at the ε-amino group of lysine (Fig. 17.5).

The species name of *Amanita muscaria* derives from the toxicity of this mushroom towards flies. Small pieces mixed with milk or honey have long

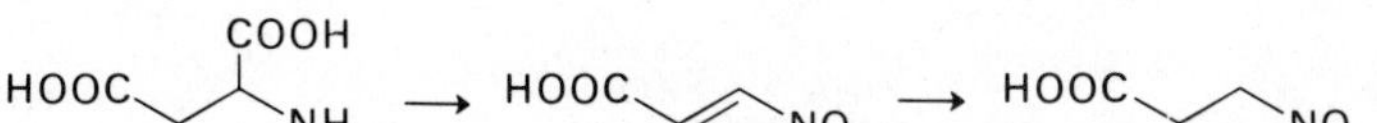

β-Nitroacrylic acid β-Nitropropionic acid

Fig. 17.4 Biogenesis of β-nitropropionic acid.

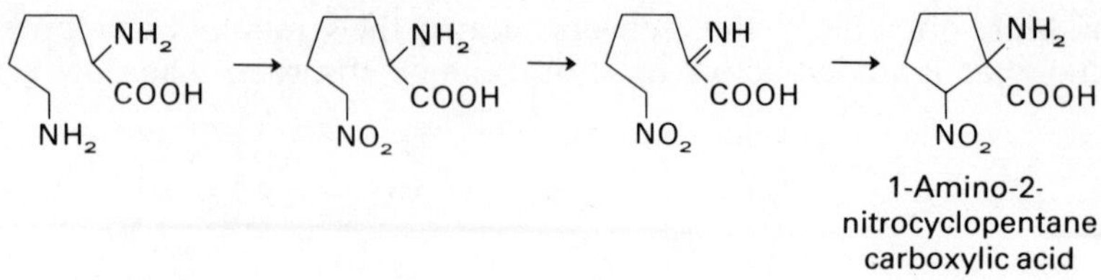

Fig. 17.5 Biosynthesis of 1-amino-2-nitrocyclopentane carboxylic acid.

been used in fly-traps, and it is commonly known as the fly agaric. Until recently its potency as an insect toxin was credited to the muscarine bases, but it is now known that the unusual oxazole and isoxazole derivatives, tricholomic acid, ibotenic acid, pantherine and muscazone, are responsible. These substances, found in a number of related Agaricaceae, induce a narcosis which can keep the insect in a state of apparent death for two days, after which it recovers with no ill-effects. Although their biogenesis has not

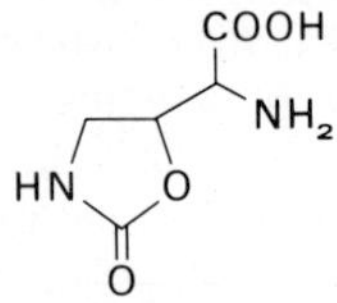

Muscazone

been explored experimentally, β-hydroxyglutamic acid has been suggested (Matsumoto, Trueb, Gwinner & Eugster, 1969) as a precursor of the isoxazoles and accompanying 4-(R)-hydroxypyrrolidone in *A. muscaria* (Fig. 17.6).

In higher plants and bacteria pyridine metabolites are derived from amino acid precursors. This is probably true in fungi as well, although firm biosynthetic evidence is lacking. Trigonelline and homarine are found in *Polyporus sulphureus* and their co-occurrence suggests that they may be formed by alternative modes of decarboxylation from quinolinic acid, followed by *N*-methylation (Fig. 17.7). Both are well-known plant alkaloids

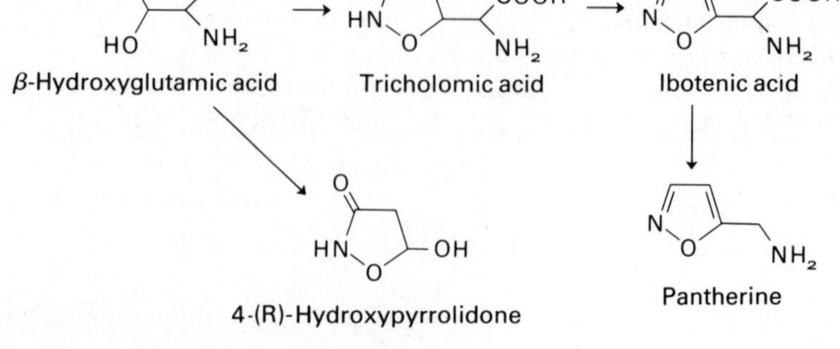

Fig. 17.6 Proposed biogenesis of Agaric fly toxins.

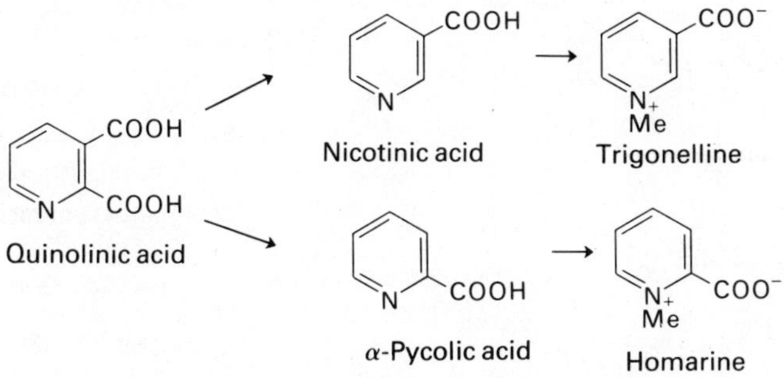

Fig. 17.7 Proposed biogenesis of α-pycolic acid, trigonelline and homarine.

and nicotinic acid is methylated to trigonelline in pea seedlings. Picolinic acid, the postulated precursor of homarine, has been isolated from *Piricularia oryzae* as α-pycolic acid, a phytotoxin involved in the pathogenic effects of this fungus on rice plants. Dipicolinic acid is a characteristic component of bacterial endospores but is also produced in *Penicillium*

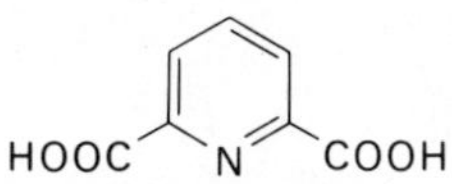

Dipicolinic acid

citreo-viride and *Isaria* species. In bacteria it is formed by a branch in the diaminopimelate pathway for lysine biosynthesis, and some radio-tracer experiments support a similar route in *P. citreo-viride* with pyruvate and aspartate as precursors. The biosynthesis of two other fungal pyridine derivatives, fusaric acid and tenellin, will be discussed in a later section.

17.4 Compounds Formed by Combining Amino Acids with Other Metabolites

In products such as fumarylalanine, where the amino acid has been combined in a simple way with a second metabolite, the biogenesis appears self-evident, though we might wonder why the compound is racemic in *Penicillium resticulosum*, but optically active in *Aspergillus indicus*. The jasmanoylisoleucines, ochratoxins, and agaritine also contain an intact amino acid in their structures. Ochratoxins are produced by *Aspergillus ochraceus* and several other fungi in stored grain and are responsible for livestock poisoning. Agaritine is found in most *Agaricus* species, including the common cultivated mushrooms, *A. bisporus* and *A. hortensis*. In *A. hortensis* it is accompanied by the *p*-hydroxyanilide of glutamic acid. Variotin, an antifungal antibiotic from *Paecilomyces varioti*, versimide from *Aspergillus versicolor* (Brown, 1970), and pencolide from *Penicillium multicolor*, seem to possess amino acid moieties that have undergone changes in addition to *N*-acylation.

An appreciable number of fungi produce compounds known as sideramines which are used for the sequestration of iron from the environment. They contain two or three *N*-acylated hydroxamic acid 'monomer

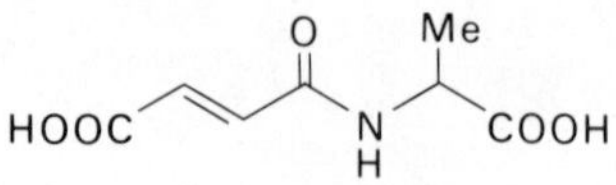

Fumarylalanine

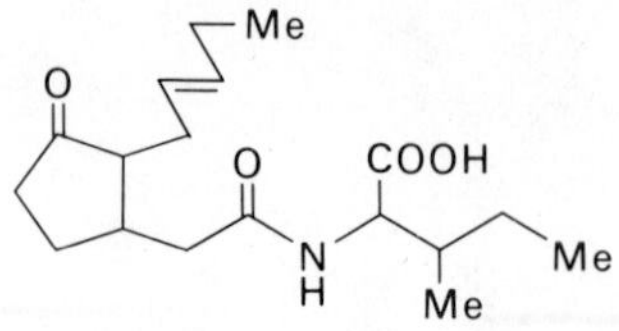

Jasmanoylisoleucines

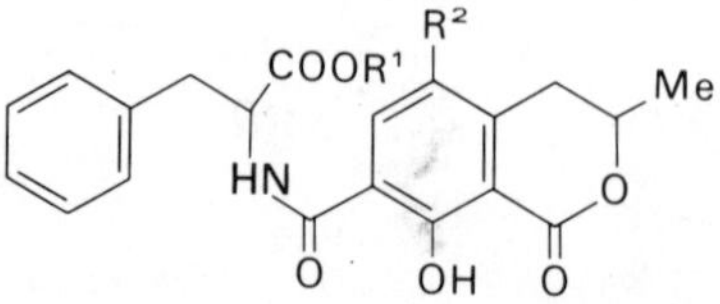

	R¹	R²
Ochratoxins	H	Cl
	H	H
	Et	Cl

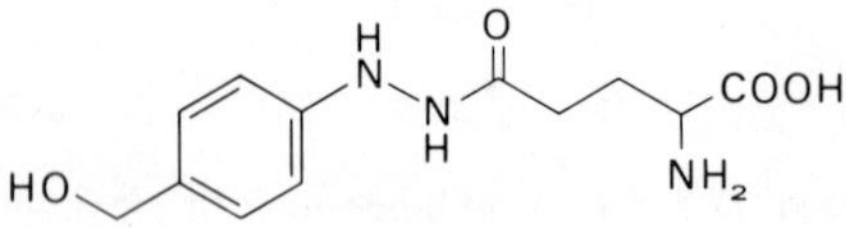

Agaritine

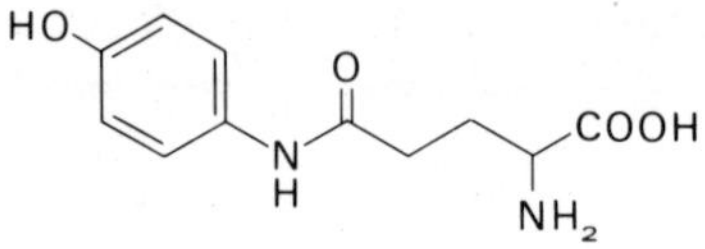

β-Hydroxyanilide of glutamic acid

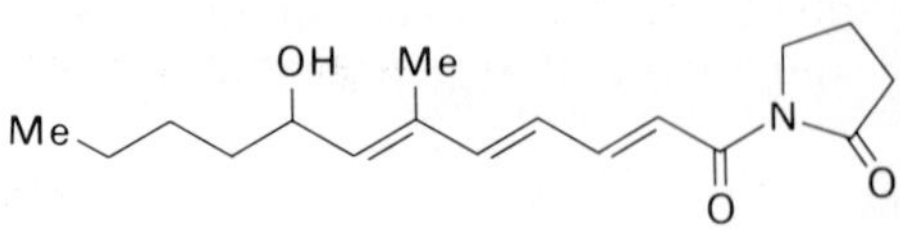

Variotin

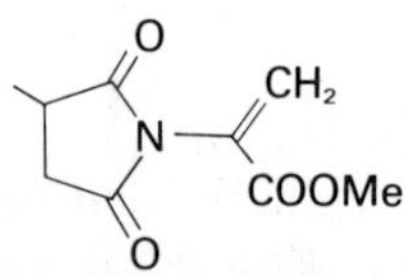

Versimide

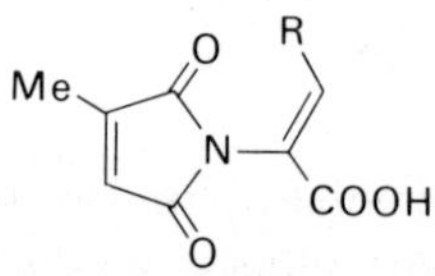

Pencolide

units' and fall into several structural types (Maehr, 1971). The most active contain three monomers. Fusarinine B, produced by *Fusarium roseum*, is a linear trimer of *cis*-fusarinine linked head-to-tail by ester bonds. In fusigen this structure is fully cyclized. Coprogen, found in numerous fungi, is a linear trimer of *trans*-fusarinine, in which the first monomer is N^2-acetylated and linked head-to-tail with the second; the second and third monomers are condensed head-to-head as a diketopiperazine, which also occurs free in *Fusarium dimerum* as dimeric acid. Ferrioxamine B is a linear trimer of N^5-hydroxy-N^5-succinylcadaverine linked head-to-tail by amide bonds. It is widely distributed in fungi and occurs in the fully cyclized form as ferrioxamine D_2. *Aspergilli* produce a related group of sideramines represented by ferrichrome. Other members of this group vary in the kind of acyl substituent attached to N^5-hydroxyornithine, and in having one or more glycines replaced by serine. The first step in the biosynthesis of sideramines containing fusarinine monomers (Fig. 17.8) is an oxidation yielding N^5-hydroxyornithine, followed by *N*-acylation (Ong & Emery, 1972). The *cis*- and *trans*-anhydrome-valonate acyl groups are derived from mevalonate. These as well as the *cis*- and *trans*-fusarinine intermediates have been isolated from *Fusarium* cultures. A cell-free extract from *F. cubense* catalyses the incorporation of *cis*-fusarinine into fusigen, and the final stage in the

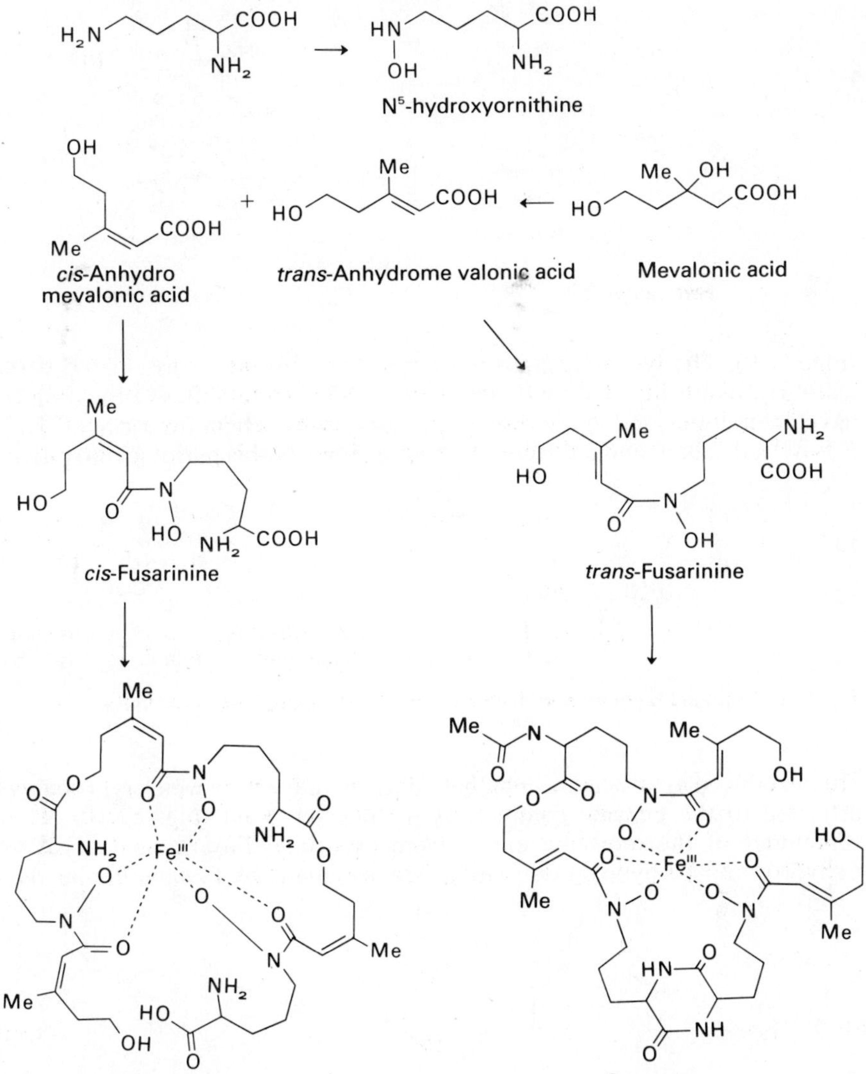

Fig. 17.8 Biogenesis of fusarinine B and coprogen.

biosynthesis of this sideramine appears to be the assembly of components in a manner similar to that used for the biosynthesis of gramicidin S and other cyclic peptides (Anke, Anke & Diekmann, 1973).

The structures of lycomarasmin and aspergillomarasmin B can readily be dissected into amino acid moieties, but the biogenesis of these substituted amines is not obvious. Studies with ^{14}C-labelled substrates in *Fusarium oxysporum* (Popplestone & Unrau, 1973) point to aspartic acid and glycine as precursors of the outer moieties of lycomarasmin with a three-carbon intermediate, possibly phosphoenolpyruvate, linking their amine groups

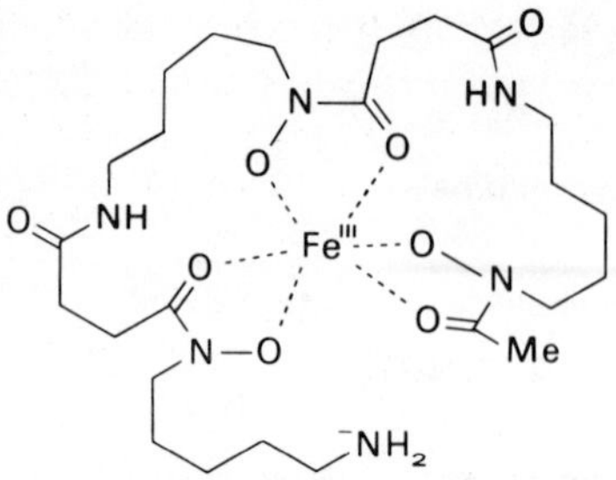

Ferrioxamine B

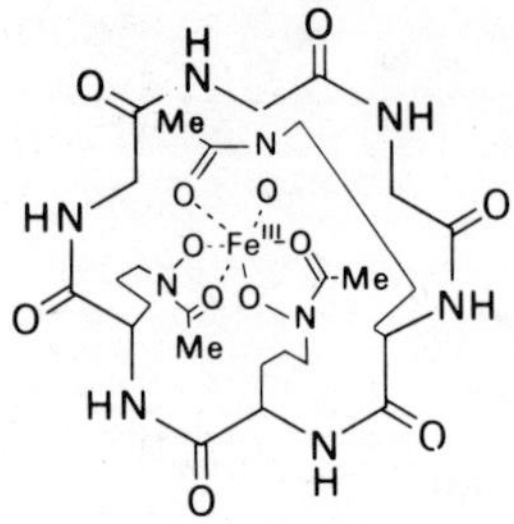

Ferrichrome

(Fig. 17.9). The lycomarasmins are phytotoxic substances associated with wilting of plants invaded by *F. oxysporum*. Other metabolites of the fungus have been implicated in the disease process, among them fusaric acid. The *n*-butyl side chain and adjacent two ring carbons of this pyridine derivative

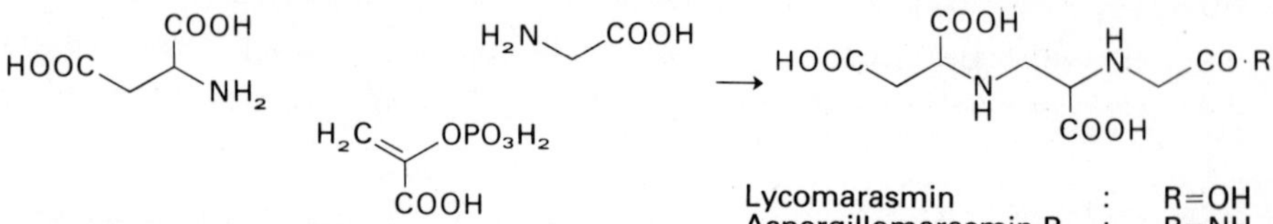

Lycomarasmin : R=OH
Aspergillomarasmin B : R=NH_2

Fig. 17.9 Proposed biogenesis of lycomarasmin and aspergillomarasmin B.

are introduced as an acetate-malonate derived polyketide which is probably attached to the enzyme complex by a thioester bond (Fig. 17.10). The remainder of the molecule enters from aspartate. Fusaric acid, its Δ^{10}-dehydro- and 10-hydroxy derivatives are also made by *F. moniliforme*, the

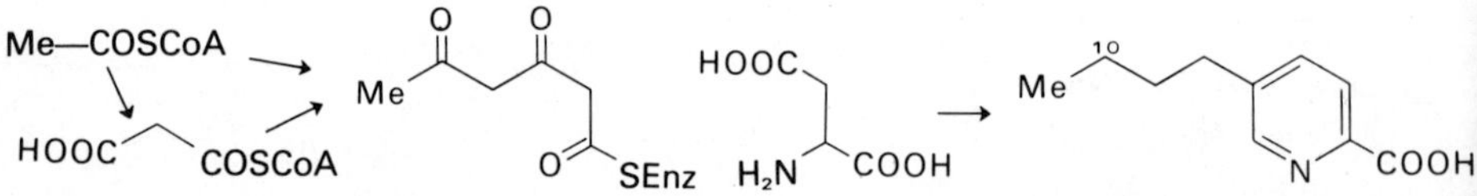

Fusaric acid

Fig. 17.10 Biosynthesis of fusaric acid.

imperfect stage of *Gibberella fujikuroi*, which is better known as a gibberellin-producer. Fusaric and dehydrofusaric acids are metabolically interconvertible, but the latter probably furnishes 10-hydroxyfusaric acid *via* an epoxide intermediate (Pitel & Vining, 1970).

An alternative mode of condensing a polyketide intermediate with an amino acid gives the tetramic acids. The first representative, tenuazoic acid, was obtained from *Alternaria tenuis* and later reisolated as an anti-tumour agent from another *Alternaria* species. It has been chemically synthesized by reacting isoleucine with ketene, and a number of related tetramic acids prepared in this way show antiviral activity (Harris, Fisher & Folkers, 1965). Biosynthetically, it is formed from isoleucine and a diketide (Fig. 17.11). In this case the ketide appears to be acetoacetylcoenzyme A since it comes from acetate without intervention of malonate and can be formed directly

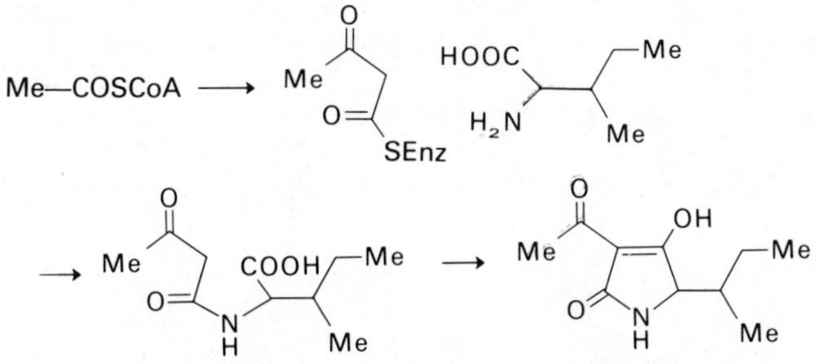

N-Acetoacetyl-L-isoleucine Tenuazonic acid

Fig. 17.11 Biosynthesis of tenuazonic acid.

from butyrate by β-oxidation (Gatenbeck & Sierankiewicz, 1973). Of the two possible ways in which initial condensation of isoleucine with the ketide could accur, *N*-acylation is favoured by the detection of *N*-acetoacetyl-L-isoleucine in the culture. The biosynthesis of cyclopiazonic acid by *Penicillium cyclopium* appears to follow a parallel course, the initial step being a condensation of acetoacetylcoenzyme A with tryptophan to form the tetramic acid, which is subsequently alkylated by dimethylallyl pyrophosphate. Some of the enzymes for this pathway have been isolated (McGrath, R. M. pers. comm.). Erythroskyrin, an orange-red pigment of *Penicillium islandicum*, is formed by condensation of valine with a decaketide (Shibata, Sankawa, Taguchi & Yamasaki, 1966).

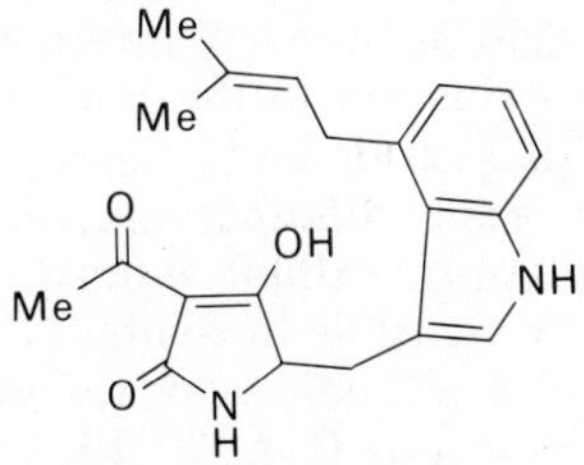

Cyclopiazonic acid

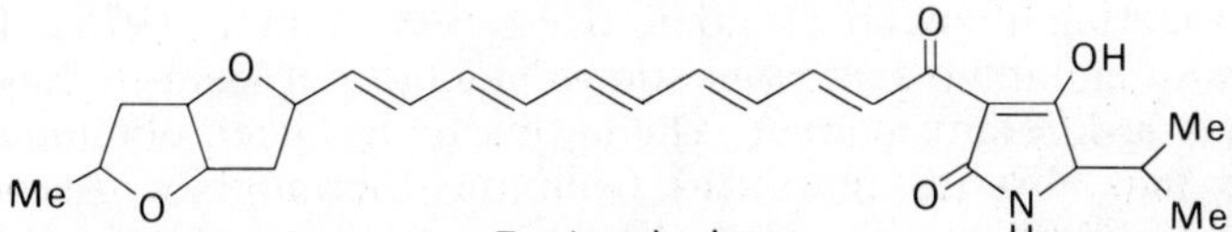

Erythroskyrin

Beauveria bassiana and *B. tenella* are insect pathogens, notorious in the past for their depredations in the silk-worm industry. Their characteristic yellow pigmentation is due to a group of *N*-hydroxylated pyridones, two of which, tenellin ($n = 1$) and bassianin ($n = 2$), have been isolated. Like fusaric and tetramic acids, they are biosynthesized by condensing an amino acid with a polyketide intermediate (McInnes, Smith, Walter, Vining & Wright, 1974). In this case the polyketide is methylated and the amino acid undergoes a rearrangement (Fig. 17.12). The timing of the rearrangement within the biosynthetic sequence is not yet known, but one possible route

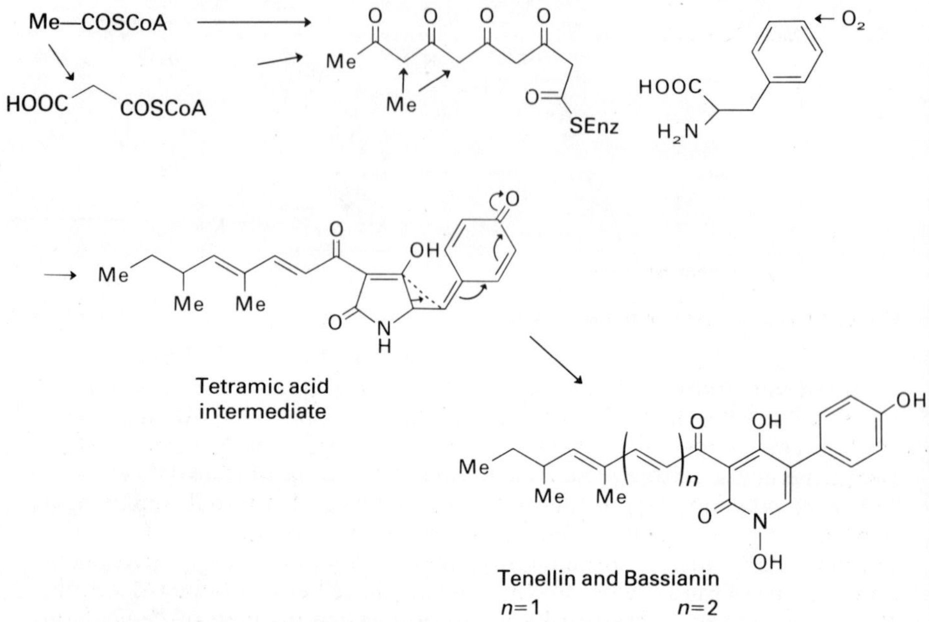

Fig. 17.12 Biogenesis of tenellin.

would give the tetramic acid as an early intermediate, ring enlargement being associated with an oxidation in the benzyl moiety. More complex condensations between a polyketide and an amino acid give rise in species of *Phoma*, *Zygosporium*, *Helminthosporium*, and *Metarrhizium* to the cytochalasins. These antibiotics exhibit unique 'cell-relaxing' effects in cultured eukaryotic cells. A plausible biosynthetic pathway would proceed via the initial formation of a tetramic acid, followed by folding back and cyclization of the polyketide tail (Fig. 17.13). Cytochalasins A and B (phomin) appear to have undergone a peroxidative cleavage of the carbon chain at the ring junction (Binder, Kiechel & Tamm, 1970). In thermozymocidin an amino acid polyketide condensation seems to have taken place in yet a different manner. The antibiotic has been obtained from a Eumycete and also (as myriocin) from the thermophilic Ascomycete, *Myriococcum albomyces*. The aliphatic chain and adjacent two hydroxylated carbons are known to be derived from acetate (Aragozzi, Beretta, Ricca, Scolasti & Wehrli, 1973). If the remainder of the molecule is introduced as serine, its α-carbon probably reacts with the coenzyme A ester of a C_{18} fatty acid as in the biosynthesis of sphingosine (Fig. 17.14), but without an accompanying loss of the carboxyl group. Whether the fatty acid chain is

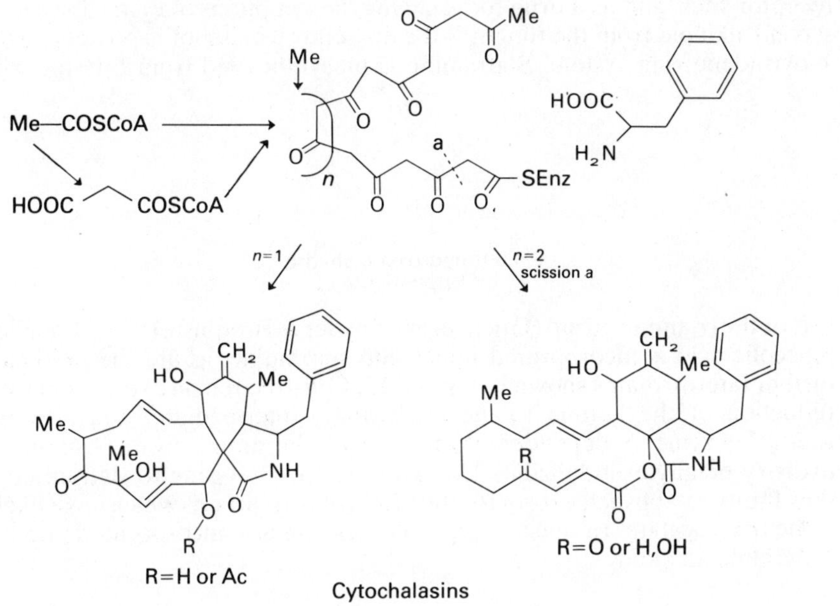

Fig. 17.13 Biogenesis of cytochalasins.

modified before or, as in sphingosine biosynthesis, after the condensation, has not yet been established.

When livestock feed on clover infected with black spot fungus, *Rhizoctonia legumicola*, they salivate excessively. The agent responsible is slaframine, a metabolite of the fungus which has become accessible through chemical synthesis. It is useful as a research tool for locating acetylcholine

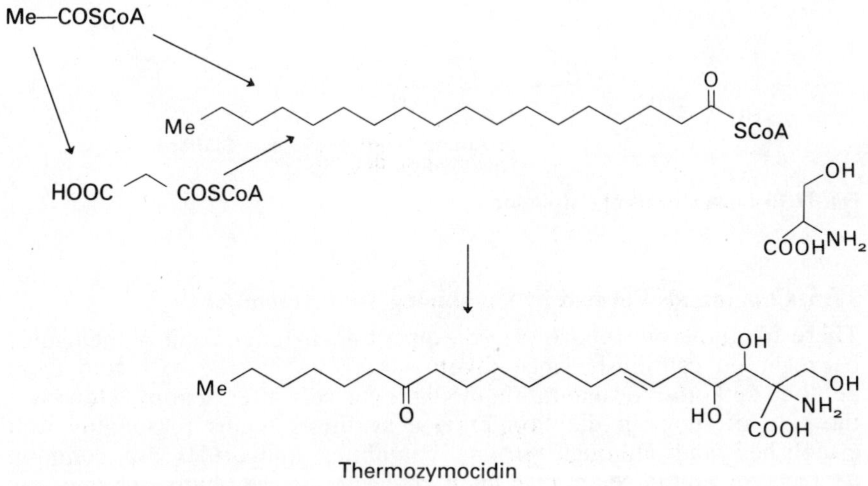

Fig. 17.14 Biogenesis of thermozymocidin.

receptor sites, and as a drug for relieving the symptoms of cystic fibrosis. A second alkaloid from the fungus is the first known metabolite possessing the 1-pyrindene ring system. Slaframine is biosynthesized from L-lysine with

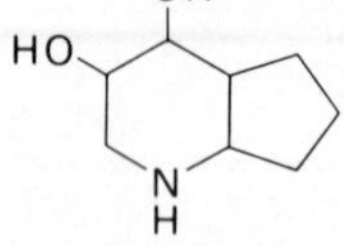

3,4,5-trihydroxyoctahydro-
1-pyrindene

loss of the α-amino group (Guengerich, Snyder & Broquist, 1973). Labelled pipecolic acid is incorporated intact into slaframine, as are the predicted further intermediates shown in Fig. 17.15. Cell-free fungal extracts catalyse reduction of the ketone to the *cis*-hydroxy intermediate, and also the acetylcoenzyme A-dependent formation of slaframine from 6-amino-1-hydroxy octahydroindolizine. The pathway from L-lysine to slaframine is thus firmly established except for the origin of two carbons which most likely come from acetate. Lysine and pipecolic acid are also incorporated into the 1-pyridine alkaloid.

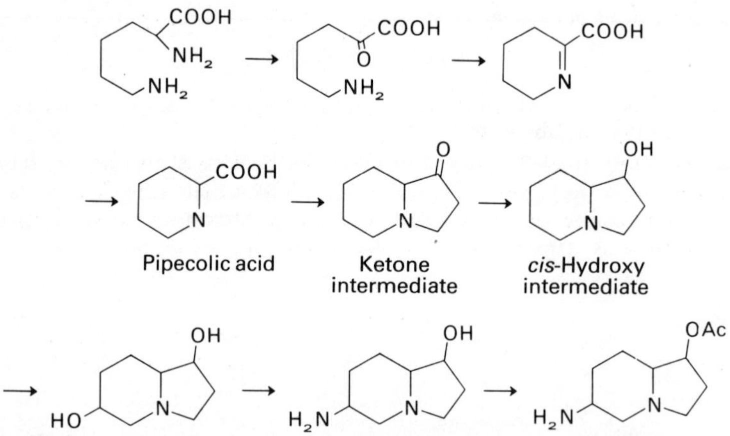

Fig. 17.15 Biosynthesis of slaframine.

17.5 Compounds Formed by Combining Two Amino Acids

There are numerous reports of cyclic dipeptides in fungi. Some of them must be regarded cautiously since diketopiperazines can be extracted from peptone and other fermentation media, especially after heating. However, the role of fungi in diketopiperazine synthesis seems reasonably well established and although proline containing anhydrides are common artifacts, several of them have been implicated in the phytotoxicity of the white root rot fungus, *Rosellina necatrix* (Chen, 1960). Of undisputed

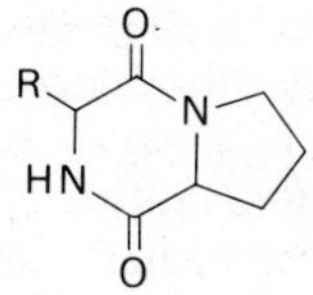

	R
Proline containing anhydrides	Bu^i
	Pr^i
	Bz

fungal origin are diketopiperazines such as echinulin where one of the amino acid components has been modified. Cyclo-L-alanyl-L-tryptophanyl is an intermediate in the biosynthesis of echinulin from alanine and tyrosine by *Aspergillus amstelodami* (Fig. 17.16) and is isoprenylated by specific

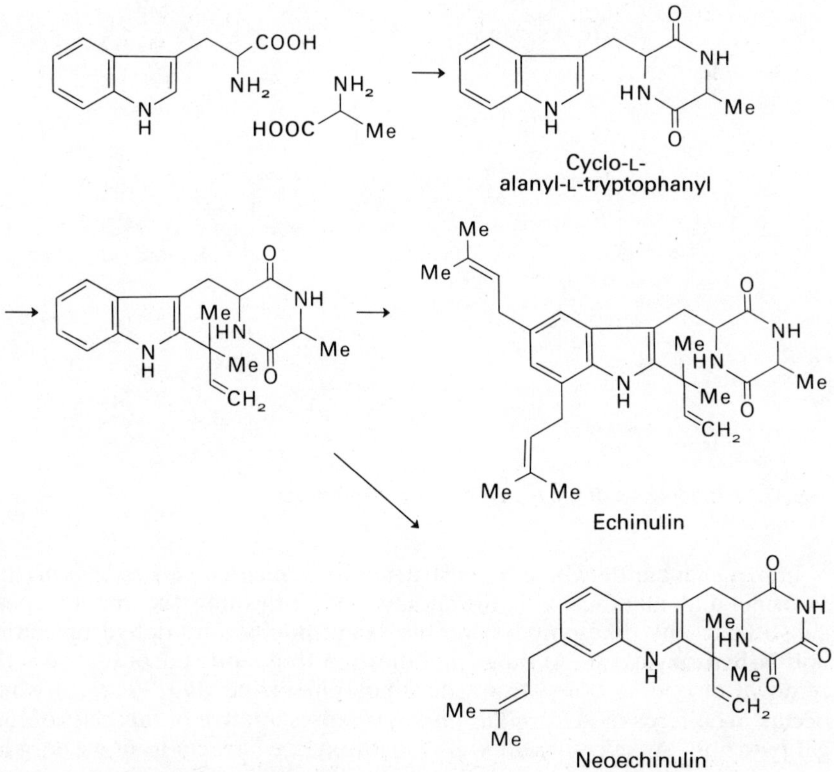

Fig. 17.16 Biosynthesis of echinulins.

enzymes (Allen, 1973). The unusual attachment of the dimethylallyl group at position-2 is believed due to rearrangement after initial substitution at the indole nitrogen (Casnati, Marchelli & Pochini, 1974). Neoechinulin, which accompanies echinulin in *A. amstelodami*, is derived from the same precursors. The group of brevianamides in *Penicillium brevi-compactum* and

Aspergillus ustus are biosynthetic half-brethren of the echinulins with proline replacing alanine as the second parent. Cyclo-L-prolyl-L-tryptophanyl is an intermediate (Baldas, Birch & Russell, 1974) and the 2-isoprenyl derivative is present in cultures (Steyn, 1973). Pathways can be proposed to account for the formation of each metabolite (Fig. 17.17).

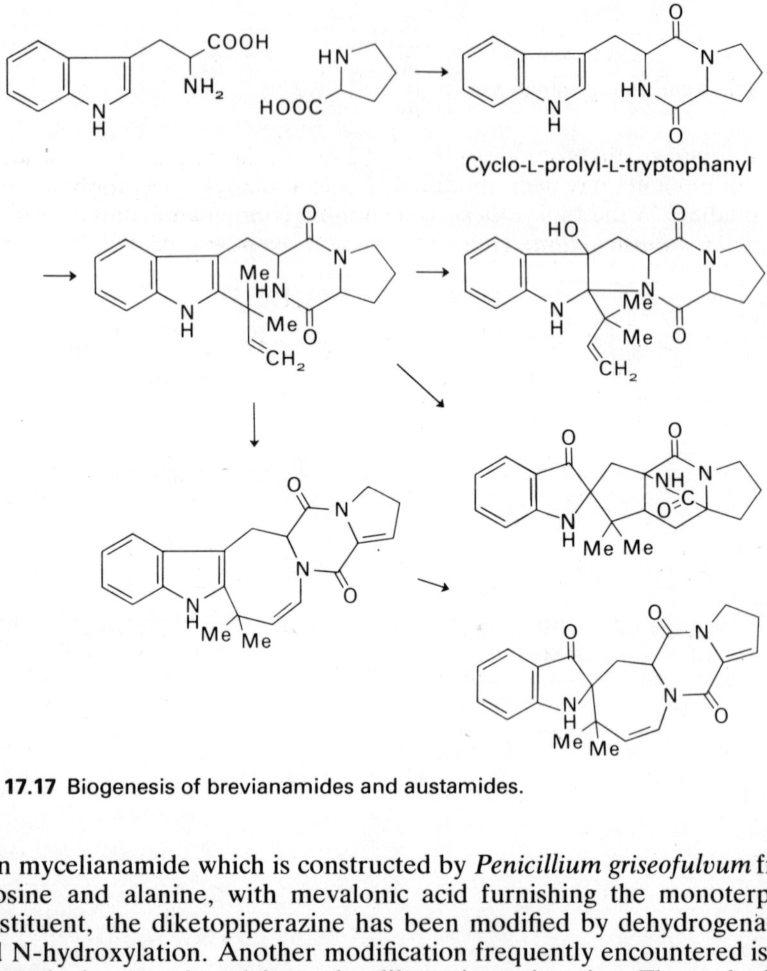

Fig. 17.17 Biogenesis of brevianamides and austamides.

In mycelianamide which is constructed by *Penicillium griseofulvum* from tyrosine and alanine, with mevalonic acid furnishing the monoterpene substituent, the diketopiperazine has been modified by dehydrogenation and N-hydroxylation. Another modification frequently encountered is the removal of one carbonyl from the diketopiperazine ring. Flavacol which occurs in cultures of *Aspergillus flavus*, is representative of this class of fungal pyrazine. *A. sclerotiorum* will *N*-hydroxylate flavacol to neoaspergillic

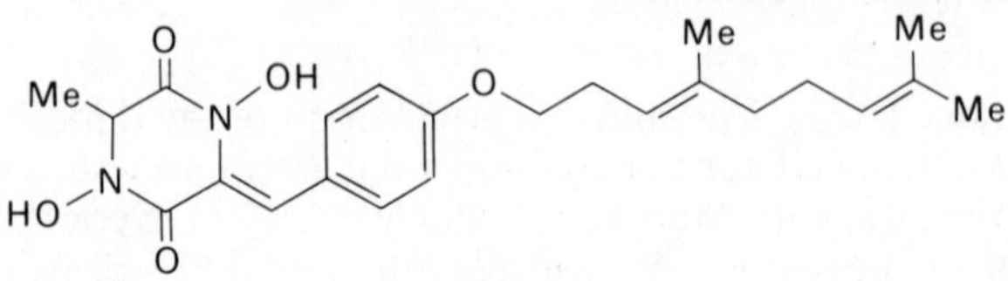

Mycelianamide

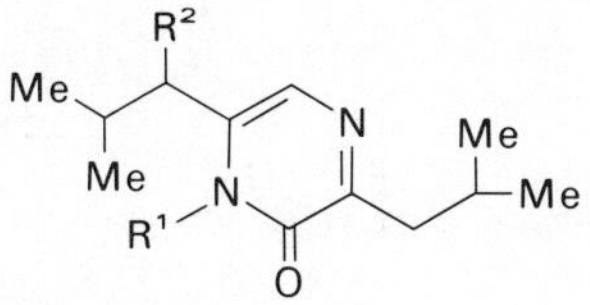

		R¹	R²
Flavacol	:	H	H
Neoaspergillic acid	:	OH	H
Neohydroxyaspergillic	:	OH	OH

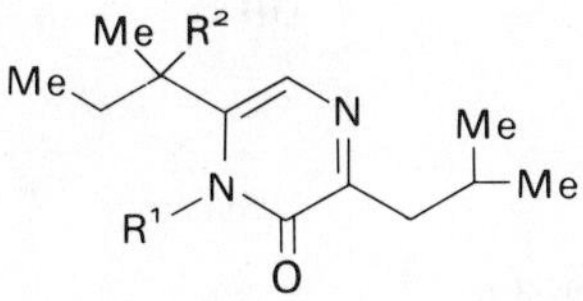

		R¹	R²
Deoxyaspergillic acid	:	H	H
Aspergillic acid	:	OH	H
Hydroxyaspergillic acid	:	OH	OH

acid, and introduce an additional hydroxyl into one of the isobutyl groups to give neohydroxyaspergillic acid. Although leucine supplies all of the carbon and nitrogen atoms of neoaspergillic acid, the diketopiperazine is not an intermediate, probably because leucylleucine is reduced at the dipeptide stage before cyclization (MacDonald, 1967). A similar sequence of reactions has been demonstrated in *A. flavus* during the synthesis of deoxyaspergillic, aspergillic, and hydroxyaspergillic acids from one molecule of leucine and one of isoleucine. Aspergillic acid and its congener are toxic antibiotics frequently encountered in *Aspergillus* species. The enzymes which catalyse their synthesis exhibit considerable tolerance in the kind of amino acids accepted as substrates, and the composition of the product can be influenced by the content of amino acids in the cytoplasmic pool. A large variety of new aspergillic acids has been made in directed biosyntheses by supplementing the nutrient medium with amino acids (MacDonald, 1970). The enzyme system in *A. flavus* shows a preference for at least one 4-carbon side chain in the product, and a reluctance to form products with unsymmetrical branched side groups in the reversed pattern from aspergillic acid.

Diketopiperazines are also modified by sulphur bridging. Gliotoxin, a venerable antibiotic discovered in 1936 in a *Trichoderma* species and widely distributed in Fungi Imperfecti is biosynthesized from phenylalanine and serine (Fig. 17.18). The aromatic ring is probably modified by an initial 2,3-epoxidation analogous to the 3,4-epoxidation step in phenylalanine hydroxylase-catalysed reactions. Attack by the α-amino nitrogen at position 2 of the epoxide would generate the hydroxylated heterocyclic ring system of gliotoxin (Bu'lock & Ryles, 1970). In support of the existence of such aromatic epoxide intermediates is the discovery in *Arachniotus aureus* and *Aspergillus terreus* of the aranotin group of antibiotics containing oxepine rings which are rearranged forms of phenyl epoxides. The aranotins are derived from two molecules of phenylalanine (Brannon, Mabe, Molloy

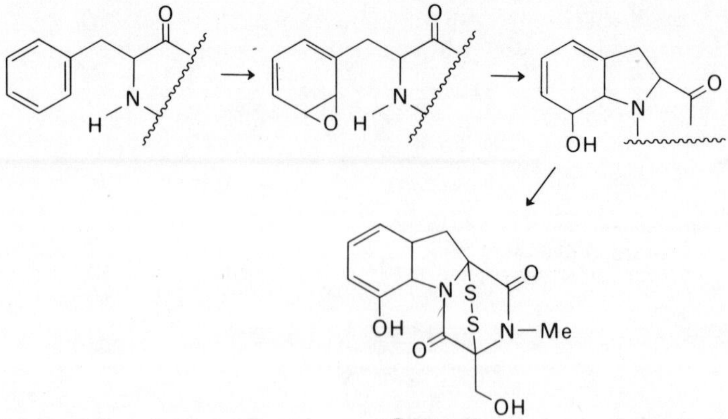

Gliotoxin

Fig. 17.18 Biogenesis of gliotoxin.

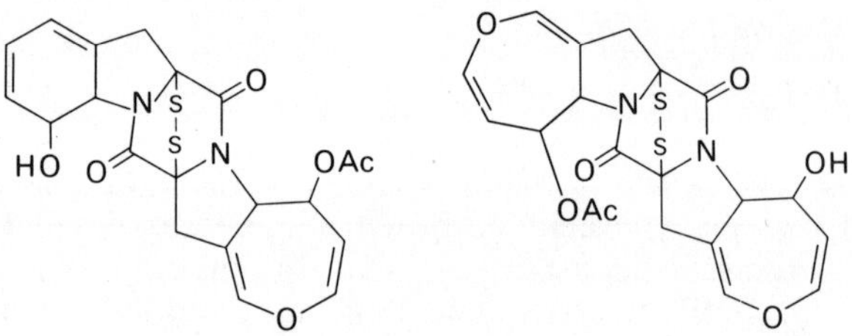

Aranotin antibiotics

& Day, 1971) and their biosynthesis according to the above scheme would require two epoxidation steps for each oxepine ring. There is no information yet on the sequence of biosynthetic reactions but discovery of the antibiotic hyalodendrin in a *Hyalodendron* species (Stillwell, Magasi & Strung, 1974)

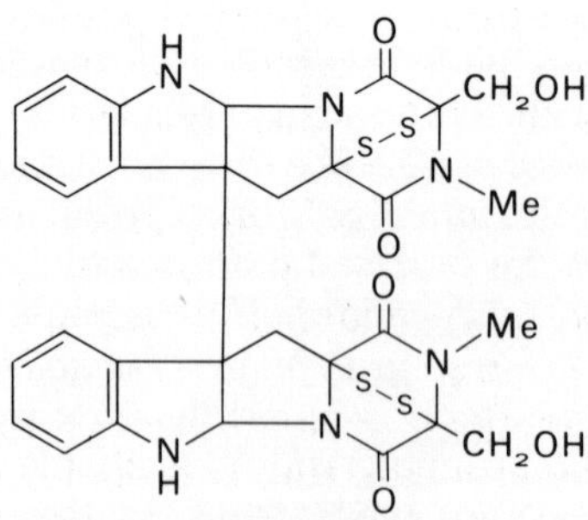

Chaetocin

suggests that the diketopiperazine might be an early intermediate with 2,3-epoxidation or *N*-methylation being alternative later events. The route by which sulphur is introduced into the diketopiperazine is also uncertain, though it does not appear to involve desaturation of the methine carbons as in neoechinulen and mycelianamide. Presumably sulphur is substituted directly for the bridgehead hydrogens, but the mechanism must account for bridging by one to four sulphur atoms and the presence of *S*-methyl groups at these locations in some members of the thiadiketopiperazine family. This diversity is exhibited in the sporidesmins, a group of powerful hepatotoxins produced by *Pithomyces chartarum.* The fungus grows on pasture detritus and large amounts may be ingested with the grass eaten by close-grazing animals such as sheep. Sporidesmins present in the fungus cause severe toxicosis in the animals. Radiotracer studies support a biogenesis from alanine and tryptophan (Towers & Wright, 1969). Chaetocin is a dimeric

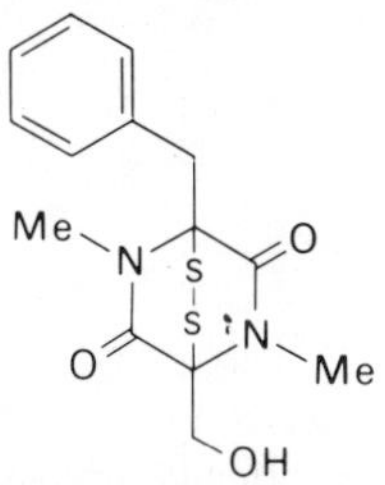

Hyalodendrin

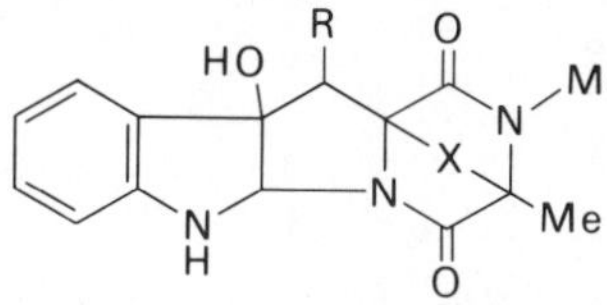

Sporidesmins

R	X
H	S_2
OH	S_2
OH	S_3
OH	$(SCH_3)_2$

relative of sporidesmin found in *Chaetomium minutum* and chetomin, a less symmetric dimer of uncertain linkage, is produced by *Chaetomium cochlioides.*

The penicillins and cephalosporins have been isolated from species of *Penicillium, Aspergillus, Cephalosporium* and even *Streptomyces.* They can be looked upon as a group of modified dipeptides, since the finished molecule in those representatives used clinically as antibiotics contains the elements of only two of the original amino acid precursors. These are L-cysteine and L-valine, which were shown in classic radiotracer experiments to have condensed and cyclized to generate the 6-aminophenicillanic acid nucleus. The remainder of the penicillin molecule, the acyl substituent, is variable, and in fact, these variations gave rise to the large family of penicillins which complicated early work on isolation and characterization of the antibiotic. The nature of the acyl group is influenced by the composition of the medium, and acyl group precursors are normally added to fermentation broths to swamp the endogenous acyl pool and steer the culture towards biosynthesis of the most desirable penicillin. This is usually penicillin G, the acyl group of which is derived from phenylalanine via phenylacetic acid. In practice large supplements of the latter, or its less toxic amide, are added. Biosynthetic studies have implicated the tripeptide δ-(L-α-aminoadipyl)-L-cysteinyl-D-valine as an intermediate in the pathway to both the penicillins and cephalosporins (Lodger & Abraham, 1971). Oxidative cyclization of this intermediate in alternative modes is believed to generate isopenicillin N or a cephalosporin intermediate. Only the former is made in *Penicillium chrysogenum,* but both are made in *Cephalosporium* species. The cephalosporin intermediate undergoes hydroxylation and acetylation at the exocyclic methyl group and epimerization at the asymmetric centre of the adipyl moiety to give cephalosporin C. In *Cephalosporium* cultures isopenicillin N is also epimerized to penicillin N. Biosynthetic

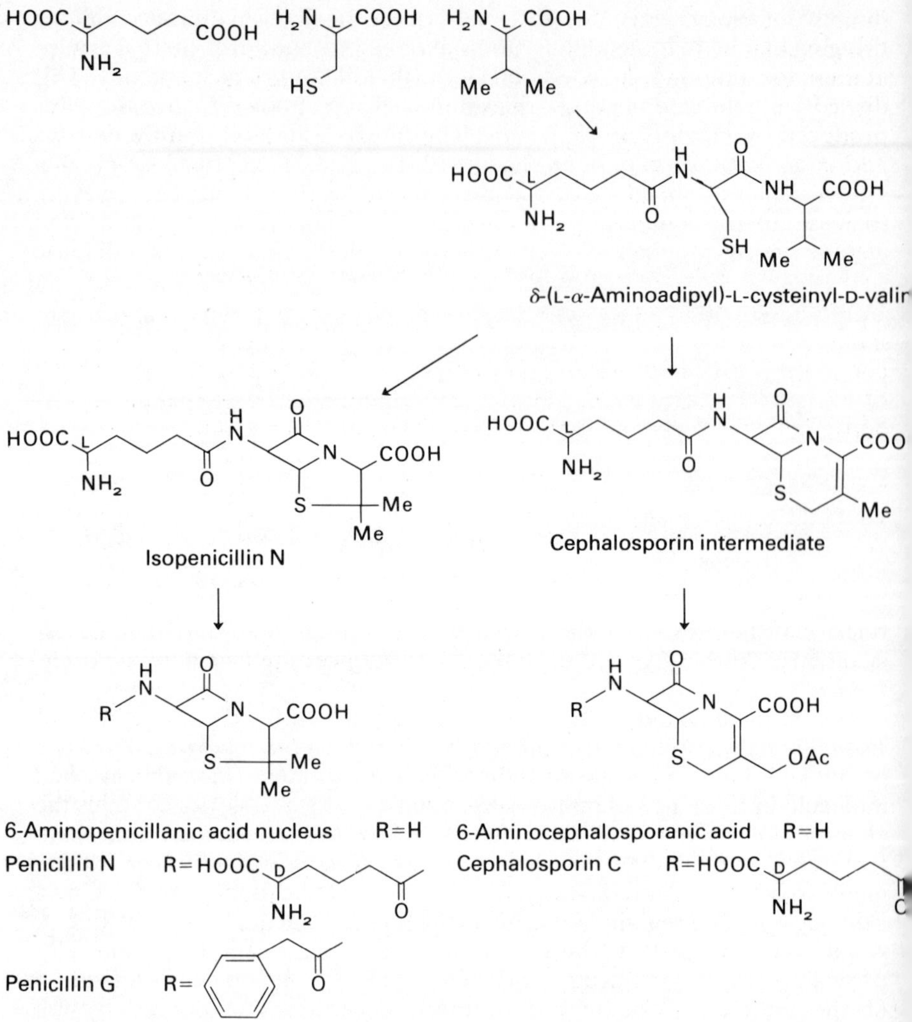

Fig. 17.19 Biosynthesis of penicillins and cephalosporins.

studies using valine chirally labelled with ^{13}C have shown that the amino acid is incorporated into both antibiotics. The two methyl groups are distinguished by *C. acremonium* enzymes, the (3*S*) methyl carbon appearing exclusively in the β-methyl group of penicillin N and in the exocyclic methylene of cephalosporin C (Kluender, Bradley, Sih, Fawcett & Abraham, 1973). Accumulation of the δ-(D-aminoadipyl) substituted products in *Cephalosporium* is probably due to a specific epimerase which is lacking in *Penicillium* species. Instead these seem to possess an acyl transferase which catalyses replacement of the L-aminoadipyl with another acyl moiety (Demain, 1966). When the phenylacetyl group is available penicillin G accumulates. The acyl side chain can be lost through the action of specific amidases to give 6-aminopenicillanic acid or 6-aminocephalosporanic acid. These normally minor reactions have been exploited on an industrial scale to obtain the penicillin and cephalosporin nuclei. Chemical substitution with acyl groups that could not be introduced biosynthetically affords the new semi-synthetic penicillins and cephalosporins that have greatly extended the scope of penicillin therapy.

17.6 Compounds Derived from More than Two Amino Acids

In these metabolites the amino acids are linked by peptide or ester bonds into cyclic structures. It is convenient to divide the group into true peptides containing only amino acids in peptide linkage, peptidolactones consisting of a peptide with one hydroxy acid generating a cyclic ester linkage, and depsipeptides which contain multiple peptide and ester sequences.

Polypeptides

Of all the fungal polypeptides only fungisporin from the spores of numerous *Aspergilli* and *Penicillia* has not been reported to have some kind of physiologic activity. None are described as antibiotics, but aspochracein from *Aspergillus ochraceus* is a potent insecticide; aspercolorin is the poisonous substance in corn infested with *Aspergillus versicolor*; islanditoxin is responsible for the toxicity of foodstuffs contaminated by *Penicillium*

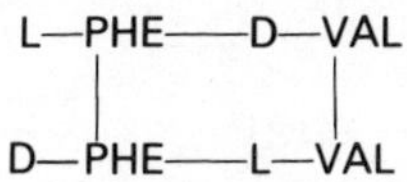

Fungisporin

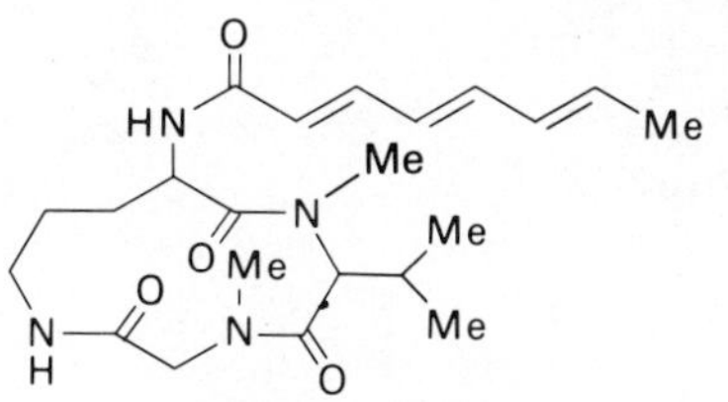

Aspochracein

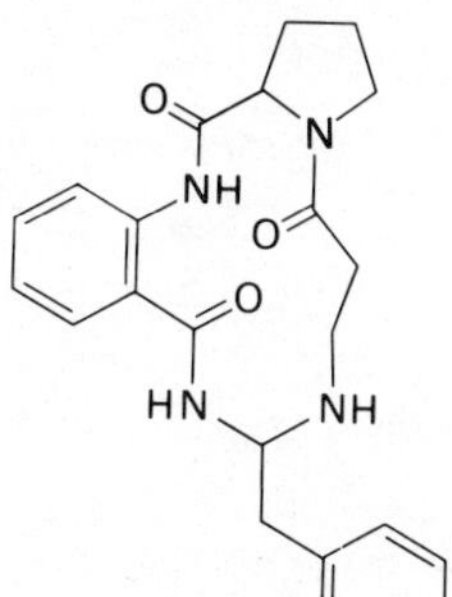

Aspercolorin

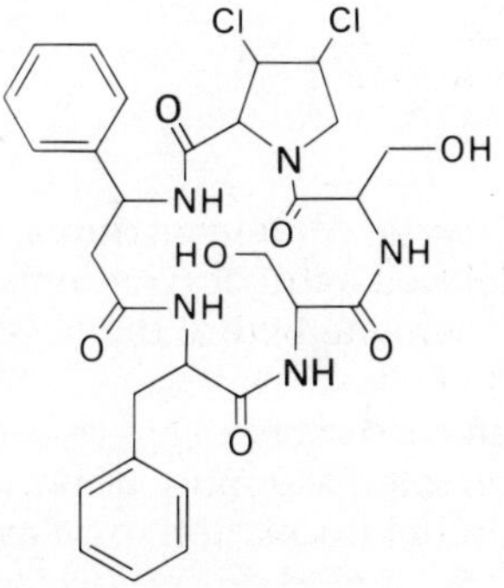

Islanditoxin

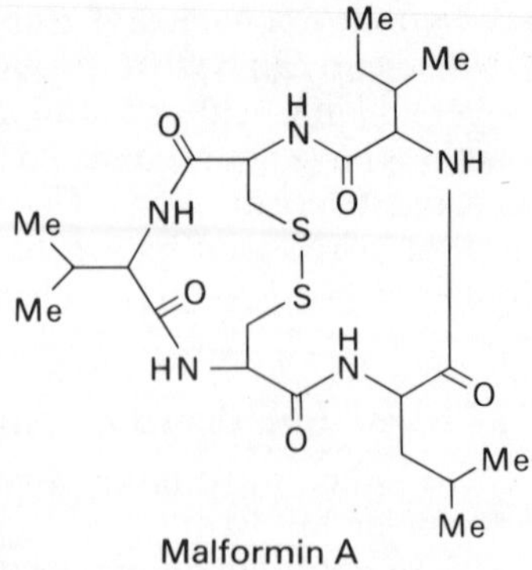

Malformin A

islandicum; malformin A which is found in a number of *Aspergillus* species, causes curvature and malformation in bean plants. The *Amanita* toxins are extremely active cell poisons which, at the molecular level, block RNA synthesis in eukaryotic cells. Two groups, amanitins and phalloidins, differing in general structure and pharmacologic activity can be recognized

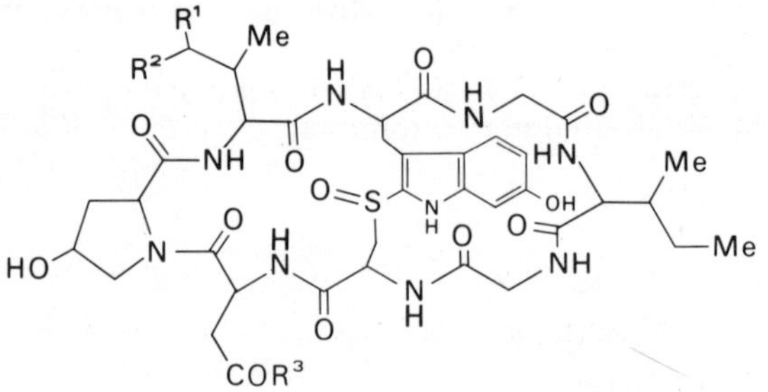

R^1	R^2	R^3
CH_2OH	OH	NH_2
CH_2OH	OH	OH
CH_3	OH	NH_2
CH_3	OH	OH
CH_3	CH_3	NH_2

Amanitins

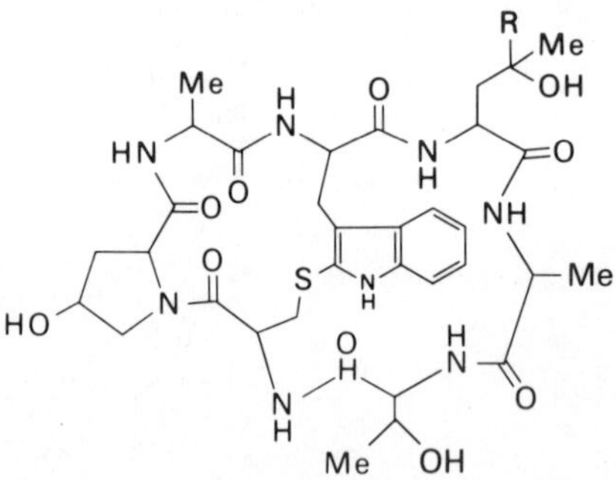

$R = CH_2OH$
$R = Me$

Phalloidins

(Wieland, 1967), but most of the lethal *Amanitas* contain mixtures. As little as 1 cm^3 of *A. verna* can be fatal, and the delayed onset of symptoms usually means that gastrointestinal absorption is complete before the poisoning is detected.

The biosynthesis of accessory peptides in bacteria is an enzyme-directed process, different from the nucleic acid-directed assembly of polypeptide chains in proteins. From the meagre amount of information so far available, fungal peptide biosynthesis proceeds by a similar mechanism. *In vivo* studies on the synthesis of malformin A in *Aspergillus niger* have shown that it is insensitive to inhibitors of protein and nucleic acid synthesis. Cell extracts which catalyse malformin A formation *in vitro* have the properties of a multi-enzyme complex with no requirement for ribosomes, transfer-RNA, or messenger-RNA. While the direction of peptide chain extension is from the amino to the carboxyl end, as in protein synthesis, the mechanism is probably more closely related to the chain-extending addition of malonate units in fatty acid or polyketide biosynthesis. Indeed, the enzyme complex catalysing gramicidin S synthesis in *Bacillus brevis* and the acyl carrier protein in the fatty acid synthetase complex both contain a phosphopantetheine prosthetic group which can accept and transfer the growing chain at each unit extension.

Many of the fungal peptides contain unusual amino acids, often the 'unnatural' D-isomer. The biogenesis may be guessed at, but little is known of the detailed reaction sequences or the timing of these conversions. In malformin A where the origin of D-leucine has been examined, the L-isomer serves as an exclusive precursor and the inversion appears to take place, as in bacterial polypeptide synthesis, after the precursor L-amino acid has become covalently attached to the enzyme complex. Epimerization does not occur via the keto acid but may involve desaturation of the bond between the α-methine carbon and nitrogen. *N*-Methylamino acids are invariably of L-configuration suggesting that *N*-methylation blocks the epimerization mechanism and that the two reactions are alternative biosynthetic processes.

Peptidolactones

The destruxins are toxic metabolites of the insect pathogen *Metarrhizium anisoplia* and contain *N*-methylamino acids. In isariin, the peptidolactone

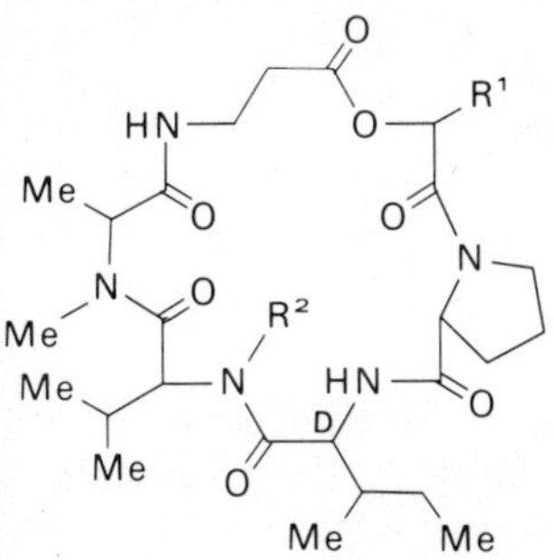

Destruxins

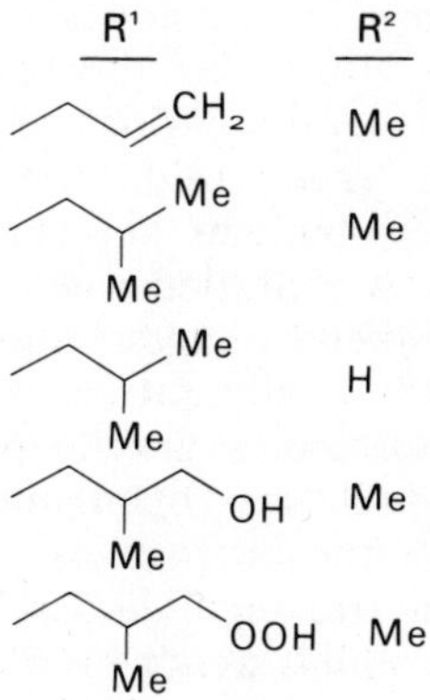

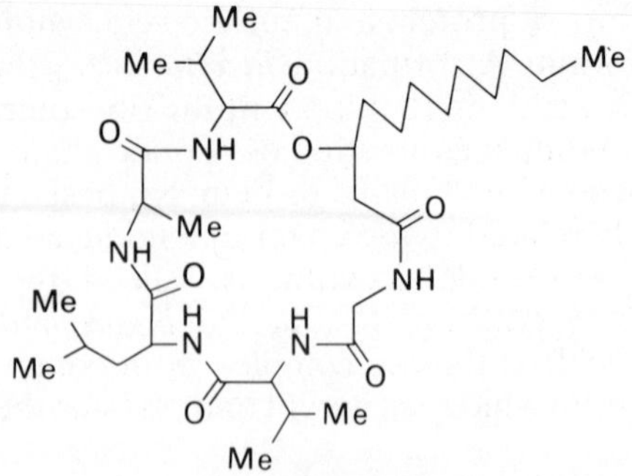

Isariin

from *Isaria cretaceae*, one of the amino acids is in the D-configuration. The β-hydroxylauric acid in isariin is replaced by β-hydroxyundecanoic acid in the isarolides which are present in another *Isaria* species (Briggs, Fergus & Shannon, 1966). There have been no biosynthetic studies on the fungal peptidolactones.

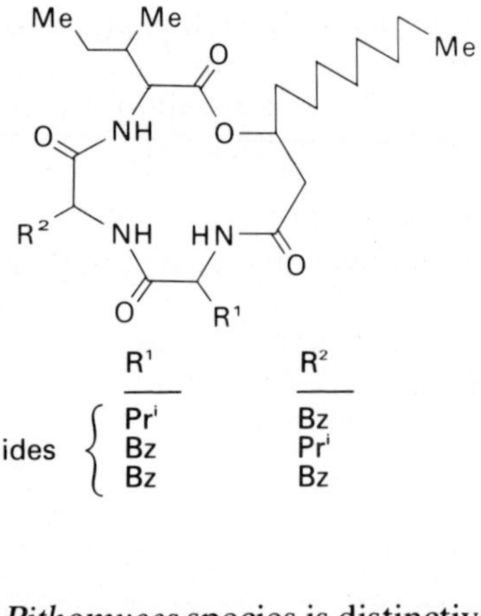

	R^1	R^2
Isarolides	Pr^i	Bz
	Bz	Pr^i
	Bz	Bz

Depsipeptides

The spore surface of *Pithomyces* species is distinctive in being covered with closely-spaced spicules which project out from the wall. Their function is uncertain, but is possibly related to water-repellancy which aids in spore dispersal. A similar role could be attributed to the water-insoluble fungisporins associated with *Penicillium* and *Aspergillus* spores. *Pithomyces* spicules consist of depsipeptides which vary with the species examined but three types can be recognized: the angolides, the sporidesmolides, and pithomycolide. Angolides consist of two dipeptolide units linked head to tail by ester bonds. A peptolide is defined here as one or more amino acids and a hydroxy acid linked by amide bonds. In angolides the component peptolides differ only in the configuration of the amino acid moiety. These relatively small cyclodepsipeptides are the main fraction in spicules of *P. sacchari* and *P. cynodontis* but the type of angolide present can be influenced by the amino acid content of the nutrient on which the fungus is grown (Okotore & Russell, 1972). The angolide derived exclusively from isoleucine in unsupplemented medium is steadily replaced by mixtures containing one isoleucyl

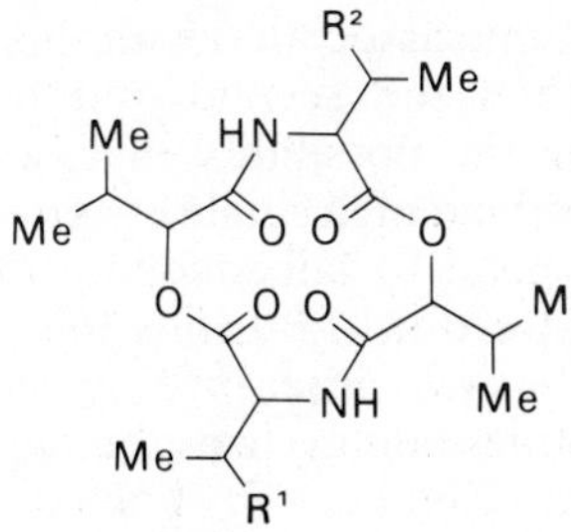

	R¹	R²
Angolides	Et (*allo*)	Et
	Et (*allo*)	Me
	Me	Et
	Me	Me

and one valyl residue, and eventually by the angolide with two valyl residues, as valine is added to the medium. The sporidesmolides from *P. chartarum* and *P. maydicus* spicules consist of two tripeptolide units linked head-to-tail, and once again their composition is influenced by the nutrient supply. Thus, sporidesmolides I and III are normally the main components, presumably because of a higher enzyme specificity for valine, but if isoleucine is

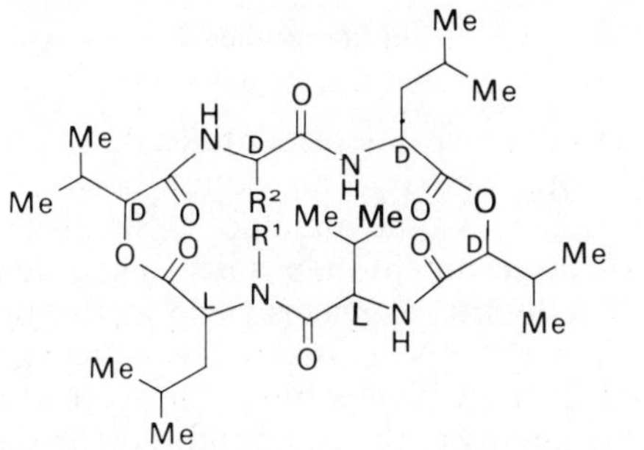

R¹	R²	
Me	Pr^i	Sporidesmolide I
Me	Bu^s (*allo*)	Sporidesmolide II
H	Pr^i	Sporidesmolide III

added to the medium, sporidesmolide II is the main product; sporidesmolides I and III, lacking an isoleucine moiety are suppressed. The hydroxy acid moieties are relatively resistant to precursor pressure. Amino acids appear to be converted to the corresponding hydroxy acids via keto acid intermediates, and although some replacement may occur this process does not compete strongly with endogenous synthesis of the keto acids. Interestingly, only the D-valine moiety in sporidesmolide I is replaced by isoleucine, the L-isomer in the other peptolide unit being unaffected. Sporidesmolides resemble other amino-acid derived metabolites in that D-amino acids are not incorporated directly but are generated during the biosynthetic sequence by α-epimerization of precursors with the normal L-configuration.

The active site on the enzyme associated with this epimerization appears to be much less specific than the second and non-epimerizing L-valine-accepting site, precluding the possibility that both peptolide units of the sporidesmolides are assembled on a single enzyme complex, with one being modified during the final head-to-tail condensation. More likely the *Pithomyces* enzyme resembles the one in *Bacillus brevis* which catalyzes gramicidin S biosynthesis; *i.e.* it has two adjacent sites on a single protein complex for assembling the peptolides and cyclizes the two completed chains *in situ* (Stoll, Froyshov, Holm, Zimmer & Laland, 1970; Russell, 1972). Pithomycolide has been obtained as a minor component of the depsipeptides from *P. chartarum* and its structure suggests that one peptolide unit of the sporidesmolides has been replaced by a novel di-ester of phenyl-D-glycollic acid.

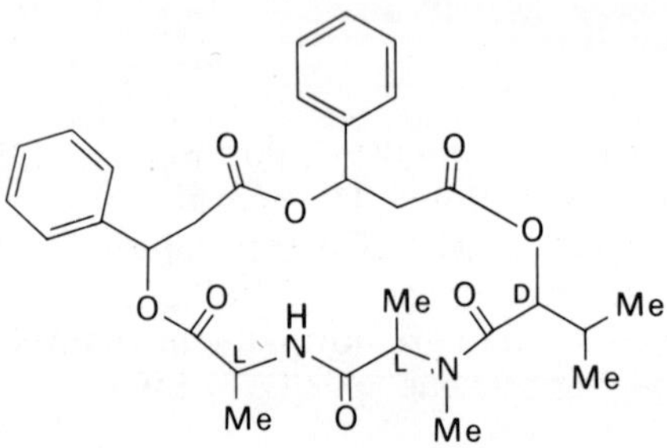

Pithomycolide

In contrast to the *Pithomyces* depsipeptides those produced by *Fusarium* species have antibiotic activity, no doubt because they are effective ionophores and complex alkaline metal ions (Shemyakin *et al.*, 1969). The enniatins consist of three dipeptolide units linked head-to-tail in a cyclic structure. All have α-hydroxyisovaleric acid as the hydroxy acid component; the amino acid may be *N*-methylated isoleucine, valine, or leucine in enniatins A, B, and C, respectively, but is the same in each peptolide unit. Again here, as with so many of the compounds discussed in this chapter, the

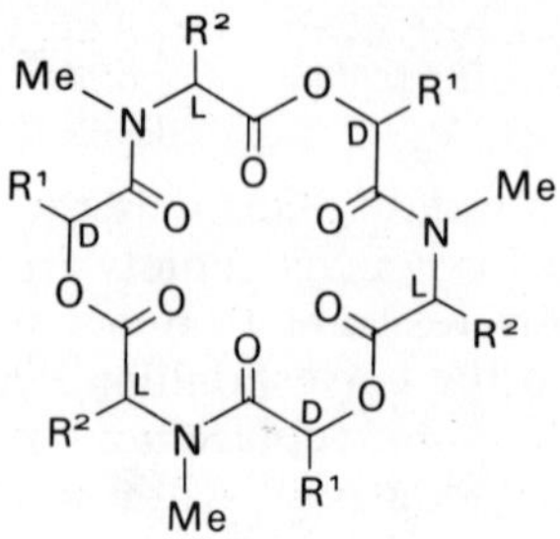

	R^1	R^2
Enniatins	Bu^s	Bu^s
	Pr^i	Pr^i
	Pr^i	Bu^s
Beauvericin	Pr^i	Bz

composition of the product is influenced by the amino acid content of the nutrient medium. However, the role of heredity in specifying the product may not be underrated. *Beauveria bassiana* produces only the congeneric depsipeptide beauvericin with *N*-methylphenylalanine in the peptolide unit, illustrating nicely the diversity as well as the unity among fungal metabolites derived from amino acids.

17.7 References

ALLEN, C. (1973). Monoisoprenylated *cyclo*-L-alanyl-L-tryptophanyl: biosynthetic precursor of echinulin. *Journal of the American Chemical Society* **95**, 2386–7.

ANKE, H., ANKE, T. & DIEKMANN, H. (1973). Biosynthesis of sideramines in fungi. Fusigen synthetase from extracts of *Fusarium cubense. FEBS Letters* **36**, 323–5.

ARAGOZZI, F., BERETTA, M. G.,RICCA, G. S., SCOLASTI, C. & WEHRLI, F. W. (1973). Biosynthesis of the antibiotic thermozymocidin. Incorporation of (1-^{13}C) acetate: ^{13}C-Nuclear magnetic resonance study. *Chemical Communications* pp. 788–9.

ASKARI, A. & MELVILLE, D. B. (1962). The reaction sequence in ergothioneine biosynthesis: hercynine as an intermediate. *Journal of Biological Chemistry* **237**,1615–8.

BALDAS, J., BIRCH, A. J. & RUSSELL, R. A. (1974). Studies in relation to biosynthesis. 46. Incorporation of *cyclo*-L-tryptophyl-L-proline into brevianamide A. *Journal of the Chemical Society, Perkins Transactions*, I, pp. 50–2.

BINDER, M., KIECHEL, J. R. & TAMM, C. (1970). Biogenesis of the antibiotic phomin. Basic building blocks. *Helvetica Chimica Acta* **53**, 1797–812.

BIRKINSHAW, J. H. (1965). Special chemical products. In *The fungi*, Vol. 1, pp. 179–228. Edited by G. C. Ainsworth and A. S. Sussman. London and New York: Academic Press.

BRANNON, D. R., MABE, J. A., MOLLOY, B. B. & DAY, W. A. (1971). Biosynthesis of dithiadiketopiperazine antibiotics: comparison of possible aromatic amino acid precursors. *Biochemical and Biophysical Research Communications* **43**, 588–94.

BRIGGS, L. H., FERGUS, B. J. & SHANNON, J. S. (1966). Chemistry of fungi. IV. Cyclodepsipeptides from a new species of *Isaria. Tetrahedron* pp. 269–78.

BROWN, A. G. (1970). Versimide, a metabolite of *Aspergillus versicolor. Journal of the Chemical Society* (C) pp. 2572–3.

BU'LOCK, J. D. & RYLES, A. P. (1970). The biosynthesis of the fungal toxin gliotoxin; the origin of the 'extra' hydrogens as established by heavy-isotope labelling and mass spectrometry. *Chemical Communications* pp. 1404–6.

CASNATI, G., MARCHELLI, R. & POCHINI, A. (1974). Rearrangement of 3-alkyl-1-allylindoles. Model reaction for biogenesis of echinulin-type compounds. *Journal of the Chemical Society, Perkins Transactions* I. pp. 754–7.

CHEN, Y. S. (1960). Studies on the metabolic products of *Rosellinia necatrix* Berlese. 1. Isolation and characterization of several physiologically active neutral substances. *Bulletin of the Agricultural Chemical Society of Japan* **24**, 372–81.

DEMAIN, A. L. (1966). Biosynthesis of penicillins and cephalosporins. In *Biosynthesis of antibiotics*, Vol. 1. pp. 30–94. Edited by J. F. Snell. New York: Academic Press.

GATENBECK, S. & SIERANKIEWICZ, J. (1973). On the biosynthesis of tenuazonic acid in *Alternaria tenuis. Acta Chemica Scandinavica* **27**, 1825–6.

GUENGERICH, F. P., SNYDER, J. J. & BROQUIST, H. (1973). Biosynthesis of slaframine, (1*S*, 6*S*, 9a*S*)-acetoxy-6-amino-octahydroindolizine, a parasympathomimetic alkaloid of fungal origin. I. Pipecolic acid and slaframine biogenesis. *Biochemistry* **12**, 426–9.

HARRIS, S. A., FISHER, L. V. & FOLKERS, K. (1965). The synthesis of tenuazonic and congeneric tetramic acids. *Journal of Medicinal Chemistry*. **8**, 478–82.

HEATH, H. & WILDY, J. (1957). Biosynthesis of ergothioneine. *Nature* **179**, 196–7.

HORNE, D. W. & BROQUIST, H. P. (1973). Role of lysine and ε-*N*-trimethyllysine in carnitine biosynthesis. 1. Studies in *Neurospora crassa. Journal of Biological Chemistry* **248**, 2170–5.

KLUENDER, H., BRADLEY, C. H., SIH, C. J., FAWCETT, P. & ABRAHAM, E. P. (1973). Synthesis and incorporation of (2*S*, 3*S*)-[4-^{13}C] valine into β-lactam antibiotics. *Journal of the American Chemical Society* **95**, 6149–50.

LASKIN, A. I. & LECHEVALIER, H. A. (eds.) (1974). *Handbook of microbiology*. Vol. III. Microbial Products. Cleveland: CRC Press Inc.

LODGER, P. B. & ABRAHAM, E. P. (1971). Biosynthesis of peptides containing α-aminoadipic acid and cysteine in extracts of a *Cephalosporium* sp. *Biochemical Journal* **123**, 477–82.

MACDONALD, J. C. (1967). Aspergillic acid and related compounds. In *Antibiotics*. Vol. II, pp. 43–51. Edited by D. Gottlieb and P. D. Shaw. New York: Springer-Verlag.

MACDONALD, J. C. (1970). Biosynthesis of compounds similar to aspergillic acid. *Canadian Journal of Biochemistry* **48**, 1165–74.

MAEHR, H. (1971). Antibiotics and other naturally occurring hydroxamic acids and hydroxamates. *Pure and Applied Chemistry* **28**, 603–36.

MATSUMOTO, T., TRUEB, W., GWINNER, R. & EUGSTER, C. H. (1969). Isolierung von (−)-*R*-4-Hydroxy-pyrrolidon-(2) und einigen weiteren Verbindungen aus *Amanita muscaria*. *Helvetica Chimica Acta* **52**, 716–20.

McINNES, A. G., SMITH, D. G., WALTER, J. A., VINING, L. C. & WRIGHT, J. L. C. (1974). New techniques in biosynthetic studies using ^{13}C nuclear magnetic resonance spectroscopy. The biosynthesis of tenellin enriched from singly and doubly labelled precursors. *Chemical Communications* pp. 282–4.

OKOTORE, R. O. & RUSSELL, D. W. (1972). Evidence for the biosynthetical equivalence of the epimeric isoleucine residues in angolide. *Canadian Journal of Biochemistry* **50**, 428–39.

ONG, D. E. & EMERY, T. F. (1972). Ferrichrome biosynthesis: enzyme catalysed formation of the hydroxamic acid group. *Archives of Biochemistry and Biophysics* **148**, 77–83.

PITEL, D. W. & VINING, L. C., (1970). Accumulation of dehydrofusaric acid and its conversion to fusaric and 10-hydroxyfusaric acids in cultures of *Gibberella fujikuroi*. *Canadian Journal of Biochemistry*, **48**, 623–30.

POPPLESTONE, C. R. & UNRAU, A. M. (1973). Biosynthesis of lycomarasmin. *Canadian Journal of Chemistry* **51**, 3943–9.

RUSSELL, D. W. (1972). Biosynthetical non-equivalence of the D-and L-valine residues in sporidesmolide I. *Biochimica et Biophysica Acta* **261**, 469–74.

SHEMYAKIN, M. M., OVCHINNIKOV, Y. A., IVANOV, V. T., ANTONOV, V. K., VINOGRADOVA, E. I., SHKROB, A. M., MALENKOV, G. G., EVSTRATOV, A. V., LAINE, I. A., MELNIK, E. I. & RYABOVA, I. D. (1969). Cyclodepsipeptides as chemical tools for studying ionic transport through membranes. *Journal of Membrane Biology*, **1**, 402–30.

SHIBATA, S., SANKAWA, U., TAGUCHI, H. & YAMASAKI, K. (1966). Biosynthesis of natural products. III. Biosynthesis of erythroskyrine, a coloring matter of *Penicillium islandicum*. *Chemical and Pharmaceutical Bulletin* (Tokyo) **14**, 474–8.

STEYN, P. S. (1973). The structures of five diketopiperazines from *Aspergillus ustus*. *Tetrahedron* **29**, 107–20.

STILLWELL, M. A., MAGASI, L. P. & STRUNZ, G. M. (1974). Production, isolation, and antimicrobial activity of hyalodendrin, a new antibiotic produced by a species of *Hyalodendron*. *Canadian Journal of Microbiology* **20**, 759–64.

STOLL, E., FROYSHOV, O., HOLM, H., ZIMMER, T. L. & LALAND, S. G. (1970). On the mechanism of gramicidin S formation from intermediate peptides. *FEBS Letters* **11**, 348–52.

TOWERS, N. R. & WRIGHT, D. E. (1969). Biosynthesis of sporidesmin from amino acids. *New Zealand Journal of Agricultural Research* **12**, 275–80.

TURNER, W. B. (1971). *Fungal metabolites*. London: Academic Press.

WIELAND, T. (1967). The toxic peptides of *Amanita phalloides*. *Progress in the Chemistry of Organic Natural Products* **25**, 214–50.

Species Index

Subject Index